The elements

Name	Symbol	Atomic number	Atomic weight
Actinium	Ac	89	[227]
Aluminum	Al	13	26.981538 (2)
Americium	Am	95	[243]
Antimony	Sb	51	121.76 (1)
Argon	Ar	18	39.948 (1)
Arsenic	As	33	74.9216 (2)
Astatine	At	85	[210]
Barium	Ba	56	137.327 (7)
Berkelium	Bk	97	[247]
Beryllium	Be	4	9.012182 (3)
Bismuth	Bi	83	208.98038 (2)
Bohrium	Bh	107	[264]
Boron	B	5	10.811 (7)
Bromine	Br	35	79.904 (1)
Cadmium	Cd	48	112.411(8)
Calcium	Ca	20	40.078 (4)
Caesium	Cs	55	132.90545 (2)
Californium	Cf	98	[251]
Carbon	C	6	12.0107 (8)
Cerium	Ce	58	140.116 (1)
Chlorine	Cl	17	35.4527 (9)
Chromium	Cr	24	51.9961 (6)
Cobalt	Co	27	58.9332 (9)
Copper	Cu	29	63.546 (3)
Curium	Cm	96	[247]
Dubnium	Db	105	[262]
Dysprosium	Dy	66	162.5 (3)
Einsteinium	Es	99	[252]
Erbium	Er	68	167.26 (3)
Europium	Eu	63	151.964 (1)
Fermium	Fm	100	[257]
Fluorine	F	9	18.9984032 (5)
Francium	Fr	87	[223]
Gadolinium	Gd	64	157.25 (3)
Gallium	Ga	31	69.723 (1)
Germanium	Ge	32	72.61 (2)
Gold	Au	79	196.96655 (2)
Hafnium	Hf	72	178.49 (2)
Hassium	Hs	108	[269]
Helium	He	2	4.002602 (2)
Holmium	Ho	67	164.93032 (2)
Hydrogen	H	1	1.00794 (7)
Indium	In	49	114.818 (3)
Iodine	I	53	126.90447 (3)
Iridium	Ir	77	192.217 (3)
Iron	Fe	26	55.845 (2)
Krypton	Kr	36	83.8 (1)

(*Continued*)

Name	Symbol	Atomic number	Atomic weight
Lanthanum	La	57	138.9055 (2)
Lawrencium	Lr	103	[262]
Lead	Pb	82	207.2 (1)
Lithium	Li	3	[6.941 (2)]
Lutetium	Lu	71	174.967 (1)
Magnesium	Mg	12	24.305 (6)
Manganese	Mn	25	54.938049 (9)
Meitnerium	Mt	109	[268]
Mendelevium	Md	101	[258]
Mercury	Hg	80	200.59 (2)
Molybdenum	Mo	42	95.94 (1)
Neodymium	Nd	60	144.24 (3)
Neon	Ne	10	20.1797 (6)
Neptunium	Np	93	[237]
Nickel	Ni	28	58.6934 (2)
Niobium	Nb	41	92.90638 (2)
Nitrogen	N	7	14.00674 (7)
Nobelium	No	102	[259]
Osmium	Os	76	190.23 (3)
Oxygen	O	8	15.9994 (3)
Palladium	Pd	46	106.42 (1)
Phosphorus	P	15	30.973762 (4)
Platinum	Pt	78	195.078 (2)
Plutonium	Pu	94	[244]
Polonium	Po	84	[210]
Potassium	K	19	39.0983 (1)
Praseodymium	Pr	59	140.90765 (2)
Promethium	Pm	61	[145]
Protactinium	Pa	91	231.03588 (2)
Radium	Ra	88	[226]
Radon	Rn	86	[222]
Rhenium	Re	75	186.207 (1)
Rhodium	Rh	45	102.9055 (2)
Rubidium	Rb	37	85.4678 (3)
Ruthenium	Ru	44	101.07 (2)
Rutherfordium	Rf	104	[261]
Samarium	Sm	62	150.36 (3)
Scandium	Sc	21	44.95591 (8)
Seaborgium	Sg	106	[266]
Selenium	Se	34	78.96 (3)
Silicon	Si	14	28.0855 (3)
Silver	Ag	47	107.8682 (2)
Sodium	Na	11	22.98977 (2)
Strontium	Sr	38	87.62 (1)
Sulfur	S	16	32.066 (6)
Tantalum	Ta	73	180.9479 (1)
Technetium	Tc	43	[98]
Tellurium	Te	52	127.6 (3)
Terbium	Tb	65	158.92534 (2)

Materials Engineering and Science

Materials Engineering and Science

Principles, Properties, and Processes

Second Edition

Brian S. Mitchell
Tulane University
New Orleans, USA

WILEY

Edition History
John Wiley & Sons, Inc. (1e, 2004)

Published by John Wiley & Sons, Inc., Hoboken, New Jersey.
Published simultaneously in Canada.

For general information on our other products and services or for technical support, please contact our Customer Care Department within the United States at (800) 762-2974, outside the United States at (317) 572-3993 or fax (317) 572-4002.

Wiley also publishes its books in a variety of electronic formats. Some content that appears in print may not be available in electronic formats. For more information about Wiley products, visit our web site at www.wiley.com.

Library of Congress Cataloging-in-Publication Data applied for:

Hardback ISBN: 9781119857105

Cover Design: Wiley
Cover Image: Courtesy Jeffrey Stolier

Set in 9.5/12.5pt STIXTwoText by Straive, Pondicherry, India

To the memory of my parents; whose
Material was loam;
Engineering was labor;
Science was lore;
And greatest product was love.

Contents

*Sections marked with an asterisk can be omitted in an introductory course.

Preface

This textbook is intended for use in a one- or two-semester undergraduate course in materials science that is primarily populated by engineering students. It is organized on two levels: by engineering subject area and by materials class:

	Metals and Alloys	Ceramics and Glasses	Polymers	Composites	Biologics
Structure	Bonding Crystal Structures Nanostructures Defects	Atomic Packing The Glassy State	Configuration Conformation Molecular Weight Crystallinity	Matrices Reinforcements Nanocomposites	Biochemistry Cells and Tissue Hard and Soft Biologics
Thermodynamics of Condensed Phases	Phase Equilibria Gibbs Phase Rule Lever Rule	Ternary Systems Surface Energy Sintering	Phase Separation Polymer Solutions Polymer Blends	Adhesion Cohesion Spreading	Cell Adhesion Cell Spreading Protein Folding
Kinetic Processes in Materials	Transformations Corrosion	Devitrification Nucleation Growth	Polymerization Degradation	Deposition Infiltration	Receptors Ligand Binding
Transport Properties of Materials	Inviscid Systems Heat Capacity Diffusion	Newtonian Flow Heat Capacity Diffusion	Non-Newtonian Flow Heat Capacity Diffusion	Porous Flow Heat Capacity Diffusion	Convection Diffusion
Mechanics of Materials	Stress-Strain Elasticity Ductility	Fatigue Fracture Creep	Viscoelasticity Elastomers	Laminates Nanocomposites	Sutures Bone Teeth
Electrical, Magnetic, and Optical Properties of Materials	Resistivity Magnetism Reflectance Quantum Dots	Dielectrics Ferrites Absorbance Thermoelectrics	Ion conductors Molecular Magnets OLEDs Solid-State Electrolytes	Dielectrics Storage Media Thin Film Devices	Biosensors MRI

(*Continued*)

	Metals and Alloys	Ceramics and Glasses	Polymers	Composites	Biologics
Processing of Materials	Casting Rolling Compaction	Pressing Casting Firing	Continuous Processing Cyclic Processing 3D Printing	Pultrusion RTM Filament Winding	Surface Modification

This organizational approach is intended to foster a deeper understanding of the subject material since it is presented in the context of courses the student has already had or is currently taking. It also means that the material can be presented either in this subject-oriented way (in rows) or in a materials-oriented sequence (in columns). The latter approach is recommended for a two-semester course. Sections marked with an asterisk (∗) are optional for an introductory-level course.

As with the first edition, there are educational objectives at the beginning of each chapter, as well as active-learning tools in the form of cooperative learning exercises (CLEs) throughout. These CLEs are intended for in-class use to help the students identify difficulties that the instructor can help to overcome in real time. They are a starting point, however, and should be used along with Example Problems to help bring difficult concepts to practice.

There are several new features to this second edition. First, "Need More Help?" callout boxes have been added that feature mostly YouTube videos on selected topics that are meant to enhance the student experience. Full links are provided in the hard copy text; active links should be available in electronic versions of the text. Exposure to alternative explanations is generally a good thing, but it can also lead to confusion. Differences in nomenclature, equation derivation, and even definitions should be addressed early and directly. No guarantees are made as to the validity or accuracy of the information provided in these supplemental videos and we all know that links can come and go. Every effort has been made to provide stable and reliable links.

Second, "Take a Peek" callout boxes have been added to direct the student to other parts of the text where a principle may be applied or a topic revisited. Again, section numbers are given in the hard text but direct links should be available in electronic versions. It is important to see the applications of certain principles, especially introductory topics such as bonding and intermolecular forces.

Third, "Historical Highlights" now (mostly) direct the reader to internet content, especially Wikipedia. (A donation has been made to Wikipedia for the use of these links.) You may have your own opinions about the accuracy of information in Wikipedia, but for our purposes of societal and professional context, it is deemed reliable enough to utilize.

Fourth, many of the chapter-end problems have been rewritten to introduce random variables and an internet-based approach to finding physical property data. The purpose of this approach is not only to provide the student with some ownership of the problem, but to dissuade them from searching for vendor-provided or AI-generated solutions, neither of which helps them learn the material. It also reinforces the concept that, because there is much physical property data available through an internet search, the role of the engineer is to critique the veracity of these data and use them in an analytical manner.

Finally, some content from the first edition has been moved to ***Advanced Topics*** which is available on the textbook support webpage. This material may be particularly useful for an introductory graduate-level course on materials engineering. You will also find updated information and corrections to the print edition on the support webpage.

Acknowledgments

In addition to those individuals who contributed to making the first edition of this textbook a success, the author wishes to thank the many contributors to this second edition. They include students and colleagues who provided corrections, reviewers who provided ideas for revisions, and the editorial staff who helped improve figures and content. The unwavering support of Bonnie, Britt, Rory, and Chelsie and their loved ones is gratefully acknowledged.

About the Companion Website

This book is accompanied by instructor and student companion websites.

www.wiley.com/go/materialsengineeringandscience2e

The instructor website includes:

- Instructor's manual
- Sample lecture slides (PowerPoint)

The student website includes:

- Advanced topics
- "Need More Help?" YouTube video links

1

The Structure of Materials

1.0 Introduction and Objectives

A wealth of information can be obtained by looking at the structure of a material. Though there are different levels of structure (atomic, nanoscale, macroscopic) many physical properties of a material can be related directly to the arrangement and types of bonds comprising that material. We will begin by reviewing some general chemical principles that will aid us in our description of material structure. Such topics as periodic structure, types of bonding, and potential energy diagrams will be reviewed. We will then use this information to look at the specific material categories in more detail: metals, ceramics, polymers, composites, and biological materials (biologics). There will be topics that are specific to each material class; there will also be some that are common to all types of materials. In subsequent chapters, we will explore not only how the building blocks of a material can significantly impact physical properties, but how the material interacts with its environment, as well.

In terms of learning outcomes, by the end of this chapter you should be able to:

- Identify trends in the periodic table for ionization energy (IE), electron affinity (EA), electronegativity, and atomic/ionic radii.
- Identify the type of bonding in a compound.
- Utilize the concepts of molecular orbital and hybridization theories to explain multiple bonds, bond angles, diamagnetism, and paramagnetism.
- Identify the seven unique crystal systems and 14 Bravais lattices.
- Calculate the volume of a unit cell from the lattice translation vectors.
- Calculate atomic density along directions, planes, and volumes in a unit cell.
- Calculate the density of a compound from its crystal structure and atomic mass.
- Locate and identify the interstitial sites in a crystal structure.
- Assign coordinates to a location, indices to a direction, and Miller indices to a plane in a unit cell.
- Use Bragg's Law to convert between diffraction angle and interplanar spacing.
- Read and interpret a simple X-ray diffraction (XRD) pattern.
- Identify types of point and line defects in solids.
- Calculate the concentration of point defects in solids.
- Draw a Burger's circuit and identify the direction of dislocation propagation.
- Identify the characteristic dimension of a nanostructured material and calculate its surface-to-volume ratio

Materials Engineering and Science: Principles, Properties, and Processes, Second Edition. Brian S. Mitchell.

Companion website: www.wiley.com/go/materialsengineeringandscience2e

- Use Pauling's rules to determine the stability of a compound.
- Predict the structure of a silicate from the Si/O ratio.
- Apply Zachariasen's rules to determine the glass-forming ability of an oxide.
- Classify polymers according to structure or formability.
- Calculate the first three moments of a polymer molecular weight distribution.
- Apply principles of glass transition and polymer crystallinity to polymer classification.
- Identify nematic, smectic, and cholesteric structures in liquid crystalline polymers (LCPs).
- Identify the components in a composite material.
- Approximate physical properties of a composite material based on component properties.
- Be conversant in terms that relate to the structure of biological materials, such as α-helix and protein folding.

1.0.1 The Elements

Elements are materials, too. Think about all the materials from our daily lives that are mostly elements: lithium in our batteries; gold for our jewelry; aluminum for our beverage cans; carbon, both as a sparkling diamond and as a mundane mechanical pencil lead; and silicon in our computer chips. Most elemental materials, however, are relatively scarce in the grand scheme of things. A table of the relative abundance of elements (Table 1.1) shows that most of our universe is made up of hydrogen and helium. A little closer to home, things are much different. A similar table of relative abundance (Table 1.2) shows that hydrogen on earth is relatively scarce (even with all the water!), while oxygen dominates the crust of our planet. Just think of how much molecular oxygen, water, and aluminosilicate rocks are contained in the earth's crust. But those are molecules—we are concentrating on atoms for the moment. Still, elements are of vital importance on earth, and the ones we use most often are primarily in the solid form.

Table 1.1 Relative Abundance of Elements in the Universe

Element	Relative Abundance (ppm)
Hydrogen (H)	750,000
Helium (He)	230,000
Oxygen (O)	10,000
Carbon (C)	5000
Neon (Ne)	1300
Iron (Fe)	1100
Nitrogen (N)	1000
Silicon (Si)	700
Magnesium (Mg)	600
Sulfur (S)	500

Source: WebElements (http://www.webelements.com).

Table 1.2 Relative Abundance of Selected Elements in the Earth's Crust

Element	Relative Abundance (ppm)
Oxygen (O)	461,000
Silicon (Si)	282,000
Aluminum (Al)	82,300
Iron (Fe)	56,300
Calcium (Ca)	41,500
Sodium (Na)	23,600
Magnesium (Mg)	23,300
Potassium (K)	20,900
Titanium (Ti)	5650
Hydrogen (H)	1400

Source: Rumble [1], CRC Press/Taylor & Francis.

Recall from your introductory chemistry course that the elements can be systematically arranged in a periodic table according to their electronic structure (see Table 1.3[1]). The heat map shows that many elements are solids above 0°C (shaded boxes) and remain solid well above ambient temperatures. As we heat to 1000°C, note that many of alkali and alkaline earth metals (columns 1 and 2) and light metals and metalloids (columns 13–15) have melted (lightly shaded), but that most of the transition metal elements (columns 3–12) are still in solid form. At 2000°C, many of the transition elements have begun to melt. There are only a few elements with melting points above 3000°C. Such elements as tungsten, platinum, molybdenum, and tantalum have exceptional high-temperature properties and are useful as *refractory metals*. Note that the element with highest melting point is carbon (in diamond form). Keep in mind that this is in an inert atmosphere. What should happen to this element in the presence of oxygen?

In addition, many elements are materials of construction. Aluminum and copper are just a few examples of elements that are used extensively for fabricating mechanical parts. Elements have special electrical characteristics, too. Silver and gold are used not just for jewelry but also for a wide variety of electrical components. We will visit all these topics in the course of this textbook.

1.0.2 Trends in the Periodic Table

A broader look at the periodic table points out some interesting trends. These trends not only help us predict how one element might perform relative to another but also give us some insight into the important properties of atoms and ions that determine their performance. For example, further examination of the melting points in Table 1.3 shows that there is a general trend to decrease melting point as we go down a column for the alkali metals and alkali earth elements (columns 1 and 2),

1 Note that the Lanthanide (atomic numbers 58–71) and Actinide (90–103) series elements, as well as the synthetic elements of atomic number greater than 87, are omitted from the tables in this text. With the exception of nuclear fuels such as uranium and plutonium, these elements are of little commercial interest.

Table 1.3 Electronic Structure and Melting Points of the Elements

	1	2	3	4	5	6	7	8	9	10	11	12	13	14	15	16	17	18
es *aw* *mp*	1 **H** $1s^1$ 1.008 −259		*es* = Electronic structure *aw* = Atomic weight (average including isotopes) *mp* = Melting point (°C) Darker shading indicates higher melting point.															2 **He** $1s^2$ 4.003 —
es *aw* *mp*	3 **Li** He$2s^1$ 6.94 180.5	4 **Be** He$2s^2$ 9.012 1287											5 **B** Be$2p^1$ 10.81 2077	6 **C** Be$2p^2$ 12.011 4440	7 **N** Be$2p^3$ 14.007 −210.0	8 **O** Be$2p^4$ 15.999 −218.8	9 **F** Be$2p^5$ 18.998 −219.7	10 **Ne** Be$2p^6$ 20.180 −249
es *aw* *mp*	11 **Na** Ne$3s^1$ 22.99 97.8	12 **Mg** Ne$3s^2$ 24.30 650											13 **Al** Mg$3p^1$ 26.98 660.3	14 **Si** Mg$3p^2$ 28.09 1414	15 **P** Mg$3p^3$ 30.974 44.2	16 **S** Mg$3p^4$ 32.06 115.2	17 **Cl** Mg$3p^5$ 35.45 −101.5	18 **Ar** Mg$3p^6$ 39.95 −189
es *aw* *mp*	19 **K** Ar$4s^1$ 39.098 63.5	20 **Ca** Ar$4s^2$ 40.08 840	21 **Sc** Ca$3d^1$ 44.96 1541	22 **Ti** Ca$3d^2$ 47.9 1670	23 **V** Ca$3d^3$ 50.94 1910	24 **Cr** K$3d^5$ 51.9 1907	25 **Mn** Ca$3d^5$ 54.93 1246	26 **Fe** Ca$3d^6$ 55.85 1538	27 **Co** Ca$3d^7$ 58.93 1495	28 **Ni** Ca$3d^8$ 58.71 1455	29 **Cu** K$3d^{10}$ 63.55 1084.6	30 **Zn** Ca$3d^{10}$ 65.37 419.5	31 **Ga** Ca$4p^1$ 69.72 29.8	32 **Ge** Ca$4p^2$ 72.59 938.3	33 **As** Ca$4p^3$ 74.92 817	34 **Se** Ca$4p^4$ 78.96 221	35 **Br** Ca$4p^5$ 79.90 −7.2	36 **Kr** Ca$4p^6$ 83.80 −157
es *aw* *mp*	37 **Rb** Kr$5s^1$ 85.47 39.3	38 **Sr** Kr$5s^2$ 87.62 777	39 **Y** Sr$4d^1$ 88.91 1522	40 **Zr** Sr$4d^2$ 91.22 1854	41 **Nb** Rb$4d^4$ 92.91 2477	42 **Mo** Rb$4d^5$ 95.94 2622	43 **Tc** Rb$4d^6$ 98.91 2157	44 **Ru** Rb$4d^7$ 101.7 2333	45 **Rh** Rb$4d^8$ 102.9 1963	46 **Pd** Kr$4d^{10}$ 106.4 1555	47 **Ag** Rb$4d^{10}$ 107.87 961.8	48 **Cd** Sr$4d^{10}$ 112.4 321.1	49 **In** Sr$5p^1$ 114.8 156.6	50 **Sn** Sr$5p^2$ 118.7 232.0	51 **Sb** Sr$5p^3$ 121.8 630.6	52 **Te** Sr$5p^4$ 127.6 449.5	53 **I** Sr$5p^5$ 126.9 113.7	54 **Xe** Sr$5p^6$ 131.3 −112
es *aw* *mp*	55 **Cs** Xe$6s^1$ 132.9 28.5	56 **Ba** Xe$6s^2$ 137.3 727	57 **La** Ba$5d^1$ 138.9 920	72 **Hf** Ba$5d^2$ 178.5 2233	73 **Ta** Ba$5d^3$ 180.9 3017	74 **W** Ba$5d^4$ 183.8 3414	75 **Re** Ba$5d^5$ 186.2 3185	76 **Os** Ba$5d^6$ 190.2 3033	77 **Ir** Xe$5d^9$ 192.2 2447	78 **Pt** Cs$5d^9$ 195.1 1768	79 **Au** Cs$5d^{10}$ 196.97 1064.2	80 **Hg** Ba$5d^{10}$ 200.6 −38.8	81 **Tl** Ba$6p^1$ 204.4 304	82 **Pb** Ba$6p^2$ 207.2 327.5	83 **Bi** Ba$6p^3$ 208.9 271.4	84 **Po** Ba$6p^4$ 210 254	85 **At** Ba$6p^5$ — 302	86 **Rn** Ba$6p^6$ 222 −71

Sources: IUPAC, 2018; Rumble [2].

but that the column trend for the transition metals (columns 3–12) appears to be different. There are some trends that are more uniform, however, and are related to the electronic structure of the element.

First Ionization Energy (IE)

The first *ionization energy*, abbreviated *IE*, is sometimes referred to as the "ionization potential." It is the energy required to remove the most weakly bound (usually outermost) electron from an isolated gaseous atom.

$$\text{atom (g)} + IE \rightarrow \text{positive ion (g)} + e^- \tag{1.1}$$

and can be calculated using the energy of the outermost electron as given by the Bohr model and Schrödinger's equation (in electron volts, eV):

$$IE = \frac{13.6Z^2}{n^2} \tag{1.2}$$

where Z is the effective nuclear charge and n is the principal quantum number. Typical values of *IE* (in kJ/mol) are listed in Table 1.4.

As shown in Figure 1.1, the general trend in the periodic table is for the IE to increase from bottom to top and from left to right (why?). A quantity related to the *IE* is the *work function*. The work function is the energy necessary to remove an electron from the metal surface in thermoelectric or photoelectric emission. We will describe this phenomenon in more detail in conjunction with electronic properties of materials in Chapter 6.

Take a Peek! Jump to Eq. (6.75) to see the definition of the work function.

Electron Affinity (EA)

Electron affinity is the reverse process to the IE; it is the energy change (often expressed in eV but also in kJ/mol) associated with an isolated gaseous atom accepting one electron:

$$\text{atom (g)} + e^- \rightarrow \text{negative ion (g)} \tag{1.3}$$

Unlike IE, *EA* can have either a negative or positive value, so it is not included in Eq. (1.3). The *EA* is positive if energy is released upon the formation of the negative ion. If energy is required, *EA* is negative. The general trend in the periodic table is again toward an increase in *EA* as we go from the bottom to top and left to right (Table 1.4, Figure 1.1), though this trend is much less uniform than for the *IE*.

Atomic and Ionic Radii

In general, positive ions are smaller than neutral atoms, while negative ions are larger (why?). The trend in ionic and atomic radii is opposite to that of *IE* and *EA* (Table 1.4, Figure 1.1); there is an increase in radius from top to bottom, right to left. In this case, the effective nuclear charge increases from left to right, the inner electrons cannot shield as effectively, and the outer electrons are drawn close to the nucleus, reducing the atomic radius. Note that the radii are only approximations because the orbitals, in theory, extend to infinity. We will discuss atomic and ionic radii in more detail when we get to chemical bonding.

Electronegativity

The IE and EA are characteristics of isolated atoms; they say very little about how two atoms will interact with each other. It would be nice to have an independent measure of the attraction an atom

Table 1.4 Ionization Energies, Electron Affinities, and Electronegativities of the Elements[a]

IE EA χ	1 **H** 1312 72.8 2.20																	2 **He** 2372 — —
IE EA χ	3 **Li** 520 59.6 0.98	4 **Be** 900 — 1.57											5 **B** 801 27.0 2.04	6 **C** 1087 122 2.55	7 **N** 1401 — 3.04	8 **O** 1314 141 3.44	9 **F** 1681 328 3.98	10 **Ne** 2081 — —
IE EA χ	11 **Na** 496 52.9 0.93	12 **Mg** 738 — 1.31											13 **Al** 578 41.8 1.61	14 **Si** 787 134 1.90	15 **P** 1012 72.0 2.19	16 **S** 1000 200 2.58	17 **Cl** 1251 349 3.16	18 **Ar** 1521 — —
IE EA χ	19 **K** 419 48.4 0.82	20 **Ca** 590 2.37 1.00	21 **Sc** 633 18.1 1.36	22 **Ti** 659 7.29 1.54	23 **V** 651 50.7 1.63	24 **Cr** 653 65.2 1.66	25 **Mn** 717 — 1.55	26 **Fe** 763 14.7 1.83	27 **Co** 760 63.9 1.88	28 **Ni** 737 112 1.91	29 **Cu** 746 119 1.90	30 **Zn** 906 — 1.65	31 **Ga** 579 41.5 1.81	32 **Ge** 762 119 2.01	33 **As** 944 77.6 2.01	34 **Se** 941 195 2.55	35 **Br** 1140 325 2.96	36 **Kr** 1351 — —
IE EA χ	37 **Rb** 403 46.9 0.82	38 **Sr** 549 5.02 0.95	39 **Y** 600 29.6 1.22	40 **Zr** 640 41.1 1.33	41 **Nb** 652 88.5 1.6	42 **Mo** 684 72.1 2.16	43 **Tc** 687 53.1 2.1	44 **Ru** 710 101 2.28	45 **Rh** 720 110 2.2	46 **Pd** 804 54.2 2.20	47 **Ag** 731 126 1.93	48 **Cd** 868 — 1.69	49 **In** 558 39.0 1.78	50 **Sn** 709 107 1.96	51 **Sb** 831 101 2.05	52 **Te** 869 191 2.1	53 **I** 1008 295 2.66	54 **Xe** 1170 — 2.6
IE EA χ	55 **Cs** 376 45.5 0.79	56 **Ba** 503 14.0 0.89	57 **La** 538 53.1 1.10	72 **Hf** 659 17.2 1.3	73 **Ta** 728 31.1 1.5	74 **W** 759 78.8 1.7	75 **Re** 756 5.83 1.9	76 **Os** 814 104 2.2	77 **Ir** 865 151 2.2	78 **Pt** 864 205 2.2	79 **Au** 890 223 2.4	80 **Hg** 1007 — 1.9	81 **Tl** 589 36.4 1.8	82 **Pb** 716 34.4 1.8	83 **Bi** 703 90.9 1.9	84 **Po** 812 183 2.0	85 **At** 899 270 2.2	86 **Rn** 1037 — —

[a] Ionization energy (*IE*) and Electron affinities (*EA*) are expressed as kilojoules per mole (kJ/mol). 1 eV = 96.49 kJ/mol.
Source: Rumble [3].

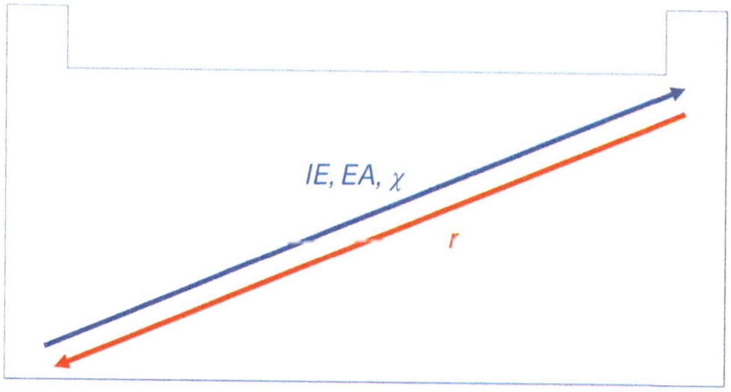

Figure 1.1 Some important trends in the periodic table for ionization energy (*IE*), electron affinity (*EA*), atomic and ionic radii (*r*), and electronegativity (χ). Increasing values are in the direction of the arrow.

has for electrons in a bond formed with another atom. *Electronegativity* is such a quantity. It is represented by the lowercase Greek letter "chi," χ. Its value can be calculated using one of the several methods discussed below. Values of χ are always relative to one another for a given method of calculation, and values from one method should not be used with values from another method.

Based upon a scale developed by Mulliken, electronegativity is the average of the IE and the EA:

$$\chi = \frac{IE + EA}{2} \tag{1.4}$$

There are other types of electronegativity scales as well, the most widely utilized of which is the one from the developer of the electronegativity concept, Linus Pauling:

$$\chi = \frac{0.31(n + 1 \pm c)}{r} + 0.5 \tag{1.5}$$

Historical Highlight
Learn more about the interesting and sometimes controversial contributions to science of two-time Nobel Prize winner ***Linus Pauling*** at https://en.wikipedia.org/wiki/Linus_Pauling

where n is the number of valence electrons, c is any formal valence charge on the atom and the sign corresponding to it, and r is the covalent radius. As with *IE* and *EA*, electronegativity values tend to increase from top to bottom and left to right in the periodic table (Table 1.4, Figure 1.1). This should make sense to you given the functional dependence shown in Eqs. (1.4) and (1.5). We will use the concept of electronegativity to discuss chemical bonding.

1.0.3 Types of Bonds

Electronegativity is a very useful quantity to help categorize bonds, because it provides a measure of the *excess binding energy* between atoms A and B, $\Delta_{A\text{-}B}$ (in kJ/mol):

$$\Delta_{A-B} = 96.5(\chi_A - \chi_B)^2 \tag{1.6}$$

The excess binding energy, in turn, is related to a measurable quantity, namely the *bond dissociation energy* between two atoms, DE_{ij}:

$$\Delta_{A-B} = DE_{AB} - [(DE_{AA})(DE_{BB})]^{1/2} \tag{1.7}$$

The bond dissociation energy is the energy required to separate two bonded atoms (see Table 1.5 for typical values). The greater the electronegativity difference, the greater the excess binding energy. These quantities give us a method for characterizing bond types. More importantly, they relate to important physical properties, such as melting point (see Table 1.6). First, let us review the bond types and characteristics, then describe each in more detail.

Primary Bonds

Primary bonds, also known as "strong bonds," are created when there is direct interaction of electrons between two or more atoms, either through transfer or from sharing. The more electrons per atom that participate, the higher the bond "order" (e.g., single, double, or triple bond) and the stronger the connection between atoms. There are four general categories of primary bonds: *ionic, covalent, polar covalent*, and *metallic*. An ionic bond, also called a *heteropolar bond*, results when electrons are transferred from the more electropositive atom to the more electronegative atom, as in sodium chloride, NaCl. Ionic bonds usually result when the electronegativity difference between two atoms in a diatomic molecule is greater than about 2.0. Because of the large discrepancy in electronegativities, one atom will generally gain an electron, while the other atom in a diatomic molecule will lose an electron. Both atoms tend to be "satisfied" with this arrangement because they often end up with noble gas electron configurations—that is, full electronic orbitals. The classic example of an ionic bond is NaCl, but CaF_2 and MgO are also examples of molecules in which ionic bonding dominates.

A covalent bond, or *homopolar bond*, arises when electrons are shared between two atoms (e.g., H—H). This means that a binding electron in a covalent diatomic molecule such as H_2 has equal likelihood of being found around either hydrogen atom. Covalent bonds are typically found

Table 1.5 Bond Dissociation Energy Values for Selected Single Bonds (in kJ/mol)

Bond	Energy	Bond	Energy	Bond	Energy	Bond	Energy
H—H	435.78	Al—Al	104.5	H—O	429.7	N—O	630.6
He—He	3.8	Si—Si	310	H—N	358.8	N—S	467
Li—Li	105.0	P—P	489.1	H—S	353.4	O—P	589
Be—Be	59	S—S	430.0	H—Si	293.3	O—S	517.9
B—B	290	Cl—Cl	242.9	C—H	338.7	O—Si	799.6
C—C	605.03	Ar—Ar	4.9	C—O	1076.6	Al—O	501.9
N—N	944.9	Ti—Ti	117.6	C—N	749.3	Al—N	368
O—O	498.5	Fe—Fe	118	C—S	713.3	S—Si	617
P—P	489.1	Ni—Ni	204	C—Si	447	P—S	442
F—F	158.7	Cu—Cu	182	C—P	507.5	Fe—H	148
Ne—Ne	4.1	Ge—Ge	264.4	C—Cl	394.9	Fe—C	385.9
Na—Na	74.8	Ag—Ag	162.9	C—F	513.8	Fe—O	407
Mg—Mg	11.3	Au—Au	226.2	C—I	253.1	Fe—S	316.3

Source: Rumble [4].

Table 1.6 Energies of Selected Types of Interatomic Bonding

Type of Bond	Example	Bond Energy[a] (kJ/mol)	Melting Point (°C)	Characteristics
Ionic		400–4000		Low electrical conductivity, transparent, brittle, high melting point
	LiF	577	848	
	CuF_2	539	836	
	Al_2O_3	1675	2053	
Covalent		100–1000		Low electrical conductivity, very hard, very high melting point
	GaAs	203	1238	
	SiC	447	2830	
	Diamond	605	4440	
Metallic		70–850		High electrical and thermal conductivity, easily deformable, opaque
	Al	140	660	
	Cu	182	1085	
	W	666	3414	
van der Waals		0.4–4.0		Weak binding, low melting and boiling points, very compressible
	Ar --- Ar	4.9	−189	
	Kr --- Kr	5.4	−157	
Hydrogen bonding		10–40		Higher melting point than van der Waals bonding, tendency to form groups of many molecules
	H_2O --- H_2O	21	0	
	NH_3 --- NH_3	13	−77.7	

[a] Bond energy values can vary greatly; ranges given are typical. For primary bonds, the bond energy is the bond dissociation energy. For secondary bonds, the bond energy is an interatomic or intermolecular energy.
Source: Rumble [4].

in homonuclear diatomics, such as O_2 and N_2, though the atoms need not be the same to have similar electronegativities. Electronegativity differences of less than about 0.4 characterize covalent bonds. For two atoms with an electronegativity difference of between 0.4 and 2.0, a polar covalent bond is formed—one that is neither truly ionic nor totally covalent. An example of a polar covalent bond can be found in the molecule hydrogen fluoride, HF. Though there is significant sharing of the electrons, some charge distribution exists that results in a polar or *partial ionic character* to the bond. The percent ionic character of the bond can again be related to the electronegativities of the individual atoms:

$$\%\,\text{ionic character} = 100\{1 - \exp[-0.25(\chi_A - \chi_B)^2]\} \tag{1.8}$$

Example Problem 1.1

What is the percent ionic character of the Al—O bond in Al_2O_3?

Answer: According to Table 1.4, the electronegativity of aluminum is 1.61 and that of oxygen is 3.44. Putting these values into Eq. (1.8) gives

$$\%\,\text{ionic character } Al_2O_3 = 100\left[1 - exp\left\{-0.25(1.61 - 3.44)^2\right\}\right] = 57\%$$

So, the classification of Al_2O_3 as an ionic compound in Table 1.6 is only partly appropriate.

Take a Peek! Jump to Section 6.0.1 to see how electronic bands determine electrical properties of materials.

The larger the electronegativity difference, the more ionic character the bond has. Of course, if the electronegativity difference is greater than about 2.0, we know that an ionic bond should result.

Finally, a special type of primary bond known as a metallic bond is found in an assembly of homonuclear atoms, such as copper or sodium. Here the bonding electrons become "decentralized" and are shared by the core of positive nuclei. Metallic bonds occur when elements of low electronegativity (usually found in the lower left region of the periodic table) bond with each other to form a class of materials we call *metals*. Metals tend to have common characteristics such as ductility, luster, and high thermal and electrical conductivity. These characteristics can to some degree be accounted for by the nature of the metallic bond. The model of a metallic bond—first proposed by Lorentz—consists of an assembly of positively charged ion cores surrounded by free electrons, or an "electron gas." We will see later, when we describe intermolecular forces and bonding, that the electron cloud does indeed have "structure" in the quantum mechanical sense, which accounts nicely for the observed electrical properties of these materials.

Historical Highlight

Learn more about Dutch physicist ***Johannes Diderik van der Waals*** at: https://en.wikipedia.org/wiki/Johannes_Diderik_van_der_Waals.

Secondary Bonds

Secondary bonds, or *weak bonds*, occur due to indirect interaction of electrons in adjacent atoms or molecules. There are three main types of secondary bonding: *hydrogen bonding, dipole–dipole interactions*, and *van der Waals forces*. The latter, named after the famous Dutch physicist who first described them, arise due to momentary electric *dipoles* (regions of positive and negative charge) that can occur in all atoms and molecules due to statistical variations in the charge density. These intermolecular forces are common, but very weak, and are found in inert gases where other types of bonding do not exist.

Hydrogen bonding is the attraction between hydrogen in a highly polar molecule and the electronegative atom in another polar molecule. In the water molecule, oxygen draws much of the electron density around it, creating positively charged centers at the two hydrogen atoms. These positively charged hydrogen atoms can interact with the negative center around the oxygen in adjacent water molecules. Although this type of bonding is of the same order of magnitude in strength as van der Waals bonding, it can have a profound influence on the properties of a material, such as boiling and melting points. In addition to having important chemical and physical implications, hydrogen bonding plays an important role in many biological and environmental phenomena. It is responsible for causing ice to be less dense than water (how many other substances do you know that are less dense in the solid state than in the liquid state?), an occurrence that allows fish to survive at the bottom of frozen lakes.

Finally, some molecules possess permanent charge separations, or dipoles, such as those found in water. The general case for the interaction of any positive dipole with a negative dipole is called dipole–dipole interaction. Hydrogen bonding can be thought of as a specific type of dipole–dipole interaction. A dipolar molecule like ammonia, NH_3, can dissolve other polar molecules like water due to dipole–dipole interactions. In the case of NaCl in water, the dipole–dipole interactions are so strong as to break the intermolecular forces within the molecular solid.

Now that the types of bonds have been reviewed, we will concentrate on the primary bond because it correlates more directly with physical properties in solids than do secondary bonds. Be aware that secondary forces exist, though, and that they play a larger role in liquids and gases than in solids.

Need More Help?
Try these videos on types of bonds: https://www.youtube.com/watch?v=p9MA6Od-zBA https://www.youtube.com/watch?v=5whtIJHW_VQ

1.0.4 Intermolecular Forces and Bonding

We have described the different types of primary bonds, but how do these bonds form in the first place? What is it that causes a sodium ion and a chloride ion to form a compound, and what is it that prevents the nuclei from fusing together to form one element? These questions all lead to the topics of intermolecular forces and bond formation. We know that atoms approach each other only to a certain distance, and then, if they form a compound, they will maintain some equilibrium separation distance known as the *bond length*. Hence, we expect that there is some *attractive energy* that brings them together, as well as some *repulsive energy* that keeps the atoms a certain distance apart.

Also known as *chemical affinity*, the attractive energy between atoms is what causes them to approach each other. This attraction is due to the electrostatic force between the nucleus and electron clouds of the separate atoms. It should make sense to you that the attractive energy (U_A) is inversely proportional to the separation distance, r; that is, the further the atoms are apart, the weaker the attraction:

$$U_A = -\frac{a}{r^m} \tag{1.9}$$

where a is a constant that we will describe in more detail in a moment, and m is a constant with a value of 1 for ions and 6 for molecules. Notice that there is a negative sign in Eq. (1.9). By convention, we will refer to the attractive energy as a "negative energy."

Once the atoms begin to approach each other, they can only come so close together due to the impenetrability of matter. The result is a repulsive energy, which we assign a positive value to, again, by convention. The primary constituents of this repulsive energy are nucleus–nucleus and electron–electron repulsions. As with the attractive energy, the repulsive energy is inversely proportional to the separation distance; the closer the atoms are, the more they repel each other:

$$U_R = \frac{b}{r^n} \tag{1.10}$$

where b and n are constants. The value of n, called the *repulsion exponent*, depends on the outer core configuration of the atom. Values of the repulsion exponent are given in Table 1.7.

The total or potential energy of the system is then the sum of the attractive and repulsive components:

$$U = U_A + U_R = \frac{-a}{r^m} + \frac{b}{r^n} \tag{1.11}$$

The result is the *potential energy well* (see Figure 1.2). The well-known Lennard-Jones potential

$$U = -\frac{a}{r^6} + \frac{b}{r^{12}} \tag{1.12}$$

Table 1.7 Values of the Repulsion Exponent

Noble Gas Ion Core	Outer Core Configuration	n
He	$1s^2$	5
Ne	$2s^22p^6$	7
Ar	$3s^23p^6$	9
Kr	$3d^{10}4s^24p^6$	10
Xe	$4d^{10}5s^25p^6$	12

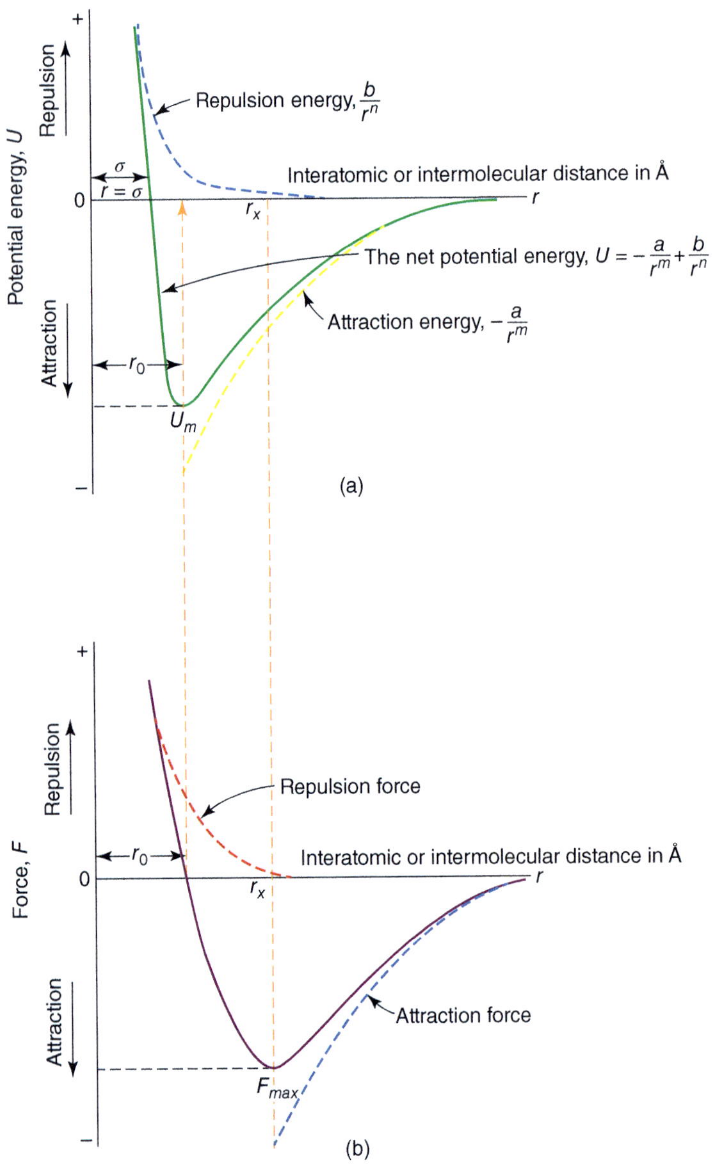

Figure 1.2 The interatomic (a) potential energy and (b) force diagrams. From Jastrzebski [5]. Copyright © 1976 by John Wiley & Sons, Inc. This material is used by permission of John Wiley & Sons, Inc.

is a common potential energy function used in numerous models, including collision theory for kinetics. It is simply a special case of Eq. (1.11) with $n = 12$ (Xe configuration) and $m = 6$ (molecules).

It is useful to know the forces involved in bonding as well as the energy. Recall that energy and force, F, are related by

$$F = -\frac{dU}{dr} \tag{1.13}$$

We will see that we can use this expression to convert between force and energy for specific types of atoms and molecules (specific values of n and m). For now, this expression helps us find the *equilibrium bond distance*, r_0, which occurs when forces are equal (the sum of attractive and repulsive forces is zero) or at *minimum potential energy* (take the derivative and set it equal to zero):

$$F = -\frac{dU}{dr} = 0, \quad \text{at } r = r_0 \tag{1.14}$$

This is not the same thing as the *maximum attractive force,* which we get by maximizing F:

$$F_{\max} = \frac{dF}{dr} = -\frac{d^2U}{dr^2} = 0 \tag{1.15}$$

The forces are equal when the potential energy is at a minimum and the separation distance is at the bond length, r_0. Differentiation of Eq. (1.11) and solving for r_0 in terms of a, b, n, and m gives

$$r_0 = \left(\frac{nb}{ma}\right)^{\frac{1}{n-m}} \tag{1.16}$$

The potential energy well concept is an important one, not only for the calculation of binding energies—as we will see in a moment—but for many important physical properties of materials. How tightly atoms are bound together in a compound has a direct impact on such properties as melting point, elastic modulus, and thermal expansion coefficient. Figure 1.3 shows a qualitative comparison of a material that has a deep and narrow potential energy well versus one in which

Take a Peek! Jump ahead to Section 5.1.1 to see how potential energy relates to modulus.

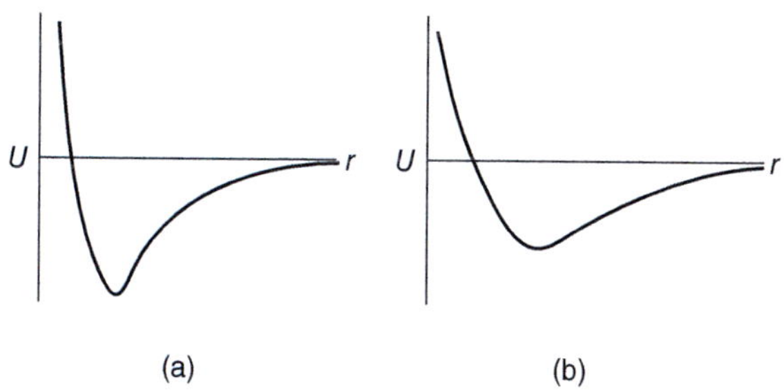

Figure 1.3 Schematic representation of the relationship between the shape of the potential energy well and selected physical properties. Materials with a deep well (a) have a high melting point, high elastic modulus, and low thermal expansion coefficient. Those with a shallow well (b) have a low melting point, low elastic modulus, and high thermal expansion coefficient. Adapted from Barrett et al. [6]. Copyright © 1973 by Prentice-Hall, Inc.

the potential energy well is wide and shallow. The deeper well represents stronger interatomic attraction; it is more difficult to melt these substances, which have correspondingly large elastic moduli and low thermal expansion coefficients.

Cooperative Learning Exercise 1.1

Work with a neighbor. Consider the Lennard-Jones potential, as given by Eq. (1.12), for which $m = 6$ and $n = 12$. You wish to determine the separation distance, r, at which the maximum force occurs, F_{max}, in terms of the equilibrium bond distance, r_0.

Person 1: Use Eq. (1.16) with the values of m and n of the Lennard-Jones potential to solve for the constant a in terms of b and the equilibrium bond distance, r_0. Now perform the determination of F_{max} as given by Eq. (1.15): substitute this value of a back into Eq. (1.12), differentiate it twice with respect to r (remember that r_0 is a constant), and set this equal to zero (determine, then maximize the force function). Solve this equation for r in terms of r_0. The other constant should drop out.

Person 2: Use Eq. (1.16) with the values of m and n of the Lennard-Jones potential to solve for the constant b in terms of a and the equilibrium bond distance, r_0. Now perform the determination of F_{max} as given by Eq. (1.15); substitute this value of b back into Eq. (1.12), differentiate it twice with respect to r (remember that r_0 is a constant), and set this equal to zero (determine, then maximize the force function). Solve this equation for r in terms of r_0. The other constant should drop out.

Compare your answers. You should both get the same result.

Answer: $r = 1.1087r_0$

Eqs. (1.12) and (1.13) and the potential energy well also form the basis for a rapidly growing area of materials engineering research called *molecular dynamics simulations* or *MD simulations*. In theory, if you knew the positions and velocities of all the atoms in an assembly, you could write a computer code to apply the equations of motion (in the simplest case, Newton's First Law $F = m \cdot a$, where a is acceleration or the second derivative of the position), minimize the energy function, and calculate a wide range of physical properties such as viscosity and diffusion coefficient, and even the spectroscopic fingerprint of the assembly. We will see the power of MD simulations as the assemblies not only become large but also when atoms can be replaced by larger structural units like monomers in a polymer or amino acids in proteins. Doing so allows us to simulate—even predict—the rheological properties of polymers and the conformation of folded proteins. For now, however, we continue to use these two equations to form bonds and build molecules from atoms.

Take a Peek! Jump ahead to Section 2.5.3 to see how MD simulations can be used to predict protein folding.

The Ionic Bond

To form an ionic bond, we must account for the complete transfer of electrons from one atom to the other. The easiest approach is to first transfer the electrons to form *ions*, then bring the two ions together to form a bond. Sodium chloride is a simple example that allows us to obtain both the bond energy and equilibrium bond distance using the potential energy approach. In this case, the potential energy, U, is not only the sum of the attractive and repulsive energies, U_A and U_R, respectively, but must also take into account the energy required to form ions from sodium and chlorine atoms, ΔE_{ions}. So, our energy expression looks like:

$$U = U_A + U_R + \Delta E_{ions} \tag{1.17}$$

which at the equilibrium bond distance gives the equilibrium potential energy, U_0:

$$U_0 = U_{A,0} + U_{R,0} + \Delta E_{ions} \tag{1.18}$$

Let us examine each of the three energies in Eq. (1.18) individually.

Energy is required to form an ion of Na^+ from elemental sodium. From Section 1.0.2, we already know that this process of removing an electron from an isolated atom is called the *IE*, which for sodium is 496 kJ/mol. Similarly, energy is given off when Cl^- is formed by adding an electron to an isolated gaseous atom of chlorine. This is the *EA* (see Section 1.0.2), which for chlorine is −349 kJ/mol. So, the energy required to form ions is given by:

$$\Delta E_{ions} = IE_{\text{Na}} + EA_{\text{Cl}} = 496 - 349 = 147\ \text{kJ/mol} \tag{1.19}$$

For a diatomic molecule, the attraction between the two ions is strictly due to opposite charges, so the attractive force is given by *Coulomb's Law*:

$$F_A = (Z_1 e \times Z_2 e)/\left(4\pi\varepsilon_0 r^2\right) \tag{1.20}$$

where ε_0 is a constant called the *electric permittivity* ($8.854 \times 10^{-12}\ \text{C}^2/\text{N}\cdot\text{m}^2$), e is the charge of an electron (1.6×10^{-19} C), Z is respective *number of charges* of positive and negative ions ($Z_1 = Z_2 = 1$), and r is the separation distance between ions in meters. (We will learn more about the electric permittivity in Chapter 6.) Substituting the values of Z for sodium and chloride ions gives

$$F_A = +e^2/\left(4\pi\varepsilon_0 r^2\right) \tag{1.21}$$

Energy is released by bringing the ions from infinite separation to their equilibrium separation distance, r_0. Recall that energy and force are related to one another by Eq. (1.13), so that the equilibrium attractive energy, $U_{A,0}$, can be found by integrating F_A:

$$U_{A,0} = -\int_{\infty}^{r_0} F_A\, dr = e^2/(4\pi\varepsilon_0 r_0) \tag{1.22}$$

Note the similarity in form of this expression for the attractive energy with that of Eq. (1.9). The exponent on r is 1, as it should be for ions ($m = 1$), and the other parameters can be grouped to form the constant, a. Recall that by definition, the attractive energy is a negative energy, so we will end up inserting a minus sign in front of the expression in Eq. (1.22) to match the form shown in Eq. (1.9).

The repulsive energy can be derived simply by using the general expression given in Eq. (1.10) and solving for the constant b by minimizing the potential energy function [Eq. (1.11)] with a knowledge of the constant a from Eqs. (1.22) and (1.9) (you should try this) to obtain:

$$U_{R,0} = e^2/(4\pi\varepsilon_0 n r_0) \tag{1.23}$$

where e and ε_0 are the same as for the attractive force, r_0 is again the equilibrium separation distance, and n is the repulsion exponent.

Inserting $U_{A,0}$ and $U_{R,0}$ into the main energy expression, Eq. (1.18), (recall that the attractive energy must be negative) gives us the equilibrium potential energy, U_0:

$$U_0 = \frac{-e^2}{4\pi\varepsilon_0 r_0} + \frac{e^2}{4\pi\varepsilon_0 n r_0} + \Delta E_{ions} \tag{1.24}$$

and simplifying gives:

$$U_0 = \left(1 - \frac{1}{n}\right)\left(\frac{-e^2}{4\pi\varepsilon_0 r_0}\right) + \Delta E_{ions} \tag{1.25}$$

We solved for the equilibrium bond distance, r_0, in Eq. (1.16), and the constants a and b have, in effect, just been evaluated. Inserting these values into Eq. (1.25), along with Eq. (1.19) and using $n = 8$ (why?), gives:

$$\begin{aligned} U_0 &= \left(1 - \frac{1}{8}\right)\left[\frac{-1(1.6 \times 10^{-19}\text{C})^2(6.02 \times 10^{23}\text{mol}^{-1})(10^{-3}\text{kJ/N} \bullet \text{m})}{4\pi(8.854 \times 10^{-12}\text{C}^2/\text{N} \cdot \text{m}^2)(2.36 \times 10^{-10}\text{m})}\right] + 147 \\ &= -367\,\text{kJ/mol} \end{aligned}$$

Similarly, we can calculate bond energies for any type of bond we wish to create. Refer to Table 1.5 for more single-bond energy values.

When we have an ordered assembly of atoms called a *lattice*, there is more than one bond per atom, and we must consider interactions with adjacent atoms that result in an increased interionic spacing compared to an isolated atom. We do this with the *Madelung constant*, α_M. This parameter depends on the structure of the ionic crystal, the charge on the ions, and the relative size of the ions. The Madelung constant fits directly into the energy expression (Eq. 1.25):

$$U_L = \left(\alpha_M - \frac{1}{n}\right)\left(\frac{-e^2}{4\pi\varepsilon_0 r_0}\right) + \Delta E_{ions} \tag{1.26}$$

> **Need More Help?**
>
> Try this video on bond energies:
>
> https://www.youtube.com/watch?v=U43-NTF-79E
> Cannot find a specific bond energy? Your library probably has free access to the CRC Handbook of Chemistry and Physics which contains over 1000 single-bond energy values.

For NaCl, $\alpha_M = 1.75$ so the *lattice energy*, U_L, is −807 kJ/mol. Typical values of the Madelung constant are given in Table 1.8 for different crystal structures (see Section 1.1.1). In general, the lattice energy increases (becomes more negative) with decreasing interionic distance for ions with the same charge. This increase in lattice energy translates directly into an increased melting point. For example, if we replace the chlorine in sodium chloride with other halogens, while retaining the cubic structure, the interionic spacing should change, as well as the melting point. We could also account for additional van der Waals interactions, but this effect is relatively small in lattices.

The Covalent Bond

Recall that covalent bonding results when electrons are "shared" by similar atoms. The simplest example is that of a hydrogen molecule, H_2. We begin by using *molecular orbital theory* to represent the bonding. Two atomic orbitals (1*s*) overlap to form two *molecular orbitals* (MOs), represented by σ: one *bonding orbital* ($\sigma 1s$), and one *antibonding orbital* ($\sigma^* 1s$), where the asterisk superscript

Table 1.8 Typical Values for the Madelung Constant

Compound	Crystal Lattice (see Section 1.1.1)	α_M
NaCl	FCC	1.74756
CsCl	CsCl	1.76267
CaF_2	Cubic	2.51939
$CdCl_2$	Hexagonal	2.244
MgF_2	Tetragonal	2.381
ZnS (wurtzite)	Hexagonal	1.64132
TiO_2 (rutile)	Tetragonal	2.408
β-SiO_2	Hexagonal	2.2197

indicates antibonding. The antibonding orbitals are higher in energy than corresponding bonding orbitals.

The shapes of the electron cloud densities for various MOs are shown in Figure 1.4. The overlap of two *s* orbitals results in one σ-bonding orbital and one σ-antibonding orbital. When two *p* orbitals overlap in an end-to-end fashion, as in Figure 1.4b, they are interacting in a manner similar to *s*–*s* overlap, so one σ-bonding orbital and one σ-antibonding orbital once again are the result. Note that

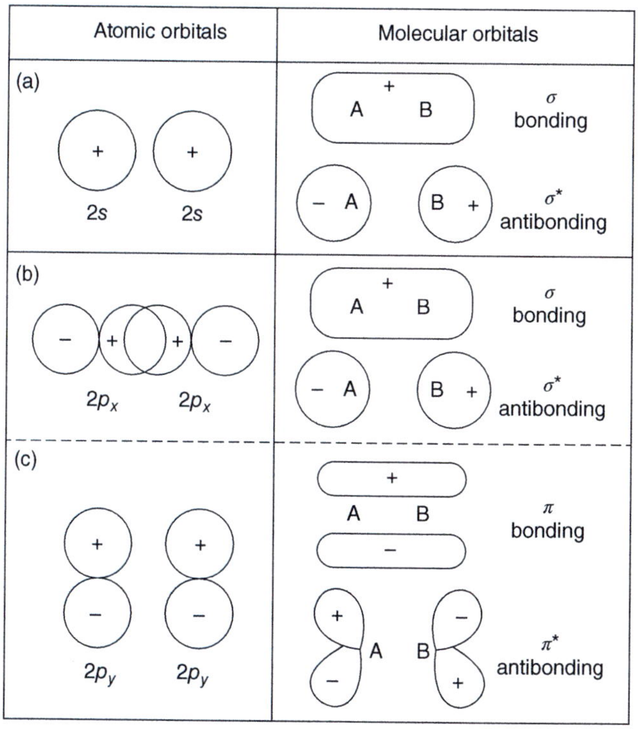

Figure 1.4 The shape of molecular orbitals formed from the overlap of (a) $2s$–$2s$, (b) $2p_x$–$2p_x$, and (c) $2p_y$–$2p_y$ atomic orbitals. From Ralls et al. [7]. Copyright © 1976 by John Wiley & Sons, Inc. This material is used by permission of John Wiley & Sons, Inc.

all σ orbitals are symmetric about a plane between the two atoms. Side-to-side overlap of p orbitals results in one π-bonding orbital and one π-antibonding orbital. There are a total of four π orbitals: two for p_x and two for p_y. Note that there is one more *node* (region of zero electron density) in an antibonding orbital than in the corresponding bonding orbital. This is what makes them higher in energy.

As in the case of ionic bonding, we use a potential energy diagram to show how orbitals form as atoms approach each other, as shown in Figure 1.5. The electrons from the isolated atoms are then placed in the MOs from bottom to top. *As long as the number of bonding electrons is greater than the number of antibonding electrons, the molecule is stable*. For atoms with p and d orbitals, diagrams become more complex, but the principles are the same. In all cases, there are the same number of molecular orbitals as atomic orbitals. Be aware that there is some change in the relative energies of the π and σ orbitals as we go down the periodic chart, particularly around O_2. As a result, you might see diagrams that have the $\pi 2p$ orbitals lower in energy than the $\sigma 2p$. Do not let this confuse you if you see some variation in the order of these orbitals in other texts or references. For our purposes, it will not affect whether the molecule is stable or not.

Let us use molecular oxygen, O_2, as an example. As shown in Figure 1.6, each oxygen atom brings six outer-core electrons to the molecular orbitals. (Note that the $1s$ orbitals are not involved in bonding and are thus not shown. They could be shown on the diagram but would be at a very low relative energy at the bottom of the diagram.). The 12 total electrons in the molecule are placed in the MOs from bottom to top; according to *Hund's rule*, the last two electrons must be placed in separate π^*2p orbitals before they can be paired.

The pairing of electrons in the MOs can manifest itself in certain physical properties of the molecule. *Paramagnetism* results when there are unpaired electrons in the molecular orbitals. Paramagnetic molecules magnetize in magnetic fields due to the alignment of unpaired electrons. *Diamagnetism* occurs when there are all paired electrons in the MOs. We will revisit these properties in Chapter 6.

We can use molecular orbital theory to explain simple heteronuclear diatomic molecules as well. A molecule such as hydrogen fluoride, HF, has molecular orbitals, but we must remember that the atomic orbitals of the isolated atoms have much different energies from each other to begin with.

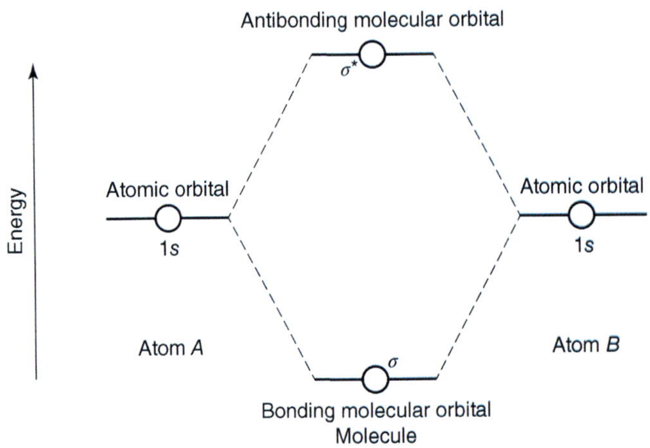

Figure 1.5 Molecular orbital diagram for the hydrogen molecule, H_2. Reproduced, by permission, from Dickerson et al. [8]. Copyright © 1979 by Pearson Education, Inc.

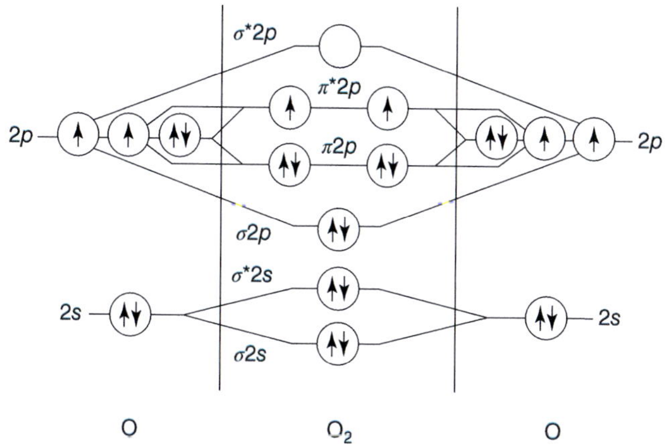

Figure 1.6 Molecular orbital diagram for molecular oxygen, O_2. From Ralls et al. [7]. Copyright © 1976 by John Wiley & Sons, Inc. This material is used by permission of John Wiley & Sons, Inc.

How do we know where these energies are relative to one another? Look back at the ionization energies in Table 1.4, and you see that the first IE for hydrogen is 1312 kJ/mol, whereas for fluorine it is 1681 kJ/mol. This means that the outer-shell electrons have energies of −1312 (1*s* electron) and−1681 kJ/mol (2*p* electron), respectively. So, the electrons in fluorine are more stable (as we would expect for an atom with a much larger nucleus relative to hydrogen), and we can construct a relative molecular energy diagram for HF (see Figure 1.7). This is a case where

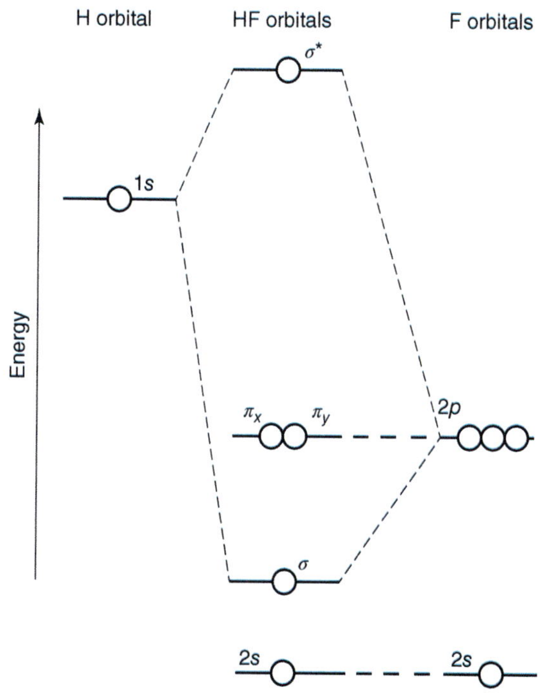

Figure 1.7 Molecular orbital diagram for HF. Reproduced, by permission, from Dickerson et al. [9]. Copyright © 1979 by Pearson Education, Inc.

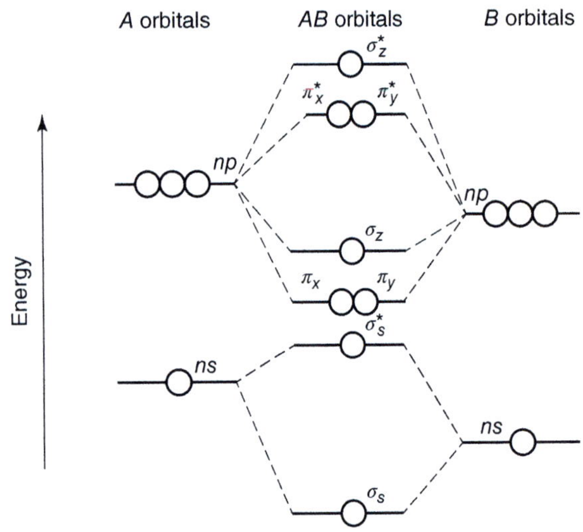

Figure 1.8 Molecular orbital diagram for the general case of a diatomic molecule AB, where B is more electronegative than A. Reproduced, by permission, from Dickerson et al. [10]. Copyright © 1979 by Pearson Education, Inc.

the electronegativity of the atoms is useful. It qualitatively describes the relative energies of the atomic orbitals and the shape of the resulting MOs. The molecular energy level diagram for the general case of molecule AB where B is more electronegative than A is shown in Figure 1.8, and the corresponding molecular orbitals are shown in Figure 1.9. In Figure 1.8, note how the B atomic orbitals are lower in energy than those of atom A. In Figure 1.9, note how the number of nodes increases from bonding to antibonding orbitals and how the electron probability is greatest around the more electronegative atom.

Example Problem 1.2

Construct the molecular orbital diagram for diatomic nitrogen, N_2. Is this molecule paramagnetic or diamagnetic?

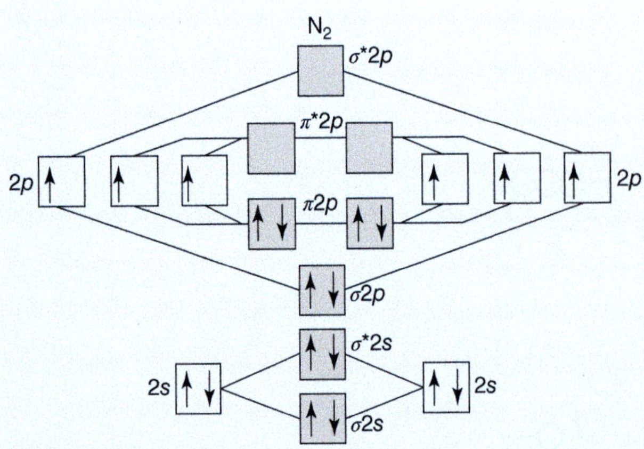

Answer: Diatomic nitrogen has all paired electrons, so it is diamagnetic.

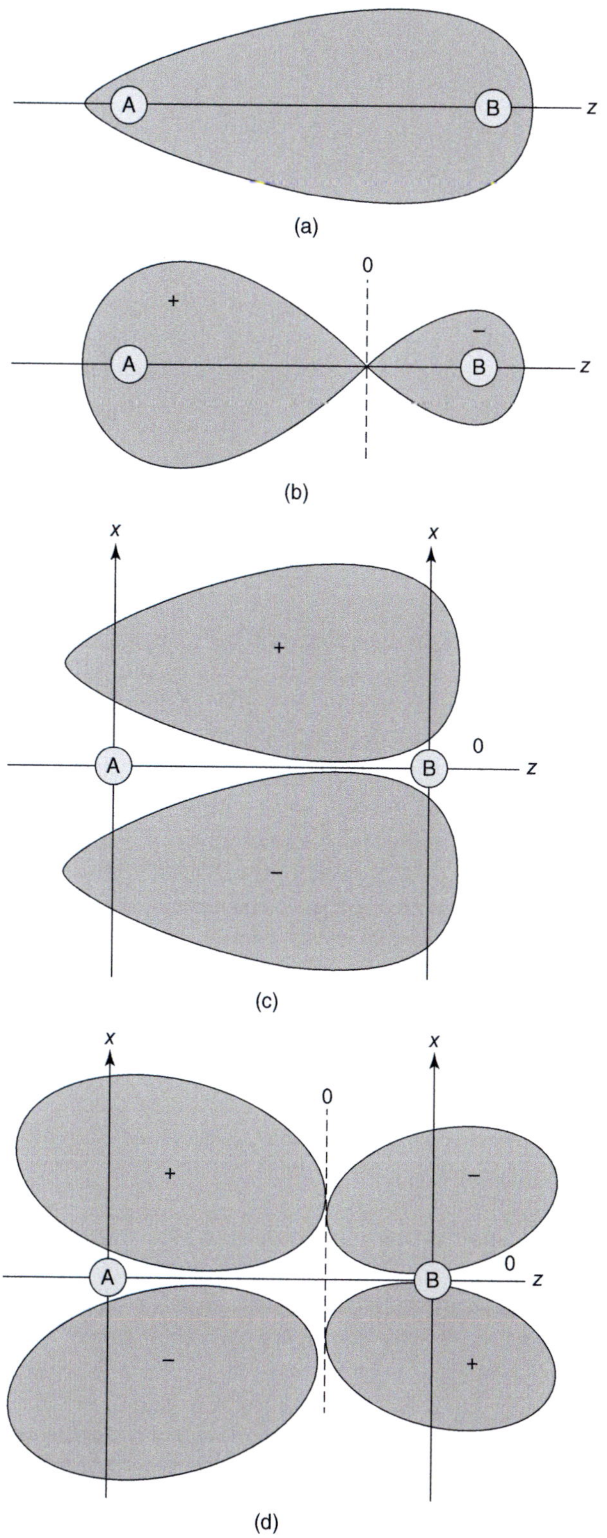

Figure 1.9 The shape of selected molecular orbitals for the diatomic molecule AB, where B is more electronegative than A: (a) σ, (b) σ^*, (c) π, and (d) π^*.

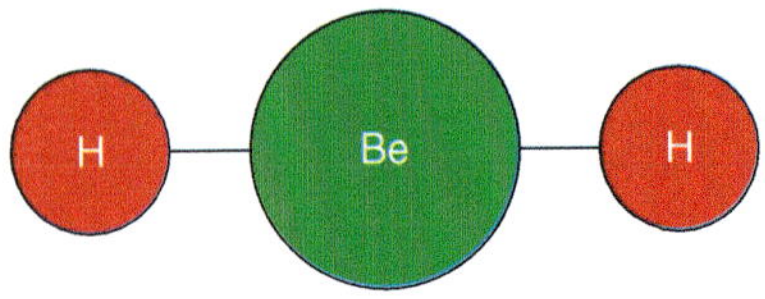

Figure 1.10 The linear structure of BeH_2.

Molecular orbitals do not explain everything and become increasingly more difficult to draw with more than two atoms. We use a model called *hybridization* to explain other effects, particularly in carbon compounds. Hybridization is a "mixing" of atomic orbitals to create new orbitals that have a geometry better suited to a particular type of bonding environment. For example, in the formation of the compound BeH_2, we aim to explain why this molecule is linear; that is, the H—Be—H bond is 180°. The hydrogen atoms only have one electron each to donate—both in their respective 1*s* orbitals—but beryllium has two electrons in the 2*s* orbital, and because its principal quantum number is 2 also has 2*p* orbitals, even though they are empty.

The goal is to make two equivalent orbitals in Be out of the atomic orbitals so that each hydrogen will see essentially the same electronic environment. We can accomplish this by "mixing" the 2*s* orbital and one of the empty 2*p* orbitals (say, the $2p_z$) to form two equivalent orbitals we call "*sp*" hybrids, since they have both *s* and *p* characteristics. As with molecular orbital theory, we must end up with the same number of orbitals we started with. The bonding lobes on the new sp_a and sp_b orbitals on Be are 180° apart, just as we need to form BeH_2. In this manner, we can mix any type of orbital we wish to come up with specific bond angles and numbers of equivalent orbitals. The most common combinations are *sp*, sp^2, and sp^3 hybrids. In *sp* hybrids, one *s* and one *p* orbital are mixed to get two *sp* orbitals, both of which are 180° apart. A linear molecule results, e.g., BeH_2, as shown in Figure 1.10. In sp^2 hybrids, one *s* and two *p* orbitals are mixed to obtain three sp^2 orbitals. Each orbital has 1/3 *s* and 2/3 *p* characteristics. A trigonal planar orbital arrangement results with 120° bond angles. An example of a trigonal planar molecule is BF_3, as shown in Figure 1.11. Finally, in sp^3 hybrids, when one *s* and three *p* orbitals are mixed, four sp^3 orbitals result, each having 1/4 *s* and 3/4 *p* characteristics. The tetrahedral arrangement of orbitals creates a 109.5° bond angle, as is found in methane, CH_4 (Figure 1.12).

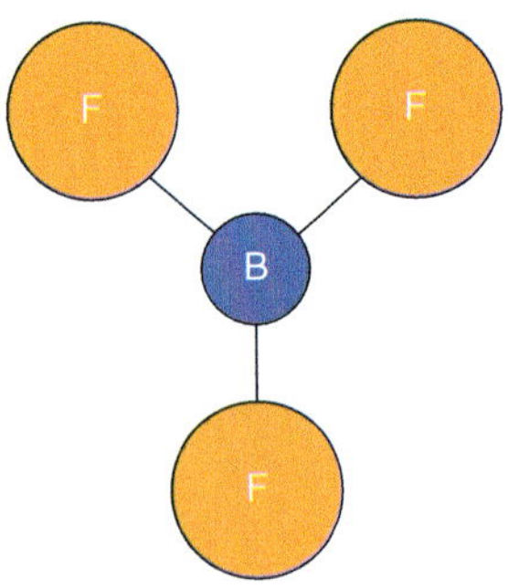

Figure 1.11 The trigonal planar structure of BF_3.

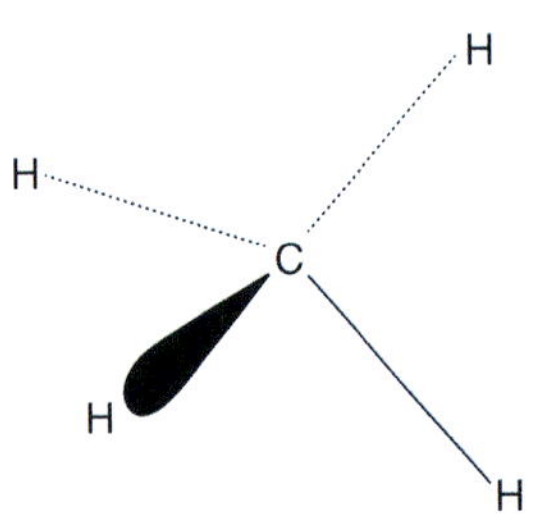

Figure 1.12 The tetrahedral structure of CH_4.

The concept of hybridization not only gives us a simple model for determining the correct geometry in simple molecules but also provides us with a rationalization for multiple bonds. A *double bond* can result from sp^2 hybridization: one sp^2—sp^2 bond and one π bond that forms between the *p* orbitals not involved in hybridization. An example is in C_2H_4 (ethylene, Figure 1.13a), where each carbon undergoes sp^2 hybridization so that it can form an sp^2—1*s* bond with two hydrogens and an sp^2—sp^2 bond with the other carbon. The remaining *p* orbitals on each carbon (say, p_z) share electrons, which form the C—C double bond. A *triple bond* can be explained in terms of *sp* hybridization. It is formed from one *sp*—*sp* bond and two π bonds which form between the two remaining *p* orbitals after hybridization. Acetylene (Figure 1.13b), C_2H_2, is such a compound

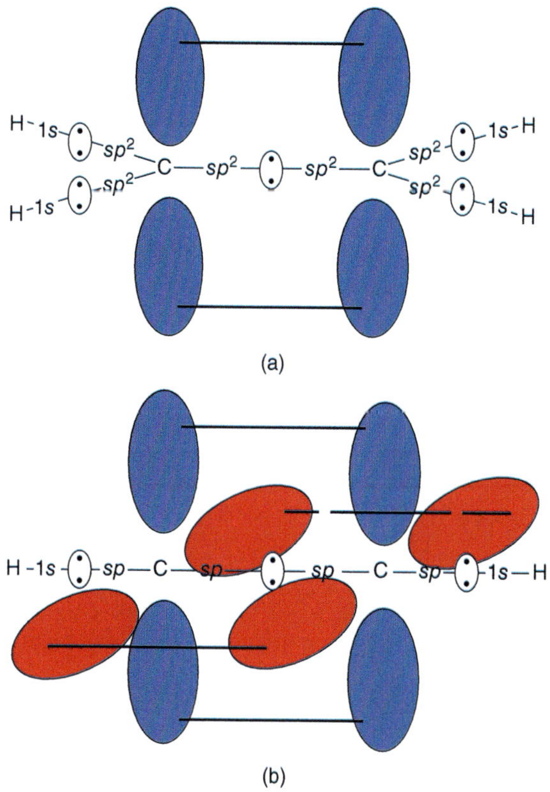

Figure 1.13 Hybridization resulting in (a) double bond and (b) triple bond.

in which both carbons undergo *sp* hybridization so that they can accommodate one bond with each other and one with hydrogen. Bonds can form between the remaining *p* orbitals, which in this case could be the p_z and p_y orbitals on each carbon, for a total of three bonds between the carbon atoms.

> **Need More Help?**
>
> Try these videos on covalent bonds:
>
> https://www.youtube.com/watch?v=nTujP4jCbsg
> https://www.youtube.com/watch?v=cPDptc0wUYI

> **Example Problem 1.3**
>
> Consider the molecule NF_3. How can we explain the observation that the F—N—F bond angles in NF_3 are 107.3° and not 120°, as we might predict?
>
> *Answer*: Nitrogen undergoes sp^3 hybridization, not sp^2, so it is tetrahedral. The additional sp^3 orbital is occupied by a lone pair of electrons from the nitrogen. This lone pair results in electron–electron repulsion that causes the other sp^3 orbitals bonded to fluorines to be closer together than the normal 109° tetrahedral bond angle, hence the 107.3 F—N—F bond angle.

The Metallic Bond

Some elements—namely those in the first two columns of the periodic table and the transition metals (e.g., Co, Mn, Ni)—not only have a propensity for two atoms to share electrons, such as in a covalent bond, but also have a tendency for large groups of atoms to come together and share valence electrons in what is called a *metallic bond*. In this case, there are now *N* atoms in the lattice, where *N* is a large number. There are *N* atomic orbitals, so there must be *N* MOs, many of which are *degenerate* or of the same energy. This leads to *bands* of electrons, as illustrated in Figure 1.14 for sodium. The characteristics of the metallic bond are that the valence electrons are not associated with any particular atom in the lattice, so they form what is figuratively referred to as an *electron gas* around the solid core of metallic nuclei. As a result, the bonds in metals are nondirectional, unlike covalent or ionic bonds, in which the electrons have a finite probability of being around a particular atom.

Take a Peek! Jump to Section 6.0.1 to see how electronic bands determine electrical properties of materials.

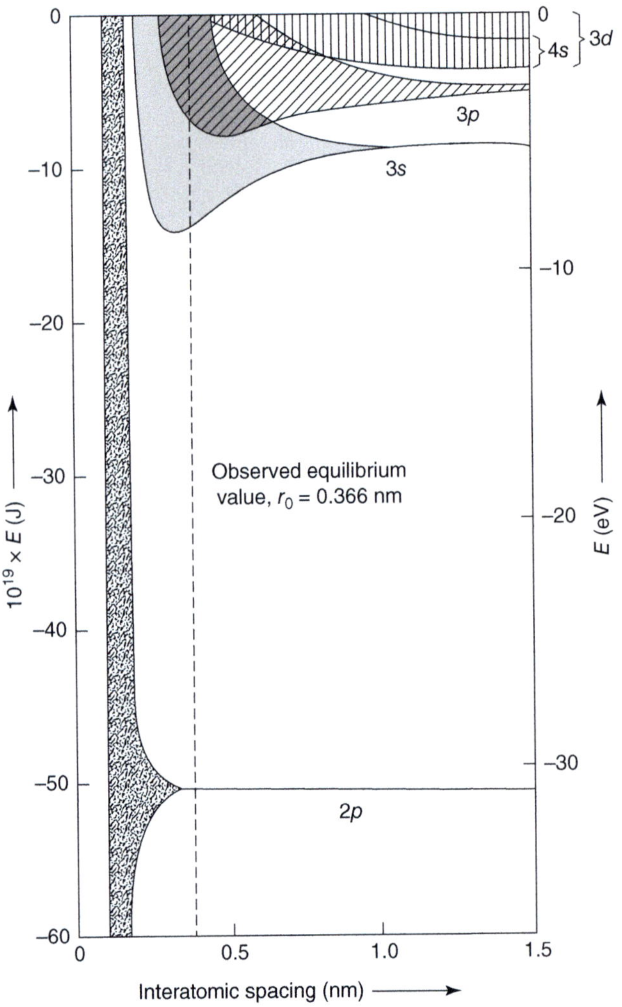

Figure 1.14 Energy band diagram for a sodium lattice. From Ralls et al. [7]. Copyright © 1976 by John Wiley & Sons, Inc. This material is used by permission of John Wiley & Sons, Inc. After Slater [11].

The electrons not involved in bonding remain in what is called the *core band*, whereas the valence electrons that form the electron gas enter the *valence band*. The remaining unfilled orbitals form higher-energy bands, called the *conduction band*. Keep in mind that even though the *d* and *f* orbitals may not be filled with electrons, they still exist for many of the heavier elements, so they must be included in the molecular orbital diagram. We will see later in Chapter 6 that the conduction band plays a very important role in the electrical, thermal, and optical properties of metals.

1.1 Structure of Metals and Alloys

Since the valence electrons in a metallic lattice are in a "gas," we must use the core electrons and nuclei to determine the structure of metals. This will be true of most solids we will describe—regardless of the type of bonding—since the electrons occupy such a small volume compared to the nucleus. For ease of visualization, we consider the atomic cores to be hard spheres. Because the electrons are delocalized, there is little in the way of electronic hindrance to restrict the number of neighbors a metallic atom may have. As a result, the atoms tend to pack in a *close-packed* arrangement, or one in which the maximum number of *nearest neighbors* (atoms directly in contact) is satisfied.

Refer to Figure 1.15. The most hard spheres one can place in the plane around a central sphere are six, regardless of the size of the spheres (remember that all of the spheres are the same size). You can then place three spheres in contact with the central sphere, both above and below the plane containing the central sphere. This results in a total of 12 nearest-neighbor spheres in contact with the central sphere in the close-packed structure.

A closer inspection of Figure 1.15a shows that there are two different ways to place the three nearest neighbors above the original plane of hard spheres. They can be directly aligned with the layer below in an ABA type of structure, or they can be rotated so that the top layer does not align core centers with the bottom layer, resulting in an ABC structure. This leads to two different types of close-packed structures. The ABAB... structure (Figure 1.15b) is called *hexagonal close-packed* (HCP) and the ABCABC... structure is called *face-centered cubic* (FCC). Remember

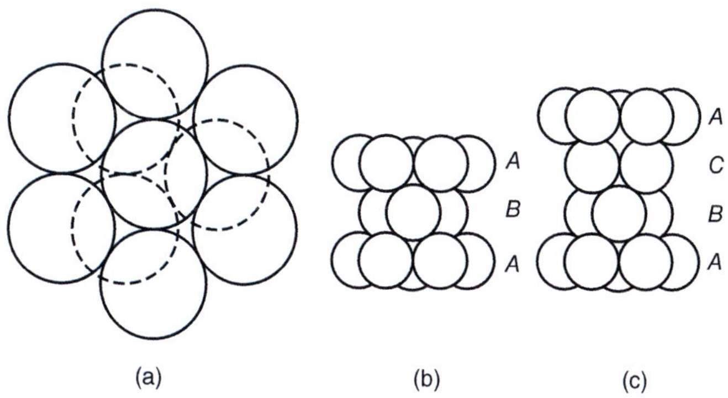

Figure 1.15 Close-packing of spheres. (a) Top view, (b) side view of ABA structure, (c) side view of ABC structure. From Jastrzebski [5]. Copyright © 1976 by John Wiley & Sons, Inc. This material is used by permission of John Wiley & Sons, Inc.

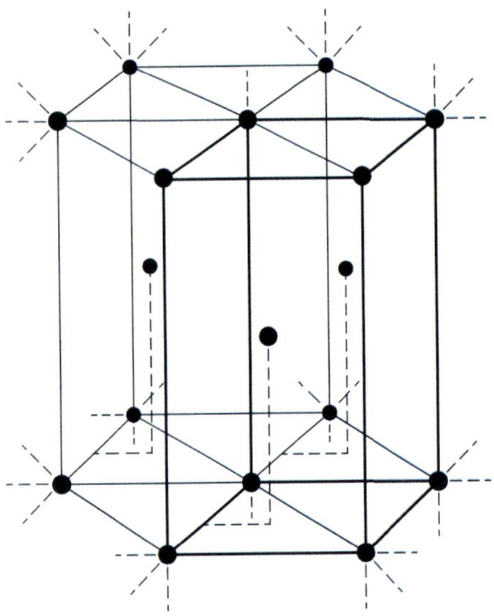

Figure 1.16 The extended unit cell of the hexagonal close-packed (HCP) structure.

that both of these close-packed arrangements have a *coordination number* (number of nearest neighbors surrounding an atom) of 12:6 in plane, 3 above, and 3 below.[2]

Keep in mind that for close-packed structures, the atoms touch each other in all directions, and all nearest neighbors are equivalent. Let us first examine the HCP structure. Figure 1.16 is a section of the HCP lattice, from which you should be able to see both hexagons formed at the top and bottom of what is called the *unit cell*. You should also be able to identify the ABA-layered structure in the HCP unit cell of Figure 1.16 through comparison with Figure 1.15. Let us count the number of atoms in the HCP unit cell. The three atoms in the center of the cell are completely enclosed. The atoms on the faces, however, are shared with adjacent cells in the lattice, which extends to infinity. The center atoms on each face are shared with one other HCP unit cell, either above (for the top face) or below (for the bottom face), so they contribute only half of an atom each to the HCP unit cell under consideration. This leaves the six corner atoms on each face (12 total) unaccounted for. These corner atoms are at the intersection of a total of *six* HCP unit cells (you should convince yourself of this!), so each corner atom contributes only one-sixth of an atom to our isolated HCP unit cell. So, the total number of whole atoms in the HCP unit cell is

$$\begin{aligned} 3 \times 1 &= 3 \text{ center atoms} \\ 2 \times (1/2) &= 1 \text{ face atom} \\ 12 \times (1/6) &= 2 \text{ corner atoms} \\ \hline & 6 \text{ total atoms} \end{aligned}$$

Counting the atoms in the FCC structure is performed in a similar manner, except that visualizing the FCC structure takes a little bit of imagination and is virtually impossible to show on a two-dimensional page. Take the ABC close-packed structure shown in Figure 1.15c and pick three atoms along a diagonal. These three atoms form the diagonal on the face of the FCC unit cell, which is shown in Figure 1.17. There is a trade-off in doing this: It is now difficult to see the close-packed layers in the FCC structure, but it is much easier to see the cubic structure (note that all the edges of the faces have the same length), and it is easier to count the total number of atoms in the FCC cell. In a manner similar to counting atoms in the HCP cell, we see that there are *zero* atoms completely enclosed by the FCC unit cell, six face atoms that are each shared with an adjacent unit cell, and eight corner atoms at the intersection of eight unit cells to give

2 You may find it useful to get some small, hard spheres to help you visualize these structures. We will use multiple perspectives, views, and diagrams to build these structures. Some will treat the atoms as solid spheres that can touch each other; some will use dots at the center of the atoms to help you visualize the larger structure. Not all descriptions will help you find the perspectives that work best for you.

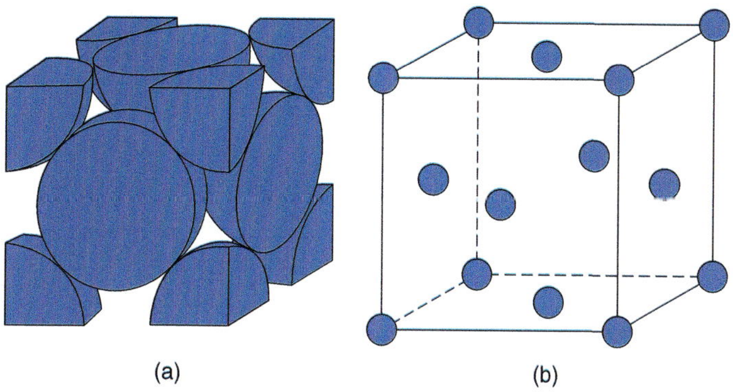

Figure 1.17 The face-centered cubic (FCC) structure showing (a) atoms touching and (b) atoms as small spheres. Reproduced, by permission, from Callister [12]. Copyright © 2000 by John Wiley & Sons, Inc.

$$\begin{aligned} 6 \times (1/2) &= 3 \text{ face atoms} \\ 8 \times (1/8) &= 1 \text{ corner atom} \\ \hline & 4 \text{ total atoms} \end{aligned}$$

Remember that both HCP and FCC are close-packed structures and that each has a coordination number of 12, but that their respective unit cells contain 6 and 4 total atoms. We will now see how these two special, close-packed structures fit into a larger assembly of crystal systems.

1.1.1 Crystal Structures

Our description of atomic packing leads naturally into *crystal structures*. While some of the simpler structures are used by metals, these structures can be employed by heteronuclear compounds, as well. We have already discussed FCC and HCP, but there are 12 other types of crystal structures, for a total of 14 *space lattices* or *Bravais lattices*. These 14 space lattices belong to more general classifications called *crystal systems*, of which there are seven.

Before describing each of the space lattices, we need to define a coordinate system. The easiest coordinate system to use depends on which crystal system we are looking at. In other words, the coordinate axes are not necessarily orthogonal and are defined by the unit cell. This may seem a bit confusing, but it simplifies the description of cell parameters for those systems that do not have crystal faces at right angles to one another. Refer to Figure 1.18. For each crystal system, we will define the space lattice in terms of three axes, x, y, and z, with interaxial angles α, β, γ. Note that the interaxial angle α is defined by the angle formed between axes z and y, and that angles β and γ are defined similarly. Only in special circumstances are α, β, γ equal to 90°. The distance along the y-axis from the origin to the edge of the unit cell is called the *lattice translation vector*, **b**. Lattice translation vectors **a** and **c** are defined similarly along the axes x and z, respectively. The magnitudes (lengths) of the lattice translation vectors are called the *lattice parameters*, a, b, and c. We will now examine each of the seven crystal systems in detail.

Crystal Systems

The cubic crystal system is composed of three space lattices, or unit cells, one of which we have already studied: *simple cubic* (SC), *body-centered cubic* (BCC), and *face-centered cubic* (FCC). The

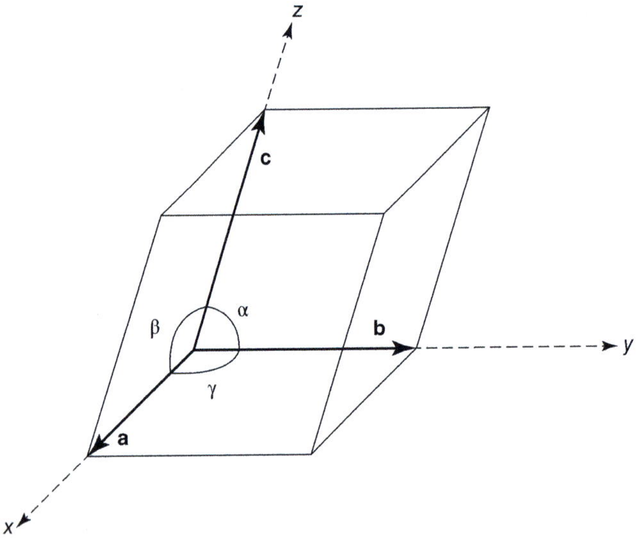

Figure 1.18 Definition of a coordinate system for crystal structures.

conditions for a crystal to be considered part of the cubic system are that the lattice parameters be the same (so there is really only one lattice parameter, a) and that the interaxial angles all be 90°.

The SC structure is the basic structural unit of many primitive cells, including NaCl. It is not, however, a close-packed structure (see Figure 1.19). In fact, it contains about 48% void space, and as a result, it is not a very dense structure. The large space in the center of the SC structure is called an *interstitial site*, which is a vacant position between atoms that can be occupied by a small impurity atom or *alloying element*. In this case, the interstitial site is surrounded by eight atoms. All eight atoms in SC are equivalent and are located at the intersection of eight adjacent unit cells, so that there are $8 \times (1/8) = 1$ total atom in the SC unit cell. Notice that Figure 1.19 shows the atoms as points, but the atoms actually occupy a larger space than that. In fact, for SC, the atoms touch along the edge of the crystal.

BCC is the unit cell of many metals and, like SC, is not a close-packed structure. The number of atoms in the BCC unit cell is calculated as follows:

$$\begin{array}{lcl} 1 \times 1 & = & 1 \text{ center atom} \\ 8 \times (1/8) & = & 1 \text{ corner atom} \\ \hline & & 2 \text{ total atoms} \end{array}$$

Take a Peek! Jump ahead to Section 2.1.3 for more information on how impurity atoms play a role in the microstructure of ferrous alloys.

Finally, FCC has already been described (Figure 1.17). Even though FCC is a close-packed structure, there are interstitial sites, just as in SC. There are two different types of interstitial sites in FCC, depending on how many atoms surround the interstitial site. A group of four atoms forms a *tetrahedral interstice*, as shown in Figure 1.20. A group of six atoms arranged in an octahedron (an eight-sided geometric figure) creates an *octahedral interstice* (Figure 1.21). Figure 1.22 shows the locations of

Crystal Structure	Lattice Parameters	Interaxial Angles
Cubic	$a = b = c$	$\alpha = \beta = \gamma = 90°$
Orthorhombic	$a \neq b \neq c$	$\alpha = \beta = \gamma = 90°$
Rhombohedral	$a = b = c$	$\alpha = \beta = \gamma \neq 90°, < 120°$
Tetragonal	$a = b \neq c$	$\alpha = \beta = \gamma = 90°$
Monoclinic	$a \neq b \neq c$	$\alpha = \gamma = 90°, \beta \neq 90°$
Triclinic	$a \neq b \neq c$	$\alpha = \beta = \gamma \neq 90°$
Hexagonal	$a = b,\ a \neq c$	$\alpha = \beta = 90°, \gamma = 120°$

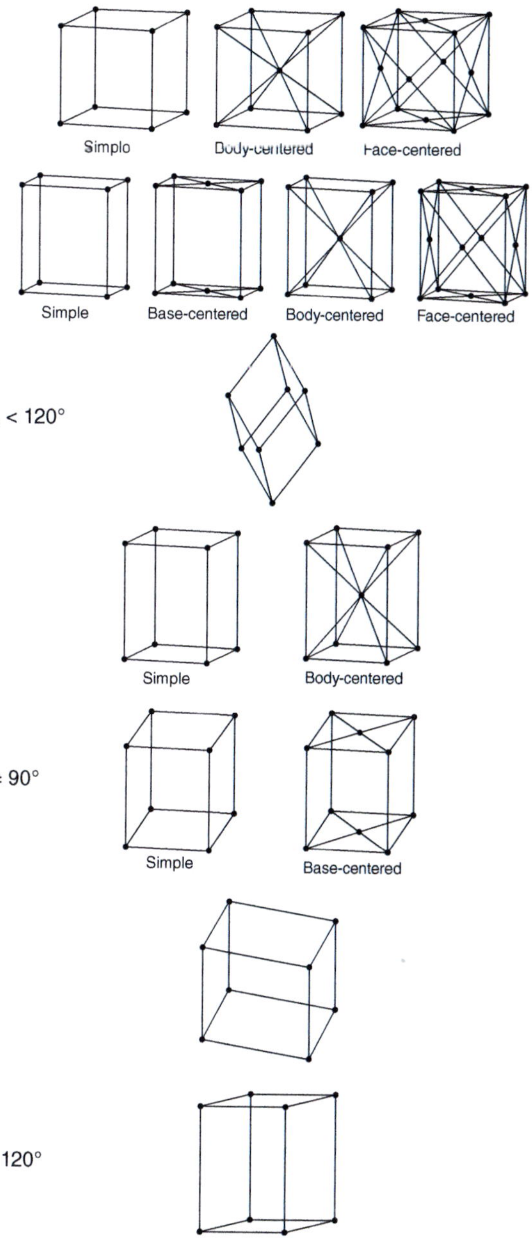

Figure 1.19 Summary of the 14 Bravais space lattices.

these interstitial sites within the FCC lattice. Note that there are eight *total* tetrahedral interstitial sites in FCC and four *total* octahedral interstitial sites in FCC (prove it!), which are counted in much the same way as we previously counted the total number of atoms in a unit cell. We will see later on that these interstitial sites play an important role in determining solubility of impurities and phase stability of alloys.

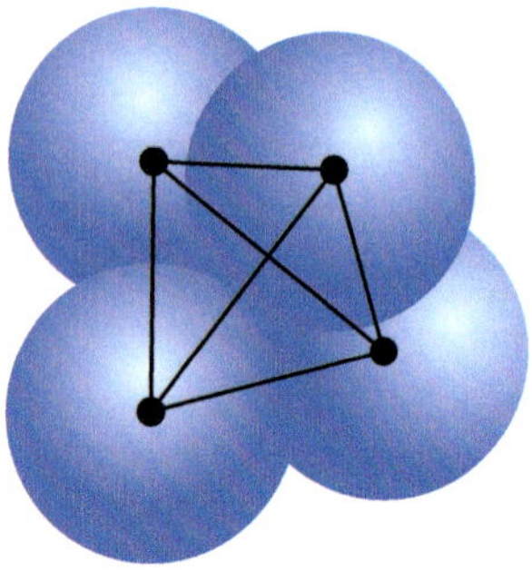

Figure 1.20 A tetrahedral interstice. From Kingery et al. [13]. Copyright © 1976 by John Wiley & Sons, Inc. This material is used by permission of John Wiley & Sons, Inc.

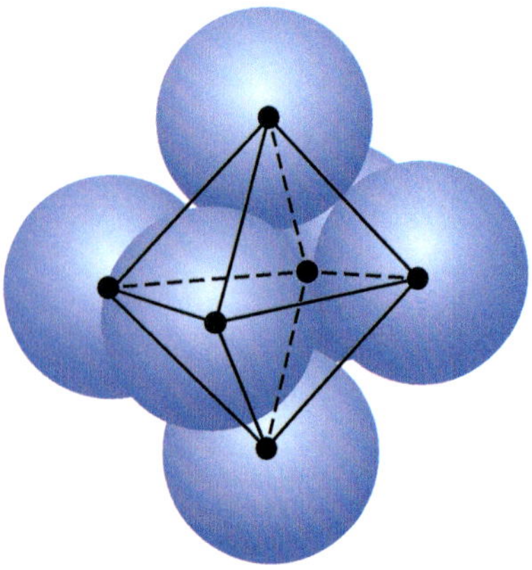

Figure 1.21 An octahedral interstice. From Kingery et al. [13]. Copyright © 1976 by John Wiley & Sons, Inc. This material is used by permission of John Wiley & Sons, Inc.

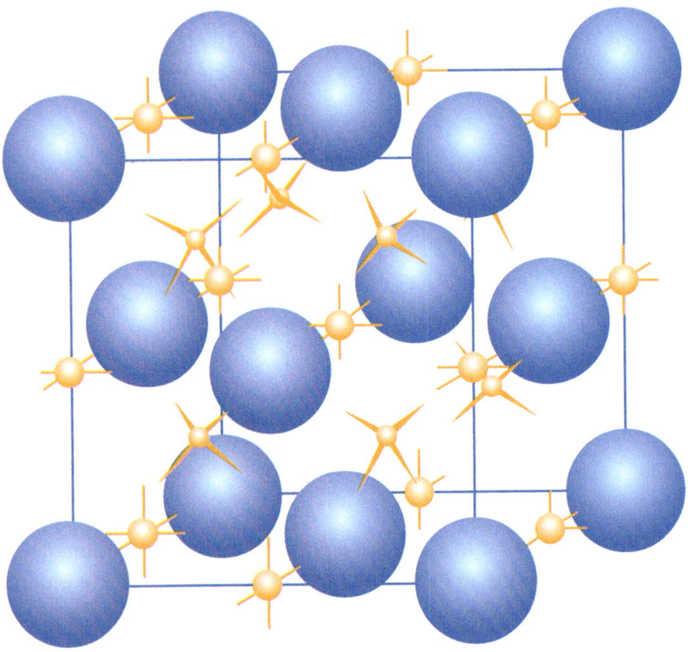

Figure 1.22 Location of interstitial sites in FCC. From Kingery et al. [13]. Copyright © 1976 by John Wiley & Sons, Inc. This material is used by permission of John Wiley & Sons, Inc.

Need More Help?

Try these videos on Bravais Lattices and Crystal Structures:

https://www.youtube.com/watch?v=qAeaHYSX0hs
https://www.youtube.com/watch?v=uKpr-9vmgsc
https://www.youtube.com/watch?v=s4os8nh4KgM

Interstitial sites are the result of packing of the spheres. Recall from Figure 1.17 that the spheres touch along the face diagonal in FCC. Similarly, the spheres touch along the body diagonal in BCC and along an edge in SC. We should, then, be able to calculate the lattice parameter, a, or the length of a face edge, from knowledge of the sphere radius. In SC, it should be evident that the side of a unit cell is simply $2r$. Application of a little geometry shows that in FCC, $a = 4r/\sqrt{2}$. The relationship between a and r for BCC is derived in Example Problem 1.4; other geometric relationships, including cell volume for cubic structures, are listed in Table 1.9. Finally, atomic radii for the elements can be found in Table 1.10. The radius of an atom is not an exactly defined quantity, and it can vary depending on the bonding environment in which it finds itself. As a result, three types of radii are listed for each element in Table 1.10: an *atomic radius* of an isolated atom, an *ionic radius*, and a *metallic radius*. Just as in Figure 1.1 for electronic structure, there are some important trends in the atomic radii. The atomic radius tends to increase as one goes down the column in a series. This is due to the addition of energy levels and more electron density. Radii tend to decrease as we move across a row because there is less shielding from inner electrons and the outer-core electrons are drawn more tightly toward the nucleus. There are some notable exceptions, or "jumps," in the row trend (why?). In general, the ionic radius is much smaller for positive ions and much larger for negative ions than the corresponding isolated atom (why?), and it follows the same general trend as for the isolated atoms. For the discussion of elemental, crystalline solids, the metallic radius is most appropriate. We will find later that the ionic values will be equally important for heteronuclear structures. There are other types of radii, such as covalent radii and van der Waals radii. The former is highly dependent upon the type of covalent bond. For example, a carbon atom in a carbon–carbon single bond has a covalent radius of 1.54 Å, whereas the same atom in a carbon–carbon triple bond has a covalent radius of only 1.35 Å.

Table 1.9 Summary of Important Parameters in the Cubic Space Lattices

	Simple Cubic	Face-Centered Cubic	Body-Centered Cubic
Unit cell side, a	$2r$	$4r/\sqrt{2}$	$4r/\sqrt{3}$
Face diagonal	$\sqrt{2}(2r)$	$4r$	$\sqrt{2/3}(4r)$
Body diagonal	$\sqrt{3}(2r)$	$\sqrt{3/2}(4r)$	$4r$
Number of atoms	1	4	2
Cell volume	$8r^3$	$\frac{32r^3}{\sqrt{2}}$	$\frac{64r^3}{3\sqrt{3}}$

r = atomic radius.

Table 1.10 Van der Waals (VDW), Covalent, and Ionic Radii* of the Elements

VDW Covalent Ionic	1+ ions **H** 1.10 0.32 —	2+ Ions											3+ ions	4+ ions	5+ ions	2– ions	1– ions	2 **He** 1.40 0.37 —
VDW Covalent Ionic	3 **Li** 1.82 1.30 0.76	4 **Be** 1.53 0.99 0.45											5 **B** 1.92 0.84 —	6 **C** 1.70 0.75 0.16	7 **N** 1.55 0.71 0.13	8 **O** 1.52 0.64 1.40	9 **F** 1.47 0.60 1.33	10 **Ne** 1.54 0.62 —
VDW Covalent Ionic	11 **Na** 2.27 1.60 1.02	12 **Mg** 1.73 1.40 0.72	3+ ions	4+ ions	5+ ions	6+ ions	4+ ions	4+ ions	3+ ions	2+ ions	1+ ions	2+ ions	13 **Al** 1.84 1.24 0.54	14 **Si** 2.10 1.14 0.40	15 **P** 1.80 1.09 —	16 **S** 1.80 1.04 1.84	17 **Cl** 1.75 1.00 1.81	18 **Ar** 1.88 1.01 —
VDW Covalent Ionic	19 **K** 2.75 2.00 1.38	20 **Ca** 2.31 1.74 1.12	21 **Sc** 2.15 1.59 0.75	22 **Ti** 2.11 1.48 0.31	23 **V** 2.07 1.44 0.54	24 **Cr** 2.06 1.30 0.55	25 **Mn** 2.05 1.29 0.53	26 **Fe** 2.04 1.24 see above	27 **Co** 2.00 1.18 0.55	28 **Ni** 1.97 1.17 0.69	29 **Cu** 1.96 1.22 0.77	30 **Zn** 2.01 1.20 0.74	31 **Ga** 1.87 1.23 0.62	32 **Ge** 2.11 1.20 0.53	33 **As** 1.85 1.20 —	34 **Se** 1.90 1.18 1.98	35 **Br** 1.85 1.17 1.96	36 **Kr** 2.02 1.16 —
VDW Covalent Ionic	37 **Rb** 3.03 2.15 1.52	38 **Sr** 2.49 1.90 1.18	39 **Y** 2.32 1.76 0.90	40 **Zr** 2.23 1.64 0.72	41 **Nb** 2.18 1.56 0.64	42 **Mo** 2.17 1.46 0.59	43 **Tc** 2.16 1.38 0.65	44 **Ru** 2.13 1.36 0.68	45 **Rh** 2.10 1.34 0.67	46 **Pd** 2.10 1.30 0.86	47 **Ag** 2.11 1.36 1.15	48 **Cd** 2.18 1.40 0.95	49 **In** 1.93 1.32 0.80	50 **Sn** 2.17 1.40 0.69	51 **Sb** 2.06 1.40 —	52 **Te** 2.06 1.37 2.21	53 **I** 1.98 1.36 2.20	54 **Xe** 2.16 1.36 —
VDW Covalent Ionic	55 **Cs** 3.43 2.38 1.67	56 **Ba** 2.68 2.06 1.35	57 **La** 2.43 1.94 1.03	72 **Hf** 2.23 1.64 0.71	73 **Ta** 2.22 1.58 0.64	74 **W** 2.18 1.50 0.60	75 **Re** 2.16 1.41 0.63	76 **Os** 2.16 1.36 0.63	77 **Ir** 2.13 1.32 0.68	78 **Pt** 2.13 1.30 0.80	79 **Au** 2.14 1.30 1.37	80 **Hg** 2.23 1.32 1.02	81 **Tl** 1.96 1.44 0.89	82 **Pb** 2.02 1.45 0.78	83 **Bi** 2.07 1.50 0.76	84 **Po** 1.97 1.42 —	85 **At** 2.02 1.48 —	86 **Rn** 2.20 1.46 —

* Ionic radii depend on the coordination number and valence state. Values are for the valence state listed in the column heading for a coordination number of 6. Some additional ionic radii at this coordination number are listed below. Values for other coordination numbers are in the reference in the table caption.

Fe^{3+}	0.55	Fe^{2+}	0.61	Mn^{3+}	0.58	Mn^{2+}	0.83
Pb^{2+}	1.19	Pt^{4+}	0.63	Co^{2+}	0.65	W^{6+}	0.60
S^{6+}	0.29	Rh^{4+}	0.60	Mo^{4+}	0.65	Cr^{3+}	0.62
Te^{4+}	0.65	V^{3+}	0.64	V^{4+}	0.58	Se^{6+}	0.42
Nb^{4+}	0.68	Tl^{+}	1.50	Ti^{3+}	0.67	Ti^{2+}	0.86

Source: Rumble [14].All values in angstroms, Å (1 Å = 0.1 nm).

Example Problem 1.4

Molybdenum has a BCC structure with a metallic atomic radius of 1.36 Å. Calculate the lattice parameter for BCC Mo.

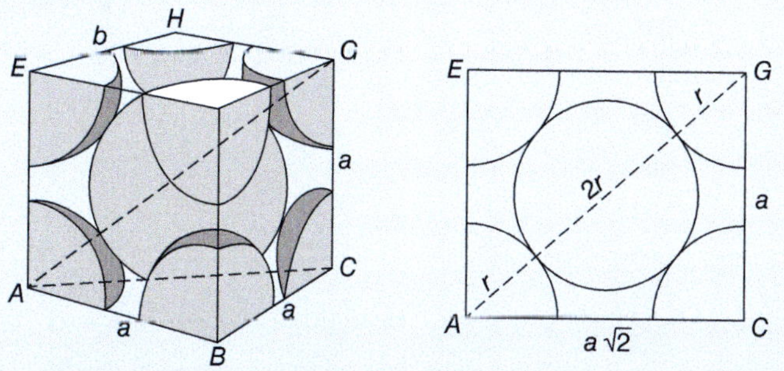

Reproduced by permission, from Z. Jastrzebski, *The Nature and Properties of Engineering Materials*, p. 47, 2nd ed. Copyright © 1976, John Wiley & Sons, Inc.

Answer: We know that the molybdenum atoms touch along the body diagonal in BCC, as shown in the projection at right. The length of the body diagonal, then, is $4r$, and is related to the lattice parameter, a (which is the length of the cube edge, not the length of the face diagonal, which is $a\sqrt{2}$) by application of the Pythagorean theorem:

$$(4r)^2 = a^2 + \left(a\sqrt{2}\right)^2 = 3a^2$$

$$a = 4r/\sqrt{3} = 4(1.36)/\sqrt{3} = 3.14$$

The lattice parameter for BCC Mo is 3.14 Å, which is consistent with the value in Table 1.12.

Continuing with our survey of the seven crystal systems, we see that the *tetragonal* crystal system is similar to the cubic system in that all the interaxial angles are 90°. However, the cell height, characterized by the lattice parameter, c, is not equal to the base, which is square ($a = b$). There are two types of tetragonal space lattices: *simple tetragonal*, with atoms only at the corners of the unit cell, and *body-centered tetragonal*, with an additional atom at the center of the unit cell.

Orthorhombic crystals are similar to both tetragonal and cubic crystals because their coordinate axes are still orthogonal, but now all the lattice parameters are unequal. There are four types of orthorhombic space lattices: *simple orthorhombic, face-centered orthorhombic, body-centered orthorhombic*, and a type we have not yet encountered, *base-centered orthorhombic*. The first three types are similar to those we have seen for the cubic and tetragonal systems. The base-centered orthorhombic space lattice has a lattice point (atom) at each corner, as well as a lattice point only on the top and bottom faces (called *basal faces*). All four orthorhombic space lattices are shown in Figure 1.19.

There is only one space lattice in the *rhombohedral* crystal system. This crystal is sometimes called *hexagonal R* or *trigonal R*, so do not confuse it with the other two similarly named crystal systems. The rhombohedral crystal has uniform lattice parameters in all directions and has equivalent interaxial angles, but the angles are nonorthogonal and are less than 120°.

The crystal descriptions become increasingly more complex as we move to the *monoclinic* system. Here, all lattice parameters are different, and only two of the interaxial angles are orthogonal.

Table 1.11 Axial Ratios for Some HCP Metals

Metal	c/a
Be, Y	1.57
Hf, Os, Ru, Ti	1.58
Sc, Zr	1.59
Tc, Tl	1.60
La	1.61
Co, Re	1.62
Mg	1.63
Zn	1.85
Cd	1.89
Ideal (sphere packing)	1.633

The third angle is not 90°. There are two types of monoclinic space lattices: *simple monoclinic* and *base-centered monoclinic*. The *triclinic* crystal, of which there is only one type, has three different lattice parameters, and none of its interaxial angles are orthogonal, though they are all equal.

Finally, we revisit the hexagonal system to provide some additional details. The lattice parameter and interaxial angle conditions shown in Figure 1.19 for the hexagonal cell refer to what is called the *primitive cell* for the hexagonal crystal, which can be seen in the front quadrant of the extended cell in Figure 1.16. The primitive hexagonal cell has lattice points only at its corners and has one atom in the center of the primitive cell, for a *basis* of two atoms. A basis is a unit assembly of atoms identical in composition, arrangement, and orientation that is placed in a regular manner on the lattice to form a space lattice. You should be able to recognize that there are three equivalent primitive cells in the extended HCP structure. The HCP *extended cell*, which is more often used to represent the hexagonal structure, contains a total of six atoms, as we calculated earlier. In the extended structure, the ratio of the height of the cell to its base, c/a, is called the *axial ratio*. Table 1.11 lists typical values of the axial ratio for some common HCP crystals.

A table of crystal structures for the elements can be found in Table 1.12 (excluding the Lanthanide and Actinide series). Some elements can have multiple crystal structures, depending on temperature and pressure. This phenomenon is called *allotropy* and is very common in elemental metals (see Table 1.13). It is not unusual for close-packed crystals to transform from one stacking sequence to the other, simply through a shift in one of the layers of atoms. Other common allotropes include carbon (graphite at ambient conditions, diamond at high pressures and temperatures), pure iron (BCC at room temperature, FCC at 912°C, and back to BCC at 1394°C), and titanium (HCP to BCC at 882°C).

Crystal Locations, Planes, and Directions

To calculate such important quantities as cell volumes and densities, we need to specify locations and directions within the crystal. *Cell coordinates* specify a position in the lattice and are indicated by the variables u, v, and w, separated by commas with no brackets:

u distance along the lattice translation vector **a**
v distance along the lattice translation vector **b**
w distance along the lattice translation vector **c**

Table 1.12 Common Crystal Structures, Solid Densities, and Lattice Parameters of the Elements

Struct. ρ, g/cc a, Å	1 **H** hcp n/a 3.78																	2 **He** fcc n/a 4.24
Struct. ρ, g/cc a, Å	3 **Li** bcc 0.534 3.51	4 **Be** hcp 1.85 2.29											5 **B** rhom 2.34 5.06	6 **C** diam 3.51 3.56	7 **N** hcp n/a 4.05	8 **O** mon n/a 5.40	9 **F** mon n/a 5.50	10 **Ne** fcc n/a 4.46
Struct. ρ, g/cc a, Å	11 **Na** bcc 0.97 4.29	12 **Mg** hcp 1.74 3.21											13 **Al** fcc 2.70 4.05	14 **Si** diam 2.33 5.43	15 **P** tricl 1.82 11.5	16 **S** orth 2.07 10.5	17 **Cl** orth n/a 6.24	18 **Ar** fcc n/a 5.32
Struct. ρ, g/cc a, Å	19 **K** bcc 0.89 5.32	20 **Ca** fcc 1.54 5.59	21 **Sc** hcp 2.99 3.31	22 **Ti** hcp 4.51 2.95	23 **V** bcc 6.0 3.02	24 **Cr** bcc 7.15 2.88	25 **Mn** cubic 7.3 8.91	26 **Fe** bcc 7.87 2.87	27 **Co** hcp 8.86 2.51	28 **Ni** fcc 8.90 3.52	29 **Cu** fcc 8.96 3.61	30 **Zn** hcp 7.13 2.66	31 **Ga** orth 5.91 4.52	32 **Ge** diam 5.32 5.66	33 **As** rhom 5.75 4.13	34 **Se** hcp 4.81 4.37	35 **Br** orth 3.10 6.68	36 **Kr** fcc n/a 5.81
Struct. ρ, g/cc a, Å	37 **Rb** bcc 1.53 5.71	38 **Sr** fcc 2.64 6.08	39 **Y** hcp 4.47 3.65	40 **Zr** hcp 6.52 3.61	41 **Nb** bcc 8.57 3.30	42 **Mo** bcc 10.22 3.15	43 **Tc** hcp 11 2.74	44 **Ru** hcp 12.1 2.71	45 **Rh** fcc 12.4 3.80	46 **Pd** fcc 12.0 3.89	47 **Ag** fcc 10.5 4.09	48 **Cd** hcp 8.69 2.98	49 **In** tetrag 7.31 3.25	50 **Sn** tetrag 7.29 5.83	51 **Sb** trig 6.68 3.38	52 **Te** trig 6.23 4.46	53 **I** orth 4.95 7.27	54 **Xe** fcc n/a 6.35
Struct. ρ, g/cc a, Å	55 **Cs** bcc 1.87 6.14	56 **Ba** bcc 3.62 5.02	57 **La** hcp 6.15 3.77	72 **Hf** hcp 13.3 3.20	73 **Ta** bcc 16.4 3.30	74 **W** bcc 19.3 3.17	75 **Re** hcp 21.0 2.76	76 **Os** hcp 22.59 2.73	77 **Ir** fcc 22.56 3.84	78 **Pt** fcc 21.5 3.92	79 **Au** fcc 19.3 4.08	80 **Hg** rhom n/a 3.01	81 **Tl** hcp 11.87 3.46	82 **Pb** fcc 11.3 4.95	83 **Bi** rhom 9.79 4.75	84 **Po** cub 9.20 3.37	85 **At** n/a n/a n/a	86 **Rn** n/a n/a n/a

Source: Rumble[15]; (1 Å = 0.1 nm).Note that many elements have allotropes. Cited structures are generally the most stable.

Table 1.13 Some Metal Allotropes

Metal	R.T. Crystal Structure	Structure at Other Temperatures
Ca	FCC	BCC (>447°C)
Co	HCP	FCC (>427°C)
Hf	HCP	BCC (>1742°C)
Fe	BCC	FCC (>912°C)
		BCC (>1394°C)
Li	BCC	BCC (< −193°C)
Na	BCC	BCC (< −233°C)
Sn	BCT	Cubic (<13°C)
Tl	HCP	BCC (>234°C)
Ti	HCP	BCC (>883°C)
Y	HCP	BCC (>1481°C)
Zr	HCP	BCC (>872°C)

Historical Highlight

Allotropy has led to some interesting public phenomena and their interpretations. Read more about "tin cry" and "tin plague" here: https://www.webelements.com/tin

For example, the center atom in the BCC space lattice (see Figure 1.19) has cell coordinates of 1/2, 1/2, 1/2. Any two points are equivalent if the fractional portions of their coordinates are equal:

$$1/2, 1/2, 1/2 \equiv -1/2, -1/2, -1/2 \text{ (center)}$$

$$0, 0, 0 \equiv 1, 0, 1 \text{ (corner)}$$

A *cell direction* is designated by the vector **r**, which is a combination of the lattice translation vectors **a**, **b**, and **c**:

$$\boldsymbol{r} = u\mathbf{a} + v\mathbf{b} + w\mathbf{c} \tag{1.27}$$

A direction can also be specified with the cell coordinates in square brackets, with commas and fractions removed:

$$[1\,1\,1] \equiv [\tfrac{1}{2}\ \tfrac{1}{2}\ \tfrac{1}{2}] \equiv 1/2\mathbf{a} + 1/2\mathbf{b} + 1/2\mathbf{c}$$

Negative directions are indicated by an overbar, e.g., $[1\bar{1}1]$, which is called the "one negative one one" direction. All directions are relative to the origin, where the three lattice translation vectors originate (see Figure 1.18).

The *cell volume*, *V*, can be calculated using the lattice translation vectors:

$$V = |\mathbf{a} \times \mathbf{b} \cdot \mathbf{c}| \tag{1.28}$$

Mathematically, this is a triple scalar product and can be used to calculate the volume of any cell with only a knowledge of the lattice translation vectors. If the lattice parameters and interaxial angles are known, the following expression for V can be derived from the vector expression:

$$V = abc\left(1 - \cos^2\alpha - \cos^2\beta - \cos^2\gamma + 2\cos\alpha\cos\beta\cos\gamma\right)^{1/2} \tag{1.29}$$

This looks complicated, but for orthogonal systems ($\alpha, \beta, \gamma = 90°$) it reduces quite nicely to the expected expression:

$$V = abc \tag{1.29a}$$

Now that we know how to find the cell volume, we can use some previous information to calculate an important property of a material, namely, its *density*, which we represent with the lowercase Greek letter rho, ρ. For example, aluminum has an FCC space lattice. Recall that there are four atoms in the FCC unit cell. We know that each aluminum atom has an atomic weight of 27 g/mol. From Table 1.12, the cubic lattice parameter for aluminum is 4.05 Å, or 0.405 nm (4.05×10^{-8} cm). This gives us a volume of $a^3 = 6.64 \times 10^{-23}$ cm^3. You should confirm that the theoretical density for aluminum is then:

$$\rho = \frac{(4\,\text{atoms/unit cell})(27\,\text{g/mol})}{(6.02 \times 10^{23}\text{atoms/mol})(6.64 \times 10^{-23}\text{cm}^3)} = 2.70\,\text{g/cm}^3 \text{ per unit cell}$$

Cooperative Learning Exercise 1.2

The actinide-series element protactinium (Pa, AW = 231.04) has a body-centered tetragonal structure with cell dimensions a = 0.3925 nm, c = 0.3238 nm.

Person 1: Determine the weight, in grams, of a single unit cell of Pa.

Person 2: Calculate the volume of a single Pa unit cell.

Combine your answers appropriately to arrive at the density.

Answer: 15.38 g/cm^3

In addition to cell coordinates and directions, *crystal planes* are very important for the determination and analysis of structure. We begin with the cell's coordinate system, with axes x, y, and z. Recall that the axes are not necessarily orthogonal, and that a, b, and c are the lattice parameters. Look at Figure 1.23. The equation of an arbitrary plane with intercepts A, B, and C, relative to the lattice parameters is given by

$$\frac{1}{A}\frac{x}{a} + \frac{1}{B}\frac{y}{b} + \frac{1}{C}\frac{z}{c} = 1 \tag{1.30}$$

We designate a plane by *Miller indices*, h, k, and l, which are simply the reciprocals of the intercepts, A, B, and C:

$$\left(\frac{1}{A}\,\frac{1}{B}\,\frac{1}{C}\right) = (hkl) \tag{1.31}$$

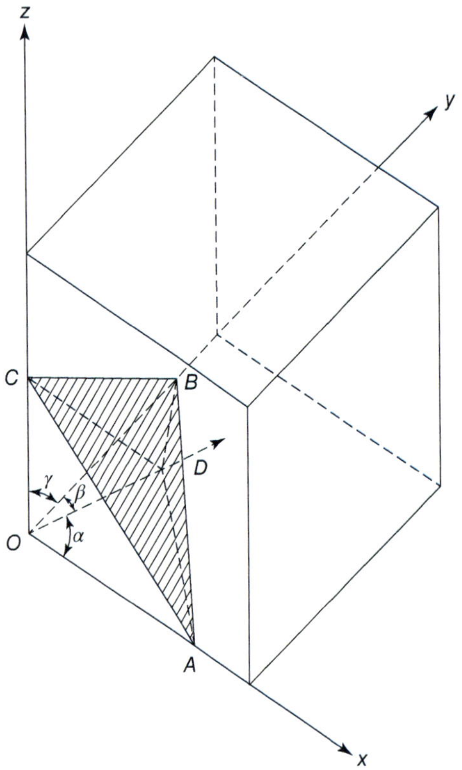

Figure 1.23 Definition of Miller indices for an arbitrary plane (shaded area). From Jastrzebski [5]. Copyright © 1976 by John Wiley & Sons, Inc. This material is used by permission of John Wiley & Sons, Inc.

Note that the Miller indices are enclosed in parentheses and **not** separated by commas. Miller indices are determined as follows:

- Remove all indeterminacy; that is, the planes should have nonzero intercepts.
- Find intercepts along three axes of the crystal system.
- Take the reciprocals of the intercepts in terms of a, b, and c.
- Multiply reciprocals by the smallest integer necessary to convert them into a set of integers.
- Enclose resulting integers in parentheses, (hkl), without commas.

Need More Help?
Try these videos on lattice directions and atomic planes: https://www.youtube.com/watch?v=n84gzYlB0BQ https://www.youtube.com/watch?v=9-us_oENGoM

Need More Help?
Try this video on linear and planar densities: https://www.youtube.com/watch?v=VAP_SozPa8M

Any planes that have common factors are parallel. For example, a (222) and a (111) plane are parallel, as are (442) and (221) planes. As with cell directions, a minus sign (in this case, indicating a negative intercept) is designated by an overbar. The $(2\bar{2}1)$ plane has intercepts at 1/2, −1/2, and 1 along the x, y, and z axes, respectively. Some important planes in the cubic crystal system are shown in Figure 1.24.

In a manner similar to that used to calculate the density of a unit cell, we can

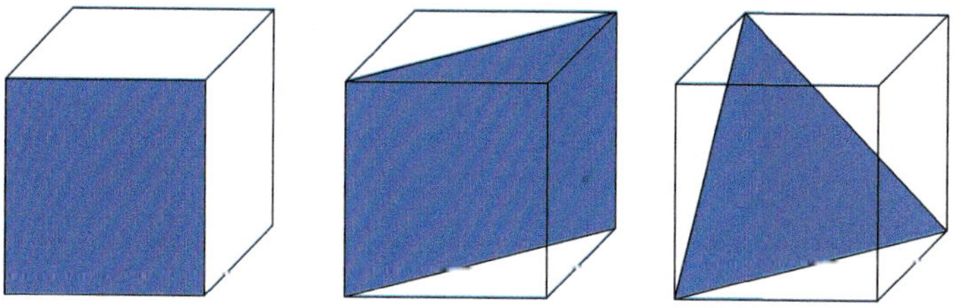

Figure 1.24 Some important planes in the cubic space lattice.

calculate the density of atoms on a plane, or *planar density*. The perpendicular intersection of a plane and sphere is a circle, so the radius of the atoms will be helpful in calculating the area they occupy on the plane. Refer again to Example Problem 1.4, where we calculated the lattice parameter for a BCC metal. The section shown along the body diagonal is the (110) plane. The body-centered atom is entirely enclosed by this plane, and the corner atoms are located at the confluence of four adjacent planes, so each contributes 1/4 of an atom to the (110) plane. So, there are a total of two atoms on the (110) plane. If we know the lattice parameter or atomic radius, we can calculate the *area of the plane*, A_p, the *area occupied by the atoms*, A_c, and the corresponding *planar density*, $PD = A_c/A_p$. We will see in Chapter 5 that an important property of metals, ductility, is affected by planar density.

Interplanar Spacings

To this point, we have concentrated on planes in an isolated cell. A crystal lattice is, of course, composed of many individual unit cells, with the planes extending in all directions. So, a real crystal lattice has many (111) planes, for example, all of which are parallel to one another. There is a uniform distance between like planes in a lattice, which we call the *interplanar spacing* and designate with d, the perpendicular distance between adjacent planes in a set. Note that even though the (111) and (222) planes are parallel to one another, they are **not** the same plane, since their planar densities may be much different depending on the lattice (for example, compare these two planes in SC). What we are calculating here is the perpendicular distance between the same plane in adjacent cells.

The general expression for d in terms of lattice parameters and interaxial angles is somewhat complicated.

$$d = V\begin{bmatrix} h^2b^2c^2\sin^2\alpha + k^2a^2c^2\sin^2\beta + l^2a^2b^2\gamma \\ + 2hlab^2c(\cos\alpha\cos\gamma - \cos\beta) + 2hkabc^2(\cos\alpha\cos\beta - \cos\gamma) \\ + 2kla^2bc(\cos\beta\cos\gamma - \cos\alpha) \end{bmatrix}^{-\frac{1}{2}} \tag{1.32}$$

where V is the cell volume as defined by Eq. (1.28). Note that the Miller indices of the plane under consideration are included here, to distinguish the distance between different planes in the same space lattice. For orthogonal systems (orthorhombic, tetragonal, and cubic), $\alpha = \beta = \gamma = 90°$

$$\frac{1}{d} = \left(\frac{h^2}{a^2} + \frac{k^2}{b^2} + \frac{l^2}{c^2}\right)^{\frac{1}{2}} \tag{1.33}$$

For a cubic system ($a = b = c$), this expression simplifies even further to

$$d = \frac{a}{\sqrt{h^2 + k^2 + l^2}} \tag{1.34}$$

Interplanar spacing is an important parameter for characterizing many types of materials. All we need to do is subject it to some X-rays.

Cooperative Learning Exercise 1.3

The (111) plane in FCC is shown at right.

Person 1: Calculate the area of the (111) plane, A_p, in the FCC cell in terms of the atomic radius, r.

Person 2: Calculate the area occupied by the atoms, A_c, in the (111) plane of the FCC cell in terms of r.

Combine your answers appropriately to arrive at the planar density, *PD*. If $r = 1.36$ Å, what is the *number density* (number of atoms per unit area) for the (111) plane in FCC?

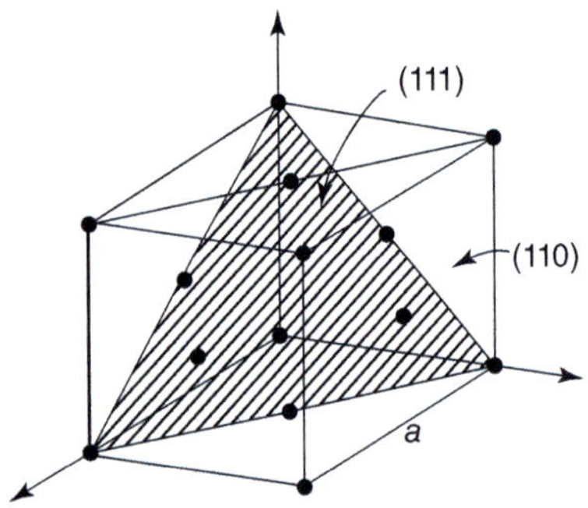

Reproduced by permission, from Z. Jastrzebski, *The Nature and Properties of Engineering Materials*, p. 37, 2nd ed. Copyright © 1976, John Wiley & Sons, Inc.

Answer: $PD = 0.19$; number density = 1.56×10^{15} atoms/cm^2

1.1.2 X-Ray Diffraction

The interplanar spacing between adjacent planes allows us to use a very powerful tool for structural determination called *X-ray diffraction (XRD)*. If we bombard a crystal with X-rays of a certain wavelength, λ, at an incident angle, θ, the X-rays can either pass directly through the crystal or, depending on the angle of incidence, interact with certain atoms in the lattice. This interaction is best visualized as a "shooting gallery" in which the incident X-rays "bounce off" the hard-core spheres in the lattice (see Figure 1.25). Really, the X-ray photons are interacting with the electron density around the atoms, leading to diffraction. The path difference for rays reflected from adjacent planes is $2d(\sin\theta)$, and *constructive interference* occurs when the path difference is an integer, n, of the wavelength:

$$n\lambda = 2d \sin\theta \tag{1.35}$$

Need More Help?

Try this video on Miller indices and *d*-spacing:

https://www.youtube.com/watch?v=LcoUFX3_A1s

Equation (1.35) is called *Bragg's law* and is a very important result. It says that if we bombard a crystal lattice with X-rays of a known wavelength and at a known

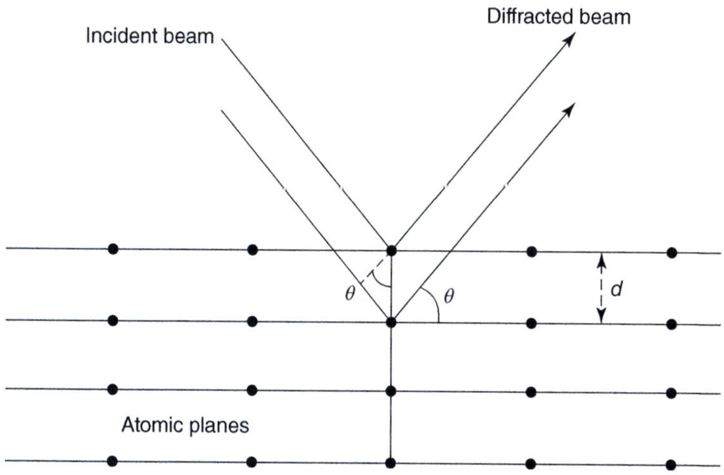

Figure 1.25 Schematic illustration of incident radiation diffraction by a crystal lattice.

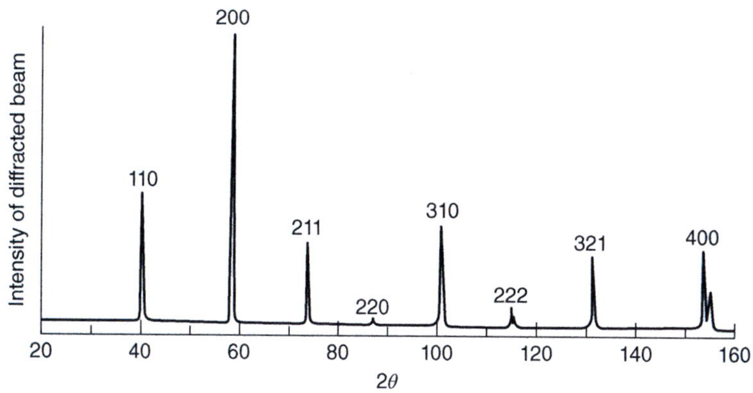

Figure 1.26 X-ray diffraction pattern for tungsten. Adapted from Guy and Hren [16]. Copyright © 1974 by Pearson Education.

angle, we will be able to detect diffracted X-rays of various intensities that represent a specific interplanar spacing in the lattice. The result of scanning through different angles at a fixed wavelength and counting the number (intensity) of diffracted X-rays is a diffraction pattern, an example of which is shown in Figure 1.26. Note how the sum of the Miller indices, or *order*, for each plane increases from left to right in the pattern, and how different planes have different intensities. Also, not all planes result in diffracted X-rays. From the abscissa (which is by convention listed as 2θ for instrumentation reasons), the interplanar spacing, d, can be calculated using Bragg's law.

The study of crystals and their X-ray patterns is both fascinating and complex. From a known crystal structure, a theoretical diffraction pattern can be constructed using vector algebra, including such considerations as *structure factors,* which account for disallowed interactions with the crystal lattice and determine the relative intensity of the diffracted beams.

For our purposes, it is sufficient to know that a diffraction pattern can be obtained experimentally in a relatively routine fashion, and the resulting pattern is characteristic of a specific crystalline material.

1.1.3 Point Defects

Now that the most important aspects of perfect crystals have been described, it is time to recognize that things are not always perfect, even in the world of space lattices. This is not necessarily a bad thing. As we will see, many important material phenomena that are based on defective structures can be exploited for very important uses. These *defects*, also known as *imperfections*, are grouped according to spatial extent. *Point defects* have zero dimensions; *line defects*, also known as *dislocations*, are one-dimensional; and *planar defects* such as *surface defects* and *grain boundary defects* have two dimensions. These defects may occur individually or in combination.

Let us first examine what happens to a crystal when we remove, add, or displace an atom in the lattice. We will then describe how a different atom, called an *impurity* (regardless of whether it is beneficial), can fit into an established lattice. As shown by Eq. (1.36), point defects have equilibrium concentrations that are determined by temperature, pressure, and composition. This is not true for all types of dimensional defects that we will study.

$$N_d = N \exp(-E_d/k_B T) \tag{1.36}$$

In Eq. (1.36), N_d is the equilibrium number of point defects, N is the total number of atomic sites per volume or mole, E_d is the activation energy for formation of the defect, k_B is Boltzmann's constant (1.38×10^{-23} J/atom · K), and T is absolute temperature. Equation (1.36) is an *Arrhenius-type expression* of which we will see a great deal in subsequent chapters. Many of these Arrhenius expressions can be derived from the Gibbs free energy, ΔG.

Need More Help?

Try this video on point defects:

https://www.youtube.com/watch?v=EuUJApDftkI
Note that Prof. Schamberger uses Kroger-Vink notation to describe point defects. You can learn more about this special type of nomenclature that allows you to write defect reactions in ***Advanced Topics***.

When an atom is missing from a lattice, the resulting space is called a *vacancy* (not to be confused with a "hole," which has an electronic connotation), as in Figure 1.27. In this case, the activation energy, E_d, is the energy required to remove an atom from the lattice and place it on the surface. The activation energy for the formation of vacancies in some representative elements is given in Table 1.14, as well as the corresponding vacancy concentration at various temperatures. Note that the vacancy concentration decreases at lower temperatures. In a nonequilibrium situation—such as rapid cooling from the melt—we would not expect the equilibrium concentration to be attained. This is indeed the case, and vacancy concentrations in rapidly quenched metals are much closer to the liquid concentration than they are to the equilibrium solid concentration.

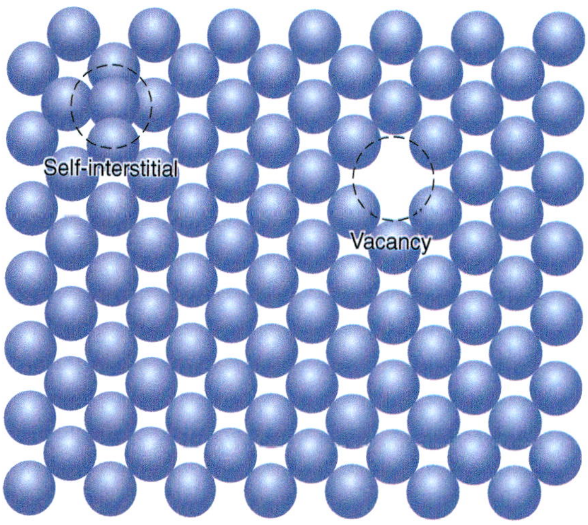

Figure 1.27 Representation of a vacancy and self-interstitial in a crystalline solid. From Ralls et al. [7]. Copyright © 1976 by John Wiley & Sons, Inc. This material is used by permission of John Wiley & Sons, Inc.

Table 1.14 Formation Energy of Vacancies for Selected Elements and Equilibrium Concentrations at Various Temperatures

			N_d (vacancies/cm^3)			
Element	**E_d (kJ/mol)**	**Melting Point, T_m (°C)**	**25°C**	**300°C**	**600°C**	**T_m**
Ag	106.1	960	1.5×10^4	1.5×10^{13}	3.0×10^{16}	7.8×10^{17}
Al	73.3	660	1.0×10^{10}	1.2×10^{16}	2.4×10^{18}	5.0×10^{18}
Au	94.5	1063	1.5×10^6	1.5×10^{14}	1.5×10^{17}	1.2×10^{19}
Cu	96.4	1083	1.1×10^6	1.4×10^{14}	1.4×10^{17}	9.0×10^{18}
Ge	192.9	958	<1	1.3×10^5	1.3×10^{11}	8.2×10^{13}
K	38.6	63	2.1×10^{15}	—	—	1.3×10^{16}
Li	39.5	186	4.7×10^{15}	—	—	1.4×10^{18}
Mg	85.8	650	4.4×10^7	6.4×101^4	3.5×10^{17}	5.7×10^{17}
Na	38.6	98	4.0×10^{15}	—	—	1.0×10^{17}
Pt	125.4	1769	8.7	2.7×10^{11}	2.0×10^{15}	4.2×10^{19}
Si	221.8	1412	<1	3.1×10^2	2.5×10^9	8.0×10^{15}

The second type of point defect is called an *impurity*. Impurities can occur in two ways: as an interstitial impurity, in which an atom occupies an interstitial site (see Figures 1.20, 1.21, and 1.28); or when an impurity atom replaces an atom in the perfect lattice (see Figure 1.28). In the first instance, either the same atom as in the lattice or an impurity atom can occupy an interstitial site, causing considerable lattice strain as the atomic planes distort slightly to accommodate the misplaced atom. The amount of strain created depends on how large the atom is relative to lattice atoms. It is also possible for a lattice atom to move off a lattice site and occupy an interstitial site. In

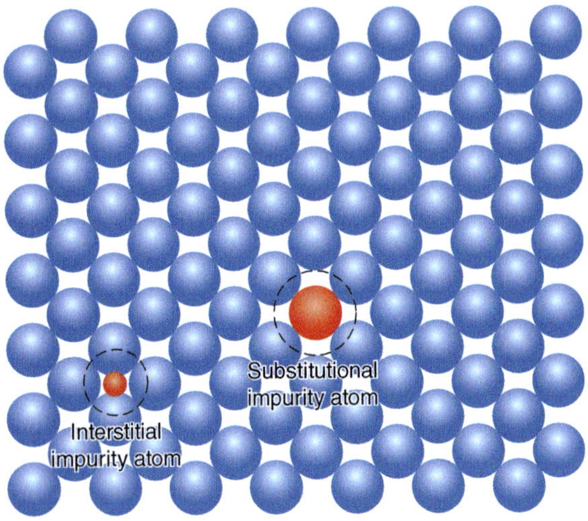

Figure 1.28 Representation of interstitial and substitutional impurity atoms in a crystalline solid. From Ralls et al. [7]. Copyright © 1976 by John Wiley & Sons, Inc. This material is used by permission of John Wiley & Sons, Inc.

this case, both defects shown in Figure 1.27 occur simultaneously, and a defect pair known as a *Frenkel defect* (or Frenkel disorder) occurs. In a pure Frenkel defect, there are always equal concentrations of interstitial impurities and vacancies.

The second type of impurity—substitution of a lattice atom with an impurity atom—allows us to enter the world of alloys and intermetallics. Let us diverge slightly for a moment to discuss how control of substitutional impurities can lead to some useful materials, and then we will conclude our description of point defects. An *alloy*, by definition, is a metallic solid or liquid formed from an intimate combination of two or more elements. By "intimate combination," we mean either a liquid or solid solution. In the instance where the solid is crystalline, some of the impurity atoms (usually defined as the minority constituent) occupy sites in the lattice that are normally occupied by the majority constituent. Alloys need not be crystalline, however. If a liquid alloy is quenched rapidly enough, an amorphous metal can result. The solid material is still an alloy, since the elements are in "intimate combination," but there is no crystalline order and hence no substitutional impurities. To aid in our description of substitutional impurities, we will limit the current description to crystalline alloys, but keep in mind that amorphous alloys exist as well.

The extent to which a lattice will allow substitutional impurity atoms depends on several things. The factors affecting the solubility of one element in another are summarized in a set of guidelines called the *Hume–Rothery rules*, though they are really not rules at all. As you can imagine, atomic size plays an important role in determining solubility. The first Hume–Rothery "rule" states that if the atomic size of the host lattice and impurity atom differ by more than about 14%, the solubility of the impurity in the lattice will be small. Refer to Table 1.10 for values of atomic size. The second rule involves electronegativity. We mentioned earlier in this chapter that electronegativity is an important concept and it plays an important role in determining not only how soluble an impurity is but also what type of bond will result. In general, the larger the electronegativity difference, $\Delta\chi$, between the host atom and the impurity, the greater the tendency to form compounds and the less solubility there is. So, elements with similar electronegativities (refer to Table 1.4) tend to alloy, whereas elements with large $\Delta\chi$ tend to have more ionic bonds (see Section 1.0.3) and form

Table 1.15 Common Crystal Structures of Alloys Based on Valences of Components

Valence Electrons/Atom	Structure
3/2	BCC, complex cubic, HCP
21/13	Complex cubic
7/4	HCP

intermetallics. Intermetallics are similar to alloys, but the bonding between the different types of atoms is partly ionic, leading to different properties than traditional alloys. The third rule deals with crystal structures. One would expect crystal structures to be more compatible, and this is generally the case. Refer to Table 1.12 for typical crystal structures, but keep in mind that the elements can have multiple structures depending on temperature, and remember that this can affect the stability of the alloy. Finally, all other things being equal, the fourth Hume–Rothery rule states that a metal of lower valency is more likely to dissolve one of higher valency than vice versa. Common valences of the elements are listed in Table 1.10. Again, elements can have multiple oxidation states. An interesting corollary to the fourth rule is that the total number of valence electrons per atom can be used as a guideline in determining the crystal structure of the alloy. As summarized in Table 1.15, by summing the valence electrons of the elements in the alloy and dividing them by the number of types of atoms (binary = 2, ternary = 3, etc.), it is sometimes possible to predict the crystal structure of an alloy. The "complex cubic" structures include cubic structures other than SC, BCC, and FCC, which we have not yet described, such as the diamond structure. As an example of this corollary, the binary alloy formed between Cu (+1 valence) and Be (+2 valence) has $(1 + 2)/2 = 3/2$ valence electrons/atom, and it turns out to have the BCC structure, which is different than either of the two-component structures.

Cooperative Learning Exercise 1.4

Person 1: Do Cu and Ni satisfy the first and second Hume–Rothery rules for complete solid solubility?

Person 2: Do Cu and Ni satisfy the third and fourth Hume–Rothery rules for complete solid solubility?

Compare your answers. Would you predict that Cu and Ni have complete, partial, or no solid solubility in each other?

Answer: Using covalent radii from Table 1.10, $r_{Ni} = 1.17$ Å, $r_{Cu} = 1.22$ Å, so $\Delta r\% = 4\%$ (<14%); $\Delta\chi = 0.01$ (small); both are FCC; Cu is lower valence than Ni (1+ compared to 2+). All four are satisfied; Cu and Ni are completely soluble.

This concludes our diversion into alloys for the time being. From this point on, we will often describe metals and alloys in similar terms, and we will make distinctions between the two classes of materials only when there are substantial dissimilarities between them. Returning now to our description of point defects, we have but one type of point defect pair left to describe. Like a Frenkel

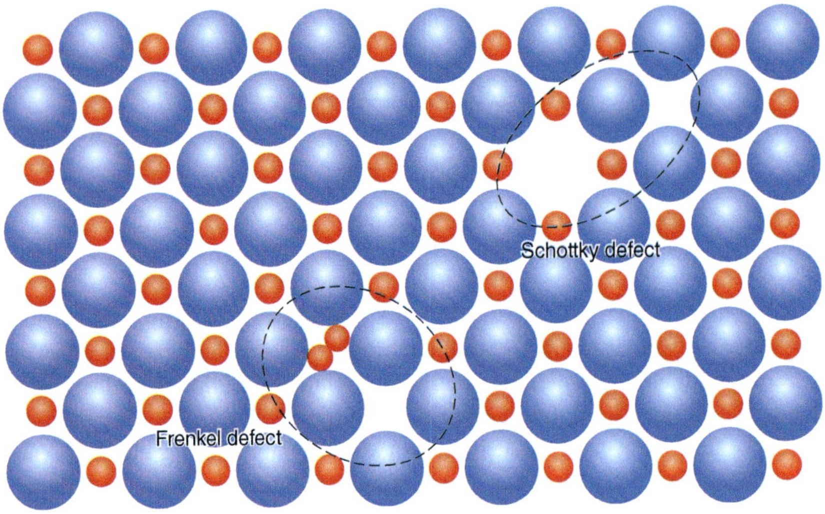

Figure 1.29 Representation of Frenkel and Schottky defects in a crystalline solid. Adapted from Moffatt et al. [17]. Copyright © 1964 by John Wiley & Sons, Inc.

defect in which both a vacancy and interstitial impurity must occur simultaneously, a *Schottky defect* (a.k.a. Schottky disorder or imperfection) arises in ionic solids when a cation–anion vacancy pair is formed. Recall that ionic compounds occur when there is a large electronegativity difference between the components, so a Schottky defect normally occurs in binary ionic compounds such as sodium chloride. Though the ionic compounds we will use as illustrations here are not technically metals or alloys, keep in mind that metallic solids such as intermetallics can have ionic bonding. In sodium chloride, the removal of one sodium ion and one chloride ion from the lattice results in a Schottky defect (see Figure 1.29). In ionic solids where the cation (positively charged ion) and anion (negatively charged ion) have the same absolute charge (e.g., $|Na^+| = 1$, $|Cl^-| = 1$), a Schottky defect arises from the same number of vacancies in both ions. For ionic solids in which the anion and cation have different absolute valencies (e.g., CaF_2), a nonstoichiometric compound must be formed to maintain charge neutrality in the lattice; that is, two fluorine ions (F^-) must leave for every calcium ion (Ca^{2+}) that is removed from the lattice. Because atoms must leave the ionic lattice, Schottky defects normally occur only at lattice perturbations such as grain boundaries or surfaces, so that the removed atoms have someplace to go. We will describe Schottky defects in more detail when we come to inorganic materials, such as oxides, where binary compounds are more prevalent.

1.1.4 Line Defects and Dislocations

We now move on to defects that have some spatial extent, even if only in one dimension. As we continue to increase the geometric complexity of these defects, you may find it more difficult to visualize them. As with crystal structures, three-dimensional models may help you with visualization. Do not limit yourself to one representation of a specific defect—look for multiple views of the same thing.

The first type of one-dimensional defect, or *line defect*, is called a *dislocation*. A dislocation is a linear disturbance of the atomic arrangement in a crystal caused by the displacement of one group of atoms from an adjacent group. There are three types of dislocations: *edge dislocations*, *screw*

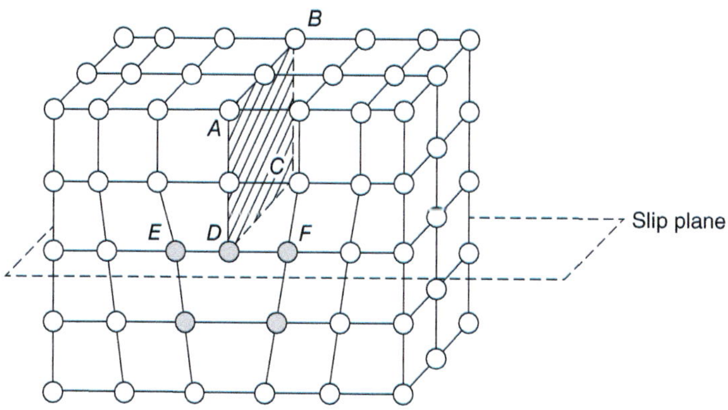

Figure 1.30 Edge dislocation DC results from introducing an extra half-plane of atoms $ABCD$, where DC is a positive dislocation designated by ⊥. From Jastrzebski [5]. Copyright © 1976 by John Wiley & Sons, Inc. This material is used by permission of John Wiley & Sons, Inc.

dislocations, and a combination of these two, termed *mixed dislocations*. An edge dislocation occurs when a single atomic plane does not extend completely through the lattice. The termination of this half-plane of atoms creates a defect line (dislocation line) in the lattice (line DC in Figure 1.30). The edge dislocation is designated by a perpendicular sign, either ⊥ if the plane is above the dislocation line or ⊤ if the plane is below the dislocation line. Edge dislocations can be quantified using a vector called the *Burger's vector*, **b**, which represents the relative atomic displacement in the lattice due to the dislocation (see Figure 1.31). The Burger's vector is determined as follows:

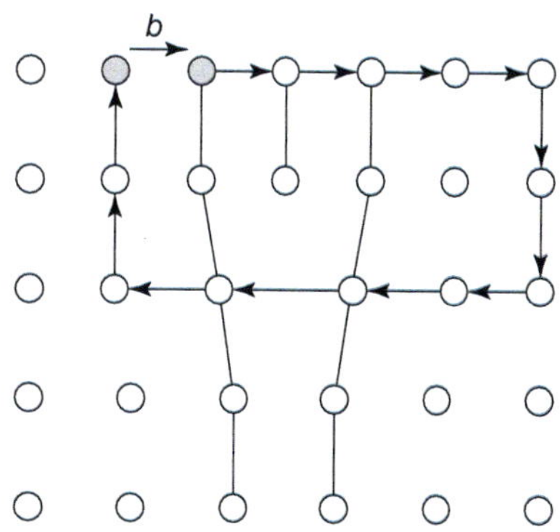

Figure 1.31 Burger's circuit around an edge dislocation. From Jastrzebski [5]. Copyright © 1976 by John Wiley & Sons, Inc. This material is used by permission of John Wiley & Sons, Inc.

- Define a positive direction along the dislocation line. This is usually done into the crystal.
- Construct a plane of atoms perpendicular to the dislocation line.
- Trace out a *clockwise* path around the dislocation line, moving n lattice vectors in each of the four mutually perpendicular directions.
- The Burger's vector is drawn from the *finish* to the *start* of the path.

Need More Help?

Try these videos on dislocations and Burger's vector:

https://www.youtube.com/watch?v=-t6btGjGKYU
https://www.youtube.com/watch?v=7qgBMKS_fyo

You can start the clockwise path—called the *Burger's circuit*—anywhere, as long as you entirely enclose the dislocation with the circuit.

The second type of line defect, the screw dislocation, occurs when the Burger's vector is parallel to the dislocation line (OC in Figure 1.32). This type of defect is called a screw dislocation because the atomic structure that results is similar to a screw. The Burger's vector for a screw dislocation is constructed in the same fashion as with the edge dislocation. When a line defect has both an edge and a screw

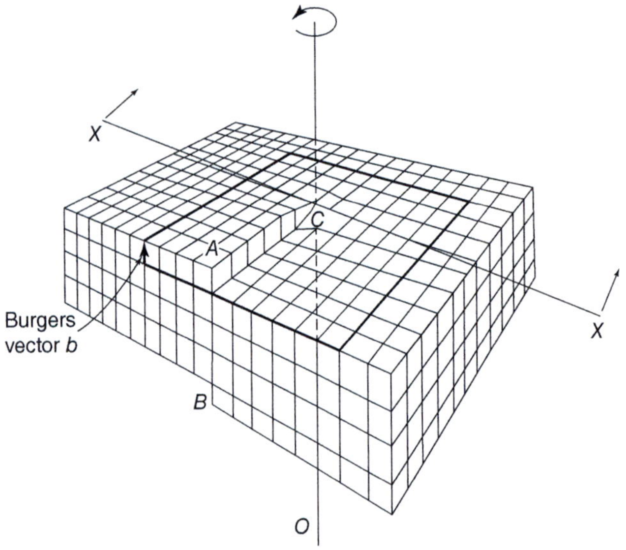

Figure 1.32 Representation of defect line (*OC*), Burger's circuit, and Burger's vector in a screw dislocation. From Jastrzebski [5]. Copyright © 1976 by John Wiley & Sons, Inc. This material is used by permission of John Wiley & Sons, Inc.

Table 1.16 Comparison of Dislocation Characteristics

Dislocation Type	Burger's Vector	Propagation Direction
Edge	⊥ to dislocation line	⊥ to dislocation line, ǁ to Burger's vector
Screw	ǁ to dislocation line	⊥ to dislocation line, ⊥ to Burger's vector
Mixed	Neither ǁ nor ⊥ to dislocation line	Neither ǁ nor ⊥ to dislocation line, Burger's vector

dislocation component, a mixed dislocation results. In this case, the Burger's vector is neither parallel nor perpendicular to the dislocation line but can be resolved into edge and screw components. A comparison of the three types of dislocations is summarized in Table 1.16.

Dislocations can move through a crystal lattice. This is most easily visualized in the edge dislocation, where the half-plane of atoms simply moves perpendicular to the defect line (line *DC* in Figure 1.30). The lattice atoms below the half-plane incorporate the atoms above them into their plane but must create a new half-plane of atoms to do so. This is how the half-plane propagates. The dislocation movement can be very rapid—approaching the speed of sound. Screw and mixed dislocations move in more complex, but similar, fashions. The propagation of all three dislocation types is summarized in Figure 1.33. We will see that dislocation propagation plays an important role in ductility and crystal slip.

1.1.5 Planar Defects

Finally, two-dimensional defects can occur in crystals. There are two categories of planar defects: *stacking faults* and *grain boundaries*.

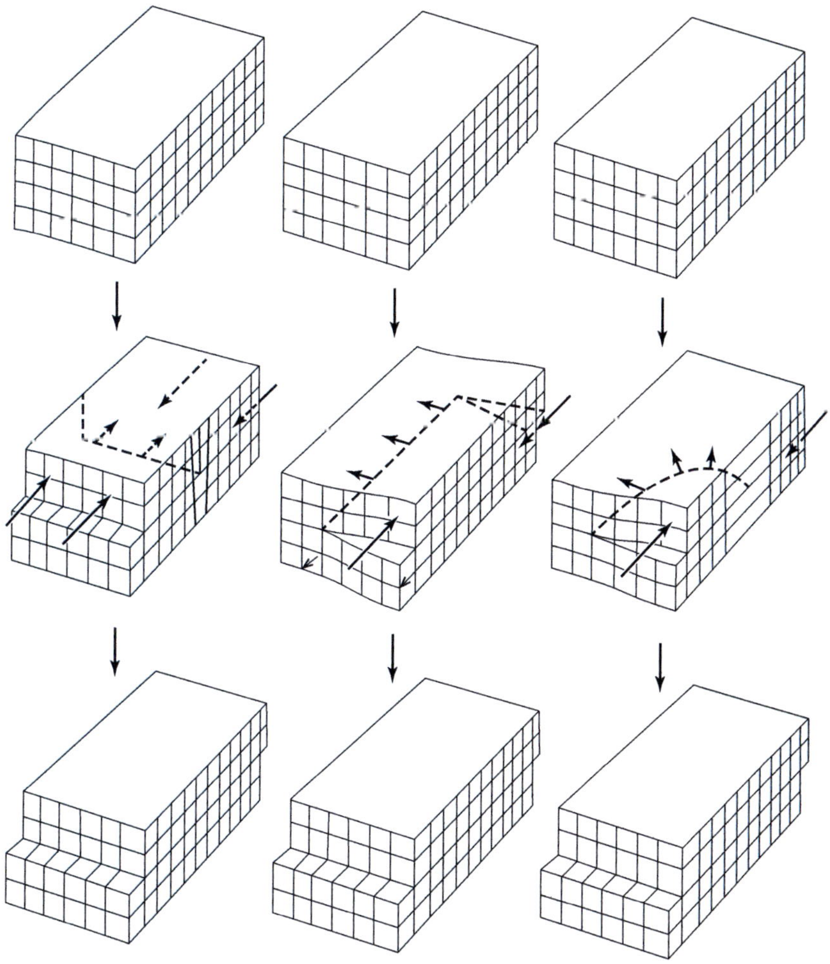

Figure 1.33 Representation of propagation modes for (a) edge, (b) screw, and (c) mixed dislocations.

Stacking faults arise due to imperfections in the stacking sequence of atomic planes. Recall that the FCC structure is the result of ABCABC... stacking of close-packed planes. If a plane of atoms is missing from this sequence, as in ABCAB|ABC, an *intrinsic stacking fault* results. If an additional plane is inserted into the sequence (e.g., an extra C plane in the sequence ABCA|C|BCABC), an *extrinsic stacking fault* results. Finally, a *twinning fault* occurs when the sequence reverses itself about a *mirror plane*, as in ABCAB$\underline{\text{C}}$BACBA, where $\underline{\text{C}}$ is the mirror plane and CBA is the reverse sequence to ABC.

To this point, we have concentrated on single crystals. Most crystalline materials are *polycrystalline*—that is, composed of many small crystals, or *grains*, that usually have random crystallographic orientations relative to each other. Unless the grains are on a surface, they are adjacent to other grains that are not necessarily of the same orientation. The region where they intersect is called a *grain boundary*. There are two general types of grain boundaries: tilt and twist.

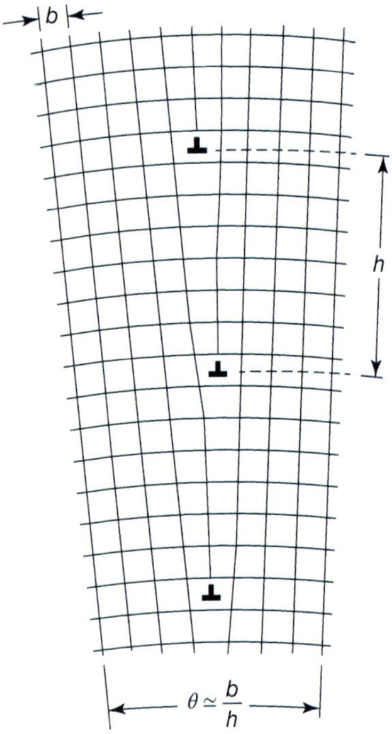

Figure 1.34 Representation of a tilt grain boundary. From Ralls et al. [7]. Copyright © 1976 by John Wiley & Sons, Inc. This material is used by permission of John Wiley & Sons, Inc.

A *tilt grain boundary* is a set of edge dislocations (see Figure 1.34). The *angle of misorientation*, θ, characterizes the tilt grain boundary and is defined as the angle between the same directions in adjacent crystals. The angle of misorientation can be calculated from the Burger's vector and the vertical separation between edge dislocations, h:

$$\tan \theta = b/h \tag{1.37}$$

A *high-angle tilt grain boundary* occurs when $\theta > 15°$. For $\theta < 10°$, a *low-angle tilt grain boundary* results, and Eq. (1.37) can be simplified to tan $\theta \approx \theta = b/h$. In a similar manner, a *twist grain boundary* is a set of screw dislocations.

Need More Help?
Try these videos on stacking faults and grain boundaries: https://www.youtube.com/watch?v=4tFKynJIwrM https://www.youtube.com/watch?v=5ecxI5YSvM0

We will have much more to say about planar defects, particularly their effect in determining the physical properties of bulk crystalline solids. In this regard, not only will defect structure play an important role, but so will crystallite size. Generally, polycrystalline materials are termed *coarse-grained*, *fine-grained*, and *ultra-fine-grained* in order of decreasing crystallite size. There are additional methods for classifying the grain size of materials, such as that used for metals by the American Society for Testing Materials (ASTM). A more recent development in this terminology has been the addition of the term *nanocrystalline* to describe materials with crystallite sizes of 10–100 nm. This term leads us naturally into the broader topic of nanostructured materials.

1.1.6 Nanostructures

The prefix "nano" means 10^{-9}, which in this case refers to a dimension in meters. A nanometer is 10 angstroms (Å, 10^{-10} m), so assemblies of a few atoms with individual diameters in angstroms (see Table 1.10) but collective dimensions in nanometers are termed *nanostructures*. From there, the prefix nano can be applied to any type of geometry to describe both the size scale and general shape of the nanostructure as in *nanosphere*, *nanorod*, *nanotube* (for a hollow cylinder), *nanosheet*, or even *nanobarrel* [18]. Although both looser and stricter definitions exist [19], most nanostructures have some characteristic dimension such as a diameter, thickness, or length that is of the order 10^{-9} m. For most technological applications that means a characteristic dimension in the range of 1–100 nm. Some examples and their relative sizes are illustrated in Figure 1.35. Such nanostructured materials or *nanomaterials* can be further classified by the dimensionality of their nanostructure. Nanostructures that are dots or spheres have zero dimensionality, tubes and cylinders are one-dimensional structures, sheets have two dimensionality, and some structures have nanoscale dimensions in all three directions (Figure 1.36).

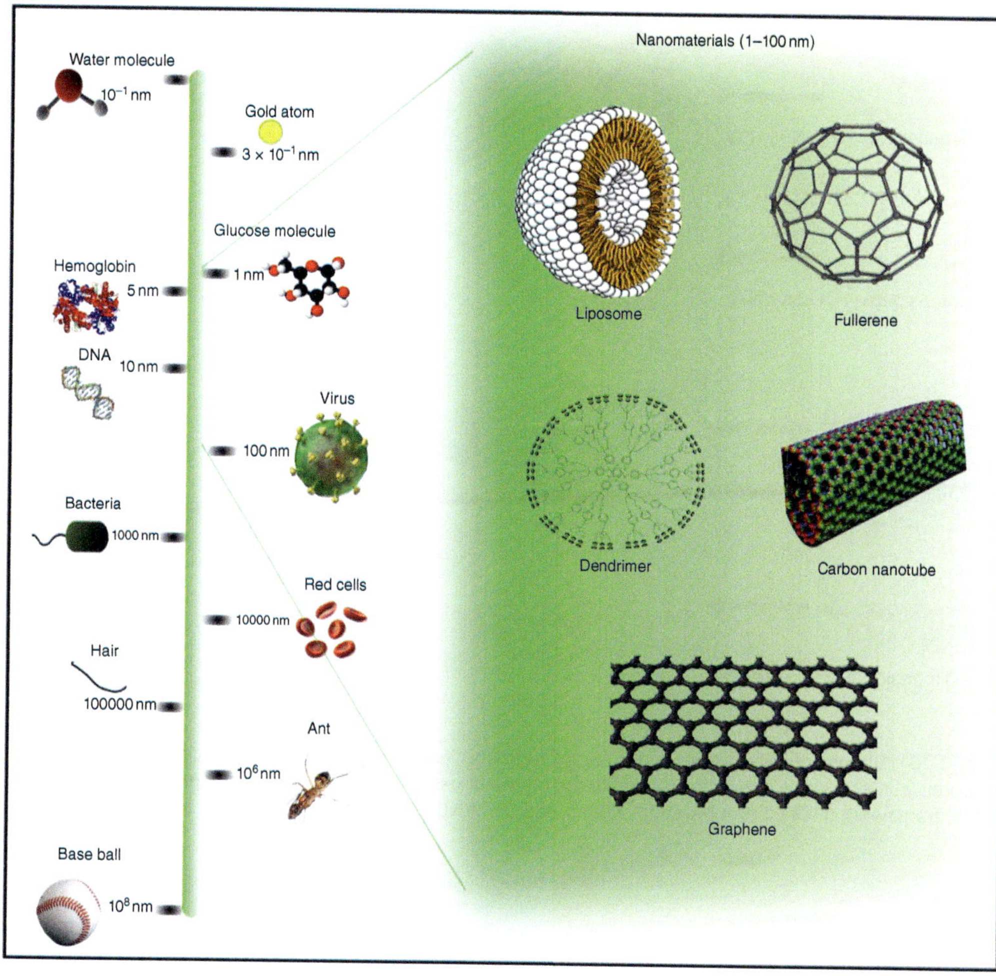

Figure 1.35 The size of nanomaterials compared to common items. From Panneerselvam et al. [20], MDPI/CC BY 3.0.

Nanomaterials may comprise any of the material classes discussed in this book. Regardless of whether they are made of an element or compound, are ceramic, polymer, or biologic, they are of particular technological interest due to some combination of their specific size, biological activity, or unique physical property that is not present at the microscale (10^{-6} m) or above. This makes nanoscience and nanostructure materials processing rapidly evolving fields. Nevertheless, there are some fundamental structural concepts that we can introduce here that will serve us well in subsequent discussions on the physical properties and processing of materials. One of those key structural components is the surface-to-volume ratio of nanomaterials. We will briefly introduce this topic as well as provide an example of industrially important nanostructured materials made from carbon. More detailed descriptions of other important nanoscale properties and the processing of nanostructure materials will be reserved for later chapters as we delve deeper into the electronic structure and properties of materials. But for now, let us see how some simple geometric arguments illustrate the fundamental importance of nanostructured materials.

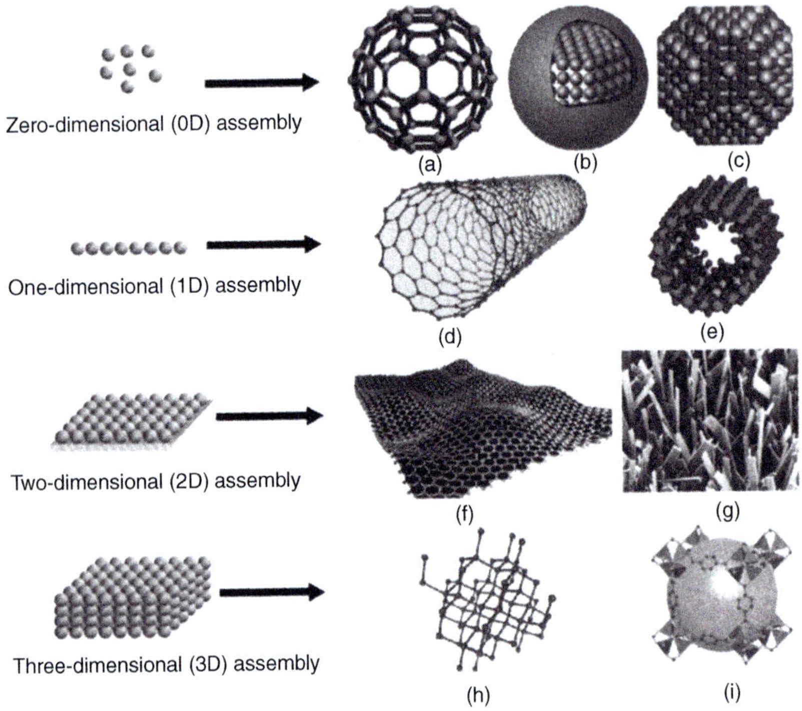

Figure 1.36 Dimensionality in nanomaterials: (a) fullerene; (b) quantum dot; (c) metal cluster; (d) carbon nanotube; (e) metal oxide nanotube; (f) graphene; (g) metal oxide nanobelts; (h) nanodiamond; and (i) metal-organic framework (MOF). From Narendra Kumar and Sunit Kumbhat [19], John Wiley & Sons, Inc.

Surface-to-Volume Ratio

For simplicity, let us consider a nanostructured sphere, or *nanosphere*. These structures are sometimes called *nanoparticles* or even *quantum dots*, depending on the context, material, properties, or application. Regardless of what it is called, a nanosphere has a surface area, S, of $4\pi r^2$, and a volume, V, of $\frac{4}{3}\pi r^3$, giving it a surface-to-volume ratio of

$$\frac{S}{V} = \frac{4\pi r^2}{\frac{4}{3}\pi r^3} = \frac{3}{r} \tag{1.38}$$

Cooperative Learning Exercise 1.5

Calculate the surface-to-volume ratio of the following nanostructures:

Person 1: nanosheet of thickness r where the length and width of the sheets are much larger than the thickness

Person 2: nanocylinder of radius r where the length of the cylinder is much greater than the radius

Person 3: nanocube of side $2r$

Compare your results with each other and that of a nanosphere of radius r. Rank them in order of lowest to highest S/V.

Answers: $2/r$; $2/r$; $3/r$; $3/r$. The nanocube and nanosphere have the largest S/V in this simplified comparison.

In other words, the surface-to-volume ratio scales with the inverse of the nanosphere size. As the sphere becomes smaller, the relative surface area exponentially increases in what is called a *scalable effect*. We will learn more about other types of size-dependent phenomena called *quantum effects* in Chapter 6, but for now, the relationship of Eq. (1.38) is important for many phenomena that depend on surface area. Which geometry maximizes relative surface area? Try CLE 1.5 for some perspective.

Carbon Nanostructures

Some of the earliest and most important applications of nanostructured materials involve carbon. As we saw earlier, carbon has a valence of 4+ and generally prefers to be tetrahedrally coordinated with surrounding atoms, resulting in sp^3 hybridization. In a single atomic layer, however, that would leave one bond unsatisfied, so carbon tends to hybridize in an sp^2 arrangement when in sheet form. These carbon sheets are called *graphene* and are so named to reflect not only the graphitic composition but also the proliferation of carbon–carbon double bonds also found in alkenes (see Fig. 1.36f).

Historical Highlight

One of the early pioneers of carbon nanostructures was ***Mildred Dresselhaus***. Learn more about the "Queen of Carbon Science" at https://en.wikipedia.org/wiki/Mildred_Dresselhaus

Take a Peek! Jump ahead to Section 5.1.1 to see the superior mechanical properties or Section 6.0.1 to see the unique electronic properties of some nanomaterials.

If we were to take this graphene sheet and curl around a central axis until the edges meet, we would have a tube, or *carbon nanotube* (CNT). When there is a single atomic layer of carbon, such tubes are termed *single-walled carbon nanotubes* (SWCNTs, see Fig. 1.36d). When there are two layers, it is a double-walled carbon nanotube (DWCNT), and more than two atomic layers of carbon form a *multi-walled carbon nanotube* (MWCNT). These CNTs have exceptional mechanical properties—especially strength—and because of their light weight, they are used as reinforcements in composites (see Section 1.4). They are readily manufactured from a variety of natural precursors such as rice hulls through a high-temperature carbonization process called *pyrolysis*. Note that there can be layered nanotubes of materials other than carbon, in which case they are generally referred to as SWNT, DWNT, or MWNTs, respectively.

Nanostructured materials need not be limited to carbon, gold, metals, or any of the elements. Note that several of the nanostructures in Figure 1.36 are metal oxides or metal-organic frameworks

(MOFs). This observation leads us naturally into the next important class of materials that include metal oxides: ceramics and glasses.

1.2 Structure of Ceramics and Glasses

Inorganic materials constitute the largest class of solids in the world. We have already described metals, and while they are not organic (they contain no biological carbon), they are also not inorganic in the strict sense of the word—they are metals due to the unique characteristics of their valence electronic structure. Inorganic materials are typically chemical compounds such as metal oxides, carbides, or nitrides. They possess many interesting properties that we will only begin to describe at this point. They can also differ structurally from metals because of their chemical heterogeneity. We begin by describing how simple physical dimensions of different elements in a compound can lead to even more complex structures.

1.2.1 Pauling's Rules

Need More Help?

Try these videos on Pauling's Rules:

https://www.youtube.com/watch?v=Lv-4xWqsbWE
https://www.youtube.com/watch?v=Fq-9w2aNWL0
https://www.youtube.com/watch?v=NQEcHhOWr_I
https://www.youtube.com/watch?v=w6xLQwpsq2g

Recall that the structure of a crystal is determined mostly by how the atoms pack together. The same is true of multi-component mixtures like alloys, and of binary compounds containing noncovalent bonds, such as ionic compounds. In addition to the concept of electronegativity, Linus Pauling also produced a set of generalizations that are used to describe most ionic crystal structures. Pauling's first rule states that *coordination polyhedra* are formed. Coordination polyhedra are three-dimensional geometric constructions such as tetrahedra and octahedra. The polyhedron that will form is related to the radii of the anions and cations in the compound. Consider the two-dimensional representation of a binary ionic compound shown in Figure 1.37. The anions (blue circles) are larger than the red cations (why?), and a central cation cannot remain in contact with the surrounding anions if the anion radius is larger than a certain value. Thus, the structure in Figure 1.37c is unstable, whereas the structures in Figures 1.37a and b are both stable. Note that the cation–anion distance is simply the sum of the cation–anion radii for the stable structures. It is also true that the coordination number is determined by the *radius ratio* of the two ions (R_{anion}/R_{cation}). The larger the central cation, the more anions that can be packed around it. It is now evident why our description of ionic radii in

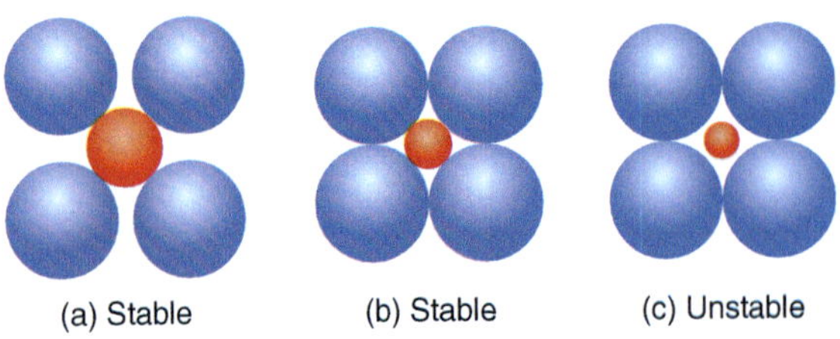

Figure 1.37 (a, b) Stable and (c) unstable coordination configurations. From Kingery et al. [13]. Copyright © 1976 by John Wiley & Sons, Inc. This material is used by permission of John Wiley & Sons, Inc.

Coordination number	*Anion-cation radius ratio*	*Coordination geometry*
2	> 6.45	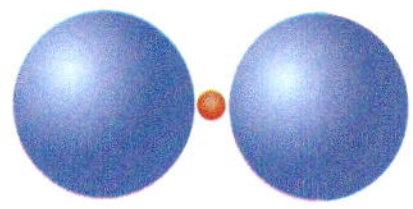
3	4.45–6.45	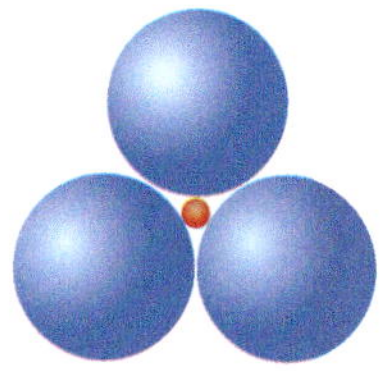
4	2.42–4.45	
6	1.37–2.42	
8	1.0–1.37	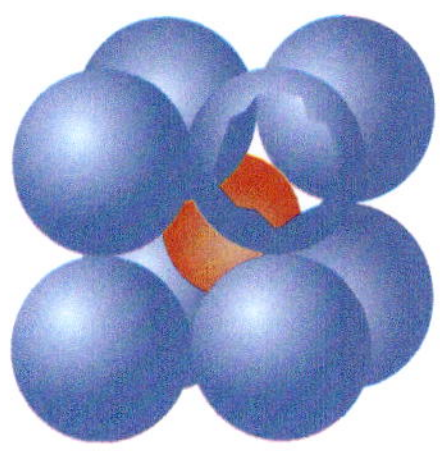

Figure 1.38 Critical radius ratios for various coordination numbers. Adapted, from Callister [21]. Copyright © 2000 by John Wiley & Sons, Inc.

Table 1.10 is so important. For each coordination number, there is some critical value of the radius ratio above which the structure will not be stable. These limits are summarized in Figure 1.38. Pauling's second rule on the packing of ions states that local electrical neutrality is maintained. We use a quantity called *bond strength* to assure electrical neutrality, where the bond strength is the ratio of formal charge on a cation to its coordination number. For example, silicon has a formal charge of +4 and a coordination number of 4, so its strength is $4/4 = 1$. For aluminum, a formal charge of +3 and a coordination number of six give a bond strength of 1/2. So, the total strength of the bonds reaching an anion from all surrounding cations is equal to the charge of the anion. Pauling's third rule tells us how to link these polyhedra together (see Figure 1.39). Corners, rather than faces or edges, tend to be shared in stable structures. This is because cation separation

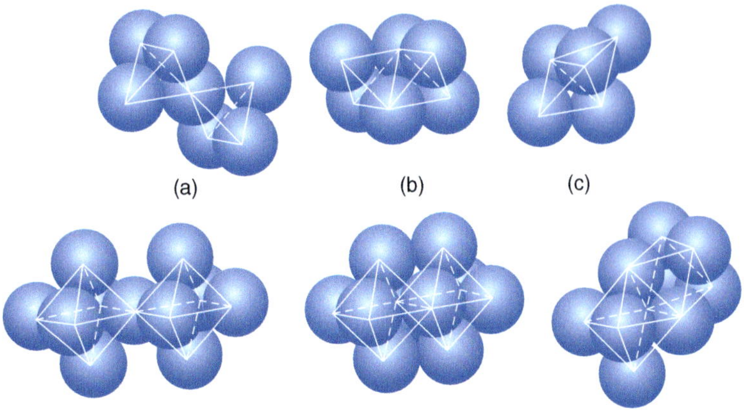

Figure 1.39 Tetrahedra (*top*) and octahedra (*bottom*) are linked by sharing (a) corners, (b) edges, and (c) faces. From Kingery et al. [13]. Copyright © 1976 by John Wiley & Sons, Inc. This material is used by permission of John Wiley & Sons, Inc.

between adjacent polyhedra decreases as edges and faces are shared, increasing repulsion and leading to unstable structures. Pauling's fourth rule is similar to the third, stating that polyhedra formed by cations of low coordination number and high charge tend to be linked by corners. The fifth and final rule states that the number of different constituents in a structure tends to be small; that is, it is difficult to efficiently pack different-sized polyhedra into a single structure.

Cooperative Learning Exercise 1.6

Magnesium silicate, Mg_2SiO_4, consists of an HCP arrangement of O^{2-} ions with the cations occupying interstitial spaces. Use the ionic radii of the elements given in Table 1.10 to answer the following questions.

Person 1: What is the coordination number for the Mg^{2+} ions and what type of interstitial site will they occupy?

Person 2: What is the coordination number for the Si^{4+} ions and what type of interstitial site will they occupy?

Combine your information. According to Table 1.17, what is the fraction of the total type of interstitial site occupied by each cation, and what is the structure name of this compound?

Answer: R_O/R_{Mg} = 1.4/0.72 = 1.94 (octahedral); R_O/R_{Si} = 1.4/0.4 = 3.5 (tetrahedral); 1/2 of octahedral sites are occupied; $\frac{1}{8}$ of tetrahedral sites are occupied; this is the olivine structure.

1.2.2 Ceramic Crystal Structures

Most crystalline inorganic compounds are based on nearly close-packing of the anions (generically referred to as O or X, though oxygen is the most common anion) with metal atom cations (generically called M or A) placed interstitially within the anion lattice. A summary of some simple ionic structures and their corresponding coordination numbers is given in Table 1.17. Some of the most common ceramic and ionic crystal structures are described in the following sections.

Table 1.17 Table of Some Simple Ionic Structures and Their Corresponding Coordination Numbers

Anion Packing	Coordination Number of M and O	Sites by Cations	Structure Name	Examples
Cubic close-packed	6:6 MO	All oct.	Rock salt	NaCl, LiF, KBr, MgO, CaO
Cubic close-packed	4:4 MO	1/2 tet.	Zinc blende	ZnS, BeO, SiC
Cubic close-packed	4:8 M_2O	All tet.	Antifluorite	Li_2O, Na_2O, K_2O, Rb_2O
Distorted cubic close-packed	6:3 MO_2	1/2 oct.	Rutile	TiO_2, GeO_2, SnO_2, PbO_2
Cubic close-packed	12:6:6 ABO_3	1/4 oct. (B)	Perovskite	$CoTiO_3$, $SrSnO_3$, $BaTiO_3$
Cubic close-packed	4 :6:4 AB_2O_4	1/8 tet. (A) 1/2 oct. (B)	Spinel	$FeAl_2O_4$, $ZnAl_2O_4$, $MgAl_2O_4$
Cubic close-packed	4:6:4 $B(AB)O_4$	1/8 tet. (B) 1/2 oct. (A, B)	Spinel (inverse)	$FeMgFeO_4$, $MgTiMgO_4$
Hexagonal close-packed	4:4 MO	1/2 tet.	Wurtzite	ZnS, ZnO, SiC
Hexagonal close-packed	6:6 MO	All oct.	Nickel arsenide	NiAs, FeS, FeSe, CoSe
Hexagonal close-packed	6:4 M_2O_3	2/3 oct.	Corundum	Al_2O_3, Fe_2O_3, Cr_2O_3, Ti_2O_3
Hexagonal close-packed	6:6:4 ABO_3	2/3 oct. (A, B)	Ilmenite	$FeTiO_3$, $NiTiO_3$, $CoTiO_3$
Hexagonal close-packed	6:4:4 A_2BO_4	1/2 oct. (A) 1/8 tet. (B)	Olivine	Mg_2SiO_4, Fe_2SiO_4
Simple cubic	8:8 MO	All cubic	CsCl	CsCl, CsBr, CsI
Simple cubic	8:4 MO_2	1/2 cubic	Fluorite	ThO_2, CeO_2, ZrO_2
Connected tetrahedra	4:2 MO_2	—	Silica types	SiO_2, GeO_2

Rock Salt Structure

The compounds NaCl, MgO, MnS, LiF, FeO, and many other oxides such as NiO and CaO have the so-called *rock salt structure*, in which the anions – such as Cl or O – are arranged in an FCC array. The cations are placed in the octahedral interstitial sites, also creating an FCC array of cations, as shown in Figure 1.40. The primitive cell of the rock salt structure, however, is SC, with anions at four corners, and cations at alternating corners of the cube, as given by one of the quadrants in Figure 1.40. As indicated in Table 1.17, the coordination number of both anions and cations is 6, so the cation–anion radius ratio is between 1.37 and 2.42, as indicated in Figure 1.38.

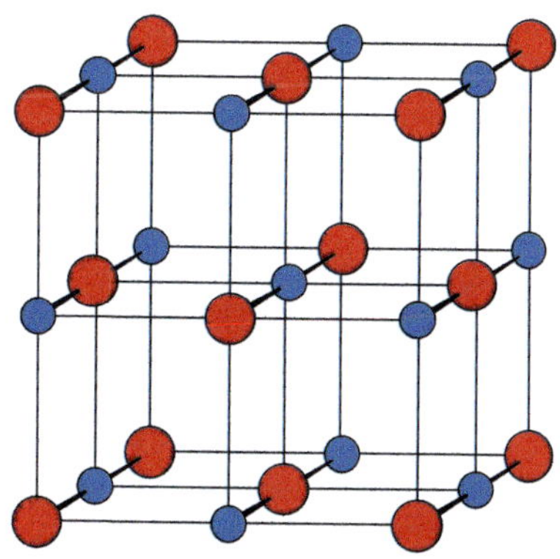

Figure 1.40 The rock salt crystal structure.

Diamond Structure

Another common ceramic structure arises when the tetrahedral sites in an FCC array of anions are occupied. For example, an FCC array of sulfur anions with Zn ions in the tetrahedral positions gives the compound ZnS and results in the *zinc blende structure* (see Figure 1.41). All the atoms can be alike also, for example, carbon, in which case diamond results. In either case, the atoms in the tetrahedral sites have a coordination number of four (by definition). Many important compounds have the zinc blende or diamond structure, including SiC.

Spinel Structure

Many compounds are formed when there is more than one metal cation in the lattice. Such is the case with the *spinel* structure, which has the general formula AB_2O_4, where A and B are different metal cations, such as in magnesium aluminate, $MgAl_2O_4$ (see Figure 1.42). This structure can be

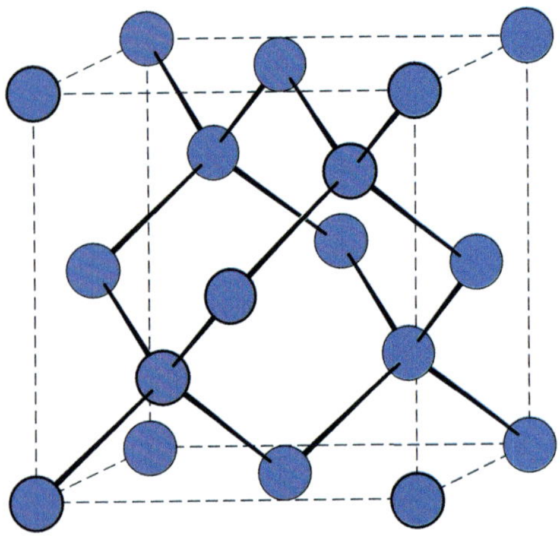

Figure 1.41 The diamond (zinc blende) crystal structure. From Ralls et al. [7]. Copyright © 1976 by John Wiley & Sons, Inc. This material is used by permission of John Wiley & Sons, Inc.

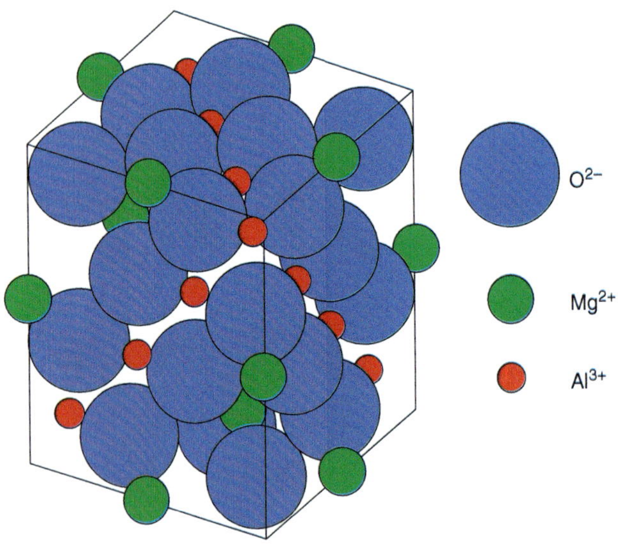

Figure 1.42 The spinel crystal structure of $MgAl_2O_4$. Reproduced, by permission, from Kittel [22]. Copyright © 1957 by John Wiley & Sons, Inc.

viewed as a combination of the rock salt and zinc blende structures. The anions, usually oxygen, are again placed in an FCC array. In a normal spinel, the divalent A ions are on tetrahedral sites and the trivalent B atoms are on octahedral sites. In an inverse spinel, divalent A atoms and half of the trivalent B atoms are on octahedral sites, with the other half of the B^{3+} atoms on tetrahedral sites. Many of the ferrites, such as Fe_3O_4 (in which iron has two different coordination states), have an inverse spinel structure.

Take a Peek! Jump ahead to Section 6.1.1 for more information on superconductors.

Other Important Ceramic Structures

There are many additional crystal structures that arise due to compounds of both simple and complex stoichiometry. These structures generally have specific names associated with them that have developed out of geology and crystallography over many years. For example, the *corundum* structure is common to Al_2O_3, the *rutile* structure comes from one of the forms of TiO_2, and a number of important ceramics, such as CaF_2, have the *fluorite* structure. One structure with current technological importance is the *perovskite* structure (see the $CaTiO_3$ perovskite structure in Figure 1.43). Many important ceramics with unique electrical and dielectric properties have the perovskite structure, including barium titanate, $BaTiO_3$, and *high-temperature superconductors* (HTS). The perovskites have the general formula ABO_3, but in the case of most superconductors, the A cation consists of more than one type of atom, such as in $Y_1Ba_2Cu_3O_{7-x}$, or the so-called "1–2–3" superconductor.

The *silicates*, made up of base units of silicon and oxygen, are an important class of ceramic compounds that can take on many structures, including some of those we have already described. They are complex structures that can contain several additional atoms such as Mg, Na, and K. What makes the silicates so important is that they can be either crystalline or *amorphous* (glassy) and provide an excellent opportunity to compare these two disparate types of structure. Let us first examine the crystalline state, which will lead us into the amorphous state.

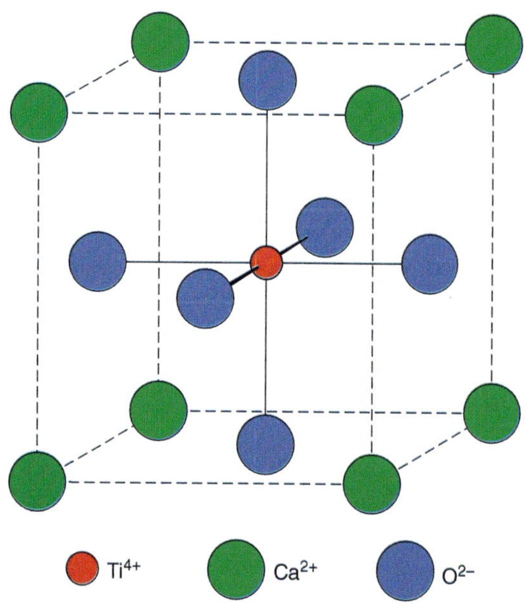

Figure 1.43 The perovskite crystal structure of $CaTiO_3$. From Kingery et al. [13]. Copyright © 1976 by John Wiley & Sons, Inc. This material is used by permission of John Wiley & Sons, Inc.

Need More Help?

Try these videos on silicate structures:

https://www.youtube.com/watch?v=lPSFt0XZlnk
https://www.youtube.com/watch?v=s4mACkEx1n8

The structural unit for the simplest silicate, SiO_2, also known as *silica*, is the tetrahedron (see Figure 1.44). This is the result of applying Pauling's principles (Section 1.2.1) to a compound between silicon and oxygen. The data in Table 1.10 indicate that the anion/cation ratio in SiO_2 is $R_O/R_{Si} = (1.4)/(0.4) = 3.5$, which, according to Figure 1.38, dictates the tetrahedron as the base structural unit. Note that the SiO_4 tetrahedron has a formal charge of −4, which must be neutralized with cations such as other Si atoms in real compounds. Pauling's second rule tells us that the bond strength in silicon is 1, and the third and fourth rules tell us that corners of the tetrahedra are generally shared. This is not always the case, and different macroscopic silicate structures result depending on how the tetrahedra are combined. Corners, edges, or faces of tetrahedra can be shared. As the nature of combination of the tetrahedra changes, so must the O/Si ratio, and charge neutrality is maintained through the addition of cations. These structures are summarized in Table 1.18.

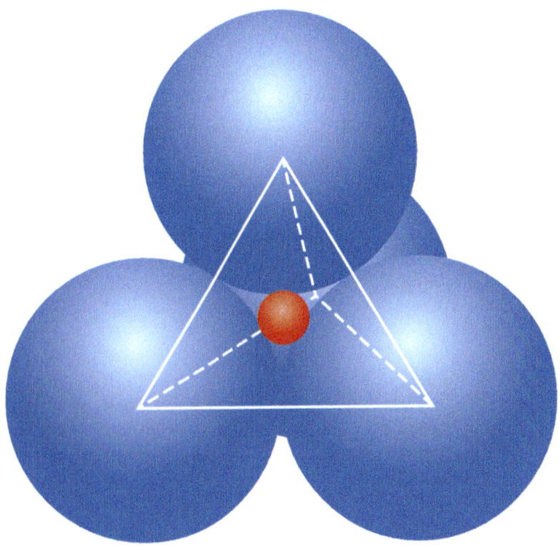

Figure 1.44 The $(SiO_4)^{4-}$ tetrahedron. The silicon atom is the red circle at the center of the tetrahedron; large blue circles are oxygens.

Table 1.18 Structural Units Observed in Crystalline Silicates

O/Si Ratio	Silicon–Oxygen Groups	Structural Units	Examples
2	SiO_2	Three-dimensional network	Quartz
2.5	Si_4O_{10}	Sheets	Talc
2.75	Si_4O_{11}	Chains	Amphiboles
3.0	SiO_3	Chains, rings	Pyroxenes, beryl
3.5	Si_2O_7	Tetrahedra sharing one oxygen ion	Pyrosilicates
4.0	SiO_4	Isolated orthosilicate tetrahedra	Orthosilicates

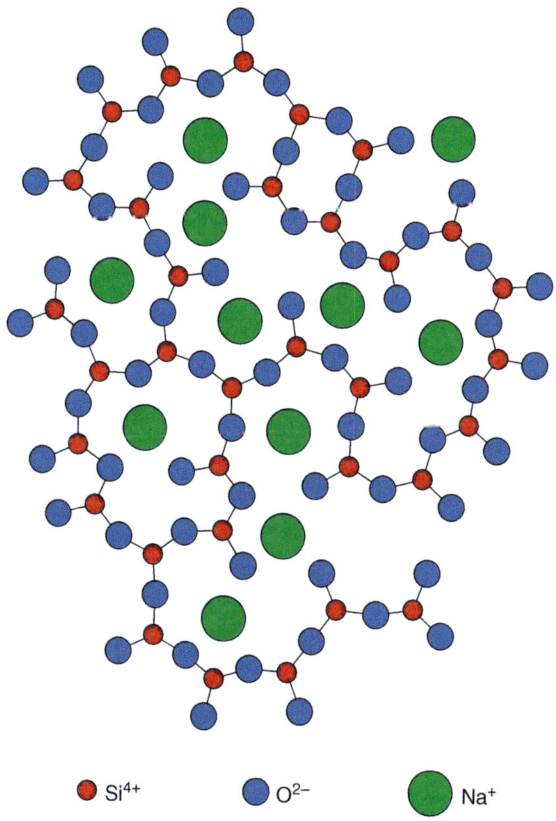

Figure 1.45 Schematic representation of a random network sodium silicate glass. From Kingery et al. [13]. Copyright © 1976 by John Wiley & Sons, Inc. This material is used by permission of John Wiley & Sons, Inc.

1.2.3 The Structure of Glasses*

The SiO_4 tetrahedra of silicates need not be arranged regularly. A perfect, three-dimensional SiO_2 lattice like quartz can be "disrupted" by introducing other cations, such as Na^+. This leads to a more random, yet still networked, structure called a *glass*. Such a random network glass is illustrated in Figure 1.45.

Historical Highlight

Some historians credit glass with such important societal advances as improving cleanliness and hygiene, to better understanding ourselves and the origins of the universe. Learn more about the history of this important substance at https://en.wikipedia.org/wiki/History_of_glass.

Before proceeding further with a description of the structural aspects of glasses, it is advisable to be aware of some pitfalls in nomenclature that abound in glass science. Technically, a glass is a type of *noncrystalline solid* that is formed from the melt. Thus, a glass need not contain silicon or oxygen at all, but it does need to be obtained by cooling a substance from its molten state. The distinction in processing condition is necessary to distinguish glasses from other types

Need More Help?

Try this video on oxide glasses:

https://www.youtube.com/watch?v=Ipioh66F9Qk

of amorphous materials that also do not contain a regular, repeating structure, but that are formed through other processing routes, such as from the vapor phase, in which case they are called *amorphous solids*, or by dehydrating a *sol* to form a *gel*. We will describe some of these processing techniques in later chapters, but for now we simply note that glasses must technically be formed from the melt and that there are no restrictions on the chemical constituents of a glass. Additional characteristics of glasses that are sometimes described in the glass literature include a rigid material, a glass transition, T_g, (see Section 1.3.7), and a viscosity greater than about 10^{15} poise. The viscosity distinction is an important one since some consider a glass to be a liquid of high viscosity. Finally, the term *vitreous* is sometimes used in connection with glasses. This term is usually reserved for glassy materials that can be crystallized through proper heat treatment in a process known as *devitrification*.

The distinction between an amorphous material and a glass is an important one. For example, an SiO_2 glass prepared from the melt has a noticeably different XRD pattern than a solid SiO_2 gel derived from dehydration of a solution (see Figure 1.46). In both cases, the glass and gel have no *long-range order* in comparison to cristobalite, which is highly crystalline and exhibits distinct XRD lines. The increase in the intensity of the gel pattern at small angles is due to microporous structures that result from the removal of water during drying—inhomogeneities that are not present in the silica glass. Both the glass and the gel do possess *short-range order*, however, as indicated

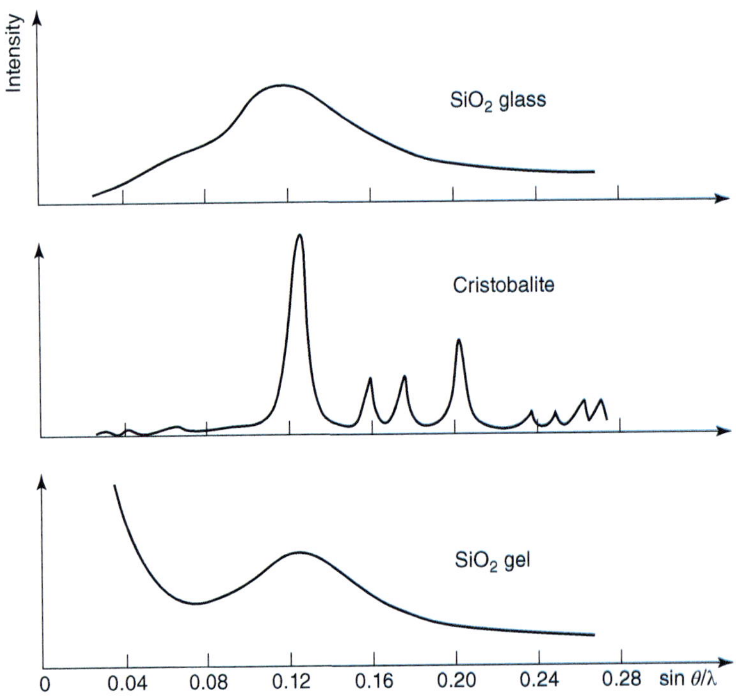

Figure 1.46 X-ray diffraction patterns of vitreous silica, crystalline silica (cristobalite), and sol-gel-derived silica. Reproduced, by permission from Scholze [23]. Copyright © 1991 by Springer-Verlag.

by the broad peak centered at a d-spacing of about 0.12 nm. This short-range order is attributed chiefly to the SiO_4 tetrahedral structural unit present in all silicates.

Though it may seem nonsensical to obtain XRD patterns from materials that have no long-range structure (i.e., crystallinity) since there is no spacing between crystal planes, the X-ray pattern provides additional information that is useful in analyzing the structure of glasses. Even though the peaks in an amorphous pattern are broad, we can extract additional information using something called the *radial distribution function* (rdf):

$$\mathrm{rdf} = 4\pi r^2 n_0 g(r) \tag{1.39}$$

where $g(r)$ is the *pair distribution function* between adjacent atoms—that is, the probability of finding another atom at a distance r from the reference atom located at $r = 0$. The pair distribution function is determined from various diffraction experiments (electron, neutron, and X-ray). The quantity n_0 in Eq. (1.39) is the average number density $= N/V$. If we plot rdf versus r, we obtain a curve similar to the one shown in Figure 1.47. The dotted line represents the parabola $4\pi r^2$, and deviations from the dotted line indicate regions of greater probability for finding an atom. The "peaks," then, correspond to likely bond distances as indicated: Si—O, O—O, and Si—Si. The radial distribution function is useful for characterizing not only glasses, but liquids and polymers as well.

So, we have seen that glasses have short-range structure, but no long-range structure, at least relative to the wavelength of the probing X-rays. But can we predict, or at least rationalize, which compounds will form glasses readily, and which ones will not? The answer is, "Yes." Once again, there are several sets of "rules," or guidelines, for describing the ability of certain cation/anion pairs to form glassy compounds. We will look at three such sets of guidelines, and you should recognize some of their components from earlier guidelines, such as Pauling's rules and the Hume–Rothery rules. Although we know that glasses by definition may consist of any types of cations and anions, oxide glasses are by far the most common and industrially most important. We will limit our discussion to oxide glasses for the moment.

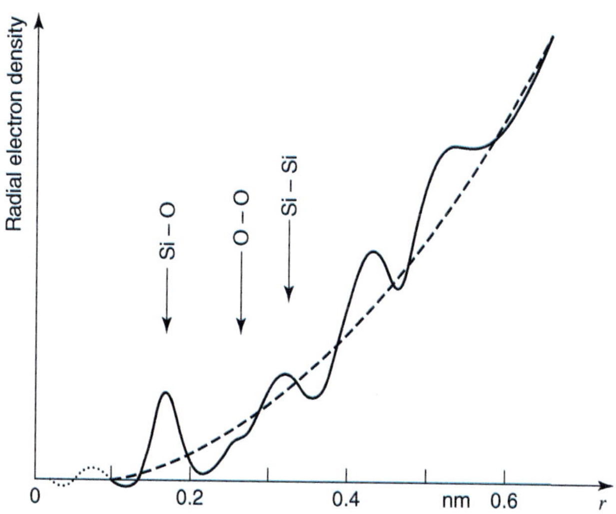

Figure 1.47 Radial distribution of electron densities of vitreous silica from X-ray exposures (Scholze). Reproduced, by permission from Scholze [24]. Copyright © 1991 by Springer-Verlag.

Zachariasen Rules

In 1932, W. H. Zachariasen considered the conditions for constructing a random network like the one shown in Figure 1.45 and proposed four rules for the formation of oxide glasses:

- An anion (oxygen atom) is linked to not more than two glass-forming cations (metal atoms).
- The coordination numbers of the glass-forming atoms (cations) are small, four or less.
- The oxygen polyhedra (structural units) share corners with each other, not edges or faces.
- The polyhedra are linked in a 3-D network (at least three corners of each polyhedra should be shared).

The "structural polyhedra" are those that we have already used: triangles, tetrahedra, and octahedra. Zachariasen's rules, as supported and modified by Warren, came to be known as the *random network theory* and, despite its limitations, are still widely used.

Stanworth Rules

In the late 1940s and early 1950s, Stanworth proposed a set of much simpler guidelines that did not rely on the formation of polyhedra. He suggested that the primary criteria for glass formation in metal oxide glasses were

- A cation valence ≥3
- An increasing tendency for glass formation with decreasing cation size
- A cation electronegativity between 1.5 and 2.1

Based on what we already know, there is circumstantial evidence to support these guidelines. For example, using the electronegativity values in Table 1.4, we see that two well-known glass-forming metal oxides B_2O_3 and SiO_2 meet the electronegativity criterion ($\chi_B = 2.04$, $\chi_{Si} = 1.90$), whereas Na_2O does not ($\chi_{Na} = 0.93$). Similarly, both B^{3+} and Si^{4+} have valencies greater than or equal to three, and they have relatively small cation sizes (0.2 and 0.40 Å, respectively) in comparison to other 3+ valence, nonnetwork-forming cations like Co and Fe (both 0.55 Å).

Oxide Glass Cations

Perhaps the single most useful guideline is Table 1.19, which classifies cations into three categories: *glass formers, intermediates*, and *modifiers*. These classifications are an extension of the random network model but also include some of Stanworth's guidelines. Glass formers are cations with a valence greater than or equal to three (this is a Stanworth rule) and a coordination number less than four (a Zachariasen rule). Note that there are many more glass-forming cations than just silicon. Intermediates are cations of lower valence and higher coordination numbers that can sometimes act as glass formers (such as aluminum) but can also act as network modifiers. A network modifier is a cation that serves to interrupt the random, glass network, partly by being of high enough valence to provide additional oxygen to the network, thereby increasing the oxygen-to-metal atom ratio and destroying the network (see Table 1.18). Note that some cations can be in several categories, such as Pb, which can have multiple oxidation states. Sodium is a well-known network modifier and is added to sand (quartz) in the form of Na_2O to form sodium silicates, which constitute a large class of glasses and, in the aqueous solution form, a large class of adhesives.

1.2.4 Glass Ceramics

There are materials that are hybrids between glasses and ceramics. *Glass ceramics* are a family of fine-grained crystalline materials achieved through controlled crystallization of glasses. They are

Table 1.19 Coordination Number and Bond Strength of Oxides

M in MO_x	Valence	Coordination Number
Glass formers		
B, Al	3	4
Si, Ge	4	4
Zr	4	6
P, V, As, Sb	5	4
Intermediates		
Zn, Pb, Cd	2	2
Be	2	4
Al	3	6
Ti	4	6
Th, Zr	4	8
Modifiers		
Li	1	4
Na	1	6
K	1	9
Rb	1	10
Cs	1	12
Pb, Zn, Cd	2	4
Mg, Cd, Hg	2	6
Ba, Ca, Sr	2	8
Sc, Ga, In	3	6
La	3	7
Y	3	8
Sn, Pb	4	6
Th	4	12

Source: Kingery et al. [13]. Copyright © 1976 by John Wiley & Sons, Inc.

nonporous and are either opaque or transparent. Their optical transparency arises from the fact that the crystals are smaller than the wavelength of visible light such that the light waves are transmitted without significant scattering (see Figure 1.48). Like glasses, glass ceramics have excellent corrosion and oxidation resistance, but have improved impact strength and dimensional stability over both glasses and ceramics. Glass ceramics are commonly made of compositions in the $MgO{-}Al_2O_3{-}SiO_2$, and $Li_2O{-}Al_2O_3{-}SiO_2$ systems, with the main crystal phases being *cordierite* ($Mg_2Al_4Si_5O_{18}$) and β-quartz and *β-spudomene* ($Li_2O \cdot Al_2O_3 \cdot 4SiO_2$) solid solutions. The residual glass phase is usually 1–10% by volume.

The desired crystalline microstructure is achieved through the addition of *nucleating agents*, often titanium dioxide, to the glass prior to heat treatment. The glass is heated to an appropriate

Take a Peek! Jump ahead to Section 3.1.2 for more information on nucleation and growth kinetics.

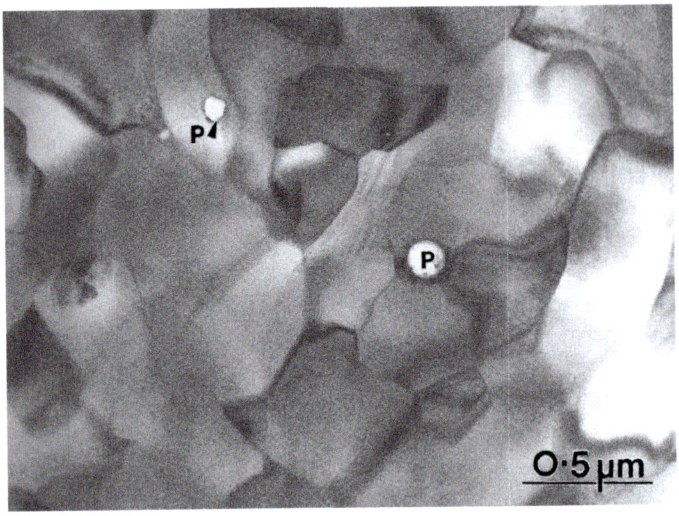

Figure 1.48 Microstructure of a typical glass ceramic with a fine grain size, minimal residual glass, and little porosity (P). Lee and Rainforth [25], © 1994, with permission from Springer Nature.

nucleating temperature that allows the maximum number of crystal nuclei to form, followed by heat treatment at a higher temperature to grow the crystals (see Chapter 3). The result is a highly crystalline material with tiny crystallites, on the order of a few nanometers. The processing of glass ceramics will be described in detail in a later chapter. Glass ceramics are used as ovenware and tableware, electrical insulators, substrates for circuit boards, architectural cladding, and heat exchangers and regenerators.

1.3 Structure of Polymers

The term *polymer* comes from "poly," meaning many, and "mer," meaning unit. Hence, polymers are composed of many units—in this case, structural units called *monomers*. A monomer is any unit that can be converted into a polymer. Similarly, a *dimer* is a combination of two monomers, a *trimer* is a combination of three monomers, and so on. Before describing the chemical composition of typical monomers and how they are put together to form polymers, it is useful to have a brief organic chemistry review. You may wish to refer to an organic chemistry text for more detailed information. We will reserve discussion of how these organic molecules are brought together to form polymers until Chapter 3, when the kinetics of polymerization are described. For the remainder of the description of polymer structure, it is sufficient to know that polymer chains are formed from the reaction of many monomers to form long-chain hydrocarbons, sometimes called *macromolecules*, but more commonly referred to as polymers.

1.3.1 Review of Organic Molecules

Alkanes, also called *paraffins*, are composed of all C—C, saturated bonds and have the general formula C_nH_{2n+2}. The naming conventions and typical properties of the first 10 alkanes are summarized in Table 1.20. The *alkenes* have one C=C double bond, and *alkynes* have a carbon—carbon triple bond. The formulae of these higher-order hydrocarbons are summarized in Table 1.21.

Table 1.20 The Alkane Series

Substance	Molecular Weight	T_m (°C)	T_v (°C)
Methane, CH_4	16.04	−182.5	−161.5
Ethane, C_2H_6	30.07	−182.8	−88.6
Propane, C_3H_8	44.10	−187.6	−42.1
Butane, C_4H_{10}	58.13	−138.2	−0.5
Pentane, C_5H_{12}	72.15	−129.7	36.1
Hexane, C_6H_{14}	86.18	−95.3	68.7
Heptane, C_7H_{16}	100.21	−90.6	98.4
Octane, C_8H_{18}	114.23	−56.7	125.6
Nonane, C_9H_{20}	128.26	−53.5	150.8
Decane, $C_{10}H_{22}$	142.29	−29.6	174.1

Source: Rumble [26].

Table 1.21 Summary of Chain and Cyclic Hydrocarbons

Hydrocarbons	Formula	Characteristics
Chain		
Alkane series	C_nH_{2n+2}	All single C—C bonds
Alkene series	C_nH_{2n}	One double C=C bond
Alkadiene series	C_nH_{2n-2}	Two double C=C bonds
Alkyne series	C_nH_{2n-2}	One triple C≡C bond
Alkadiyne series	C_nH_{2n-6}	Two triple C≡C bonds
Cyclic		
Cycloalkane series	C_nH_{2n}	All single C—C bonds, cyclic
Cycloalkene series	C_nH_{2n-2}	One double C=C bond, cyclic
Aromatic	Various	Ring structures, based on the benzene ring, in which single and double carbon bonds alternate

Many chemical reactions involving organic molecules, particularly polymerization reactions, involve *functional groups*. A functional group is an atom or groups of atoms that show a relative constancy of properties when attached to different carbon chains. For example, the attachment of a hydroxyl group, —OH, to hydrocarbons, leads to a group of compounds called *alcohols*, which, taken as a class, have similar properties, such as a more polar nature than the parent hydrocarbon. Some of the more important functional groups in polymer chemistry are listed in Table 1.22.

Another important concept from organic chemistry that has an impact on polymer structure is *isomerism*. Recall that *structural isomers* are molecules that have the same chemical formula, but different molecular architectures. For example, there are two different types of propyl alcohols,

Table 1.22 Some of the More Important Functional Groups in Polymer Chemistry

Group	General Formula	Compound Class	Example
—Cl	R—Cl	Chlorides	Methyl chloride, CH_3Cl
—OH	R—OH	Alcohols	Ethyl alcohol, CH_3CH_2OH
—C(=O)OH	R—C(=O)OH	Acids (or carboxylic acids)	Acetic acid, CH_3COOH
>C=O	R(R)C=O	Ketones	Acetone, CH_3COCH_3
H(—)C=O	H(R)C=O	Aldehydes	Propionaldehyde, CH_3CH_2CHO
—O—	R—O—R	Ethers	Ethyl ether, $CH_3CH_2OCH_2CH_3$
$—NH_2$	$R—NH_2$	Amines	Propylamine, $CH_3(CH_2)_2NH_2$
$—C(=O)NH_2$	$R—C(=O)NH_2$	Amides	Acetamide, CH_3CONH_2
—C(=O)O—R′	R—C(=O)O—R′	Esters	Ethyl acetate, $CH_3COOC_2H_6$

both with the same formula, depending on where the —OH functional group is placed on the carbon backbone (see Figure 1.49). Another type of isomerism results in *stereoisomers*, in which the functional groups are in the same position on the chain but occupy different geometric positions. For example, the *cis* and *trans* forms of butene are stereoisomers (see Figure 1.50). An *asymmetric carbon* is required for stereoisomers to form. We will see that asymmetric carbons play an important role in determining the tacticity of polymers. Finally, two terms related to the structure and chemistry of the repeat units are *conformation* and *configuration*. These terms are sometimes incorrectly used interchangeably, but they are quite different. The conformation of a molecule refers to how the bonds – such as those found in the backbone of a polymer – are rotated to cause the chain to fold or

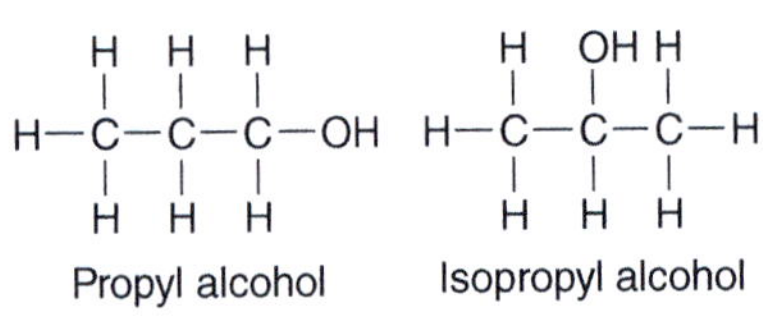

Figure 1.49 An example of structural isomers.

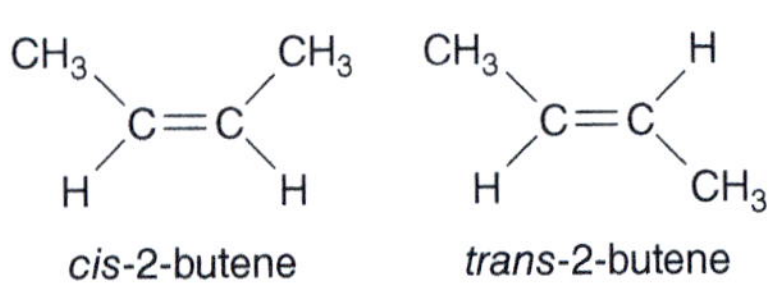

Figure 1.50 An example of stereoisomers.

straighten out. Changes in conformation involve bond rotation only. For example, some stereoisomers can convert between *cis* and *trans* conformations through bond rotation (except for those shown in Figure 1.50, since bond rotation about a double bond is not allowed). Changes in configuration, however, require that bonds be broken and reformed. For example, to change between the isopropyl and propyl alcohol configurations in Figure 1.50, the hydroxyl functional group would have to be moved from the center to an end carbon.

1.3.2 Polymer Classification

Need More Help?
Try this video on polymer classification:
https://www.youtube.com/watch?v=L4kd4DzNFqY

It is useful to classify polymers according to their chemical makeup, physical properties, formability, and how they are synthesized. The appropriate classification scheme depends on the context of the application. The first scheme groups polymers according to their chain chemistry. *Carbon-chain polymers* have a backbone composed entirely of carbon atoms. In contrast, *heterochain polymers* have other elements in the backbone, such as oxygen in a polyether, —C—O—C—. We can also classify polymers according to their macroscopic structure—that is, independent of the chemistry of the chain or functional groups. According to this scheme, there are three broad categories of polymers: *linear*, *branched*, and *networked* (crosslinked) polymers. There are exceptions to these categories of polymer architecture, such as *star polymers* and *dendrimers*, but we will discuss these specialty polymers in the appropriate context. For now, refer to Figure 1.51 for a schematic representation of the linear, branched, and networked classification schemes. Finally, polymers can be classified according to their formability. Polymers that can be repeatedly shaped and reshaped are called *thermoplastics*, whereas those that cannot be reshaped at any temperature once they are set are termed *thermosets*. Often, network polymers are thermosets, and linear and branched polymers are thermoplastics. Hence, the thermoplastic/thermoset distinction is worthy of some elaboration.

Thermoplastic Polymers

Most thermoplastic polymers are used in high-volume, widely recognized applications, so they are often referred to as *commodity plastics*. (We will elaborate upon the distinction between a polymer and a plastic in Chapter 7, but for now, we simply note that a plastic is a polymer that contains other additives and is usually identified by a variety of commercial trade names.) There are numerous databases [27, 28] that identify the primary polymer components of most plastics. With a few notable exceptions, we will refer to most polymers by their generic chemical name. The most common commodity thermoplastics are polyethylene (PE), polypropylene (PP), polyvinyl chloride (PVC), and polystyrene (PS). These thermoplastics all have in common the general repeat unit —($CHX—CH_2$)—, where —X is —H for PE, —CH_3 for PP, —Cl for PVC, and a benzene ring for PS. When we discuss polymerization reactions in Chapter 3, we will see that all of these thermoplastics can be produced by the same type of reaction.

In their simplest forms, the thermoplastics are linear, carbon-chain polymers. There are methods for creating branches—especially in PE—while still maintaining thermoplasticity. Increased branching tends to decrease the density, melting point, and certain mechanical properties of the polymer, but increases transparency and impact toughness. Thus, branched PE is important for many packaging applications. Other special types of PE include low-density PE (LDPE), high-density PE (HDPE), and linear low-density PE (LLDPE). HDPE is used when

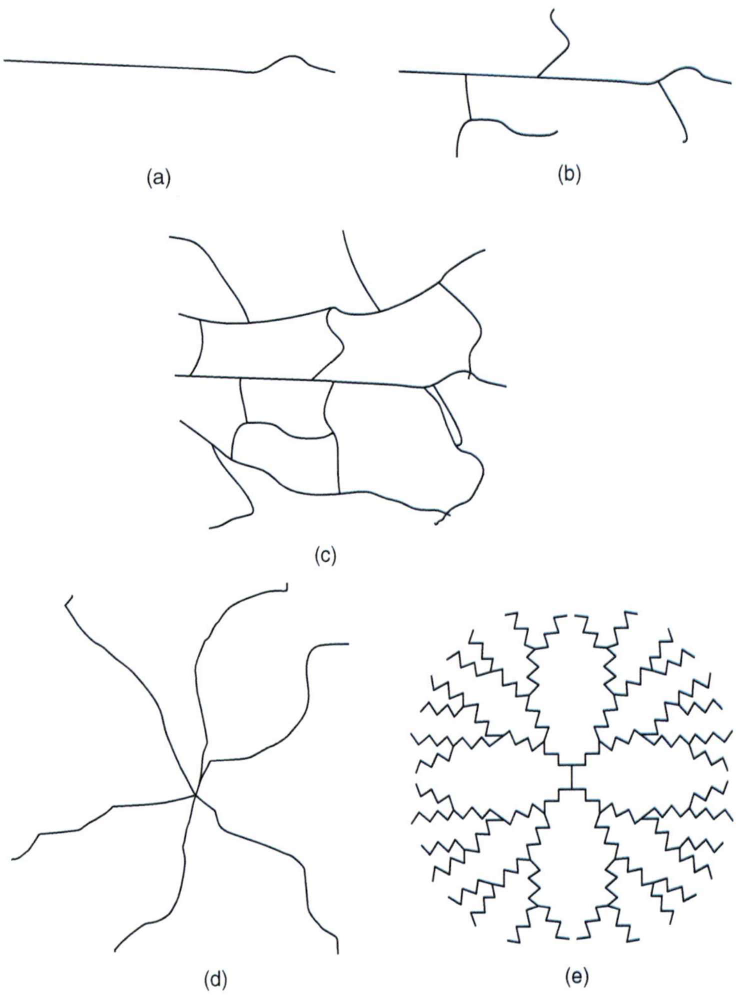

Figure 1.51 Classification of polymers according to macroscopic structure: (a) linear, (b) branched, and (c) networked, as well as specialty structures like (d) star polymers and (e) dendrimers.

greater stiffness is required, such as in milk, water, and detergent bottles. LDPE is used for many standard piping applications. LLDPE has a low density, like LDPE, but a linear structure, much like HDPE. It is less expensive to produce than LDPE, and it generally has better mechanical properties than LDPE.

PP, PVC, and PS have in common an asymmetric carbon atom in their backbone, a fact that leads to interesting structural properties as elaborated upon in the next section. They all have specific advantages leading to a variety of applications, which will also be discussed in subsequent chapters. For the moment, it is important to recognize that they are linear thermoplastics, whose properties can be tailored through blending, branching, and additives. It is important to note at this point that the ability of thermoplastics to soften when heated and harden upon cooling is what leads to the principle of *recycling*. The structure of some of these common thermoplastics can be found in Appendix 1.

Thermoset Polymers

There are many important examples of thermoset polymers, a subset of which is sometimes referred to as *resins*. The thermoplastic PE can be treated by electron radiation or chemical means to form chemical bonds between adjacent chains called *crosslinks*. In this process, some of the carbon–hydrogen or carbon–carbon bonds in the linear chain are broken—creating free radicals—which react with free radicals on other chains to form bridges between the chains. When the crosslinking is brought about by chemical means, the term *curing* is often employed. The result is that unlike linear PE, crosslinked PE has no distinct melting point due to limited chain mobility and will eventually degrade upon heating. This is a common characteristic of thermoset polymers. Due to their three-dimensional structure, they cannot be reshaped upon heating. Instead, they tend to remain rigid, or soften only slightly, until the backbone begins to break, leading to *polymer degradation*. While this may not seem like a desirable quality, it means that thermoset polymers are generally stiffer than thermoplastics. In fact, the crosslinks tend to increase the long-term thermal and dimensional stability of the polymer, such that thermosets find wide use in very large and complex parts where thermal stability is important.

The chemical structures of thermosets are generally much more diverse than those of commodity thermoplastics. The most common types of thermosets are the phenolformaldehydes (PF), urea-formaldehydes (UF), melamine-formaldehydes (MF), epoxies (EP), polyurethanes (PU), and polyimides (PI). Appendix 1 shows the chemical structure of these important thermosetting polymers.

A related yet distinctly different class of crosslinked polymers are the *elastomers*. Though they are structurally different than the thermosets, we include them here since they tend to decompose when heated rather than flow. The presence of crosslinks in some polymers allows them to be stretched, or elongated, by large amounts. Polymers that have more than 200% elastic elongation (three times the original length) and can be returned to their original form are termed elastomers. They are like thermoplastics in that they readily elongate, but the presence of crosslinks limits the elongation prior to breakage and allows the polymer to return to its original shape. Natural rubber is an important elastomer. Crosslinks are added to an emulsion of rubber, called *latex*, through the addition of heat and sulfur. The sulfur creates chemical bonds between the rubber chains in a process known as *vulcanization*. Other common elastomers are polyisoprene, butadiene rubber (BR), styrene-butadiene rubber (SBR), silicones, and fluoroelastomers.

1.3.3 Tacticity

Need More Help?
Try this video on polymer tacticity: https://www.youtube.com/watch?v=jHIZJrPlpik

An important phenomenon in some but not all industrially significant polymers arises from the ability of organic monomers with asymmetric carbons to form stereoisomers. When the repeat units (monomer units) are placed together in a long chain, monomers such as the one shown in Figure 1.52 can add to the growing chain in one of two ways: with the R group (which may be a chlorine, methyl group, or some other functional group) sticking out of the plane, or with it pointing back into the plane of the text. The variation in how the functional groups are arranged leads to differences in *tacticity*. Tacticity is variation in the configuration (not conformation!) of polymer chains because of the sequence of asymmetric carbon

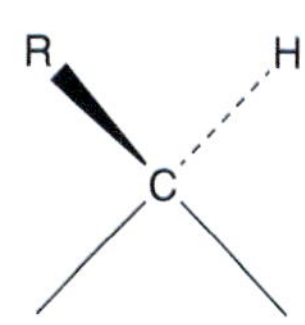

Figure 1.52 An example of an asymmetric carbon atom.

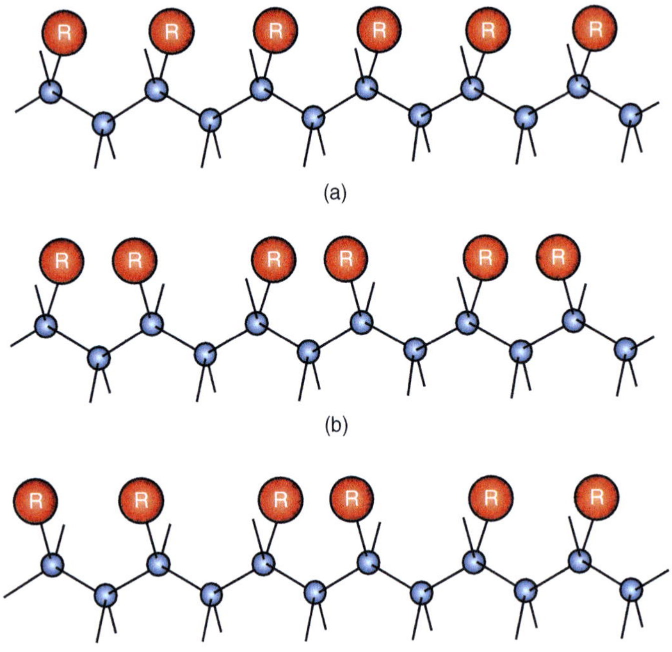

Figure 1.53 Summary of different types of tacticity, where R represents an atom other than hydrogen: (a) isotactic; (b) syndiotactic; and (c) atactic. Reproduced, by permission, from Hiemenz [29]. Copyright © 1984 by Taylor & Francis.

centers in the repeat units. There are three types of tacticity in polymers: isotactic, syndiotactic, and atactic. In *isotactic* polymers, the functional groups are all arranged on the same side of the asymmetric carbon, as in Figure 1.53a. Figure 1.53b shows the *syndiotactic* arrangement, in which the R groups perfectly alternate between sticking in and out of the plane of the carbon backbone. Finally, it is possible (and common) to have a perfectly random arrangement of functional groups, as in Figure 1.53c, in which case an *atactic* polymer results. Notable examples of polymers that exhibit tacticity are polyvinyl chloride (PVC, R = Cl), polystyrene (PS, R = Ph, phenyl group), and polypropylene (PP, R = CH_3).

1.3.4 Copolymers

Need More Help?
Try this video on copolymers: https://www.youtube.com/watch?v=CCiBMz2ig5E

Polymers can be made that contain more than one type of repeat unit. For example, the R group on the asymmetric carbon in Figure 1.52 could be chlorine in some of the monomer units and fluorine in the rest. Such polymers are called *copolymers*. The ratio of the two types of monomers can vary from 0 to 1, and there can be more than two types of monomers in a copolymer. The presence of more than one type of repeat unit opens many possibilities for variation in the structure of the polymer, or *chain architecture,* as it is sometimes called. We will not describe the myriad of possible variations and the important consequences in terms

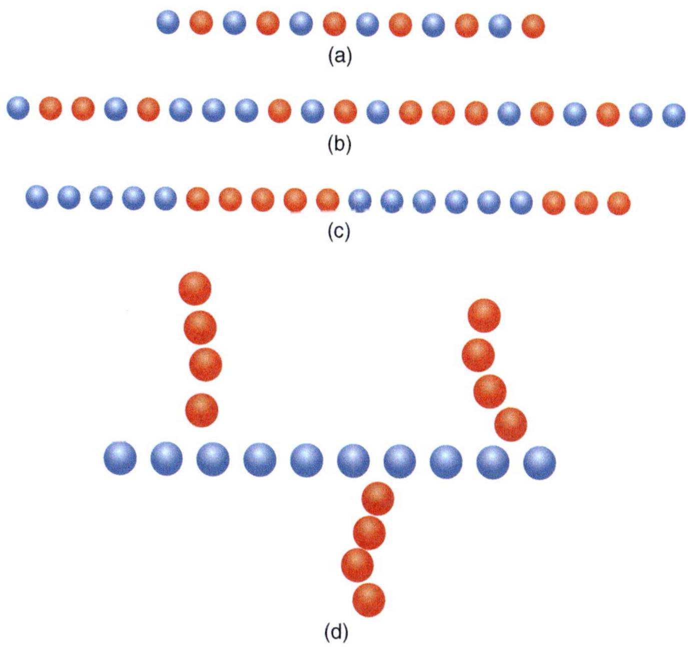

Figure 1.54 Summary of copolymer classifications: (a) alternating, (b) random, (c) block, and (d) graft.

of polymer physical properties, but here merely categorize copolymers in some broad, structural terms.

If the two monomers of a bi-component copolymer are perfectly alternating, an *alternating copolymer* results, as shown in Figure 1.54a. If the monomer units alternate in random fashion, a *random copolymer* results, as in Figure 1.54b. Figure 1.54c shows a *block copolymer*, in which the different monomers prefer to add to each other rather than themselves, resulting in large segments of one type of monomer followed by large segments of the other type of monomer. A second type of monomer can be added as a branch to a polymer backbone, as shown in Figure 1.54d, resulting in a *graft copolymer*. If a polymer has three distinct repeat units, it is called a *terpolymer*. There are also copolymers in the star and networked configurations (see Fig. 1.51). The naming conventions for copolymers vary from the historical, such as "SBS" for the commercially important poly(styrene-butadiene-styrene) copolymer, to the more systematic poly(styrene-*co*-butadiene) for the same copolymer. A summary of some copolymer naming conventions is given in Table 1.23. The different types of copolymers can have markedly different physical properties and are the basis for many important materials in the healthcare and consumer product industries.

1.3.5 Molecular Weight

We have described some of the general characteristics of polymers, and how they can be grouped according to structure, but we have not addressed any of the more quantitative aspects of polymer structures. For instance, we have stated that a polymer is made up of many monomer (repeat) units, but how many of these repeat units do we typically find in a polymer? Do all polymer chains have the same number of repeat units? These topics are addressed in this section on polymer molecular weight. Again, the kinetics of polymer formation are not discussed until Chapter 3—we merely

Table 1.23 Selected Copolymer Naming Conventions

Type	Connector	Example
General	*-co-*	poly(A-*co*-B)
Random	*-ran-*	poly(A-*ran*-B)
Alternating	*-alt-*	poly(A-*alt*-B)
Block	*-block-*	poly A-*block*-poly B
Graft	*-graft-*	poly A-*graft*-poly B
Starblock	*-star-...-block...*	*star*-poly A-*block*-poly B
Crosslinked	*-net-*	poly(A-*net*-B)

assume here that the polymer chains have been formed and that we can count the number of repeat units in each chain.

Degree of Polymerization

The number of repeat units in an isolated polymer chain is called the degree of polymerization, x_n. If the monomer from which our polymer is formed has an initial molecular weight of M_0, then the molecular weight of an isolated polymer chain is $x_n \cdot M_0$. For example, PE, $-(H_2C-CH_2)-$, is made up of ethylene repeat units, each of which has a molecular weight $M_0 = 2(12) + 4(1) = 28$ g/mol. Let us assume that an isolated PE chain has 1000 repeat units, so $x_n = 1000$. The molecular weight of this isolated PE chain would be 1000 (28 g/mol) = 28,000 g/mol. Notice that we are neglecting the contribution of the terminal hydrogens located at both ends of the PE chain. Since typical molecular weights are of the order 10^6 and larger, terminal groups do not make a significant contribution to the molecular weight, and neglecting them does not introduce substantial error in calculations. We will also see that the molecular weight of the repeat unit in the polymer chain is not necessarily numerically the same as the molecular weight of the monomer. Small molecules can sometimes be produced during polymerization reactions, such as condensation reactions in which water is formed as a by-product. Be sure to use the molecular weight of the repeat unit in your calculations.

Average Molecular Weight

In a typical polymer, not all chains have the same length. Consequently, there is a distribution of x_n values, and the molecular weight, M_i, varies from chain to chain. We could construct a histogram of the distribution of the number of chains, N_i, with molecular weight, M_i, as shown in Figure 1.55. Molecular weights have discrete values, of course, and it is possible to find some mathematical function, $f(M_i)$ that appropriately describes the curve represented by the dotted line in Figure 1.55. For molecular weights, this function is simply the *number fraction*, n_i:

$$n_i = \frac{N_i}{\sum N_i} \tag{1.40}$$

Recall from calculus (or physics or statistics) that distributions can have k different moments about the origin, μ'_k, which for a discrete random variable, M_i, are given by

$$\mu'_k = \sum_i n_i (M_i)^k \tag{1.41}$$

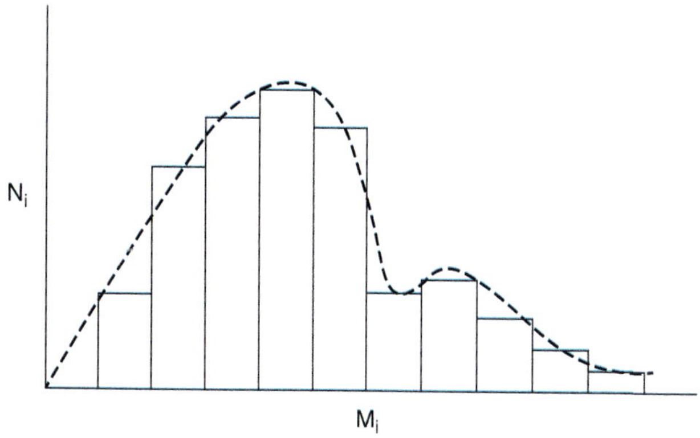

Figure 1.55 Histogram showing the number of molecules, N_i, having the molecular weight M_i. Reproduced, by permission, from Heimenz [30]. Copyright © 1984 by Taylor & Francis.

The first moment of the distribution ($k = 1$) is simply the average molecular weight, called the *number-average molecular weight*, $\overline{M}_n$,

$$\mu_1' = \overline{M}_n = \sum n_i M_i = \frac{\sum N_i M_i}{\sum N_i} \tag{1.42}$$

or, in terms of the average degree of polymerization, $\bar{x}_n$,

$$\overline{M}_n = \bar{x}_n M_0 \tag{1.43}$$

where M_0 is the molecular weight of a repeat unit.

The second moment of the molecular weight distribution is then

$$\mu_2' = \sum n_i M_i^2 \tag{1.44}$$

A ratio of the second moment to the first moment (number-average molecular weight) is called the *weight-average molecular weight*, $\overline{M}_w$, which is the average based on the *weight* of each polymer chain:

$$\overline{M}_w = \frac{\sum N_i M_i^2}{\sum N_i M_i} = \frac{\sum n_i M_i^2}{\sum n_i M_i} \tag{1.45}$$

or, in terms of the mass (in grams) of species with molecular weight M_i, $m_i = N_i M_i$

$$\overline{M}_w = \frac{\sum m_i M_i}{\sum m_i} = \sum w_i M_i \tag{1.46}$$

where w_i is the *weight fraction*, $m_i/\Sigma m_i$.

There are other types of molecular weights based on higher-order moments in the distribution (see Table 1.24), but $\overline{M}_n$ and $\overline{M}_w$ are the most widely used quantities. In general, $\overline{M}_n \neq \overline{M}_w$, but

Need More Help?

Try these videos on polymer molecular weight distributions:

https://www.youtube.com/watch?v=CMylb58vd4Q
https://www.youtube.com/watch?v=vtltTzP-bNU

Table 1.24 Summary of the Most Common Molecular Weight Averages

Average	Definition	Methods of Determination
$\bar{M}_n$	$\frac{\sum N_i M_i}{\sum N_i}$	Osmotic pressure, colligative properties, end group analysis
$\bar{M}_w$	$\frac{\sum N_i M_i^2}{\sum N_i M_i}$	Light scattering, sedimentation velocity
$\bar{M}_z$	$\frac{\sum N_i M_i^3}{\sum N_i M_i^2}$	Sedimentation equilibrium
$\bar{M}_v$	$\left(\frac{\sum N_i M_i^{1+a}}{\sum N_i M_i}\right)^{1/a}$	Intrinsic viscosity (a is characteristic of a system and lies between 0.5 and 1.0)

taken together as a ratio, they provide a measure of the breadth of the molecular weight distribution. This ratio is called the *polydispersity index*, and it takes on values greater than or equal to 1.0:

$$\frac{\overline{M}_w}{\overline{M}_n} \geq 1.0 \tag{1.47}$$

If all chains have exactly the same weight and number of repeat units, then the system is termed *monodispersed*, and the polydispersity index is exactly 1.0. Most real polymers have rather large polydispersity indexes, but some standards used for chromatography calibration, such as PS, can have values approaching unity. Calculating molecular weights and determining which form of Eqs. (1.42), (1.45), and (1.46) to use requires some practice and patience. A Cooperative Learning Exercise is provided below, but you are encouraged to consult a textbook on polymer science for further problems on molecular weight calculations and determination.

Cooperative Learning Exercise 1.7

Consider the following collection of polymer chains:

10 molecules	2800 MW
5 molecules	3000 MW
4 molecules	1200 MW
2 molecules	3600 MW
1 molecule	1000 MW

Person 1: Calculate the number average molecular weight.

Person 2: Calculate the weight average molecular weight.

Combine your answers to determine the polydispersity index.

Answer:

$$\overline{M}_w = \frac{28,000(2800) + 15,000(3000) + 4800(1200) + 7200(3600) + 1000(1000)}{56,600}$$

$$= \frac{1.56 \times 10^8}{56,000} = 2787$$

$$\overline{M}_n = \frac{10(2800) + 5(3000) + 4(1200) + 2(3600) + 1(1000)}{22}$$

$$= \frac{56,000}{22} = 2545$$

$$\overline{M}_w/\overline{M}_n = 2787/2545 = 1.095$$

1.3.6 Polymer Crystallinity

An interesting and important structural characteristic of many polymers is that they are easily transformed from the amorphous to the crystalline state. Unlike inorganic glasses, which generally need a great deal of energy to devitrify (crystallize), or metals that require nonequilibrium conditions to form amorphous structures, many polymers have amorphous to crystalline transition temperatures that are near room temperature. Not all polymers crystallize readily, however. Intuitively, we would expect that a jumbled mass of spaghetti-like strands that make up a polymer solution or melt would tend to be amorphous, and this is indeed the case. But there are several structural factors that contribute to the ability of the amorphous polymer chains to rearrange themselves into an ordered structure. These factors include chain architecture (the chemical constituents and bond angles of the backbone and side groups), order and regularity (e.g., tacticity), intermolecular forces (both within an individual chain and between adjacent chains, such as hydrogen bonding), and steric effects (the size of side groups and branches).

Types of Bulk Polymer Crystallinity

Polymer crystallinity is a complicated subject, to which numerous books and symposia are devoted; but for our purposes, we can classify crystallinity in bulk polymers into two general categories: extended chain and folded chain. *Extended chain crystallinity* arises in many polymers with highly regular structures, such as PE, poly(vinyl alcohol), syndiotactic polymers of poly(vinyl chloride) and poly(1,2-butadiene), most polyamides, and cellulose. In these molecules, the so-called "planar zigzag" structure shown in Figure 1.56a possesses the minimum energy for an isolated section of the chain and is therefore the thermodynamically favored conformation. Side groups, if they are small enough and arranged in a regular fashion, as in the syndiotactic structure, need not prevent crystallinity (see Figure 1.56b), but as the bulkiness and irregularity of the side groups grow, crystallization becomes more and more difficult. As a result, highly branched molecules, such as branched PE, do not crystallize, even though PE itself is easily crystallized. Similarly, networked polymers do not have the freedom to move in a way such that extended chain crystallinity can occur.

As with the other classes of materials, polymers can be either single crystals or polycrystalline. Polycrystalline polymers are more appropriately termed *semicrystalline polymers*, since the region between crystalline domains in polymers can be quite large and result in a significant amorphous component to the polymer. The crystalline regions in semicrystalline polymers are called *crystallites*, which have dimensions of several hundred angstroms, but the length of polymer chains is

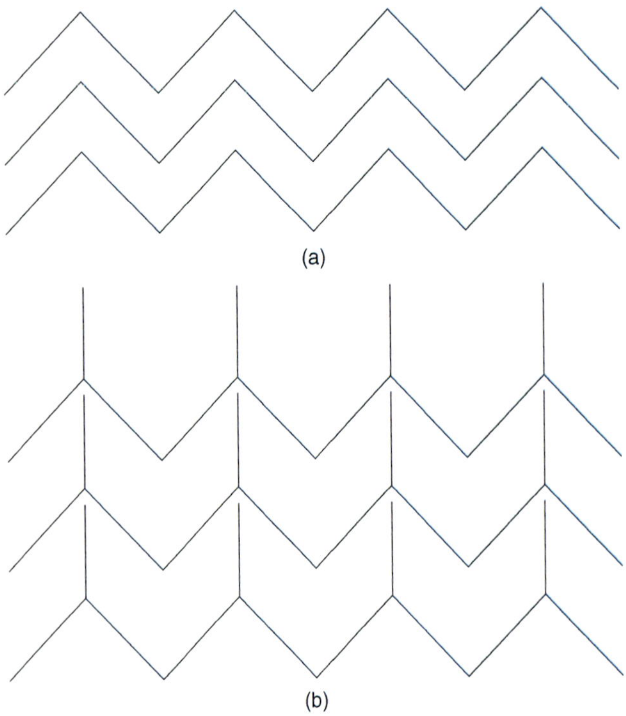

Figure 1.56 Schematic illustration of extended chain crystallinity in polymers (a) polyethylene and (b) polypropylene.

generally much larger than this. For example, a PE chain with the extended chain structure shown in Figure 1.56a with a molecular weight of 50,000 has an end-to-end length of about 4500 Å. How can this be? The second type of crystalline structure in polymers is chain folding, which gives an explanation. Polymer chains can fold in a regular fashion to form plate-like crystallites called *lamellae*, as shown schematically in the insert of Figure 1.57. Notice that the polymer chains not only fold but can extend from one lamella to another to form amorphous regions. In polymers crystallized from the melt, these lamellae often radiate from a central nucleation site, forming three-dimensional spherical structures called *spherulites* (see Figures 1.57 and 1.58). In cross-polarized light, the spherulites form a characteristic *Maltese cross* pattern due to birefringent effects associated with the lamellar structures.

A polymer crystal structure related to chain folding is called the *fringed micelle model*, in which the polymer chains do not fold in a regular fashion but extend from one crystalline region to another, again forming amorphous regions between the crystallites (see Figure 1.59). While the fringed micelle model is no longer the preferred one for describing crystallinity in most polymers, it is still used to describe the structure of highly oriented polymers, such as occurs in the stretching of rubber, viscose rayon, and Kevlar®.

The amorphous and crystalline regions each have different densities, with the crystalline density ρ_c being higher than the amorphous density ρ_a due to a more compact structure. The mass fraction crystallinity, w_c, in a semicrystalline polymer with bulk density ρ can then be calculated from the respective crystalline and amorphous densities, ρ_a and ρ_c, respectively:

$$w_c = \frac{\rho_c(\rho - \rho_a)}{\rho(\rho_c - \rho_a)} \tag{1.48}$$

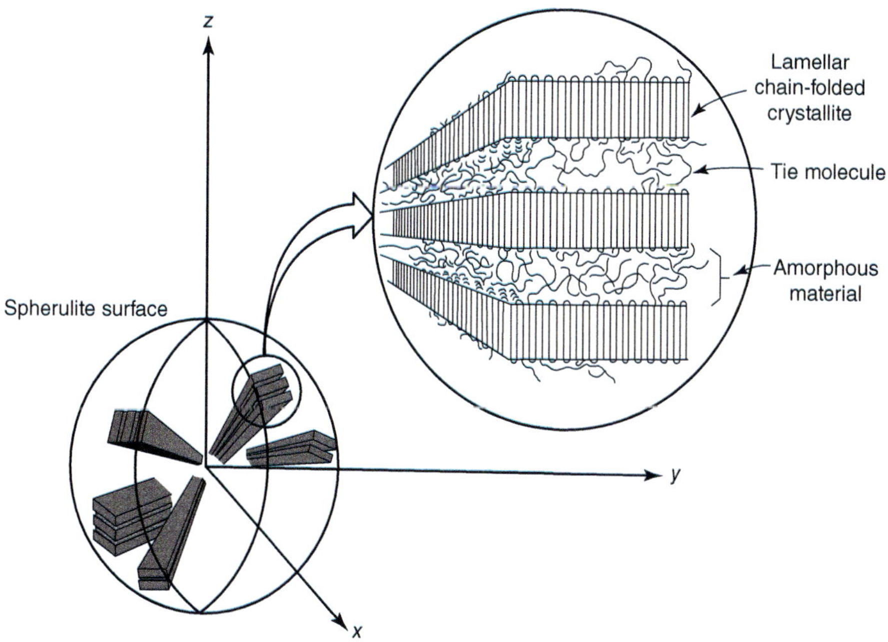

Figure 1.57 Schematic illustration of chain folding leading to lamellar crystallites (*inset*) and lamellar stacking to form spherulites.

Figure 1.58 Cross-polarized micrograph of polyethylene showing spherulitic structure. Ralls et al. [7], © 1976, with permission from John Wiley & Sons, Inc.

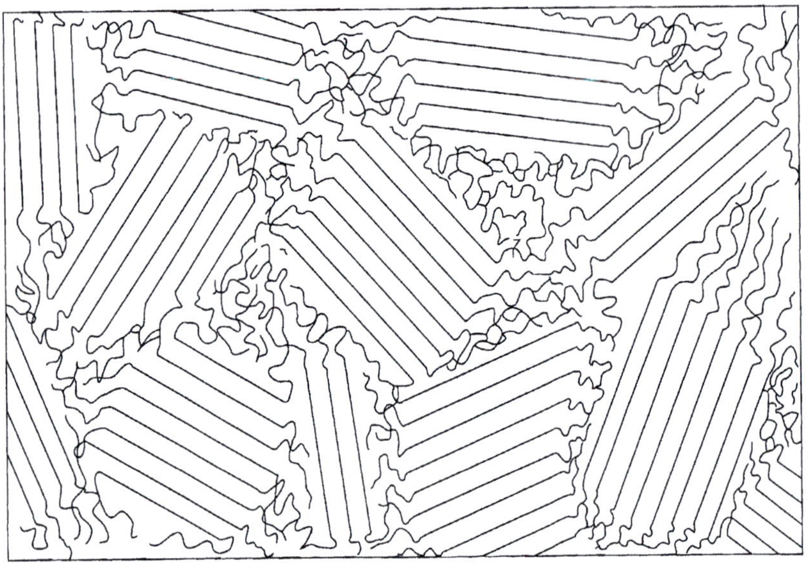

Figure 1.59 The fringed-micelle model of polymer crystallinity. From Ralls et al. [7]. Copyright © 1976 by John Wiley & Sons, Inc. This material is used by permission of John Wiley & Sons, Inc.

Need More Help?

Try these videos on polymer crystallinity:

https://www.youtube.com/watch?v=wmyd3DYqZWA
https://www.youtube.com/watch?v=uI7vXMqkM3s

Polymer single crystals possess the density of the crystal, ρ_c. Though polymer single crystals do not usually form in bulk but rather from more carefully controlled formation techniques such as vapor deposition, we will describe them here since they generally have a folded-chain crystal structure. An example of a polymer single crystal is shown in Figure 1.60. One would expect that the regular array of folded chains in a single crystal, or in a semicrystalline polymer for that matter, leads to diffraction phenomena when bombarded with X-rays, in much the same way that lower-molecular-weight materials like sodium chloride or aluminum do. This is indeed the case, and the study of XRD phenomena in polymers is a large field of interest. The primary difference in the diffraction patterns of polymer crystals is due to the fact that molecules—in this case, polymer chains— rather than atoms make up the lattice points of the unit cells. For example, the unit cell of PE is orthorhombic, with PE chains forming the lattice points, as illustrated in Figure 1.61. As a result, the dimensions of the unit cells in polymer crystals tend to be much bigger than those for ceramics and metals. The orthorhombic unit cell for PE in Figure 1.61 has dimensions $a = 7.42$ Å, $b = 4.95$ Å, $c = 2.55$ Å, with two chains per unit cell: one in the center and $4 \times (1/4)$ at each corner where four unit cells come together. The larger lattice parameters and increased interplanar spacings mean that the diffraction angles for polymers are generally much smaller than for ceramics and metals [recall that $\theta \propto 1/d$, see Eq. (1.35)]. Hence, *small-angle X-ray scattering* (SAXS) is often used for polymer structural characterization instead of the traditional *wide-angle X-ray scattering* (WAXS).

As with ceramics and metals, polymer crystals can have multiple crystal forms. PE has a metastable monoclinic form and an orthohexagonal high-pressure form. A list of some of the more

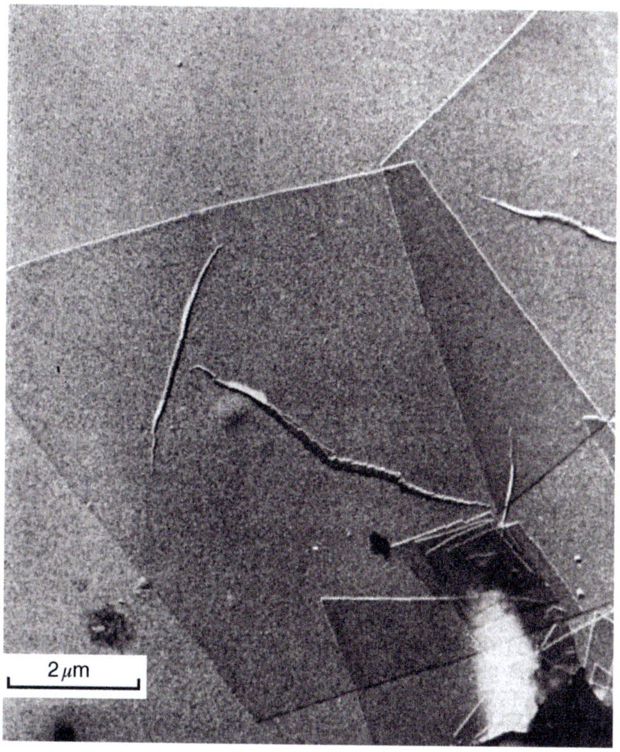

Figure 1.60 Polyethylene single crystals. From Heimenz [31], © 1984, Taylor and Francis.

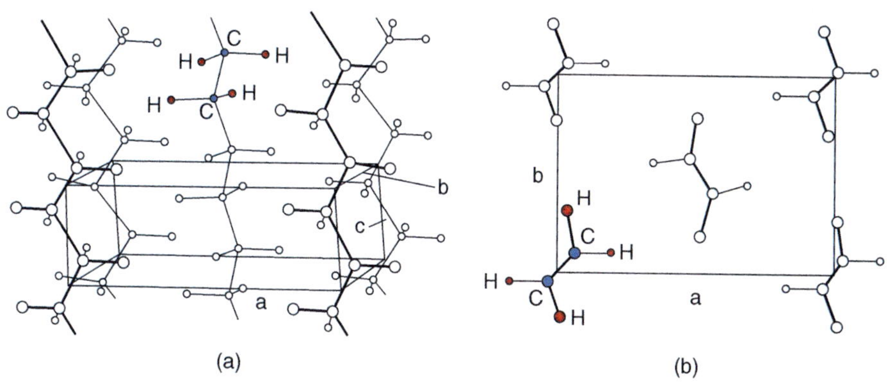

Figure 1.61 (a) Crystal structure of polyethylene unit cell shown in relation to chains. (b) View of unit cell perpendicular to the chain axis. Reproduced, by permission, from Heimenz [32]. Copyright © 1984 by Taylor & Francis.

common polymers and their corresponding crystal structures is given in Table 1.25. Finally, XRD can be used to determine the amorphous to crystalline ratio in semicrystalline polymers in much the same way that Eq. (1.48) can be used. Figure 1.62 shows a schematic illustration of the XRD patterns for semicrystalline and amorphous PE. The estimation of crystalline content is based on a ratio of the peak areas in the two samples.

More so than in metals, glasses, and ceramics, the microstructure in polymers is easily altered, and within the operating temperature and pressure of many industrial and biological processes,

Table 1.25 The Crystal Structures of Selected Polymers

Polymer	Crystal System	Lattice Constants (Å) *a*	*b*	*c*	Crystal Density (g/cm^3)
Polyethylene	Orthorhombic	7.417	4.945	2.547	1.00
Polytetrafluoroethylene	Trigonal (>19°C)	5.66	—	19.50	2.30
Isotactic polypropylene	α-Monoclinic	6.65	20.96	6.50	0.936
	β-Hexagonal	19.08	—	6.49	0.922
Syndiotactic polypropylene	Orthorhombic	14.50	5.60	7.40	0.93
Isotactic polystyrene	Trigonal	21.90	—	6.65	1.13
Poly(vinyl chloride)	Orthorhombic	10.6	5.4	5.1	1.42
Poly(vinyl alcohol)	Monoclinic	7.81	2.25	5.51	1.35
Poly(vinyl fluoride)	Orthorhombic	8.57	4.95	2.52	1.430
Poly(vinylidene fluoride)	α-Monoclinic	4.96	9.64	4.62	1.925
	β-Orthorhombic	8.58	4.91	2.56	1.973
Isotactic poly(methyl methacrylate)	Orthorhombic	20.98	12.06	10.40	1.26
trans-1,4-Polybutadiene	Monoclinic	8.63	9.11	4.83	1.04
cis-1,4-Polybutadiene	Monoclinic	4.60	9.50	8.60	1.01
Poly(ethylene oxide)	Monoclinic triclinic	8.05	13.04	19.48	1.228
		4.71	4.44	7.12	1.197
Isotactic poly(propylene oxide)	Orthorhombic	10.46	4.66	7.03	1.126
Nylon 66	α-Triclinic	4.9	5.4	17.2	1.24
	β-Triclinic	4.9	8.0	17.2	1.248
2,6-Polyurethane	Triclinic	4.93	4.58	16.8	1.27
	Triclinic	4.59	5.14	13.9	1.33
3,6-Polyurethane	Monoclinic	4.70	8.66	33.9	1.34
Polyketone	Orthorhombic	7.97	4.76	7.57	1.296
Poly(ethylene sulfide)	Orthorhombic	8.50	4.95	6.70	1.416
Polyisobutylene	Orthorhombic	6.88	11.91	18.60	0.972
Poly(isobutylene oxide)	Orthorhombic	10.76	5.76	7.00	1.10
Poly(ethylene sulfide)	Orthorhombic	8.50	4.95	6.70	1.60
Isotactic poly(vinyl methyl ether)	Trigonal	16.25	—	6.50	1.168

Source: Tadokoro et al. [33]/Krieger Publishing Company.

transitions between the amorphous and crystalline states, and the ratio between the amorphous and crystalline components, can easily take place. As we mentioned in the previous section, the amorphous component of a polymer is less dense than the crystalline component. Conversely, the *specific volume*, or volume per mole in cubic centimeters, is lower for polymer crystals than it is for amorphous polymers. This distinction is best understood by observing the volume change of a polymer melt as it cools. Consider a molten polymer at point *A* in Figure 1.63. If we cool the polymer melt slowly, as in path *ABG*, the polymer chains have sufficient time to rearrange, fold, and

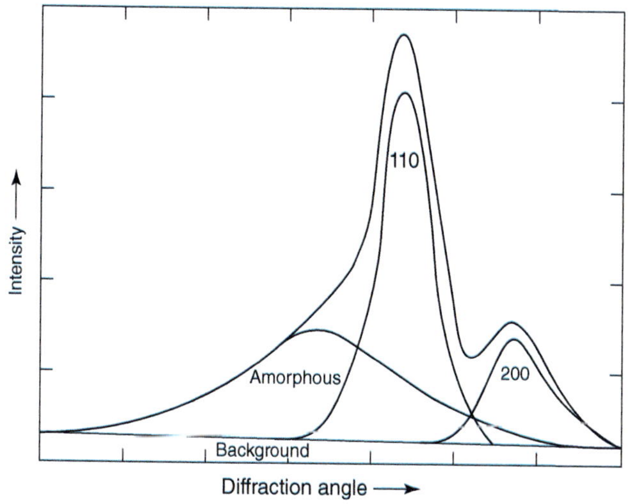

Figure 1.62 Resolution of the X-ray scattering curve of a semicrystalline polyethylene sample into contributions from crystalline (110 and 200 planes) and amorphous components. From Billmeyer [34]. Copyright © 1984 by John Wiley & Sons, Inc. This material is used by permission of John Wiley & Sons, Inc.

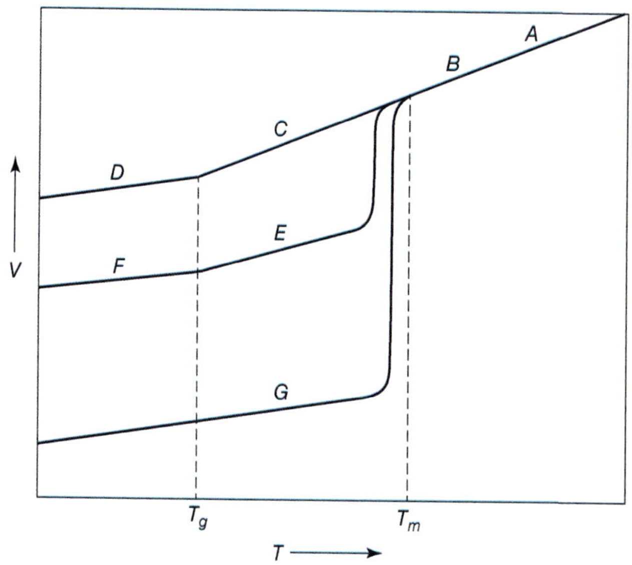

Figure 1.63 Specific volume as a function of temperature on cooling from the melt for a polymer that tends to crystallize. Region *A* is liquid, *B* is liquid with elastic response, *C* is supercooled liquid, *D* is glass, *E* crystallites in a supercooled liquid matrix, *F* crystallites in a glassy matrix, and *G* is completely crystalline. Paths *ABCD*, *ABEF*, and *ABG* represent fast, intermediate, and very slow cooling rates, respectively. From Ralls et al. [7]. Copyright © 1976 by John Wiley & Sons, Inc. This material is used by permission of John Wiley & Sons, Inc.

form lamellar structures, resulting in a crystalline polymer, provided, of course, that they have the propensity to fold in the first place, and are not prevented from doing so by such factors as steric effects. The point at which the melt solidifies in the form of crystals is called the *crystalline melting point*, T_m, and is characterized by a sharp decrease in the specific volume and an increase in the density. If the same polymer melt is cooled rapidly, as in the path *ABCD*, a supercooled liquid is first obtained at point *C*, followed by an amorphous solid, or *glassy polymer*, at point *D*, due to insufficient time for the large molecules to arrange themselves in an ordered structure. The point at which the slope of the specific volume with temperature curve decreases (between *C* and *D*), representing solidification, is called the *glass transition temperature*, T_g. The reasoning behind the term "glass transition temperature" becomes more apparent if we turn around and begin slowly heating the glassy polymer. At some point, there is sufficient mobility in the polymer chains for them to begin to align themselves in a regular array and form crystallites. The polymer is not yet molten—there is simply short-range chain movement that results in an amorphous to crystalline transformation in the solid state. This point is also T_g. As we continue to heat the sample, the now-crystalline polymer eventually reaches T_m and melts. Most polymers are a combination of amorphous and crystalline components and are the result of intermediate cooling paths such as *ABEF* in Figure 1.63. Keep in mind that some polymers, no matter how slowly we cool them, cannot crystallize, and they follow the glassy path *ABCD* all the time. We will have more to say about the glass transition temperature in Section 1.3.7.

Need More Help?

Try this video on liquid crystalline polymers:

https://www.youtube.com/watch?v=vzVjbd8_Ur0

Liquid Crystalline Polymers

One class of polymers that requires some special attention from a structural standpoint is *liquid crystalline polymers*, or LCPs. LCPs are nonisotropic materials that are composed of long molecules parallel to each other in large clusters and that have properties intermediate between those of crystalline solids and liquids. Because they are neither completely liquids nor solids, LCPs are called *mesophase* (intermediate phase) materials. These mesophase materials have liquid-like properties so that they can flow; but under certain conditions, they also have long-range order and crystal structures. Because they are liquid-like, LCPs have a translational degree of freedom that most solid crystals we have described so far do not. That is, crystals have three-dimensional order, whereas LCPs have only one- or two-dimensional order. Nevertheless, they are called "crystals," and we shall treat them as such in this section.

In many cases, these polymer chains take on a rod-like (*calamitic* LCPs) or even disc-like (*discotic* LCPs) conformation, but this does not affect the overall structural classification scheme. There are many organic compounds —though not polymeric in nature—that exhibit liquid crystallinity and play important roles in biological processes. For example, arteriosclerosis is possibly caused by the formation of a cholesterol-containing liquid crystal in the arteries of the heart. Similarly, cell wall membranes are generally considered to have liquid crystalline properties. As interesting as these examples of liquid crystallinity in small organic compounds are, we must limit the current discussion to polymers only.

There are three categories of LCPs, grouped according to the arrangement of the molecules: smectic, nematic, and cholesteric. *Nematic* (from the Greek term meaning "thread-like") LCPs have their molecules aligned along the chain axis, as shown schematically in Figure 1.64. Nematic liquids have low viscosity and tend to be turbid, or "cloudy." *Smectic* (from the Greek term for "soap") LCPs have an additional level of structure, in that the polymer chains are also aligned along the chain axis, but they also segregate into layers. Smectic liquids are also turbid but tend to be highly viscous. Finally, *cholesteric* (from the Greek term for "bile") LCPs have layered structures, but the aligned polymer chains in one layer are rotated from the aligned polymer chains in adjacent layers. Cholesteric LCPs are also highly viscous but often possess novel photochromic, optical, thermochromic, and electro-optical properties.

Clearly, not all polymeric molecules possess the ability to form LCPs. Generally, LCPs have a molecular structure in which there are two regions with dissimilar chemical properties. For example, chains that consist of aliphatic–aromatic, dipolar–nonpolar, hydrophobic–hydrophilic,

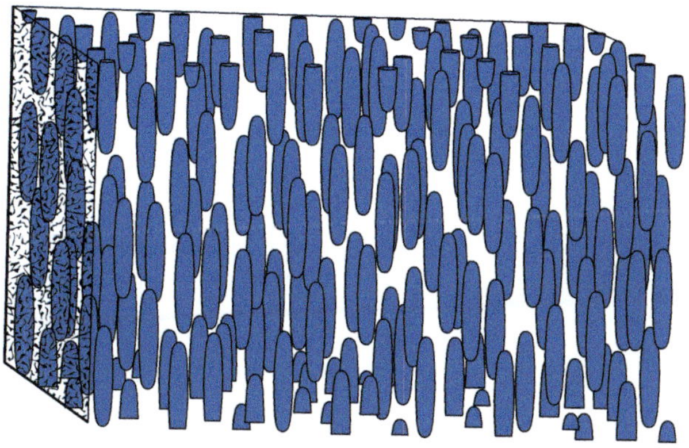

(a)

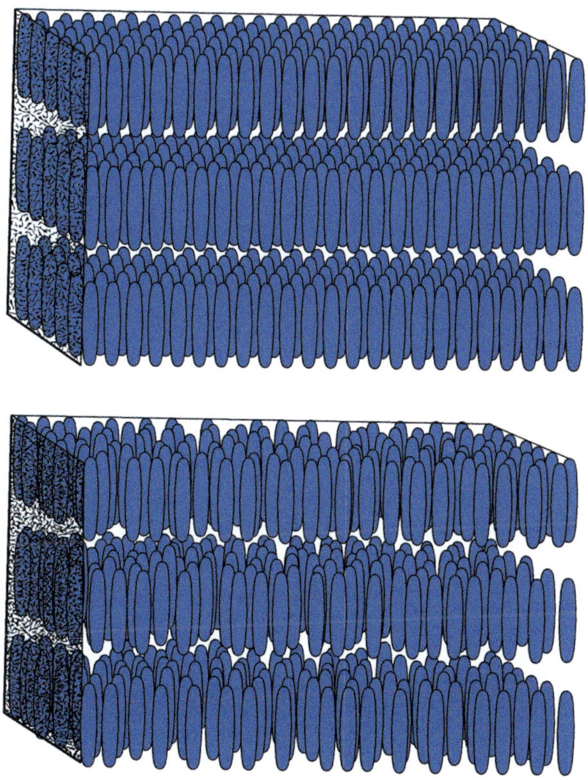

(b)

Figure 1.64 The structure of liquid crystalline polymers (a) nematic, (b) smectic, and (c) cholesteric. Reproduced, by permission, from Fergason [35]. Copyright © 1964 by Springer Nature.

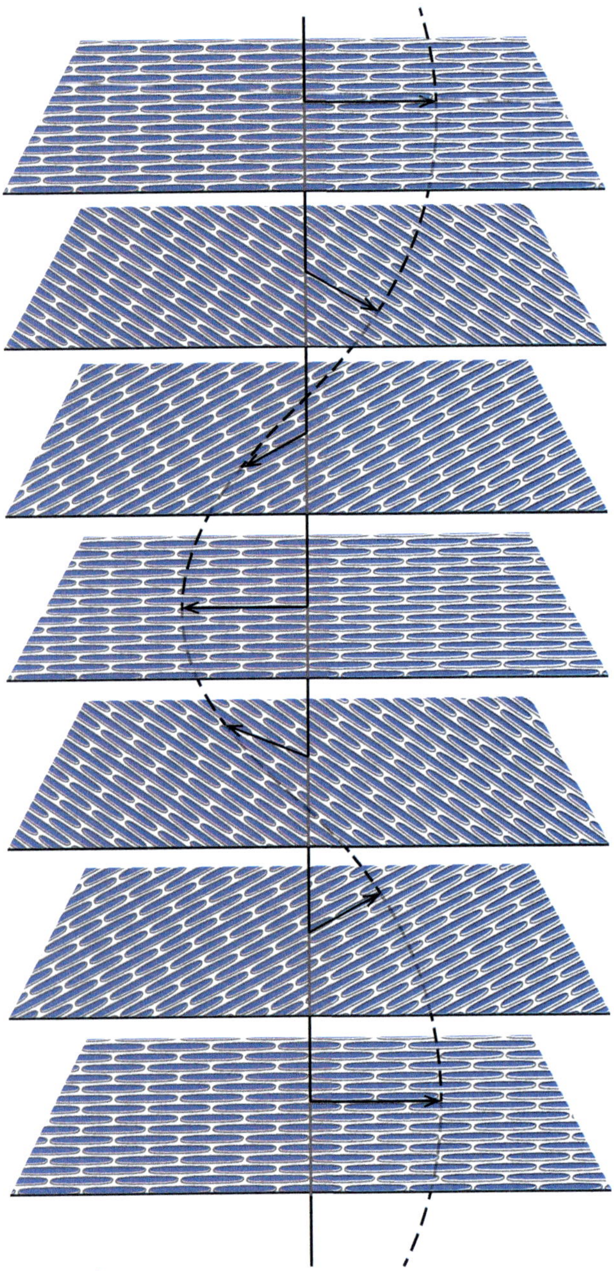

Figure 1.64 (Continued)

flexible–rigid, or hydrocarbon–fluorocarbon combinations of substantial size have a propensity to form LCPs. The two different portions of the chains can interact locally with similar regions of adjacent chains, leading to ordering. LCPs that rely on structural units in the backbone, or main chain, to impart their crystallinity are called *main-chain LCPs*. The building blocks of typical main-chain LCPs are shown in Figure 1.65. Side-chain LCPs crystallize due to the interaction of the side chains,

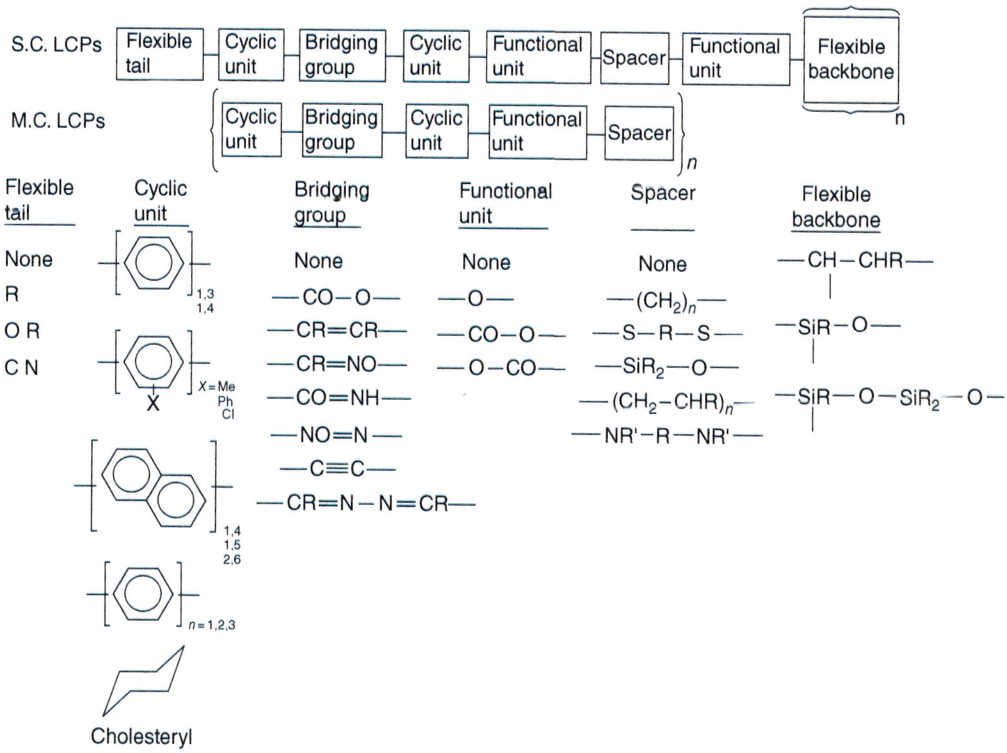

Figure 1.65 General structure of main-chain (M.C.) and side-chain (S.C.) LCPs. Adapted from Chung [36]. Copyright © 1986, John Wiley & Sons.

or branches, as shown in Figure 1.65. Side chains play an important role in determining not only the liquid crystalline activity of polymers but the type of structure that will form as well. For example, in Table 1.26, we see that the number of hydrocarbon units in the side chain (not the backbone repeat unit) affects whether the resulting LCP is nematic or smectic.

We have used the general term "liquid" to describe this special class of polymers, but we know that a liquid can be either a melt or a solution. In the case of LCPs, both types of liquids can occur.

Table 1.26 Effect of the Flexible Tail on the Structure of a Side-Chain LCP

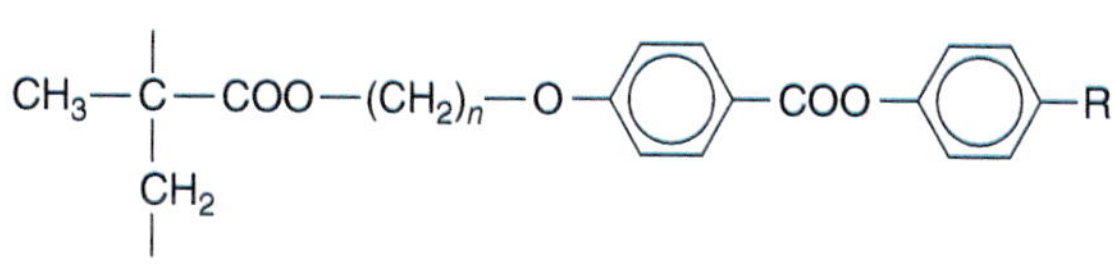

Number	n	R	Structure
1	2	OCH_3	Nematic
2	2	OC_3H_7	Smectic
3	6	OCH_3	Nematic
4	6	OC_6H_{13}	Smectic

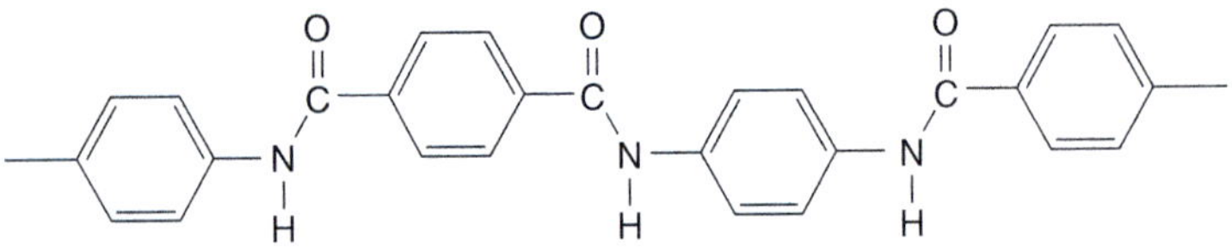

Figure 1.66 The repeat structure of Kevlar.

A polymer that exhibits crystallinity in the melt and that undergoes an ordered–disordered transformation because of thermal effects is called a *thermotropic* LCP. A polymer requiring a small-molecule solvent to exhibit crystallinity is termed a *lyotropic* LCP. All three types of LCP structures can occur in either thermotropic or lyotropic polymers, and both are industrially relevant materials. Perhaps the most widely utilized (and studied) lyotropic LCP is poly *p*-phenylene terephthalamide (PPTA), more commonly known as Kevlar (see Figure 1.66). Kevlar belongs to the class of aramids that are well known for their LCP properties. Because these polymers are crystalline in solution, they are often spun into filaments, from which the solvent is subsequently removed in order to retain the aligned polymer structure. The result is a highly oriented, strong filament that can be used for a wide variety of structural applications. Most thermotropic LCPs are polyesters or copolymers that can be melted and molded into strong, durable objects.

1.3.7 The Glass Transition

The glass transition is an important parameter since many physical properties change when polymer chains gain mobility. This is an important temperature because it marks the point where the polymer becomes more easily deformed, or more ductile. As summarized in Table 1.27, totally noncrystalline polymers are very brittle below the glass transition, whereas polymers with intermediate crystallinity are very tough above the glass transition. Most homopolymers have glass transition temperatures that are 0.5–0.75T_m, even though the crystalline melting point can vary by several hundred degrees between polymer types (see Figure 1.67). Most plastics are designed to be used between T_m and T_g, as shown in Figure 1.68.

To conclude this section on polymers, we should note that we have used the term polymer almost exclusively to refer to organic macromolecules. The term *plastic* refers not only to organic substances of high molecular weight but also to such substances that at some point in their manufacture have been shaped by flow. Thus, the term plastic is more specific than the term polymer, and this term carries with it an indication of its processing history. As we will see in Chapter 7, there are many materials that can be considered polymers, yet are formed by routes other than melt processing.

Table 1.27 Predominant Properties of Crystalline Polymers

	Degree of Crystallinity		
Temperature Range	**Low (5–10%)**	**Intermediate (20–60%)**	**High (70–90%)**
Above T_g	Rubbery	Leathery, tough	Stiff, hard, brittle
Below T_g	Glassy, brittle	Hornlike, tough	Stiff, hard, brittle

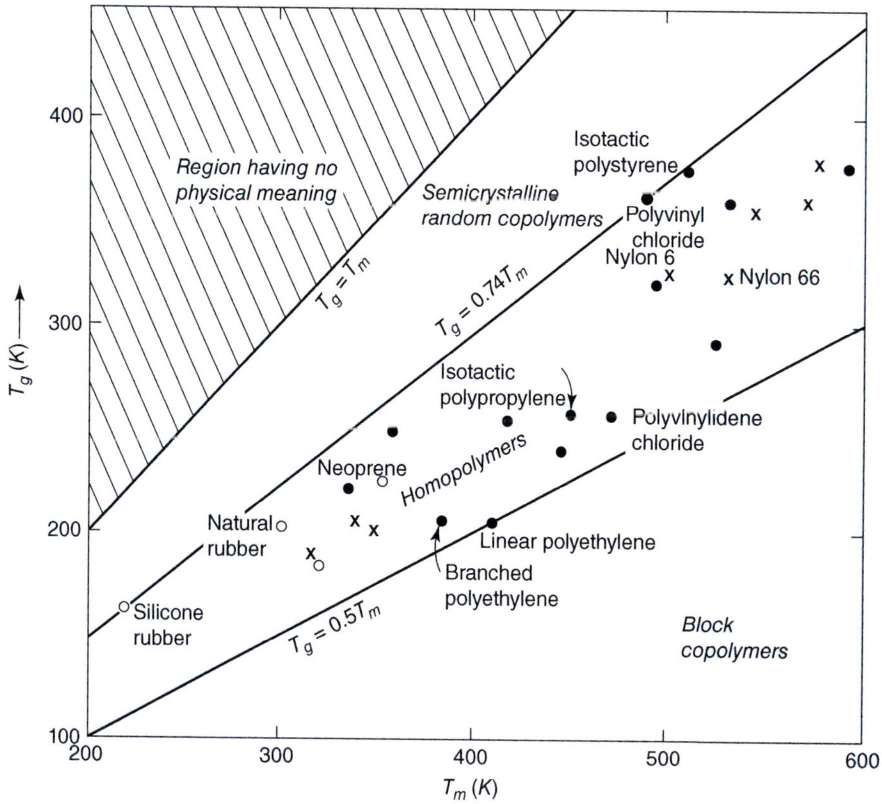

Figure 1.67 The glass transition temperature, T_g, as a function of crystalline melting point, T_m, for homopolymers. Filled circles are addition homopolymers, open circles are elastomers, and crosses are condensation homopolymers. From Ralls et al. [7]. Copyright © 1976 by John Wiley & Sons, Inc. This material is used by permission of John Wiley & Sons, Inc.

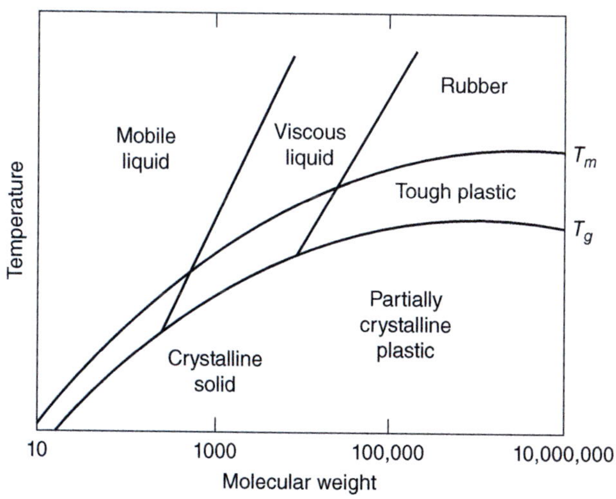

Figure 1.68 Approximate relations among molecular weight, T_g, T_m, and polymer properties. From Billmeyer [34]. Copyright © 1984 by John Wiley & Sons, Inc. This material is used by permission of John Wiley & Sons, Inc.

1.4 Structure of Composites

The preceding three sections of this chapter described the unique categories of materials: metals, ceramics, and polymers. There is an increasing emphasis on combining materials from these different categories—even different materials within each category—in such a way as to achieve properties and performance that are unique. Such materials are called *composites*. A composite can mean different things depending on the level of definition we use. In the most basic sense, all materials except elements are composites. For example, a binary mixture of two elements, like an alloy, can be considered a composite structure on an atomic scale. More practically, composite systems begin at the nanoscale where the components are identifiable. These material systems are called *nanocomposites*. At the microstructure level, composites are composed of crystals, phases, and compounds. With this definition, steel, which is a suspension of carbon in iron, is a composite, but brass, a single-phase alloy, is not a composite. If we move up one more level on the size scale, we find that there are macrostructural composites: materials composed of fibers, matrices, and particulates. These are the traditional composite materials and include such common materials as concrete (a composite of cement and particulates) and fiber-reinforced plastics. Despite the multitude of size scales and compositional variations available for composites, they all have one thing in common: they are *materials systems*. That is, they comprise two or more components that are distinguishable and perform complementary functions in the material system. This broad definition is the one we will use, so a composite is this: *a material system brought about by combining materials differing in composition or form for the purpose of obtaining specific characteristics and properties*. We will add qualifiers and clarifiers to this definition as the components and applications dictate, but this definition is the starting point for identifying the constituents of a composite.

1.4.1 Composite Constituents

The constituents in a composite retain their identity such that they can be physically identified and exhibit an interface with one another. This concept is graphically summarized in Figure 1.69. The body constituent gives the composite its bulk form; it is called the *matrix* or *body component*. The other component is a *structural constituent*, sometimes called the *reinforcement* or *filler*, which determines the internal structure of the composite. Though the structural component in Figure 1.69 is a fiber, there are other geometries that the structural component can take on, as

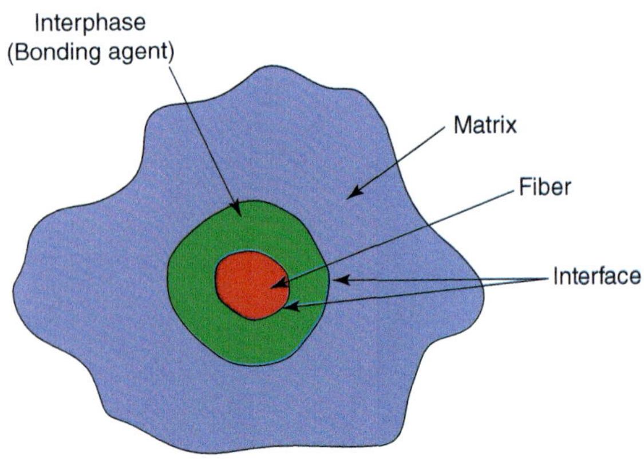

Figure 1.69 Schematic illustration of composite constituents. Reproduced, by permission, from Schwartz [37]. Copyright © 1992 by McGraw-Hill.

we will discuss in a subsequent section. The region between the body and structural constituents is called the *interphase*. It is quite common (even in the technical literature), but incorrect, to use the term *interface* to describe this region. An interface is a two-dimensional construction—an area having a common boundary between the constituents—whereas an interphase is a three-dimensional phase between the constituents and, as such, has its own properties. It turns out that these interphase properties play a very important role in determining the ultimate properties of the bulk composite. For instance, the interphase is where mechanical stresses are transferred between the matrix and the reinforcement. The interphase is also critical to the long-term stability of a composite. It will be assumed that there is always an interphase present in a composite, even though it may have a thickness of only an atomic dimension.

The chemical composition of the composite constituents and the interphase is not limited to any one material class. There are metal–matrix, ceramic–matrix, and polymer–matrix composites, all of which find industrially relevant applications. Similarly, reinforcements in important commercial composites are made of such materials as steel, E-glass, and Kevlar. Often, a bonding agent is added to the fibers prior to compounding to create an interphase of a specified chemistry. We will describe specific component chemistries in subsequent sections.

1.4.2 Composite Classification

We have already seen that there are many ways to classify composites, including the size scale of the constituents as described earlier, but other classification schemes are based upon (1) specific material combinations, such as metal–matrix, or glass-fiber-reinforced composites; (2) bulk-form characteristics, such as laminar composites or matrix composites; (3) distribution of constituents, such as continuous or discontinuous; or (4) function, like structural or electrical composites. Scheme (2) is the most general, so we will utilize it here, but other classification schemes will be useful in later sections of this chapter.

As shown in Figure 1.70, there are five general types of composites when categorized by bulk form. *Fiber composites* consist of fibers— with or without a matrix. By definition, a fiber is a particle

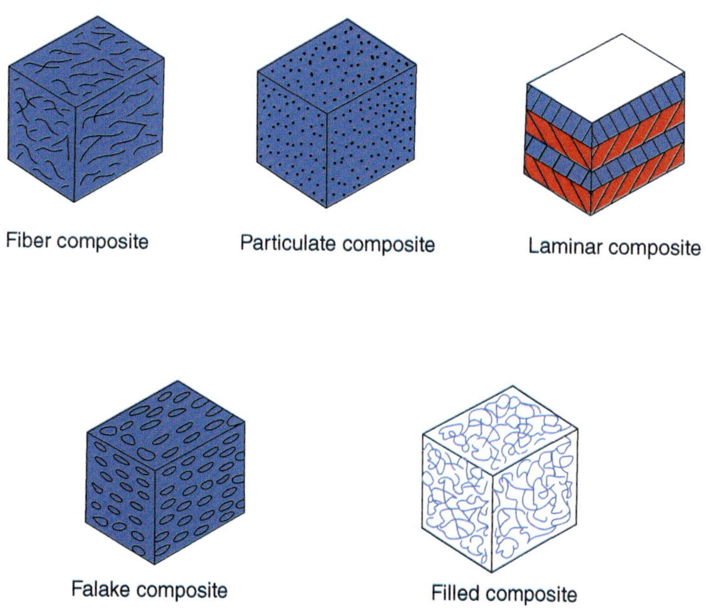

Figure 1.70 Classes of composites. Reproduced, by permission, from Schwartz [38]. Copyright © 1992 by McGraw-Hill.

longer than 100 μm with a length-to-diameter ratio (*aspect ratio*) greater than 10 : 1. *Flake composites* consist of flakes, with or without a matrix. A flake is a flat, plate-like material. *Particulate composites* can also have either a matrix or no matrix along with the particulate reinforcement. Particulates are roughly spherical in shape in comparison to fibers or flakes. In a *filled composite*, the reinforcement—which may be a three-dimensional fibrous or porous structure—is continuous and often considered the primary phase, with a second material added through such processes as *chemical vapor infiltration* (CVI). Finally, *laminar composites* are composed of distinct layers. The layers may be of different materials or the same material with different orientations. There are many variations on these classifications, of course, and we will see that the components in fiber, flake, and particulate composites need not be distributed uniformly and may even be arranged with specific orientations.

Cooperative Learning Exercise 1.8

We will see in Section 5.4.2 that the elastic modulus of a unidirectional, continuous-fiber-reinforced composite depends on whether the composite is tested along the direction of fiber orientation (parallel) or normal to the fiber direction (transverse). In fact, the elastic modulus parallel to the fibers, E_1, is given by Eq. (1.49), whereas the transverse modulus, E_2, is given by Eq. (1.50). Consider a composite material that consists of 40% (by volume) continuous, uniaxially aligned, glass fibers (E_f = 76 GPa) in a polyester matrix (E_m = 3 GPa).

Person 1: Calculate the elastic modulus parallel to the fiber direction, E_1.

Person 2: Calculate the elastic modulus transverse to the fiber direction, E_2.

Compare your answers. What would you expect the composite modulus to be if the same volume fraction of fibers were randomly oriented instead of uniaxially aligned?

Answer: $E_1 = (0.4)76 + (0.6)(3) = 32.2$ GPa; $E_2 = (76)(3)/[(0.6)(76) + (0.4)(3)] = 4.87$ GPa

1.4.3 Combination Effects in Composites

There are three ways that a composite can offer improved properties over the individual components, collectively called *combination effects*. A *summation effect* arises when the contribution of each constituent is independent of the others. For example, the density of a composite is, to a first approximation, simply the weighted average of the densities of its constituents. The density of each component is independent of the other components. Elastic modulus is also a summation effect, with the upper limit $E_c(u)$ given by

$$E_c(u) = V_p E_p + V_m E_m \tag{1.49}$$

and the lower limit, $E_c(l)$ given by

$$E_c(l) = \frac{E_m E_p}{V_m E_p + V_p E_m} \tag{1.50}$$

where E_m and E_p are the elastic moduli of the matrix and particulate, respectively, and V_m and V_p are their respective volume fractions. We will discuss mechanical properties in more detail in Chapter 5, but the point of Eqs. (1.49) and (1.50) is that summation properties can be added appropriately to give an estimate of the composite properties. A *complementation effect* occurs when each constituent contributes separate properties. For example, laminar composites are sandwich-type composites composed of several layers of materials. Sometimes the outer layer is a protective coating, such as a polymeric film, that imparts corrosion resistance to the composite. This outer layer serves no structural purpose and contributes only a specific property to the overall composite—in this case, corrosion resistance. Finally, some constituent properties are not independent of each other, and an *interaction effect* occurs. In this case, the composite property may be higher than either of the components, and the effect may be synergistic rather than additive. For example, it has been observed that the strength of some glass-fiber-reinforced plastic composites is greater than either the matrix or the reinforcement component by itself. In the subsequent sections, we look at the individual components of a composite and see what each can contribute and how it helps to improve the properties of the composite.

Take a Peek! Jump ahead to Section 5.4.2 for more information on combination effects in the mechanical properties of composites.

1.4.4 The Composite Matrix

The matrix serves two primary functions: to hold the reinforcement phase in place and to deform and distribute the stress under load to the reinforcement phase. In most cases, the matrix material for a fiber composite has an elongation at break greater than the fiber; that is, it must deform more before breaking. It is also beneficial to have a matrix that encapsulates the reinforcement phase without excessive shrinkage during processing. A secondary function of the matrix is to protect the surface of the reinforcement. Many reinforcements tend to be brittle, and the matrix protects them from abrasion and scratching, which can degrade their mechanical properties. The matrix can also protect the reinforcement component from oxidation or corrosion. In this way, many fibers with excellent mechanical properties, such as graphite fibers, can be used in oxidizing environments at elevated temperatures due to protection by the matrix constituent.

Most *fiber-matrix composites* (FMCs) are named according to the type of matrix involved. *Metal-matrix composites* (MMCs), *ceramic-matrix composites* (CMCs), and *polymer-matrix composites* (PMCs) have completely different structures and completely different applications. Often, the temperature at which the composite must operate dictates which type of matrix material is to be used. The maximum operating temperatures of the three common types of FMCs are listed in Table 1.28.

Table 1.28 Approximate Upper-Temperature Limits for Continuous Operation of Composites

Composite	Maximum Operating Temperature (°C)
Polymer matrix	400
Metal matrix	1000
Ceramic matrix	1300

Table 1.29 Some Typical Polymers Used as Matrices in PMCs

Matrix Material	Process Temperature (°C)	Upper Use Temperature (°C)
Thermosetting		
Polyester (phthalic/maleic type)	RT	70
Vinyl ester	RT	125
Epoxy	150	125
Epoxy	200	175
Phenolic	250	200
Cyanates (triazines)	250	200
Bismaleimides	250	225
Nadic end-capped polyimides (e.g., PMR-15)	316	316
Thermoplastic		
Polysulfone	325	180
Polyamide	250	100
Polycarbonate	280	100
Polyphenylene oxide (PPO)	280	100
Polysulfides (PPS)	300	150
Polyether ether ketone (PEEK)	370	175
Polyether sulfone (PES)	350	175
Polyamide-imides	325	200
Polyetherimide	400	275
Polyimide	370	316
Polyarylate	400	300
Polyester (liquid crystalline)	300	150

Most structural PMCs consist of a relatively soft matrix, such as a thermosetting plastic of polyester, phenolic, or epoxy, sometimes referred to as *resin-matrix composites*. Some typical polymers used as matrices in PMCs are listed in Table 1.29. The list of metals used in MMCs is much shorter. Aluminum, magnesium, titanium, and iron- and nickel-based alloys are the most common (see Table 1.30). These metals are typically utilized due to their combination of low density and good mechanical properties. Matrix materials for CMCs generally fall into four categories: glass ceramics like lithium aluminosilicate; oxide ceramics like aluminum oxide (alumina) and mullite; nitride ceramics such as silicon nitride; and carbide ceramics such as silicon carbide. The processing techniques used for CMCs can be quite exotic (and expensive), such as CVI, or through pyrolysis of polymeric precursors. Their maximum use temperatures are theoretically much higher than most MMCs or PMCs, exceeding 1800°C, although the practical use temperature is often much lower due to creep deformation (see Table 1.31). The attraction of CMCs, then, is for use in high-temperature structural applications.

1.4.5 The Composite Reinforcement

In Section 1.4.2, we described several classification schemes for composites, including one that is based on the distribution of the constituents. For reinforced composites, this scheme is quite useful,

Table 1.30 Common Metals for Metal-Matrix Composites

Matrix Material	Fabrication Method	Typical Composite Density (g/cm^3)	Use Temperature (°C)
Aluminum	Diffusion bonding		350
	Hot molding	2.62–3.45	
	Powder metallurgy		
	Liquid processing		
Magnesium	Liquid processing	1.82–2.80	300
	Diffusion bonding		
Titanium	Diffusion bonding	3.76–4.00	650
Iron- and nickel-based alloys	Diffusion bonding	5.41–11.7	800–1150

Table 1.31 Common Ceramics for Ceramic-Matrix Composites

Matrix Material	Density (g/cm^3)	Use Temperature (°C)
Alumina, Al_2O_3	4.0	~1000
Glass ceramics	2.7	900
Si_3N_4	3.1	~1300
SiC	3.2	~1300

as shown in Figure 1.71. In reinforced composites, the reinforcement is the structural constituent and determines the internal structure of the composite. The reinforcement may take on the form of particulates, flakes, lamina, or fibers. Fibers are the most common type of reinforcement, resulting in FMCs. Let us examine some of these reinforcement constituents in more detail.

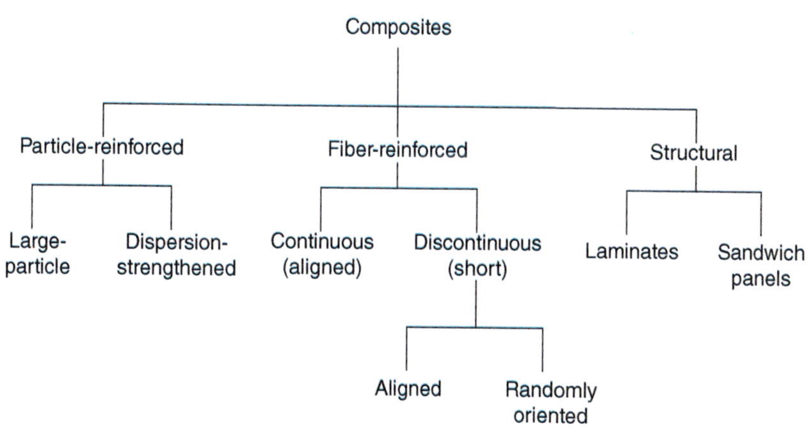

Figure 1.71 Types of reinforced composites.

Fiber-Matrix Composites

As shown in Figure 1.71, there are two main classifications of FMCs: those with continuous fiber reinforcement and those with discontinuous fiber reinforcement. *Continuous-fiber-reinforced composites* are made from fiber *rovings* (bundles of twisted filaments) that have been woven into two-dimensional sheets resembling a cloth fabric. These sheets can be cut and formed to a desired shape, or *preform*, that is then incorporated into a composite matrix, typically a thermosetting resin such as epoxy. Metallic, ceramic, and polymeric fibers of specific compositions can all be produced in continuous fashion, and the properties of the resulting composite are highly dependent not only on the type of fiber and matrix used but also on the processing techniques with which they are formed. We will discuss continuous-fiber-reinforced composites in more detail in Sections 5.4 and 7.4.

Discontinuous-fiber-reinforced composites are much more widely used, and there are some general underlying principles that affect their overall properties. Four main factors contribute to the performance level of a fiber in discontinuously reinforced FMCs. The first factor is fiber orientation. As shown in Figure 1.72, there are several ways that short fibers can be oriented within the matrix. *One-dimensional reinforcement* occurs when the fibers are oriented along their primary axis. This offers maximum mechanical strength along the orientation axis, but results in *anisotropic composites*; that is, the mechanical and physical properties are not the same in all directions. *Planar reinforcement* occurs with two-dimensional orienting of the fibers, as often occurs with woven fabrics. The fabric, as is common in woven glass fibers, is produced in sheets, and it is laid down (much like a laminate) to produce a *two-dimensional reinforcement* structure. *Three-dimensional reinforcement* results from the random orientation of the fibers. This creates an isotropic composite, in which the properties are the same in all directions, but the reinforcing value is often decreased compared to the aligned fibers.

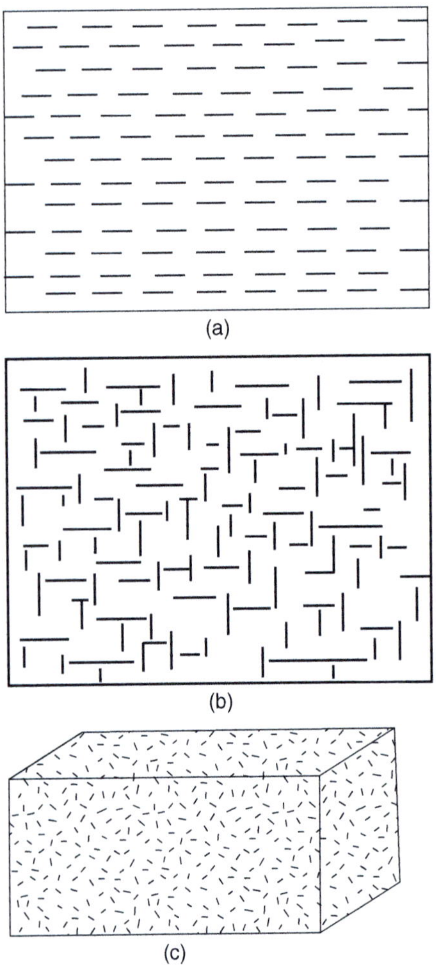

Figure 1.72 Types of fiber reinforcement orientation: (a) one-dimensional, (b) two-dimensional, and (c) three-dimensional.

The second factor that affects performance in discontinuously reinforced FMCs is fiber length. This has an effect primarily on the ease with which the composite can be manufactured. Very long fibers can create difficulties with methods used to create discontinuously reinforced FMCs and can result in nonuniform mechanical properties. The third factor is also related to fiber geometry, namely, the fiber shape. Recall that the definition of a fiber is a particulate with length greater than 100 μm and an aspect ratio greater than 10 : 1. This definition allows for a great deal of flexibility in the geometry of the fiber. For example, aspect ratios can vary widely; many reinforcement filaments have aspect ratios much larger than 10 : 1. It is also not necessary that the fiber cross section be exactly circular. Hexagonal, ellipsoidal, and even annular (hollow fiber) cross sections are quite common. Finally, it is also not necessary that the fiber, even those with circular cross sections, be exactly cylindrical. "Dumbbell"-shaped fibers are very common—even preferred—since mechanical stresses tend to concentrate at the fiber ends. The dumbbell shape helps distribute these stresses.

The final factor affecting the reinforcement's performance is its composition. Chemistry affects properties, and strength is usually the most important property of a reinforcing fiber. Though we will concentrate on the mechanical properties of materials in Chapter 5, it is useful at this point to familiarize yourself with some of the common fiber reinforcements, as summarized in Table 1.32. It is worth noting that often the important design consideration for reinforcement materials (or the matrix, for that matter) is not the absolute value of a particular design criterion, such as tensile strength or modulus, but the value per unit weight, such as *specific strength* or *specific modulus*.

Table 1.32 Selected Properties of Some Common Reinforcing Fibers

Fiber	Density (g/cm^3)	Melting Point (°C)	Specific Modulus ($GPa \cdot cm^3/g$)	Specific Strength ($MPa \cdot cm^3/g$)
Aluminum	2.70	660	27	230
Steel	7.87	1621	25	530
Tantalum	16.6	2996	12	37
Titanium	4.72	1668	24	410
Tungsten	19.3	3410	21	220
Boron	2.30	2100	192	1500
Beryllium	1.84	1284	165	710
Molybdenum	10.2	2610	35	140
Aluminum oxide	3.97	2082	132	170
Aluminosilicate	3.90	1816	26	1060
Asbestos	2.50	1521	60	550
Beryllium carbide	2.44	2093	127	420
Beryllium oxide	3.03	2566	116	170
Carbon	1.76	3700	114	1570
Graphite	1.50	3650	230	1840
E-glass	2.54	1316	28	1360
S-glass	2.49	1650	34	1940
Quartz	2.20	1927	32	407
SiC (on tungsten)	3.21	2316	140	1000
Si_3N_4	2.50	1900	100	1344
BN	1.91	3000 (decomp.)	47	722
ZrO_2	4.84	2760	71	427
Wood	0.4–0.8	–	17	–
Polyamide (Kevlar)	1.14	249	2.5	730
Polyester	1.40	249	2.9	490
Polypropylene	0.9	154	1.8	77

These values are listed in Table 1.32 rather than as absolute values to illustrate this point. This fact is extremely important for many applications, such as automotive and aerospace composites, for which weight savings are paramount. Note also that reinforcing fibers come from all material classes. Let us examine some of the more common fibers in more detail.

Particulate Composites

Particulate composites encompass a wide range of materials, from cement reinforced with rock aggregates (concrete) to mixtures of ceramic particles in metals, called *cermets*. In all cases, the particulate composite consists of a reinforcement that has similar dimensions in all directions (roughly spherical), and all phases in the composite bear a proportion of an applied load. The percentage of particulates in this class of composites ranges from a few percent to over 70%.

One important class of particulate composites is *dispersion-hardened alloys*. These composites consist of a hard particle constituent in a softer metal matrix. The particle constituent seldom exceeds 3% by volume, and the particles are very small, below micrometer sizes. The characteristics of the particles largely control the property of the alloy, and a spacing of 0.2–0.3 μm between particles usually helps optimize properties. As particle size increases, less material is required to achieve the desired interparticle spacing. Refractory oxide particles are often used, although intermetallics such as $AlFe_3$ also find use. Dispersion-hardened composites are formed in several ways, including surface oxidation of ultrafine metal powders, resulting in trapped metal oxide particles within the metal matrix. Metals of commercial interest for dispersion-hardened alloys include aluminum, nickel, and tungsten.

A *cermet* is a particulate composite similar to a dispersion-hardened alloy but consists of larger ceramic grains (*cer-*) held in a metal matrix (*-met*) (see Figure 1.73). The refractory particulates can be from the oxide category, such as alumina (Al_2O_3), magnesia (MgO), or thoria (ThO_2), or from the carbide category, like tungsten carbide (WC), chromium carbide (Cr_3C_2), or titanium carbide (TiC). Cermets are formed by traditional powder-metallurgical techniques, although bonding agents must sometimes be added to improve bonding between the ceramic particulate and the metallic matrix. Cermets are used in many applications, including fuel elements and control rods in the nuclear industry, pulse magnetrons, and cutting tools.

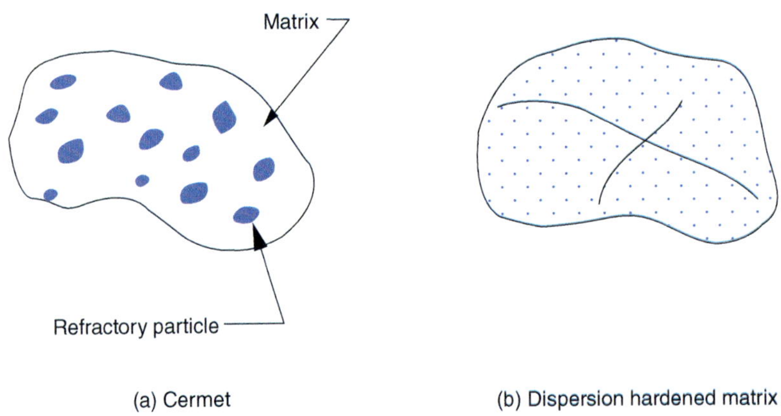

Figure 1.73 Comparison of (a) cermet and (b) dispersion-hardened alloy. Reproduced, by permission, from Schwartz [39]. Copyright © 1992 by McGraw-Hill.

Finally, metal- and resin-bonded composites are also classified as particulate composites. *Metal-bonded composites* include structural parts, electrical contact materials, metal-cutting tools, and magnet materials and are formed by incorporating metallic or ceramic particulates such as WC, TiC, W, or Mo in metal matrixes through traditional powder metallurgical or casting techniques. *Resin-bonded composites* are composed of particulate fillers such as silica flour, wood flour, mica, or glass spheres in phenolformaldehyde (Bakelite), epoxy, polyester, or thermoplastic matrixes.

1.4.6 The Composite Interphase

As mentioned in the introduction to the section on composites, the interphase, or the region between the matrix and the reinforcement, is often the primary determinant of the mechanical properties of the composite. It can affect other properties as well, but since it serves to transfer loads from the matrix to the reinforcement, its primary impact is on the strength of the composite.

There are several factors that affect the composition and spatial extent of the interphase. The first is *wettability*, or the ease with which the liquid matrix wets the reinforcing constituent prior to solidification. (There are other ways to form composites, of course, but incorporation of a solid particle into a liquid matrix is common and is germane to the description of wetting.) The ability of a liquid to wet a solid is measured by an interfacial *contact angle*, as shown in Figure 1.74. Without going into the thermodynamics of wetting at this point, it is important to the current discussion only to know that low contact angles are representative of "good" wetting (i.e., favorable liquid–solid interactions), whereas high contact angles (greater than about 90°) are indicative of "poor" wetting and unfavorable liquid–solid interactions. Obviously, a favorable liquid–solid interaction is desirable to obtain good matrix–fiber interactions. Sometimes, good wetting is not possible due to the disparate chemistries of the constituents, such as is found between silicate glass fibers and polymeric matrixes. In such cases, *wetting agents* are sometimes added to improve wettability. Another important factor in determining the makeup of the interphase is the inherent reactivity between the components. It is possible for the reinforcement and matrix to react with one another, either through direct chemical reaction or through longer-term diffusion of components from one phase to another. For example, Figure 1.75 shows a $CaO—Al_2O_3$ fiber embedded in an aluminum matrix with traces of silicon [27]. Elemental analysis of the interphase (the dark region between the fiber and the matrix) shows it contains silicon, which has diffused into the fiber surface during melt processing of the aluminum matrix to form an aluminum–silicon oxide compound.

Despite the emphasis on favorable interactions between the matrix and reinforcement and compound formation between them, it may be beneficial in certain circumstances for the interaction between the two primary constituents to be relatively weak. This is especially true in

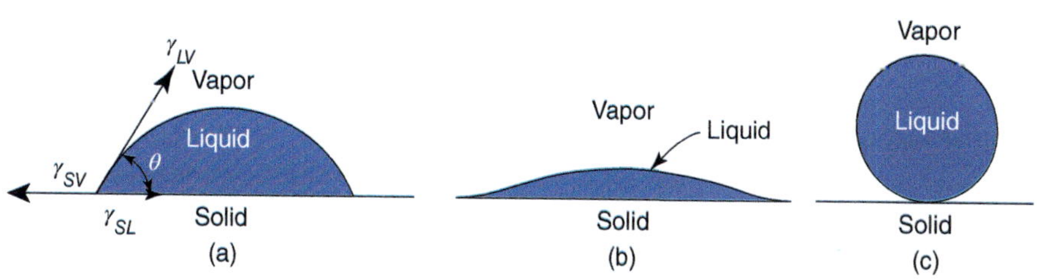

Figure 1.74 Interaction of a liquid droplet with a solid substrate: (a) Partially wetting ($0° < \theta < 180°$), (b) completely wetting ($\theta \approx 0°$), and (c) completely nonwetting ($\theta = 180°$). Interfacial energies between the solid–liquid, solid–vapor, and vapor–liquid are represented by γ_{SL}, γ_{SV}, and γ_{LV}, respectively. From Ralls et al. [7]. Copyright © 1976 by John Wiley & Sons, Inc. This material is used by permission of John Wiley & Sons, Inc.

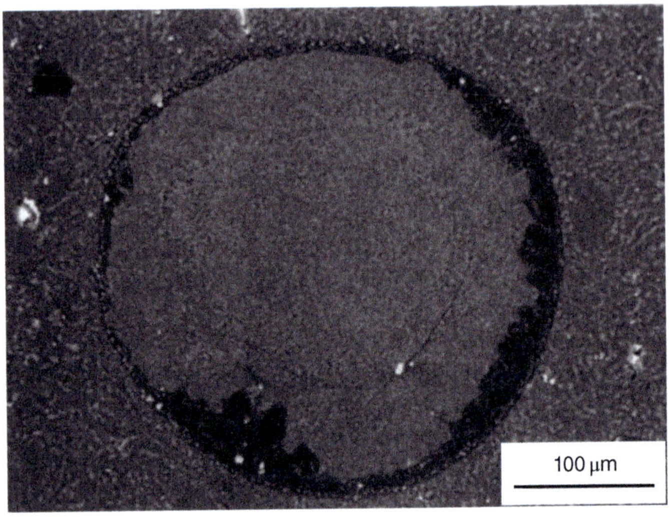

Figure 1.75 SEM photomicrograph of a $CaO—Al_2O_3$ fiber embedded in 4032 aluminum alloy. Sung et al. [40], © 1994, Reproduced with permission from Springer Nature.

ceramic–ceramic composites, where both constituents are brittle, and the only way to impart some ductility to the composite is for the interphase to fail "gracefully"—that is, for the fibers to actually "pull out" of the matrix in a controlled manner. Optimization of the interphase properties is currently the focus of much research. In fact, there is one class of composites in which the interphase can actually dominate the composite properties. This arises in nanoscale reinforcements that are mostly surface, termed nanocomposites.

1.4.7 Nanocomposites

Just as nanostructured materials are a subset of all materials, *nanocomposites* are a subset of composite materials. They are so named because one or more of the identifiable components has a characteristic dimension of the order 10^{-9} m (1–100 nm). This definition leaves room for a wide variety of materials, so it is beyond the scope of this introductory text to survey them all. A few important examples will suffice to illustrate the variety and importance of nanocomposites.

Exfoliated Clay

Recall from Section 1.2.2 that silicates are a broad class of inorganic materials that comprise a silicon-oxygen network balanced in charge by a variety of possible metallic counterions. Although *clay* is a broad term denoting a natural material of variable composition, most clays contain some type of hydrous aluminum phyllosilicate, including *kaolin*, $Al_2Si_2O_5(OH)_4$, and *montmorillonite*, which have variable levels of hydration and counterions such as Na and Ca. Table 1.18 tells us that silicates with Si/O = 2.5 should form sheets, and indeed kaolin-based clays have molecularly thin silicate sheets separated by aluminum counterions (see Fig. 1.76) and held together by hydrogen bonds from the hydroxyl groups. These hydrogen bonds are relatively weak and can be easily manipulated through the insertion of ions or molecules between the silicate layers in a process known as *intercalation* (Fig. 1.77). When the hydrogen bonds are completely broken, individual sheets are the results, and the process is known as *exfoliation*. *Exfoliated clay* is industrially important and is the result of the mixing of clays with such exfoliating molecules as hydrophilic polymers.

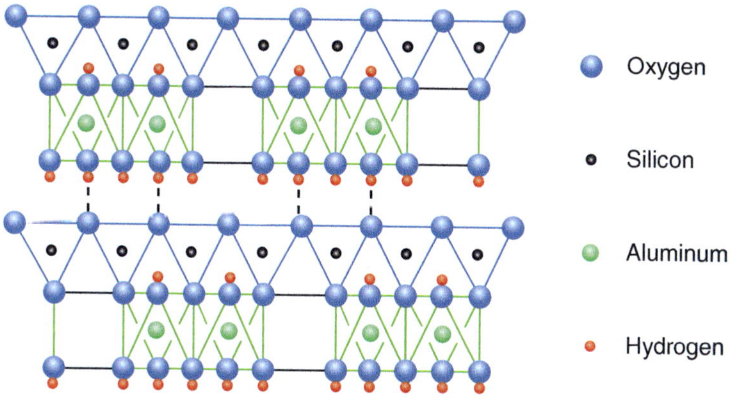

Figure 1.76 Sheet-like chemical structure of kaolin. Chris P. Schaller/https://employees.csbsju.edu/cschaller/Principles%20Chem/network/NWalumina.htm.

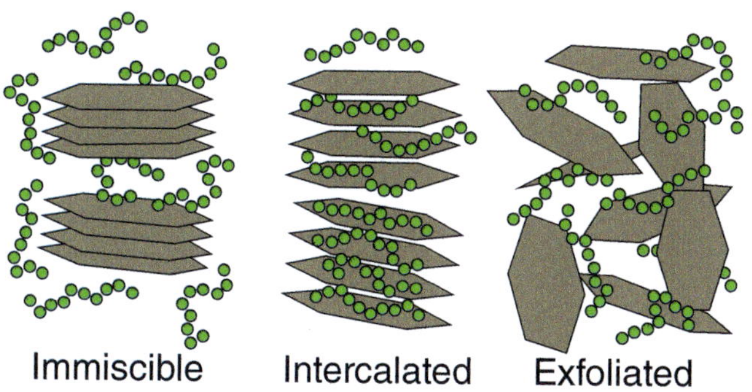

Figure 1.77 Intercalated and exfoliated clay structures from mixing of clays with polymers. James et al. 2015/American Chemical Society/CC BY 4.0.

With the polymer as the matrix and exfoliated clay as the reinforcement, the resulting nanocomposite is lightweight with improved mechanical properties over the polymer alone. These nanocomposites find use in the automotive industry for interior panels where weight and formability are important considerations. Food packaging films of clay-polymer nanocomposites improve content freshness through reduced diffusion of oxygen through the packaging and allow for the incorporation of nontoxic bioactive agents that have antibacterial, antifungal, and antioxidant properties [41].

Take a Peek! Jump ahead to Section 5.4.4 to see how the mechanical properties of nanocomposites are superior to those of their constituents.

CNT Reinforced Composites

Recall from Section 1.1.6 that CNTs can come in the single-walled, double-walled, or multi-walled form. In any case, their combination of excellent mechanical properties and light weight makes them ideal for use as reinforcement in lightweight composites. Unlike nanoclay composites, which are mostly limited to polymer matrix composites because of the need for exfoliation, CNT composites are found in most matrix classes. In metal matrix composites, for example, CNTs not only give rise to improved mechanical properties and decreased density but can also lead to interesting

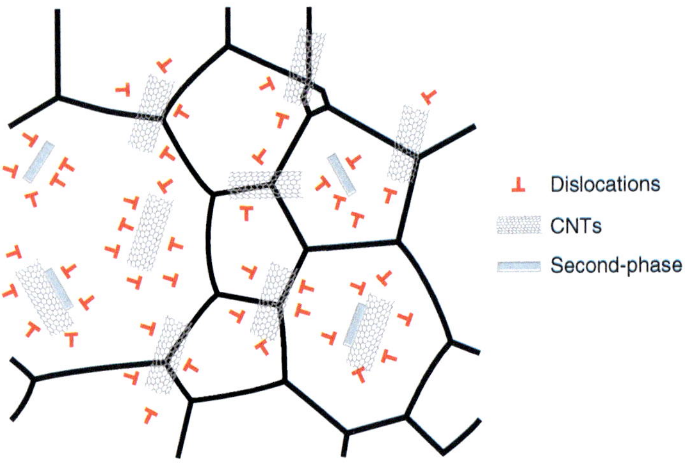

Figure 1.78 Interaction of CNTs with metal matrix to form dislocations and secondary phases. Íris Carneiro et al. 2021/MDPI/CC BY 4.0.

phenomena such as increased hardness due to mismatches in the coefficient of thermal expansion between the fiber and matrix phases (see Fig. 1.78). The formation of secondary phases and interphases due to the reaction of the CNTs with the metal matrix described earlier also affects the performance of the composite. For example, Al_3C_4 secondary phases form when CNTs are incorporated into aluminum matrix composites, leading to strain hardening and strengthening. These are the topics of Chapter 5, but also an example of interaction effects described earlier that are fundamental to the benefits composites can provide.

1.5 Structure of Biologics

The boundary between the disciplines of biology and materials engineering is becoming increasingly blurred. Molecules that were once thought of as being biological "macromolecules," such as DNA, are now being used as templates to produce molecules with desired gene sequences, and live tissue cultures are being used to reconstruct human body parts using 3-dimensional printers. More importantly, however, the scientific and engineering principles behind the application and development of biological materials are becoming more well understood, so it is useful to provide some information on the structure of these materials in the context of their use as materials of (human) construction. We begin with some simple biochemistry and then describe how nature uses these building blocks to create complex composite structures of tissue, both hard and soft.

1.5.1 Review of Biological Molecules

Just as we began our description of polymer structure with an organic chemistry review, let us begin our introduction to biological materials with some simple biochemistry.

Amino Acids and Proteins

Proteins are the molecules that perform the functions of life. They can be enzymes that catalyze biological reactions, or they can be the receptor site on a membrane that binds a specific substance.

Proteins are important parts of both bones—the so-called *hard biologics*—and the *soft biologics* such as muscle and skin. Any discussion of the structure of living organisms must begin with the structure of proteins.

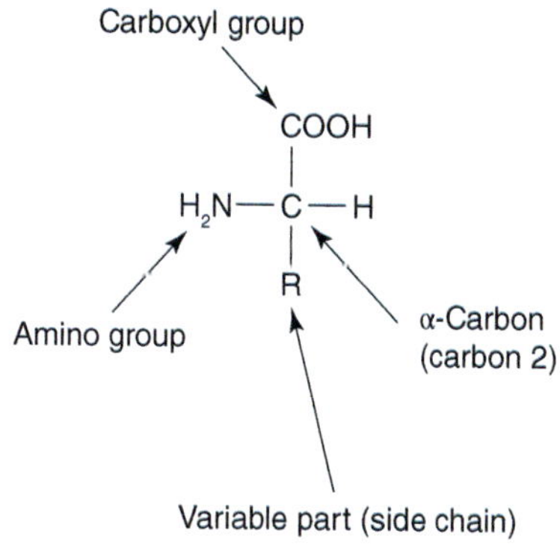

Figure 1.79 General structure of an amino acid.

Proteins are composed of *amino acids*. As shown in Figure 1.79, an amino acid has a carboxyl group, an amino group (refer to Table 1.22 for a summary of functional groups), a central atom identified as the alpha (α) carbon, and a variable part known as the side chain (R). Normally, the amino acid is in its dissociated state, such that the terminal hydrogen on the carboxyl group moves to the amino group, thereby creating a carboxylate group (COO^-) and an ammonium group (NH_3^+) known as a *zwitterion*. The molecule remains neutral overall, and there may be several stable zwitterion forms, depending on the amino acid. Note also that the α-carbon is an asymmetric carbon, resulting in a chiral molecule for all but one of the naturally occurring amino acids. All the amino acids are found in only one of the stereoisomer configurations (the L-configuration).

There are only 20 amino acids that make up proteins (technically, 19 amino acids and one imino acid), differing only in the type of R group they contain (see Table 1.33). Each amino acid has a specific name and three-letter designation. Most amino acids decompose instead of melting due to their strong intermolecular electrostatic attractions, with decomposition temperatures ranging from 185°C to 315°C, and they are only sparingly soluble in water, with the exceptions of glycine, alanine, proline, lysine, and arginine.

To form proteins, the carboxylic acid group on one amino acid reacts with the amine group on another molecule in a condensation reaction that forms one water molecule and a —CO—NH—CHR— linkage known as a *peptide bond*. A molecule containing more than about 100 amino acid sequences is called a *polypeptide*, and a protein is composed of one (or more) polypeptide chains. Thus, the number of possible proteins from the 20 amino acids is enormous (20^{100}). Replacing even one amino acid in the sequence of a protein can change its function completely. Sickle cell anemia, for example, is the result of replacing only one valine amino acid with a glutamic acid unit in one protein chain of the hemoglobin molecule. The peptide sequence is named by starting at the *N*(−amino) terminus of the polypeptide.

As with polymers, peptides and proteins can possess complex conformations by rotation of bonds in the backbone and interaction between side chains. The stability of these structures is strongly dependent upon the R groups and, hence, the specific amino acid sequence, as well as the environment in which the protein finds itself. Figure 1.80 illustrates a few of the common conformations found in proteins. The *α-helix* occurs when the chain coils like a right-hand screw to form a cylinder and is the result of hydrogen bonding between the C=O and N—H in adjacent turns of the helix. Only the right-handed α-helix occurs in nature, and its presence results in an electric dipole with excess positive charge at one end and excess negative charge at the other. In the β-sheet, the peptide chain is much more extended, with 0.35 nm between adjacent peptide groups, in comparison to 0.15 nm for the α-helix. These sheet structures are also the result of hydrogen bonding. The α-helix and the β-sheet are examples of *secondary structures* in proteins. Tertiary and quaternary structures also exist but are beyond the scope of this text. Depending on the nature of the side chains, there can also be hydrophobic interactions within the chain, leading to chain extension. Finally, disulfide bonds can occur when two cysteine residues react to form a covalent —S—S— bond. The breaking

Table 1.33 Common Amino Acids

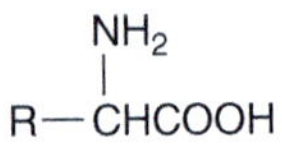	**Name**	**Abbreviation**
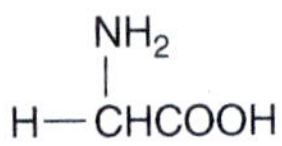	Glycine	Gly
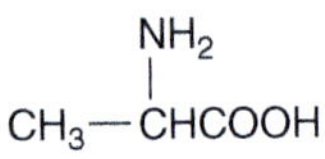	Alanine	Ala
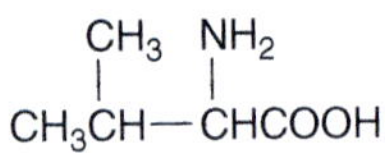	Valine	Val
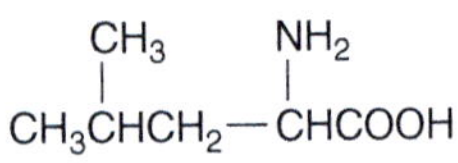	Leucine	Leu
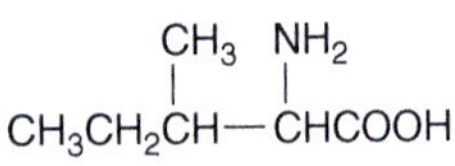	Isoleucine	Ile
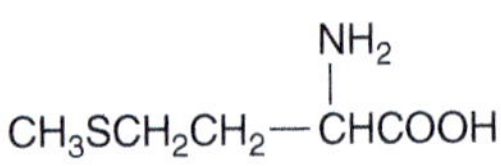	Methionine	Met
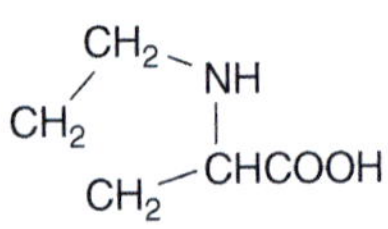	Proline	Pro
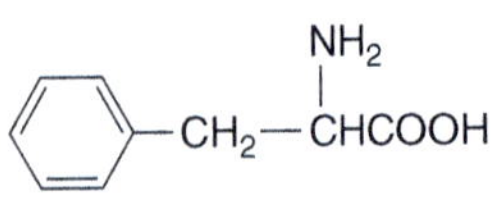	Phenylalanine	Phe
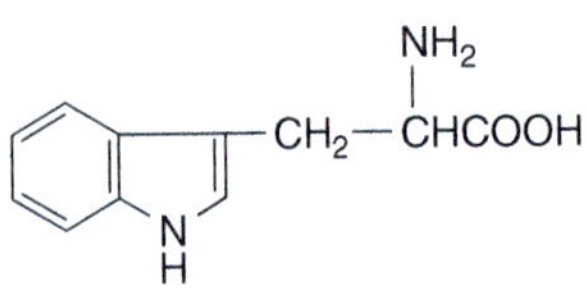	Tryptophan	Trp
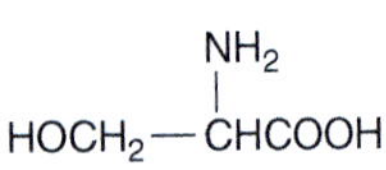	Serine	Ser
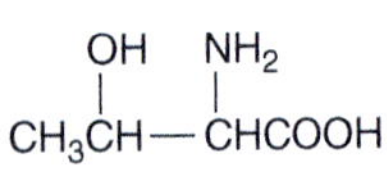	Threonine	Thr
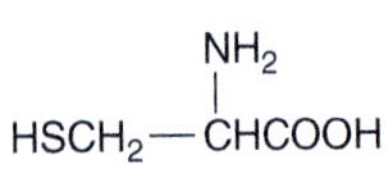	Cysteine	Cys
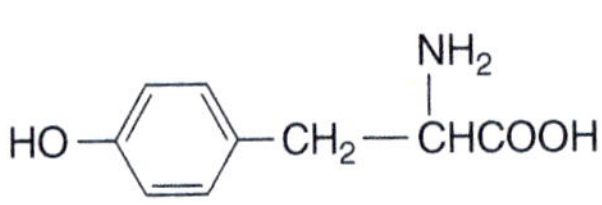	Tyrosine	Tyr

Table 1.33 (Continued)

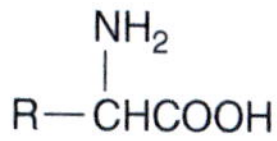	Name	Abbreviation
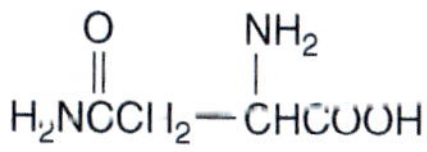	Asparagine	Asn
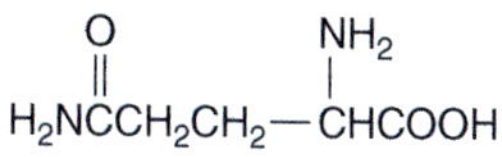	Glutamine	Gln
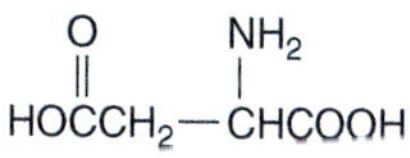	Aspartic acid	Asp
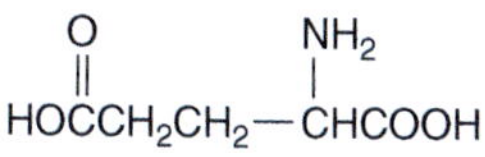	Glutamic acid	Glu
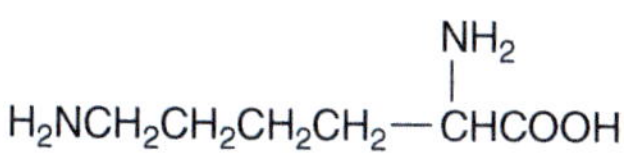	Lysine	Lys
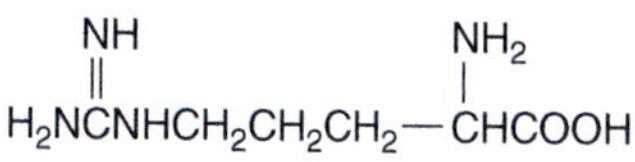	Arginine	Arg
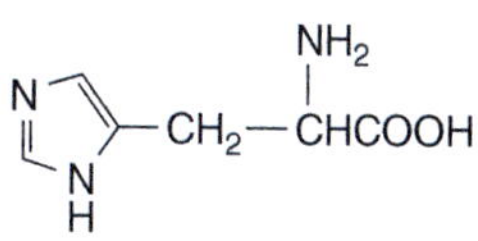	Histidine	His

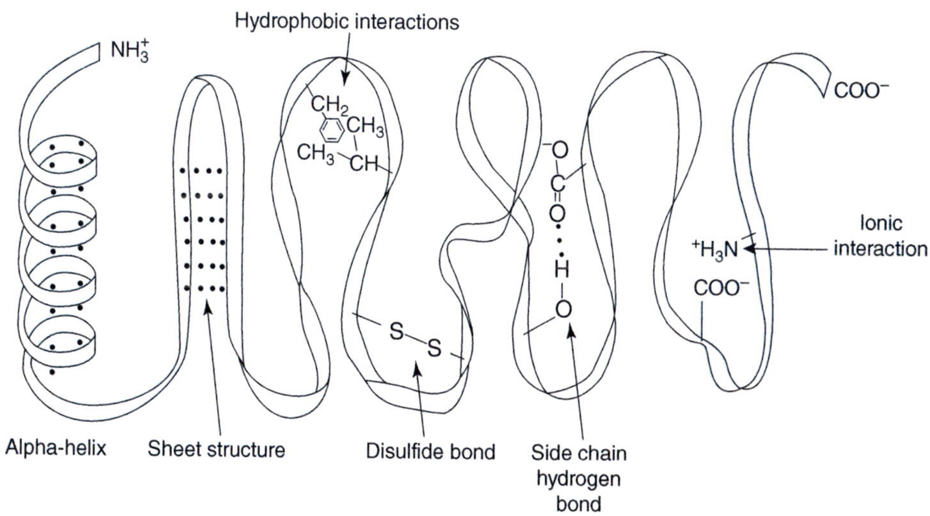

Figure 1.80 Some common structural elements in a hypothetical protein molecule. Reproduced, by permission, from Houston [42]. Copyright © 2001 by Michael E. Houston.

of disulfide bonds, or any action that leads to an alteration in the structure of a protein so as to render it inactive, leads to *denaturation*. Denatured proteins also tend to have decreased solubility.

DNA and RNA

Like proteins, *deoxyribonucleic acid* (DNA) and *ribonucleic acid* (RNA) are polymers, but instead of amino acids as repeat units, they are composed of a chain of *nucleotides*. Each nucleotide is composed of three basic structural units: a *base*, a *sugar*, and a *phosphate group* (see Figure 1.81). One base with its sugar (and without the phosphate group) is called a *nucleoside*. The sugar in RNA is called *ribose*, which is reduced in DNA by a loss of oxygen at the 2′ carbon to form *deoxyribose* (see Figure 1.82). There are only five primary bases found in polynucleotides: two *purines* represented by *adenine* (A) and *guanine* (G); and three *pyrimidines* represented by *cytosine* (C), *thymine* (T), and *uracil* (U). Thymine is found only in DNA nucleotides and uracil only in RNA nucleotides, which results in four DNA nucleotides (see Figure 1.83) and four RNA nucleotides. The four DNA nucleotides are 2′-deoxyadenosine monophosphate (dAMP), 2′-deoxyguanosine monophosphate (dGMP), 2′-deoxycytidine monophosphate (dCMP), and thymidine monophosphate (TMP), the latter of which is already assumed to have 2′-deoxyribose as the sugar since it occurs only in DNA and not in RNA.

The nucleic acid polymer is formed when the nucleotides attach to one another through phosphodiester bonds, which connect the 3′-OH group of one nucleotide to the 5′-OH group of another nucleotide through the phosphate group. The order of the nucleotides in the chain is the primary structure of the DNA or RNA molecule, and it can be represented in shorthand notation with only the base pair designation (A, G, C, T, or U). Some sequences are highly repetitive, such as the *Alu sequence*, which is a sequence of 123 base pairs that occurs millions of times in the chain. These

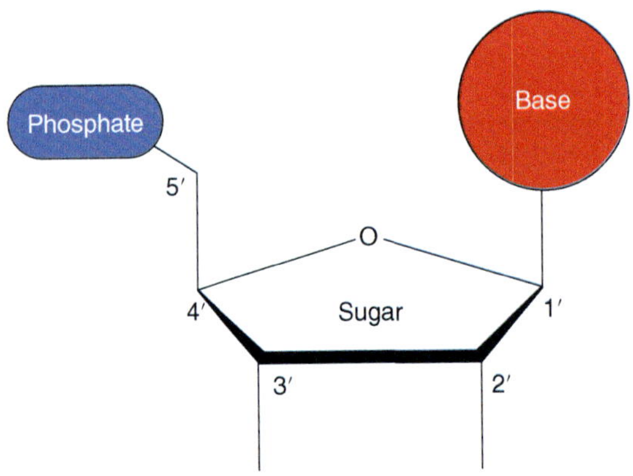

Figure 1.81 The chemical structure of DNA and RNA nucleotides. From Matthews et al. [43]. Copyright © 1997 by John Wiley & Sons, Inc. This material is used by permission of John Wiley & Sons, Inc.

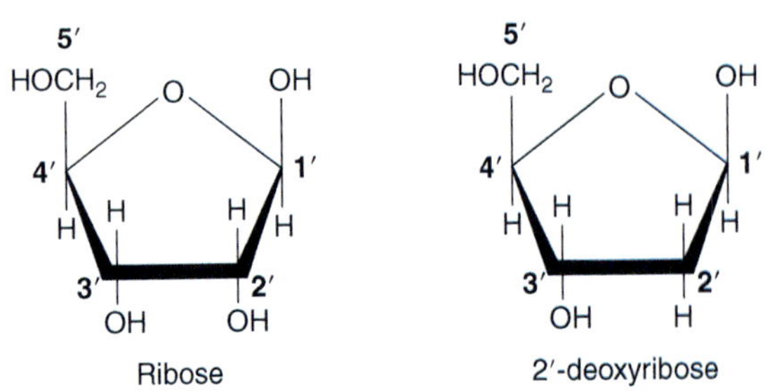

Figure 1.82 The chemical structure of the sugars in RNA (ribose) and DNA (deoxyribose). Reproduced, by permission, from Houston [44]. Copyright © 2001 by Michael E. Houston.

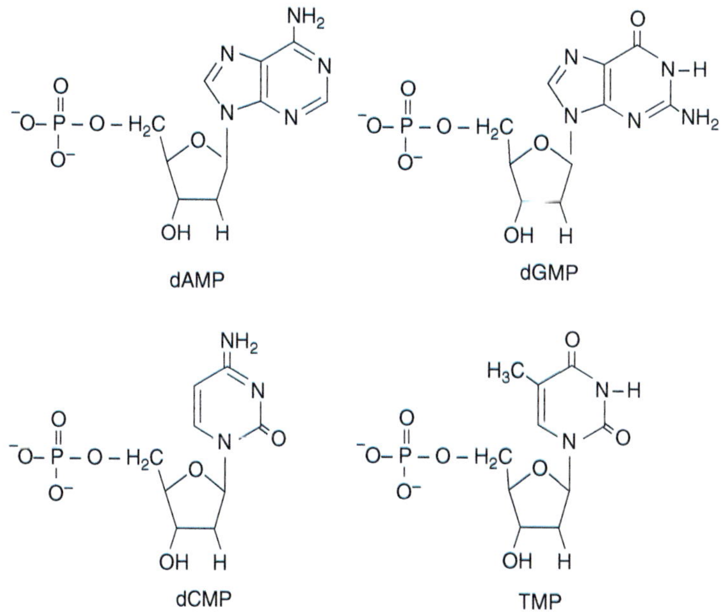

Figure 1.83 The four DNA nucleotides. Reproduced, by permission, from Schumm [45]. Copyright © 1995 by Hachette Book Group.

highly repetitive sequences are also known as short interspersed elements, or SINEs. Some sequences are specific binding sites for certain proteins, as in the TATAA sequence found near the start of many DNA molecules.

As with proteins, the nucleic acid polymers can denature, and they have secondary structure. In DNA, two nucleic acid polymer chains are twisted together with their bases facing inward to form a *double helix*. In doing so, the bases shield their hydrophobic components from the solvent, and they form hydrogen bonds in one of only two specific patterns, called *base pairs*. Adenine hydrogen bonds only with thymine (or uracil in RNA), and guanine pairs only with cytosine. Essentially, every base is part of a base pair in DNA, but only some of the bases in RNA are paired. The double-helix structure formed by DNA is comprised not only of two nucleic acid polymer chains, but the chains have complementary sequences such that when they are wound around each other in an antiparallel fashion, each base is opposite its appropriate partner, and a base pair is automatically formed. As with proteins, this double helix is also right-handed, with the phosphate groups on the outside of the structure, where they can interact with solvent ions. Thus, there are two "grooves" formed between the phosphate chains (see Figure 1.84): a *major groove* and a *minor groove*. The edges of the base pairs are accessible to the solvent in the grooves and provide regions where specific protein binding can occur. The double helix undergoes further conformational changes, called *supercoiling*, which allows a single molecule of human DNA, which is nearly one meter long if stretched out, to fit into the nucleus of a cell. Due to supercoiling, the DNA strand is only about 2 nm in diameter, making it a nanostructured biologic.

DNA is purely a molecular code: The molecule itself executes no function. A specific section of a DNA molecule, known as a *gene*, is used only as a blueprint to produce a nucleic acid chain called *messenger RNA* (mRNA) that carries out the synthesis of proteins. From a sequence of four bases in

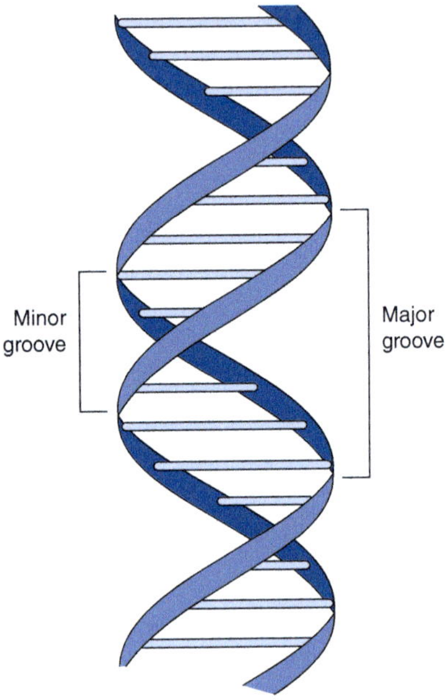

Figure 1.84 Major and minor grooves in the DNA double helix. Reproduced, by permission, from Schumm [46]. Copyright © 1995 by Hachette Book Group.

DNA, mRNA specifies the correct sequence of the 20 amino acids required to produce a specific protein. Interestingly, only about 5% of all human DNA is used to produce protein molecules. Nonetheless, this is still roughly 100,000 gene sequences.

Cells

We finally come to what are the direct building blocks of biological materials: *cells*. Cells are assemblies of molecules enclosed within a *plasma membrane* that carry out specific functions. The human body contains over 10^{14} cells, all of which take in nutrients, oxidize fuels, and excrete waste products. Despite their varied functions, all cells have a similar internal organization. We will concentrate on this internal organization for now and leave the topics of cell reproduction, energy production, and related concepts to the molecular biologist.

Surrounding the outside of all cells is the plasma membrane (see Figure 1.85). It is composed primarily of *lipids* and is selectively permeable, limiting the exchange of molecules between the inside and

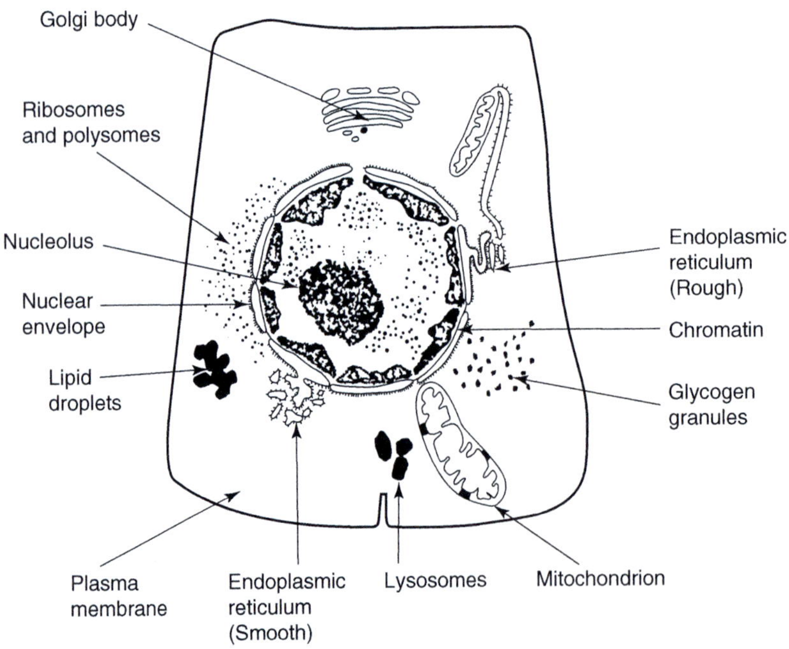

Figure 1.85 Diagram of a typical human cell showing some of the subcellular structures. Reproduced, by permission, from Schumm [47]. Copyright © 1995 by Hachette Book Group.

outside of the cell. The outside of the plasma membrane contains all the carbohydrates and receptor sites. The *cytoplasm* includes everything inside the plasma membrane except for the nucleus. Energy is generated in the cell by the *mitochondria*, where cellular fuel is oxidized. There may be several mitochondria per cell. The *Golgi complex* is composed of parallel membrane sacs and is used to secrete proteins. The *endoplasmic reticulum* (ER) is a system of membranes that store, segregate, and transport substances within the cell. The ER is continuous with the Golgi complex and the *nuclear membrane*. The nuclear membrane, or envelope, surrounds the nucleolus, which contains RNA as well as DNA and other proteins. Normally, the DNA is a diffuse tangle of fine threads called *chromatin*. The rest of the cytoplasm is composed of *lysosomes,* which degrade nucleic acids, proteins, and complex carbohydrates; *peroxisomes*, which contain a variety of oxidation enzymes; *polysomes*, which are engaged in protein synthesis using the mRNA; and *glycogen*, which is a polymer of glucose, used for energy.

Cells with similar structure and function group together to form *tissue*. Despite the astounding diversity of cell types and functions, there are only four major types of tissue: *epithelial, connective, muscular,* and *nervous*. Epithelial tissues are usually delicate cells that form linings of internal structures and organs. They also form the outer covering of our bodies, called *skin*. Connective tissues are found in the walls of organs, where they provide structural support. *Bone, cartilage, ligaments,* and *tendons* are all types of connective tissue. Muscle tissue enables the body to move, and it is characterized by its ability to contract. Nervous tissue is composed of highly specialized cells called *neurons*, and it is characterized by its ability to translate stimulation into electrochemical nervous impulses. The region between cells in a tissue is equally important. It is often termed the *extracellular matrix*, and it contains proteins and ions that perform vital functions. We will see that the extracellular matrix components have a profound effect on how well foreign materials are accepted (or rejected) by the body. For the time being, however, we will concentrate on the tissue itself by further generalizing biological materials as either "hard" or "soft" materials.

1.5.2 Hard Biologics

Historical Highlight

The basic structural features of apatite lattices were first worked out independently in the 1920s by de Jong, [48] Mehmen, [49] and st. Naray-Szabo [50]. de Jong showed for the first time—with the then relatively new technique of X-ray diffraction—that the mineral in bone bore a close resemblance to naturally occurring hydroxyapatite (HA). In this pioneering X-ray diffraction study, de Jong also observed that the apatite crystals in bone were extremely minute and ill-defined. However, a detailed spatial arrangement of the constituent ions in the apatite structure was not firmly established. It was only after 25 years that Posner et al. [51] could arrive at these structural aspects from X-ray diffraction studies on synthetically prepared singled crystals of HA. That animal bones and teeth contain HA as an ingredient was proved by identification of XRD patterns of the former with those of naturally occurring HA, as shown at right. Pattern (a) is from the synthetic crystalline sample, pattern (b) is from the synthetic amorphous sample, and pattern (c) is from the bone sample.

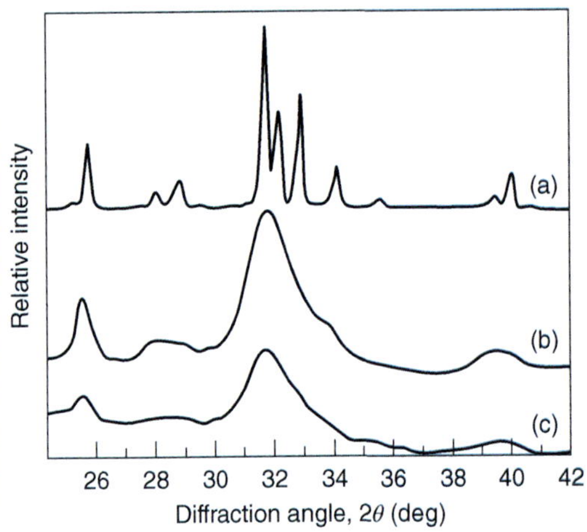

Source: Narasaraju and Phebe [52].

Though most (by volume) of the human body is composed of soft materials such as skin and muscle, we would have limited abilities to eat, move, and protect ourselves were it not for hard biologics such as bone and teeth. From an industrial standpoint, hard tissue repair accounts for roughly 20% of the multi-billion dollar biomaterials industry worldwide, [53] with demand for implants and repairs continually growing as the world's population ages.

The delineation between hard and soft biologics is not always clear, such as in the case of the connecting tissue between bone and muscle called *tendons*, but for our purposes, hard biologics contain some significant fraction of inorganic components. For both bone and teeth, the primary inorganic constituent belongs to the *calcium phosphate* family, $CaO—P_2O_5$. Actually, most calcium phosphates found in biological environments have water associated with them, so the calcium phosphates we will be discussing belong to the ternary system $CaO—P_2O_5—H_2O$. The specific calcium phosphate phase depends on such factors as the temperature and pH of the environment, as well as the chemical reactivity of that phase with its surroundings. The most prevalent form of calcium phosphate in the human body is *hydroxyapatite* (sometimes called *hydroxyl* apatite and abbreviated HA or HAp), $Ca_{10}(PO_4)_6(OH_2)$. Hydroxyapatite can occur both naturally and synthetically and has a hexagonal crystal structure (see Figure 1.86), with Ca^{2+} and $(PO_4)^{3-}$ ions arranged about columns of $(OH)^-$ ions [40, 41, 48, 49, 53]. Other biologically important calcium phosphates include

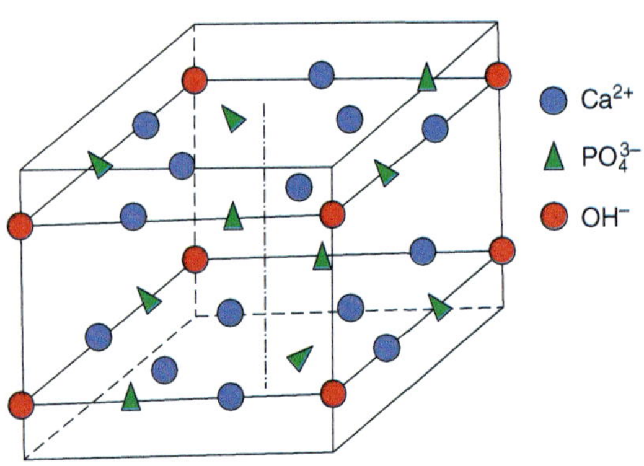

Figure 1.86 The unit cell perspective of hydroxyapatite. Reproduced, by permission, from Narasaraju and Phebe [54]. Copyright © 1996, Taylor & Francis.

dicalcium phosphate, $CaHPO_4$, and tricalcium phosphate (TCP), $Ca_3(PO_4)_2$. The remaining chemical constituents in teeth and bone are listed in Table 1.34.

The primary components of a human tooth are shown in Figure 1.87. The outer surface of the tooth is composed entirely of *enamel*. Enamel is a smooth, white, semi-transparent material that varies in thickness from 0.01 mm to about 2.5 mm over the grinding surfaces of molars. It is extremely hard and brittle and not very strong. The main constituent inorganic phase (95 wt%) is HA, with an organic phase (1 wt%) made up of protein, carbohydrates, lipids, and other matter. The remaining phase (about 4 wt%) is water, mostly present as a shell surrounding the HA crystallites. *Dentin*, the most important part of the tooth structure, is covered by the enamel in the crown and the *cementum* in the root. Dentin is a calcified mesh of collagen fibrils surrounding the cellular pathways of the dentin-forming cells. The resultant tubular structure contains about 50% HA (but

Table 1.34 Composition and Selected Properties of Inorganic Phases in Adult Human Enamel, Dentin, and Bone

	Enamel	**Dentin**	**Bone**
Composition[a]			
Calcium, Ca^{2+}[b]	36.5	35.1	34.8
Phosphorus, as P	17.7	16.9	15.2
(Ca/P) molar[b]	1.63	1.61	1.71
Sodium, Na^{+}[b]	0.5	0.6	0.9
Magnesium, Mg^{2+}[b]	0.44	1.23	0.72
Potassium, K^{+}[b]	0.08	0.05	0.03
Carbonate, as CO_3^{2-}[c]	3.5	5.6	7.4
Fluoride, F^{-}[b]	0.01	0.06	0.03
Chloride, Cl^{-}[b]	0.30	0.01	0.13
Pyrophosphate, $P_2O_7^{4-}$	0.022	0.10	0.07
Total inorganic (mineral)	97.0	70.0	65.0
Total organic[d]	1.5	20.0	25.0
Absorbed H_2O	1.5	10.0	10.0
Trace elements: Sr^{2+}, Pb^{2+}, Zn^{2+}, Cu^{2+}, Fe^{3+}, etc.			
Crystallographic properties			
Lattice parameters (±0.003 Å)			
a axis	9.441	9.42	9.41
c axis	6.880	6.88	6.89
Crystallinity index[e]	70–75	33–37	33–37
Crystallite size (average), Å	1300 × 300	200 × 40	250 × 30
Ignition products (800°C)	β-TCP + HAp	β-TCP + HAp	HAp + CaO

Source: Suchanek et al. [53]/Cambridge University Press.
[a] Weight %.
[b] Ashed sample.
[c] Unashed sample, IR method.
[d] Principal organic component: enamel, noncollagenous; dentin and bone, collagenous.
[e] Calculated from ratio of coherent/incoherent scattering, mineral, HAp = 100.

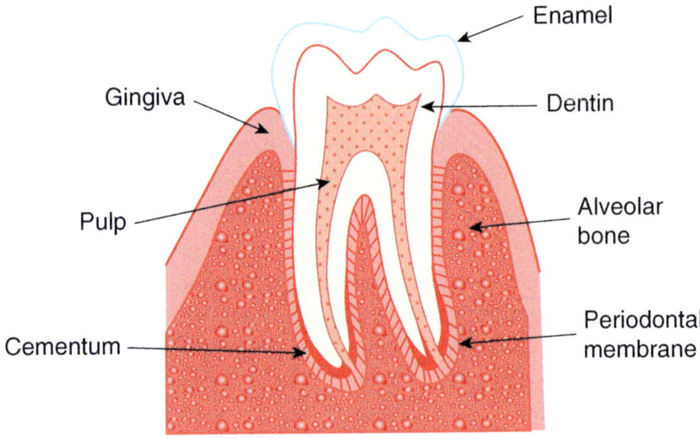

Figure 1.87 Schematic diagram of a human tooth. Reproduced, by permission, from Lewis [55]. Copyright © 1990 by Pearson Education.

the crystal size is smaller and the orientation different from that in enamel), 32% collagen, 8% mucopolysaccharides, and 10% water.

Bone is also a composite material composed of about 70% HA, as well as collagen (see Section 1.5.3), but in contrast to enamel, bones also contain blood vessels that help the bone regenerate itself (see Figure 1.88). The collagen fibers are a bundled array of crosslinked helical peptides. Bone tissue replaces itself through the action of cells called *osteoclasts*. Osteoclasts produce acids that resorb HA and break down collagen. The resulting calcium and proteins cause other cells called *osteoblasts* to lay down new matrix that mineralizes and forms HA and collagen. Bone cells produce growth factors, such as bone *morphogenetic proteins*, to increase or decrease bone regrowth. The blood vessels navigate through the porous bone structure, carrying important compounds such as *calcitonin* (a thyroid hormone that prevents bone resorption), *parathyroid hormone* and *calcitrol* (which regulate calcium and phosphate metabolism), and *prostaglandins* (fatty acids that perform hormone-like functions).

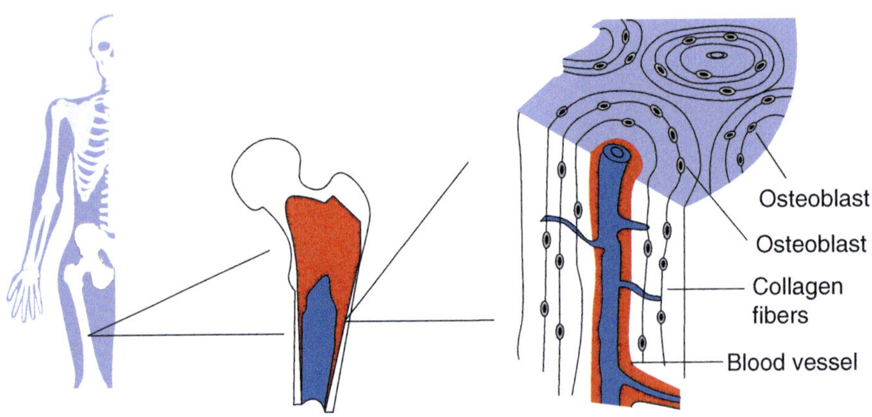

Figure 1.88 The structure of human bone. Reproduced with permission from Ritter [56]. Copyright © 1997 American Chemical Society.

As you can see, bone is a very complex composite material and is difficult to artificially replicate. Nonetheless, there are materials of all types that are used as bone replacements. Polymethylmethacrylate (PMMA), titanium, graphite/polyethyletherketone (PEEK) composites, and tricalcium phosphate have all been used to repair or replace bone in humans. The use of 3D printing in the fabrication of bone scaffolds is an emerging technological area that brings together material structure, properties, and fabrication principles.

Need More Help?

Try these videos on bone scaffold fabrication:

https://www.youtube.com/watch?v=reVOFOJk1pE
https://www.youtube.com/watch?v=Yq8jXXI7w0M

Cooperative Learning Exercise 1.9

The crystal structure of hydroxyapatite, shown in Figure 1.86, is a hexagonal unit cell with $a = 9.42$ Å and $c = 6.88$ Å. The relationship between interplanar diffraction spacing, d, and the lattice parameter for the HCP structure, analogous to Eqs. (1.33) and (1.34), is

$$d = \frac{1}{\sqrt{\frac{4}{3}\left(\frac{h^2 + hk + k^2}{a^2} + \frac{l^2}{c^2}\right)}}$$

The diffraction pattern for various forms of hydroxyapatite is shown in the Historical Highlight on page 110. Use this information to calculate the following.

Person 1: Use Bragg's Law [Eq. (1.35)] to calculate the d-spacing (in nm) for the first diffraction peak in hydroxyapatite. Assume a first-order diffraction and an X-ray source of $\lambda = 0.1537$ nm.

Person 2: Derive a relationship in the simplest terms for the d-spacing of hydroxyapatite in terms of the Miller indices only (h, k, and l). Use the cell parameters in nm.

Combine your information to determine the Miller indices of the first diffraction peak for hydroxyapatite.

Answer: $d = 0.355$ nm; $\frac{1}{d^2} = 1.51\left(h^2 + hk + k^2\right) + 2.82l^2$; $(hkl) = (111)$ for $2\theta = 25°$.

1.5.3 Soft Biologics

Unlike hard biological materials that are composed primarily of one compound (calcium phosphate) and perform a limited number of functions (primarily structural), soft biological materials are highly diversified, performing a myriad of highly specialized and/or combinatorial functions. Earlier, we classified tissue into four general categories (muscle, nervous, epithelial, and connective). Although connective tissue is primarily a hard material, there are a number of connective tissues that are more like the other soft tissues, inasmuch as they do not have a significant fraction of the calcium phosphate-based inorganic phase. Instead of taking the traditional biology-based approach to classifying and characterizing soft materials, let us instead concentrate on four important proteins found in the extracellular matrix that allow these tissue types to execute their intended functions, as well as control our ability to introduce foreign objects—biomaterials—into the human body. These proteins are *collagen*, *elastin, fibronectin,* and *laminin*.

Collagen is one of the most important and abundant substances in the human body. It is not a single protein, but rather a group of at least five different proteins that have a similar structure. Collagen contains 30% glycine, 20% proline and hydroxyproline, and a modified version of hydroxylysine. The secondary structure of collagen is a triple helix (see Figure 1.89), but not an α-helix, because the high proline content prevents the formation of the α-helix. The three chains in the helix may be identical or different. There are at least 10 different types of collagen found in connective tissue (see Table 1.35), with types I–III having the ability to form fibers called *fibrils*. Type I collagen is the principal structural component of most tissues. Type II and III collagens are found in cartilage (II) and cardiovascular tissue (III), among other places. Type IV collagen is found in basement membranes, which are sheet-like structures found beneath epithelial cells or blood vessel linings (*endothelial cells*).

Elastin is a protein also found in connective tissue that imparts an ability for these tissues to undergo large shape and size changes without permanent damage to the tissue. Elastin is the amorphous component (up to 90%) of elastic fibers that are found in the extracellular matrix of most tissue. For example, the aorta has an elastin content as high as 30–60% of the dry weight—higher than that of any other tissue. Skin, in contrast, contains less than 5% elastin by weight. Elastin is like collagen in that it is composed of about 33% glycine and 13% proline, but it contains no hydroxyproline, and contains 10–14% valine and 21–24% alanine. The latter two amino acids are nonpolar and do not form hydrogen bonds with water molecules. There are many crosslinks between the elastin chains, as one would expect for an elastic substance.

Fibronectin and *laminin* are adhesion proteins responsible for linking the outer surface of cells to collagens and other components in the extracellular matrix. Both fibronectin and laminin are

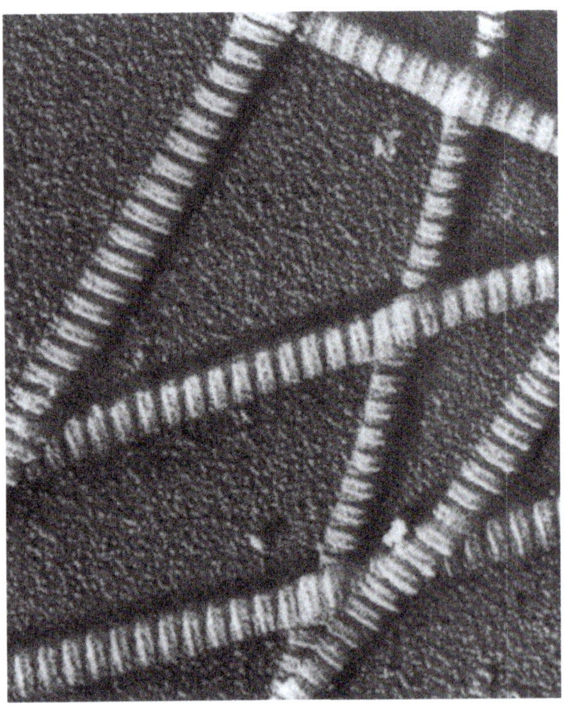

Figure 1.89 Electron photomicrograph of collagen. Reproduced with permission from L. V. Interrante and M. J. Hampden-Smith [57], © 1998, John Wiley & Sons, Inc.

Table 1.35 Types of Collagen Found in Tissue[a]

Collagen Type	Tissue or Organ	Location
I	Tendon, skin, bone, and fascia	Thick extracellular fibrils and fibers
II	Cartilage	Thin fibrils around cartilage cells
III	Cardiovascular tissue	Intermediate-size extracellular fibrils
IV	Basement membranes	Network-forming component
V	Tendon, skin, and cardiovascular tissue	Pericellular matrix around cells
VI	Cardiovascular tissue, placenta, uterus, liver, kidney, skin, ligament, and cornea	Extracellular matrix
VII	Skin	Anchoring fibrils
VIII	Cardiovascular tissue	Around endothelial cells
IX	Cartilage	Extracellular matrix
X	Cartilage	Extracellular matrix
1α, 2α, 3α	Cartilage	Extracellular matrix

[a] See Martin et al. [58].

glycoproteins—proteins that contain polysaccharide residues. Fibronectin binds to cells through a *tripeptide sequence* (—Arg—Gly—Glu—) called *RGD*, which binds to the *fibronectin receptor*, one of a family of cellular transmembrane proteins called *integrins* (see Figure 1.90). Integrins link the extracellular matrix to the cytoskeleton. We will discuss integrins in more detail in Chapter 3.

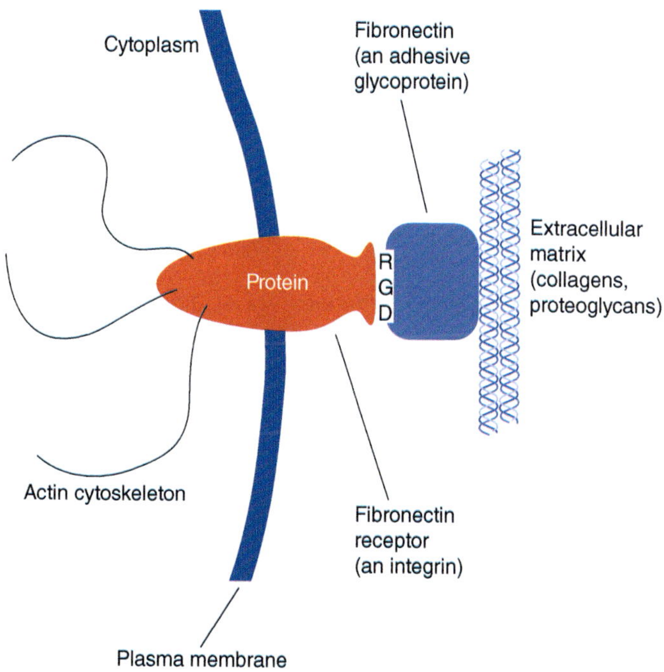

Figure 1.90 The function of integrin. From Matthews et al. [43]. Copyright © 1997 by John Wiley & Sons, Inc. This material is used by permission of John Wiley & Sons, Inc.

Need More Help?

Try these videos on soft tissue scaffolded fabrication:

https://www.youtube.com/watch?v=-Boc_TPwwYo
https://www.youtube.com/watch?v=NiwrtPoKHlg

Finally, as we saw with hard biologics, soft biologics can be used to fabricate artificial organs and assist in soft tissue repair. The advent of 3D printing and nanofabrication techniques is the future of biologics applications for the benefit of humanity.

References

1 John R. Rumble, ed., Interactive table: abundance of elements in the Earth's crust and in the sea, in *CRC Handbook of Chemistry and Physics*, 102nd Edition (Internet Version 2021), CRC Press/Taylor & Francis, Boca Raton, FL.

2 John R. Rumble, ed., Interactive table: melting, boiling, triple, and critical point temperatures of the elements, in *CRC Handbook of Chemistry and Physics*, 102nd Edition (Internet Version 2021), CRC Press/Taylor & Francis, Boca Raton, FL.

3 John R. Rumble, ed., Interactive tables: electron configuration and ionization energy of neutral atoms in the ground state; electron affinities; electronegativity values on the Pauling scale, in *CRC Handbook of Chemistry and Physics*, 102nd Edition (Internet Version 2021), CRC Press/Taylor & Francis, Boca Raton, FL.

4 John R. Rumble, ed., Interactive table: bond dissociation energies in diatomic molecules, in *CRC Handbook of Chemistry and Physics*, 102nd Edition (Internet Version 2021), CRC Press/Taylor & Francis, Boca Raton, FL.

5 Jastrzebski, Z., *The Nature and Properties of Engineering Materials*, 2nd ed., John Wiley & Sons, Inc, 1976.

6 Barrett, C. R., W. D. Nix, and A. S. Tetelman, *The Principles of Engineering Materials*, Prentice-Hall, Inc, 1973.

7 Ralls, K. M., T. H. Courtney, and J. Wulff, *Introduction to Materials Science and Engineering*, John Wiley & Sons, Inc, 1976.

8 Dickerson, R. E., H. B. Gray, and G. P. Haight Jr., *Chemical Principles*, 3rd ed., Pearson Education, Inc, 1979, p. 446.

9 Dickerson, R. E., H. B. Gray, and G. P. Haight Jr., *Chemical Principles*, 3rd ed., Pearson Education, Inc, 1979, p. 461.

10 Dickerson, R. E., H. B. Gray, and G. P. Haight Jr., *Chemical Principles*, 3rd ed., Pearson Education, Inc, 1979, p. 464.

11 Slater, J. C., Electronic energy bands in metals, *Phys. Rev.*, 45, 794 (1934).

12 Callister, W., *Materials Science and Engineering: An Introduction*, 5th ed., John Wiley & Sons, Inc, 2000, p. 32.

13 Kingery, W. D., H. K. Bowen, and D. R. Uhlmann, *Introduction to Ceramics*, John Wiley & Sons, Inc, 1976.

14 Rumble, J. R., ed., Interactive tables: "Van der Waals and Covalent Radii of the Elements" and "Ionic Radii for Ions with Various Coordination Numbers", in *CRC Handbook of Chemistry and Physics*, 102nd Edition (Internet Version 2021), CRC Press/Taylor & Francis, Boca Raton, FL.

15 John R. Rumble, ed., Interactive table: Van der Waals and covalent radii of the elements, in *CRC Handbook of Chemistry and Physics*, 103rd Edition (Internet Version 2022), CRC Press/Taylor & Francis, Boca Raton, FL.

16 Guy, G., and J. J. Hren, *Elements of Physical Metallurgy*, 3rd ed., Pearson Education, 1974, p. 208.
17 Moffatt, W. G., G. W. Pearsall, and J. Wulff, *The Structure and Properties of Materials*, Vol. 1, John Wiley & Sons, Inc, 1964.
18 Lee, K. S., S. Park, W. Lee, and Y. S. Yoon, Fe_2O_3 on reduced graphene oxide as high-performance anode for lithium-ion batteries, *ACS Appl. Mater. Interfaces*, 8(3), 2027–2034 (2016) https://doi.org/10.1021/acsami.5b10342.
19 Kumar, N., and S. Kumbhat, eds., *Essentials in Nanoscience and Nanotechnology*, Chapter 1 "Introduction," 1st ed., John Wiley & Sons, Inc, 2016, p. 15.
20 Panneerselvam, S., and S. Choi, *Nanoinformatics: emerging databases and available tools*, *Int. J. Mol. Sci.*, 15(5), 7158–7182 (2014) https://doi.org/10.3390/ijms15057158.
21 Callister, W., *Materials Science and Engineering: An Introduction*, 5th ed., John Wiley & Sons, Inc, 2000, p. 384.
22 Kittel, C., *Introduction to Solid State Physics*, John Wiley & Sons, Inc, 1957, p. 447.
23 Scholze, H., *Glass*, Springer-Verlag, 1991, p. 97.
24 Scholze, H., *Glass*, Springer-Verlag, 1991, p. 98.
25 Lee, W. E., and W. M. Rainforth, *Ceramic Microstructures*, Springer Nature, p. 59.
26 John R. Rumble, ed., Interactive table: physical constants of organic compounds, in *CRC Handbook of Chemistry and Physics*, 102nd Edition (Internet Version 2021), CRC Press/Taylor & Francis, Boca Raton, FL.
27 https://polymerdatabase.com/main.html.
28 http://www.matweb.com/search/SearchTradeName.asp.
29 Hiemenz, P., *Polymer Chemistry: The Basic Concepts*, Taylor & Francis, 1984, p. 27.
30 Hiemenz, P., *Polymer Chemistry: The Basic Concepts*, Taylor & Francis, 1984, p. 35.
31 Hiemenz, P., *Polymer Chemistry: The Basic Concepts*, Taylor & Francis, 1984, p. 239.
32 Hiemenz, P., *Polymer Chemistry: The Basic Concepts*, Taylor & Francis, 1984, p. 236.
33 Hiroyuki, Tadokoro, *Structure of Crystalline Polymers*, Krieger Publishing Company, 1990.
34 Billmeyer, F. W., *Textbook of Polymer Science*, 3rd ed., John Wiley & Sons, Inc, 1984.
35 Fergason, J. L., Liquid Crystals, *Sci. Am.*, 211(2), pp. 78, 80 (1964).
36 Chung, T. S., The recent developments of thermotropic liquid crystalline polymers, *Polym. Eng. Sci.*, 26(13), 903 (1986).
37 Schwartz, M., *Composite Materials Handbook*, 2nd ed., McGraw-Hill, 1992, p. 1.4.
38 Schwartz, M., *Composite Materials Handbook*, 2nd ed., McGraw-Hill, 1992, p. 1.7.
39 Schwartz, M., *Composite Materials Handbook*, 2nd ed., McGraw-Hill, 1992, p. 1.32.
40 Sung, Y. M., K.-Y. Yon, S. A. Dunn, and J. A. Koutsky, Wetting behavior and mullite formation at the interface of inviscid melt-spun CaO—Al_2O_3 fibre-reinforced Al—Si (4032) composite, *J. Mater. Sci.*, 29, 5583–5588 (1994).
41 Cheikh, D., H. Majdoub, and M. Darder, An overview of clay-polymer nanocomposites containing bioactive compounds for food packaging applications, *Appl. Clay Sci.*, 216, 106335 (2022) https://doi.org/10.1016/j.clay.2021.106335.
42 Houston, M. E., *Biochemistry Primer for Exercise Science*, 2nd ed., Michael E. Houston, 2001, p. 9.
43 Matthews, H. R., R. Freedland, and R. L. Miesfeld, *Biochemistry: A Short Course*, John Wiley & Sons, Inc, 1997.
44 Houston, M. E., *Biochemistry Primer for Exercise Science*, 2nd ed., Michael E. Houston, 2001, p. 30.
45 Schumm, D. E., *Essentials of Biochemistry*, 2nd ed., Hachette Book Group, 1995, p. 17.
46 Schumm, D. E., *Essentials of Biochemistry*, 2nd ed., Hachette Book Group, 1995, p. 18.
47 Schumm, D. E., *Essentials of Biochemistry*, 2nd ed., Hachette Book Group, 1995, p. 4.
48 de Jong, W. F., Le substance mineral dans les os, *Rec. Trav. Chem. Pays-Bas*, 45(6), 445 (1926).

49 Mehmen, M., Über die Struktur des Apatits. I, *Z. Kristallogr.*, 75(1), 323–331 (1930) https://doi.org/10.1515/zkri-1930-0122.

50 Naray-Szabo, S., The structure of apatite $(CaF)Ca_4(PO_4)_3$, *Z. Kristallogr.*, 75, 387 (1930) https://doi.org/10.1515/zkri-1930-0129.

51 Posner, A., A. Perloff, and A. F. Diorio, Refinement of the hydroxyapatite structure, *Acta Crystallogr.*, 11, 308 (1958).

52 Narasaraju, T. S., and D. E. Phebe, Some physico-chemical aspects of hydroxyapatite, *J. Mater. Sci.*, 31, 1 (1996).

53 Suchanek, W., and M. Yoshimura, Processing and properties of hydroxyapatite-based biomaterials for use as hard tissue replacement implants, *J. Mater. Res.*, 13(1), 94 (1998).

54 Narasaraju, T. S. B., and D. E. Phebe, Some physico-chemical aspects of hydroxylapatite, *J. Mater. Sci.*, 1, 1 (1996).

55 Lewis, G., *Selection of Engineering Materials*, Pearson Education, 1990, p. 401.

56 Ritter, S. K., *Chemical & Engineering News*, American Chemical Society, 1997, p. 27.

57 Interrante, L. V., and M. J. Hampden-Smith, eds., *Chemistry of Advanced Materials, 507*, John Wiley & Sons, 1998.

58 Martin, G. R., R. Timpl, R. K. Muller, and K. Kuhn, The genetically distinct collagens, *Trends Biochem. Sci.*, 9, 285 (1985).

Problems

Level I

1.I.1 What is the relative abundance of elements in the human body? List the top three in order and cite the source of your information.

1.I.2 Find at least one other method for estimating the electronegativity of an element and use it to calculate the electronegativity of your assigned element. Show your own work and cite the source of your information.

1.I.3 Find a method for calculating the Madelung constant and use it to calculate the value of "m" for one of the compounds in Table 1.8. How does it compare? Show your own work and cite the source of your information.

1.I.4 Li^+ and O^{2-} ions are separated by a center-to-center distance of 2.0 nm. Calculate the attractive and repulsive forces between these two ions.

1.I.5 What is a van Arkel–Ketelaar triangle, and how does it classify different types of chemical bonds? Use Eq. (1.8) to place a simple binary compound on the chart you find. Interpret your result. Give a description in your own words and cite the source of your information.

1.I.6 An amino acid has three ionizable groups, the α-amino and α-carbonyl groups, and a side chain that can be positively charged. The pH values are 3, 9, and 11, respectively. Which of the following pH values is nearest to the isoelectric point (the point at which the overall net charge is zero) for this amino acid: 1.1, 5.3, 12.2? Explain your answer.

1.I.7 Calculate the radius of a palladium atom using its crystal structure, density, and atomic weight. Cite the source of your information and show your work.

1.I.8 Calculate the number of atoms contained in a nanoparticle 4 nm in diameter of (a) silicon and (b) germanium.

1.I.9 What is the direction that results from the intersection of the (110) and $(1\bar{1}0)$ planes in a cubic crystal structure? Show your work and make a sketch of the direction.

1.I.10 (a) Can fully cured Bakelite be ground up and reused? Explain. (b) Can polyethylene be ground up and reused? Explain.

1.I.11 (a) Calculate the molecular weight of polystyrene having $\bar{x} = 100{,}000$. (b) Calculate the approximate extended chain length of one of the molecules.

1.I.12 In the formaldehyde molecule, H_2CO, a double bond exists between the carbon and oxygen atoms, (a) What type of hybridization is involved? (b) The molecule is found to be planar; one bond between the C and O atoms is a σ bond, and the other is a π bond. With a simple sketch, show the atomic orbital overlap that is responsible for the π bond.

1.I.13 If the equilibrium number of vacancies for aluminum at 500°C is $7.57 \times 10^{23}\ m^{-3}$, what is the energy of vacancy formation? State your assumptions and show your work.

1.I.14 Draw an orthorhombic cell, and within that cell draw a $[\bar{2}11]$ direction and a $(0\bar{2}1)$ plane.

1.I.15 Which of the following molecules is (are) paramagnetic: O_2^{2+}; Be_2^{2+}; F_2^{2+}?

1.I.16 Estimate the coordination number for the cation in each of these ceramic oxides: Al_2O_3, B_2O_3, CaO, MgO, SiO_2, and TiO_2.

1.I.17 Which ions or atoms of the following pairs have the greatest radius: K/K^+; O/O^{2-}; H/He; Co/Ni; Li/Cl; Li^+/Cl^-; Co^{2+}/Ni^{2+}?

1.I.18 Draw structural formulas comparing starch with cellulose.

1.I.19 Show the centers of positive and negative charge in (i) CCl_4, (ii) $C_2H_2Cl_2$, and (iii) CH_3Cl. Which of these molecules can have two or more forms?

1.I.20 Which of the following substitutions in an α-helical part of a protein is most likely to affect the function of the protein: Glu → Asp; Lys → Arg; Val → Phe; Ser → Cys; or Gln → Pro?

Level II

1.II.1 Pick an element in Table 1.3 from a column corresponding to your birth month. Use a scholarly article search engine to find a peer-reviewed journal article from a year specified by your instructor that contains this element in the title. In your own words, write a 100-word summary on how this element is important to the topic of the journal article. You need not understand all of the results or discussion presented in the article, but you should be able to discern how your element is integral to the topic. Cite your article and include a copy of the abstract in your answer.

1.II.2 A somewhat inaccurate, but geometrically convenient way of visualizing carbon bonding is to consider the carbon nucleus at the center of a tetrahedron with four valence electron clouds extending to corners of the tetrahedron. In this scheme, a carbon–carbon single bond represents tetrahedra joined tip-to-tip, a double bond represents tetrahedra joined edge-to-edge, and a triple bond represents tetrahedra joined face-to-face. Calculate the expected ratio of single, double, and triple bond lengths according to this geometrical interpretation and compare it with the measured bond lengths shown below. Comment on your results.

Bond Type	Bond Length (nm)
C—C single bond	0.154
C—C double bond	0.134
C—C triple bond	0.120

1.II.3 A journal article [James, K., and J. Kohn, New biomaterials for tissue engineering, *MRS Bull.*, **21**(11), 22–26 (1996)] describes the use of tyrosine-derived polycarbonates for tissue engineering, specifically as a resorbable substrate for small bone fixation. Three similar polycarbonates (DTE, DTH, and DTO) were considered, all having the general structure shown below, where the size of the "pendant" chain (side chain in the circle) can be varied during synthesis. If $x = 2$, the pendant chain contains an ethyl group, and the polymer is called DTE ("E" for ethyl). Similarly, $x = 6$ for DTH ("hexyl") and $x = 8$ for DTO ("octyl"). If the weight average molecular weight of DTH is 350,000, what is its number average degree of polymerization, assuming that it is monodispersed?

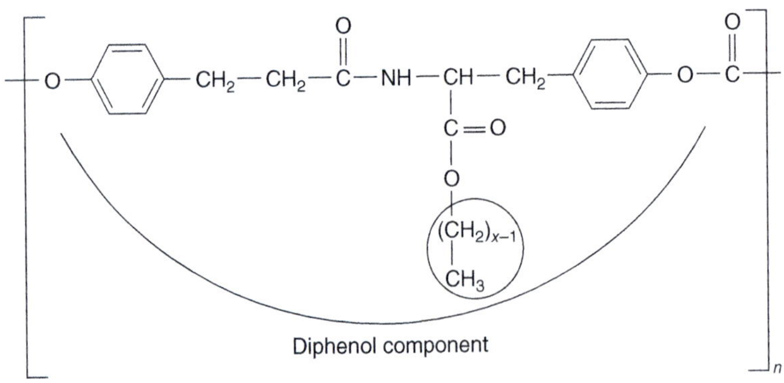

1.II.4 Find a relationship between the magnitude of the Burger's vector, |**b**|, and cell parameters in a cubic system for a plane of dislocation given by the appropriate Miller indices h, k, and l. Cite the source of your information and use this relationship to calculate the magnitude of the Burger's vector for one of the closest-packed planes. Show your work.

1.II.5 Bragg's Law [Eq. (1.35)] is a necessary but not sufficient condition for diffraction by real crystals. It specifies when diffraction will occur for unit cells having atoms positioned only at cell corners. However, atoms situated at other sites (e.g., face and interior positions in FCC or BCC) act as extra scattering centers, which can produce out-of-phase scattering at certain Bragg angles. The net result is the absence of some diffracted beams that, according to Eq. (1.35), should be present. For example, for the BCC crystal structure, $h + k + l$ must be even if diffraction is to occur, whereas for FCC, h, k, and l must all be either odd or even. Use this information to determine the Miller indices for the first five reflections that are present for a single-atom BCC and FCC unit cell. The first reflection is defined to be the one closest to $2\theta = 0$. (Contributed by Brian Grady)

1.II.6 Indicate which of the following pairs of metals would not be likely to form a continuous series of solid solutions: Ta—W; Pt—Pb, Co—Ni, Co—Zn, and Ti—Ta. Check your predictions in the *Metals Handbook*.

1.II.7 In Eq. (1.48), the mass fraction crystallinity in a polymer can be related to the volume crystallinity, ϕ_c, through the relation $w_c = \phi_c \rho_c / \rho$. Derive an expression for ϕ_c in terms of ρ, ρ_c, and ρ_a.

1.II.8 An article related to acrylic bone cements [Abboud, M. et al., PMMA-based composite materials with reactive ceramic fillers: IV. Radiopacifying particles embedded in PMMA beads for acrylic bone cements, *J. Biomed. Mater. Res.*, **53**(6), 728 (2000)] provides the following information on the PMMA matrix used in these cements: $\overline{M}_w = 295{,}000$; $\overline{M}_w / \overline{M}_n = 2.2$. Calculate the number average degree of polymerization for the PMMA used in this study.

Level III

1.III.1 Read the section on Kroger-Vink notation in Advanced Topics. Al_2O_3 will form a limited solid solution in MgO. At a specific temperature called the "eutectic temperature" (1995°C), approximately 18 wt% of Al_2O_3 is soluble in MgO. Predict the change in density resulting from (a) interstitial Al^{3+} ions and (b) substitutional Al^{3+} ions.

1.III.2 The three materials listed in the table below are available in either fiber or sheet form. Each material may also be used as a matrix. The individual physical and chemical characteristics listed in the table are independent of geometry.

	Strength (kpsi)	Density (g/cm^3)	Oxidation Resistance
Polymer	1	1	Poor
Metal	97	7	Poor
Ceramic	21	3	Excellent

Design a composite that has good oxidation resistance, a density of less than 3.0 g/cm^3, and an *isotropic* strength of at least 30 kpsi. You need not use all three materials in your design.

Assume:

- Density is a summation effect; the total density is a weight average of the components.
- Oxidation resistance is a complementary effect.
- Strength is either an interactive or a summation effect, depending on the form of the material. The total strength of the composite is three times the strength of the matrix for one-dimensional fiber orientation in a fiber-matrix composite (FMC). The total strength is two times the strength of the matrix for two-dimensional fiber orientation in an FMC. Three-dimensional (random) fiber orientation, or a nonfibrous composite, causes the total strength to be a weight average of all the components.

Describe the form (e.g., fiber, matrix, layer, etc.) of each material in your composite and the weight fraction of each component. Also, indicate the composite density and strength. Make a diagram of your composite, indicating the different components and any important features.

1.III.3* As an oxide modifier (such as Na_2O) is added to silica glass, the oxygen-to-silicon ratio increases, and it is empirically observed that the limit of glass formation is reached when O/Si is about 2.5–3. Explain, in terms of structure, why a soda–silica mixture such that $2 < O/Si < 2.5$ will form a glass, whereas a soda–silica mixture such that $O/Si = 3$ will crystallize rather than forming a glass.

1.III.4 In an article by the "Queen of Carbon Science," Mildred Dresselhaus et al. [Hydrogen adsorption in carbon materials, *MRS Bull.*, **24**(11), 45 (1999)] on the storage of molecular hydrogen, the use of carbon as an economical, safe, hydrogen storage medium for a hydrogen-fueled transportation system is discussed. Use the following excerpts from this article to provide answers to the following questions:

"To gain insight into the hydrogen adsorption problem, it is first necessary to review a few basic facts about hydrogen molecules and the surfaces to which they might bind. In the ground state, the hydrogen molecule is nearly spherical ... and the intermolecular interaction between H_2 molecules is weak. Experimentally, solid hydrogen at 4.2 K forms a hexagonal close-packed structure, with lattice parameters $a = 3.76$ Å and $c = 6.14$ Å." (a) What is the axial ratio for the hexagonal cell of solid hydrogen molecules? (b) What is the theoretical axial ratio for a hexagonal cell? (c) Compare your answers to parts (a) and (b). What does the difference between them, if any, mean physically?

The article continues:

"Using purely geometric arguments, we can thus gain a simple geometric estimate for the close-packing capacity of hydrogen molecules above a plane of graphite. Graphite has a honeycomb structure, with an in-plane lattice parameter, $a_g = 2.46$ Å and an interplanar separation of 3.35 Å. Since the value of the ... diameter for the hydrogen molecule is greater than a_g, the closest packing of hydrogen molecules would have to be incommensurate with the (graphite surface). Commensurate H_2 adsorption on a two-dimensional ... superlattice would yield a lattice constant of $a = 4.26$ Å." (See figure below.)

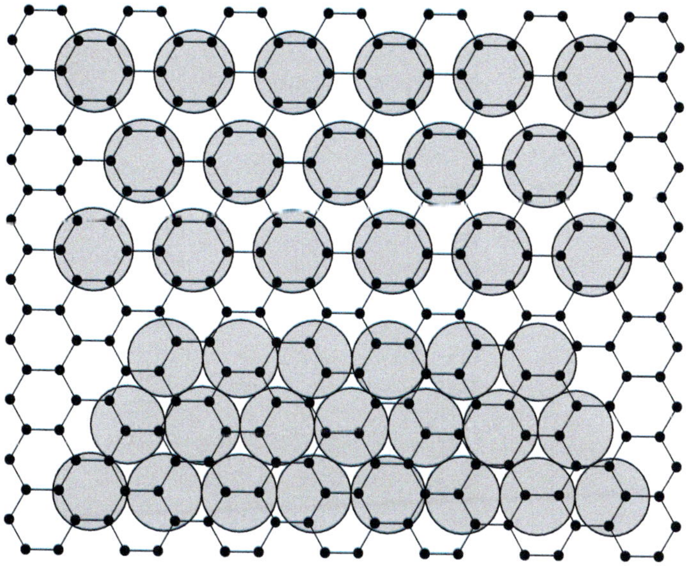

Relative density of a $\sqrt{3} \times \sqrt{3}$ commensurate (top) and an incommensurate (bottom) monolayer of H_2 on a graphite surface. Reprinted, by permission, from M. S. Dresselhaus, K. A. Williams, and P. C. Eklund, *MRS Bulletin*, **24**(11), p. 47. Copyright © 1999 by Materials Research Society.

(d) What is the number of nearest neighbors that a hydrogen molecule contained in a single, incommensurate, adsorbed layer on a surface will have? (e) What is the ratio of the number of hydrogen atoms to carbon atoms (H:C) for commensurate packing? (f) What is the weight percentage (g H/g C × 100) of atomic hydrogen that could be adsorbed onto a graphite surface using commensurate adsorption? (g) At what diffraction angle (2θ) would you expect to see a peak due to the interplanar separation of graphite layers? Assume you are using Cu Kα radiation with wavelength $\lambda = 1.54$ Å. (h) Give at least two reasons why carbon (graphite) would be desirable as a hydrogen storage medium in fuel cells.

2

Thermodynamics of Condensed Phases

2.0 Introduction and Objectives

Thermodynamics is the foundation of all engineering and many of the natural science disciplines; materials engineering and science is no exception. A certain familiarity with the tenets of thermodynamics—First and Second Laws, enthalpy, and entropy—will be assumed since you have probably been exposed to them already in chemistry and physics, and possibly in an introductory thermodynamics course. A review of these concepts is certainly in order—not only to reacquaint you with them but also to establish certain conventions (sign conventions, which letters represent which quantities) that can vary between disciplines, instructors, and books. As we traverse this review and begin applying thermodynamic principles to real material systems, you should be delighted that some concepts you may have found confusing in other courses (e.g., Carnot cycles, compressibility factors, and fugacities) are not of concern here. There is even little need to distinguish between internal energy and enthalpy. This stems from the fact that we are dealing (almost) exclusively with condensed systems—that is, liquids and solids. Though gases can certainly behave as material systems, we will not find much utility in treating them as such.

By the end of this chapter, you should be able to

- Identify the number of components present, the number of phases present, the composition of each phase, and the quantity of each phase from unary, binary, and ternary-phase diagrams—that is, apply the Gibbs Phase Rule.
- Apply the lever rule to a two-phase field in a binary-phase diagram.
- Identify three-phase reactions in binary-component systems.
- Calculate the free energy and heat of mixing for a simple binary mixture.
- Determine relative composition from a ternary-phase diagram by applying the center of gravity rule.
- Define surface energy and relate it to thermodynamic quantities such as free energy.
- Apply the Laplace equation to determine the pressure across a curved surface.
- Apply the Young equation to relate contact angle with surface energies.
- Identify the three stages of sintering and describe how surface energy drives each process.
- Differentiate between binodal and spinodal decomposition in polymer mixtures.
- Differentiate between adhesion, cohesion, and spreading and calculate the work associated with all three.
- Describe cell–cell adhesion in terms of free energy concepts.

Materials Engineering and Science: Principles, Properties, and Processes, Second Edition. Brian S. Mitchell.

Companion website: www.wiley.com/go/materialsengineeringandscience2e

2.0.1 Internal Energy

The letter U will be used to represent the *internal energy* of a system. Recall that the internal energy of a system is the sum of the microscopic kinetic and potential energies of the particles. *Kinetic energy* is the energy due to motion of the particles, including translation, rotation, and vibration. The *potential energy* is the energy due to composition. We saw in Chapter 1 how there is potential energy stored in chemical bonds and how the potential energy between two atoms or ions changes as a function of separation distance. The quantity ΔU represents the change of internal energy for the system from an initial state to a final state.

The letter E will be used to represent the *total energy* of a system. The quantity ΔE represents the change of total energy from an initial state to a final state. The total energy comprises the internal energy, U, the kinetic energy of the system, and the potential energy of the system. Do not confuse the *macroscopic* kinetic and potential energies, ΔE_k, and ΔE_p, respectively, with the *microscopic* kinetic and potential energy contributions to the internal energy just described above. By way of analogy, the former has to do with a rock being pushed off a 10-m-high ledge, while the latter has to do with the bonds between molecules in the rock and the movement of the individual atoms within the rock.

Already, you should be thinking to yourself, "But the particles in solids really don't move that much!" and you are certainly correct. They do move or "translate" in the liquid state of that same solid; however, do not forget about rotation and vibration, which we will see in subsequent chapters can be very important in solids. But along this line of thinking, we can simplify the First Law of Thermodynamics, which in general terms can be written for a closed system (no transfer of matter between the system and surroundings) as

$$\Delta U + \Delta E_k + \Delta E_p = \pm Q \pm W \tag{2.1}$$

where Q is the *heat* transferred between the surroundings and the system and W is the *work* performed. The signs used in front of Q and W are a matter of convention. Each is taken as positive when the exchange is from the surroundings to the system—that is, *by* the surroundings *to* the system. This is the convention we will use. This convention is universally accepted for Q, but not so for W. In this way, *exothermic processes* in which heat is transferred from the system to the surroundings have a negative sign in front of Q, and *endothermic processes*, in which heat is transferred from the surroundings to our system, have a positive sign. The sign convention for W is less problematic. We will see that in condensed, closed systems, the work term will not be of interest to us and will be ignored, since it mostly arises from pressure–volume (PV) work. Similarly, the macroscopic kinetic and potential energies, ΔE_k, and ΔE_p, respectively, are not of importance, so that the functional form of the First Law is

$$\Delta U = Q \tag{2.2}$$

or, in differential form for infinitesimal changes of state,

$$dU = dQ \tag{2.3}$$

Both U and Q have units of joules, J, in the SI system, ergs in cgs, and calories (cal) in American engineering units.

2.0.2 Enthalpy

A closely related quantity to the internal energy is the *enthalpy*, H. It, too, has SI units of joules and is defined as the internal energy plus the pressure–volume product, PV. As in most cases, we are concerned with changes in internal energy and enthalpy from one state to another, so that the definition of enthalpy for infinitesimal changes in state is

$$dH = dU + d(PV) \tag{2.4}$$

Substitution of Eq. (2.3) into Eq. (2.4) and recognition that for condensed systems the $d(PV)$ term is negligible leads to

$$dH = dQ \tag{2.5}$$

which in integrated form is

$$\Delta H = Q \tag{2.6}$$

2.0.3 Entropy

Entropy is represented by the letter S. Entropy is a measure of randomness or disorder in a system and has SI units of J/K. Recall that the Second Law of Thermodynamics states that *the entropy change of all processes must be positive*. We will see that the origins of entropy are best described from statistical thermodynamics, but for now let us concentrate on how we can use entropy to describe real material systems.

The differential change in entropy for a closed system from one state to another is, by definition, directly proportional to the change in reversible heat, dQ_{rev}, and inversely proportional to the absolute temperature, T:

$$dS = \frac{dQ_{rev}}{T} \tag{2.7}$$

Reversible processes will be studied in this chapter, so it is not necessary to retain the subscript on Q. Also, equilibrium transformations will be the primary focus of this chapter, such that we will be concerned mostly with constant temperature processes. These facts allow us to simplify Eq. (2.7) by integration

$$\Delta S = \frac{Q}{T} \tag{2.8}$$

Substitution of Eq. (2.6) into Eq. (2.8) gives a useful relationship between entropy and enthalpy for constant pressure and temperature processes:

$$\Delta S = \frac{\Delta H}{T} \tag{2.9}$$

Entropy also plays a role in the Third Law of Thermodynamics, which states that *the entropy of a perfect crystal is zero at zero absolute temperature.*

2.0.4 Free Energy

The most useful quantity for this chapter is the *Gibbs free energy*, G. The Gibbs free energy for a closed system is defined in terms of the enthalpy and entropy as

$$G = H - TS \tag{2.10}$$

which in differential form at constant pressure and temperature is written

$$dG = dH - TdS \tag{2.11}$$

or in the integrated form

$$\Delta G = \Delta H - T\Delta S \tag{2.12}$$

Historical Highlight
Learn more about Dutch physicist ***Josiah Willard Gibbs*** at https://en.wikipedia.org/wiki/Josiah_Willard_Gibbs

Take a Peek! Jump ahead to Section 3.2.1 to see how the kinetics of phase transformations takes place.

Recall that the important function of the free energy change from one state to another is to determine whether the process is *spontaneous*—that is, thermodynamically favored. The conditions under which a process is considered spontaneous are summarized in Table 2.1. The processes described here in a generic sense are the topic of this chapter, as are the implications of equilibrium for material systems. The issue of spontaneity and what this means to the rate at which a process occurs is the subject of Chapter 3.

2.0.5 Chemical Potential

The final thermodynamic quantity for review is the *chemical potential*, which is represented by the Greek letter mu, μ. The chemical potential can be defined in terms of the partial derivative of any of the previous thermodynamic quantities with respect to the number of moles of species i, n_i, at constant n_j (where j indicates all species other than i) and thermodynamic quantities as indicated:

$$\mu_i = \left(\frac{\partial U}{\partial n_i}\right)_{S,V,n_j} = \left(\frac{\partial H}{\partial n_i}\right)_{P,S,n_j} = \left(\frac{\partial G}{\partial n_i}\right)_{T,P,n_j} \tag{2.13}$$

The advantage of the chemical potential over the other thermodynamic quantities, U, H, and G, is that it is an *intensive quantity*—that is, independent of the number of moles or quantity of species present. Internal energy, enthalpy, free energy, and entropy are all *extensive variables*. Their values

Table 2.1 Summary of Free Energy Effects on Process Spontaneity

$\Delta G < 0$	Process proceeds spontaneously
$\Delta G > 0$	Process not spontaneous
$\Delta G = 0$	Process at equilibrium

depend on the extent of the system—how much there is. We will see in the next section that intensive variables such as μ, T, and P are useful in defining equilibrium.

2.1 Thermodynamics of Metals and Alloys

In Chapter 1, the assertion was made that "Elements are materials, too!" We will use this fact to begin our description of equilibrium transformation processes in materials. Not all elements are metals, of course, but most of the elements that are of interest from a material application standpoint are.

2.1.1 Phase Equilibria in Single-Component Systems

Gibbs Phase Rule

The goal of this section is to predict what will happen to our element when it is subjected to changes in those variables that we typically can manipulate: temperature and pressure. For example, what happens when we heat a sample of pure sulfur? It will probably melt at some point. What happens when we subject carbon to very high pressures? We predict that diamond will form. We seek quantitative explanations of these phenomena and an ability to predict under what conditions they will occur.

Consider a single *component*, A. This component is a chemical constituent (element or compound) that has a specified composition. For simplicity, we will assume that component A is an element, but we will see in subsequent sections that it can be anything that does not undergo a chemical change, including compounds. But for now, we have an element, A, that can exist in two *phases*, which we will designate α and β. A phase is defined as a homogeneous portion of a system that has uniform physical and chemical characteristics. It need not be continuous. For example, a carbonated beverage consists of two phases: the liquid phase, which is continuous, and the gas phase, which is dispersed in the liquid phase as discrete bubbles. For the current discussion, it is easiest to visualize the two phases as a solid and a liquid, respectively, since we are familiar with the processes of melting and solidification—but these phases could be two solids, two liquids, a liquid and a gas, or even a solid and a gas.

Both phase α and phase β are in *equilibrium* with one another. What does this mean? First, it means that even though there is an exchange of atoms between the two phases; (some of the solid phase α is melting to form β, and some of the liquid phase β is solidifying to form α) these processes are occurring at essentially equal rates such that the relative amounts of each phase are unchanged. This is known as a *dynamic equilibrium*. In terms of intensive variables, equilibrium means that

$$T_\alpha = T_\beta \tag{2.14}$$

$$P_\alpha = P_\beta \tag{2.15}$$

$$\mu_\alpha = \mu_\beta \tag{2.16}$$

where T_α and P_α are the temperature and pressure of the solid phase α; T_β and P_β are the temperature and pressure of the liquid phase β, and μ_α and μ_β are the chemical potentials of each phase. Thus, we have six intensive quantities that establish equilibrium between the two phases. However, these six variables are not all independent. Changing any one of them can affect the others. This can be shown mathematically by assuming that the chemical potential is a function of temperature and pressure:

$$\mu_\alpha = \mu_\alpha(T_\alpha, P_\alpha) \tag{2.17}$$

$$\mu_\beta = \mu_\beta(T_\beta, P_\beta) \tag{2.18}$$

So, for a one-component system containing two phases in equilibrium, we have three thermodynamic conditions of equilibrium [Eqs. (2.14)–(2.16)] and four unknown parameters, T_α, P_α, T_β, and P_β. If we arbitrarily assign a value to one of the parameters, we can solve for the other three (three equations, three unknowns).

Let us extend this analysis to the general case of C independent, nonreacting components, so that we might arrive at a very useful, general conclusion. Instead of one component, we now have an arbitrary number 1, 2, up to C; and instead of two phases (liquid and solid), we now have an arbitrary number, ϕ. The conditions of equilibrium are now analogous to Eqs. (2.14)–(2.16):

$$T_\alpha = T_\beta = T_\gamma = \dots T_\phi \tag{2.19}$$

$$P_\alpha = P_\beta = P_\gamma = \dots P_\phi \tag{2.20}$$

$$\begin{aligned} \mu_{1\alpha} &= \mu_{1\beta} = \mu_{1\gamma} = \dots \mu_{1\phi} \\ \mu_{2\alpha} &= \mu_{2\beta} = \mu_{2\gamma} = \dots \mu_{2\phi} \\ &\vdots \\ \mu_{C\alpha} &= \mu_{C\beta} = \mu_{C\gamma} = \dots \mu_{C\phi} \end{aligned} \tag{2.21}$$

There are a total of $(\phi - 1)$ equalities for temperature in Eq. (2.19), $(\phi - 1)$ equalities for pressure in Eq. (2.20), and $(\phi - 1)$ equalities for chemical potential of each of the components, C in Eq. (2.21). There are C equations in (2.21), so the total number of equations is $(C + 2)$. Thus, we have $(\phi - 1)(C + 2)$ total equalities, or restrictions, to be satisfied for equilibrium. The total number of intensive variables (excluding composition) in the system is $\phi(C + 2)$. The difference between the number of intensive variables and the number of independent restrictions is known as the *degrees of freedom* of the system, F. The degrees of freedom are the number of variables (including composition) that must be specified for the system to be defined in a mathematical sense. So,

$$\begin{aligned} \text{degrees of freedom} &= \text{number of intensive variables} \\ &\quad - \text{number of independent restrictions} \\ F &= \phi(C + 1) - (\phi - 1)(C + 2) \end{aligned} \tag{2.22}$$

$$F = C - \phi + 2 \tag{2.23}$$

Eq. (2.23) is a very important result. It is known as the *Gibbs Phase Rule*, or simply the "phase rule," and relates the number of components and phases to the number of degrees of freedom in a system. It is a more specific case of the general case for N independent, noncompositional variables

$$F = C - \phi + N \tag{2.24}$$

We utilized only temperature and pressure as independent, noncompositional variables in our derivation ($N = 2$), which are of most practical importance.

Return now to the case of a single component and two phases, $C = 1$ and $\phi = 2$, so that the phase rule for our element, A, is

$$F = 1 - 2 + 2 = 1 \tag{2.25}$$

This means that when phases α and β coexist at equilibrium, only one variable may be changed independently. For example, if temperature is changed, pressure cannot be changed simultaneously without affecting the balance of equilibrium. In those cases where $F = 2$, both temperature and pressure can be changed without affecting the balance of equilibrium (number of phases present); and when $F = 0$, none of the intensive variables can be changed without altering equilibrium between phases.

Need More Help?

Try this video on phase diagrams:

https://www.youtube.com/watch?v=MJoYwtX_zFA

Unary Phase Diagrams

The phase rule provides us with a powerful tool for connecting the thermodynamics of phase equilibria with graphical representations of the phases known as *phase diagrams*. Phase diagrams tell us many things, but at a minimum, describe the number of components present, the number of phases present, the composition of each phase, and the quantity of each phase. There are several ways in which this information can be presented. We start with the free energy presentation.

Take, for example, the plot of G versus temperature for elemental sulfur, represented by the bottom diagram in Figure 2.1. We know from experiments and observation that there are four phases to consider for sulfur: two solid forms (a low-temperature orthorhombic form, R, and high-temperature monoclinic form, M), liquid (L), and vapor (or gas, V). The lines of G versus T for each phase—which are partially solid and continue as dashed lines—are constructed at constant pressure using Eq. (2.10), where each phase has a different enthalpy and entropy associated with it. The solid line in this figure represents the minimum free energy, regardless of which phase it represents, at a specified temperature. Hence, at this pressure (which is unspecified, but fixed at P'), the orthorhombic form, R, is the most stable at low temperatures. As temperature is increased, monoclinic (M) becomes more stable at point a, followed by the liquid, L, at point c, and finally the vapor phase (V) at point e.

Plots of free energy as a function of temperature can be made at any pressure (in theory), but they are not particularly useful in and of themselves. Free energy is something we cannot generally measure directly, but temperature and pressure are. Hence, plots of two intensive variables, such as pressure and temperature, are much more practical. Such plots are produced by simply translating the information in the G versus T plot at various pressures onto a P versus T plot, as shown in the top diagram of Figure 2.1. Point a in the bottom figure now becomes a point on the solid–solid equilibrium line between the orthorhombic and monoclinic forms of sulfur. Point b is distinctly in the monoclinic phase field, M, since this form has the lowest free energy at this temperature and pressure. Point c is also on an equilibrium phase boundary between the monoclinic and liquid, L, forms of sulfur. Note that this is the melting (or solidification) point of sulfur at pressure P'. Point d is distinctly in the liquid region, and point e is on the liquid–vapor equilibrium line. These plots of pressure versus temperature for a single-component system are called *unary phase diagrams*. There is no limit to the number of phases that may be present on a unary diagram. The only constraint is that there be only one component—that is, no chemical transformations are taking place. Although our discussion here is limited to elements and metals, keep in mind that we are not limiting the number of atoms that make up the component, only the number of components. For example, water is a single component that is composed of hydrogen and oxygen. In a unary phase diagram for water, only the phase transformations of that component are shown—for example, ice, water, and steam. Two hydrogens are always bound to one oxygen. For now, we will continue to limit our discussion to metals.

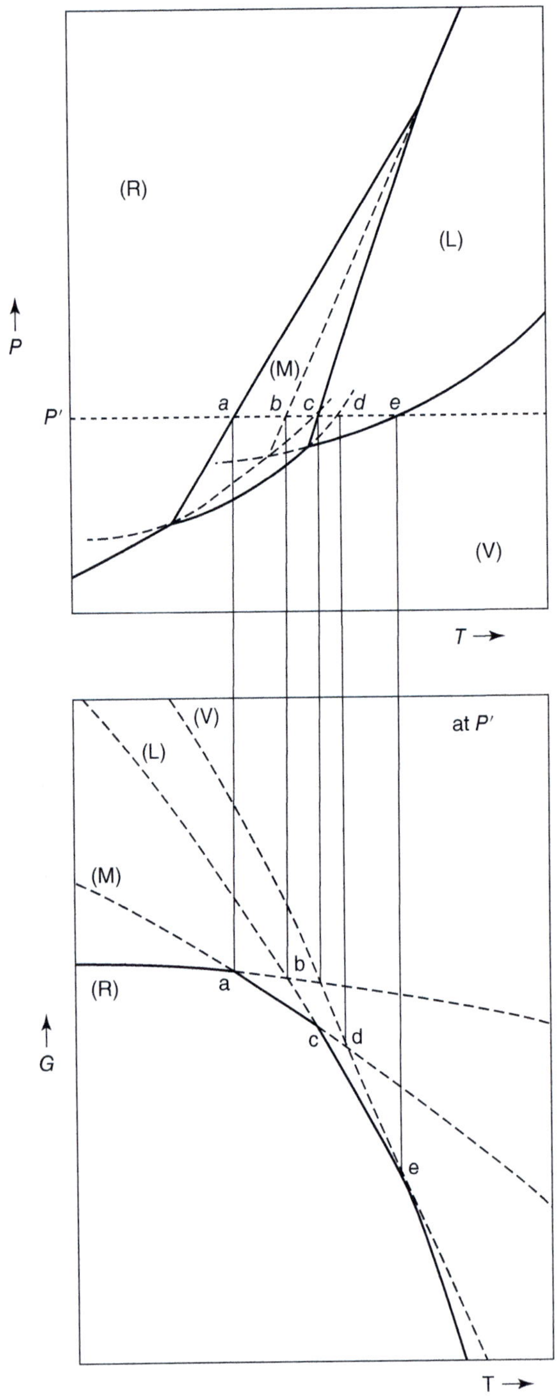

Figure 2.1 Unary phase diagram (*top*) and Gibbs free energy plot (*bottom*) for elemental sulfur. Reproduced by permission, from Gaskell [1]. Copyright © 1981 by Hemisphere Publishing Corporation.

The power of the phase rule is illustrated in a second example. Consider the *T–P* phase diagram for carbon shown in Figure 2.2. First, notice that pressure is plotted as the independent variable instead of the usual temperature variable, but given what we know about the effects of pressure on carbon, this makes sense. Let us then examine the equilibrium between two solid forms of

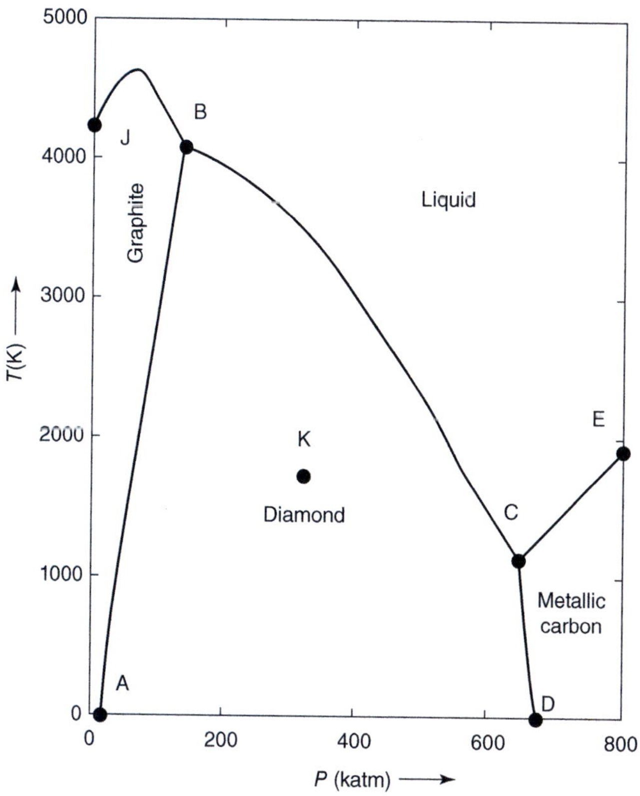

Figure 2.2 Temperature–pressure unary phase diagram for carbon. From Ralls et al. [2]. Copyright © 1976 by John Wiley & Sons, Inc. This material is used by permission of John Wiley & Sons, Inc.

carbon: graphite and diamond. This equilibrium is shown graphically by the line A–B in Figure 2.2. Application of Eq. (2.23) to any point on this line, or *phase boundary*, results in $F = 1$ ($C = 1$ component, carbon; $\phi = 2$, graphite and diamond). This means that any change in temperature requires a corresponding change in pressure to maintain this equilibrium. Temperature and pressure are not independent. Similarly, the phase boundaries represented by curves B–C (diamond–liquid), C–D (diamond–metallic carbon), C–E (metallic carbon–liquid), and J–B (graphite–liquid) represent lines of $F = 1$.

Any point within a single-phase region ($\phi = 1$) results in $F = 2$. Point K is such a point. It is located in the diamond *phase field*, where both temperature and pressure can be changed independently without creating or destroying the phase. No appreciable dynamic equilibrium exists here—there is only one phase.

Since $F = 2$ results in an area (phase field), and $F = 1$ results in a line (phase boundary), we can predict that $F = 0$ should occur at a point, and indeed it does. Point B is such a point ($\phi = 3$; graphite, diamond, and liquid; $C = 1$, carbon), as is point C. There are no degrees of freedom at this point—any variation in temperature or pressure will result in movement into a distinctly separate phase field, and at least one of the phases must necessarily be lost. In the cases where the three phases in coexistence are solid, liquid, and vapor, this *invariant point* is known as the *triple point*.

2.1.2 Phase Equilibria in Binary-Component Systems

Most metallic materials of practical importance are mixtures of two or more metals. Recall from Section 1.1.3 that these "intimate mixtures" of metals are called alloys, and when the bond between the metals is partially ionic, they are termed *intermetallics*. For the purposes of this chapter, and especially this section, we will not need to distinguish between an intermetallic and an alloy, except to note that when a compound is indicated on a phase diagram (e.g., $CuAl_2$), it represents an intermetallic compound. We are concerned only with the thermodynamics that describe the intimate mixing of two species under equilibrium conditions. The factors affecting how two metal atoms mix have already been described in Section 1.1.3. Recall that the solubility of one element in another depends on the relative atomic radii, the electronegativity difference between the two elements, the similarity in crystal structures, and the valences of the two elements. Thermodynamics does not yet allow us to translate these properties of atoms directly into free energies, but these factors are what contribute to the free energy of the metallic mixture and determine whether the species are soluble in one another, whether they form intermetallic compounds, or whether they phase separate into two distinct phases of different composition.

The conditions for equilibrium have not changed; application of the phase rule is conducted as in the previous section. The difference now is that composition must be included as an intensive variable because it can change. Composition is accounted for through direct introduction into the thermodynamic quantities of enthalpy and entropy. The free energy of a mixture of two pure elements, A and B, is still given by the definition

$$G = H - TS \tag{2.10}$$

but the enthalpy and entropy for the mixture are now taken as a combination of the enthalpies and entropies of components A and B, plus parameters that account for the interaction of the two species due to differences in crystal structure, electronegativity, valence, and atomic radii, known as the *enthalpy* and *entropy of mixing*, ΔH_{mix} and ΔS_{mix}, respectively:

$$H = X_A H_A^\circ + X_B H_B^\circ + \Delta H_{mix} \tag{2.26}$$

$$S = X_A S_A^\circ + X_B S_B^\circ + \Delta S_{mix} \tag{2.27}$$

Here, X_A and X_B are the mole fractions of components A and B, respectively, and are related by $(X_A + X_B) = 1$. We have used a superscript circle on the enthalpies and entropies of pure components A and B to indicate that these are *standard state* enthalpies and entropies of the pure components. The standard state of a component in a condensed system is its stable state at the temperature and pressure of interest. So, depending on the temperature and pressure of the system, the standard state could be either a liquid or a solid for either of components A and B.

Substitution of Eqs. (2.26) and (2.27) into Eq. (2.10), and using a similar definition for standard state free energies, leads to the following relationship for the free energy of the mixture

$$\begin{aligned} G &= X_A\left(H_A^\circ - TS_A^\circ\right) + X_B\left(H_B^\circ - TS_B^\circ\right) + \Delta H_{mix} - T\Delta S_{mix} \\ G &= X_A G_A^\circ + X_B G_B^\circ + \Delta G_{mix} \end{aligned} \tag{2.28}$$

where

$$\Delta G_{mix} = \Delta H_{mix} - T\Delta S_{mix} \tag{2.29}$$

Though Eq. (2.28) is rigorously correct, it contains a rather ill-defined parameter that has been introduced specifically to account for the mixing of the two species, namely, ΔG_{mix}. To proceed further, we must explore this quantity and its definition [given in Eq. (2.29)] more thoroughly.

The entropy of mixing can be thought of as a measure of the increase in the number of spatial configurations that become available to the system due to the mixing process, ΔS_{conf}, which can be shown with statistical thermodynamic arguments to be

$$\Delta S_{conf} = \Delta S_{mix} = -R(X_A \ln X_A + X_B \ln X_B) \tag{2.30}$$

where R is the gas constant. Eq. (2.30) shows that the entropy of mixing in a binary-component solution is dependent only on the composition (relative number of moles of components) in the solution and is independent of temperature.

Similarly, it can be shown that the enthalpy of mixing can be approximated by the internal energy of mixing [see Eq. (2.4)] for condensed phases, which in turn can be related to the mole fractions of the two components, and an *interaction energy*, α, which also has units of joules:

$$\Delta H_{mix} = \alpha X_A X_B \tag{2.31}$$

If $\alpha < 0$, *exothermic mixing* occurs ($\Delta H_{mix} < 0$) because an A—B bond is stronger (more negative) than either the A—A or B—B bonds and unlike nearest neighbors are favored in the system. *Endothermic mixing* ($\Delta H_{mix} > 0$) occurs when like nearest neighbors are favored, and A—A and B—B bonds are stronger (more negative) than A—B bonds, resulting in $\alpha > 0$. For an ideal solution, the bond energies of A—A, B—B, and A—B bonds are identical, there is no preference for either atom as a nearest neighbor in solution, and both the interaction energy and the heat of mixing are zero:

$$\alpha^{ideal} = \Delta H_{mix}^{ideal} = 0 \tag{2.32}$$

The free energy of mixing is found by substituting Eqs. (2.31) and (2.30) into Eq. (2.29) to obtain

$$\Delta G_{mix} = \alpha X_A X_B + RT(X_A \ln X_A + X_B \ln X_B) \tag{2.33}$$

which is then substituted into Eq. (2.28) for ΔG_{mix} to obtain

$$G^{reg} = X_A G_A^{\circ} + X_B G_B^{\circ} + \alpha X_A X_B + RT(X_A \ln X_A + X_B \ln X_B) \tag{2.34}$$

Solutions whose free energies follow Eq. (2.34) are said to be *regular* (so we use G^{reg}) to distinguish them from irregular solutions, wherein the entropy of mixing is governed by a relationship other than Eq. (2.30). The free energy for a mixture that behaves as an *ideal solution* ($\alpha = 0$) is indicated by G^{ideal} and reduces to

$$G^{ideal} = X_A G_A^{\circ} + X_B G_B^{\circ} + RT(X_A \ln X_A + X_B \ln X_B) \tag{2.35}$$

Keep in mind that at this point, the mixtures can be either solid or liquid, depending on the temperature, and that both solid and liquid mixtures may coexist at certain temperatures and compositions. Thus, when performing free energy calculations using Eqs. (2.34) or (2.35), the standard state free energies for both components must be carefully selected. This process is illustrated in Example Problem 2.1.

The conditions for equilibrium between two phases in coexistence are still the same as for the case of a single component; that is, Eqs. (2.19)–(2.21) still apply. The phase rule is also still applicable. Before applying these principles to a binary system, let us first examine how a binary-phase diagram can be constructed from the free energy equations.

Binary-Phase Diagrams

Just as in the case of the unary phase diagram in the previous section, we can construct graphical representations of the equilibrium phases present in a binary-component system by first plotting—with the assistance of Eq. (2.33)—free energy of mixing versus composition for a system consisting of two pure elements A and B. Such a plot is shown in Figure 2.3. It should be noted that the free energy of mixing is selected as the ordinate in these plots, since we are concerned only with the change in free energy that results when the two components are taken from separate, standard states into the mixture. The mixture may be either a liquid or a solid.

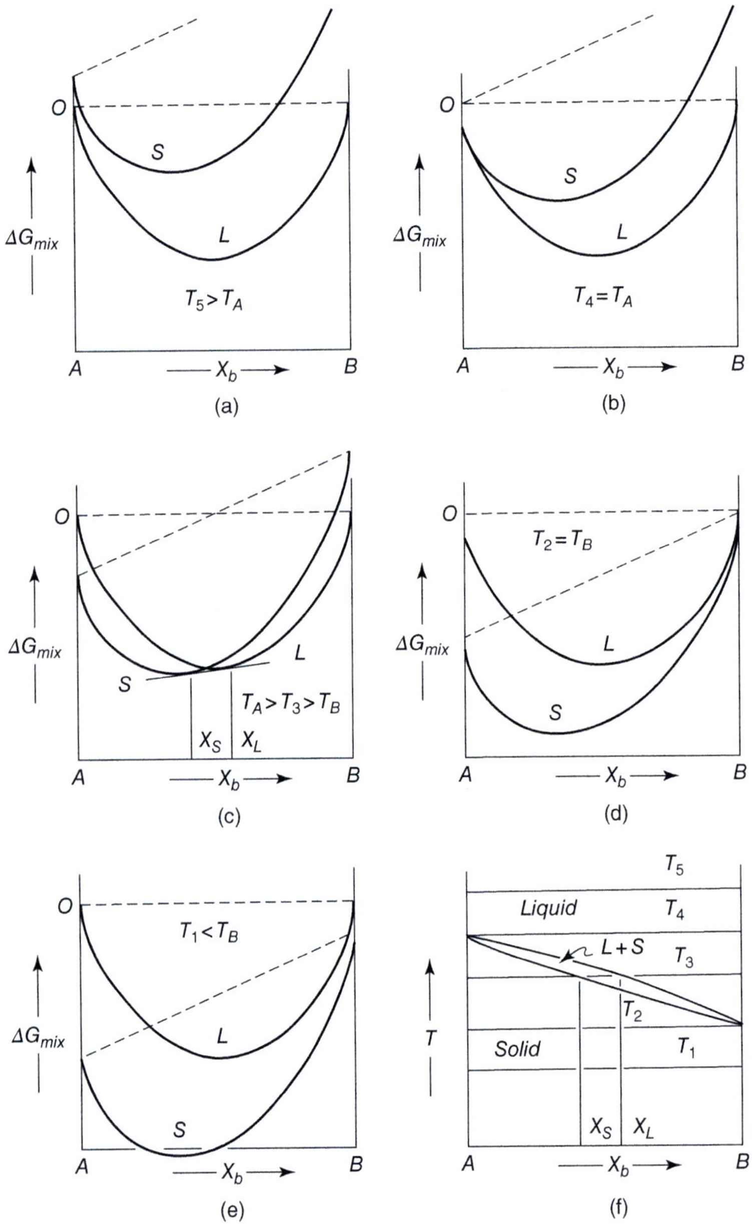

Figure 2.3 Free energy of mixing curves for solid and liquid phases at various temperatures (a–e) and resulting temperature–composition phase diagram for a completely soluble binary-component system (f). From Devereux [3]. Copyright © 1983 by John Wiley & Sons, Inc. This material is used by permission of John Wiley & Sons, Inc.

Example Problem 2.1

Cesium, Cs, and rubidium, Rb, form ideal solutions in the liquid phase and regular solutions in the solid phase. Their standard state Gibbs free energy changes of melting as a function of temperature, $\Delta G^{\circ}_{m,Cs} = \left(G^{\circ}_{Cs,L} - G^{\circ}_{Cs,S}\right)$ and $\Delta G^{\circ}_{m,Rb} = \left(G^{\circ}_{Rb,L} - G^{\circ}_{Rb,S}\right)$, respectively, are given by the following empirical relationships (in J):

$$\Delta G^{\circ}_{m,Cs} = 2100 - 6.95T \quad \Delta G^{\circ}_{m,Rb} = 2200 - 76.05T$$

1) Determine the melting temperatures for Cs and Rb.
2) Determine the relationship for Gibbs free energy of mixing as a function of composition at 9.7°C (282.7 K) for both liquid and solid solutions.

Solution

1) The melting temperatures are found from the free energy expressions by setting them equal to zero, since for each pure component, the liquid and solid free energies must be equal for both phases to co-exist at the melting temperature. For Cs

$$\Delta G^{\circ}_{m,Cs} = 2100 - 6.95T_{m,Cs} = 0$$
$$T_{m,Cs} = 302\ K = 29.0°C$$

Similarly, $T_{m,Rb} = 312$ K (38.9°C).

2) First, we note that the temperature under consideration (282.7 K) is below the melting point of both elements, so that the solid-state free energies of both components should be selected as the reference state for our free energy calculations. For the liquid solution, which is ideal, the free energy change of mixing relative to the solid components is found by subtracting the weighted average (in terms of mole fraction) of the standard state free energies of the components from the free energy of the ideal mixture, as given by Eq. (2.35):

$$\Delta G_L^{ideal} = \left[X_{Cs}G^{\circ}_{Cs,L} - X_{Rb}G^{\circ}_{Rb,L} + RT(X_{Cs}\ln X_{Cs} + X_{Rb}\ln X_{Rb})\right] - \left[X_{Cs}G^{\circ}_{Cs,S} - X_{Rb}G^{\circ}_{Rb,s}\right]$$

which, upon combining terms and using the definitions of free energy changes of melting, simplifies to

$$\Delta G_L^{ideal} = X_{Cs}\Delta G^{\circ}_{m,Cs} - X_{Rb}\Delta G^{\circ}_{m,Rb} + RT(X_{Cs}\ln X_{Cs} + X_{Rb}\ln X_{Rb})$$

Substitution of the empirical free energy relationships provided for each component, simplification using $X_{Cs} + X_{Rb} = 1$, and evaluation at $T = 282.7$ K gives

$$\Delta G_L^{ideal} = 205X_{Rb} + 133\,(1 - X_{Rb}) + (8.314)(282.7)\left[(1 - X_{Rb})\ln(1 - X_{Rb}) + X_{Rb}\ln X_{Rb}\right]$$

This is the free energy change of mixing for the liquid as a function of composition at 282.7 K.

A similar evaluation for the solid solution starts with Eq. (2.34), since the solid solution is regular and not ideal. This equation simplifies rapidly, since the standard state free energies of the components are in the solid phase in both the separate and mixed states, such that

$$\Delta G_S^{reg} = \alpha X_A X_B + RT(X_A\ln X_A + X_B\ln X_B)$$

which, when evaluated at 282.7 K, becomes

$$\Delta G_S^{reg} = \alpha X_{Rb}(1 - X_{Rb}) + (8.314)(282.7)[X_{Rb}\ ln X_{Rb} + (1 - X_{Rb})\ ln(1 - X_{Rb})]$$

Cooperative Learning Exercise 2.1

Using the data in Example Problem 2.1 and assuming that $\alpha = 668$ J, determine the composition of a mixture of Cs and Rb that will melt *congruently* at 282.7 K—that is, will go from a solid solution to a liquid solution without changing composition (going through phase separation).

Person 1: Calculate the Gibbs free energy of mixing for the liquid solution from $X_{Rb} = 0$ to 1, in increments of 0.1. (*Note*: You will have to use approximations for $X_{Rb} = 0.0$ and 1.0. Why?)

Person 2: Calculate the Gibbs free energy of mixing for the solid solution from $X_{Rb} = 0$ to 1, in increments of 0.1. (*Note:* You will have to use approximations for $X_{Rb} = 0.0$ and 1.0. Why?)

Plot both of your values on one graph and estimate the value of X_{Rb} at which the free energies of the liquid and solution solutions are equivalent.

Answer: $X_{Rb} \approx 0.45$. This is also easily accomplished using a spreadsheet and equation solver.

Figure 2.3 requires careful consideration. Figures 2.3a through 2.3e are free energy of mixing curves as a function of concentration (here plotted as X_B, where $X_B = 1.0$ at pure B and $X_B = 0$ at pure A) at decreasing temperatures ($T_5 > T_4 > T_3 > T_2 > T_1$). Temperatures T_1 and T_2 are selected such that they are at or below the melting point (T_B) of the lowest melting point species, B. Temperatures T_4 and T_5 are selected such that they are at or above the melting point (T_A) of the higher melting species, A. Temperature T_3 is between the melting points of the individual components. Figure 2.3f results from application of the equilibrium criteria at all temperatures, including T_1 through T_5. Notice that the binary-component phase diagram is a plot of temperature versus composition, and not pressure versus temperature as was the case for a unary phase diagram. This has to do with the relative importance of our intensive variables. For the unary diagram, we only had one component so that composition was fixed. For the binary diagram, we have three intensive variables (temperature, pressure, and composition), so to make an $x-y$ diagram, we must fix one of the variables. Pressure is normally selected as the fixed variable. Moreover, pressure is typically fixed at 1 atm. This allows us to plot the most commonly manipulated variables in a binary-component system: temperature and composition.

In Figure 2.3a, at a temperature T_5, the liquid phase free energy of mixing is lower than the solid phase free energy of mixing at all compositions. This means that the liquid is more stable at this temperature for all compositions, and indeed this is the case in Figure 2.3f. Any composition at temperature T_5 will be located in the single-phase, liquid region of the phase diagram. A similar situation applies to temperature T_4, shown in Figure 2.3b. Any changes in composition of the system at these temperatures, e.g., adding more B to the mixture, will be reflected in a commensurate change in composition of the liquid phase. At temperatures T_2 and T_1 (Figure 2.3d and e, respectively), the solid solution is the most stable at all compositions. Since A and B mix completely at all compositions, any change in composition at these temperatures, e.g., adding more A to the mixture, will result in a corresponding change in the solid alloy.

The situation shown in Figure 2.3c is a little more complex. Here, the solid phase has the lowest free energy at A-rich compositions ($X_B < X_S$), and the liquid phase has the lowest free energy at

B-rich compositions ($X_B > X_L$). In between the two minima of the solid and liquid free energy curves, $X_S < X_B < X_L$, both the liquid and solid phases can coexist. Why is this? Recall that the stability criteria are as follows for a two-phase system:

$$T_\alpha = T_\beta \tag{2.14}$$

$$P_\alpha = P_\beta \tag{2.15}$$

$$\mu_\alpha = \mu_\beta \tag{2.16}$$

Pressure has already been fixed, as has temperature for the diagram in Figure 2.3c. The only criterion of concern here is that the chemical potentials of the two phases be equal. Recall that the chemical potential is defined as the derivative of the free energy with respect to composition

$$\mu_i = \left(\frac{\partial G}{\partial n_i}\right)_{T,P,n_j} \tag{2.13}$$

or, for the two phases under consideration here,

$$\mu_S = \left(\frac{\partial \Delta G_{mix,S}}{\partial X_B}\right)_{T,P,n_L} = \mu_L = \left(\frac{\partial \Delta G_{mix,L}}{\partial X_B}\right)_{T,P,n_S} \tag{2.36}$$

Graphically, this condition can only occur when the two free energy of mixing curves share the same tangent, which represents the derivative of the curve at that point. This tangent is shown in Figure 2.3c. This is the situation that exists at all temperatures between T_4 and T_2, so there exists a two-phase region at all compositions between these two temperatures, as reflected in Figure 2.3f. Recall that the minima of the two free energy of mixing curves at any given temperature are located by taking the second derivative and setting it equal to 0:

$$\left(\frac{\partial^2 \Delta G_{mix,S}}{\partial X_B^2}\right) = 0, \quad \left(\frac{\partial^2 \Delta G_{mix,L}}{\partial X_B^2}\right) = 0 \tag{2.37}$$

The locus of points generated by these conditions at various temperatures constructs the phase boundary lines between the solid and liquid phases shown in Figure 2.3f. That is, there is a different solid and liquid composition, X_S and X_L, resulting from the tangent between the free energy of mixing curves, at all temperatures between T_2 and T_4, as illustrated in Figure 2.3c. We will return to the conditions stipulated by Eq. (2.37) in subsequent sections.

So, for this binary solution of components A and B, which mix perfectly at all compositions, there is a two-phase region at which both solid and liquid phases can coexist. The uppermost boundary between the liquid and liquid + solid phase regions in Figure 2.3f is known as the *liquidus*, or the point at which solid first begins to form when a melt of constant composition is cooled under equilibrium conditions. Similarly, the lower phase boundary between the solid and liquid + solid phase regions is known as the *solidus*, or the point at which solidification is complete upon further equilibrium cooling at a fixed composition.

Let us examine a real alloy phase diagram. Figure 2.4 is the binary-phase diagram for the system Cu—Ni. First, note that the composition is indicated in terms of weight fraction of one of the components, in this case Ni, and not mole fraction. The conversion between the two is straightforward

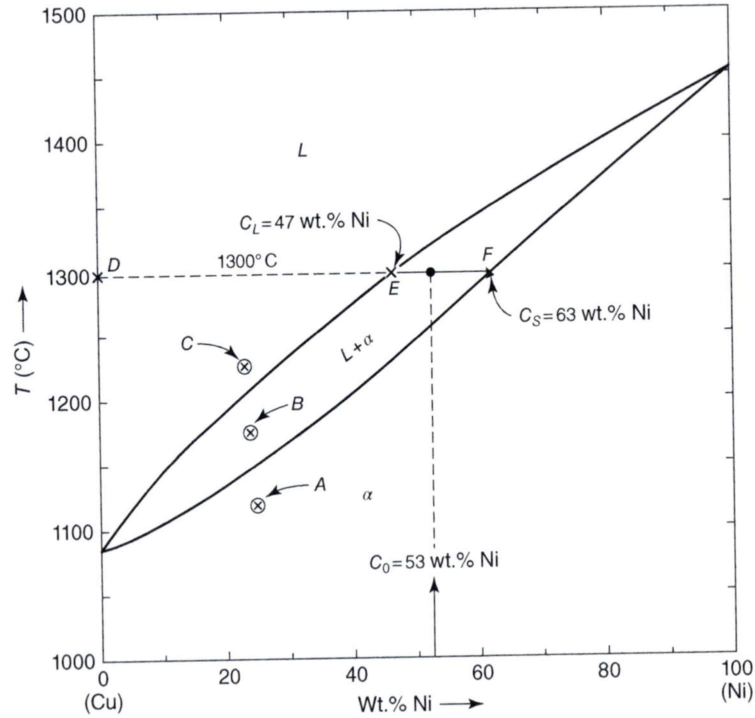

Figure 2.4 The Cu–Ni phase diagram, illustrating the use of the lever rule. From Ralls et al. [2]. Copyright © 1976 by John Wiley & Sons, Inc. This material is used by permission of John Wiley & Sons, Inc.

if the molecular weights are known, and weight fraction is used only because it is more useful experimentally than mole fraction.

In terms of the phase rule, we must utilize Eq. (2.24) instead of Eq. (2.23), since we have only one noncompositional intensive variable, T (pressure is fixed), so $N = 1$. Application of Eq. (2.24) to points A and C in Figure 2.4 indicates that the liquid and solid regions ($\phi = 1$; $C = 2$) correspond to $F = 2$, which means that both temperature and composition can be altered independently in these regions. Point B is in the liquid + solid two-phase region ($\phi = 2$; $C = 2$), so that $F = 1$. Temperature and composition are not independent in this region; any change in temperature necessarily results in a change in composition of both the liquid and the solid phases. The liquidus and solidus phase boundaries are included in this two-phase region.

More involved phase diagrams result when the two elements, A and B, are only slightly soluble in each other, or when more than one solid alloy phase can be formed from the two components. The principles of generating the phase diagram from free energy of mixing versus composition diagrams still apply, however, as do the application of the stability requirements represented by Eq. (2.36). For example, when A and B are insoluble in the solid phase, a phase diagram such as that shown in Figure 2.5f can result. As in the previous example, the phase boundaries are constructed from the free energy of mixing curves as a function of composition at various temperatures. The primary difference here is that the solid free energy curve is a straight line instead of a curve with a minimum, such that at intermediate temperatures, as in Figure 2.3c and d, tangents can be drawn between the free energy curves of the pure solids (which occur at the pure component free energies) and the liquid solution, sometimes involving both pure components, as in Figure 2.3d. When the

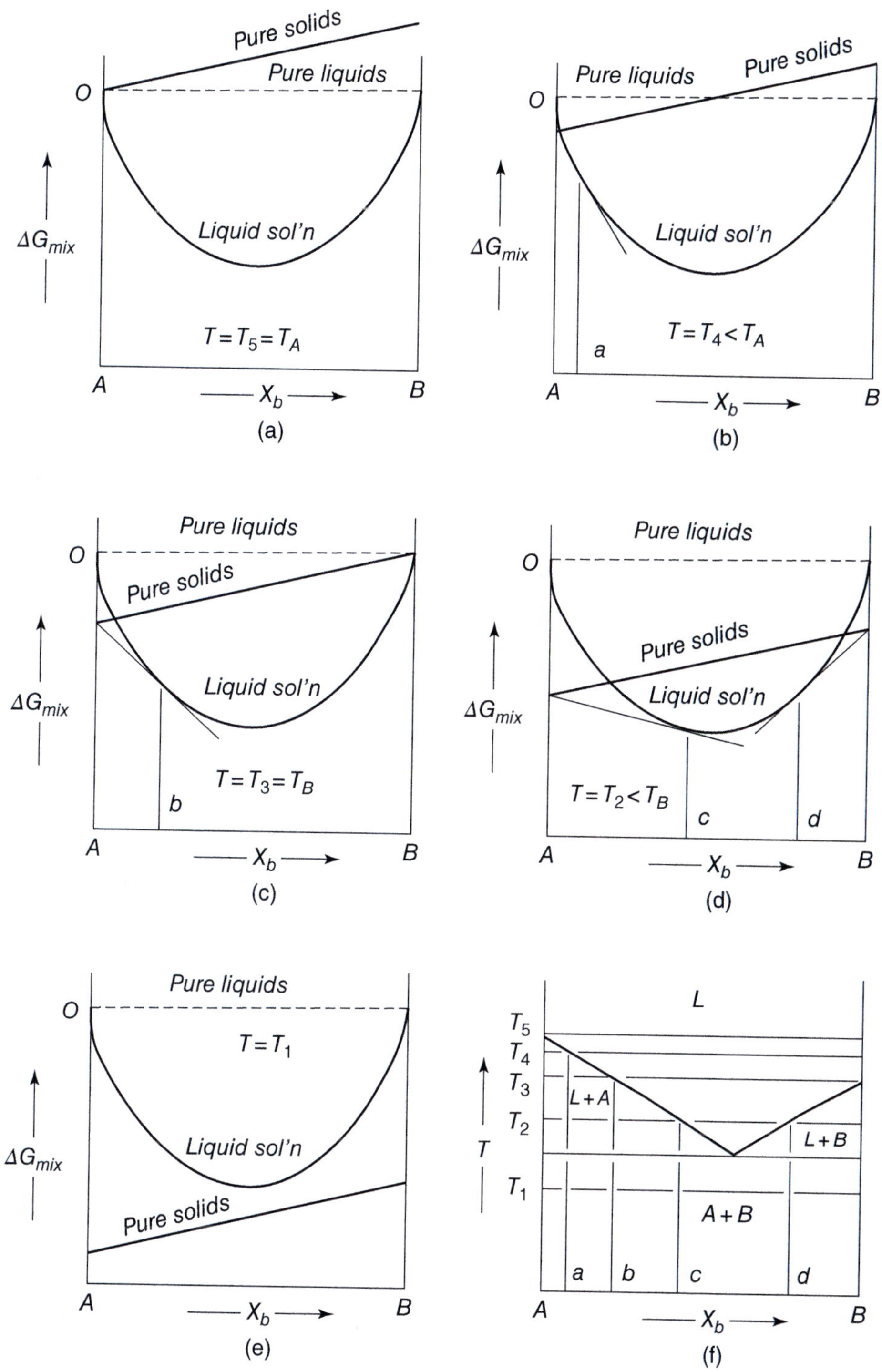

Figure 2.5 Free energy of mixing for solid and liquid phases at various temperatures (a–e) and resulting temperature–composition phase diagram for a slightly soluble eutectic binary-component system (f). From Devereux [3]. Copyright © 1983 by John Wiley & Sons, Inc. This material is used by permission of John Wiley & Sons, Inc.

free energy of the pure solid segment is the tangent to the liquid solution (located exactly at the minimum of the liquid solution free energy curve), an interesting phenomenon occurs. It is at this temperature–composition combination, called the *eutectic point*, at which the liquid, solid A, and solid B coexist. According to the phase rule, there should be zero degrees of freedom, and indeed this is the case ($\phi = 3$, $C = 2$, $N = 1$). We will learn more about the eutectic and other three-phase transformations in Section "Three-Phase Transformations in Binary Systems".

Another typical binary-phase diagram that results when two different solid solutions of A and B can form, here labeled α and β, is shown in Figure 2.6f. Again, the phase diagram is constructed

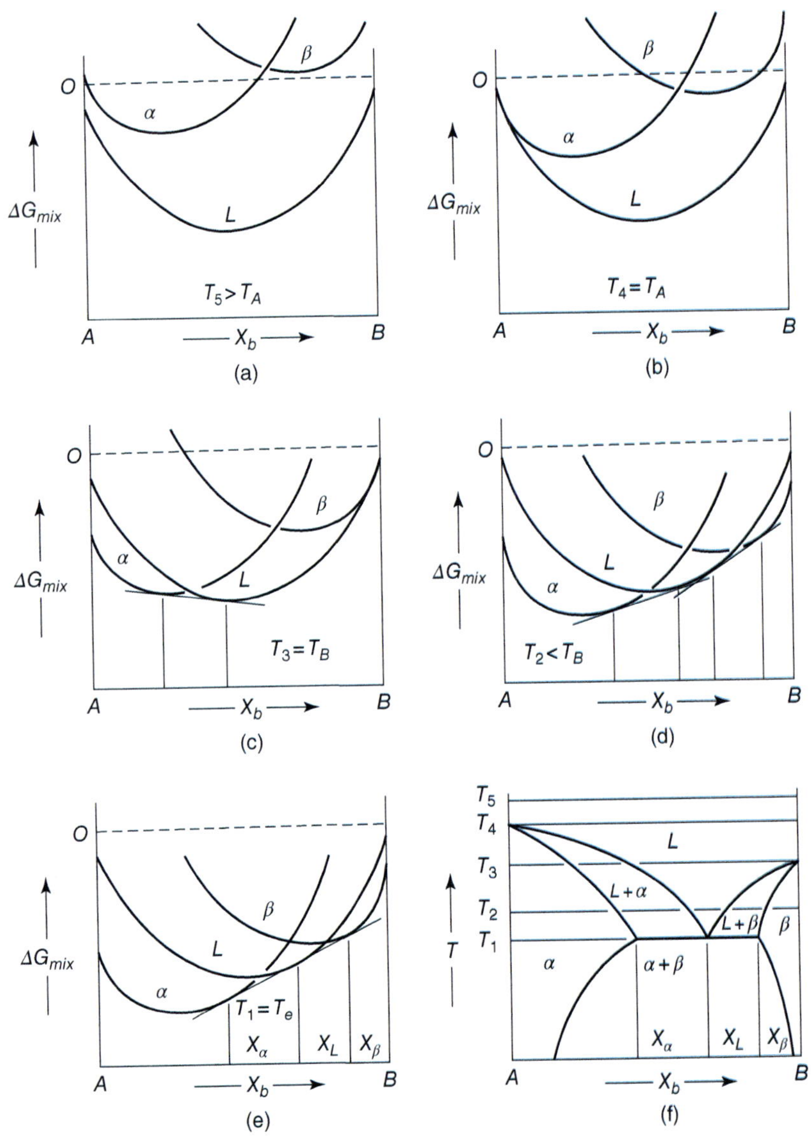

Figure 2.6 Free energy of mixing for solid and liquid phases at various temperatures (a–e) and resulting temperature–composition phase diagram for a mostly insoluble binary-component system (f). From Devereux [3]. Copyright © 1983 by John Wiley & Sons, Inc. This material is used by permission of John Wiley & Sons, Inc.

Table 2.2 Summary of Phase Relationships in a Phase Diagram.

- One-phase regions may touch each other only at single points, never along boundary lines.
- Adjacent one-phase regions are separated from each other by two-phase regions involving the same two phases.
- Three two-phase regions must originate upon every three-phase isotherm; that is, six boundary lines must radiate from each three-phase reaction horizontally.
- Two three-phase isotherms may be connected by a two-phase region, provided that there are two phases which are common to both of the three-phase equilibria.

from the free energy of mixing as a function of concentration curves (Figure 2.6a-e). In this case, it is always possible to draw a tangent between the free energy curves of the α and β phases at low temperatures, and in some cases, one tangent will be sufficient to touch all three curves for the α, β, and liquid phases, as in Figure 2.6d and e. Note that unlike the binary-component system in Figure 2.5 in which A and B were never soluble in the solid phase, in this system, A and B are always soluble in each other to some extent. Thus, the composition of the solid α phase will vary, as will that of the β phase. This can be thought of as each major component having an ability to dissolve small amounts of the impurity phase, either as interstitial or substitutional impurities (see Section 1.1.3). This phase diagram also has a eutectic point, found at the intersection of the X_L composition line and the T_1 isotherm.

Close examination of a variety of binary-component phase diagrams allows us to draw several generalizations regarding the spatial relationship between phases in a diagram. These relationships are summarized in Table 2.2.

The Lever Rule

The phase rule tells us that there is only one degree of freedom in a two-phase region of a binary-component phase diagram—for example, the solid + liquid region of the Cu – Ni phase diagram in Figure 2.4. If we change temperature in this region, the compositions of the two phases must also change. This means that the relative amounts of the two phases must, in turn, be adjusted. If there is an equal amount (in terms of weight) of the solid and liquid phase, regardless of their compositions, removing some of component B from the liquid and moving it to the solid reduces the weight of the liquid phase and increases the weight of the solid phase. The relationship between composition of the two phases and the relative amount of each phase in a phase diagram is given by the *lever rule*. We will develop the lever rule by way of example.

Consider the initial alloy composition C_0 = 53 wt% Ni, 47 wt% Cu indicated by the vertical arrow and dashed line in Figure 2.4. At 1300°C, this composition will exist as both liquid and solid (called the α phase), according to the two-phase region in the diagram. Let W_α and W_L be the fractional amounts by weight of solid and liquid, respectively, such that

$$W_\alpha + W_\beta = 1 \tag{2.38}$$

Also, the mass of each component, in this case, Ni and Cu, distributed between the two phases must equal their masses in the original composition:

$$W_\alpha C_S + W_\beta C_L = C_0 \tag{2.39}$$

Solution of Eqs. (2.38) and (2.39) gives the weight fraction of liquid in terms of the compositions:

$$W_\beta = \frac{C_S - C_0}{C_S - C_L} \tag{2.40}$$

and the weight fraction of solid in terms of compositions:

$$W_\alpha = \frac{C_0 - C_L}{C_S - C_L} \tag{2.41}$$

Eqs. (2.40) and (2.41) are the lever rule and can be used to determine the relative amounts of each phase in any two-phase region of a binary-component phase diagram. For the example under consideration, the amount of liquid present turns out to be

$$W_\beta = \frac{63 - 53}{63 - 47} = 0.625 \tag{2.42}$$

So, 62.5% by weight of the alloy is present as liquid, and 37.5% is solid. The compositions of the two phases are determined by extending the horizontal line, or *tie line*, to the phase boundaries (line EF in Figure 2.4) and then reading the compositions of these phases off the abscissa. The liquid phase has a composition of 47 wt% Ni/53 wt% Cu (point E), and the solid phase is 63 wt% Ni/37% Cu (point F).

There are a few important points regarding application of the lever rule. First, it can only be used in the form of Eqs. (2.40) and (2.41) when concentrations are expressed in terms of weight percentages. Mole percentages must first be converted to weight percentages when concentrations are so

Cooperative Learning Exercise 2.2

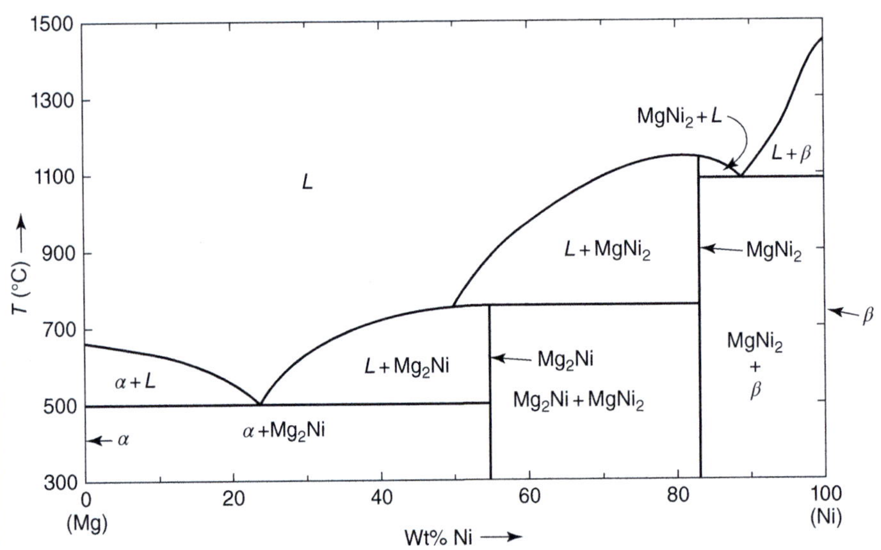

Consider the Mg—Ni binary-phase diagram shown above. Note that in this system, intermetallic compound formation is common; for example, Mg_2Ni and $MgNi_2$ are stoichiometric compounds.

Person 1: Determine the phases present, the relative amounts of each phase, and the composition of each phase when the only *incongruently melting* intermetallic compound in this system is heated to 800°C.

Person 2: Determine the phases present, the relative amounts of each phase, and the composition of each phase when the liquid phase from Person 1's problem is cooled to 600°C.

Answer:

Person 1: 88% L (51 wt% Ni, 49 wt% Mg); 12% $MgNi_2$ (83 wt % Ni, 17 wt% Mg).
Person 2: 14% L (27 wt% Ni, 73 wt% Mg); 86% Mg_2Ni (55 wt% Ni, 45 wt % Mg).

expressed. Second, Eqs. (2.40) and (2.41) can be used by simply measuring the distances of the corresponding line segments directly off the diagram. For example, the line EF in Figure 2.4 represents the quantity $C_S - C_L$. If it is measured using a ruler in, say, millimeters, then $C_O - C_L$ could be measured similarly, and W_α determined from a ratio of these two measurements. In fact, this method is much more accurate than extrapolating lines from the phase boundaries down to the abscissa (which can introduce error) and then interpolating the composition from markings that may, in some cases, be given only every 20 wt% or so. Finally, as you begin to work problems with the lever rule, you may find that the ratio of line segments given by Eqs. 2.40 and 2.41 is a little bit counterintuitive. That is, the line segment in the numerator for the weight fraction of solids [Eq. (2.41)] is that *opposite* the solid–liquid + solid phase boundary, and the line segment in the numerator for the liquid fraction [Eq. (2.42)] is opposite the liquid–solid + liquid phase boundary. A little practice will eliminate this confusion.

Need More Help?

Try this video on the lever rule and eutectics:

https://www.youtube.com/watch?v=h5dwpTqacqc

Three-Phase Transformations in Binary Systems

Although this chapter focuses on the equilibrium between phases in binary-component systems, we have already seen that in the case of a eutectic point, phase transformations that occur over minute temperature fluctuations can be represented on phase diagrams as well. These transformations are known as *three-phase transformations* because they involve three distinct phases that coexist at the transformation temperature. Their characteristic shapes as they occur in binary-component phase diagrams are summarized in Table 2.3. Here, the Greek letters α, β, γ, and so on designate solid phases, and L designates the liquid phase. Subscripts differentiate between immiscible phases of different compositions. For example, L_I and L_{II} are immiscible liquids, and α_{I} and α_{II} are allotropic solid phases (different crystal structures).

The Cu—Zn system (see Figure 2.7) displays several *intermediate solid solutions* that arise due to limited solubility between the two elements. For example, at low wt% Zn (which incidentally is the composition of alloys known as *brass*), the relatively pure copper α phase can accommodate small

Table 2.3 Common Three-Phase Transformations in Condensed Binary Systems

Name of Reaction	Equation	Phase Diagram Characteristic
Monotectic	$L_I \xrightarrow{\text{cooling}} \alpha + L_{II}$	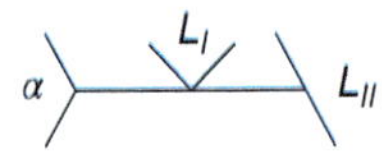
Monotectoid	$\alpha_1 \xrightarrow{\text{cooling}} \alpha_2 + \beta$	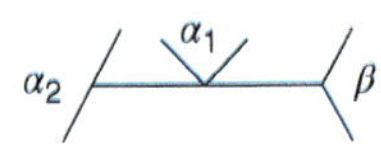
Eutectic	$L \xrightarrow{\text{cooling}} \alpha + \beta$	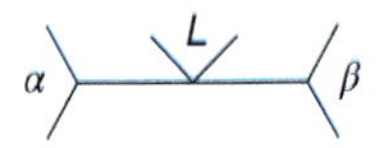
Eutectoid	$\alpha \xrightarrow{\text{cooling}} \beta + \gamma$	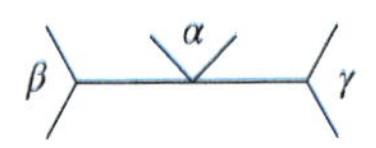
Syntectic	$L_I + L_{II} \xrightarrow{\text{cooling}} \alpha$	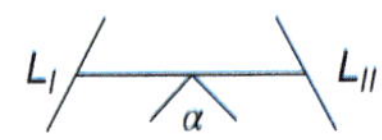
Peritectic	$\alpha + L \xrightarrow{\text{cooling}} \beta$	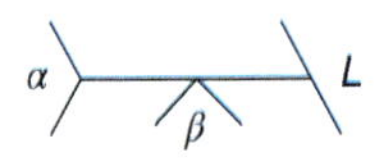
Peritectoid	$\alpha + \beta \xrightarrow{\text{cooling}} \gamma$	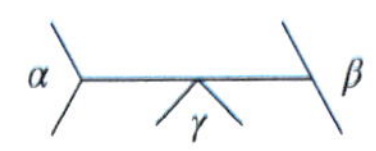

Source: K. M. Ralls, T. H. Courtney and J. Wulff, *Introduction to Materials Science and Engineering*, p. 331. Copyright © 1976 by John Wiley & Sons, Inc.

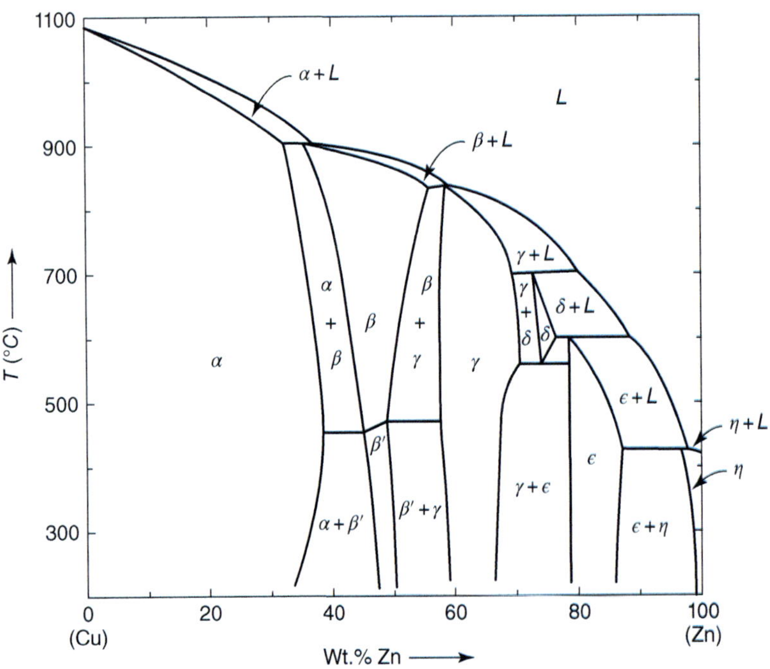

Figure 2.7 Cu–Zn phase diagram, illustrating several three-phase reactions. From Ralls et al. [2]. Copyright © 1976 by John Wiley & Sons, Inc. This material is used by permission of John Wiley & Sons, Inc.

amounts of Zn as an impurity in the crystal structure. This is known as a *terminal solid phase*, and the solubility limit where intermediate solid solutions (such as $\alpha + \beta$) begin to occur is called the *solvus* line. Some of the three-phase transformations that are found in this diagram include a peritectic ($\delta + L \rightarrow \varepsilon$) and a eutectoid ($\delta \rightarrow \gamma + \varepsilon$). Remember that these three-phase transformations are defined for equilibrium cooling processes, not heating or nonequilibrium conditions.

2.1.3 The Iron–Carbon Phase Diagram*

Iron and its alloys continue to be the most widely utilized metallic systems for construction. The process by which *iron-based alloys*, including *steels*, are produced from iron ore involves the reduction of iron oxide by means of carbon to form a product known as *pig iron*, which contains approximately 4% carbon. As a result, the iron–carbon phase diagram is an industrially important one. The compounds formed between iron and carbon are sufficiently complex in their microstructures and properties to warrant intensive investigation and more detailed description. However, the iron–carbon phase diagram is just another binary phase diagram, and, as such, we have all the tools we need to fully understand it.

Historical Highlight
Learn more about the metallurgical development of iron as a material of construction at https://en.wikipedia.org/wiki/Iron#Development_of_iron_metallurgy

The iron–carbon phase diagram at low weight percentages of carbon is shown in Figure 2.8. The phase boundary at 6.69 wt% carbon represents the compound iron carbide, Fe_3C, known as *cementite*, so that the phase diagram in Figure 2.8 is more appropriately that of Fe_3C—C. The allotropy of elemental iron plays an important role in the formation of iron alloys. Upon solidification from the melt, iron undergoes two allotropic transformations (see Figure 2.9). At 1539°C, iron assumes a BCC structure, called *delta-iron* (δ-Fe). Upon further cooling, this structure transforms to the FCC structure at 1400°C, resulting in *gamma-iron* (γ-Fe). The FCC structure is stable down to 910°C, where it transforms back into a low-temperature BCC structure, *alpha-iron* (α-Fe). Thus, δ-Fe and α-Fe are actually the same form of iron but are treated as distinct forms due to their two different temperature ranges of stability.

Carbon is soluble to varying degrees in each of these allotropic forms of iron. The solid solutions of carbon in α-Fe, γ-Fe, and δ-Fe are called, respectively, *ferrite, austenite*, and *δ-ferrite*. So, for example, the single-phase region labeled as γ in Figure 2.8 could be (and is in many diagrams) labeled as the austenite phase region. Similarly, the α and δ phases can be called the ferrite and δ-ferrite regions, respectively. Keep in mind, however, that these single-phase regions are simply a solid solution between one of the allotropic forms of elemental iron and carbon. Note that the ferrite region is particularly limited in both temperature and composition, as is δ-ferrite, and that austenite has a particularly large range of stability at higher temperatures. All alloys with carbon contents between 0.5 and 2 wt% C solidify as austenite. Alloys containing more than 2% carbon are subject to a eutectic transformation that forms austenite and cementite. At 0.8% carbon, what has solidified as austenite undergoes a eutectoid transformation at 723°C to an intimate mixture of ferrite and cementite. This mixture is given a special name, *pearlite*. Recall

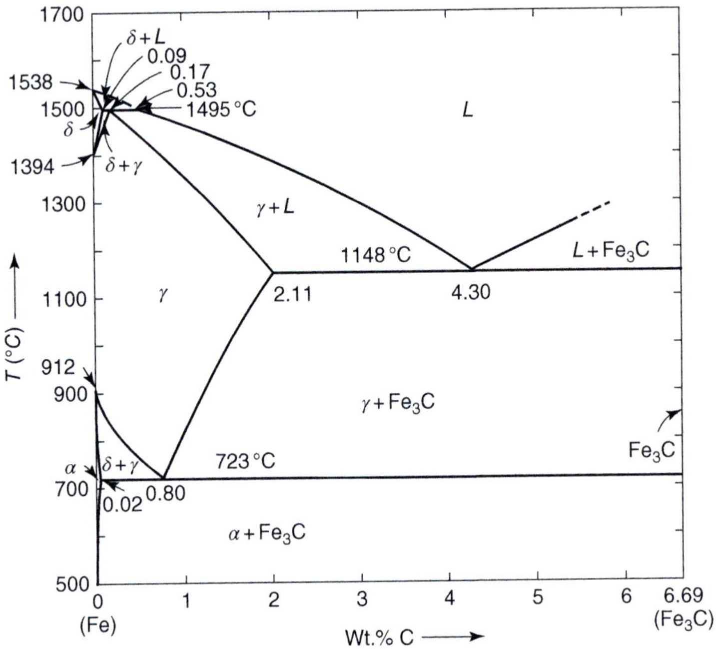

Figure 2.8 The Fe—C phase diagram (low weight % C). From Ralls et al. [2]. Copyright © 1976 by John Wiley & Sons, Inc. This material is used by permission of John Wiley & Sons, Inc.

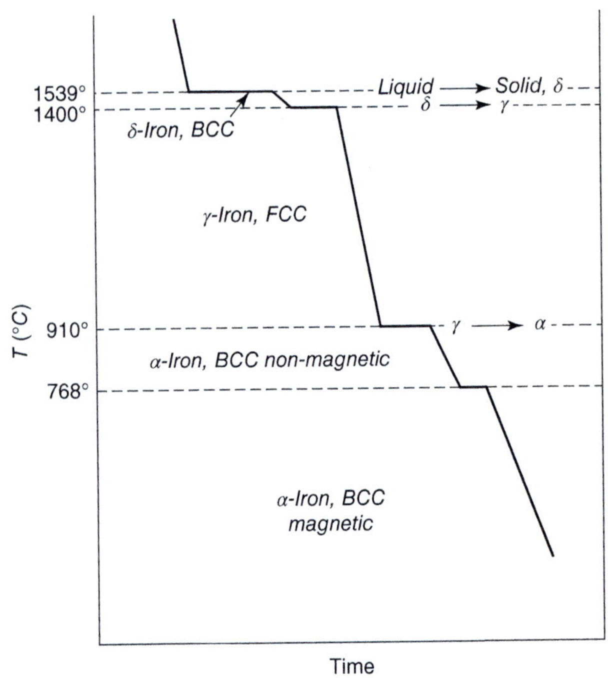

Figure 2.9 Cooling curve for pure iron. Reproduced by permission, from Committee on Metallurgy [4]. Copyright © 1957 by Pitman Publishing.

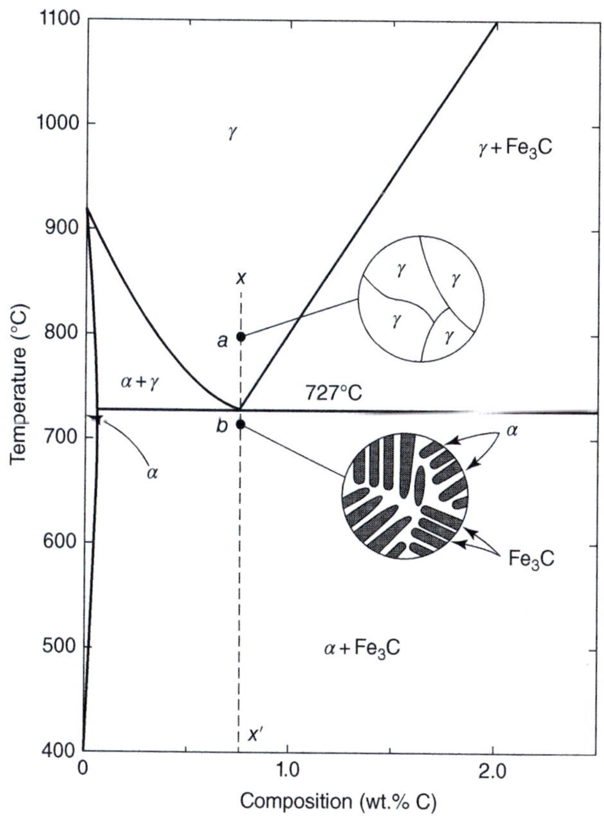

Figure 2.10 Schematic representation of the microstructures for a eutectoid transformation in the Fe–C system. Reproduced by permission, from Callister [5]. Copyright © 2000 by John Wiley & Sons, Inc.

that phases need not be continuous, and such is the case in pearlite where the cementite (Fe_3C) forms lamellar structures within the austenite phase, which can also have some carbon dissolved in it (see Figure 2.10).

The eutectoid transformation is an important one not only for this specific carbon composition but also for classifying all types of steels. *Carbon steels* have carbon contents between 0.1 and 1.5 wt %. Those with carbon contents less than 0.8% are termed *hypoeutectoid steels*, and those with greater than 0.8% C are called *hypereutectoid steels*. Further classifications of steels are given in Table 2.4. (Most ferrous alloys including stainless steel contain many alloying elements, not just carbon. See Appendix 2 for the composition of some common ferrous alloys.)

When austenite is cooled under more rapid conditions, a compound called *bainite* is produced. Bainite is a nonequilibrium product that is similar to pearlite but consists of a dispersion of very small Fe_3C particles between the ferrite plates. Bainite formation is favored at a high degree of supercooling from the austenite phase, whereas pearlite forms at low degrees of supercooling, or more equilibrium cooling.

Finally, if an austenite steel is rapidly quenched, *martensite* steel can form. Steel martensite possesses a body-centered tetragonal structure (see Figure 2.11), which is only slightly distorted from the BCC structure of austenite. The distortion arises from interstitial carbon atoms that occupy one set of preferred octahedral interstitial sites. During the martensitic transformation upon rapid

Table 2.4 Classification of Ferrous Alloys

Component (wt%)	Classification	Examples
C < 0.01	Iron	
0.1 < C < 1.5	Carbon steel	1000 series—ferrite/pearlite
		2000 series—ferrite/pearlite or bainite
		3000 series—martensitic 4000 series
2 < C < 4.5	Cast iron	Gray iron
		White iron
Various	Alloy steels	Ni, Mn, Cu, Co, C, N—austenite stabilizers
		Cr, Mo, Si, W, V, Sn, Cb, Ti—ferrite stabilizers
Cr > 12	Stainless steels	200/300—austenitic
		400—ferritic
		400/500—martensitic

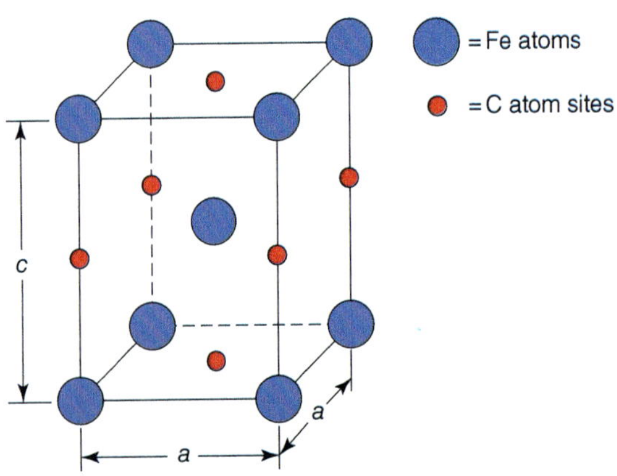

Figure 2.11 The body-centered tetragonal unit cell of steel martensite. From Ralls et al. [2]. Copyright © 1976 by John Wiley & Sons, Inc. This material is used by permission of John Wiley & Sons, Inc.

cooling, carbon atoms residing at equivalent octahedral sites in BCC change position slightly to only one (along the *c*-axis) of the three possible sets of octahedral sites in the body-centered tetragonal structure, thus creating the distortion. The extent of tetragonal distortion, as measured by the axial ratio, increases with increasing carbon content (see Figure 2.12). The transformation process from austenite to martensite is very rapid because it involves only a distortion of the BCC unit cell, as opposed to the BCC–FCC transition to pearlite or bainite, which is a diffusion-controlled process. This type of diffusionless transformation is termed a *martensitic transformation* and can occur in alloys other than steel.

As you can see, the process by which the iron–carbon alloy is processed and solidified is just as important as the overall stoichiometry. Although a discussion regarding phase transformations is more the realm of kinetic processes (Chapter 3), it is nonetheless pertinent to summarize here the types of important ferrous alloys, particularly those in the *cast iron* categories. This is done in Figure 2.13.

As indicated in Table 2.4, cast irons contain more than 2% carbon by weight, compared to steels which usually have less than 1.5% carbon. Cast irons also differ microstructurally from steels, in that a separate graphite phase typically exists. Graphite is seldom found in steels because the

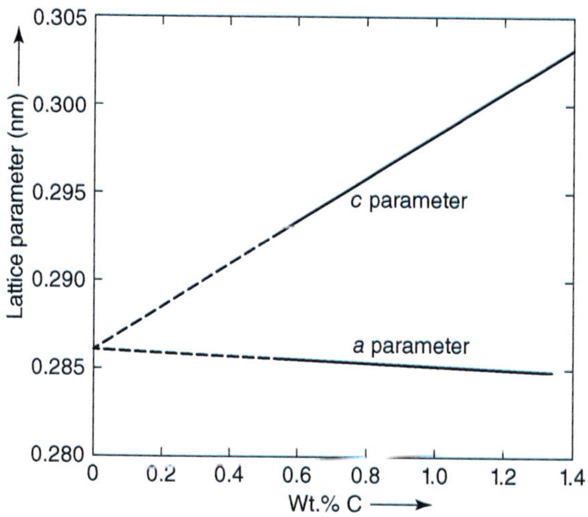

Figure 2.12 Variation of tetragonality in steel martensite with axial ratio. From M. Ralls et al. [2]. Copyright © 1976 by John Wiley & Sons, Inc. This material is used by permission of John Wiley & Sons, Inc.

solid-state transformations that give rise to it are so slow that transformations yielding cementite always predominate. In cast irons, however, graphite can form directly from eutectic solidification and by solid-state transformations. Whether carbon is present as graphite or is tied up in cementite in cast irons depends on whether the solidification and cooling processes are carried out under close to equilibrium conditions, which favor graphite formation, or under highly nonequilibrium conditions, which favor Fe_3C formation. There is an analogy here with austenite decomposition that we have already described: pearlite forms under near-equilibrium cooling conditions in preference to the less stable bainite, which forms under rapid cooling conditions. As shown in Figure 2.13, slow cooling rates favor graphite formation from the eutectic reaction. The graphite phase continues to grow with decreasing temperature as the carbon content of the iron-rich phase (γ or α) decreases. Under these conditions, the phases present in zones I, II, and III are austenite and liquid, austenite and graphite, and ferrite and graphite, respectively. Under normal conditions, the graphite is present as flakes, but if the melt is *desulfurized* (where sulfur is an impurity in the ore), the graphite will be present as nodules. The normal structure is called *ferritic gray cast iron*, and that for the desulfurized material is called *ferritic ductile cast iron*.

Under moderate cooling rates, graphite still forms in the eutectic reaction, but Fe_3C is one product of the eutectoid reaction. The final structure, consisting of locally interconnected graphite flakes dispersed in a pearlitic matrix, is called *pearlitic gray cast iron* or, if desulfurized, *pearlitic ductile cast iron*. White cast irons are formed under rapid cooling and have similar structures to steel insofar as there is no graphite present. Thus, austenite and Fe_3C are present in zone II, and pearlite and Fe_3C are present in zone III. White casts can be converted into more usable structures by heating into zone II for longer time periods. This treatment causes graphite, in the shape of *rosettes*, to precipitate and the cementite is eliminated. Keep in mind that the cooling classifications of "slow," "medium," and "rapid," are arbitrary and that commercial processes can produce materials with a mixture of the various microstructures shown in Figure 2.13. We will discuss transformation processes and the effect of heating and cooling rates on them in Chapter 3.

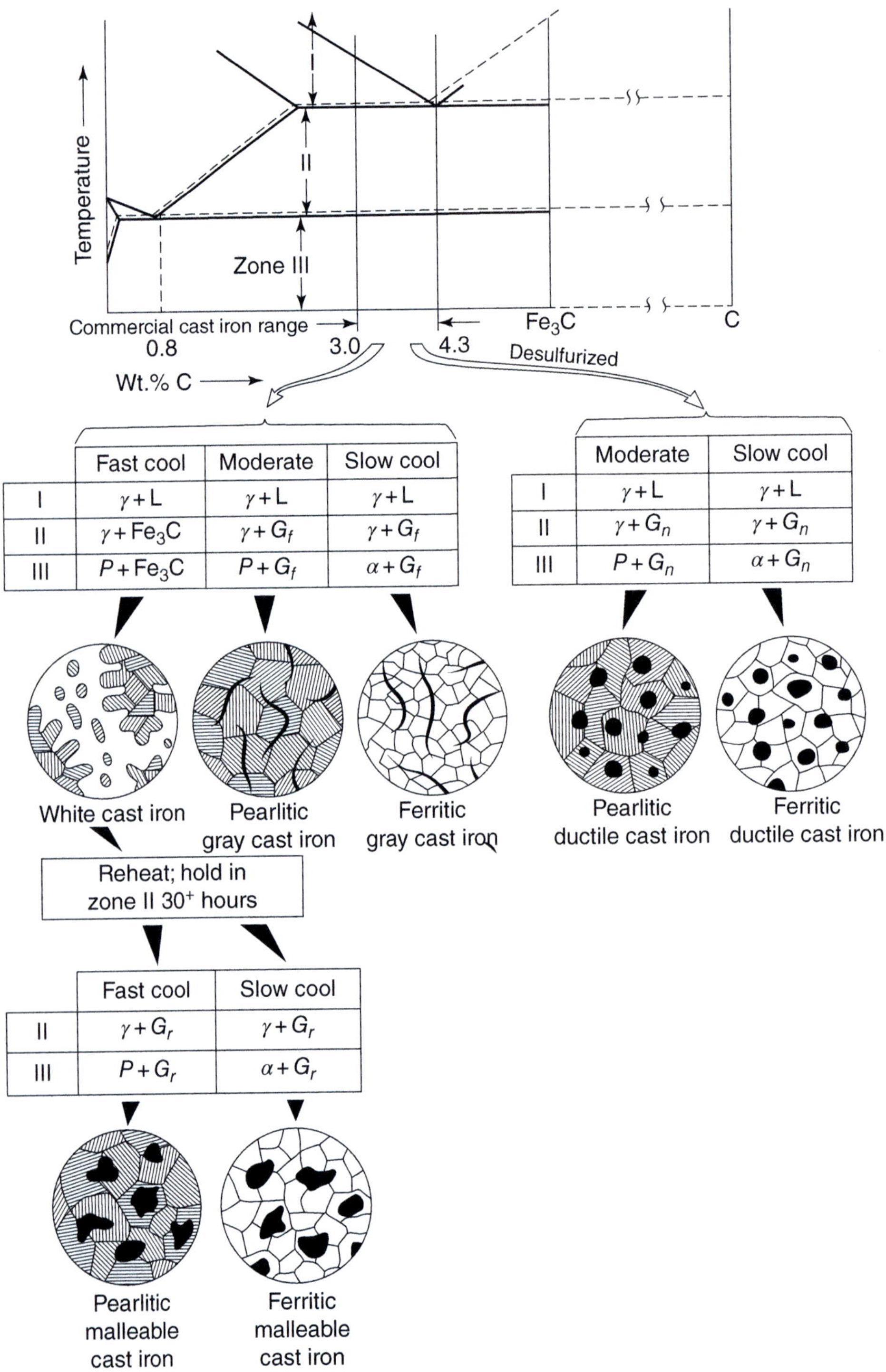

Figure 2.13 Microstructures obtained by varying thermal treatments in cast irons (G_f = graphite flakes; G_r = graphite rosettes; G_n = graphite nodules; P = pearlite). From Ralls et al. [2]. Copyright © 1976 by John Wiley & Sons, Inc. This material is used by permission of John Wiley & Sons, Inc.

2.2 Thermodynamics of Ceramics and Glasses

2.2.1 Phase Equilibria in Ternary-Component Systems

Three-component systems, or *ternary systems*, are fundamentally no different from two-component systems in terms of their thermodynamics. Phases in equilibrium must still meet the equilibrium criteria [Eqs. (2.14)–(2.16)], except that there may now be as many as five coexisting phases in equilibrium with each other. The phase rule still applies to ternary systems, and we will see that we can use an adaptation of the lever rule to three-component phase diagrams.

As with unary and binary-component systems, the three components of a ternary system can be anything—even elements—as in our preceding descriptions. In fact, there are many important alloy systems that consist of three primary components, such as iron–carbon–silicon and aluminum–zinc–magnesium. However, the addition of one important element, oxygen, to any metallic binary-component system suddenly transforms the discussion from that which is based primarily in metallurgy to one which is more the realm of ceramics. For example, the aluminum–silicon binary metal alloy system becomes a ternary system of *aluminosilicates*, and an important class of ceramics, when oxygen is introduced as the third component, and compounds such as Al_2O_3 and SiO_2 can form. Thus, the discussion here will be limited to ceramics, but keep in mind that the principles of ternary-phase diagrams can be applied to any three-component system, including three compounds. Again, let us consider only condensed phases.

Ternary-Phase Diagrams

In a ternary system, it is necessary to specify temperature, pressure, and two composition parameters to completely describe the system. Typically, pressure is fixed, so that there are three independent variables that are needed to fix the system: temperature and two compositions. The third composition is, of course, fixed by the first two. We could create a three-dimensional plot with three mutually perpendicular axes, as is usually the case in mathematics; however, it is more convenient, and graphically more appealing, to establish two compositional axes 60° apart from each other, with a third, redundant compositional axis, as in the form of an equilateral triangle (see Figure 2.14). The temperature axis is then constructed perpendicular to the plane of the triangle, if desired.

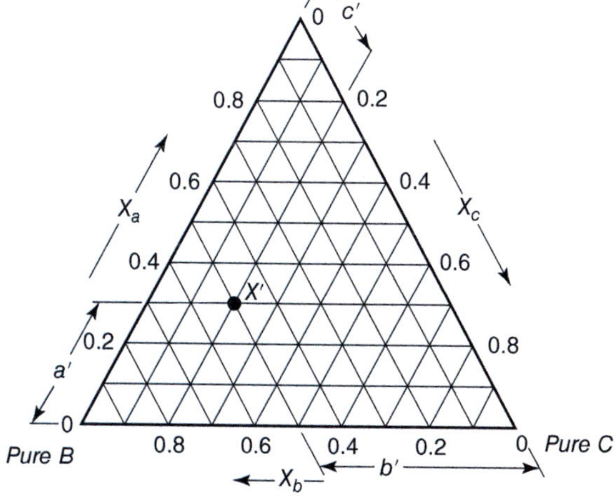

Figure 2.14 Illustration of how to express compositions on a three-component diagram. From Devereux [3]. Copyright © 1983 by John Wiley & Sons, Inc. This material is used by permission of John Wiley & Sons, Inc.

Need More Help?

Try these videos on ternary-phase diagrams:

https://www.youtube.com/watch?v=h8tFnC4pNGc
https://www.youtube.com/watch?v=okctrfb3v1o
https://www.youtube.com/watch?v=3QUI1RhSuqU

Figure 2.14 requires a bit of instruction and practice before it can be used correctly. First, the pure components—in this case, A, B, and C—are located at the vertices of the equilateral triangle. The base of the triangle opposite the corresponding vertex refers to 0% of that component, so that the edge of the triangle connecting vertices B and C is 0 wt% A. Lines parallel to each edge constitute lines of constant composition for one of the components. For example, lines parallel to BC—which are spaced a uniform distance apart and become shorter in length as they approach the vertex A—are lines of constant wt% A. Sometimes these lines are every 10 wt%, sometimes every 20 wt%, and sometimes they are not there at all. Similarly, lines parallel to AC and AB represent lines of constant composition for components B and C, respectively. The triangular grid generated by these lines of constant composition helps us determine the overall composition of a point in the diagram. For example, point X′ in Figure 2.14 is located at 30% A, 50% B, and 20% C. Obviously, the overall weight percentages of the components must add up to 100%. Not all compositions fall neatly on the intersection of three grid lines, of course, but when the grid is present, it is relatively easy to interpolate between gridlines.

Gridlines can clutter the phase diagram, however, especially when phases and boundaries start appearing in real systems, so that it is convenient to have a method for determining the composition of any point on the diagram without the assistance of gridlines. This is accomplished using the *center of gravity rule*, which states that the sum of the perpendicular distances from any point within an equilateral triangle to the three sides is constant and equal to the altitude of the triangle. With reference to Figure 2.15, this means that the quantity $(a + b + c)$ is a constant, regardless of where the point is located inside the triangle, and it represents the length from any edge to its opposite vertex.[1] Moreover, the relative amounts of the components are then given by the ratio of their respective line segments to the altitude:

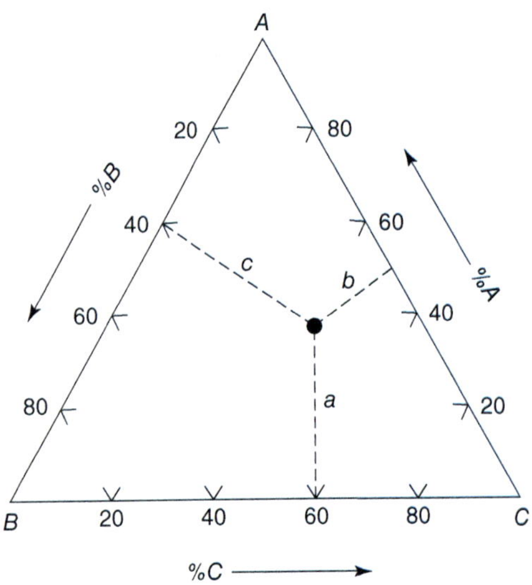

Figure 2.15 Illustration of center-of-gravity rule for determining compositions in ternary system.

$$\%\text{A} = \frac{100a}{(a + b + c)} \tag{2.43a}$$

$$\%\text{B} = \frac{100b}{(a + b + c)} \tag{2.43b}$$

$$\%\text{C} = \frac{100c}{(a + b + c)} \tag{2.43c}$$

1 Do not let the different ways the axes are labeled between Figures 2.14 and 2.15 confuse you. For example, the labels for the gridlines of component A are on the left in Figure 2.14, but on the right in Figure 2.15. Similar adjustments are made to the labels for components B and C. Both methods and both figures are entirely correct. You should convince yourself of this fact.

Just as with the lever rule in binary-phase diagrams, segments can be measured in any consistent units of length.

Cooperative Learning Exercise 2.3

Using the diagram in Figure 2.15:

Person 1: Lightly sketch in grid lines on the diagram and use them to determine the composition at the indicated point.

Person 2: Use the center-of-gravity Eq. (2.43) to calculate the composition at the indicated point.

Compare your answers.
Answer: 37% A; 21% B; 42% C

Now that we can determine overall compositions on a ternary diagram, we can return to the issue of representing temperature in the diagram. As mentioned earlier, the temperature axis is perpendicular to the compositional, triangular plane, as shown in Figure 2.16 for the case of three components that form a solid solution, α. Recall that to form solid solutions, the components must generally have the same crystal structure. Such substances are said to be *isomorphous*.

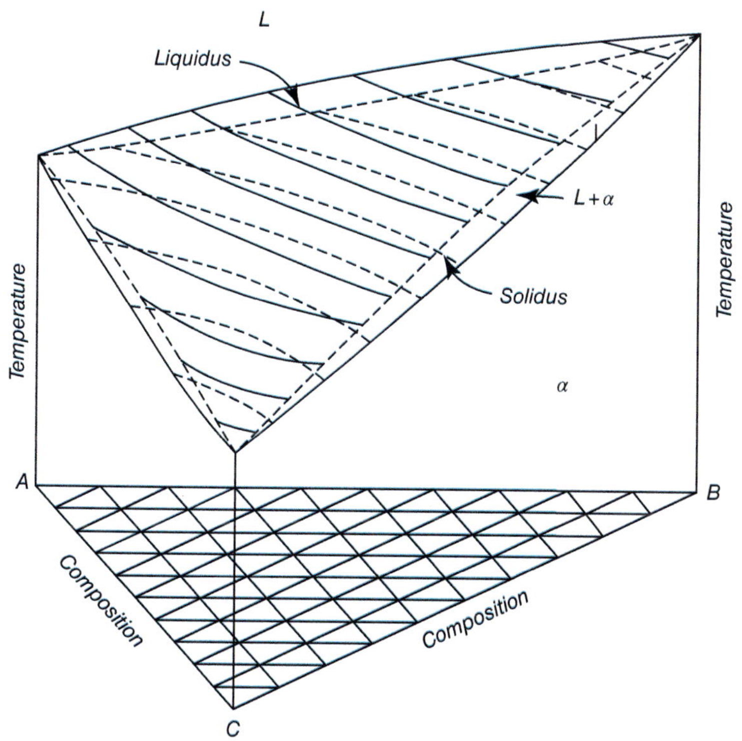

Figure 2.16 Temperature–composition space diagram of a ternary-isomorphous system. Reproduced by permission, from Rhines [6]. Copyright © 1956 by McGraw-Hill Book Co.

There are several important things to note in Figure 2.16. First, we are assuming a fixed pressure, as is the case in most condensed-phase diagrams. Second, the phase rule still applies. The single-phase regions, α or liquid, have three degrees of freedom, as given by the phase rule

$$F = C - \phi + N = 3 - 1 + 1 = 3 \tag{2.44}$$

so that composition of all three components can be varied continuously in these regions. In Figure 2.16, there is only one two-phase region, $L + \alpha$. Here, there are only $F = 3{-}2 + 1 = 2°$ of freedom. The upper surface of the two-phase region is the liquidus, and the lower surface of the region is the solidus. These phase boundary surfaces are indicated by selected solid and dashed lines, respectively. Finally, you should be able to visualize that each vertical face of the diagram represents a complete binary-component phase diagram. For example, the C–B face in Figure 2.16 occurs only at $A = 0$ for all temperatures, and the resulting diagram is like the binary-component phase diagram for isomorphous substances shown in Figure 2.4. Similarly, the faces represented by A–B ($C = 0$) and A–C ($B = 0$) are independent binary-phase diagrams. Thus, a ternary-phase diagram contains not only all of the phase information we will be describing in this section, but three independent binary-component phase diagrams as well.

Before moving on to more complicated ternary-phase diagrams, it should be noted that the complete temperature–composition ternary-component phase diagram can be "sectioned" in two important ways to simplify the representation. First, we can take horizontal "slices" of the diagram. Such sections are parallel to the base of the diagram and are, as such, necessarily at a fixed temperature, or *isotherm*. Examples of isothermal sections from a temperature–composition ternary-phase diagram are shown in Figure 2.17. Note that at temperatures above the melting point of the highest-melting compound, the resulting diagram consists completely of a single-phase liquid region. As we

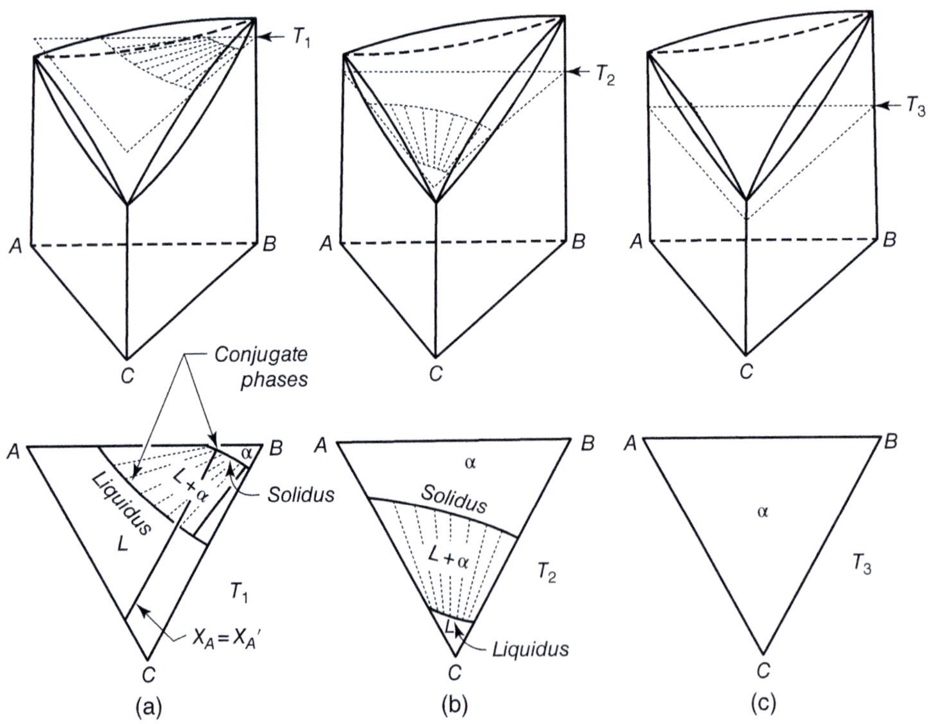

Figure 2.17 Isotherms through the ternary-isomorphous phase diagram. Reproduced by permission, from Rhines [6]. Copyright © 1956 by McGraw-Hill Book Co.

decrease temperature and take a section below the liquidus, as in Figure 2.17a, part of the diagram contains liquid, part contains solid, and part contains the liquid + solid two-phase region. The liquidus and solidus lines can now be identified, and what was once a two-phase region (L + α) with two degrees of freedom in Figure 2.16 is now a two-phase region with one degree of freedom, just as in a binary-component diagram, since the temperature has now been fixed and one degree of freedom has been lost. The one degree of freedom in this region is a composition, with the other compositional variables being fixed when the first is determined. Unlike true binary-component phase diagrams, the tie lines connecting the two *conjugate phases* (in this case liquid and α) that coexist at temperature T_1 cannot be drawn as completely horizontal or vertical lines but must be determined experimentally. These tie lines are represented as dashed lines in the two-phase region. A similar type of diagram results from an isothermal section taken at T_2, as in Figure 2.17b, which is slightly lower than T_1. Finally, at T_3, the isotherm is completely within the single-phase solid region.

The second type of "slice" through the temperature-composition plot is a vertical section, known as an *isopleth*. Isopleths are sections of constant relative composition. For example, in Figure 2.18a,

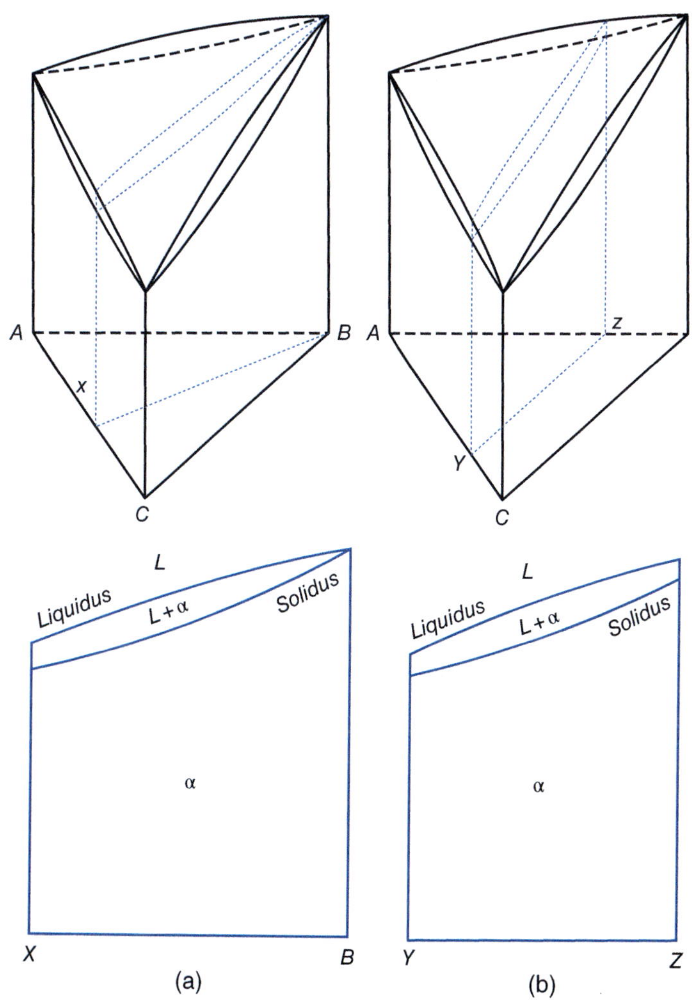

Figure 2.18 Examples of isopleths and pseudobinary-phase diagrams, (a) constant A/C; (b) constant A. Reproduced by permission, from Rhines [6]. Copyright © 1956 by McGraw-Hill Book Co.

the line X–B is drawn on the base of the diagram and extended upward along the temperature axis to generate the diagram shown below, known as a *pseudobinary-phase diagram*, since it looks similar to a binary-component phase diagram, but is obviously different. Note, for example, that the liquidus and solidus lines do not converge at X on the pseudobinary-phase diagram. This section has been selected such that A and C are always in the ratio of 1:1 at any point on the diagram. Obviously, the absolute amount of B can change, from 0% at X to 100% at B, thus changing the absolute amounts of A and C, but A and C always stay in the same proportion relative to one another. Vertical sections need not bisect a vertex. They are, however, normally taken in a logical manner—for example, a constant composition of one component or constant relative composition of two components. Figure 2.18b shows a section taken at a constant composition for A, as indicated by the line Y–Z, extended upward along the temperature axis. The resulting pseudobinary diagram is for a fixed composition of A, where the relative amounts of C and B are allowed to vary between pure C at Y and pure B at Z.

Just as in the case of binary systems, ternary systems can contain components that have only partial solubility on one another. This leads to eutectic diagrams, such as the one shown in Figure 2.19. This is the three-component analog to Figure 2.5. Again, each face is a binary-phase diagram, with binary eutectic melting points located at e_1, e_2, and e_3. The ternary eutectic melting point is located at point E, which is a "well" in the middle of the diagram. The red lines connecting

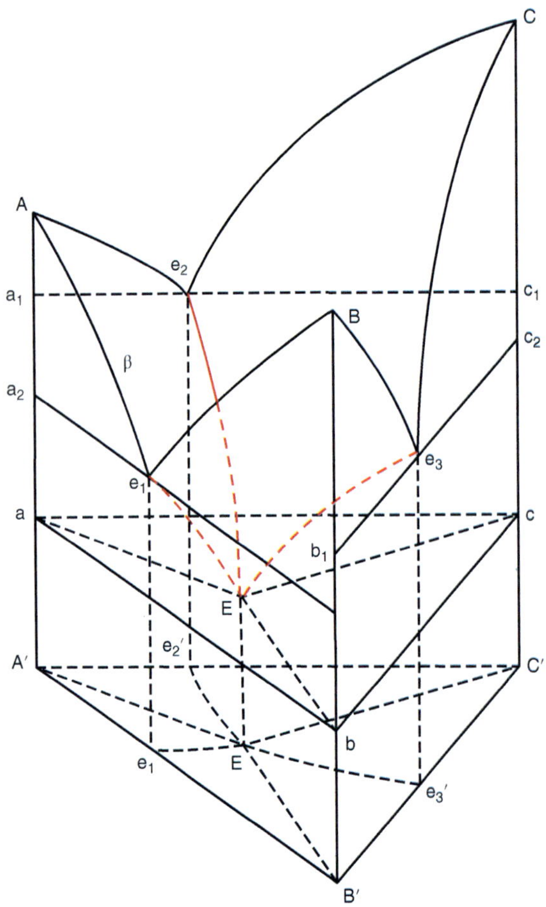

Figure 2.19 Perspective drawing of a ternary system with a simple eutectic and no ternary compound. Reproduced by permission, from Ref. [7]. Copyright © 1964, The American Ceramic Society.

the binary eutectics with the ternary eutectics, e_1–E, e_2–E, and e_3–E, are part of the liquidus surface, separating the single-phase liquid region from the solid solution regions.

There are, of course, numerous combinations of solid solutions, phases, and transformations possible in these diagrams, and we will not attempt to address more than a few of the basic ones here. But keep in mind that these phase diagrams are generated in the same way that the unary and binary-phase diagrams were generated: from free energy versus composition curves at various temperatures. The calculation of free energies of mixing for three components is directly analogous to that for two components, and the analogs for the minimization of free energy criteria as given by Eqs. (2.37) are appropriate.

Cooperative Learning Exercise 2.4

Work in groups or three. The melting points of components A, B, and C are 1000°C, 900°C, and 750°C, respectively. The constituent binary systems of the ternary system ABC each show complete solubility in both the liquid and the solid states. The following data refer to three binary alloys:

Composition (wt%)			Temperature (°C)	
A	**B**	**C**	**Liquidus**	**Solidus**
50	50	—	975	950
50	—	50	920	850
—	50	50	840	800

Person 1: Sketch and label the AB binary temperature–composition phase diagram.

Person 2: Sketch and label the BC binary temperature–composition phase diagram.

Person 3: Sketch and label the AC binary temperature–composition phase diagram.

Combine your information to sketch the ABC temperature–composition ternary-phase diagram, assuming that the liquidus and solidus surfaces do not exhibit a maximum or minimum. Now take a vertical section joining 100% A to the midpoint of the BC binary face. Sketch the resulting pseudobinary-phase diagram.

Answer: Each binary-phase diagram will look like Figure 2.4, with the melting point of each component at the appropriate end of each diagram. The ternary-phase diagram is shown below (right), along with the vertical section joining the vertex at A to the 50% BC composition (left).

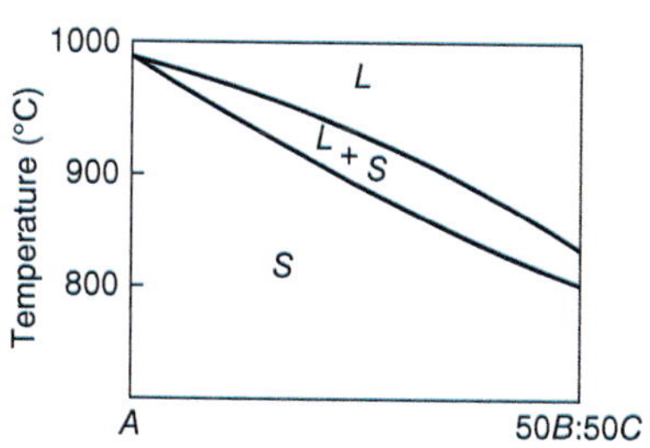

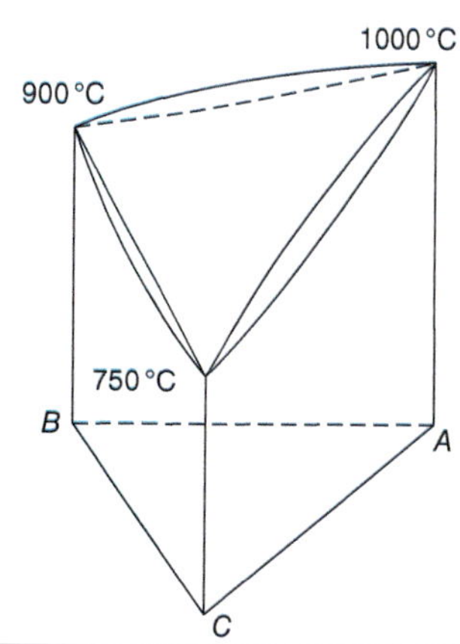

Ceramic Phase Diagrams

Let us now examine a real three-component phase diagram. This could be three elements—for example, copper—with any of several other metallic elements that lead to an important class of alloys called *bronzes*. Copper–zinc–manganese alloys are *manganese bronzes*, copper–tin–nickel alloys are *nickel-tin bronzes*, copper–tin–lead alloys are *leaded tin bronzes*, and so on. But we are not limited only to metals in ternary-phase diagrams. When we introduce nonmetallic elements, such as carbon, nitrogen, and oxygen, we enter a new realm of materials engineering and science—that of ceramics. The compounds formed in these systems are carbides, nitrides, and oxides. For example, consider the copper–iron–oxygen (Cu—Fe—O) phase diagram shown in Figure 2.20. We know that this must be an isothermal section of a more extended diagram, such as that in Figure 2.19, and in this case turns out to be a range of temperatures at which all the phases indicated are stable, namely, above 675°C.

Along the Fe—O binary, we find common ferrous oxides such as *hematite* (Fe_2O_3), *magnetite* (Fe_3O_4), and *wüstite* (FeO). The lines that connect these intermediate binary compounds with the third component are called *alkemade lines*, or alkemades. The alkemades divide ternary systems into composition triangles. For example, the alkemade connecting magnetite with Cu in Figure 2.20 could serve as the edge of a Fe—Cu—Fe_3O_4 phase diagram. In practice, any compound can be used as the vertex of the triangle. Such diagrams that illustrate the phase relationships between important compounds are called *subsystem* diagrams.

If we were to isolate the portion of the diagram along the Fe—O binary and expand the phase diagram along the temperature axis, we would end up with a binary-component phase diagram

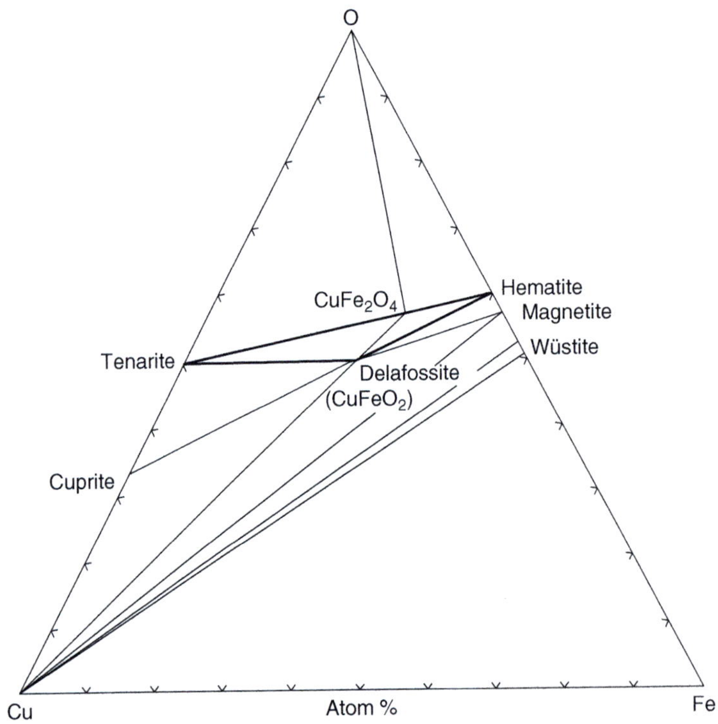

Figure 2.20 Stable assemblages above 675°C for the system Cu—Fe—O. Reproduced by permission, from Ref. [8]. Copyright © 1969, The American Ceramic Society.

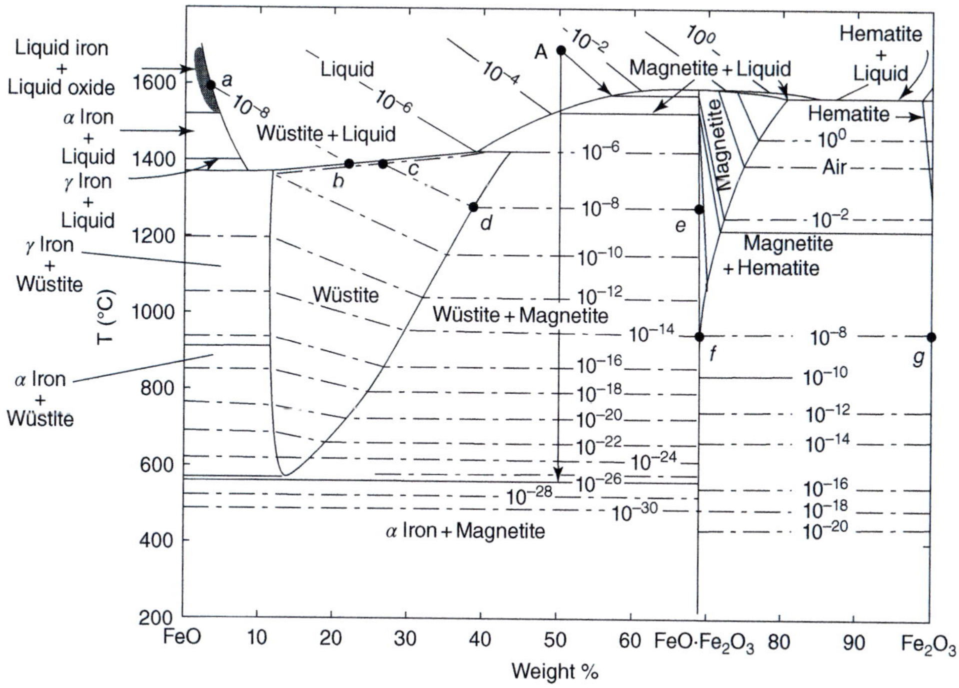

Figure 2.21 The FeO–Fe_2O_3 phase diagram showing oxygen isobars (in atm). Adapted from Kingery et al. [9]. Copyright © 1976 by John Wiley & Sons, Inc. This material is used by permission of John Wiley & Sons, Inc.

like that shown in Figure 2.21, which begins at the compound FeO (the phases are mostly the same from pure Fe to the compound FeO), then continue up to Fe_2O_3, which corresponds to about 30 wt% O. This is in fact how many oxide ceramic phase diagrams are presented—that is, not with oxygen as a continuously varying independent element, but as an integral part of metal oxide compounds that form when it reacts with the metallic components—in this case, ferrous oxides. After all, pure oxygen is in the gaseous form at the temperatures under consideration here. As a result, phase diagrams such as that in Figure 2.21 are often presented at a total pressure of 1 atm, with oxygen partial pressures superimposed on it in the form of *isobars*, or lines of constant partial pressure. The isobars help us follow the variation in equilibrium compositions with temperature under a fixed oxygen partial pressure in the system. This is important from an industrial standpoint, where the partial pressure can be controlled in manufacturing operations. For example, consider a small quantity of liquid iron at 1635°C, saturated with oxygen in a gas reservoir of oxygen partial pressure, $p_{O_2} = 10^{-8}$atm. The volume of the gas reservoir is sufficiently large that any oxygen consumed by oxidation or produced by reduction of the iron or oxides has no effect on the total pressure of the system. The oxygen-saturated liquid iron, located at point *a* in Figure 2.21, is cooled slowly, such that equilibrium is maintained with the gas reservoir. Since we carry out cooling at a constant oxygen partial pressure, we must follow along the isobar, represented by the dotted line *a*–*b*. We are still in the single-phase liquid region, but the dissolved oxygen is now beginning to react with the iron to form FeO:

$$Fe(l) + \frac{1}{2}O_2(g) \rightarrow FeO(l) \tag{2.45}$$

At point *b*, the liquid has mostly turned to wüstite, FeO, which is in the solid phase. From point *b* to point *c*, the remaining liquid becomes solid wüstite. A further decrease in temperature leads through the wüstite solid-phase field along the 10^{-8} atm isobar, until *d* is reached, at which point magnetite, Fe_3O_4 begins to form from the following oxidation reaction:

$$3\,FeO(s) + \frac{1}{2}O_2(g) \rightarrow Fe_3O_4(s) \tag{2.46}$$

This reaction proceeds at 1275°C along the isobar to point *e* on the phase diagram. At this point, only magnetite remains. Magnetite remains a stable compound upon further cooling, to point *f*. Here, the formation of hematite, Fe_2O_3, becomes thermodynamically favorable at this partial pressure:

$$2\,Fe_3O_4(s) + \frac{1}{2}O_2(g) \rightarrow 3Fe_2O_3(s) \tag{2.47}$$

Take a Peek! Learn more about the thermodynamics of metal oxide formation reactions using graphical representations called *Ellingham Diagrams* in ***Advanced Topics***.

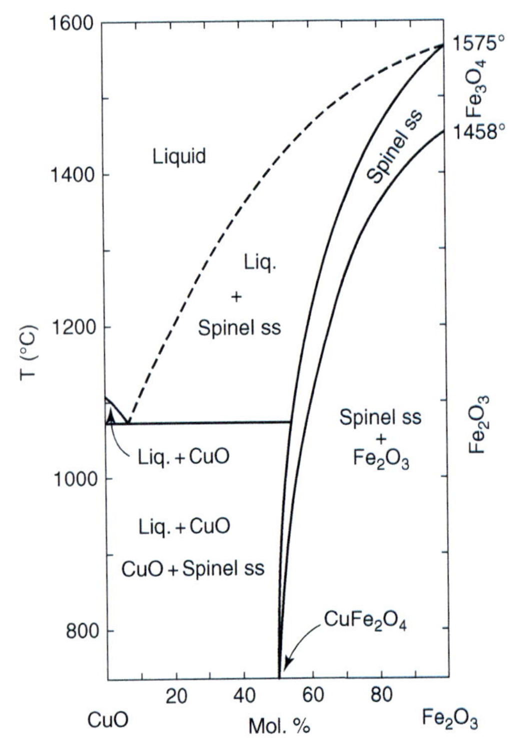

Figure 2.22 Binary-component phase diagram for the system $CuOFe_2O_3$. Reproduced by permission, from Ref. [10]. Copyright © 1969 by The American Ceramic Society.

This reaction proceeds at 875°C until hematite is the only remaining component at point *g*.

Eqs. (2.45) through (2.47) are typical oxidation reactions for the formation of metal oxides. The reverse reactions (steps *g*–*a*) are the reduction of metal oxides to form metal, such as is found in *smelting operations*. It should be apparent that the control of the oxygen partial pressure during heating is an important parameter in determining which phases will form.

As mentioned before, the free energy of mixing of the various components in the system is calculated, and the equilibrium criteria are applied to determine which phases are stable at the indicated temperature. An analysis like the one just performed for the iron–oxygen system can be performed for the copper–oxygen binary system in Figure 2.20, for which there are two stable copper oxides, *cuprite* (cuprous oxide, Cu_2O) and *tenarite* (cupric oxide, CuO). Further inspection of Figure 2.20 shows that copper oxide and iron oxide form two stable compounds: $CuFe_2O_4$ and $CuFeO_2$, respectively.

Recall that we can take vertical "slices" of the ternary-phase diagrams, as well as isothermal (horizontal) slices. If we take, for example, a slice that begins at the tenorite composition (CuO) and extends across to the hematite composition (Fe_2O_3), we would end up with a pseudobinary-phase diagram, which, when plotted on the appropriate temperature–composition axes, would look like Figure 2.22. Note that the compound $CuFe_2O_4$ is present, here labeled as *spinel* (see Section 1.2.2), but there is much more phase and temperature information available to us. This is, in fact, how many metal oxide phase diagrams are presented. The most stable forms of the oxides, or at least the oxides of interest, are selected for the end compositions of the phase diagram, rather than plotting the elemental metals, along with oxygen, at the vertices of an equilateral triangle. In this way, a ternary-phase diagram composed of three elements (Figure 2.20) can be represented as a binary-phase diagram of two compounds (Figures 2.21 and 2.22).

This type of representation allows us to do something even more powerful. We can now take what would normally be a *quaternary*-component system, composed of four elements, and represent it on a ternary-phase diagram system by representing three of the elements as important compounds of the fourth. Again, this is done primarily with metal oxide systems, in which three of the components are metals and the fourth is oxygen. One of the most important ternary-component phase diagram systems represented in this way is for the system $MgO—Al_2O_3—SiO_2$, shown in Figure 2.23.

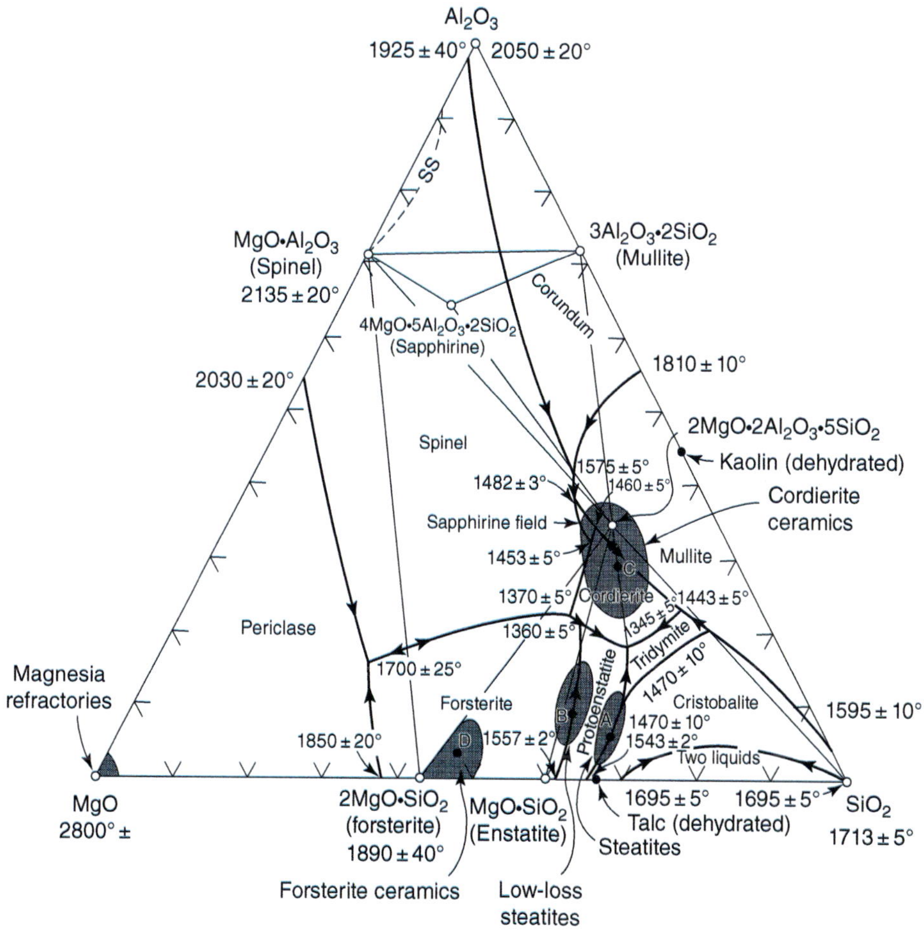

Figure 2.23 The $MgO—Al_2O_3—SiO_2$ phase diagram. Temperatures are in °C. From Kingery et al. [9]. Copyright © 1976 by John Wiley & Sons, Inc. This material is used by permission of John Wiley & Sons, Inc.

This system is of great industrial importance because of the wide variety of naturally occurring compounds it contains. *Silica*, SiO_2, is found in nature in many forms, including sand and quartz. *Alumina*, Al_2O_3, is found as *bauxite*. Notice that we are using a new naming convention. Compounds that contain metallic elements as their root and end in "-a" are common industrial terms for oxides of that metal. *Magnesia* (MgO), *yttria* (Y_2O_3), and *zirconia* (ZrO_2) are a few more examples. This terminology is not universal, especially when the metal can have more than one common oxidation state. Many of the compounds along the Al_2O_3—SiO_2 binary, termed *aluminosilicates*, are the building blocks of most clays. *Kaolin*, for example, is a hydrated aluminosilicate of nominal composition $Al_2O_3{\cdot}2SiO_2{\cdot}2H_2O$, sometimes expressed as the compound $Al_2Si_2O_5(OH)_4$. It is used for making porcelain for valves, tubes, and fittings; as a refractory for bricks and furnace linings; and as a pigment filler in paints. Recall, also, from Chapter 1 that it is also a type of exfoliated clay in polymer–clay nanocomposites. *Talc* is a hydrated form of magnesium silicate, $3MgO{\cdot}4SiO_2{\cdot}H_2O$, and can be dehydrated to form *protoenstatite* crystals, $MgSiO_3$, in a silica matrix. *Cordierite* ceramics have variable compositions as shown in the shaded region of Figure 2.23 but have a nominal composition of $2MgO{\cdot}2Al_2O_3{\cdot}5SiO_2$. Cordierite ceramics are particularly useful due to their low coefficient of thermal expansion (see Section 5.1.3) and thermal shock resistance. Many of the ceramics in this system are termed *refractories*, since they have very high melting points, generally above 1580°C, and can bear loads and resist corrosion at elevated temperatures.

Example Problem 2.2

Yttrium–barium–copper oxide (YBCO) superconductors were one of the first class of metal oxides to be found to superconduct (see Section 6.0.2) at liquid nitrogen temperatures (above about 77 K). Not all YBCO will superconduct, however, owing to the complex dependence of this phenomenon on the copper oxide lattice. For example, the "2–1–1" phase, named after the relative molar quantities of the three metals, Y_2BaCuO_5, does not superconduct, but the "1–2–3" phase, $YBa_2Cu_3C_7$, does. A portion of the Y_2O_3—BaO—CuO ternary-phase diagram is shown below, with the corresponding pseudobinary-phase diagram connecting the 1–2–3 compound with point F in the ternary-phase diagram next to it. Determine the amount of 1–2–3 formed when a liquid of composition shown in the hatched region of the binary diagram (about 88 mol% $BaCu_3O_4$) is cooled to the vicinity of 960°C.

Answer: Application of the lever rule in the 1–2–3 + liquid region shows that even in the best case (at the left edge of the hatched region), only about 3% of the 1–2–3 phase is formed, the rest being liquid of different composition. As the liquid is cooled further, more yttria-poor compounds are formed, such as barium cuprate (BC), that are difficult to remove. Thus, melt processing of pure 1–2–3 superconductors is virtually impossible via this route. One solution has been to use nonequilibrium processing, which renders the equilibrium phase diagrams shown below useless.

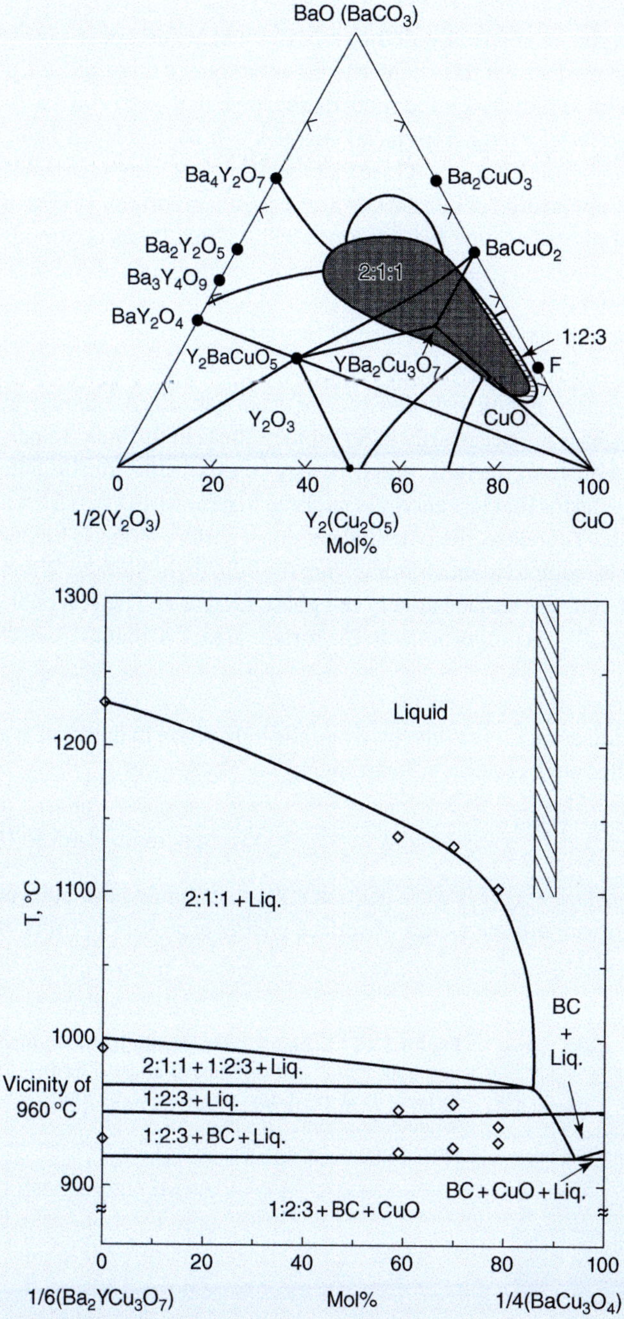
BaO ($BaCO_3$)
$Ba_4Y_2O_7$
Ba_2CuO_3
$Ba_2Y_2O_5$
$BaCuO_2$
$Ba_3Y_4O_9$
2:1:1
BaY_2O_4
1:2:3
Y_2BaCuO_5
$YBa_2Cu_3O_7$
F
Y_2O_3
CuO
0
20
40
60
80
100
1/2(Y_2O_3)
$Y_2(Cu_2O_5)$
Mol%
CuO
1300
1200
1100
1000
900
T, °C
Liquid
2:1:1 + Liq.
BC
+
Liq.
2:1:1 + 1:2:3 + Liq.
Vicinity of
960 °C
1:2:3 + Liq.
1:2:3 + BC + Liq.
BC + CuO + Liq.
1:2:3 + BC + CuO
0
20
40
60
80
100
1/6($Ba_2YCu_3O_7$)
Mol%
1/4($BaCu_3O_4$)

2.2.2 Interfacial Thermodynamics

Thermodynamics plays a fundamental role in practically all the processes that are described in this book. Phase equilibria, which has been the focus of this chapter so far, is but one of these areas. Let us diverge from phase equilibria for a bit and discuss a thermodynamic topic that will be of use in many of the subsequent chapters: *interfacial energy*. Interfacial energies will be developed using thermodynamic arguments primarily for liquids, but we will see that the results are general to solids as well. This point will be illustrated by application of interfacial energies to an important process in the densification of commercial materials called *sintering*. Later, we will see that it has enormous utility in the description of many surface-related phenomena in materials science, such as fracture, adhesion, lubrication, and reaction kinetics.

Surface Energy

A *surface* is an inhomogeneous boundary region between two adjacent phases. As shown in Figure 2.24, atoms on the surface of a phase are necessarily different than those in the bulk. In particular, they have fewer nearest neighbors than the bulk, and they may be exposed to constituents from an adjacent phase. This generally means that less energy is required to remove an atom from a surface than to remove it from the bulk. Therefore, the potential energy of surface atoms is higher than that of bulk atoms. In turn, work is required to move atoms from the bulk to the surface. When this is done, a new surface is created, and the surface area of the phase increases. The reversible work required to form the new surface, dW_s, is proportional to the surface area, dA, that is created:

$$dW_S = \gamma dA \tag{2.48}$$

The proportionality constant, γ, is called the *surface energy*. When the bulk phase in question is a liquid, the surface energy is often called *surface tension*. Rearranging Eq. (2.48) gives

$$\gamma = \frac{dW_S}{dA} \tag{2.49}$$

We can see from this relation that surface energy is work per unit area, so it should have units of J/m^2 in SI and $ergs/cm^2$ in cgs. Often you will see values of surface energy expressed in units of dyne/cm, where 1 dyne = 1 erg/cm.

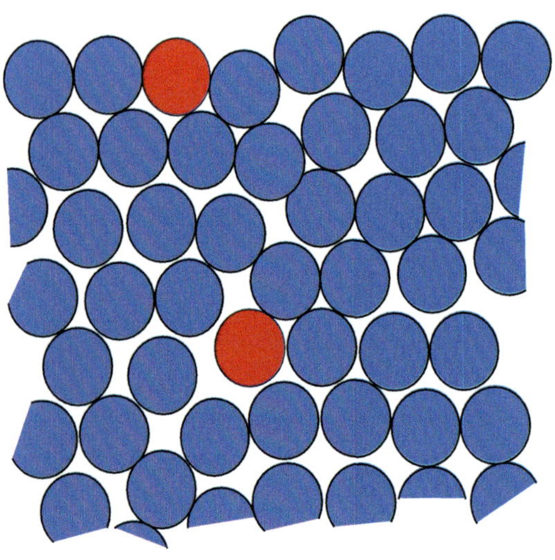

Figure 2.24 Schematic representation of surface and bulk atoms in a condensed phase. From Kingery et al. [9]. Copyright © 1976 by John Wiley & Sons, Inc. This material is used by permission of John Wiley & Sons, Inc.

With the aid of Eqs. (2.1), (2.7), (2.9), and (2.11), you should be able to prove to yourself that the reversible, nonpressure–volume work, dW_s, is equivalent to the free energy change, dG, so that Eq. (2.49) becomes (with proper use of partial differentials)

$$\gamma = \left(\frac{\partial G}{\partial A}\right)_{T,P,N_i} \tag{2.50}$$

This relationship identifies the surface energy as the increment of the Gibbs free energy per unit change in area at constant temperature, pressure, and number of moles. The path-dependent variable dW_s in Eq. (2.49) has been replaced by a state variable, namely, the Gibbs free energy. The energy interpretation of γ has been carried to the point where it has been identified with a specific thermodynamic function. As a result, many of the relationships that apply to G also apply to γ:

$$\gamma = H_S - TS_S \tag{2.51}$$

where the subscript s on the enthalpy and entropy indicates that these are surface properties. As mentioned earlier, the surface atoms are fundamentally different than bulk atoms, so that they have different enthalpies and entropies associated with them. Differentiation of Eq. (2.51) with respect to temperature at constant pressure gives

$$\left(\frac{\partial \gamma}{\partial T}\right)_P = -S_S \tag{2.52}$$

Substitution of Eq. (2.51) in Eq. (2.52) gives

$$\gamma = H_S + T\left(\frac{\partial \gamma}{\partial T}\right)_P \tag{2.53}$$

Eq. (2.53) is useful from an experimental standpoint because the measurements of surface energies at various temperatures can, in principle, provide a measurement of the surface enthalpy. The surface enthalpy, H_S, can also be determined directly because it is equivalent to the heat of sublimation or vaporization.

Since the surface energy is a direct result of intermolecular forces, its value will depend on the type of bond and the structural arrangement of the atoms. Normally, densely packed planes would have lower surface energies. Liquid hydrocarbons having only weak van der Waals forces have values of surface tension of 15–30 dyn/cm, while liquids with polar forces and hydrogen bonds, like water, have surface tensions in the range of 3–72 dyn/cm. Fused salts and glasses with additional ionic bonds have surface energies from 100 to 600 dyn/cm, and molten metals have surface tensions from 100 to 3000 dyn/cm. Even higher surface energies can be found in covalent solids (see Example Problem 2.3). See Appendix 3 for values of solid and liquid surface energies for a variety of materials.

The Laplace Equation

The concept of surface energy allows us to describe several naturally occurring phenomena involving liquids and solids. One such situation that plays an important role in the processing and application of both liquids and solids is the pressure difference that arises due to a curved surface, such as a bubble or spherical particle. For the most part, we have ignored pressure effects, but for the isolated surfaces under consideration here, we must take pressure into account.

Consider a generic curved surface such as that found in a sphere or cylinder (see Figure 2.25). The curved surface has two principal radii of curvature, R_1 and R_2. The front of this surface is indicated

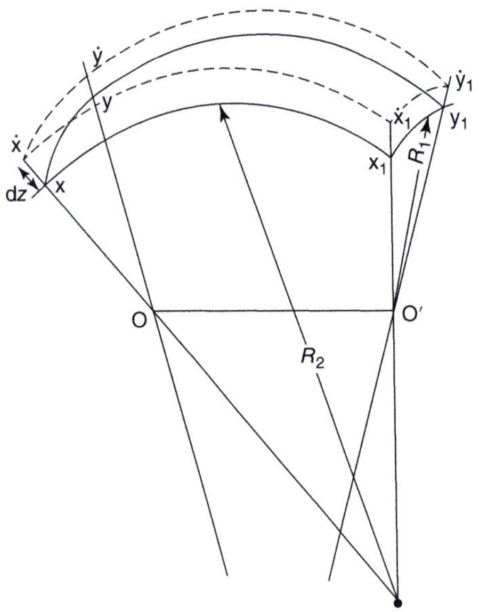

Figure 2.25 An element of a curved surface with principal radii R_1 and R_2. Reproduced by permission, from Padday [11]. Copyright © 1969 by John Wiley & Sons, Inc.

by the line xx_1, and the back is represented by the line yy_1. Let us now move the surface out by a differential element, dz. The front of the new surface is now given by $\dot{x}\dot{x}_1$, and the back is indicated by $\dot{y}\dot{y}_1$. The work required to displace the surface this amount is supplied by a pressure difference, ΔP. The pressure acts on an area given by $(xx_1)(x_1y_1)$ moving through the differential element dz, such that the total pressure–volume work associated with extending the surface through dz is:

$$W = \Delta P(xx_1)(x_1y_1)dz \tag{2.54}$$

The pressure–volume work must be counterbalanced by surface tension forces. The work required to move against surface tension forces is best calculated by breaking it into two parts, W_1 and W_2. W_1 is the work required to move side xx_1 away from yy_1 a distance $(x_1y_1/R_1)\,dz$ during the expansion:

$$W_1 = \frac{\gamma(xx_1)(x_1y_1)dz}{R_1} \tag{2.55}$$

Example Problem 2.3

An alternative method for estimating surface energies is to calculate the work required to separate two surfaces of a crystal along a certain crystallographic plane. At 0 K, this work can be approximated as the energy required to break the number of bonds per unit area, or the energy of cohesion, E_{coh}. Since two surfaces are being formed in this cleavage process, the surface energy of a single surface is then

$$\gamma \approx \frac{1}{2}E_{coh}$$

Let us consider, for example, separation along the (111) plane in diamond. The lattice constant for diamond is a = 3.56 Å (see Table 1.12), so the number of atoms per square centimeter on the (111) surface is

$$\frac{2\text{ atoms per plane}(111)}{\sqrt{3}a^2/2} = \frac{4}{\sqrt{3}(3.56\times10^{-8})^2}$$
$$= 1.82\times10^{15}\text{atoms/cm}^2$$

From Table 1.5, the bond energy for C—C bonds is 605.03 kJ/mol, so that the cohesive energy is:

$$E_{coh} = \frac{(1.82\times10^{15}\text{bond/cm}^2)(605{,}030\text{ J/mol})(10^7\text{ erg/J})}{6.02\times10^{23}\text{ atoms/mol}}$$
$$= (18{,}290\text{ erg/cm}^2)$$

so that the surface energy is

$$\gamma = 1/2(18{,}290\text{ erg/cm}^2) = 9{,}145\text{ erg/cm}^2$$

Similarly, the work required to move side xy away from x_1y_1 through a distance $(xx_1/R_2)\,dz$ is W_2:

$$W_2 = \frac{\gamma(x_1y_1)(xx_1)dz}{R_2} \tag{2.56}$$

Adding Eqs. (2.55) and (2.56) together to arrive at the work against surface tension, equating them with the pressure–volume work in Eq. (2.54) and simplifying leads to the *Laplace equation*:

$$\Delta P = \gamma\left(\frac{1}{R_1} + \frac{1}{R_2}\right) \tag{2.57}$$

The Laplace equation in this form is general and applies equally well to geometrical bodies whose radii of curvature are constant over the entire surface to more intricate shapes for which the Rs, are a function of surface position. In the instance of constant radii of curvature across the surface, Eq. (2.57) reduces for several common cases. For spherical surfaces, $R_1 = R_2 = R$, where R is the radius of the sphere, and Eq. (2.57) becomes

$$\Delta P = \frac{2\gamma}{R} \tag{2.58}$$

For a cylindrical surface, R_1 (or R_2) is infinity, so the remaining radius, R, is the radius of the cylinder and:

$$\Delta P = \frac{\gamma}{R} \tag{2.59}$$

Finally, for a planar surface, both radii of curvature become infinity, and

$$\Delta P = 0 \tag{2.60}$$

The pressure difference may also be numerically 0 in the instance where the two principal radii of curvature lie on opposite sides of the surface, such as in the case of a saddle.

Need More Help?

Try these videos on surface tension, and the Laplace and Young equations:

https://www.youtube.com/watch?v=YdLeLvNfE-w
https://www.youtube.com/watch?v=PezcU4hHML4
https://www.youtube.com/watch?v=9WigcUFS3eQ

The Young Equation

The principle of balancing forces used in the derivation of the Laplace equation can also be used to derive another important equation in surface thermodynamics, the *Young equation*, sometimes called *Young's equation*. Consider a liquid droplet in equilibrium with its own vapor and a flat, solid surface at constant temperature, as shown in Figure 2.26. The liquid–solid, liquid–vapor, and solid–vapor *interfacial surface energies* are defined as γ_{LS}, γ_{LV}, and γ_{SV}, respectively. Technically, the

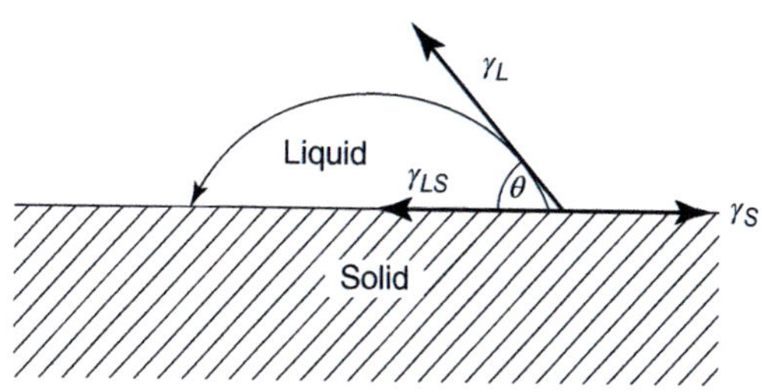

Figure 2.26 Schematic diagram of liquid droplet on solid surface. From Jastrzebski [12]. Copyright © 1976 by John Wiley & Sons, Inc. This material is used by permission of John Wiley & Sons, Inc.

liquid–vapor and solid–vapor interfacial energies should be for the liquid and solid in equilibrium with their respective vapors, which are probably not the same. In practice, however, the vapor is usually a nonreactive gas, so γ_{LV} and γ_{SV} become the surface tension and solid surface energy in the gas of interest and are simply labeled γ_L and γ_S, respectively, as shown in Figure 2.26. These are both approximations. The surface tensions and surface energies of some common liquids and solids are listed in Appendix 3.

At equilibrium, an angle θ, called the *contact angle*, is formed at the three-phase solid–liquid–gas junction. The contact angle can have values from 0 to 180°. When $\theta = 0$, the liquid completely spreads off the solid surface, forming a thin monolayer (see Figure 2.27). Such a condition is termed *wetting*. The other extreme is called *nonwetting*, and it occurs when the entire liquid droplet sits as a sphere on the solid. All values of contact angle in between these two extremes are theoretically possible. Obviously, the chemical nature of the liquid, solid, and vapor determines the extent to which the liquid will wet the solid, and temperature has an important influence. Keep in mind that the contact angle under consideration here is an *equilibrium contact angle*. There are also *dynamic*, *receding*, and *advancing contact angles* associated with droplets as they spread and move on substrates. For now, the term contact angle will refer to the equilibrium condition.

The three interfacial surface energies, as shown at the three-phase junction in Figure 2.26, can be used to perform a simple force balance. The liquid–solid interfacial energy plus the component of the liquid–vapor interfacial energy that lies in the same direction must exactly balance the solid–vapor interfacial energy at equilibrium:

$$\gamma_L \cos\theta + \gamma_{SL} = \gamma_S \tag{2.61}$$

This equation is *Young's equation*; it is named after Thomas Young, who first proposed it in 1805. The derivation presented here in terms of force balances is simplistic, but there are more rigorous thermodynamic arguments to support its development.

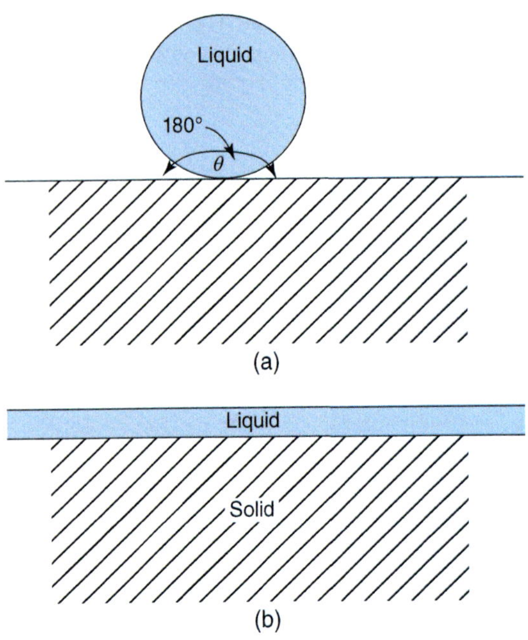

Figure 2.27 Schematic illustration of (a) nonwetting and (b) wetting of a liquid on a solid. From Jastrzebski [12]. Copyright © 1976 by John Wiley & Sons, Inc. This material is used by permission of John Wiley & Sons, Inc.

Figure 2.28 A water droplet on a lotus leaf illustrating a superhydrophobic surface in nature. keetareel0/74 images/Pixabay.

In practice, the contact angle can be experimentally determined in a rather routine manner, as can the liquid surface tension and even the solid surface energy. The interfacial energy for the liquid–solid system of interest, γ_{SL}, can then be calculated using Young's equation. Alternatively, if γ_{SL}, γ_L, and γ_S are known as a function of temperature, the contact angle can be predicted at a specified temperature.

An important application of spreading and wetting on surfaces arises when the contact angles are very high (>150°) and the difference between the advancing and receding contact angles—called *hysteresis*—is very low. When these conditions are met, the surface is said to be *superhydrophobic*. Superhydrophobicity occurs in nature like on the lotus leaf (Figure 2.28) but can also be produced artificially on surfaces through both structural and chemical modifications. This is a growing area of materials and nanoscience research as superhydrophobic surfaces have many potential industrial and commercial applications in the areas of self-cleaning and antifouling surfaces.

Need More Help?
Try this video on superhydrophobic surfaces: https://www.youtube.com/watch?v=FLegmQ8_dHg

Sintering and Densification

Curved surfaces and surface energies play a large role in the description of an important phenomenon in ceramic processing called *sintering*. Sintering is a process in which a particulate material is consolidated during heat treatment. Consolidation implies that within the material, particles have joined together into an aggregate that has strength. *Densification*, or the increase in density through the accompanying reduction in interparticle voids, often accompanies sintering, but there are some highly porous insulation products that can actually be less dense after they have been sintered. Nonetheless, we will examine sintering and densification simultaneously. We will also limit the discussion to solid-state sintering.

The driving force for sintering is the reduction in the total free energy of the particulate system, ΔG, which is composed of free energy changes of volume, ΔG_V, boundaries, ΔG_b, and surfaces, ΔG_s:

$$\Delta G = \Delta G_V + \Delta G_b + \Delta G_S \tag{2.62}$$

In conventional sintering, the dominant term in the free energy reduction is that due to surface area reduction. Recall from Eq. (2.50) that the surface energy is the increment of the Gibbs free energy per unit change in area at constant temperature, pressure, and total number of moles

$$\gamma = \left(\frac{\partial G}{\partial A}\right) \tag{2.63}$$

such that if A_s is the surface area of the particles, then the free energy change due to surface area is

$$\Delta G_S = \gamma \Delta A_s \tag{2.64}$$

Three stages of solid-state sintering are recognized, although there is not a clear distinction between them, and they overlap to some extent.

- *Initial sintering* involves rearrangement of the powder particles and formation of a strong bond or neck at the contact points between particles (see Figure 2.29). The relative density of the powder may increase from 0.5 to 0.6, due mostly to an increase in particle packing.
- *Intermediate sintering* occurs as the necks grow, the number of pores decreases substantially, and shrinkage of the particle assembly occurs. Grain boundaries form, and some grains grow while others shrink. This stage continues while the pore channels are connected, called *open porosity*, but is considered over when the pores are isolated or in closed porosity. The majority of shrinkage occurs during intermediate sintering, and the relative density at the end of this stage is about 0.9.
- *Final-stage sintering* occurs as the pores become closed and are slowly eliminated by diffusion of vacancies from the pores along the grain boundaries. Little densification occurs in this stage, although grain sizes continue to grow.

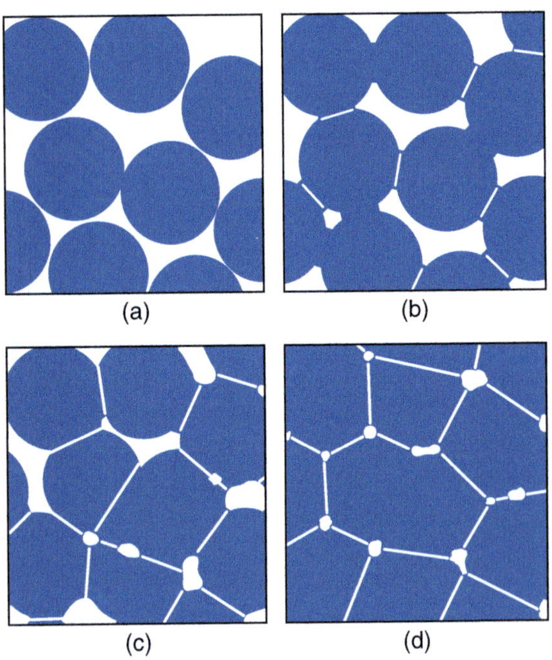

Figure 2.29 Development of ceramic microstructure during sintering: (a) loose powder particles; (b) initial stage; (c) intermediate stage; and (d) final stage. German, R.M., 1996. Sintering theory and practice (p. 568) / with permission of John Wiley & Sons.

During the initial stage, pore rounding causes a decrease in ΔG_s in proportion to the reduction of the surface area, as dictated by Eq. (2.64), but the formation of a grain boundary causes an increase in ΔG_b. The angle of intersection of the pore at the pore–grain boundary juncture (see Figure 2.30) is called the *dihedral angle*, ϕ, and is related to the relative interfacial energies of the grain boundary, γ_{gb}, and the solid surface, γ_s

$$\cos\left(\frac{\phi}{2}\right) = \frac{\gamma_{gb}}{2\gamma_s} \tag{2.65}$$

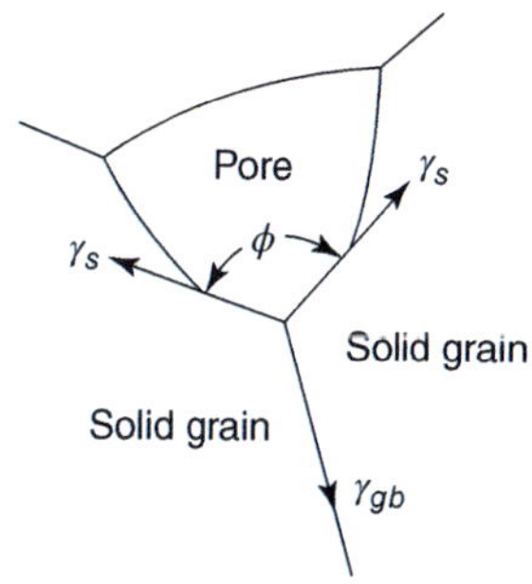

Figure 2.30 Schematic illustration of dihedral angle for solid–pore interaction.

Values of ϕ range from about 105° to 113° in ceramics, implying that γ_{gb}/γ_s is 1.1–1.2, whereas in metals this ratio ranges from 0.25 to 0.5.

There is much more to be said concerning sintering, especially the role that diffusion plays. Many of the models that describe the sintering processes are based on the diffusivity of solid species, which are addressed in Section 4.3. The practical aspects of the sintering of metals and ceramics are described in Sections 7.1 and 7.2, respectively.

2.3 Thermodynamics of Polymers

Polymers do not behave like the atoms or compounds that have been described in the previous sections. We saw in Chapter 1 that their crystalline structure is different from that of metals and ceramics, and we know that they can in many cases form amorphous structures just as easily as they crystallize. In addition, unlike metals and ceramics, whose thermodynamics can be adequately described in most cases with theories of mixing and compound formation, the thermodynamics of polymers involves solution thermodynamics—that is, the behavior of the polymer molecules in a liquid solvent. These factors contribute to a thermodynamic approach to describing polymer systems that is necessarily different from that for simple mixtures of metals and compounds. Rest assured that free energy will play an important role in these discussions, just as it has in previous sections, but we are now dealing with highly inhomogeneous systems that will require some new parameters.

2.3.1 Solution Thermodynamics and Phase Separation

Most polymers are synthesized in solution. The solvent could be water, carbon tetrachloride, benzene, or any number of solvents, both polar and nonpolar. Many polymers are also processed into usable shapes in solution during and after synthesis. They can be cast as solutions wherein the solvent is allowed to evaporate to leave only the polymer. So, it is critically important to understand how a polymer molecule interacts with the solvent. Again, intermolecular forces will play an important role here, but as always free energy will be used to determine the stability of any system we can create at any temperature.

Recall that the free energy of mixing for a binary-component system is

$$\Delta G_{mix} = \Delta H_{mix} - T\Delta S_{mix} \tag{2.29}$$

and that for regular solutions, this relation becomes

$$\Delta G_{mix} = \alpha X_A X_B + RT(X_A \ln X_A + X_B \ln X_B) \tag{2.33}$$

The first term in Eq. (2.33) is the enthalpy of mixing, and the second term is the entropy of mixing. Let us examine each of these terms independently, beginning with the entropy of mixing.

For solutions of a polymer in a solvent, it can be shown—again through statistical thermodynamic arguments—that the entropy of mixing is given by

$$\Delta S_{mix} = -R(X_A \ln \nu_A + X_B \ln \nu_B) = -k_B(N_A \ln \nu_A + N_B \ln \nu_B) \tag{2.66}$$

Here k_B is Boltzmann's constant, or the gas constant per molecule, R/N_0, where N_0 is Avogadro's number (or $N_A + N_B$ for one mole of solution) and ν_A and ν_B are the volume fractions of the solvent and polymer. Comparison of Eq. (2.66) with the entropy of mixing presented earlier in the chapter by Eq. (2.30) shows that they are similar in form, except that now the volume fractions of the components, υ_A and υ_B, are found to be the most convenient way of expressing the entropy change for polymers, rather than the mole fraction used for most small molecules. This change arises from the differences in size between the large polymer molecules and the small solvent molecules which would normally mean mole fractions close to unity for the solvent, especially when dilute solutions are being studied.

Need More Help?

Try these videos on ideal polymer solutions:

https://www.youtube.com/watch?v=sPnQpTMPVfQ

https://www.youtube.com/watch?v=BLIpMfRalkg

and the Flory-Huggins interaction parameter:

https://www.youtube.com/watch?v=dB74U8idZOs

It can also be shown with related statistical arguments that the enthalpy of mixing for a regular polymer–solvent solution is given by

$$\Delta H_{mix} = k_B T \chi N_A \nu_B = RT\chi \nu_A \nu_B \tag{2.67}$$

where N_A is the number of solvent molecules, and ν_A and ν_B are the volume fraction of solvent and polymer molecules, respectively. Compare the second form of Eq. (2.67) with Eq. (2.31) and you will see that they have the same general form, except that the mole fractions have been replaced with volume fractions, just as for the entropy of mixing, and the interaction energy, α, is composed of the gas constant, R, temperature, T, and a new parameter, χ, which is called the *Flory–Huggins interaction parameter*, or just interaction parameter (do not confuse the interaction parameter with electronegativity, which goes by the same symbol!). Note that the interaction parameter is the interaction energy per solvent molecule per $k_B T$ and is inversely proportional to temperature. It is 0 for ideal mixtures (zero enthalpy of mixing), positive for endothermic mixing, and negative for exothermic mixing. The interaction parameter is an important feature of polymer solution theory. It is an indication of how well the polymer interacts with the solvent. "Poor" solvents have values of χ close to 0.5, while "good" solvents have lower, or negative, values. The typical range for most synthetic polymer solutions is $0.3 < \chi < 0.6$ (see Table 2.5). A linear temperature dependence for χ is also predicted of the form $\chi = a + b/T$, which suggests that as the temperature increases, the solvating power of the liquid should increase.

Table 2.5 Some Polymer–Solvent Interaction Parameters at 25°C

Polymer	Solvent	Interaction Parameter, χ
cis-Polyisoprene	Toluene ($V_1 = 106$)[a]	0.391
	Benzene ($V_1 = 89.0$)	0.437
Polyisobutylene	Toluene	0.557
	Cyclohexane ($V_1 = 108$)	0.436
Butadiene–styrene	Benzene	0.442
71.5:28.5	Cyclohexane	0.489
Butadiene–acrylonitrile		
82:18	Benzene	0.390
70:30	Benzene	0.486
61:39	Benzene	0.564

Source: F. Rodriguez, *Principles of Polymer Systems*, 2nd ed. Copyright © 1982 by McGraw-Hill Book Company.
[a] Note that solvent molar volume, V_1, is in cubic centimeters per mole.

Cooperative Learning Exercise 2.5

It can be shown that the volume fraction of polymer in solution, v_B, can be related to the mole fraction of polymer in solution, X_B, as follows:

$$v_B = \frac{X_B}{(1/x_n) + X_B[1 - (1/x_n)]}$$

where n is the number of repeat units in a polymer chain or the degree of polymerization (cf. Section 1.3.5). [This is a good thing to sit down and prove to yourself sometime. Just recall that $X_B = N_B/(N_A + N_B)$ and $v_B/v_A = x_n N_B/N_A$.] For simplicity, let us assume a monodispersed polymer of $x_n = 500$; that is, all the polymer chains have exactly 500 repeat units.

First, work together to create a table of mole fractions in increments of 0.1 from 0.1 to 0.9; for example, $X_B = 0.1$, 0.2, and so on, and their corresponding volume fractions using the equation above. Then, perform the following calculations separately.

Person 1: Use Eq. (2.30) to calculate values of $\Delta S_{mix}/R$ for the nine values of X_B and X_A. These values are the entropy of mixing for an ideal solution.

Person 2: Use Eq. (2.66) to calculate values of $\Delta S_{mix}/R$ for the nine values of v_B and v_A. [*Hint*: Use the first form of Eq. (2.66).] These values are the free energy of mixing for a real "athermal" polymer solution.

Combine your information to make a single, rough plot of $\Delta S_{mix}/R$ versus X_B. At what mole fraction is the difference in the entropy of mixing between an ideal and athermal polymer solution the greatest for this system?

Answer:
$X_B = 0.1$; $X_A = 0.9$; $v_B = 0.982$; $v_A = 0.018$; max difference occurs about $X_B = 0.2$. Polymer solutions are highly nonideal.

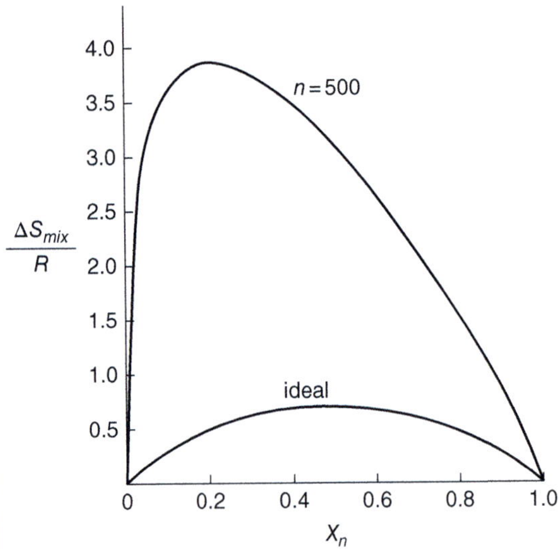

Adapted from P. Hiemenz, *Polymer Chemistry*, p. 520. Copyright © 1984 by Taylor & Francis.

Combining the enthalpy of mixing and the entropy of mixing, along with substituting them into Eq. (2.29), gives us the free energy of mixing for a polymer solution:

$$\Delta G_{mix} = k_B T[\chi N_A v_B + N_A \ln v_A + N_B \ln v_B] \tag{2.68}$$

From this expression, one can derive many useful relations involving experimentally obtainable quantities, such as osmotic pressure. We will use it to develop a phase diagram for a polymer solution.

Consider a binary-component system consisting of a liquid, A, which is a poor solvent for a polymer, B. We already know that complete miscibility occurs when the Gibbs free energy of mixing is less than the Gibbs free energies of the components, and the solution maintains its homogeneity only as long as ΔG_{mix} remains less than the Gibbs free energy of any two possible coexisting phases. This situation is shown in Figure 2.31. As with our other phase diagrams, the free energy is plotted as a function of composition in the top part of the curve, and the resulting temperature–composition plot is shown in the bottom of the diagram. At high temperatures, T_5, the polymer and solvent are miscible, and a single phase is formed (Region I). As T decreases, the solution begins to separate into two phases. Phase separation, or *immiscibility*, begins to occur at the critical temperature, T_c. Here, the free energy curve begins to have maxima, minima, and inflection points, and the condition of phase stability as given by Eq. (2.36), namely, that the chemical potentials of the two phases (here labeled L1 and L2) be equal, is no longer met:

$$\mu_{L1} = \left(\frac{\partial \Delta G_{mix}}{\partial v_B}\right)_{T,P,n_{L2}} = \mu_{L2} = \left(\frac{\partial \Delta G_{mix}}{\partial v_B}\right)_{T,P,n_{L1}} \tag{2.36}$$

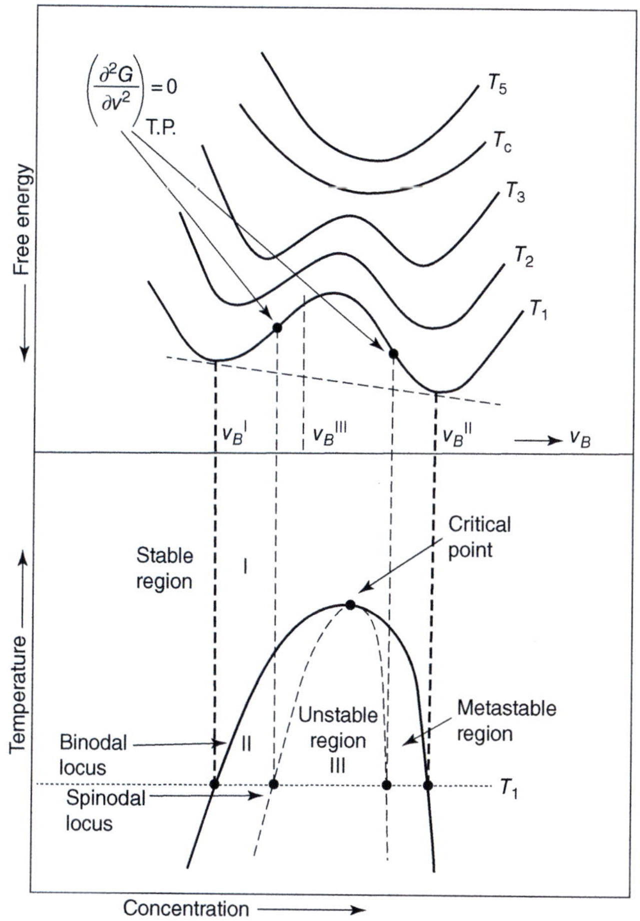

Figure 2.31 Polymer–solvent phase diagram showing binodal, spinodal, and miscibility gap. Reproduced, by permission, from Cowie [14]. Copyright © 1991 by Taylor & Francis.

(You should prove to yourself that it makes no difference if we differentiate with respect to component A or B, nor whether we differentiate with respect to mole or volume fraction, since the derivatives are set equal to 0.) The locus of points swept out by the stability condition shown in Eq. (2.36) at various temperatures is called the *binodal* or *cloud point curve*. For example, just as we drew tangents to the minima of the free energy curves in Figures 2.3 and 2.6 to map out the phase boundaries in solid–liquid equilibria, we can use tangents to the minima (line A–B) in a single free energy–composition plot for our polymer solution to establish the phase boundary—in this case, the binodal. A similar mapping of the inflection points, mathematically described by the second derivative of the free energy versus composition curve given in

$$\left(\frac{\partial^2 \Delta G_{mix}}{\partial v_B^2}\right) = 0 \tag{2.37}$$

generates the *spinodal*. The region between the binodal and spinodal curves (Region II) represents the region of two-phase metastability. Below the spinodal (Region III), the solution is unstable, phase separation occurs, and two liquid solutions are formed whose relative amounts are given by the lever rule. Note that as temperature increases, the solvating power of the solvent increases, and a single polymer–solvent phase becomes stable.

We can substitute our expression for the free energy of mixing in Eq. (2.68) into the stability criteria of Eqs. (2.36) and (2.37) to solve for the critical interaction parameter at the onset of phase separation, χ_c:

$$\chi_c = \frac{1}{2v_{A,c}^2} \tag{2.69}$$

where $v_{A,c}$ is the volume fraction of the solvent at the critical point.

Take a Peek! Learn more about cohesive energy density and the thermodynamics of polymer blends in ***Advanced Topics***.

The instance we have considered here, that of a polymer in a poor solvent, results in an *upper critical solution temperature* (UCST) as shown in Figure 2.31. This occurs due to (a) decreased attractive forces between like molecules at higher temperatures and (b) increased solubility. For some systems, however, a decrease in solubility can occur, and the corresponding critical temperature is located at the minimum of the miscibility curve, resulting in a *lower critical solution temperature* (LCST). This situation is illustrated in Figure 2.32.

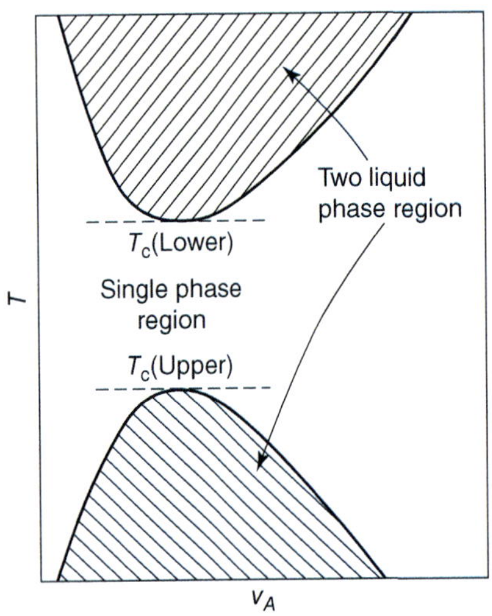

Figure 2.32 Schematic diagram of two-phase regions resulting in UCST (*bottom*) and LCST (*top*). Reproduced, by permission, from Cowie [15]. Copyright © 1991 by Taylor & Francis.

2.4 Thermodynamics of Composites

Much of what we need to know about the thermodynamics of composites has been described in the previous sections. For example, if the composite matrix is composed of a metal, ceramic, or polymer, its phase stability behavior will be dictated by the free energy considerations of the preceding sections. Unary, binary, ternary, and even higher-order phase diagrams can be employed as appropriate to describe the phase behavior of both the reinforcement or matrix component of the composite system. At this level of discussion on composites, there is really only one topic that needs some further elaboration: a thermodynamic description of the interphase. As we did back in Chapter 1, we will reserve the term "interphase" for a phase consisting of three-dimensional structure (e.g., with a characteristic thickness) and will use the term "interface" for a two-dimensional surface. Once this topic has been addressed, we will briefly describe how composite phase diagrams differ from those of the metal, ceramic, and polymer constituents that we have studied so far.

Adhesion, Cohesion, and Spreading

We will see in subsequent chapters—particularly those concerning mechanical properties—that the physical properties of a composite are dictated to a large degree by the interphase between the matrix and reinforcement phases. The interphase plays an increasingly larger role as its proportion relative to the other constituents increases; for example, the surface-to-volume ratio of the reinforcement increases as its characteristic dimension decreases. As a result, it is important to understand the energetics of interphase formation. This is typically done by dividing the formation process into a combination of three possible processes: *adhesion*, *cohesion*, and *spreading*.

Let us consider two hypothetical phases in our composite, A and B, without specifying their physical state. They could be a polymer melt and a glass fiber reinforcement during melt infiltration processing, a metal powder and ceramic powder that are being subjected to consolidation at elevated temperature and pressure, or two immiscible polymer melts that will be coextruded and solidified into a two-phase, three-dimensional object. In any case, the surface that forms between the two phases is designated AB, and their individual surfaces that are exposed to their own vapor, air, or inert gas (we make no distinction here) are labeled either A or B. The following three processes are defined as these surfaces interact and form

Cohesion	No surface $\rightarrow$ 2A (or 2B) surfaces
Adhesion	1AB surface $\rightarrow$ 1 A + 1 B surface
Spreading (B on A)	1 A surface $\rightarrow$ 1 AB + 1 B surface

These processes are schematically illustrated in Figure 2.33. We are interested in the free energy change that accompanies each process.

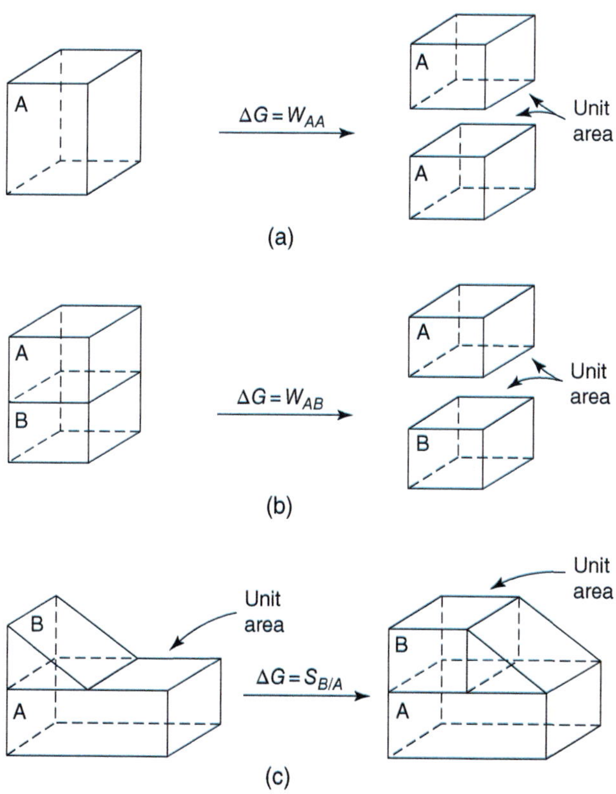

Figure 2.33 Schematic illustrations of the processes for which ΔG equals (a) the work of cohesion, (b) the work of adhesion, and (c) the work of spreading. Reproduced by permission, from Heimenz [16]. Copyright © 1986 by Taylor & Francis.

The separation process shown in Figure 2.33a consists of the formation of two new interfaces, each of unit cross-sectional area, at a location where no interface previously existed. The free energy change associated with the separation process comes directly from the definition of surface energy [Eq. (2.50)] where two surfaces of unit surface area are formed. With appropriate assumptions regarding constant temperature, pressure, and incompressible fluids, we can equate this free energy change with the work required to form the two surfaces, which is called the *work of cohesion* and is labeled W_{AA}:

$$\Delta G = 2\gamma_A = W_{AA} \tag{2.70}$$

Here γ_A is the surface energy of component A.

For the case where A is separated from B, as shown in Figure 2.33b, we once again have the formation of two free surfaces, A and B, but must subtract from the energy associated with their formation the energy that was associated with the interface AB, γ_{AB}, to obtain the free energy change of this process, which is called adhesion. Again, equating work with the free energy change leads to the *work of adhesion*, W_{AB}.

$$\Delta G = \gamma_A + \gamma_B - \gamma_{AB} = W_{AB} \tag{2.71}$$

The work of adhesion measures the degree of attraction between the two phases. For the specific case where A is a solid (S) and B is a liquid (L), γ_A becomes the solid surface energy, γ_S, γ_B become the liquid surface tension, γ_L, and Young's equation, Eq. (2.61), can be inserted into Eq. (2.71) to obtain a relationship between the liquid–solid contact angle, θ, liquid surface tension, and the work of adhesion:

$$W_{SL} = \gamma_L(1 + \cos\theta) \tag{2.72}$$

Finally, for the case of spreading of B on A shown in Figure 2.33c, the free energy change is

$$\Delta G = \gamma_{AB} + \gamma_B - \gamma_A \tag{2.73}$$

As usual with free energies, a negative value means that the process occurs spontaneously. We define a quantity called the *spreading coefficient*, $S_{B/A}$, which is the negative of the free energy of spreading, such that a thermodynamically favorable spreading process (negative free energy) leads to a positive value of the spreading coefficient. By combining Eqs. (2.70), (2.71), and (2.73), we note that the spreading coefficient can be related to the work of adhesion and work of cohesion:

$$S_{B/A} = W_{AB} - W_{BB} \tag{2.74}$$

If $W_{AB} > W_{BB}$, the A–B interaction is sufficiently strong to promote the wetting of A by B. Conversely, no wetting occurs if $W_{BB} > W_{AB}$, since the work required to overcome the attraction between two B molecules is not compensated for by the attraction between A and B. Hence, a negative spreading coefficient means that B will not spread over A.

Cooperative Learning Exercise 2.6

Consider the case of an adhesive joint created by bonding two poly(vinyl chloride) (PVC, component A) sheets together with an epoxy adhesive (component B). The surface energies for the PVC solid surface, liquid epoxy, and PVC/epoxy interface are 47, 41.7, and 4.0 erg/cm^2, respectively.

Person 1: Calculate the work of cohesion within the epoxy adhesive. This is the work required to separate epoxy bonds from one another.

Person 2: Calculate the work of adhesion between one surface of the PVC sheet and the liquid epoxy. This is the work required to separate the epoxy from the PVC.

Combine your results to calculate the spreading coefficient, $S_{B/A}$. Will the epoxy wet the PVC? Assuming that complete wetting can occur (and that a sufficiently thin layer of adhesive is applied), how will the adhesive joint fail when stressed; i.e., will the epoxy or the PVC/epoxy interface fail first?

Answer: $W_{BB} = 2(41.7) = 83.4$ ergs/cm^2: $W_{AB} = 47 + 41.7 - 4 = 84.7$ erg/cm^2: $S_{B/A} = 84.7 - 83.4 = 1.3$ erg/cm^2; the epoxy will wet the PVC. Since $W_{AB} > W_{BB}$, cohesive failure will occur prior to adhesive failure, and the epoxy will fail before the interface.

2.5 Thermodynamics of Biologics

Recall from Section 1.5 that tissue is composed not only of cells but also of an extracellular matrix, of which proteins are an important constituent. In this section, we are concerned primarily with the thermodynamics that dictates protein behavior in biological systems. Let us begin by considering the insertion of a synthetic material into a tissue sample—for example, a metal catheter into the skin or a metal–ceramic composite artificial hip into a human body. What happens? Does the foreign material stay isolated from the body? Obviously, it interacts with the tissue cells in some fashion. It can be argued that in the case of a hip implant, what we would like to see is cell adhesion—cells sticking to the material in an attempt to integrate it into the body. It turns out that one of the extracellular matrix proteins we described in Chapter 1, fibronectin, plays an important role in cell attachment and spreading. In this process, fibronectin spreads on the foreign surface, after which point the cell attaches to the fibronectin via a specific type of interaction called *receptor–ligand binding* (see Figure 1.90). What we are concerned with here is the process that allows a specific chemical constituent in a cell or a protein, such as fibronectin, called the *ligand*, to interact, or bind, to the corresponding specific chemical constituent in another cell membrane, called the *receptor*. In this process, the ligand, which is sort of a chemical "key," fits into the chemical "lock" of the receptor, which leads to a *receptor–ligand complex*. Receptor–ligand binding is a type of *specific binding*; that is, there is one receptor that is intended to interact with one ligand. There are other types of *nonspecific binding* that can occur and contribute to a small extent to the overall binding of the cell to the protein. Let us examine these two situations independently, by first describing the adhesion and spreading of a cell on a surface and then examining specific cell–cell interactions. The discussion at this point will be purely thermodynamic in nature. The kinetics of the process—for example, how fast or slow receptor–ligand binding occurs—is the subject of Chapter 3.

Take a Peek! Jump ahead to Section3.5 to see how the kinetics of receptor–ligand binding are described.

2.5.1 Cell Adhesion and Spreading on Surfaces

For cells, the processes of adhesion and spreading on surfaces are two distinctly separate processes. To date, only adhesion has been described in a thermodynamic way. Cell spreading, on the other hand, is difficult to describe in this manner, due to cellular activity, such as protein production and cytoskeleton transport which complicate the spreading process. As a result, the strength of adhesion does not correlate well with the amount of spreading the cell has undergone. Let us look first at the thermodynamics of cell adhesion on a substrate and then separately at the spreading process.

The free energy change for adhesion of cells from a liquid suspension onto a solid substrate is given by

$$G_{adh} = \gamma_{cs} - \gamma_{cl} - \gamma_{sl} \tag{2.75}$$

where the interfacial free energies are for cell–solid (γ_{cs}), cell–liquid (γ_{cl}), and solid–liquid (γ_{sl}) interfaces, respectively. If the free energy change of adhesion is negative, the adhesion process is spontaneous. If it is positive, the process is thermodynamically unfavorable. As a result, neither extremely hydrophilic nor extremely hydrophobic ($\gamma_{sl} < 40$ dyn/cm) substrates promote adhesion of cells such as fibroblasts. This trend, while still the subject of much experimental investigation and verification, is illustrated in Figure 2.34. Here, area A represents a nonadhesive zone between 20 and 30 dyn/cm critical surface energy. Area B represents biomaterials with good adhesive properties. Other factors such as surface charge, surface topography, porosity, and mechanical forces can greatly influence the ability of a cell to adhere to a surface.

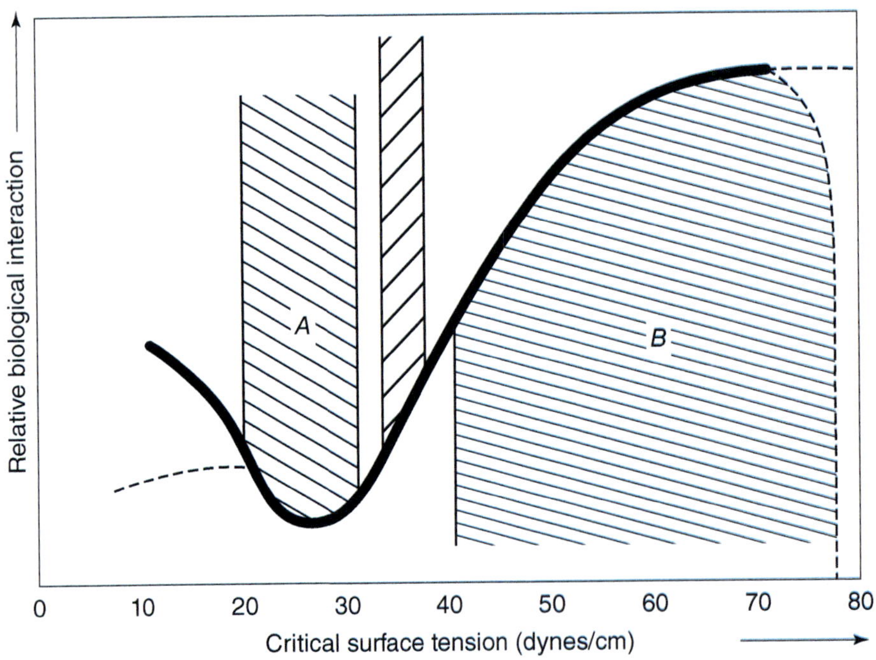

Figure 2.34 Relationship between the biological interactiveness and the critical surface tension of a biomaterial. Reproduced, by permission, from Schakenraad [17]. Copyright © 1996 by Elsevier.

Cooperative Learning Exercise 2.7

It has been observed that receptor–ligand binding processes are generally exothermic ($\Delta H < 0$). Protein adsorption on surfaces can be described with thermodynamic arguments similar to those used for cell adhesion—that is, with free energy arguments similar to those in Eqs. (2.12) and (2.75). For example, the enthalpies of adsorption for two proteins, human serum albumin (HSA) and RNase, on α-Fe_2O_3 and negatively charged polystyrene (PS-H), respectively, at various pH levels are given below.

Protein	Substrate	pH	Enthalpy (mJ/m^2)
HSA	α-Fe_2O_3	5	+1.9
		7	+7.0
RNase	PS-H	5	+4
		11	−2

Discuss with your neighbor how protein adsorption can still be spontaneous ($\Delta G < 0$), even though some of these processes are endothermic ($\Delta H > 0$). Look up values for the solid surface energies of the two substrates in Appendix 3 (you may have to select similar materials, but this is sufficient for these purposes). How do these enthalpies compare in magnitude to the respective surface energies?

Answer: According to Eq. (2.12), the entropy must also be considered. In fact, many of the binding/adhesion processes are entropy-driven—even the exothermic ones. The surface energy of Fe_2O_3 is probably similar to FeO–about 500 erg/cm^2, and PS is around 33 erg/cm^2. Thus, the enthalpy of adsorption is much larger relative to the surface energy for the RNase/PS system than for the HSA/Fe_2O_3 system. Note that 1 mJ/m^2 = 1 erg/cm^2.

Once the cell is adhered to the surface, it may undergo spreading. Cell spreading is the combined process of continued adhesion and cytoplasmic contractile meshwork activity. This activity involves the formation of *microfilaments* and *lamellar protrusions*. When a cell meets the membrane of another cell, further spreading is prohibited. Due to these complexities in the spreading process, it is difficult to give a purely thermodynamic explanation for cell spreading. Nonetheless, some correlation between the surface energy of the substrate and the ability of a cell to spread upon it has been noted (see Figure 2.35). Here, the presence of preadsorbed proteins (solid line in Figure 2.35) can affect the ability of the cell to spread relative to a surface free of preadsorbed proteins (dashed line in Figure 2.35).

2.5.2 Cell–Cell Adhesion

Consider two cells, capable of binding to one another (see Figure 2.36). The surface of a cell consists of a 100-Å-thick layer called the *glycocalyx*, which contains the surface receptors for binding. There are R_{1T} and R_{2T} total surface receptors, respectively, where R_1 is the ligand for R_2 and R_2 is the ligand for R_1. Let B represent the number of *receptor–ligand complexes* acting to bridge one cell to another, called *bridges*. The number of free receptors within and outside the cell–cell contact area is R_{ic} and

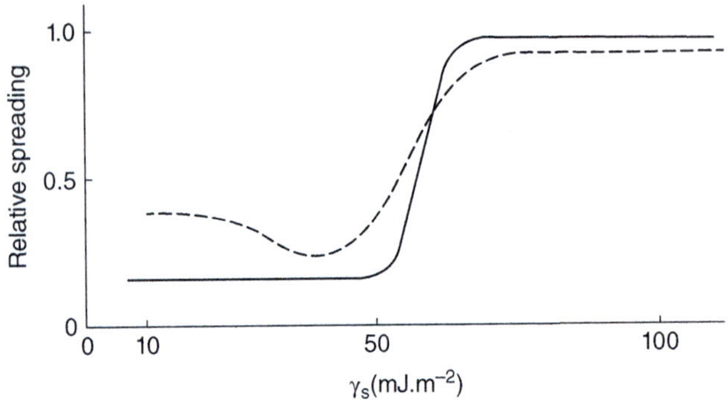

Figure 2.35 Influence of substrate surface energy on cell spreading with preadsorbed serum proteins (solid line) and without (dashed line). Reproduced, by permission, from Schakenraad [18]. Copyright © 1996 by Elsevier.

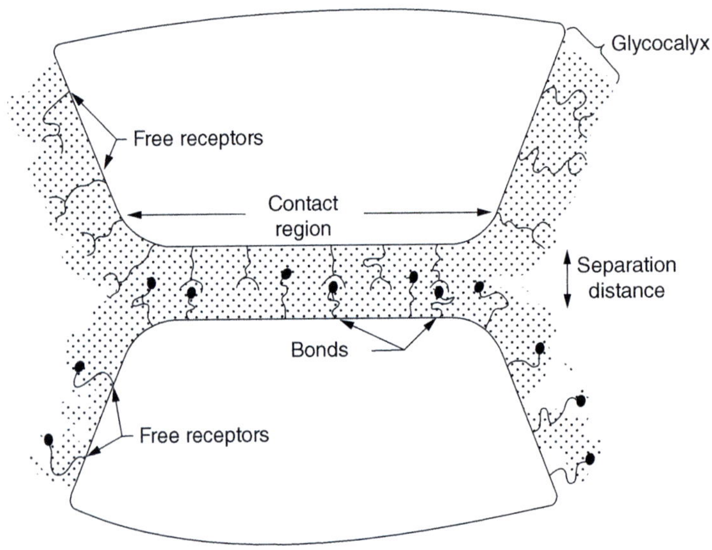

Figure 2.36 Schematic illustration of cell–cell and cell–substrate adhesion. Reproduced, by permission, from Bell et al. [19].

R_{io}, respectively, where $i = 1, 2$. Assuming a constant number of total surface receptors on each cell, the conservation of receptors for cell i will be

$$R_{iT} = B + R_{ic} + R_{io} \tag{2.76}$$

The corresponding receptor densities are $n_{iT} = R_{iT}/A_i$, $n_b = B/A_c$, $n_{ic} = R_{ic}/A_c$, and $n_{io} = R_{io}/(A_i - A_c)$, where A_i is the total surface area of cell i and A_c is the contact area. We will assume that the distance between the cell membranes is constant and equal to s.

The Gibbs free energy for the system can be expressed as

$$\Delta G = \sum_j \left(N_j n_j \mu_j\right) + A_c \Gamma(s) \tag{2.77}$$

where N_j is the number of receptor species (either within, R_{ic}, or outside, R_{io}, of the contact area or as a bridge, B), n_j is the density of each of the species (n_{ic}, n_{io}, or n_b), and μ_j is the chemical potential of each receptor state. The quantity $\Gamma(s)$ is called the *nonspecific interaction energy* and is the mechanical work that must be done against nonspecific repulsive forces to bring a unit area of membrane from an infinite separation to a cell–cell separation distance of s. The repulsive forces that must be overcome are steric and electrostatic in nature, whereas the attractive forces are of the van der Waals type (see Figure 2.37). At small separation distances, repulsive forces dominate the cell–cell interaction and diminish at values of s greater than 200 Å. This is important because cell–cell and cell–surface adhesion occur with separation distances between 100 and 300 Å. As a result, repulsive forces dominate, and van der Waals forces are often neglected. So, the nonspecific interaction energy can be modeled empirically as a separation-distance-dependent net repulsion:

$$\Gamma(s) = \frac{\sigma}{s} \exp\left(\frac{-s}{\delta_g}\right) \tag{2.78}$$

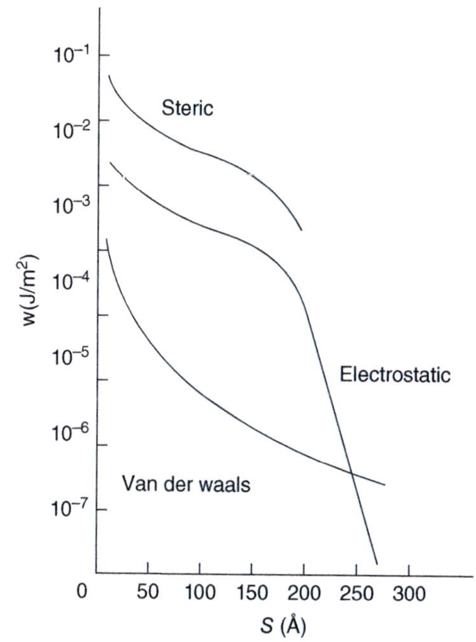

Figure 2.37 Nonspecific work of adhesion required to separate two like cells as a function of cell–cell separation distances. Reproduced, by permission, from Bongrand and Bell [20]. Copyright © 1984 by Taylor & Francis.

where σ is a measure of the glycocalyx stiffness and δ_g is its thickness. Substitution of Eq. (2.78) into Eq. (2.77) and expansion of the individual free energy terms of the receptor species as described above yields the final form of the free energy change for cell–cell adhesion:

$$\Delta G = R_{ic}n_{ic}\mu_{ic} + R_{io}n_{io}\mu_{io} + Bn_b\mu_b + \frac{A_c\sigma}{s} \exp\left(\frac{-s}{\delta_g}\right) \tag{2.79}$$

As with all free energy expressions in this chapter—minimization with respect to concentration, in this case—the concentration of receptor sites provides the information for the phase diagram, an example of which is shown in Figure 2.38. There are three regions for cell–cell adhesion in this diagram, according to the values R_{1T}, the total number of receptors on cell 1, and a parameter ξ, which is a dimensionless quantity that characterizes a modified ratio of nonspecific binding repulsion to binding. In region I, no stable adhesive interaction occurs between the cells. The attractive forces offered by the receptor–ligand bonds are not strong enough to overcome the repulsive forces resulting from cell–cell contact. In region II, stable cell–cell adhesion exists with a contact area less than A_{max}. Here, a balance exists between attractive and repulsive forces. In region III, stable cell–cell adhesion results and the contact area is A_{max}. Here, attractive forces outweigh repulsive forces, and the area of contact is limited only by geometric and morphological constraints.

2.5.3 Protein Folding*

The balance of attractive and repulsive forces is also a key contributor to one of the most important processes in living systems: *protein folding*. Recall from Section 1.5.1 that proteins have a primary

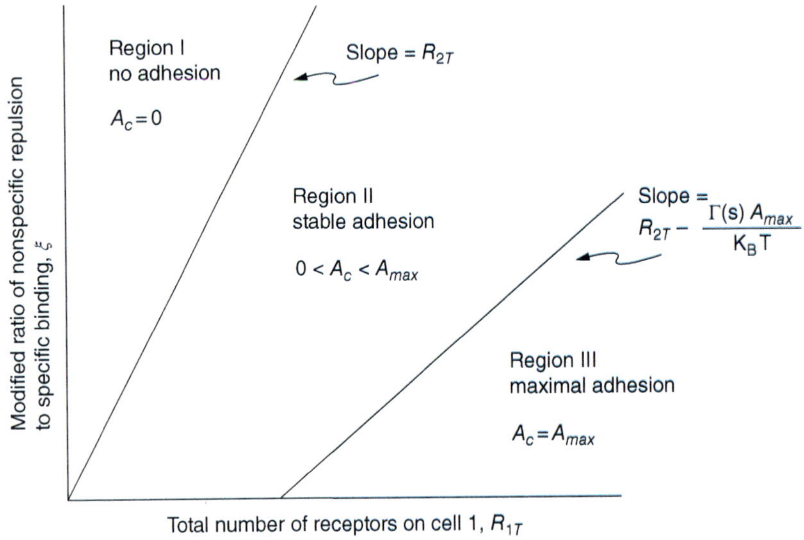

Figure 2.38 Predicted phase diagram for cell–cell adhesion. Reproduced, by permission, from Lauffenburger and Linderman [21]. Copyright © 1993 by Oxford University Press.

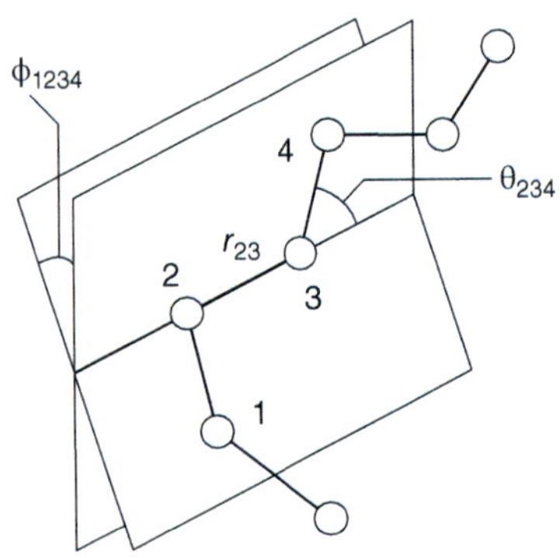

Figure 2.39 Representation of a simple protein where circles represent amino acid units, r_{ij} is the distance between adjacent units, θ is the bond angle formed by three consecutive units, and ϕ is called the torsional angle formed by four consecutive units. Ref. [22]/John von Neumann Institute for Computing.

structure (their chemical sequences) and a secondary structure as a result of chain conformation changes. These conformational changes are the result of not only hydrophobic/hydrophilic interactions but also the contribution of *conformational entropy*—the thermodynamic "penalty" for changing conformations. As with all thermodynamic processes, protein folding is only spontaneous when the enthalpic and entropic contributions lead to a negative free energy change. Only when the protein is in the proper folder conformation does it become biologically active to perform the function for which it is designed.

The use of computer simulations to perform these complex thermodynamic calculations has led to the ability to not only understand the proteins that currently exist but also design new proteins that could be used to cure diseases or improve the environment. Recall from Section 1.0.4 that we introduced MD simulations as a tool for calculating both the energy and forces in an assembly of particles. If we take those "particles" to be not atoms but those amino acid groups that constitute a protein, we can begin to use MD simulations to minimize the interparticle energy and obtain an equilibrium structure for the protein that describes its folded conformation. The mathematical and computer coding details of this procedure are beyond the scope of this text, but we can see from Figure 2.39 how this simulation might be accomplished.

Instead of the intermolecular potential of Eq. (1.12), the bending potential energy function of this protein could be represented by the following function:

$$U(r) = -k\cos\theta_{234} = -k\left(\vec{\boldsymbol{r}}_{23}\cdot\vec{\boldsymbol{r}}_{23}\right)^{-\frac{1}{2}}\left(\vec{\boldsymbol{r}}_{34}\cdot\vec{\boldsymbol{r}}_{34}\right)^{-\frac{1}{2}}\left(\vec{\boldsymbol{r}}_{23}\cdot\vec{\boldsymbol{r}}_{34}\right) \tag{2.80}$$

where the interunit distances, $\boldsymbol{r}_{ij}$, are now vector quantities because we have a multiparticle system which makes their product a vector dot product. This potential energy function can be inserted into Eq. (1.13), where it is differentiated with respect to each interunit distance to give the force. The same thing can be done to calculate the torsional potential, ϕ_{1234}. You can see how a computer program can simplify these repeated calculations as they are executed across time increments. As computational power increases and the algorithms used to simplify the calculations are improved, the ability to visualize the folded structure of proteins also improves.

Need More Help?

Try these videos on protein folding:

https://www.youtube.com/watch?v=KpedmJdrTpY
https://www.youtube.com/watch?v=qkSBWRlHKfE

References

1 Gaskell, D. R., *Introduction to Metallurgical Thermodynamics*, 2nd ed., Hemisphere Publishing Corporation, 1981, p. 178.
2 Ralls, K. M., T. H. Courtney, and J. Wulff, *Introduction to Materials Science and Engineering*, John Wiley & Sons, Inc, 1976.
3 Devereux, O. F., *Topics in Metallurgical Thermodynamics*, John Wiley & Sons, Inc, 1983.
4 Committee on Metallurgy, *Engineering Metallurgy*, Pitman Publishing, 1957, p. 245.
5 Callister, W., *Materials Science and Engineering: An Introduction*, 5th ed., John Wiley & Sons, Inc, 2000, p. 277.
6 Rhines, F. N., *Phase Diagrams in Metallurgy*, McGraw-Hill Book Co, 1956.
7 "Phase Diagrams for Ceramists," is authored by Ernest M Levin, Carl R. Robbins, Howard F. McMurdie, and edited by Margie K. Reser, The American Ceramic Society, 1964, p. 15.
8 "Phase Diagrams for Ceramists," is authored by Ernest M Levin, Carl R. Robbins, Howard F. McMurdie, and edited by Margie K. Reser, The American Ceramic Society, 1969, p. 30.
9 Kingery, W. D., H. K. Bowen, and D. R. Uhlmann, *Introduction to Ceramics*, John Wiley & Sons, Inc, 1976.
10 "Phase Diagrams for Ceramists," is authored by Ernest M Levin, Carl R. Robbins, Howard F. McMurdie, and edited by Margie K. Reser, The American Ceramic Society, 1969, p. 31.
11 Padday, J. F., "Surface Tension, Part I. The Theory of Surface Tension.", in *Surface and Colloid Science*, E. Matijevic, ed., Vol. 1, John Wiley & Sons, Inc, 1969, p. 79.
12 Jastrzebski, Z., *The Nature and Properties of Engineering Materials*, 2nd ed., John Wiley & Sons, Inc, 1976.
13 Lee, W. E., and W. M. Rainforth, *Ceramic Microstructures*, Kluwer Academic Publishers, 1994, p. 37.
14 Cowie, J. M. G., *Polymers: Chemistry & Physics of Modern Materials*, 2nd ed., Taylor & Francis, 1991, p. 167.
15 Cowie, J. M. G., *Polymers: Chemistry & Physics of Modern Materials*, 2nd ed., Taylor & Francis, 1991, p. 175.
16 Heimenz, P., *Principles of Colloid and Surface Chemistry*, 2nd ed., Taylor & Francis, 1986, p. 316.

17 Schakenraad, J. M., Cells: their surfaces and interactions with materials, in *Biomaterials Science: An Introduction to Materials in Medicine*, B. D. Ratner, A. S. Hoffman, F. J. Schoen, and J. E. Lemons, eds., Elsevier, 1996, p. 146.

18 Schakenraad, J. M., Cells: their surfaces and interactions with materials, in *Biomaterials Science: An Introduction to Materials in Medicine*, B. D. Ratner, A. S. Hoffman, F. J. Schoen, and J. E. Lemons, eds., Elsevier, 1996, p. 144.

19 Bell, G. I., M. I. C. A. H. Dembo, and P. I. E. R. R. E. Bongrand, Cell adhesion. Competition between nonspecific repulsion and specific bonding, *Biophysical Journal*, **45**, 1051 (1984).

20 Bongrand, P., and G. I. Bell, Cell-cell adhesion: parameters and possible mechanisms, in *Cell Surface Dynamics: Concepts and Models*, A. S. Perelson, C. DeLisi, and F. W. Wiegel, eds., Taylor & Francis, 1984, pp. 459–493.

21 D. A. Lauffenburger and J. J. Linderman, *Receptors*, p. 281. 1993, Oxford University Press.

22 Allen, M. P., Introduction to molecular dynamics simulations, in *Computational Soft Matter: From Synthetic Polymers to Proteins, Lecture Notes*, N. Attig, K. Binder, H. Grubmüller, and K. Kremer, eds., Vol. 23, ISBN 3-00-012641-4, John von Neumann Institute for Computing, Jülich, NIC Series, 2004, pp. 1–28.

Problems

Level I

2.I.1 Draw schematic ΔG_{mix} versus X_{Ni} curves for the Cu—Ni system shown in Figure 2.4 at 1455, 1300, 1100, and 300°C.

2.I.2 Read about Ellingham Diagrams in Advanced Topics Section A2.1. Compare the relative efficiencies of H_2 and CO as reducing agents for metal oxides; for example, if cobalt oxide, CoO, were to be reduced to the pure metal, which would do a better job and at what temperatures?

2.I.3 At 200°C, a 50:50 Pb—Sn alloy exists as two phases, a lead-rich solid and a tin-rich liquid. Calculate the degrees of freedom for this alloy and comment on its practical significance.

2.I.4 For 1 kg of eutectoid steel at room temperature, calculate the amount of each phase (α and Fe_3C) present.

2.I.5 Calculate the amount of each phase present in 1 kg of a 50:50 Ni—Cu alloy at (a) 1400°C, (b) 1300°C, and (c) 1200°C.

2.I.6 Calculate the amount of each phase present in 50 kg of free-cutting brass at 200°C. Neglect the presence of alloying elements less than 3% by weight.

2.I.7 Find at least two sources for the surface energy of the (111) plane in diamond. State the values and cite your sources. Do they agree with one another? Are they different from the estimated value in Example Problem 2.3? Briefly describe why these three values may differ.

Level II

2.II.1 We will see in Chapter 5 that the strength of a material, σ, is the work required to separate surfaces divided by the separation distance. In this way, the cohesive strength, σ_c, is the work of cohesion, W_{BB}, divided by some characteristic intermolecular distance, x_{BB}. Similarly, the adhesive strength, σ_a, is the work of adhesion, W_{AB}, divided by some characteristic intermolecular distance, x_{AB}. (a) Assuming that $x_{BB} = x_{AB} \approx 5.0$ Å for both the epoxy adhesive and the PVC–epoxy interface, calculate the cohesive strength and adhesive strength of the PVC–epoxy composite joint in Cooperative Learning Exercise 2.6. Which has a higher strength, the epoxy with itself or the PVC–epoxy interface? Recall that 1erg/cm^2 = 1 dyn/cm. (b) Now assume that the thickness of the interface, x_{AB}, is increased. At what value of x_{AB} does mechanical failure switch from cohesive to adhesive? What does this tell you about the optimal thickness of an adhesive layer if it is to be stronger than the materials it binds together?

2.II.2 Consider 1 kg of a brass with composition 30 wt% Zn. (a) Upon cooling, at what temperature would the first solid appear? (b) What is the first solid phase to appear, and what is its composition? (c) At what temperature will the alloy completely solidify? (d) Over what temperature range will the microstructure be completely in the α-phase?

2.II.3 You have supplies of kaolin, silica, and mullite as raw materials. Using kaolin plus either silica or mullite, calculate the batch composition (in weight percent) necessary to produce a final microstructure that is equimolar in silica and mullite.

2.II.4 Hot isostatic pressing (HIP), in which an inert gas is applied at high pressure, can be used to consolidate and densify ceramic components. The driving force for pore shrinkage in such cases is the sum of the applied pressure, P_e, and the surface energy-driven component, as given by Eq. 2.58. Calculate the ratio of driving forces for HIPing at 35 MPa compared to conventional sintering. Assume a pore diameter of 5 μm and γ_s = 700 mN/m at the densification temperature.

2.II.5 The melting points of a series of poly(α-olefin) crystals were studied. All of the polymers were isotactic and had chain substituents of different bulkiness. The results are listed below. Use Eq. (2.12) to derive a relationship between the melting point, T_m, and the enthalpy and entropy of fusion, ΔH_f and ΔS_f, respectively. Use this relationship, plus what you know about polymer crystallinity and structure from Chapter 1, to rationalize the trend in melting point.

Substituent	T_m (°C)
$-CH_3$	165
$-CH_2CH_3$	125
$-CH_2CH_2CH_3$	75
$-CH_2CH_2CH_2CH_3$	−55
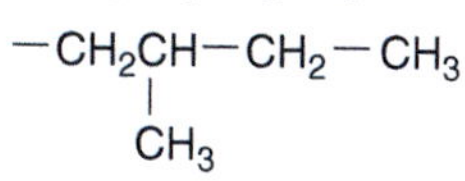	196
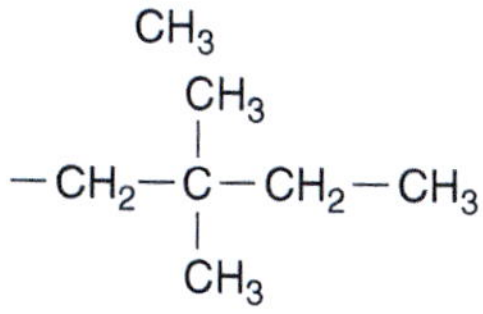	350

2.II.6 The diagram shown below qualitatively describes the phase relationships for Nylon-66, water, and phenol at $T > 70°C$. Use this diagram to draw arrows that trace the following procedures: (1) Water is added to solution A until the system separates into solutions B and C; (2) solution C is removed and water is added to solution B until the system separates into solutions D and E; (3) phenol is added to solution C until a solution A′ results which is close to A in concentration. Would you expect solutions C and E, or B and D, to contain the higher-molecular-weight polymer? Briefly explain. Outline a strategy for Nylon fractionation based on steps 1–3. Some steps may be repeated as necessary.

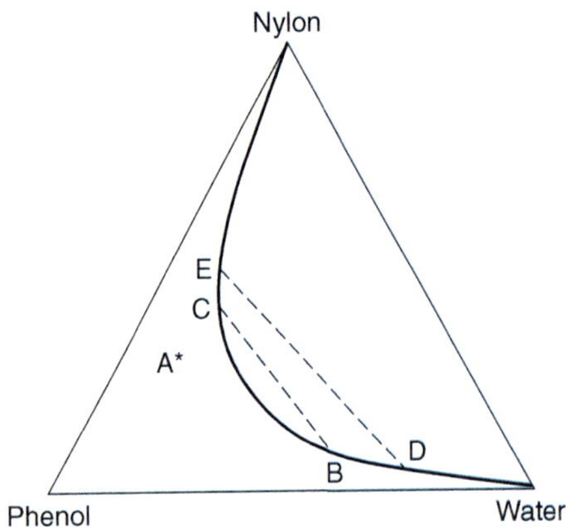

2.II.7 The following data were obtained for the work of adhesion between the listed liquids and carbon black. Use these data, together with the surface tensions of pure liquids from Appendix 3 (or from additional sources as necessary), to calculate the spreading coefficients for the various liquids on carbon black.

Liquid	W_{AB} (erg/cm^2)
Benzene	109.3
Toluene	110.2
CCl_4	112.4
CS_2	122.1
Ethyl ether	76.4
H_2O	126.8

Level III

2.III.1 The Maxwell relations in thermodynamics are obtained by treating a thermodynamic relation as an exact differential equation. Exact differential equations are of the form

$$dz = \left(\frac{dz}{dx}\right)_y dx + \left(\frac{dz}{dy}\right)_x dy = Mdx + Ndy$$

where $z = z(x, y)$ and M and N correspond to the partial derivatives of z with respect to x and y, respectively. A useful relation in exact differential equations is that

$$\left(\frac{\partial M}{\partial y}\right)_x = \left(\frac{\partial N}{\partial x}\right)_y$$

(Can you prove why this is so?) When x, y, and z are thermodynamic quantities, such as free energy, volume, temperature, or enthalpy, the relationship between the partial differentials of M and N as described above are called Maxwell relations. Use Maxwell relations to derive the Laplace equation for a sphere at constant temperature using the following thermodynamic expression for free energy, G:

$$dG = VdP - SdT + \gamma dA$$

where V is the volume of the sphere, P is the pressure, T is temperature, γ is the surface energy, and A is the spheres surface area.

2.III.2 Why does pore coalescence in a glassy material cause an increase in porosity and negative densification? (*Hint*: Consider two pores containing n moles of gas that become one pore during isothermal coalescence. The driving force for pore shrinkage is $2\gamma/r$. Use this information with the ideal gas law to calculate the change in pore radius.)

2.III.3 Calculate the three components of the bending force on the central three units, $\boldsymbol{F}_{234}$, of the simple protein shown in Figure 2.39. Hint: Use the vector form of Eq. (1.13) where the potential energy function is given by Eq. 2.80. If you do this by hand, you may find it useful to break up each of the three partial derivatives [e.g., $(\partial/\partial \boldsymbol{r}_2)\cos\theta_{234}$] into its components before applying the chain rule. Alternatively, you could set up a Matlab (or similar) program if your instructor allows you to solve these equations and calculate the forces.

3

Kinetic Processes in Materials

3.0 Introduction and Objectives

Thermodynamics tells us whether a process is favorable; kinetics tells us how quickly that process will take place. The rate at which chemical and biological processes occur is of tantamount importance. For example, we all know that an exposed piece of iron will rust (oxidize) over time, but is this important if the iron piece will only be used once in the near future?

There are three categories of kinetic processes that are of concern to us: the rate at which materials are formed, the rate at which they are transformed, and the rate at which they decompose. In general, formation and decomposition are chemical processes, involving the reaction of two or more chemical species. Transformations, on the other hand, are usually physical processes, such as the melting of ice, and do not involve chemical reactions. We can also have situations in which both physical and chemical kinetic processes occur simultaneously. In all cases, our goal is to be able to describe the rate at which the process is occurring. There are two important concepts that help us accomplish our task: the thermodynamic concept of free energy, ΔG, which was introduced in Chapter 2; and a related quantity called the activation energy, E_a. We will see how one—or both—can be modified to help determine the rate of processes in the formation, transformation, and decomposition of materials.

By the end of this chapter, you should be able to:

- Apply the law of mass action to a reaction to determine the equilibrium constant at a specified temperature.
- Calculate the activation energy and preexponential factor from kinetic data.
- Identify the dimensionality and calculate rates of a nucleation and growth process from kinetic data.
- Distinguish between the different types of corrosion in all material classes.
- Calculate the electromotive force (EMF) of a galvanic cell from the half-cell potentials.
- Use the Nernst equation to calculate the potential of a half-cell not at unit activity.
- Distinguish between activation and concentration polarization.
- Distinguish between homogeneous and heterogeneous nucleation processes.
- Distinguish between stepwise and radical chain polymerizations.
- Calculate the degree of polymerization, number-average, and weight-average molecular weights given kinetic data for a polymerization reaction.
- Identify the different types of copolymers and use the copolymer equation with appropriate kinetic data to determine which type of copolymer will form.

Materials Engineering and Science: Principles, Properties, and Processes, Second Edition. Brian S. Mitchell.

Companion website: www.wiley.com/go/materialsengineeringandscience2e

- Distinguish between the different types of polymerization processes and identify the equipment and kinetics that are specific to each.
- Distinguish between the different types of polymer degradation.
- Distinguish between the different types of vapor deposition processes and calculate the rate of deposition for each from kinetic data.
- Determine the rate of receptor–ligand binding from kinetic data.

3.0.1 The Law of Mass Action

We saw in Chapter 2 that an important thermodynamic quantity is the Gibbs free energy, ΔG. The specific functional relationship we use to describe the free energy will depend on whether we are studying a physical or a chemical transformation. For physical processes, such as phase transformations, the most useful form of the Gibbs free energy is its definition given in Chapter 2:

$$\Delta G = \Delta H - T\Delta S \tag{2.12}$$

In the case of reacting systems, however, the number of moles is not constant, and it is more convenient to consider the total free energy as a sum of individual free energy contributions. For this case, let us begin by considering the generalized reaction shown below, in which reactants A and B go to products C and D:

$$aA + bB \rightleftarrows cC + dD \tag{3.1}$$

where a, b, c, and d are the *stoichiometric coefficients* of chemical species A, B, C, and D, respectively, and can be represented generically by v_i. Recall that stoichiometric coefficients are positive for products and negative for reactants. There can, of course, be more reactants and/or products than those listed here, and the reaction may be reversible, hence arrows in both directions. The *Law of Mass Action* states that the velocity of the reaction at a given temperature is proportional to the product of the active masses of the reacting substances. In other words, the *forward reaction rate*, r_1, is a function of the concentrations of A and B raised to their corresponding exponents:

$$r_1 = k_1[A]^a[B]^b \tag{3.2}$$

Similarly, the *reverse reaction rate* is related to the amount of products, C and D, present

$$r_2 = k_2[C]^c[D]^d \tag{3.3}$$

In Eqs. (3.2) and (3.3), k_1 and k_2 are proportionality constants—or *rate constants*—for the forward and reverse reactions, respectively, and have units of inverse time (typically seconds) and an appropriate inverse concentration, depending on the way the concentration of chemical species are represented and the magnitude of the stoichiometric coefficients.

At equilibrium (recall $\Delta G = 0$) the rate of the forward reaction must be equal to the rate of the reverse reaction, $r_1 = r_2$, and we can equate Eqs. (3.2) and (3.3) to give:

$$K = \frac{k_2}{k_1} = \frac{[C]^c[D]^d}{[A]^a[B]^b} \tag{3.4}$$

The constant K is the *equilibrium constant* for the entire reaction at constant temperature.

We can relate the equilibrium constant, K, to the free energy of the system. Recall that the free energy change for a reaction must be less than zero for it to proceed spontaneously, and any system whose overall free energy change is zero is in equilibrium. The free energy of a system is simply the sum of the free energy contributions of each of the components, which are measured using their chemical potentials, μ_i [see Eq. (2.13) for the definition of chemical potential], and the stoichiometric coefficients

$$\Delta G = \sum \nu_i \mu_i \tag{3.5}$$

When all the components are in their standard states (see Section 2.1.2), the standard Gibbs free energy, ΔG^0, is given by:

$$\Delta G^0 = \sum \nu_i \mu_i^0 \tag{3.6}$$

The chemical potentials in Eq. (3.5), μ_i, and the standard chemical potentials in Eq. (3.6), μ_i^0, are related to each other by the *activity* of each species, a_i:

$$\mu_i = \mu_i^0 + RT\, ln\, a_i \tag{3.7}$$

where R is the gas constant and T is the absolute temperature. The activity of a species, then, is an indication of how far away from standard state that species is.

If we subtract Eq. (3.6) from Eq. (3.5) and substitute Eq. (3.7) into the result, we obtain a very useful relationship between the activity of the reactants and products and the free energy for a reacting system:

$$\Delta G - \Delta G^0 = RT \sum \nu_i \ln a_i = RT \ln\left(\prod_i a_i^{\nu_i}\right) \tag{3.8}$$

At equilibrium, then, $\Delta G = 0$, and the activities of the products and reactants, a_i, are related to their respective concentrations:

$$\Delta G^0 = -RT \ln\left\{\frac{[A]^a[B]^b}{[C]^c[D]^d}\right\} \tag{3.9}$$

and we can identify the term in the logarithm from Eq. (3.4) as the equilibrium constant, K:

$$\Delta G^0 = -RT \ln K \tag{3.10}$$

Rearranging Eq. (3.10) to express K in terms of the standard free energy change gives:

$$K = \exp\left(\frac{-\Delta G^0}{RT}\right) \tag{3.11}$$

Equation (3.11) gives the effect of temperature on the equilibrium constant. Note that a plot of ln K versus $1/T$ should give a straight line of slope $-\Delta G^0/R$.

3.0.2 The Activation Energy

Since the equilibrium constant, K, is a ratio of the forward to reverse reaction rate constants, k_1 and k_2, respectively, it is reasonable to assume that the forward and reverse rate constants also follow a relationship similar in form to Eq. (3.11). If we replace the standard free energy with some general energy term for activation, E_a, and include a proportionality constant called the *preexponential factor*, k_0, we get what is called an *Arrhenius expression* for the rate constant:

$$k = k_0\, exp(-E_a/RT) \tag{3.12}$$

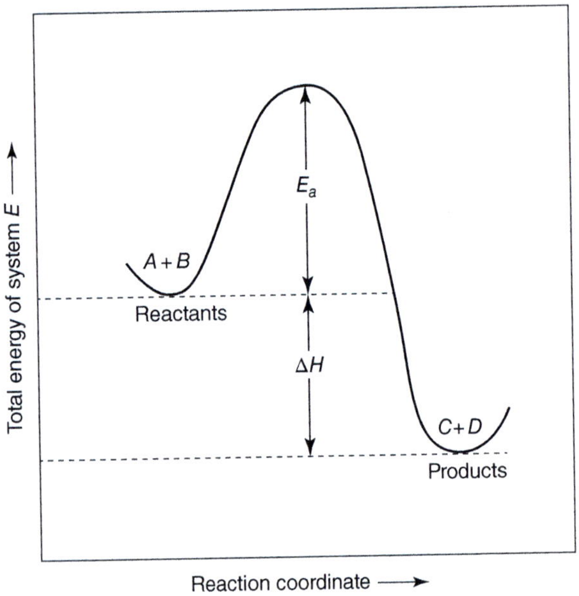

Figure 3.1 Activation energy barrier for a chemical reaction.

We will not prove the Arrhenius relationship here, but it falls out nicely from statistical thermodynamics by considering that all molecules in a reaction must overcome an activation energy before they react and form products. The Boltzmann distribution tells us that the fraction of molecules with the required energy is given by exp.($-E_a/RT$), which leads to the functional dependence shown in Eq. (3.12).

> **Need More Help?**
>
> Try these videos on the Law of Mass Action:
>
> https://www.youtube.com/watch?v=cmVuUFIaLWO
>
> and activation energy and the Arrhenius equation:
>
> https://www.youtube.com/watch?v=jX4dEOFwaLQ

Many of the processes we will study in this and other chapters have parameters that appear to follow the very general Arrhenius exponential relationship. The quantity E_a is called the *activation energy* and represents the energy barrier for the reaction under consideration. As shown in Figure 3.1, the reactants can be thought of as having to overcome an activation energy barrier of height E_a in order to achieve the activated complex state and proceed to products. The *heat of reaction*, ΔH, can be either negative (exothermic reaction) or positive (endothermic reaction), in which case the products C and D actually have a higher final energy than the reactants. Let us now see how a wide variety of kinetic processes in material formation, transformation, and degradation can be described using these relationships.

3.1 Kinetic Processes in Metals and Alloys

We begin by examining the kinetics processes of formation reactions in metals and alloys. Although many of these substances are elements or binary mixtures of elements, they can be formed through complex reactions, such as the reduction of metal oxides. We will first examine the formation of an intermetallic compound.

3.1.1 Kinetics of Intermetallic Formation

We made a distinction between alloys and intermetallics in Chapter 1, indicating that intermetallics are more like compounds than alloys (which are more like mixtures), even though both contain two or more metallic components. Nonetheless, both are composed entirely of metallic elements, and a discussion of the formation of either would be appropriate here. We focus on the formation of an intermetallic, $TiAl_3$, to illustrate how kinetic parameters can be obtained from experimental observations.

Consider the dissolution of elemental titanium, Ti, in molten aluminum, Al, at various temperatures that leads to the formation of intermetallic $TiAl_3$ layers. Figure 3.2 shows the experimental data for the decrease in thickness of the titanium as a function of time, and Figure 3.3 shows the increase in thickness of the titanium aluminide ($TiAl_3$) layer with time. From these plots, the rate of dissolution and rate of formation can be determined, respectively, for each temperature. These values are presented in Cooperative Learning Exercise 3.1. Application of the Arrhenius Eq. (3.12) in both instances allows us to calculate the activation energy for both processes by plotting ln k_i versus $1/T$ (do not forget that T is in Kelvin!), with the slope of each line equal to $-E_a/R$. The fact that the two activation energies calculated in Cooperative Learning Exercise 3.1 are essentially equivalent tells us that the rate of the chemical reaction between these two metals—not the diffusion of either aluminum or titanium through the intermetallic layer formed on the surface of the solid titanium—determines the overall rate of $TiAl_3$ formation. As we will see, diffusion is also an activated process and can be described using an Arrhenius-type relationship.

Take a Peek! Jump ahead to §3.2.1 to see how diffusion is an activated process.

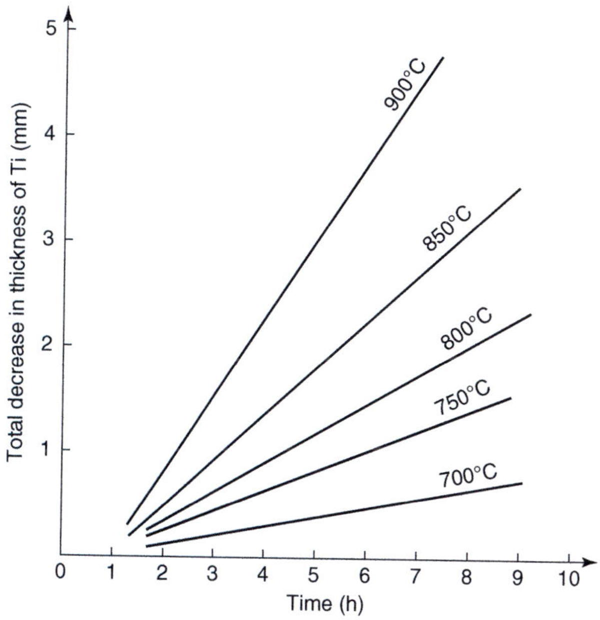

Figure 3.2 Rate of dissolution of Ti at various temperatures.

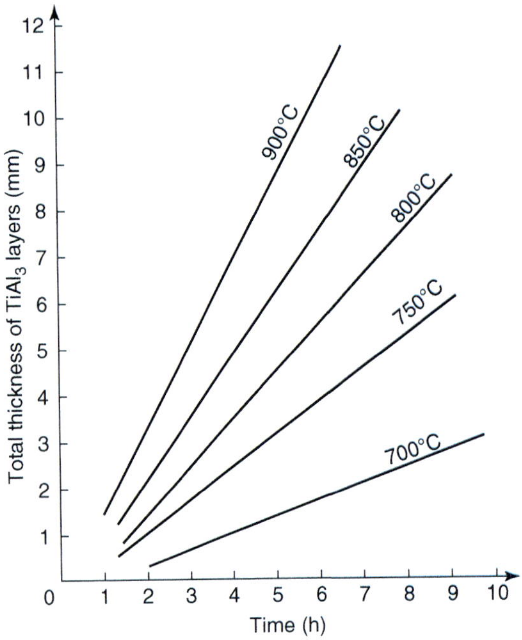

Figure 3.3 Rate of formation of $TiAl_3$ at various temperatures.

Cooperative Learning Exercise 3.1

Consider the following data for the dissolution of Ti in molten aluminum to form the intermetallic $TiAl_3$.

Temperature (°C)	Rate of Ti Dissolution, k_{Ti} (cm/s)	Rate of $TiAl_3$ Formation, k_{TiAl3} (cm/s)
700	1.51×10^{-6}	1.042×10^{-5}
750	2.605×10^{-6}	1.98×10^{-5}
800	4.17×10^{-6}	2.777×10^{-5}
850	6.13×10^{-6}	3.798×10^{-5}
900	9.36×10^{-6}	6.805×10^{-5}

Person 1: Determine the activation energy, $E_{a,Ti}$, and preexponential factor, $k_{0,Ti}$, for the rate of dissolution of Ti.

Person 2: Determine the activation energy, $k_{a,TiAl_3}$, and preexponential factor, $k_{0,TiAl_3}$, for the rate of growth of $TiAl_3$.

Group: Compare your results and suggest a reason for your findings. What other processes could affect the rate of either of these reactions?

Answers: $E_{a,Ti}$ = 86 kJ/mol; $k_{0,Ti}$ = 0.06 $E_{a,TiAl_3}$ = 84 kJ/mol; $E_{0,TiAl_3}$ = 5.3.

3.1.2 Kinetics of Phase Transformations in Metals and Alloys

Nucleation and Growth (Round 1)

Phase transformations, such as the solidification of a solid from a liquid phase, or the transformation of one solid crystal form to another (remember allotropy?), are important for many industrial processes. We have investigated the thermodynamics that leads to phase stability and the establishment of equilibrium between phases in Chapter 2, but we now turn our attention toward determining what factors influence the rate at which transformations occur. In this section, we will simply look at the phase transformation kinetics from an overall rate standpoint. In Section 3.2.1, we will look at the fundamental principles involved in creating ordered, solid particles from a disordered, solid phase, termed *crystallization* or *devitrification*.

Consider the isothermal transformation of a disordered (amorphous) solid to an ordered (crystalline) solid. Obviously, there must be enough thermal energy to allow individual atoms to move around and reorient themselves, but assuming this is possible, it is generally found that *at constant temperature*, the amount of amorphous material transformed to crystalline material, dx (on a volume basis) per unit time, dt, is given by

$$\frac{dx}{dt} = nk(1-x)t^{n-1} \tag{3.13}$$

where the quantity $(1-x)$ is the fraction of amorphous material remaining, n is the reaction order, t is time, and k is the reaction rate constant as expressed by Eq. (3.12). The solution to this equation for the amount of amorphous materials transformed, x, at time t, is known as the *Johnson–Mehl–Avrami (JMA) equation* [1–3]:

$$x = 1 - \exp[-kt^n] \tag{3.14}$$

Taking the natural log of this equation twice yields a useful form of the JMA equation:

$$\ln[-\ln(1-x)] = \ln k + n \ln t \tag{3.15}$$

Thus, a plot of $\ln[-\ln(1-x)]$ versus $\ln t$ yields a straight line of slope n. The value of n is important not only because it indicates the reaction order but also because it has some physical interpretation, depending on whether the crystals are growing at the surface of a material or in the bulk (sometimes called volume crystallization). See Table 3.1 for a description of the dimension of the crystals for various values of n. The intercept

Need More Help?

Try these videos on the Johnson–Mehl–Avrami equation:

https://www.youtube.com/watch?v=3SWMOHjbUVk
https://www.youtube.com/watch?v=w4gTOsuzwPs

Table 3.1 Values of the Growth Dimension, n (aka Reaction Order), for Different Crystallization Processes

Mechanism	n
Bulk nucleation, 3-dimensional growth	4
Bulk nucleation, 2-dimensional growth	3
Bulk nucleation, 1-dimensional growth	2
Surface nucleation	1

of this plot gives the rate constant, k. Obtaining k at different isothermal temperatures and applying Eq. (3.12) gives the activation energy for the amorphous–crystalline transformation.

Most industrially relevant transformation processes are not isothermal; even in a controlled laboratory environment, it is difficult to perform experiments that are completely isothermal. The kinetics of nonisothermal phase transformations are more complex, of course, but there are some useful relationships that have been developed that allow for the evaluation of kinetic parameters under nonisothermal conditions. One such equation includes the heating rate, ϕ usually in K/min, used in the experiment [4]:

$$\ln[-\ln(1-x)] = n\ln\phi + \ln k + \text{constant} \tag{3.16}$$

An equation developed by Kissinger and modified by others can be used to determine the activation energy using the temperature at which the transformation rate is a maximum, T_p, for various heating rates [5]:

$$\text{In}\left(\frac{\phi}{T_p^2}\right) = \text{constant} - \frac{E_a}{RT_p} \tag{3.17}$$

The temperature of maximum transformation rate is easily determined using either of two similar techniques called *differential scanning calorimetry* (DSC) or *differential thermal analysis* (DTA). These techniques are extremely useful in the kinetic study of both isothermal and nonisothermal phase transformations.

Martensitic Transformations

Some phase transformations in the solid phase are so rapid that they approach the speed of sound. These *congruent phase transformations* (no change in composition) that involve the displacement of atoms over short distances, rather than diffusion (such as in the previous section), are termed *displacive* or *martensitic transformations*. The term "martensitic" comes from this type of transformation that is observed in the cooling of austenitic steel (see Section 2.1.3). Martensite is a supersaturated solution of carbon in ferrite with interstitial carbon atoms in the body-centered tetragonal structure that forms when face-centered cubic iron with carbon, called austenite, is rapidly cooled to about 260°C. The amount of austenite transformed to martensite is a function of the temperature only, varying from 0% to 100% between the temperatures M_s and M_f, as indicated in the accompanying time–temperature–transformation (TTT) curve shown in Figure 3.4.

Martensitic transformations occur in a variety of nonferrous alloy systems, particularly for allotropes that transform at low temperatures. The reasons for this are that (a) the free energy difference between the high-temperature and low-temperature phases becomes more negative with decreasing temperature, and (b) the crystal structures of allotropes are relatively simple and similar to one another. Examples of some common martensitic transformations are given in Table 3.2.

Need More Help?
Watch these fascinating videos on the shape memory effect: https://www.youtube.com/watch?v=wl-qAxKJoSU https://www.youtube.com/watch?v=Pn-6bGORyOU

The martensitic transformation plays an important role in a special class of alloys called *shape memory alloys*. These materials have the unique ability to "remember" their dimensions before deformation, and they return to their original shape upon undergoing the martensitic transition.

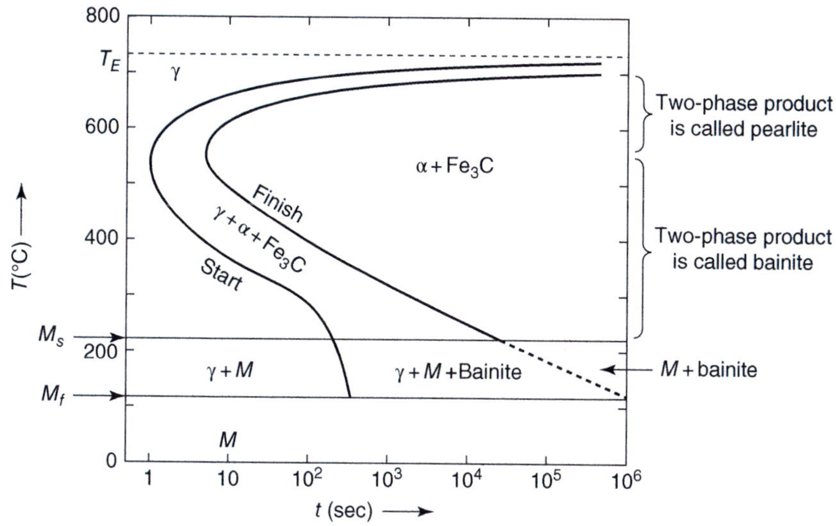

Figure 3.4 TTT diagram for eutectoid steel composition (0.80 wt% C). From Ralls et al. [19]. Copyright © 1976 by John Wiley & Sons, Inc. This material is used by permission John Wiley & Sons, Inc.

Table 3.2 Some Common Martensitic Transformations

Alloy	Transition Temperature (°C)
fcc Co → hcp Co	427
Cu–Zn–Al	−200 to +120
βTi → αTi	883
Cu–Al–Ni	−200 to +170
50% Ni–50% Ti	−200 to +100

This process is shown schematically in Figure 3.5 for the *one-way shape memory effect* (SME). There is also a *two-way* SME that we will not describe here. The SME is the result of a partial ordering of the structure as the BCC to CsCl structure transition takes place (see Figure 3.6a and b). The martensitic phase (CsCl structure in this case) is easily deformed, and a distorted multidomain martensite phase results (Figure 3.6c). When the object is reheated, the austenitic phase (BCC in this case) reforms and returns to its original structure. The SME has been studied for applications ranging from drive shafts that stiffen in response to increased shaft vibration [6] to orthodontic wires and other potential biomedical uses [7, 8].

3.1.3 Kinetics of Corrosion in Metals and Alloys

Corrosion is the deterioration of a material by reaction with its environment. Although the term is used primarily in conjunction with the deterioration of metals, the broader definition allows it to be used in conjunction with all types of materials. We will limit the description to corrosion of metals and alloys for the moment and will save the degradation of other types of materials such as polymers for a later section. In this section, we will see how corrosion is perhaps the clearest example of

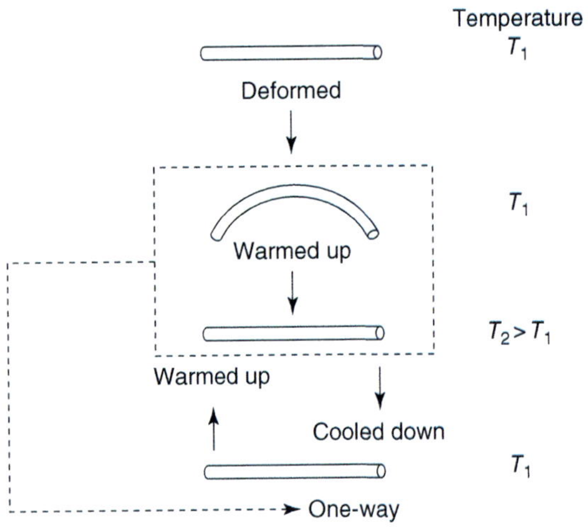

Figure 3.5 Illustration of the one-way shape memory effect.

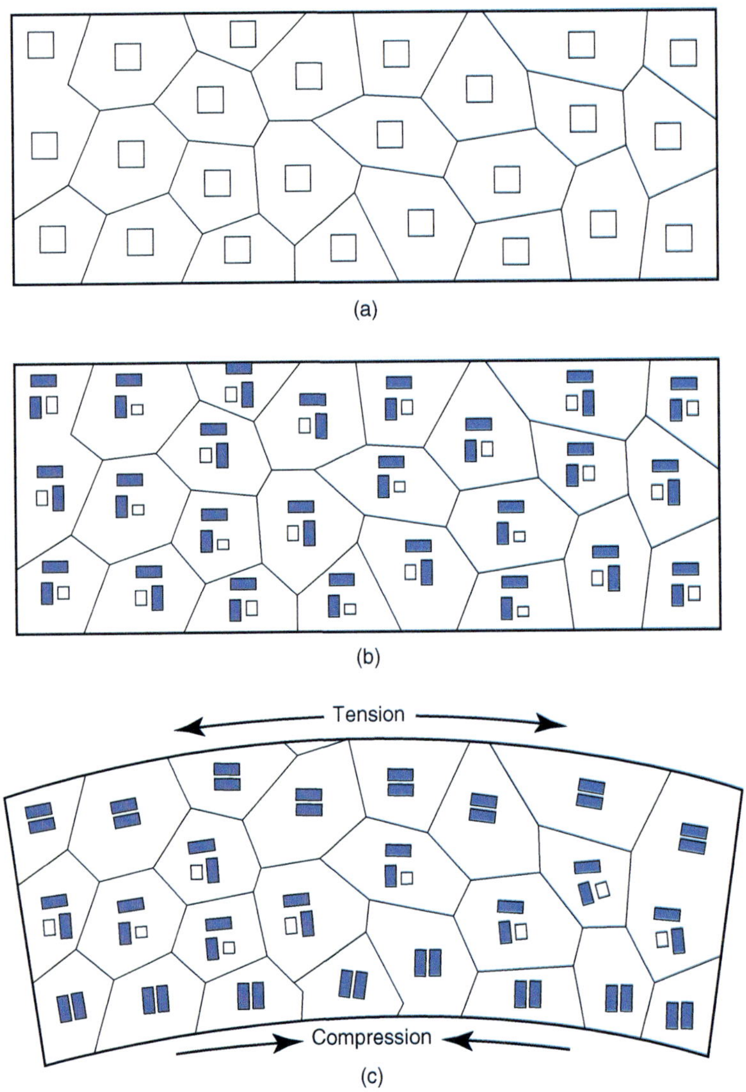

Figure 3.6 Shape-memory alloys transform from (a) a partially ordered, high-temperature austenitic phase to (b) a mixed austenite–martensite low-temperature state to (c) an ordered mixed-phase state under deformation.

the battle between thermodynamics and kinetics for determining the likelihood of a given reaction occurring within a specified time period. We will also see how important this process is from an industrial standpoint. For example, a 2013 assessment estimated the global costs of corrosion at $2.5 trillion each year and that 15–35% of this cost could be prevented by using modern corrosion control techniques [9]. It is important to understand the mechanisms of corrosion before we can attempt to control it.

Types of Corrosion

There are many ways in which to categorize corrosion. One system classifies corrosion into two general categories: macroscopic and microscopic (see Figure 3.7). *Macroscopic corrosion* is basically "outside-in" corrosion and includes such phenomena as *pitting* and *galvanic* (two-metal) *corrosion*. *Microscopic corrosion* is more of an "inside-out" process, where such environmental influences as mechanical stress can cause intergranular corrosion and eventually cracking deep within the sample. Another way to classify corrosion is according to the type of environment—that is, wet versus dry corrosion. This is a useful classification scheme for describing why ordinary steel is relatively resistant to dry chlorine gas, but moist chlorine gas, and especially hydrochloric acid solutions, can cause severe damage to steel.

Need More Help?
Try this video on types of corrosion: https://www.youtube.com/watch?v=aUKBO06_ymw

For our purposes, the classification system is not as important as (a) recognizing that many different forms of corrosion exist and (b) understanding the fundamental kinetic processes behind these different types. To this end, we will first look at the important principles of corrosion and then see how they can be applied to some important types of corrosion.

Electrochemical Reactions

Consider a simple *galvanic cell*, composed of two metal *electrodes*, zinc and copper, immersed in two different aqueous solutions of unit activity—in this case, 1.0 M $ZnSO_4$ and 1.0 M $CuSO_4$, respectively, connected by an electrical circuit, and separated by a *semipermeable membrane* (see Figure 3.8). The membrane allows passage of ions, but not bulk flow of the aqueous solutions from one side of the cell to the other. Electrons are liberated at the *anode* by the *oxidation* (increase in the oxidation number) of the zinc electrode:

$$Zn \rightarrow Zn^{2+} + 2e^- \tag{3.18}$$

Electrons flow through the circuit connecting the two electrodes, and they are consumed at the *cathode* by the *reduction* (decrease in oxidation number) of copper ions in solution:

$$Cu^{2+} + 2e^- \rightarrow Cu \tag{3.19}$$

The overall reaction, then, is the sum of these two *half-reactions*:

$$Zn + Cu^{2+} \rightarrow Zn^{2+} + Cu \tag{3.20}$$

The sulfate ions that remain as a result of reaction (3.19) pass from right to left through the semipermeable membrane in order to balance the charges arising from the zinc ions forming at the anode. Current continues to pass through the circuit, and the zinc electrode slowly dissolves while the copper electrode slowly grows through the precipitation of copper. If we place a high-resistance voltmeter between the copper and zinc electrodes, we measure a potential difference of approximately 1.1 V.

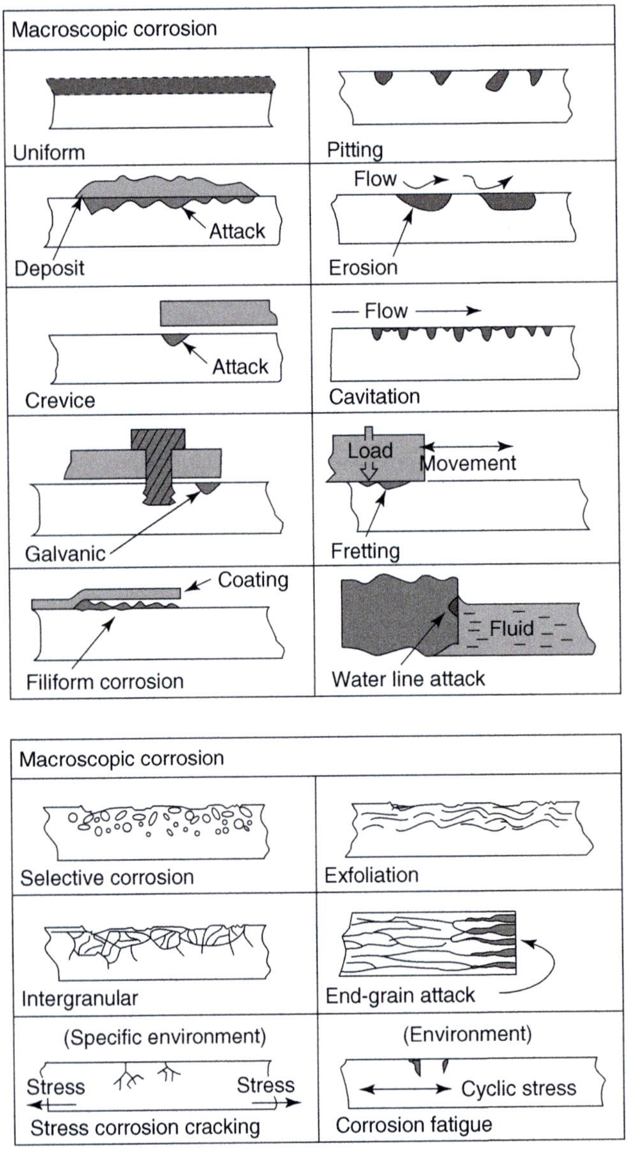

Figure 3.7 Types of corrosion. Reproduced, by permission, from Elliott [20]. Copyright © 1998 by the American Institute of Chemical Engineers.

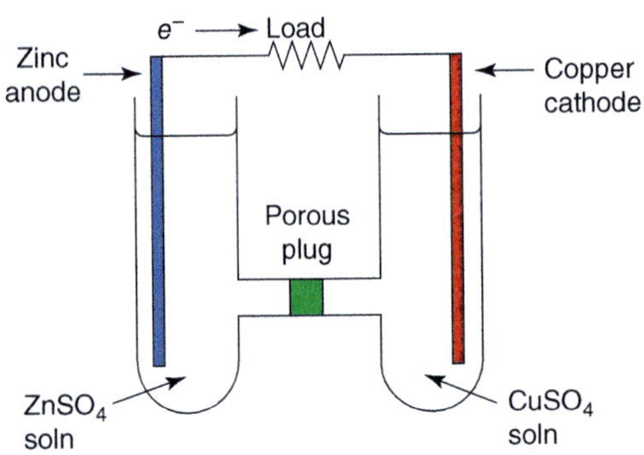

Figure 3.8 Schematic diagram of an electrochemical cell.

The voltage we measure is characteristic of the metals we use. As an additional example, unit activity solutions of $CuCl_2$ and AgCl with copper and silver electrodes, respectively, give a potential difference of about 0.45 V. We could continue with this type of measurement for all the different anode–cathode combinations, but the number of galvanic cells needed would be very large. Fortunately, the half-reactions for most metals have been calculated relative to a *standard reference electrode*, which is arbitrarily selected as the reduction of hydrogen:

$$2H^{+} + 2e^{-} \rightarrow H_2(g) \tag{3.21}$$

The hydrogen EMF is defined as exactly zero, and all other half-cell EMFs are listed relative to the hydrogen electrode at 25°C (see Table 3.3). By convention, the half-cell EMFs are listed as reduction reactions with electrons on the left side of the reaction. All reactions must be reversible, of course, and the corresponding oxidation potential is found by reversing the reaction and reversing the sign on the voltage. For our Zn/Cu galvanic cell example, the EMF we predict for the cell is the sum of

Table 3.3 Standard EMF Potentials for Selected Metals

	Electrode Reaction	Standard Potential, *E* (volts), 25°C
Corrosion less likely ↑	$Au^{3+} + 3e^{-} \rightarrow Au$	1.50
	$Pt^{2+} + 2e^{-} \rightarrow Pt$	1.2
	$Pd^{2+} + 2e^{-} \rightarrow Pd$	0.987
	$Hg^{2+} + 2e^{-} \rightarrow Hg$	0.854
	$Ag^{+} + e^{-} \rightarrow Ag$	0.800
	$Cu^{+} + e^{-} \rightarrow Cu$	0.521
	$Cu^{2+} + 2e^{-} \rightarrow Cu$	0.337
	$\mathbf{2H^{+} + 2e^{-} \rightarrow H_2}$	**0.000**
	$Pb^{2+} + 2e^{-} \rightarrow Pb$	−0.126
	$Sn^{2+} + 2e^{-} \rightarrow Sn$	−0.136
	$Ni^{2+} + 2e^{-} \rightarrow Ni$	−0.250
	$Co^{2+} + 2e^{-} \rightarrow Co$	−0.277
	$Fe^{2+} + 2e^{-} \rightarrow Fe$	−0.440
	$Cr^{3+} + 3e^{-} \rightarrow Cr$	−0.74
	$Zn^{2+} + 2e^{-} \rightarrow Zn$	−0.763
	$Mn^{2+} + 2e^{-} \rightarrow Mn$	−1.18
	$Zr^{4+} + 4e^{-} \rightarrow Zr$	−1.53
	$Ti^{2+} + 2e^{-} \rightarrow Ti$	−1.63
	$Al^{3+} + 3e^{-} \rightarrow Al$	−1.66
Corrosion more likely ↓	$Mg^{2+} + 2e^{-} \rightarrow Mg$	−2.37
	$Na^{+} + e^{-} \rightarrow Na$	−2.71
	$Ca^{2+} + 2e^{-} \rightarrow Ca$	−2.87
	$K^{+} + e^{-} \rightarrow K$	−2.93
	$Li^{+} + e^{-} \rightarrow Li$	−3.05

Source: Craig [24]. Copyright © 1991 by Plenum Publishing.

the two half-cell reactions. From Table 3.3, this should be the EMF for the reduction of copper (0.337 V) plus the EMF for the oxidation of zinc, which we obtain by reversing the reduction reaction and changing the sign (0.763). The sum of these two EMFs is 1.1 V, as measured earlier. It is important to note that the EMF potentials *do not* need to be multiplied by a factor when balancing half-reactions with different oxidation numbers (why?)—for example, a 2+ cathodic reaction with a 1+ anodic reaction. Test this with the Cu/Ag cell example above, where the measured EMF was about 0.45 V, using the data from Table 3.3.

Most environmental conditions do not allow for calculation of these potentials at 25°C; in this case, temperature effects cannot be easily neglected. Electric vehicle (EV) batteries are less efficient, and driving range is compromised in cold weather, even before taking cabin heating into account. Temperature effects can be accounted for by returning to the free energy, which already accounts for temperature fluctuations. Equation (3.8) developed at the beginning of this chapter is where we start.

$$\Delta G - \Delta G^{\circ} = RT \ln\left(\prod_i a_i^{\nu_i}\right) \tag{3.8}$$

For electrochemical systems, we do not have pressure–vapor work that is normally associated with gaseous systems, nor do we have significant thermal effects like heat of fusion or formation. The result is that our usual description of free energy using enthalpies and entropies is not the best representation of free energy here. Instead, the work being generated is electrochemical work that can be best represented by the following relationship:

$$\Delta G = -nEF \tag{3.22}$$

where n is the number of electrons taking part in the reaction, F is a constant called the Faraday constant (96,500 coulombs/mol), and E is the potential of the half-reaction in volts. The corresponding standard state free energy is

$$\Delta G^{\circ} = -nE^{\circ} F \tag{3.23}$$

Substitution of Eqs. (3.22) and (3.23) into Eq. (3.8) results in the well-known *Nernst equation*:

$$E - E^{\circ} = \frac{-RT}{nF} \ln\left(\prod_i a_i^{\nu_i}\right) \tag{3.24}$$

The Nernst equation tells us the electrical potential of a half-cell when the reactants are not at unit activity.

Exchange Current Density

Let us now return to our electrochemical cell shown in Figure 3.8. This cell is a combination of two half-cells, with the oxidation reaction occurring at the anode and the reduction reaction occurring at the cathode resulting in a net flow of electrons from the anode to the cathode. Equilibrium conditions dictate that the rate of oxidation and reduction, r_{oxid} and r_{red}, be equal, where both rates can be obtained from *Faraday's Law*:

$$r_{\text{oxid}} = r_{\text{red}} = \frac{i_0}{nF} \tag{3.25}$$

As in Eq. (3.22), F is the Faraday constant, n is the number of electrons taking part in the reaction, but i_0 is a new quantity called the *exchange current density*. These rates have units of mol/cm$^2 \cdot$ s, so the exchange current density has units of A/cm^2. Typical values of i_0 for some common oxidation

Table 3.4 Some Exchange Current Densities

Reaction	Electrode	Solution	i_0, A/cm^2
$2H^+ + 2e^- \rightleftarrows H_2$	Al	$2\,N\ H_2SO_4$	10^{-10}
$2H^+ + 2e^- \rightleftarrows H_2$	Au	$1N$ HCl	10^{-6}
$2H^+ + 2e^- \rightleftarrows H_2$	Cu	$0.1N$ HCl	2×10^{-7}
$2H^+ + 2e^- \rightleftarrows H_2$	Fe	$2N\ H_2SO_4$	10^{-6}
$2H^+ + 2e^- \rightleftarrows H_2$	Hg	$1N$ HCl	2×10^{-12}
$2H+ + 2e^- \rightleftarrows H_2$	Hg	$5N$ HCl	4×10^{-11}
$2H^+ + 2e^- \rightleftarrows H_2$	Ni	$1N$ HCl	4×10^{-6}
$2H^+ + 2e^- \rightleftarrows H_2$	Pb	$1N$ HCl	2×10^{-13}
$2H^+ + 2e^- \rightleftarrows H_2$	Pt	$1N$ HCl	10^{-3}
$2H^+ + 2e^- \rightleftarrows H_2$	Pd	$0.6N$ HCl	2×10^{-4}
$2H^+ + 2e^- \rightleftarrows H_2$	Sn	$1N$ HCl	10^{-8}
$O_2 + 4H^+ + 4e^- \rightleftarrows 2H_2O$	Au	$0.1N$ NaOH	5×10^{-13}
$O_2 + 4H^+ + 4e^- \rightleftarrows 2H_2O$	Pt	$0.1N$ NaOH	4×10^{-13}
$Fe^{3+} + e^- \rightleftarrows Fe^{2+}$	Pt		2×10^{-3}
$Ni \rightleftarrows Ni^{2+} + 2e^-$	Ni	$0.5\,N\ NiSO_4$	10^{-6}

Source: Fontana [35]. Copyright © 1986 by McGraw-Hill Book Co.

and reduction reactions of various metals are shown in Table 3.4. Like reversible potentials, exchange current densities are influenced by temperature, surface roughness, and such factors as the ratio of oxidized and reduced species present in the system. Therefore, they must be determined experimentally.

Need More Help?

Try this video on the Nernst Equation:

https://www.youtube.com/watch?v=jousNNceCXs

and the Butler–Volmer and Tafel Equations:

https://www.youtube.com/watch?v=NMgfJehxPNE

Polarization

The net current flow produced in a cell results in a deviation of each half-cell potential from the equilibrium value listed in Table 3.3. This deviation from equilibrium is termed *polarization*, the magnitude of which is given the lowercase Greek symbol eta, η, and is called the *overpotential*, E–E°. There are two primary types of polarization: *activation polarization* and *concentration polarization*.

Activation polarization is an activated process and possesses an activation energy barrier, just as described earlier in Figure 3.1. There are two parts to the activation process, either one of which can determine the rate of the reaction at the electrode. The slowest step, or *rate-determining step*, can be either (a) electron transfer at the electrode–solution interface or (b) formation of atoms at the electrode surface. The activation polarization component of the overpotential, η_a, is related to the actual rate of oxidation or reduction, i, and the exchange current density:

$$\eta_a = \pm \beta \log(i/i_0) \tag{3.26}$$

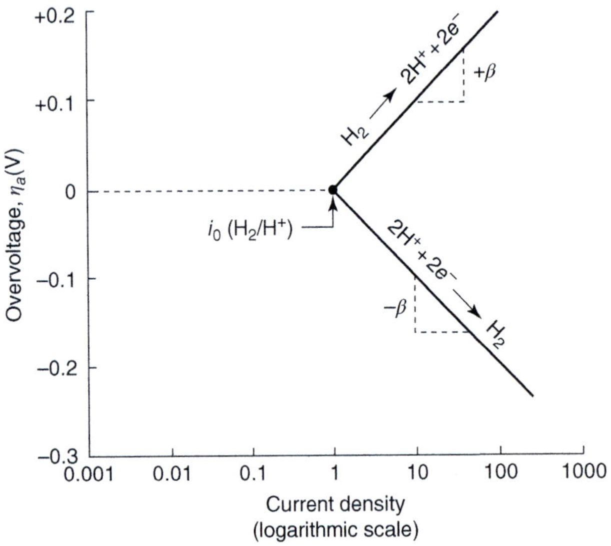

Figure 3.9 Evans diagram showing effect of activation polarization on overpotential for a hydrogen electrode. Reproduced, by permission, from Callister [21]. Copyright © 2000 by John Wiley & Sons, Inc.

This equation is often called the *Tafel equation*. Here, β is a constant, representing the expression $2.3RT/\alpha nF$, where R, T, n, and F have been defined previously, and α is the *symmetry coefficient*, which describes the shape of the rate-controlling energy barrier. Equation (3.26) is represented graphically in Figure 3.9, the resulting plot of which is called an *Evans diagram*. The linear relationship between overpotential and current density is normally valid over a range of about ±50 mV, but linearity is assumed for most Evans diagrams. The reaction rate, as represented by the current density, changes by one order of magnitude for each 100 mV change in overpotential. Note, also, that at $\eta_a = 0$, the rates of oxidation and reduction are equal, and the current density is given by the exchange current density, $i = i_0$.

The second type of polarization, concentration polarization, results from the depletion of ions at the electrode surface as the reaction proceeds. A concentration gradient builds up between the electrode surface and the bulk solution, and the reaction rate is controlled by the rate of diffusion of ions from the bulk to the electrode surface. Hence, the *limiting current* under concentration polarization, i_L, is proportional to the *diffusion coefficient* for the reacting ion, D:

$$i_L = \frac{DnFC_0}{x} \tag{3.27}$$

Take a Peek! Jump ahead to Section 4.3.1 to see how the diffusion coefficient is defined.

Cooperative Learning Exercise 3.2

Iron metal corrodes in acidic solution according to the following reaction:

$$Fe + 2H^+ \rightarrow Fe^{2+} + H_2$$

Given that the rates of oxidation and reduction of the half-reactions are controlled by activation polarization only, that $\beta_{Fe} = +0.07$ and $\beta_{H_2} = -0.08$, and that the exchange current densities for both the oxidation of Fe and reduction of hydrogen in acidic solution are identical, use the data in Tables 3.3 and 3.4 to determine the following quantities. Recall that the potential for each half-cell is the sum of the equilibrium potential and the corresponding overpotential, in this case, η_a. The exchange current density for iron in acidified solution is 10^{-8} A/cm^2.

Person 1: Derive an expression for the electrical potential, E_H for the reduction of hydrogen as a function of the current density, i.

Person 2: Derive an expression for the electrical potential for the oxidation of iron, E_{Fe} as a function of the current density, i.

Group: Recognize the fact that at equilibrium, $E_H = E_{Fe}$ and combine your results to determine the current density, i, for the cell, and use this value to calculate the corrosion rate of iron, r_{Fe}.

Answer: E_H = 0.000 V − 0.08 log $(I/10^{-6})$; E_{Fe} = −0.440 V + 0.07 log $(i/10^{-8})$; i_c = 10^{-4} A/cm^2; r_{Fe} = 5.18 × 10^{-10} mol/cm^2

where n and F have been defined previously, C_0 is the concentration of reacting ions in the bulk solution, and x is the thickness of the *depleted region* near the electrode, or the distance through which the ions must travel from the bulk to the electrode, termed the *Nernst diffusion layer*. The overpotential for concentration polarization, η_c, is:

$$\eta_c = \frac{2.3RT}{nF} \log\left(1 - \frac{i}{i_L}\right) \tag{3.28}$$

Both activation and concentration polarization typically occur at the same electrode, although activation polarization is predominant at low reaction rates (small current densities) and concentration polarization controls at higher reaction rates (see Figure 3.10). The combined effect of activation and concentration polarization on the current density can be obtained by adding the contributions from each [Eqs. (3.26) and (3.28)], with appropriate signs for a reduction process only to obtain the *Butler–Volmer equation*:

$$\eta_{\text{reduction}} = -\beta \log\left(\frac{i}{i_0}\right) + \frac{2.3RT}{nF} \log\left(1 - \frac{i}{i_L}\right) \tag{3.29}$$

Eq. (3.29) allows us to determine the kinetics of most corrosion reactions from only three parameters: β, i_0, and i_L.

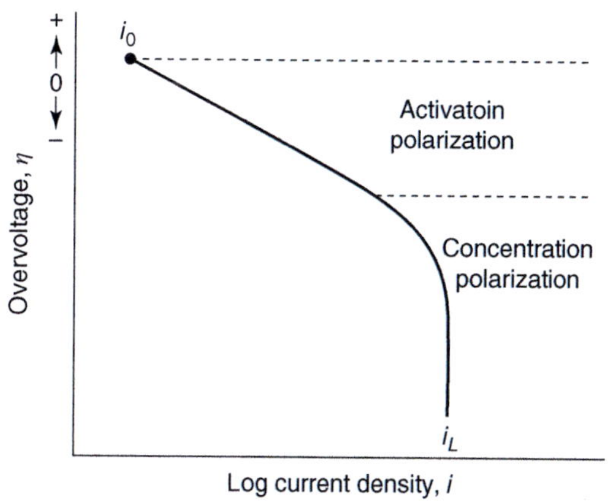

Figure 3.10 Effect of activation and concentration polarization on overpotential. Reproduced, by permission, from Callister [22]. Copyright © 2000 by John Wiley & Sons, Inc.

Corrosion Rate

The current densities give us an estimate of the rate of the electrochemical reactions involved in corrosion, but they do not give us a direct measurement of the corrosion rate. A simple, empirical expression can be used to quantify the rate of corrosion in terms of a quantity called the *corrosion penetration rate* (*CPR*):

$$CPR = \frac{KW}{\rho A t} \tag{3.30}$$

where W is the weight loss of the sample per unit time, t. The variables ρ and A represent the density of the material and the exposed surface area of the sample, respectively. The constant K varies with the system of units being used; $K = 534$ to give *CPR* in mils per year (*mpy*, where 1 mil = 0.001 in.), and $K = 87.6$ to give *CPR* in mm/y.

3.2 Kinetic Processes in Ceramics and Glasses*

3.2.1 Nucleation and Growth (Round 2)

We described spinodal decomposition of polymers in Chapter 2 and martensitic-type transformations in metals in Section 3.0.2, where we also initiated a description of crystallization processes. We wish to elaborate on crystallization by identifying two primary phenomena called *nucleation* and *growth*. Nucleation and growth kinetics comprise two separate processes. Nucleation is the initial formation of small particles of the product phase from the parent phase. The resulting nuclei are often composed of just a few molecules. The growth step involves the increase in size of the nucleated particles. Many natural phenomena follow nucleation and growth kinetics, such as ice formation. Unlike martensitic transformations, both spinodal decomposition and nucleation and growth processes involve diffusion, but there are some subtle differences. Figure 3.11 shows a comparison based on composition and spatial extent between a spinodal transformation and a

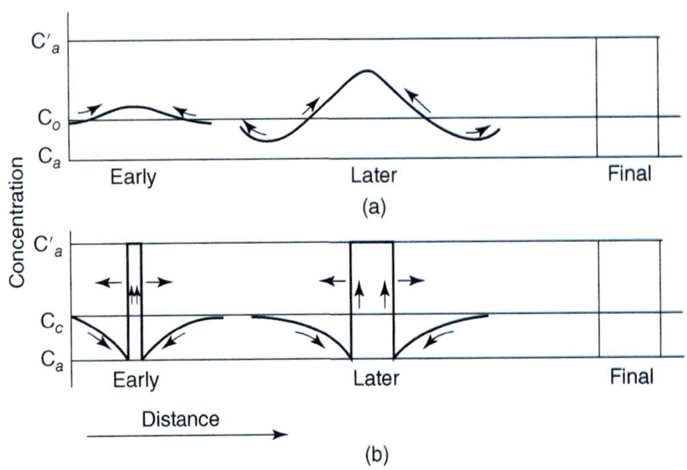

Figure 3.11 Schematic comparison of dimensional changes that occur in (a) spinodal and (b) nucleation and growth transformation processes. From Kingery et al. [23]. Copyright © 1976 by John Wiley & Sons, Inc. This material is used by permission of John Wiley & Sons, Inc.

nucleation and growth transformation. Spinodal transformations are large in spatial extent but involve relatively small concentration differences throughout the sample, whereas the nucleation process involves the formation of small domains of composition very different from the parent phase, which then grow in spatial extent via concentration-gradient-driven diffusion.

Let us first consider the liquid–solid phase transformation. At the melting point (or more appropriately, fusion point for a solidification process), liquid and solid are in equilibrium with each other. At equilibrium, we know that the free energy change for the liquid–solid transition must be zero. We can modify Eq. (2.12) for this situation:

$$\Delta G_v = \Delta H_f - T_m \Delta S_f = 0 \tag{3.31}$$

where ΔG_v is the free energy change per unit volume for fusion, ΔH_f is the enthalpy of fusion and is related to the strength of interatomic or intermolecular forces of the solid, ΔS_f is the entropy of fusion, representing the degree of disorder in the solid, and T_m is the melting point. We can solve Eq. (3.31) for the entropy change of fusion:

$$\Delta S_f = \frac{\Delta H_f}{T_m} \tag{3.32}$$

If we assume that the entropy and enthalpy are relatively independent of temperature, we can drop the subscript "m" on temperature in Eq. (3.32), thereby obtaining a more general relationship between the entropy and enthalpy of fusion as a function of temperature. Substitution of this modified form of Eq. (3.32) into Eq. (3.31) gives the following relation for the free energy:

$$\Delta G_v = \Delta H_f \left[\frac{(T_m - T)}{T_m} \right] \tag{3.33}$$

This expression will be very useful in the discussion of nucleation and growth kinetics.

In the nucleation step, there must be sites upon which the crystals can form. This is similar to "seeding" the clouds to cause water to precipitate (rain). There are two sources for these nucleating particles: homogeneous and heterogeneous agents.

Homogeneous Nucleation

In *homogeneous nucleation*, critical nuclei (nuclei with enough molecules to initiate growth) form without the aid of a foreign agent. This happens very rarely in practice but is useful for theoretical discussions. Consider the formation of a small, solid, spherical particle from the parent (liquid) phase. There is a free energy change, ΔG, associated with this process. Much like the potential energy function between two atoms, there are two competing forces that contribute to the overall free energy change: (1) the free energy change due to a volume change (a negative contribution) and (2) the free energy change due to formation of an interphase (a positive contribution). These forces are quantified in Eq. (3.34):

$$\Delta G = \left[\frac{4\pi r^3}{3} \right] \Delta G_v + 4\pi r^2 \gamma \tag{3.34}$$

where ΔG_v is the free energy change per unit volume for the phase change (which must be negative); r is the nucleated particle radius; γ is the *interfacial energy* (often called "surface tension" for a liquid–gas interface; refer back to Section 2.2.2), which has units of energy/area (erg/cm^2) or force/length (dyne/cm). The first term in Eq. (3.34) is the free energy change associated with the volume change, and the second term is the free energy change associated with the interphase formation.

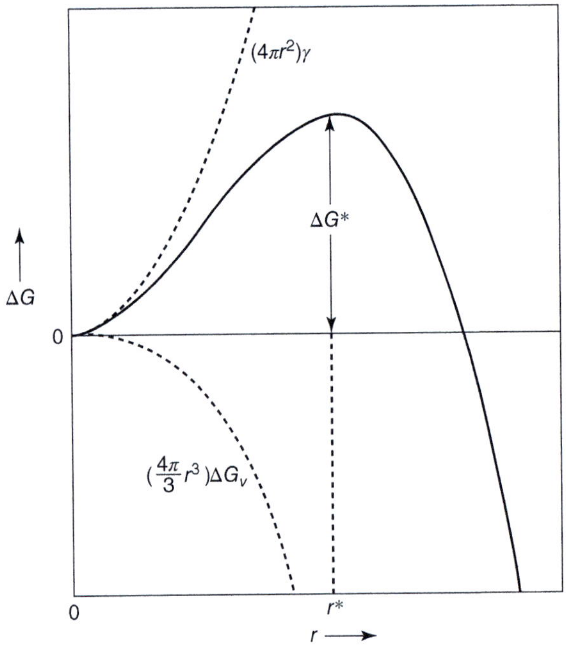

Figure 3.12 Representation of activation energy barrier for nucleation. From Ralls et al. [19]. Copyright © 1976 by John Wiley & Sons, Inc. This material is used by permission John Wiley & Sons, Inc.

As with any activated process, there is a free energy barrier, ΔG^*, or "critical free energy" that must be surmounted to nucleate a particle. This is illustrated in Figure 3.12 as a plot of free energy versus the particle size, r. Once the overall free energy change decreases past ΔG^*, the nucleation process becomes spontaneous. This occurs as enough molecules gather to form a nucleus with critical radius, r^*. We can solve for r^* by taking the derivative of ΔG in Eq. (3.34) with respect to r, setting it equal to zero (the slope of ΔG versus r in Figure 3.12 is zero at $\Delta G = \Delta G^*$), replacing r by r^* to get:

$$r^* = \frac{-2\gamma}{\Delta G_v} \tag{3.35}$$

Cooperative Learning Exercise 3.3

Work with a neighbor. Consider Eq. (3.34), which describes the free energy change associated with the homogeneous nucleation of a solid, spherical particle of radius, r, from a parent liquid phase:

$$\Delta G = \left(\frac{4\pi r^3}{3}\right)\Delta G_v + 4\pi r^2\gamma$$

Recall that this equation could be minimized with respect to particle radius to determine the critical particle size, r^*, as given by Eq. (3.35). This critical radius could then be used to determine the height of the free energy activation energy barrier, ΔG^*, as given by Eq. (3.36). A similar derivation can be performed for a cubic particle with edge length, a.

Person 1: What is the equivalent expression for the first term in Eq. (3.34) for the cube? Recall that ΔG_v is the free energy change per unit volume.

Person 2: What is the equivalent expression for the second term in Eq. (3.34) for the cube? Recall that γ is the interfacial energy for the liquid–solid interface.

Combine your results to arrive at an equivalent expression for Eq. (3.34) for the cube. Minimize this equation to determine the critical cube length, a^*. Substitute this value back into your expression to determine the critical free energy barrier for the cube, $\Delta G_v{}^*$. Why is $\Delta G_v{}^*$ higher for a cube than for a sphere? What does this tell you about the likelihood of cubes nucleating rather than spheres?

Answer: $a^3\Delta G_v$; $6a^2\gamma$; $a^* = -4\gamma/\Delta G_v$; $\Delta G_v{}^* = 32\gamma^3/\Delta G_v{}^2$. The barrier is lower for spherical particles so their formation is more likely.

The height of the activation energy barrier, ΔG^*, can now be found by evaluating Eq. (3.34) at r by r^* with the aid of Eq. (3.35)

$$\Delta G^* = \frac{16\pi\gamma^3}{3\Delta G_v^2} \tag{3.36}$$

To summarize the nucleation stage, then, particles that achieve a critical radius $r > r^*$ may enter the growth stage of this phase transformation process.

Heterogeneous Nucleation

In *heterogeneous nucleation*, the critical nuclei form with the assistance of nucleating sites. These sites can be impurities, container walls, cracks, or pores and may either be naturally present or intentionally added to promote nucleation. The effect of nucleating agents is to decrease ΔG^*.

Unlike homogeneous nucleation, heterogeneous nucleation involves an interface between two compositionally different materials, so we must account for the interaction of the parent phase with the nucleating particle. This is accomplished through the introduction of the contact angle, θ, at the three-phase interphase of the parent liquid, its solid phase, and the nucleating agent (see Figure 3.13 and refer back to Eq. 2.60). Without derivation, we present the free energy barrier for heterogeneous nucleation, ΔG_s^*, as

$$\Delta G_s^* = \Delta G^* \frac{(2 + \cos\theta)(1 - \cos\theta)^2}{4} \tag{3.37}$$

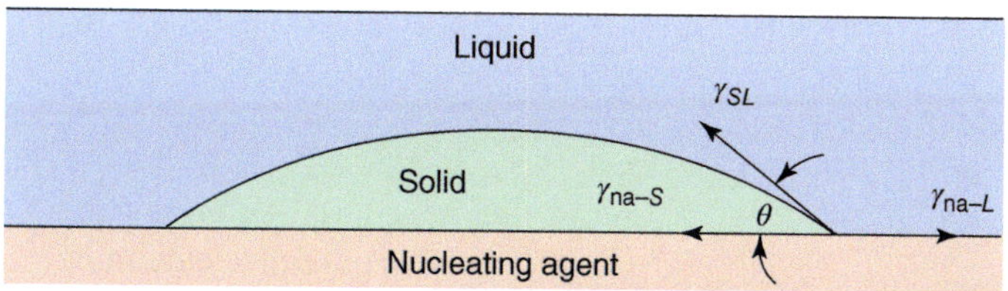

Figure 3.13 Schematic diagram of contact angle formed at three-phase intersection during heterogeneous nucleation. From Ralls et al. [19]. Copyright © 1976 by John Wiley & Sons, Inc. This material is used by permission John Wiley & Sons, Inc.

where ΔG^* is the free energy barrier for homogeneous nucleation and the subscript "s" on ΔG_s^* indicates heterogeneous or "surface" nucleation.

Nucleation Rate

We can now define an overall nucleation rate:

$$\mathbb{N} = \nu n_s n^* \tag{3.38}$$

where $\mathbb{N}$ is the number of critical nuclei that form per unit volume per unit time, n_s is the number of molecules in contact with critical nucleus, n^* is the number of critical size clusters per unit volume, and ν is the collision frequency of single molecules with the nuclei (see Figure 3.14). The number of critical nuclei, n^* can be related to the free energy barrier, ΔG^*, through an Arrhenius-type expression:

$$n^* = n_0 \exp(-\Delta G^*/k_B T) \tag{3.39}$$

where n_0 is the number of single molecules per unit volume, k_B is Boltzmann's constant, and T is absolute temperature. Similarly, the collision frequency, ν, can be expressed by an Arrhenius relation:

$$\nu = \nu_0 \exp(-\Delta G_m/k_B T) \tag{3.40}$$

where ν_0 is the molecular jump frequency, and ΔG_m is the activation energy for transport across the nucleus–matrix interface, which is related to short-range diffusion. Putting these expressions back into Eq. (3.38) gives:

$$\mathbb{N} = \nu n_s n_0 \exp(-\Delta G^*/k_B T) \exp(-\Delta G_m/k_B T) \tag{3.41}$$

For heterogeneous nucleation, we simply use ΔG_s^* instead of ΔG^*:

$$\mathbb{N} = \nu_0 n_s n_0 \exp\left(-\Delta G_s^*/k_B T\right) \exp(-\Delta G_m/k_B T) \tag{3.42}$$

Though Eqs. (3.41) and (3.42) are somewhat cumbersome, they describe the rate at which nuclei form as a function of temperature. Note that in both equations, n_s and n_0 have dimensions of number per unit volume, and ν_0 has units of time^{-1}.

What would be more useful are forms of Eqs. (3.41) and (3.42) that contain some measurable physical properties of the system. Free energies hardly fit this description. The danger in doing this

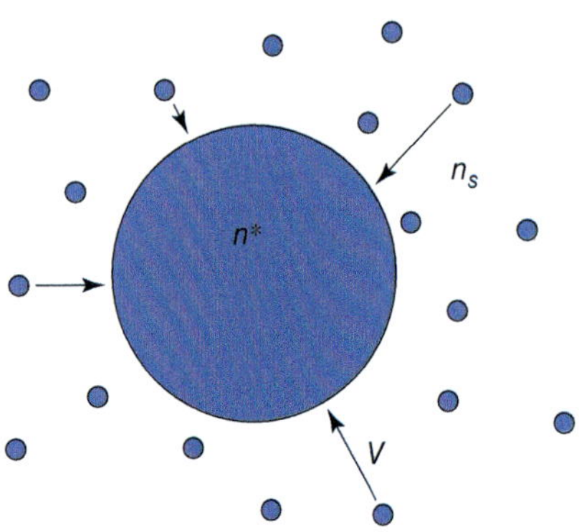

Figure 3.14 Schematic illustration of nucleation site formation.

is that we must make some simplifications and assumptions. Let us begin by recalling that the free energy for homogeneous nucleation is given by Eq. (3.36):

$$\Delta G^* = \frac{16\pi\gamma^3}{3(\Delta G_v)^2} \tag{3.36}$$

Recall that for liquid–solid transitions at temperatures close to T_m, ΔG_v is given by Eq. (3.33):

$$\Delta G_v = \frac{\Delta H_v(T_m - T)}{T_m} \tag{3.33}$$

where ΔH_v is the heat of transformation per unit volume. Substitution of Eqs. (3.33) and (3.36) into the first exponential in Eq. (3.42) gives us the temperature dependence of the nucleation rate for homogeneous nucleation:

$$\mathbb{N} = \nu_0 n_s n_0 \exp\left[\frac{-16\pi\gamma^3 T^2}{3k_B T \Delta H_v^2 (T_m - T)^2}\right] \exp\left(\frac{-\Delta G_m}{k_B T}\right) \tag{3.43}$$

While this relation may appear even more cumbersome than Eq. (3.41), Eq. 3.43 contains some parameters that are directly measurable such as the interfacial surface energy, γ, and the heat of fusion, ΔH_v, but more importantly, it contains the temperature difference $(T_m - T)$, which is the degree of undercooling—that is, how far the temperature is below the melting (solidification) point!

The two exponentials in Eq. (3.43) compete against each other. As T decreases below the melting point:

- $(T_m - T)$ becomes larger (remember that this term is squared, so it is always a positive value), the term inside the first exponential becomes smaller, but because it is negative, the entire exponential gets bigger and the *nucleation rate increases* due to this exponential—*the driving force for nucleation becomes greater*, and
- the term in the second exponential becomes more negative, the exponential gets smaller and the *nucleation rate decreases* due to this exponential term—*diffusion becomes more difficult as the temperature decreases and particles cannot migrate to the nucleation surface.*

These generalizations are true for both homogeneous and heterogeneous nucleation. As a result, we would expect that the two competing exponentials give rise to a maximum nucleation rate at some temperature below the melting point, and this is indeed the case, as illustrated in Figure 3.15.

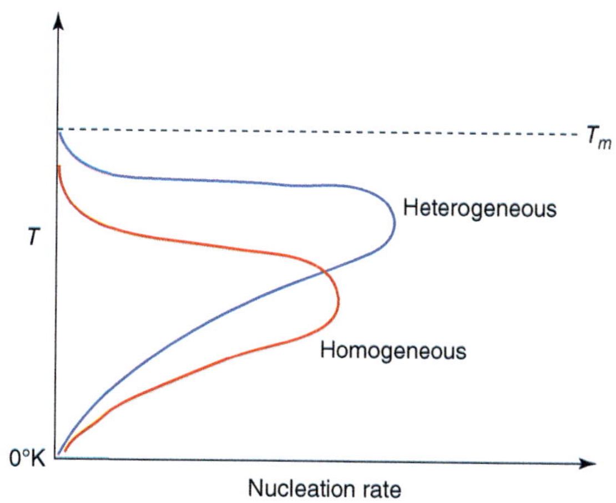

Figure 3.15 Effects of temperature and undercooling on homogeneous and heterogeneous nucleation rates.

It is also logical that heterogeneous nucleation has a higher absolute nucleation rate than homogeneous nucleation and that heterogeneous occurs at a higher temperature (lower degree of undercooling) than homogeneous nucleation.

Growth

The second portion of nucleation and growth is the growth process. Here, the tiny nuclei grow into larger crystals through the addition of molecules to the solid phase. There are two primary types of crystal growth: thermally activated (diffusion controlled) and diffusionless (martensitic). We have already discussed martensitic transformations, so only *thermally activated crystal growth* will be considered here. The development of the proper growth rate expression is highly dependent upon the type of phase transformation—that is, crystallization from the melt, vapor phase, or dilute solution. We will simply use a general form of the growth rate expression which is based upon an Arrhenius-type expression:

$$\mathbb{R} = A\left[1 - \exp\left(\Delta G / k_{\mathrm{B}} T\right)\right] \tag{3.44}$$

where $\mathbb{R}$ is the growth rate, ΔG is the molar free energy difference between product and parent phase, and A is a preexponential factor. The form of the preexponential factor, A, depends on the type of theory one wishes to employ. It can be directly related to the liquid phase viscosity, as in $A = \nu a_0$, where ν is the *frequency factor* for transport across the interphase

$$\nu = \frac{k_{\mathrm{B}} T}{3\pi a^3 \eta} \tag{3.45}$$

in which η is the viscosity of the liquid phase, and a_0 is the molecular diameter; or A can be related to the diffusivity of atoms across the interphase, as in $A = KD$, where K is a constant and the diffusivity of atoms jumping across interphase, D, is given by

$$D = D \exp\left(\frac{-\Delta H^0}{k_{\mathrm{B}} T}\right) \tag{3.46}$$

These two expressions are similar since D and η are interrelated (see Section 4.3). Other expressions exist as well for such phenomena as surface growth and screw dislocation growth.

Overall Phase Transformation Rate

Finally, the overall transformation rate, $\mathbb{P}$ is given by the product of the nucleation and growth expressions [Eqs. (3.41) or (3.42), and (3.44), respectively]:

$$\mathbb{P} = \mathbb{N} \times \mathbb{R} \tag{3.47}$$

The overall transformation rate is shown qualitatively in Figure 3.16. Notice that the maximum nucleation rate occurs at a lower temperature than the maximum growth rate and that the maximum transformation rate may not be at either of these two rate maxima. Note also that there is some finite transformation rate, even at very low temperatures which is why some glasses can crystallize over very long periods of time. As long as there is some molecular motion, there is a probability of crystallization taking place.

Finally, the breaking down of the liquid to solid phase transformation into two separate steps—nucleation and growth—is not entirely artificial. An interesting application of the distinct nature of the nucleation and growth processes is found in the formation of glass ceramics. As shown in Figure 3.17, certain glass-forming inorganic materials can be heat-treated in very controlled ways

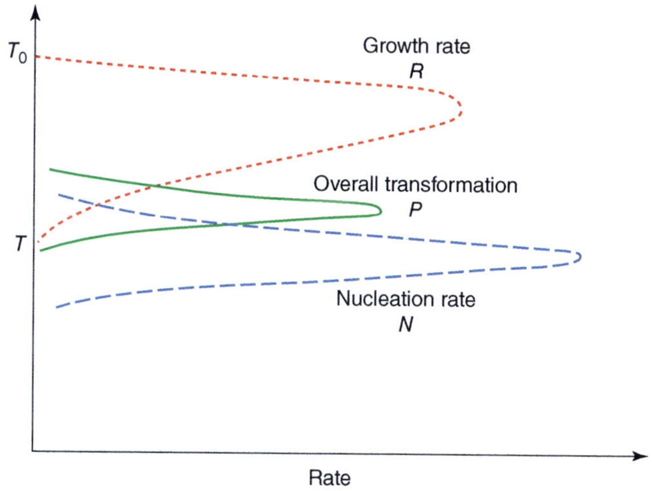

Figure 3.16 Schematic representation of transformation rates involved in crystallization by nucleation and growth kinetics.

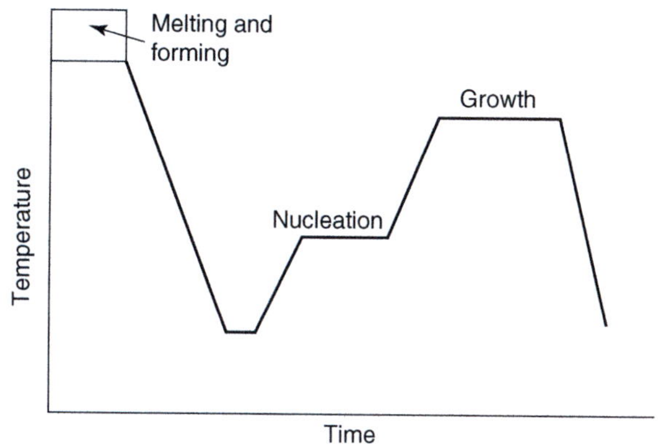

Figure 3.17 Schematic time–temperature cycle for the controlled crystallization of a glass–ceramic body.

to affect the structure of the crystals that form. For instance, a glass can be rapidly cooled below its melting point, then heated to the maximum nucleation rate, which (recall) is below both the melting point and the maximum growth rate. At this temperature and with sufficient time, many small nuclei will be formed. The glass, now replete with many tiny nuclei, is heated to the maximum growth rate temperature and held there while the crystals grow. The crystals can only grow so large, though, since they soon run into another growing crystal from a neighboring nucleation site. In this way, the size of the crystals, or *crystallites*, in the glass can be controlled, and a *glass ceramic* is formed. Glass ceramics are unique in that they are crystalline materials, yet they are transparent in many cases (see Section 1.2.4). They also possess unique physical properties, such as low thermal expansion, due to the large amount of interphase relative to the same material made of larger crystals. In the extreme, crystallites on the order of 10^{-9} m in diameter can be formed, resulting in so-called *nanostructured ceramics*. These materials, like glass ceramics, have unique physical properties, which may one day lead to improved ductility and thermal properties in ceramics.

3.2.2 Kinetics of Corrosion in Ceramics and Glasses

Although the term corrosion is normally associated with metals, the definition presented in Section 3.1.3 is equally applicable to ceramic materials. The cost to industry due to ceramic corrosion is similarly large, over $1 billion annually. For example, the cost to replace the refractory insulation in a glass-melting furnace might be about $10 million, nearly 60% of which is related to replacing the insulation itself, with the remainder due to labor costs. As with metallic corrosion, types of ceramic corrosion can be grouped in different ways, but classification according to the type of attacking medium (i.e., liquid versus gaseous corrosion) is a useful distinction. We will investigate these two methods of attack separately.

Kinetics of Ceramic and Glass Corrosion by Liquids

The corrosion of a solid, crystalline ceramic material by a liquid occurs through the formation of a reaction product between the solid and the solvent. This reaction product is generally less soluble than the bulk ceramic solid, and it may form an attached surface layer on the solid. This type of corrosion in a liquid is called *indirect dissolution*, *incongruent dissolution*, or *heterogeneous dissolution*. The rate-limiting step in this type of ceramic corrosion can be the interfacial layer formation, diffusion through the interfacial layer, or diffusion through the liquid. *Direct dissolution*, *congruent dissolution*, or *homogeneous dissolution* occur when the solid, crystalline ceramic material dissolves directly into the liquid through dissociation or reaction. The concentrations of the ceramic species in solution, along with the corresponding diffusion coefficients of the species, determine the rate of direct dissolution.

The type of attack that occurs in liquid media is highly dependent on the chemical nature of the liquid—that is, molten metal, molten ceramic, or aqueous solution. We will consider two industrially important cases: attack by molten metals and attack by aqueous media. The attack of most metal oxide ceramics by molten metals involves a simple exchange of one metal ion for another. For example, silicon dioxide in contact with molten aluminum is susceptible to the following corrosion reaction:

$$3\,\mathrm{SiO_2}(s) + 2\,\mathrm{Al}(l) \rightarrow 2\mathrm{Al_2\,O_3}(s) + 3\mathrm{Si}(l) \tag{3.48}$$

The ease with which this exchange can occur is given by the free energy of reaction, ΔG, which is the summation of the free energy of formation for the individual oxides, presented here at 300 K in kJ/mol:

$$\begin{array}{ll} 3\,\mathrm{SiO_2}(s) \rightarrow 3\mathrm{O_2}(g) + 3\,\mathrm{Si}(l) & \Delta G_f = 2568 \\ \underline{3\mathrm{O_2}(g) + 4\mathrm{Al}(l) \rightarrow 2\mathrm{Al_2O_3}(s)} & \Delta G_f = -3164 \\ 3\,\mathrm{SiO_2}(s) + 4\mathrm{Al}(l) \rightarrow 2\mathrm{Al_2O_3}(s) + 3\,\mathrm{Si}(l) & \\ & \Delta G_f = -596 \end{array} \tag{3.49}$$

Values for formation reactions can be found in the Joint Army-Navy-Air Force (JANAF) thermochemical tables [10]. Another way in which a molten metal can attack an oxide is through compound formation, such as the formation of spinel:

$$4\mathrm{Al_2O_3}(s) + 3\mathrm{Mg}(l) \rightarrow 3\mathrm{MgAl_2O_4}(s) + 2\mathrm{Al}(l) \tag{3.50}$$

This reaction has a lower free energy of formation (−218 kJ/mol at 1000°C) compared to the simple oxidation–reduction reaction (−118 kJ/mol) analogous to Eq. (3.48) above. It is also possible that the molten metal can completely reduce the metal oxide ceramic, forming an alloy and gaseous oxygen:

$$M_xO_y(s) \xrightarrow{\text{liquid metal}} xM(\text{solution}) + \frac{y}{2}O_2(g) \tag{3.51}$$

Ceramic materials in contact with aqueous liquids are also susceptible to attack. This is of particular importance in the realm of civil engineering, where mineral-containing materials of construction routinely contact groundwater and soil. Minerals dissolve into aqueous solutions through the diffusion of leachable species into a stationary thin film of water, about 110 μm thick:

$$\text{Mineral } A + nH^+ + mH_2O \rightleftarrows \text{Mineral } B + qM^+ \tag{3.52}$$

where M^+ is the soluble species. The equilibrium constant for this equation is given by Eq. (3.4):

$$K = \frac{[M^+]^q}{[H^+]^n[H_2O]^m} \tag{3.53}$$

where the solids species (minerals) have unit activity. In dilute solutions and at one-atmosphere pressure, the activity of water is also approximately unity. At higher pressures, the activity of water is approximately proportional to the pressure. It is evident from this relationship that the dissolution of ceramics is highly dependent upon the pH ($[H^+]$) of the aqueous medium. Generally, at low pH values (acidic environments) the equilibrium shifts to the right, and corrosion is promoted, whereas high pH values (basic environments) retard corrosion by shifting the equilibrium back to the left.

Not all ceramic materials behave the same at a given pH, however. As the material begins to dissolve, ions form at the surface, water molecules orient themselves accordingly, and an *electrical double layer* is established, as shown in Figure 3.18. The first layer of charged ions and oriented

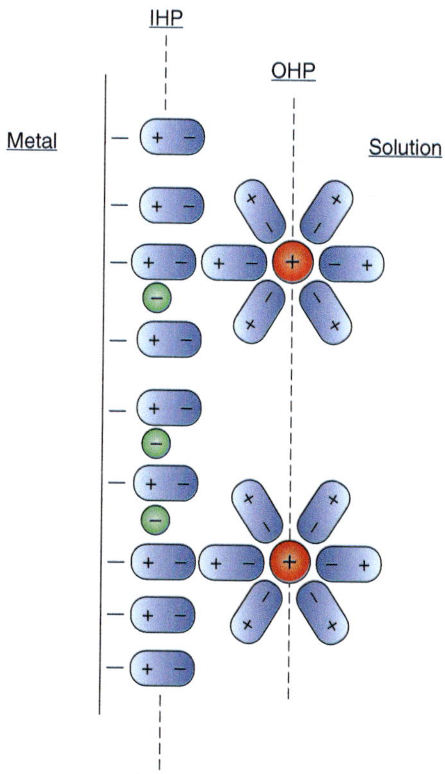

Figure 3.18 Formation of the electrical double layer of a surface in solution, showing the inner Helmholtz plane (IHP) and outer Helmholtz plane (OHP). Reproduced, by permission, from Craig [24]. Copyright © 1991 by Plenum Press.

water molecules is called the *inner Helmholtz plane* (IHP), and the second layer of oppositely charged particles is called the *outer Helmholtz plane* (OHP). This arrangement should remind you of a capacitor; and indeed, this is how this system is often modeled. The potential established between these two layers decreases as we move away from the surface into the solution. The extent to which this double layer is established and the magnitude of the charge on the surface is directly related to the pH of the surrounding solution. The pH value for which there is no net surface change is called the *isoelectric point* (IEP). An empirical formula for determining the IEP of oxides has been established [11]:

$$\mathrm{IEP} = 18.6 - \frac{11.5z}{d} \tag{3.54}$$

where z is the oxidation state of the cation in the film, and d is the distance (in angstroms) of an adsorbed proton from the cation via the oxygen ion. Additional values of IEPs for some typical oxides and hydroxides are given in Table 3.5.

Related to the attack of polycrystalline ceramic materials by aqueous media is the hydrolysis of silicate glasses. The following relationship has been developed to describe the effect of time and temperature on the acid corrosion (10% HCl) of silicate glasses [12]:

$$W = at^{b_1} \exp(-b_2/T) \tag{3.55}$$

where W is the weight loss, t is time, T is temperature, and a, b_1, and b_2 are experimentally determined coefficients. Like aqueous attack of crystalline ceramics, glass dissolution rates are a strong function of pH. Unlike crystalline ceramics, however, there are two competing mechanisms in glass dissolution: *ion exchange* and *matrix dissolution* (see Figure 3.19). In ion exchange, the metal–oxygen bond is coordinated with hydrogen ions, whereas in matrix dissolution the metal–oxygen bond is coordinated with hydroxyl ions. Ultimately, the corrosion of glasses affects their mechanical properties and determines their useful service lifetime on an industrial scale.

Table 3.5 Some Isoelectric Points (IEP) for Selected Oxides and Hydroxides

Material	IEP	Material	IEP
α-Al_2O_3	5–9.2	ZnO	8.8
γ-Al_2O_3	8.0	TiO_2	4.7–62
γ-$Al(OH)_3$	9.25	$Pb(OH)_2$	11.0
α-$Al\ (OH)_3$	5.0	$Mg(OH)_2$	12
CuO	9.5	SiO_2	2.2
Fe_2O_3	8.6	ZrO_2	6.7
Fe_3O_4	6.5	SnO_2	4.3
CdO	12	WO_3	0.4
Ta_2O_5	2.9	M_2O_5, MO_3	<0.5
MO_2	0–7.5	M_2O_3	6.5–10.4
MO	8.5–12.5	M_2O	>11.5

M = metal cation.

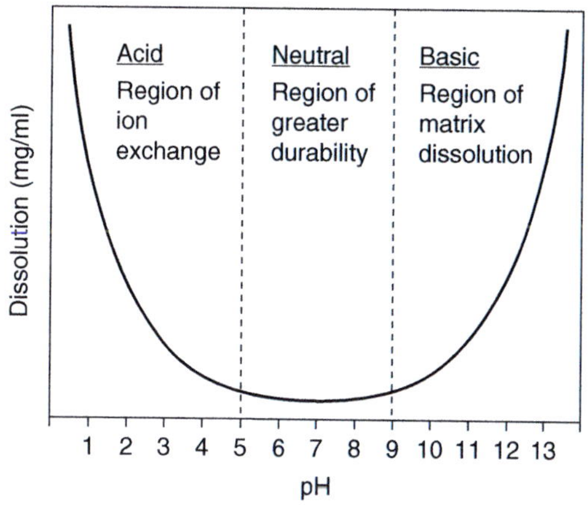

Figure 3.19 Effect of pH upon glass dissolution.

Kinetics of Ceramic and Glass Corrosion by Gases

The attack of ceramics and glasses by gaseous reactants is much more prevalent than corrosion due to liquids. In addition to the chemical composition of the attacking medium, the geometry of the ceramic solid is of tantamount importance to its corrosion resistance. That is, a porous or highly permeable substrate will be much more susceptible to attack by gases than a theoretically dense one. Pore size, pore size distribution, surface-to-volume ratio, and diffusivity all play important roles in this process.

Gas or, more appropriately, vapor attack can result in the formation of either a solid, liquid, or gas, depending on the reaction conditions and the type of product formed. For example, Na_2O vapors can attack SiO_2 to form liquid sodium silicate. In some cases, the vapor may enter the ceramic due to a thermal gradient, then condense and dissolve the ceramic by liquid attack.

The corrosion rate can be limited by any number of steps in the dissolution process:

- Diffusion of the gas to the solid.
- Adsorption of the gas molecule onto the solid surface.
- Surface diffusion of the adsorbed gas.
- Decomposition of reactants at surface sites.
- Reaction at the surface.
- Removal of products from reaction sites.
- Surface diffusion of products.
- Desorption of gas molecules from the surface.
- Diffusion of products away from the solid.

Notice that most of the possible rate-limiting steps involve movement of atoms—diffusion, adsorption, and desorption—and only one direct chemical reaction is involved. In general, the total pressure of the system will determine which step is rate-limiting. At low total pressures ($<10^{-4}$ atm), gas transport is "line-of-sight," and the surface reaction may be rate-controlling. At intermediate pressures (10^{-4}–10^{-1} atm), bulk diffusion is usually rate-limiting, while diffusion through the reaction boundary layer, formed as products accumulate at the surface, controls the overall rate at high pressures ($>10^{-1}$ atm). We will reserve most of our description of diffusion in solids to Chapter 4; but for now, recall that the diffusion coefficient is given by D and has units of cm^2/s.

For gases, the concentration is often represented by the partial pressure of component A, P_A, which gives the functional dependence of corrosion rate on pressure, as described previously.

When the rate of the chemical reaction occurring at the surface is the rate-limiting step, the principles we have described to this point apply. The reaction rate can have any order, and the gas reacts with the ceramic substrate to produce products. Although our discussion to this point has focused on oxide ceramics, there are several nonoxide ceramics, such as carbides, nitrides, or borides, that are of importance and that undergo common decomposition reactions in the presence of oxygen. These ceramics are particularly susceptible to corrosion since they are often used at elevated temperatures in oxidizing and/or corrosive environments. For example, metal nitrides can be oxidized to form oxides:

$$2\mathrm{M}_x\mathrm{N}_y(s) + \mathrm{O}_2(g) \rightarrow 2\mathrm{M}_x\mathrm{O}(s) + y\mathrm{N}_2(g) \tag{3.56}$$

In this case, the reaction rate, r, is a function of the partial pressure of the gaseous species, P_i, since solids are taken to have unit activity:

$$r = k\frac{P_{\mathrm{N}_2}^{y}}{P_{\mathrm{O}_2}} \tag{3.57}$$

Note from reaction (3.56) that the chemical composition of the surface is changing. As this surface layer builds over the course of the reaction, diffusion of species back through this *surface reaction layer*, or *boundary layer*, can become the rate-limiting step (see Figure 3.20). Again, the intricacies of boundary layer diffusion will be addressed in Chapter 4, but for now, we can look at the kinetics associated with the growth of this boundary layer. For thin reaction films (<5 nm in thickness), the thickness of the film, x, changes with time, t, according to one of three general forms: logarithmic, inverse logarithmic, or asymptotic

$$x = k_1 \log t \tag{3.58a}$$

$$\frac{1}{x} = k_2 - k_3 \log t \tag{3.58b}$$

$$x = k_4[1 - \exp(-k_5 t)] \tag{3.58c}$$

For thick films (>5 nm), the film thickness follows either a parabolic or rectilinear relationship with time

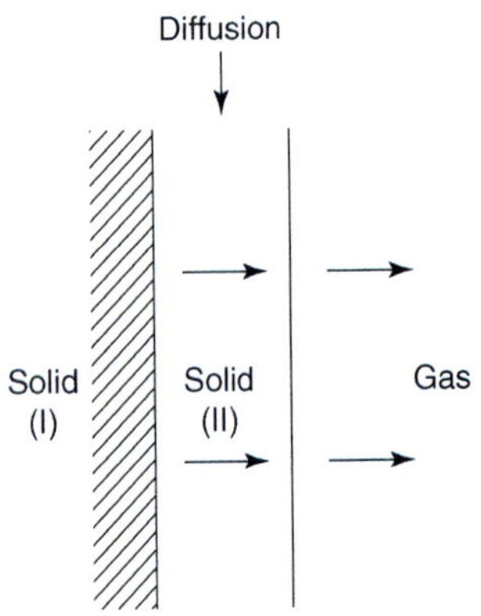

Figure 3.20 Diffusion of reaction products through boundary layer.

$$x^2 = k_6 t \tag{3.59a}$$

$$x = k_7 \tag{3.59b}$$

In Eqs. (3.58a)–(3.58c) and (3.59a) and (3.59b), the k_i are the reaction rate constants. We will see in Chapter 4 that many solid-state ceramic processes involve simultaneous mass transport (diffusion), thermal transport, and reaction.

3.3 Kinetic Processes in Polymers

The previous sections dealt primarily with phase transformations and corrosion in materials. Polymers also undergo phase transformations. For example, there are many polymers that utilize nucleation and growth kinetics to transform from amorphous to crystalline polymers. The kinetics of these transformations are very similar, in principle, to the preceding descriptions for glasses, so it is not necessary to duplicate that material here. Polymers also are susceptible to corrosion, but the term *degradation* is more widely utilized for deterioration in polymers. The kinetics of polymer degradation will be discussed in Section 3.3.3. Before we get to degradation, we should first discuss formation of polymers. Here, we find kinetic processes that are unique to polymers. These are primarily formation reactions whose rates can vary widely depending on such variables as temperature and concentration. This is not the case with most metals. Thermodynamics tells us whether an alloy will form between two metals; and if that process is favorable, it generally proceeds spontaneously at a very rapid rate. The same is true of most ceramics and glasses: when the temperature is high enough, the reactants go to products—not only rapidly, but sometimes explosively. Polymer systems are more complex in that the reacting molecules and products are much larger than those in metallic and ceramic systems, leading to mass- and heat-transport limitations that can affect the kinetics. It is now not enough to know that a reaction is thermodynamically favorable—we must look closely at the actual rate at which polymers are formed. This is the subject of Section 3.3.1. The effect of different types of processing conditions on these reactions will then be discussed in Section 3.3.2.

3.3.1 Kinetics of Polymerization

Polymerization reactions are generally grouped into *stepwise* (also known as *condensation*) reactions and *radical chain* (sometimes called *addition*) reactions, depending upon the nature in which the monomers react with each other or with growing polymer chains to form higher-molecular-weight structures. Condensation reactions are characterized by the reaction of two functional groups that result in the loss of a small molecule. Thus, the repeat unit has a different molecular formula than the monomer from which it was formed. The small molecule formed during condensation is often water, though the specific formation of water is not a necessity for a condensation reaction. In contrast, addition polymerizations are those in which no loss of a small molecule occurs. This type of polymerization reaction usually requires an active center, such as a free radical or an ion, to proceed. Analysis of the kinetics for these two classes of reactions reveals some interesting differences not only in how the polymer is formed but in how the reactions can be controlled, as summarized in Table 3.6.

Table 3.6 Characteristics of Stepwise and Addition Polymerization Reactions

Stepwise Polymerization	Addition Polymerization
Any two molecular species present can react	Only growth reaction adds repeating units one at a time to the chain
Monomer disappears early in the reaction	Monomer concentration decreases steadily throughout the reaction
Polymer molecular weight rises steadily throughout the reaction	High polymer is formed immediately, polymer molecular weight changes little throughout the reaction
Long reaction times necessary to obtain high molecular weight	Long reaction times do not affect molecular weight much but do increase yield

Stepwise Polymerization

Consider the reaction of hexamethylene diamine and adipic acid:

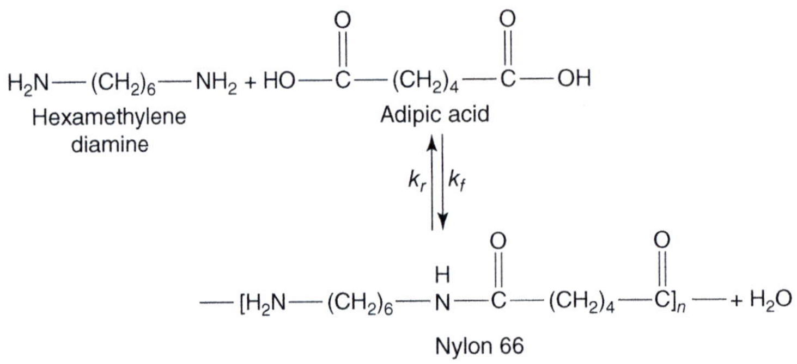

(3.60)

Note first of all that there is the characteristic loss of a small molecule for stepwise reactions, which in this case is water, and that the loss of water results in an amide (HN–CO) linkage; hence, the resulting polymer is called a *polyamide*. This particular polyamide is very common and is better known by its trade name *nylon*. There are many types of nylons, depending on the exact formula of the diamine and diacid used to form the polyamide, but in this case, there are six carbon units between the nitrogen in the diamine and there are six carbon units (including the carbons which contain the double-bonded oxygens) in the diacid, so this polyamide is called *Nylon 66*.

The kinetics of this reaction are analyzed by monitoring the disappearance of one of the reactants, or the formation of the product. If we choose to follow the disappearance of the adipic acid, the rate expression for a catalyzed reaction* is

$$\frac{-d[\mathrm{COOH}]}{dt} = k[\mathrm{COOH}][\mathrm{NH}] \tag{3.61}$$

If we start with equal amounts of diamine and diacid, then for the duration of the reaction we have [COOH] = [NH] = c, and we have a second-order differential equation:

$$\frac{-dc}{dt} = kc^2 \tag{3.62}$$

---

* In the absence of an added strong acid, the reaction is self-catalyzed by a second molecule of the adipic acid, so the reaction is second order in acid: $\frac{-d[\mathrm{COOH}]}{dt} = k[\mathrm{COOH}]^2[\mathrm{NH}]$.

that is easily separated and integrated:

$$\frac{1}{c} = kt + \text{constant} \tag{3.63}$$

A convenient parameter for following the progress of these reactions is the fraction of functional groups that reacted at any time, called the *extent of reaction*, $p = (c_0 - c)/c_0$. Here, c_0 is the initial concentration. In terms of the extent of reaction, the rate expression becomes

$$c_0 kt = \frac{1}{1-p} + \text{constant} \tag{3.64}$$

The quantity $1/(1-p)$ is called the *number-average degree of polymerization*, $\bar{x}_n$, which represents the initial number of structural units present in the monomer relative to the total number of molecules, both monomer and polymer chains, present at any time t:

$$\bar{x}_n = c_0 kt + \text{constant} \tag{3.65}$$

Eq. (3.65) shows us that the degree of polymerization increases linearly with time, an observation that has been verified experimentally. It can be shown that the weight-average degree of polymerization for condensation reactions is similarly related to the extent of reaction:

Need More Help?

Try these videos on stepwise polymerization:

https://www.youtube.com/watch?v=XTleyli6yhk
https://www.youtube.com/watch?v=fp2ISQD-8V4
https://www.youtube.com/watch?v=GoTYtPREvxQ

$$\bar{x}_w = \frac{1+p}{1-p} \tag{3.66}$$

Addition Polymerization

As the name suggests, addition polymerizations proceed by the addition of many monomer units to a single active center on the growing polymer chain. Though there are many types of active centers, and thus many types of addition polymerizations, such as anionic, cationic, and coordination polymerizations, the most common active center is a radical, usually formed at a double bond in the monomer, such as in vinyl chloride. As a result, addition polymerizations are sometimes called *free radical*, *chain growth*, or even *vinyl polymerizations*, although it should be noted that each of these terms is increasingly narrow in its definition. However, analysis of vinyl polymerization provides a good example of the kinetics of addition polymerizations.

In the 1930s, Paul Flory showed that radical polymerizations generally consist of three distinct steps: initiation, propagation, and termination. The free radical must first be initiated, then must propagate through the addition of monomers to the chain, and must eventually be terminated, either through consumption of the monomer or through addition of an agent that kills the free radicals. There are many ways to *initiate* radicals, such as ultraviolet radiation, electrochemical initiation, or oxidation–reduction reactions, but the most common is through addition of an initiator that transfers its radicals to the monomer to begin the polymerization. Organic peroxides, such as benzoyl peroxide, are common initiators that can transfer their radical (represented as a dot, •) to a monomer unit:

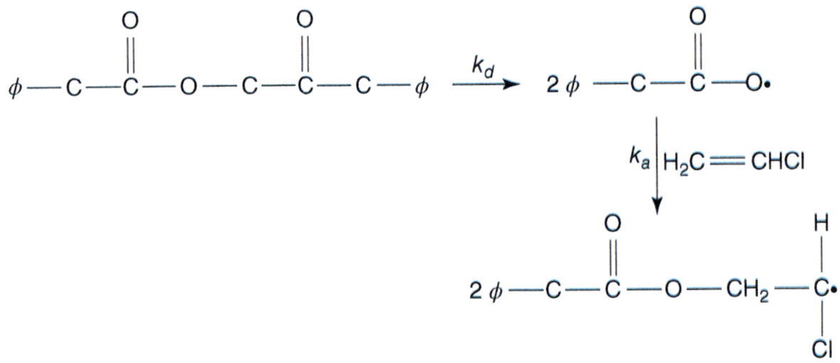

(3.67)

where ϕ is a phenol group. Most of these initiators have efficiencies of between 60% and nearly 100%, with recombination of the radical pairs the most common cause of low efficiency. If the fraction of radicals formed in the dissociation step that leads to successful radical chains is represented by f, then the rate of initiation is given by the change in initiated monomer concentration with time:

$$\frac{d[\mathrm{M}\bullet]}{dt} = 2f\, k_d[\mathrm{I}] \tag{3.68}$$

where [I] is the concentration of the initiator, in this case, benzoyl peroxide, and [M•] is the radical concentration, in this case, the initiated vinyl chloride monomer.

Once the monomer radical has been initiated, high polymer is formed through addition of monomer units to the radical in the *propagation* phase of the polymerization. Each time a monomer unit is added, the radical transfers to the end of the chain to allow the polymerization to continue:

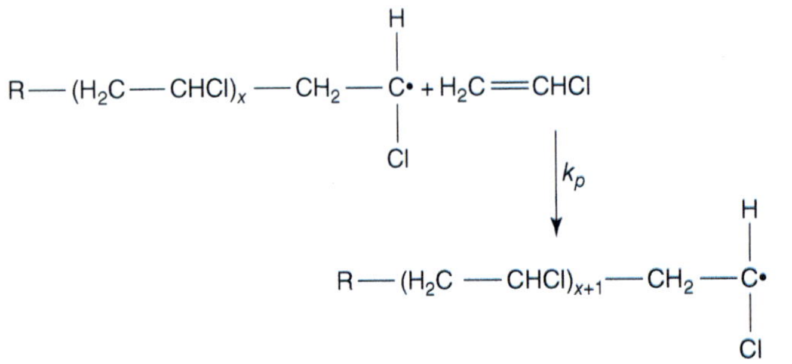

(3.69)

The rate expression for propagation is given by the rate of monomer disappearance:

$$\frac{-d[\mathrm{M}]}{dt} = k_p[\mathrm{M}][\mathrm{M}\bullet] \tag{3.70}$$

where [M] is the instantaneous monomer concentration. It is assumed that the rate constant for propagation, k_p, is the same for all propagation reactions in which a monomer is added to a growing radical chain, regardless of the molecular weight of the chain.

The propagation reactions proceed at multiple radical sites until the monomer is exhausted, or until one of two common types of *termination* reactions occur: either *combination* or *disproportionation*.

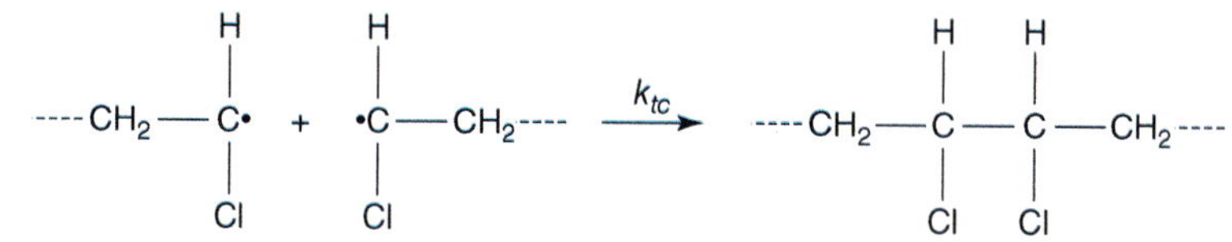

Termination by combination (3.71)

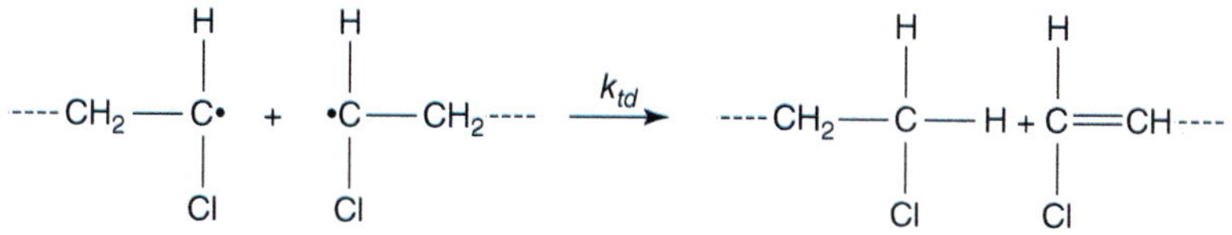

Termination by disproportionation (3.72)

Each of these termination reactions has a similar rate expression that is second order in radical concentration, so the overall termination rate can be expressed as the rate of disappearance of the radicals due to both types of termination:

$$\frac{-d[\mathrm{M}\bullet]}{dt} = 2k_t[\mathrm{M}\bullet]^2 \tag{3.73}$$

> **Need More Help?**
>
> Try these videos on addition polymerization:
>
> https://www.youtube.com/watch?v=o3Hq4gXuKB0
> https://www.youtube.com/watch?v=kMHYNuyKQ2Q
> https://www.youtube.com/watch?v=URT8Dc_feP0
> https://www.youtube.com/watch?v=a9K8D5YI-JE
> https://www.youtube.com/watch?v=XNJOViATGPI

Other reactions can also take place, such as chain transfer reactions, in which the radical is transferred from a growing polymer chain to a small molecule such as the initiator or a solvent molecule. These so-called *chain transfer agents* are often deliberately added to the reaction mix to either control the rate of the reaction or create branched polymers. The rate of the reaction can also be controlled with *inhibitors* or *retarders*. These additives also serve to help control the molecular weight of the polymer chain. In this instance, the rate of disappearance of monomer is given by an equation analogous to (3.70), except that the rate constant is now $k_{tr,i}$; and instead of adding monomer to continue chain propagation, M• now transfers its radical to some agent, A, which may be an initiator, a solvent, a monomer, or a chain-transfer agent:

$$\frac{-d[\mathrm{M}]}{dt} = \sum k_{tr,i}[\mathrm{A}][\mathrm{M}\bullet] \tag{3.74}$$

In the steady-state condition, the rate of initiation is equivalent to the rate of radical termination, and the radical concentration, [M•], remains essentially constant with time. Reactions (3.68) and (3.73) can be set equal at steady state to solve for [M•],

$$[\mathrm{M}\bullet] = \left(\frac{fk_d[\mathrm{I}]}{k_t}\right)^{1/2} \tag{3.75}$$

which is then substituted back into Eq. (3.70) to obtain the overall rate expression for addition polymerization:

$$\frac{-d[\mathrm{M}]}{dt} = k_p\left(\frac{fk_d[\mathrm{I}]}{k_t}\right)^{1/2}[\mathrm{M}] \tag{3.76}$$

As was the case for condensation-type reactions, there is a number-average degree of polymerization, $\bar{x}_n$, where

$$\bar{x}_n = \frac{\text{rate of growth}}{\sum \text{rate of reactions leading to dead polymer}} \quad (3.77)$$

The rate of growth is simply given by Eq. (3.76), and the sum of the reaction rates leading to dead polymer comes from termination [Eq. (3.73)] and chain transfer [Eq. (3.74)] reactions

$$\bar{x}_n = \frac{k_p[\text{M}\bullet][\text{M}]}{\Upsilon k_t[\text{M}\bullet]^2 + \sum_i k_{tr,i}[\text{A}][\text{M}\bullet]} \quad (3.78)$$

where $\Upsilon = 1$ for termination by coupling and 2 if termination is by disproportionation, since two dead polymer molecules are formed in the latter reaction but only one dead polymer molecule is formed in coupling. Eq. (3.78) is more easily simplified by inverting it to obtain an expression for $1/\bar{x}_n$:

$$\frac{1}{\bar{x}_n} = \frac{\Upsilon k_t[\text{M}\bullet]^2 + \sum_i k_{tr,i}[\text{A}][\text{M}\bullet]}{k_p[\text{M}\bullet][\text{M}]} = \frac{\Upsilon k_t[\text{M}\bullet]}{k_p[\text{M}]} + \sum_i \left(\frac{k_{tr,i}[\text{A}]}{k_p[\text{M}]}\right) \quad (3.79)$$

Cooperative Learning Exercise 3.4

Work in groups of three. Equation (3.76) can be rewritten to reflect the fact that experimental determination of polymerization rates typically yields a single rate constant, called the *apparent rate constant*, k_{app}, which is a composite of the three different rate constants, k_d, k_t, and k_p. If we call the rate of disappearance of monomer concentration, $-d[\text{M}]/dt$ the rate of polymerization, R_p, then we can rewrite Eq. 3.76 as

$$R_p = k_{app}[\text{M}][\text{I}]^{1/2} \text{ or } \ln R_p = \ln k_{app} + \ln[\text{M}] + 0.5\ln[\text{I}]$$

This relationship will allow us to estimate the variation in polymerization rate with temperature. Taking the derivative and treating the monomer and initiator concentrations as constants with respect to temperature gives

$$d\ln R_p = d\ln k_{app} \text{ or } (dR_p)/R_p = d\ln k_{app}$$

From here, we can introduce temperature dependence by assuming that k_{app} is given by an Arrhenius-type expression (Eq. 3.12).

1) Work together to derive the following expression for the temperature-dependence of the polymerization rate from the previous expression:

$$\frac{dR_p}{R_p} = \frac{E_{a,p} + \frac{E_{a,d}}{2} - \frac{E_{a,t}}{2}}{RT^2} dT$$

where $E_{a,p}$, $E_{a,d}$, and $E_{a,t}$ are the activation energies for propagation, initiator decomposition, and termination, respectively.

2) Use the data below for the activation energies of some typical addition polymerization reactions at 60°C to calculate an average activation energy for propagation, termination, and initiator decomposition. Each person should calculate one of the three average activation energies.

Monomer	$E_{a,p}$ (kJ/mol)	$E_{a,t}$ (kJ/mol)
Acrylonitrile	16.2	15.5
Methyl acrylate	29.7	22.2
Methyl methacrylate	26.4	11.9
Styrene	26.0	8.0
Vinyl acetate	18.0	21.9
2-Vinyl pyridine	33.0	21.0

Initiator	T (°C)	$E_{a,d}$ (kJ/mol)
2,2′ Azobisisobutyronitrile in benzene	70	123.4
t-Butyl peroxide in benzene	100	146.9
Benzoyl peroxide in benzene	70	123.8
t-Butyl hydroperoxide in benzene	169	170.7

3) Combined your answers to estimate the percent rate of change in the polymerization rate ($dR_p/R_p \times 100$) for each 1°C change in temperature ($dT = 1$) at 50°C.

Answers: $E_{a,p} = 24.9$ kJ/mol; $E_{a,t} = 16.8$ kJ/mol; $E_{a,d} = 141$ kJ/mol; $dR_p/R_p = \{[24.9 + 141/2 - 16.8/2](10^3)(1)/(8.314)(323)^2\} \times 100 = 10\%°C^{-1}$

Once again, elimination of [M•] using Eq. (3.75) gives a general expression for the instantaneous degree of polymerization as a function of rate constants and concentrations:

$$\frac{1}{\bar{x}_n} = \frac{\Upsilon(f\,k_d k_t[\mathrm{I}])^{1/2}}{k_p[\mathrm{M}]} + \sum_i \left(\frac{k_{tr,i}[\mathrm{A}]}{k_p[\mathrm{M}]}\right) \tag{3.80}$$

The first term in Eq. (3.80) is the inverse of the degree of polymerization that would occur in the absence of chain transfer, $(\bar{x}_n)_0$. The second term represents the effect of different chain transfer agents on the molecular weight.

Polymerization of Copolymers

Recall from Chapter 1 that a copolymer is a polymer composed of two or more different repeat units—the result of the simultaneous polymerization of two or more monomers. Though copolymers can be formed by any of the reaction mechanisms described above, the most interesting copolymerizations are of the free radical type. This is because the reactivity of the radicals can vary depending on the structure of the monomer to which they are attached. Such changes in reactivity between monomers are not generally found in condensation-type reactions; that is, a carboxylic acid group is a carboxylic acid group, independent of the chain or monomer to which it is attached. In chain polymerizations, however, the monomers can add to the growing chain in different ways and at different rates. Let us examine the simple system of two monomers M_1 and M_2 that are initiated to form radicals $M_1\bullet$ and $M_2\bullet$. A growing radical chain will then either terminate in a $M_1\bullet$ or $M_2\bullet$ radical. There are four ways in which the next monomer can add to a growing radical chain (of any size) and four corresponding rate expressions:

$$\cdots\cdots\cdots M_1\bullet + M_1 \xrightarrow{k_{11}} \cdots\cdots\cdots M_1\bullet \tag{3.81a}$$

$$\cdots\cdots\cdots M_1\bullet + M_2 \xrightarrow{k_{12}} \cdots\cdots\cdots M_2\bullet \tag{3.81b}$$

$$\cdots\cdots\cdots M_2\bullet + M_1 \xrightarrow{k_{21}} \cdots\cdots\cdots M_1\bullet \tag{3.81c}$$

$$\cdots\cdots\cdots M_2\bullet + M_2 \xrightarrow{k_{22}} \cdots\cdots\cdots M_2\bullet \tag{3.81d}$$

The rate of disappearance of each monomer is then given by

$$\frac{-d[M_1]}{dt} = k_{11}[M_1\bullet][M_1] + k_{21}[M_2\bullet][M_1] \tag{3.82}$$

$$\frac{-d[M_2]}{dt} = k_{12}[M_1\bullet][M_2] + k_{22}[M_2\bullet][M_2] \tag{3.83}$$

If we define the monomer *reactivity ratio* for monomers 1 and 2, r_1 and r_2, respectively, as the ratio of rate constants for a given radical adding to its own monomer to the rate constant for it adding to the other monomer ($r_1 = k_{11}/k_{12}$ and $r_2 = k_{22}/k_{21}$ see Table 3.7 for typical values), then we arrive at the following relationship known as the *copolymer equation*:

$$\frac{d[M_1]}{d[M_2]} = \frac{[M_1](r_1[M_1] + [M_2])}{[M_2]([M_1] + r_2[M_2])} \tag{3.84}$$

Table 3.7 Typical Monomer Reactivity Ratios in the Addition Polymerization of Copolymers

Monomer 1	Monomer 2	r_1	r_2	T (°C)
Acrylonitrile	1,3-Butadiene	0.02	0.3	40
	Methyl methacrylate	0.15	1.22	80
	Styrene	0.04	0.4	60
	Vinyl acetate	4.2	0.05	50
	Vinyl chloride	2.7	0.04	60
1,3-Butadiene	Methyl methacrylate	0.75	0.25	90
	Styrene	1.35	0.58	50
	Vinyl chloride	8.8	0.035	50
Methyl methacrylate	Styrene	0.46	0.52	60
	Vinyl acetate	20	0.015	60
	Vinyl chloride	10	0.1	68
Styrene	Vinyl acetate	55	0.01	60
	Vinyl chloride	17	0.02	60
Vinyl acetate	Vinyl chloride	0.23	1.68	60

[a]Data from Young [36].

Though this is a useful relationship, there are some special cases that are more easily described by introducing two types of ratios. The first ratio is F_1, the rate of change of monomer M_1 mole fraction:

$$F_1 = \frac{d[M_1]/dt}{d[M_1]/dt + d[M_2]/dt} \tag{3.85}$$

Similarly,

$$F_2 = \frac{d[M_2]/dt}{d[M_1]/dt + d[M_2]/dt} \tag{3.86}$$

The second useful ratio is the instantaneous concentration of a given monomer, f_1 or f_2:

$$f_1 = \frac{[M_1]}{[M_1] + [M_2]}, \; f_2 = \frac{[M_2]}{[M_1] + [M_2]} \tag{3.87}$$

It is important to distinguish between the concentration ratio, f_i, and the rate of change of concentration, F_i, since if monomers M_1 and M_2 are consumed at different rates, then $F_i \neq f_i$. Substitution of these ratios F_1, F_2, f_1, and f_2 into the copolymer equation gives:

$$\frac{F_1}{F_2} = \frac{(r_1 f_1/f_2) + 1}{(r_2 f_2/f_1) + 1} \tag{3.88}$$

or

$$\frac{F_1}{1 - F_1} = \frac{[r_1 f_1/(1 - f_1)] + 1}{[r_2(1 - f_1)/f_1] + 1} \tag{3.89}$$

This form of the copolymer equation allows us to identify several simplifying cases:

- k_{12} and f_{21} are negligible; that is, a radical prefers to react with its own monomer, and no copolymer is formed—only homopolymer.

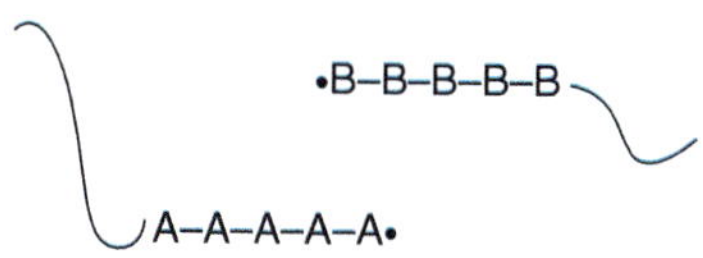

- k_{11} and k_{22} are negligible; that is, a radical prefers to react with the other monomer, $F_1 = F_2 = 0.5$,

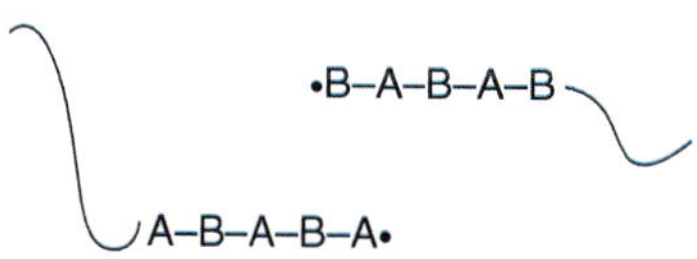

and a perfectly alternating copolymer is formed.

- $k_{11} \approx k_{12}$ and $k_{22} \approx k_{21}$; $r_1 = r_2 = 1$; $F_1 = f_1$, and a perfectly random copolymer is formed.

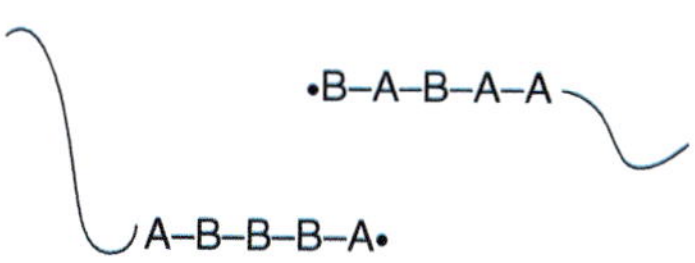

- $r_1 r_2 = 1$, so that Eq. (3.89) becomes

$$\frac{F_1}{1-F_1} = \frac{r_1 f_1}{1-f_1} \tag{3.90}$$

Need More Help?

Try this video on co-polymerization:

https://www.youtube.com/watch?v=lCzhDSHkfkg
https://www.youtube.com/watch?v=vr-eMRWDYzM

3.3.2 Polymerization Processes*

For the engineer, a knowledge of the reaction kinetics is merely a piece of what is usually a much larger problem involving the design of reactor components and the production and processing of large quantities of product—in this case, polymers. As a result, it is useful to discuss not only how the kinetic principles developed so far are applied to different methods of polymerization, but how problems associated with real processes, such as heat generation and removal, can, in turn, affect the polymerization kinetics (see Table 3.8). We will briefly look at four common polymerization processes: *bulk polymerization, solution polymerization, suspension polymerization,* and *emulsion polymerization*. Though most of these processes are used primarily for addition-type polymerizations, applications involving condensation-type polymerizations will be noted where appropriate.

Bulk Polymerization

Monomer and polymer (with traces of initiator) are the only constituents in bulk polymerizations. Obviously, the monomer must be soluble in the polymer for this type of process to effectively

Table 3.8 Comparison of Polymerization Processes

Type	Advantages	Disadvantages
Batch bulk	Minimum contamination Simple equipment	Poor heat control Broad MW distribution
Continuous bulk	Better heat control Narrower MW distribution	Requires stirring, separation, and recycle
Solution	Good heat control Direct use of solution possible	Not useful for dry polymer Solvent removal difficult
Suspension	Excellent heat control Direct use of suspension possible	Requires stirring Contamination by stabilizer possible Additional processing (washing and drying) required
Emulsion	Excellent heat control Narrow MW distribution High MW attainable Direct use of emulsion possible	Contamination by emulsifier likely Additional processing (washing, drying) required

Source: Adapted from Billmeyer [37].

proceed. Bulk polymerization, also called *mass* or *block polymerization*, can occur in stirred-tank reactors or can be unstirred, in which instance it is called *quiescent bulk polymerization*. The primary difficulty with bulk polymerizations is that as the polymerization proceeds and more polymer is formed, the viscosity increases, thermal conductivity decreases, and heat removal becomes difficult.

There is an increase in temperature as heat removal slows, along with a corresponding increase in the reaction rate. This phenomenon is known as the *autoacceleration* or *Trommsdorff effect* and can lead to catastrophic results if not properly controlled. Even with successful control of the reaction, it is difficult to remove the traces of remaining monomer from the polymer due to decreased diffusion. Similarly, it is difficult to get the reactions to proceed to completion due to limited monomer mobility.

A bulk polymerization reactor can be as simple as a tube into which the reactants are fed and from which the polymer mixture emerges at the end; it can be more of a traditional, continuous stirred-tank reactor (CSTR), or even a high-pressure autoclave-type reactor. A bulk polymerization process need not be continuous, but it should not be confused with a batch reaction. There can be batch bulk polymerizations just as there are continuous bulk polymerization processes.

Solution Polymerization

By adding a solvent to the monomer–polymer mixture, heat removal can be improved dramatically over bulk reactions. The solvent must be removed after the polymerization is completed, however, which leads to a primary disadvantage of *solution polymerization*. Another problem associated with radical chain polymerizations carried out in solution is associated with chain transfer to the solvent. As we saw earlier, chain transfer can significantly affect the molecular weight of the final polymer. This is particularly true in solution polymerization, where there are many solvent molecules present. In fact, chain transfer to solvent often dominates over chain transfer to other types of molecules, so that Eq. (3.80) reduces to

$$\frac{1}{\bar{x}_n} = \left(\frac{1}{\bar{x}_n}\right)_0 + \frac{C_S[\mathrm{S}]}{[\mathrm{M}]} \tag{3.91}$$

where the first term once again describes the contributions of propagation to the molecular weight, [S] and [M] are concentrations of the solvent and monomer, respectively, and C_s is the ratio of rate constants $k_{tr,s}/k_p$. This equation shows that a plot of $1/\bar{x}_n$ versus [S]/[M] should yield a straight line, the intercept for which gives the degree of polymerization in the absence of chain transfer to solvent. This relationship has been borne out experimentally for polystyrene and several solvent systems (see Figure 3.21).

The reactor equipment used for solution polymerizations is typically glass-lined stainless steel. An example of solution polymerization is the reaction of ethylene in isooctane with a chromia–silica–alumina catalyst initiator to form polyethylene. Typical reaction conditions for this polymerization are 150–180°C and 2.1–4.8 MPa (300–700 psi).

Suspension Polymerization

When water is the solvent and the monomer is insoluble in water, *suspension polymerization* is often employed. The monomer and initiator form small droplets within the aqueous phase, and polymerization proceeds via bulk polymerization. Droplet sizes are typically between 10 and 1000 μm in diameter, so heat transfer to the suspending medium and subsequently to the reactor

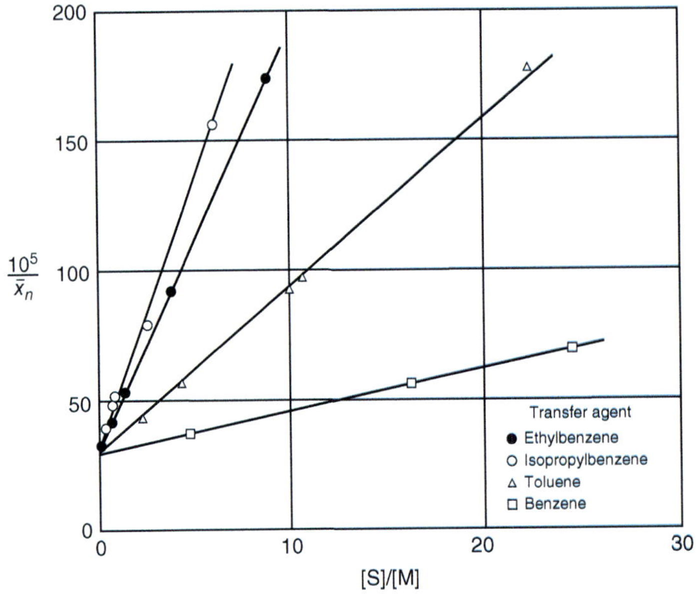

Figure 3.21 Effect of chain transfer to solvent on number average degree of polymerization for polystyrene. [Data from Gregg and Mayo [25]] From Billmeyer [26]. Copyright © 1984 by John Wiley & Sons, Inc. This material is used by permission of John Wiley & Sons, Inc.

walls is very efficient throughout the course of the reaction. The result of the polymerization within each of the droplets is a polymer "bead." Though some agglomeration of particles takes place, the addition of a suspending agent such as poly(vinyl alcohol) minimizes coalescence, and the final polymer bead diameters are typically 100–1000 μm. The polymer beads are recovered by filtration, which is followed by a washing step. Suspension polymerization can be carried to nearly 90% completion at moderate pressures of 600 kPa (88 psi) and 80°C with stirring at about 150–300 rpm.

Need More Help?
Try this video on the emulsion polymerization: https://www.youtube.com/watch?v=ghuJDO5XVcM and suspension polymerization: https://www.youtube.com/watch?v=a_pF6Oz3EIg

Emulsion Polymerization

Emulsion polymerization is like suspension polymerization in that the monomer is insoluble in the solvent, water. The similarity ends there. In emulsion polymerization, the initiator is water-soluble, the particles in which the propagation step occurs are smaller, typically 0.05–5 μm in diameter, and the final beads are not usually filtered from the aqueous solution. The *emulsifier*, or *surfactant*, plays a very important role in emulsion polymerization. It is usually a long-chained hydrocarbon with a hydrophilic "head" and hydrophobic "tail." This charge polarization causes the tails to surround the monomer molecules, with the heads pointing outward into the water phase, forming a spherical droplet called a *micelle* (see Figure 3.22). As radicals initiate in the aqueous phase, they react with what few monomer molecules are dissolved in the water. As the radicals slowly grow and become

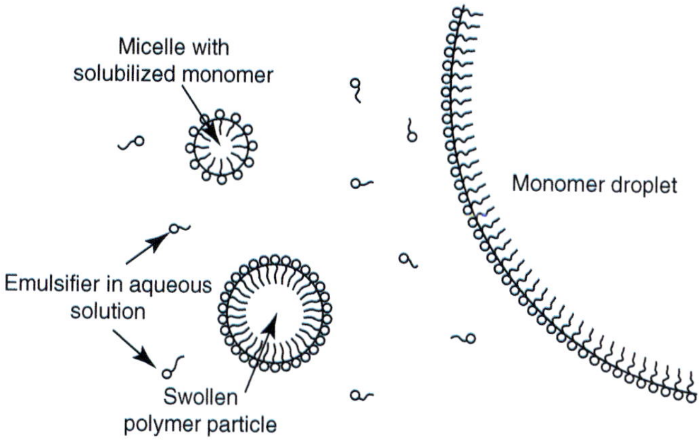

Figure 3.22 Schematic representation of micelle formation in emulsion polymerization.

more hydrophobic, they find it more stable to be inside of the micelles. Upon entering a micelle, the radicals cause rapid propagation, and the micelle expands.

It can be shown that the degree of polymerization in emulsion polymerization can be derived from general polymerization kinetics [Eq. (3.80)] by neglecting chain transfer:

$$\frac{1}{\bar{x}_n} = \frac{2f\,k_d[I]}{k_p[\mathrm{M}^*][\mathrm{N}^*]} \tag{3.92}$$

where the superscript "*" on the monomer concentration indicates that this is the concentration of monomer in the micelle, not the solution. Since the initiator is soluble in water, its concentration remains based upon the total volume of water present. In Eq. (3.92), $[\mathrm{N}^*]$ is the concentration of micelle-swollen polymer particles.

The final product of emulsion polymerization is an *emulsion*—a stable, heterogeneous mixture of fine polymer beads in an aqueous solution, sometimes called a *latex emulsion*. Water-based paints, for example, can be formed from the emulsion polymerization of vinyl acetate. In this process, 1 m^3 of water containing 3% poly(vinyl alcohol) and 1% surfactant are heated to 60°C in a reaction vessel. The temperature rises to around 80°C over a 4- to 5-hour period as monomer and an aqueous persulfate solution are added. The rate at which heat can be removed limits the rate at which monomers can be added.

Copolymerizations can also be carried out in an emulsion reaction. For example, *nitrile rubber* (NR) can be made from the emulsion polymerization of acrylonitrile and butadiene:

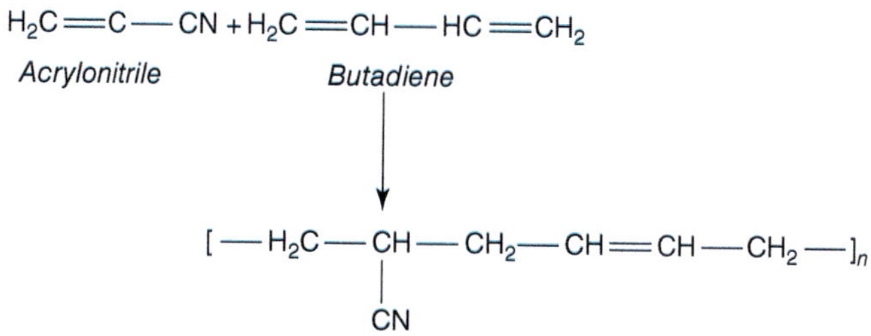

(3.93)

As the acrylonitrile content in NR increases, so does the resistance to nonpolar solvents.

Gas-Phase Polymerization

Not all polymerization reactions are carried out in the liquid phase. Polyethylene, for example, can be polymerized not only in solution as discussed previously, but also through gas-phase polymerization. Significant economic advantage is gained by eliminating costly solvent and catalyst recovery equipment. A fluidized bed reactor is used, into which purified ethylene and a modified chromia on silica powdered catalyst are fed. The reactor pressure is about 2 MPa (290 psi) and the temperature is 85–100°C. Heat is removed as gas is circulated using external coolers. The growing particles remain in the reactor for three to five hours; emerging with an average diameter of about 500 μm. Small amounts of catalyst are used, so catalyst removal is often not necessary. Hydrogen is used as a chain transfer agent, and branching can be introduced by the addition of 1-butene.

3.3.3 Kinetics of Polymer Degradation*

Degradation of polymers may be considered as any type of undesired modification of a polymer chain backbone, side groups, or both. These modifications are usually chemical in nature; that is, they involve the breaking of primary bonds, leading to reduced molecular weight, and altered physical properties. The causes of degradation may be environmental (such as heat, light, or atmospheric pollutants), induced, (as in rubber mastication), or a result of processing due to heat and/or air. We will be concerned primarily with *environmental degradation*. Even within environmental degradation, there are six primary classes of polymer degradation: thermal, oxidative, radiative, chemical, mechanochemical, and biological. We will look briefly at the mechanisms of each of these types of degradative processes and present without derivation their related kinetic expressions.

Thermal Degradation

In *thermal degradation*, a polymer chain—in this case, a vinyl polymer chain—*depolymerizes* or *unzips*. Just as in the formation reactions, there is initiation, chain transfer, and termination. Once a free radical has been initiated through some sort of *random* or *chain-end scission* process, the *depropagation* step proceeds, and monomer fragments split off from the polymer chain in a stepwise manner (see Eq. 3.94). The monomer fragments are usually assumed to be volatile.

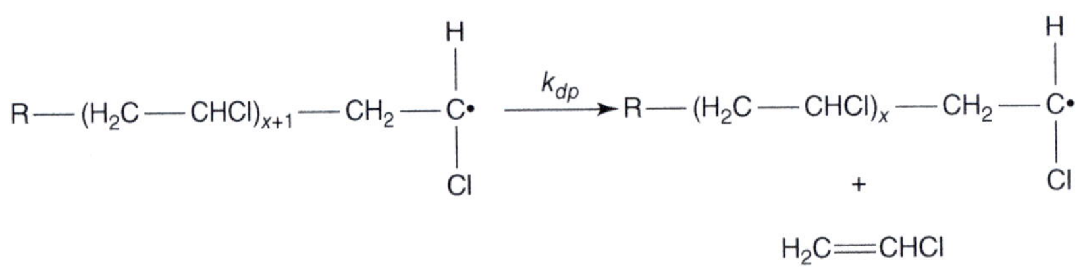

(3.94)

It can be shown that the change in polymer weight, W, with time, t, for thermal degradation is given by

$$\frac{-dW}{dt} = \frac{2k_{ir}b(\overline{x}_n)_0 W^{(2b-1)/b}}{W_0^{(b-1)/b}} \tag{3.95}$$

where k_{ir} is the rate constant for random initiation of radicals, W_0 is the initial weight of the polymer, $(\overline{x}_n)_0$ is the initial degree of polymerization, and b is a parameter that depends on the degree of polydispersity ($b = 1$ for a monodispersed system). The value of b is a measure of the breadth of the molecular weight distribution curve. For example, when the molecular weight distribution follows the exponential distribution, $b = 2$. According to Eq. (3.95), this corresponds to a reaction order of 3/2. As b increases, the reaction order approaches a limiting value of 2.

Oxidative Degradation

There is a wealth of literature available on the *oxidative degradation* of polymers. Unfortunately, there is no single mathematical treatment that is applicable to all polymer types. As a result, the development of an appropriate kinetic expression is highly dependent upon the type of polymer being considered and the presence of additives, inhibitors, and antioxidants. For the sake of simplicity, we will consider a general case of autoxidation in a polyolefin in the absence of additives. It will be assumed that the diffusion of oxygen is rapid and is therefore not the rate limiting kinetic step.

As in thermal degradation, the process begins with an initiation step, in this case, the formation of radicals where X denotes a source of free radicals, such as polymer chains, which then react with atmospheric oxygen:

$$\mathrm{X} \xrightarrow{k_1} \mathrm{R}\bullet \tag{3.96a}$$

$$\mathrm{R}\bullet + \mathrm{O}_2 \xrightarrow{k_2} \mathrm{RO}_2\bullet \tag{3.96b}$$

The oxidized radicals can then react with another polymer chain, RH, to form peroxides:

$$\mathrm{RO}_2\bullet + \mathrm{RH} \xrightarrow{k_3} \mathrm{RO}_2\mathrm{H} + \mathrm{R}\bullet \tag{3.97}$$

Reactions (3.96b) and (3.97) are the propagation steps, although the former is very rapid, with a negligible activation energy, whereas the latter is slower, with an activation energy of about 1.7 kJ/mol (7 kcal/mol). Three types of termination reactions ensue, all of which are very rapid, with activation energies of less than 1 kJ/mol (3 kcal/mol):

$$2\mathrm{R}\bullet \xrightarrow{k_4} \text{Products} \tag{3.98a}$$

$$\mathrm{RO}_2 + \mathrm{R}\bullet \xrightarrow{k_5} \text{Products} \tag{3.98b}$$

$$2\mathrm{RO}_2\bullet \xrightarrow{k_6} \text{Products} + \mathrm{O}_2 \tag{3.98c}$$

In the presence of oxygen, the first two termination reactions can be neglected. The oxidation rate can then be calculated from the rate-limiting propagation step, Eq. (3.97):

$$\frac{d[\mathrm{RO}_2\mathrm{H}]}{dt} = k_3[\mathrm{RO}_2\bullet][\mathrm{RH}] \tag{3.99}$$

A steady-state approximation allows us to solve for the intermediate free-radical concentration, $[\mathrm{RO}_2\bullet]$, to obtain

$$\frac{d[\mathrm{RO}_2\mathrm{H}]}{dt} = \frac{k_3 k_1^{1/2}[\mathrm{X}]^{1/2}[\mathrm{RH}]}{k_6^{1/2}} \tag{3.100}$$

Radiative Degradation

Radiative degradation in polymers can give rise to both low-molecular-weight species, as in thermal degradation or crosslinked structures, which are insoluble and infusible. Though crosslinking leads to an increase in the molecular weight of the sample, it is still considered a degradative process since the resulting product has unfavorably altered physical properties. There are generally two types of radiative degradation processes in polymers, grouped according to the amount of radiant energy involved and which result in different types of degradation. *Photolysis* occurs when ultraviolet light (wavelength $\lambda = 10^2$–10^4 Å) imparts energy on the sample of the order 10^2–10^3 kJ/mol. *Radiolysis* occurs when higher radiant waves ($\lambda = 10^{-3} - 10^2$ Å) impart energy of the order 10^5–10^{10} kJ/mol to the sample. Each of these types attacks the polymer chain in a different fashion.

Chemical Degradation

For most applications in the chemical process industry, *chemical degradation* is the most important type of polymer degradation. It is also the largest class overall, encompassing degradation due to both gaseous and liquid species on all polymer classes. Even within a polymer class, such as polyolefins, slight changes in the chain chemistry can lead to enormous changes in chemical compatibility. As a result, the presentation here must be severely limited. The intent will be to provide some insight into how the chemical degradation occurs, and to present practical information in guiding the selection of an appropriate polymer for a specific chemical environment.

In addition to oxygen, the most important gaseous agents that lead to polymer chemical degradation are nitrogen dioxide (NO_2), sulfur dioxide (SO_2), and ozone (O_3). Nitrogen dioxide attacks hydrocarbons by removing a tertiary hydrogen atom:

$$NO_2(g) + RH \rightarrow HNO_2 + R\bullet \tag{3.101}$$

The resulting radical then reacts further with nitrogen dioxide,

$$NO_2(g) + R\bullet \rightarrow RNO_2 \tag{3.102}$$

which then leads to products via chain scission and/or crosslinking. Polystyrene undergoes such a degradation process, the degree of polymerization for which changes according to the following empirical relationship

$$\frac{1}{\overline{x}_n} - \frac{1}{(\overline{x}_n)_0} = kt \tag{3.103}$$

where the rate constant, k, follows an Arrhenius-type relationship of the form

$$k = P_{NO_2}[41.4\exp(-3870/RT)]\text{h}^{-1} \tag{3.104}$$

Here, P_{NO_2} is the partial pressure of nitrogen dioxide in pascals (1 Pa = 0.000145 psi) and the activation energy is 3870 kJ/mol. Other saturated polymers are less susceptible to attack by NO_2 than most unsaturated polymers such as synthetic rubbers (polyisoprene, polybutadiene, and butyl rubber). The presence of oxygen also tends to accelerate degradation by NO_2. The reaction of sulfur dioxide with saturated polymers is complex but appears to be activated by ultraviolet radiation.

Mechanochemical Degradation

Mechanochemical degradation occurs in polymers as the result of an applied mechanical force. This type of degradation is quite common in machining processes such as *grinding*, *ball milling*, and *mastication*. In general, the degradation processes are similar to the other types of degradation, in that radicals are formed and cause subsequent reactions, such as oxidation and chain transfer, that can lead to either crosslinking or a reduction in molecular weight, though molecular weight reduction is the most common result. There are some unique aspects of mechanochemical degradation, however. In contrast to thermal degradation, radical formation is nonrandom, and the location of chain scission depends on the manner in which the forces are applied to the polymer. Also, the molecular weight decreases rapidly from an initial degree of polymerization, $(\bar{x}_n)_0$, for polymers under mechanical shear, then level off to a limiting degree of polymerization, $(\bar{x}_n)_\infty$ (Figure 3.23).

This exponential decrease in molecular weight and degree of polymerization with time can be expressed by an empirical relationship of the form

$$\frac{(\bar{x}_n)_0 - (\bar{x}_n)_t}{(\bar{x}_n)_0 - (\bar{x}_n)_\infty} = 1 - \exp(-kt) \tag{3.105}$$

where $(\bar{x}_n)_t$ is the degree of polymerization at time t. Typical values of k for common polymers of various molecular weights are given in Table 3.9.

In addition to a reduction in molecular weight, mechanical forces generally lead to an increase in solubility, a narrowing of the molecular weight distribution, decreased tensile strength, decreased

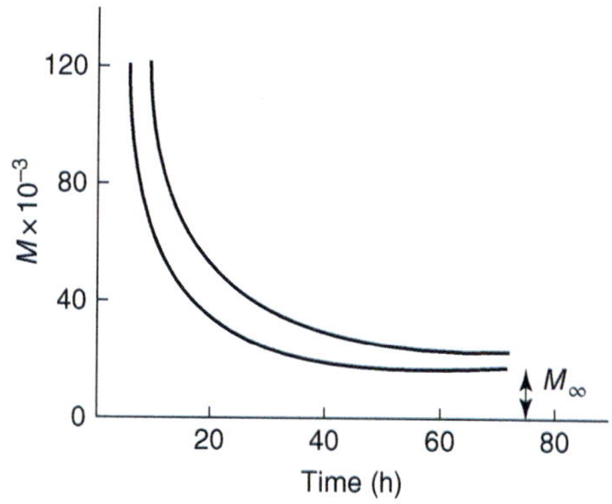

Figure 3.23 Effect of machining time on molecular weight of polymer.

Table 3.9 Rate Constants for Mechanochemical Chemical Degradation of Polymers

Polymer	Molecular Weight	k, h^{-1}
Poly(vinyl alcohol)	4000	0.0237
Polystyrene	7000	0.0945
Poly(methyl methacrylate)	9000	0.1200
Poly(vinyl acetate)	11 000	0.0468

crystallinity, and increased plasticity. These effects are highly dependent upon the machining conditions. In particular, higher machining temperatures result in non-Arrhenius behavior since there is a corresponding decrease in viscosity. The presence of radical scavengers and the type of machining equipment employed also play significant roles in determining the extent of mechanochemical degradation during machining.

Biological Degradation

In contrast to the degradation processes discussed to this point, *biological degradation* in polymers is oftentimes advantageous. In fact, there is a class of polymers that have been developed to enhance their susceptibility to attack by biological agents in order to improve their biodegradability and reduce their long-term impact on the environment. Other important applications of biodegradable polymers are as biomaterials, such as *resorbable sutures*, and *controlled drug delivery* devices.

There are generally two types of polymeric materials that are susceptible to biological agents such as water, enzymes, and microbes: (a) natural materials like collagen, cellulose, and starch and (b) synthetic polymers. Though natural polymers represent a large fraction of all biodegradable polymers, particularly for biomedical uses like catgut sutures, we will concentrate on synthetic polymers, since they represent the class of materials where most of the advancement is still possible. It is the synthetic polymers that can be optimized in terms of mechanical properties and rate of degradation for biological and biomedical applications.

Substitutes for petroleum-based polymers used in packaging (e.g., polystyrene, polyethylene, and polypropylene) represent one of the largest potential markets for biodegradable polymers. Some natural polymers have been used as substitutes, but their processability and mechanical properties have limited their use. In addition, factors such as cost, recycling, and alternative strategies like incineration have limited the need for the development of synthetic biodegradable polymers. Work continues in these areas, but it is the realm of synthetic polymers for biomedical applications that represents one of the more important areas of new material development. It is also the area where the kinetics of the degradation processes have been most widely studied. Though several classes of biodegradable synthetic polymers have been developed, including polyamides based on 4, 4′-spirobibutyrolactone [13] and poly(hydroxylalkanoates), it is the poly(α-esters) that have been most widely utilized for biomedical applications. Let us look at a specific example of synthetic polymers for resorbable suture material.

The most common synthetic biodegradable polymers for suture material and their corresponding weight loss in aqueous solution are listed in Table 3.10. Of these, poly(glycolic acid) (PGA), poly(lactic acid) (PLA), and copolymers of these two polyesters are the most widely used for resorbable suture material. PGA is a tough, crystalline, inelastic material with a T_m of 230°C and T_g of 36°C. There are two types of PLA: poly(L-lactic acid), PLLA, which is a tough, crystalline, and inelastic material with $T_m = 180°C$, $T_g = 67°C$; and poly(D-lactic acid), PDLA, which is a tough, amorphous, and inelastic material with $T_g = 57°C$. The *in vitro* and molecular weight loss data for PLLA show that over a 16-week period, only 10–15% of the mass is lost, and the molecular weight is reduced to 50% of its original value [14]. Similar studies for PGA show that mass loss begins in the fourth week and is completed in 10–12 weeks (see Figure 3.24 for two different types of commercial PGA sutures), and the molecular weight loss data (Figure 3.25) shows a steady decrease in the molecular weight of the primary peak.

Table 3.10 Comparative Weight Loss for Some Common Biodegradable Polymers, Given as Time for 10% Weight Loss at 37°C and pH 7.4

Polymer	t_{10} (h)
Poly(L-lactic acid) (PLLA 8%)-*co*-Poly(glycolic acid) (PGA 92%)	450
Poly(glycolic acid) (PGA)	550
Poly(*p*-dioxanone) (PDS)	1200
Poly(3-hydroxybutyrate-*co*-3-valerate) PHBV (20% hydroxyvalerate, HV)	4.7% @ 5500
PHBV (12% HV)	5.6% @ 5500
Poly(3-hydroxybuturate) HB (0% HV)	1.8% @ 2500

Source: Amass et al. [38].

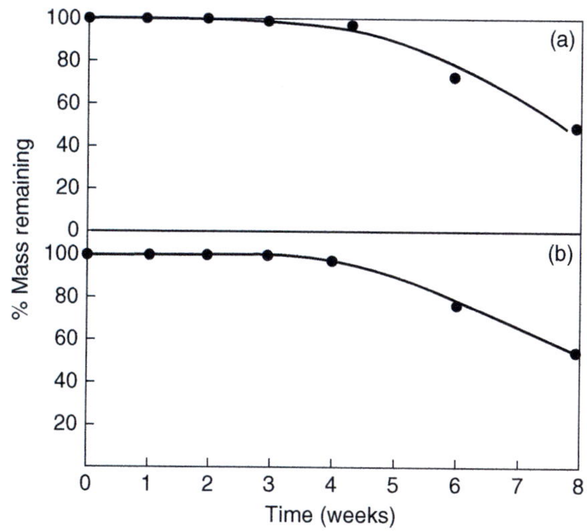

Figure 3.24 Percent weight loss for two types of PGA sutures: (a) Dexon and (b) Dexon-s, as function of exposure time to a buffered solution (pH 7, 37°C). Reproduced, by permission, from Feijin [27], in *Polymeric Biomaterials*, p. 68. Copyright © 1986 by Springer Nature.

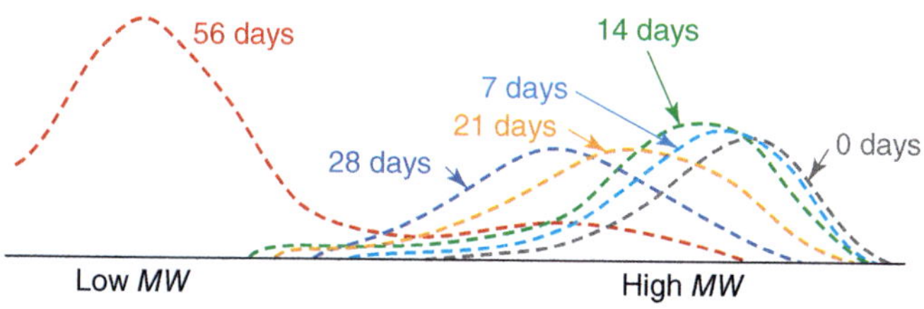

Figure 3.25 Shift in molecular weight distribution for a commercial PGA suture material (Dexon) as a function of exposure time to a buffered solution (pH 7, 37°C). Reproduced, by permission, from Feijin [27]. Copyright © 1986 by Springer Nature.

The degradation mechanisms of these polymers are quite complex. Factors influencing the biological stability include the molecular weight (high MW gives stability), the presence of residual monomer and low-molecular-weight compounds, temperature, % crystallinity (high crystallinity decreases the rate of biodegradation), and extent of sterilization by β- and γ-radiation (both lead to degradation). Polyesters generally undergo a two-stage hydrolysis process. In the first stage, water diffuses into the amorphous regions, producing random hydrolytic scission at the susceptible ester linkage. This leads to an apparent increase in the percent crystallinity of the polymer during this stage. After most of the amorphous regions have been eroded, hydrolysis of the crystalline regions commences. A method for determining the mode of hydrolysis has been developed by Shih [15], in which the mole fraction of monomer, m_1, and the degree of polymerization, α, are determined by a technique like nuclear magnetic resonance (NMR) spectroscopy. If completely random scission of the backbone bonds occurs, $m_1 = \alpha^2$, and if exclusive chain-end unzipping occurs, $m_1 = \alpha$. In addition to hydrolysis, these polymers can undergo degradation due to the effect of enzymes, chemicals, and radiation.

Take a Peek! Jump ahead to Section 4.3.2 to see how diffusion in porous solids is relevant to the kinetics of CVD and CVI processes.

3.4 Kinetic Processes in Composites*

Industrial composites are made up of combinations of metals, polymers, and ceramics such that the kinetic processes involved in their formation, transformation, and degradation are often the same as those of the individual components. Most of the processes we have described to this point have involved condensed phases—liquids or solids—but there are two gas-phase processes, widely utilized for composite formation, that require some individualized attention. *Chemical vapor deposition* (CVD) and *chemical vapor infiltration* (CVI) involve the reaction of gas-phase species with a solid substrate to form a heterogeneous, solid-phase composite. Because this discussion must necessarily involve some of the concepts of transport phenomena, namely diffusion, you may wish to refresh your memory from your transport course or refer to the specific topics in Chapter 4 as they come up in this description.

3.4.1 Kinetics of Chemical Vapor Deposition

CVD involves the reaction of gaseous reactants to form solid products. There are two general types of CVD processes: (a) the thermal decomposition of a homogeneous gas to form a solid and (b) the chemical reaction of two or more gaseous species to form a solid. Both types of CVD reactions are used industrially to form a variety of important elements and compounds for semiconductor, superconductor, and ceramic coating applications.

An example of the first kind can be found in the decomposition of trichloromethylsilane (TMS) to give silicon carbide and hydrogen chloride:

$$(CH_3)SiCl_3(g) \rightarrow SiC(s) + 3HCl(g) \tag{3.106}$$

The reaction of titanium tetrachloride with boron trichloride and hydrogen to give solid titanium diboride is an example of the second type of CVD reaction:

$$TiCl_4(g) + 2BCl_3(g) + 5H_2(g) \rightarrow TiB_2(s) + 10HCl(g) \tag{3.107}$$

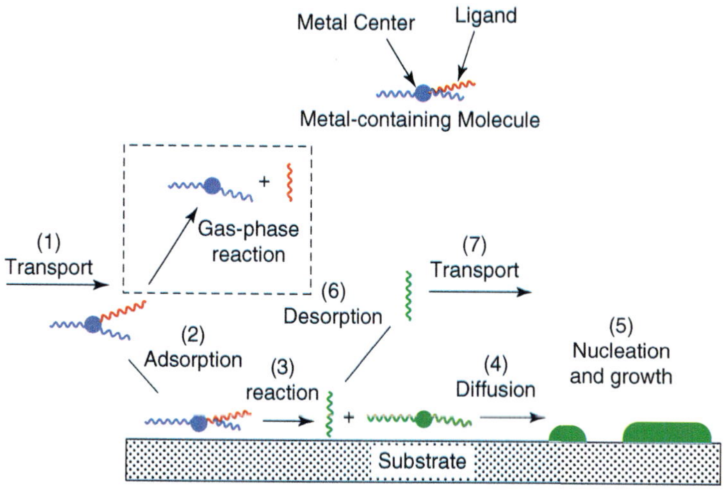

Figure 3.26 Processes involved in chemical vapor deposition. From Ref. [28]. This material is used by permission of John Wiley & Sons, Inc.

Reaction (3.107) illustrates the primary advantage of a CVD process. Solid titanium diboride, TiB_2, melts at 3325°C, yet it can be produced via reaction (3.107) at 1027°C or lower on a suitable substrate by the CVD process. Additionally, the diffusion of gas phase reactants is generally rapid, allowing for the efficient coating of high-surface-area (i.e., porous), substrates. The mechanisms of these CVD reactions can be highly complex, often involving interactions with solid surface sites, and the combination of heat and momentum transport coupled with reaction kinetics can lead to complex models (see Figure 3.26). Let us look at some simple, yet important, systems to gain a better understanding of the kinetics of CVD processes.

Consider the growth of a germanium (Ge) film in the production of a semiconducting thin film [16]. The germanium can be deposited onto a substrate through the reaction of $GeCl_4$ with hydrogen gas:

$$GeCl_4(g) + H_2(g) \rightarrow Ge(s) + 2HCl(g) + Cl_2(g) \tag{3.108}$$

The proposed mechanism for this overall reaction involves a gas-phase dissociation step

$$GeCl_4(g) \leftrightarrow GeCl_2(g) + Cl_2(g) \tag{3.109}$$

two adsorption processes, where S is a surface site on the solid

$$GeCl_2(g) + S \xleftrightarrow{K_A} GeCl_2{\bullet}S \tag{3.110}$$

$$H_2(g) + 2S \xleftrightarrow{K_H} 2H{\bullet}S \tag{3.111}$$

and a surface reaction

$$GeCl_2S + 2HS \xrightarrow{k_s} Ge(s) + 2HCl(g) + 2S \tag{3.112}$$

The rate-limiting step is believed to be the surface reaction between adsorbed molecular hydrogen and germanium dichloride (3.112):

$$\frac{d[Ge]}{dt} = k_s[GeCl_2{\bullet}S][H{\bullet}S]^2 \tag{3.113}$$

where the concentration of adsorbed surface species is often represented by their *fractional coverage*, f_i, or the fraction of the surface occupied by species i:

$$\frac{d[\mathrm{Ge}]}{dt} = k_S f_{\mathrm{GeCl_2 \cdot S}} f^2_{\mathrm{HS}} \tag{3.114}$$

The *deposition rate*, or *film growth rate*, is usually expressed in nm/s, as is the reaction rate constant, k_S. Deposition rates can be expressed in mol/m^2·s by multiplying by the molar density of solid germanium.

It can be shown that through appropriate substitution of the rate expressions for Eqs. (3.109) through (3.111) and use of the fractional area balance $f_v + f_{\mathrm{GeCl_2}} + f_{\mathrm{H}} = 1$, where f_v is the fraction of vacant surface sites, the deposition rate can be expressed in terms of the partial pressures (concentrations) of the reactants as

$$\frac{d[\mathrm{Ge}]}{dt} = \frac{k' P_{\mathrm{GeCl_2}} P_{\mathrm{H_2}}}{\left(1 + K_A P_{\mathrm{GeCl_2}} + \sqrt{K_{\mathrm{H}} P_{\mathrm{H_2}}}\right)^3} \tag{3.115}$$

where k' is the product of the two adsorption equilibrium constants, K_A and K_H, and the reaction rate constant k_S.

Another example of an industrially relevant CVD process is the production of tungsten (W) films on various substrates for (a) wear and corrosion protection and (b) diffusion barriers in electronic devices. One CVD method for depositing tungsten is through the reduction of WF_6 with hydrogen, which follows the overall reaction:

$$\mathrm{WF_6}(g) + 3\mathrm{H_2}(g) \rightarrow \mathrm{W}(s) + 6\mathrm{HF}(g) \tag{3.116}$$

This reaction is slightly endothermic ($\Delta H = 88$ kJ/mol), but the large net increase in entropy and the nonequilibrium nature of most CVD processes lead to significant tungsten deposition. As with the Ge example, the deposition mechanism involves adsorption steps and surface reactions. At low pressures and under conditions of excess hydrogen gas, the deposition rate follows the general form:

$$\frac{d[\mathrm{W}(s)]}{dt} \propto p^{1/2}_{\mathrm{H_2}} p^0_{\mathrm{WF_6}} \tag{3.117}$$

Need More Help?
Try these videos on chemical vapor deposition: https://www.youtube.com/watch?v=9XKGVHPXXho https://www.youtube.com/watch?v=k27ukETADD8

Ideally, the functional dependence of the growth rate on the pressure (concentration) of the gaseous species would be determined by analysis of all the mechanistic steps, as was the case for our Ge example and as could be done for the W example. This is not always possible, however, due to insufficient information on such factors as reaction rate constants, adsorption energies, and preexponential factors. As a result, deposition rate data are often determined experimentally and are used to either support or refute suspected rate-limiting steps. Arrhenius-type plots are often made by plotting the logarithm of the measured deposition rate versus inverse temperature (see Figure 3.27 for the formation of Si_3N_4 from SiF_4 and NH_3), sometimes for various total gas pressures (see Figure 3.28 for the formation of B_4C from BCl_3, CH_4, and H_2).

Figure 3.27 Arrhenius plot for the deposition of Si_3N_4. Reproduced, by permission, from Galasso [29]. Copyright © by CRC Press.

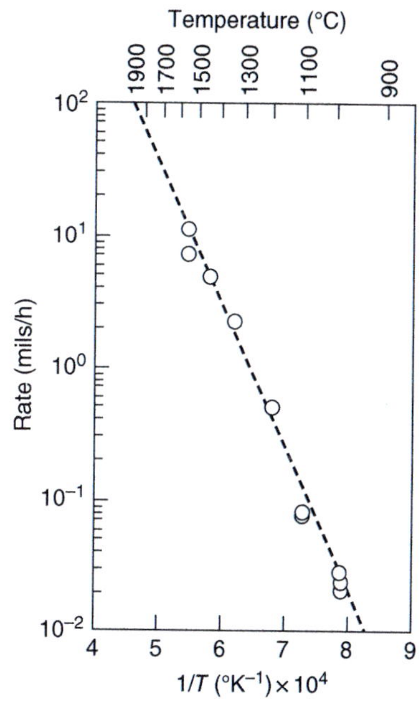

Figure 3.28 Arrhenius plot for the deposition of B_4C at various total pressures. Reproduced, by permission, from Naslain [30]. Copyright © 1992 by Chapman & Hall, London.

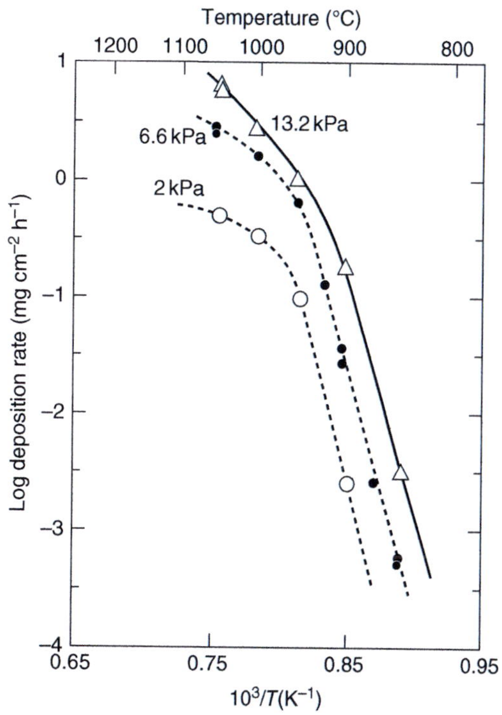

There are numerous materials, both metallic and ceramic, that are produced via CVD processes, including some exciting new applications such as CVD diamond, but they all involve deposition on some substrate, making them fundamentally composite materials. There are equally numerous modifications to the basic CVD processes, leading to such exotic-sounding processes as *vapor-phase epitaxy* (VPE), *atomic-layer epitaxy* (ALE), *chemical-beam epitaxy* (CBE), *plasma-enhanced CVD* (PECVD), *laser-assisted CVD* (LACVD), and *metal-organic compound CVD* (MOCVD). You can learn more about specific CVD processes for structural materials and optical fiber in ***Advanced Topics***.

3.4.2 Kinetics of Chemical Vapor Infiltration

CVI is similar to CVD in that gaseous reactants are used to form solid products on a substrate, but it is more specialized in that the substrate is generally porous, instead of a more uniform, nominally flat surface, as in CVD. The porous substrate introduces an additional complexity with regard to transport of the reactants to the surface, which can play an important role in the reaction as illustrated earlier with CVD reactions. The reactants can be introduced into the porous substrate by either a diffusive or convective process prior to the deposition step. As infiltration proceeds, the deposit (matrix) becomes thicker, eventually (in the ideal situation) filling the pores and producing a dense composite.

Need More Help?
Try this video on chemical vapor infiltration: https://www.youtube.com/watch?v=x6_vtq5RD4g

There are five general categories of CVI processes, depending on which type of transport process they utilize (diffusion or convection) and whether there is an imposed thermal gradient or not (see Figure 3.29). Type I, known as *isothermal*

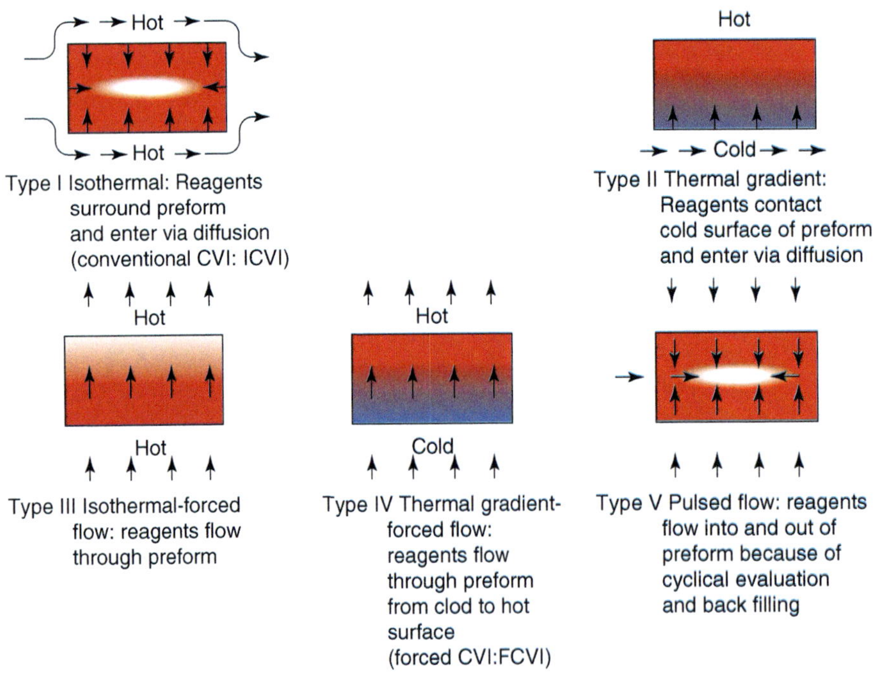

Figure 3.29 Classification of chemical vapor infiltration processes. From Ref. [31]. Copyright © 1997 by Chapman & Hall, London, UK, with kind permission of Kluwer Academic Publishers.

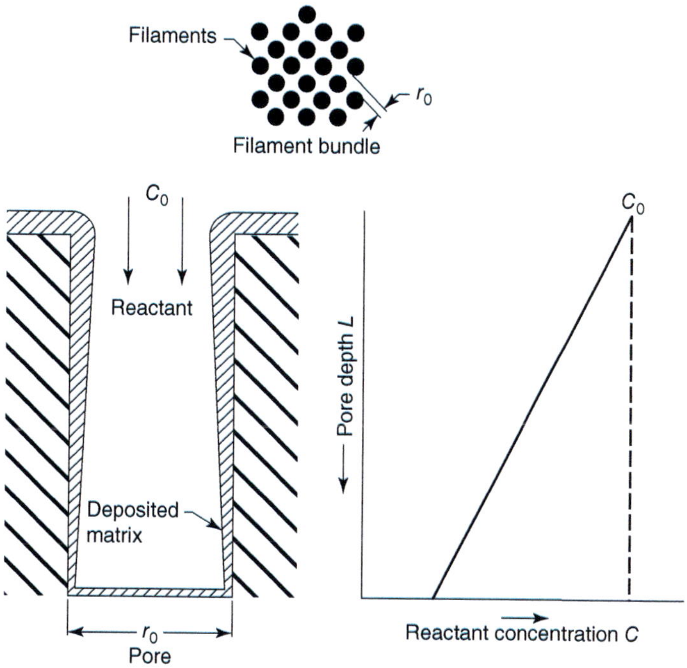

Figure 3.30 Schematic diagram of pore-filling process in the chemical vapor infiltration of a fiber bundle. From Ref. [32]. Copyright © 1997 by Chapman & Hall, London, with kind permission of Kluwer Academic Publishers.

CVI (ICVI) has no thermal gradients imposed, and reagents enter the substrate via diffusion. The disadvantage of this process is that the entire substrate is at the reaction temperature, so that pores on the surface can close off before the inner pores are completely filled. This problem is circumvented with a *thermal gradient*, as in Type II CVI. Here, the reactants contact the substrate in a cold region and diffuse to the heated reaction zone. Type III and Type IV CVI are the forced convection analogs of Type I and Type II, respectively, where the reactants are now forced into the substrate by convection instead of being allowed to diffuse. Type V CVI is specialized, in that not only are the reactants forced into the substrate, but gaseous products are also forcefully removed in a cyclical manner.

The isothermal reaction/diffusion system (Type I) is the easiest to model and provides us with some insight into the competition between kinetics and diffusion. Consider the case of CVI in a filament bundle, as shown schematically in Figure 3.30. The regions between the fibers can be approximated as cylindrical pores with diameter r_0 and depth L. Ideally, reactants will diffuse into the pore and create a uniform concentration throughout the pore length before they react to form product, thereby filling the pore uniformly from bottom to top. This is often not the case, and there

Historical Highlight

Read the story of how the first synthetic diamonds were made by the General Electric Company in the 1950s:

https://www.sciencefriday.com/articles/diamond-fake-excerpt/

and how they are now made more easily by chemical vapor deposition:

https://www.gia.edu/news-research-CVD-grown-part1

can be a concentration gradient, as shown in Figure 3.30, leading to nonuniform deposition. Eventually, the top of the pore will close off, and reactants will not be able to find their way to the bottom of the pore to complete the densification process.

The competition between kinetics and transport processes can be characterized using one of two dimensionless groups. The first is called the *Dahmköhler number, Da*, $k_1 r_0/D$, where k_1 is the reaction rate constant for the rate-limiting step, r_0 is the pore radius, and D is the diffusivity of the gaseous reactants. A related dimensionless group is the *Thiele modulus*, $\Phi^2 = 2k_1L^2/r_0D$, where the additional parameter, L, is the pore length. Both groups represent a ratio of the kinetic deposition rate to the gas phase mass transport rate, or, similarly, the diffusion time to the reaction time. For very small values of either dimensionless group, gas transport is very rapid and the reaction rate is very slow. This is the ideal situation, with uniform deposition along the pore length the result. In contrast, deposition is rapid for large values of Da and Φ^2, but diffusion is slow, so there is a decreasing concentration of reactants down the pore length, and a nonuniform density results. Mixed control occurs when the reaction and transport rates are comparable ($Da \approx \Phi^2 \approx 1$).

The CVI procedure can lead to some very interesting composite materials. Lackey et al. [17] have produced composites of carbon fibers in a laminated SiC/carbon matrix (see Figure 3.31). The alternating SiC/carbon layers were produced by CVI of two different reactions: the decomposition of methyltrichlorosilane (MTS) in the presence of hydrogen; and the decomposition of propylene in hydrogen, respectively. There is also an excellent opportunity to model these processes with the intent of gaining better control of the CVI process [18]. You can learn more about specific CVI processes for composite materials in ***Advanced Topics***.

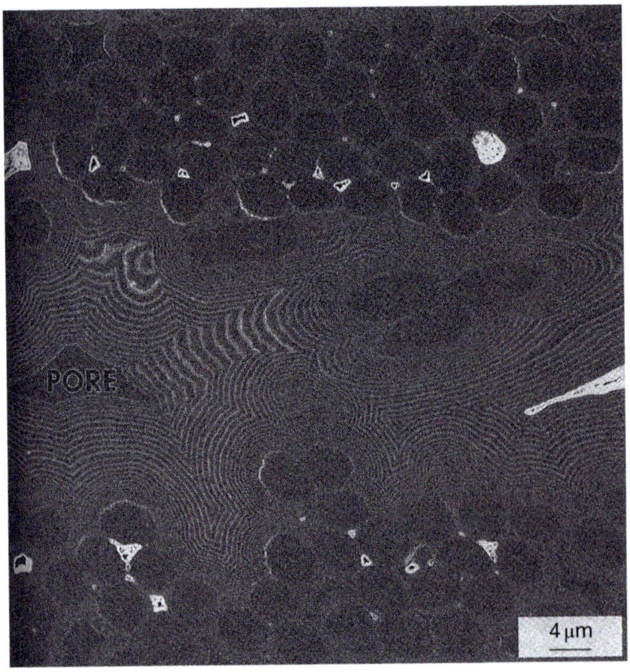

Figure 3.31 CVI-laminated SiC (light) and carbon (dark) matrix layers surrounding carbon fibers. Reproduced with permission from Lackey et al. [33]. © 1997, The American Ceramic Society.

Cooperative Learning Exercise 3.5

Consider the thermal decomposition of TMS in the presence of hydrogen gas to give SiC in a porous substrate:

$$(CH_3)SiCl_3(g) + H_2(g) \xrightarrow{k_1} SiC(s) + 3HCl(g) + H_2(g)$$

The following operating conditions are typically used for this CVI process:

Pore diameter	1–10 μm
Pore length	100–1000 μm
Ambient temperature	800–1200°C
Inlet pressure	5–200 Torr
Molecular diffusivity	726 cm²/s@1000°C and 5 Torr
Preexponential reaction constant	262 cm/s
Activation energy	120 kJ/mol

Person 1: Estimate the Dahmköhler number, *Da*.

Person 2: Estimate the Thiele modulus, Φ^2.

Compare your results. Is this CVI process reaction-controlled, diffusion-controlled, or a combination of both?

Answers: For $T = 1273$ K, $r_0 = 0.0005$ cm, and $L = 0.1$ cm, Da is about 10^{-9}, and Φ^2 is about 10^{-4}, both of which indicate a reaction-controlled CVI process.

3.5 Kinetic Processes in Biologics*

To attempt to describe the kinetics of formation, adaptation, and decomposition of tissue is to try and detail the processes inherent to life itself. This presents a rather daunting challenge, even if we limit ourselves to human tissue. Luckily, other disciplines such as biology have performed this task for us. The factors involved in the reproduction and growth of living tissue, as well as the kinetics of energy production, respiration, and excretion in cells, are not of concern to us here. What we are hoping to understand are the kinetic processes that will eventually allow us to seamlessly integrate biological and synthetic systems. We aim to describe not only the building blocks of biological systems (as we did in Chapter 1), and whether these building blocks are thermodynamically compatible with each other (as we did in Chapter 2), but also how rapidly the incorporation of synthetic materials into living systems can proceed. We will describe the kinetics of one such process here, namely, the kinetics of receptor–ligand binding in cells.

3.5.1 Kinetics of Cell Surface Receptor–Ligand Binding

Recall from Sections 1.5 and 2.5 that there are types of specific binding in and between cells called receptor–ligand binding (see Figure 3.32). The receptor–ligand interaction phenomenon is an extremely important one–one that goes far beyond simple cell adhesion. Such other cellular

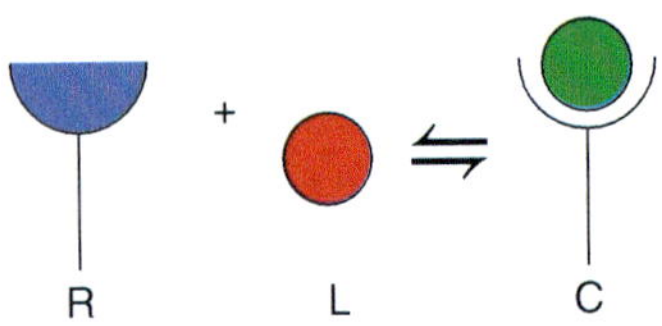

Figure 3.32 Schematic representation of receptor (R)–ligand (L) binding to form a receptor–ligand complex (C).

functions as growth, secretion, contraction, and motion are known to be receptor-mediated cell behaviors. Moreover, an understanding of the receptor–ligand interaction affords us the potential to control these cellular functions. For example, cell biologists can chemically alter the structure of the receptor and/or the ligand in order to modify the association and dissociation rate constants.

Figure 3.32 suggests that we can write a reversible reaction for the formation of the receptor–ligand complex

$$R + L \overset{k_f k_r}{\rightleftarrows} C \tag{3.118}$$

and that the time rate of change of the complex density is given by

$$\frac{d[C]}{dt} = k_f[R][L] - k_r[C] \tag{3.119}$$

Let us continue by considering the simplest case, where the ligand concentration remains essentially constant at its initial value of $[L]_0$, and the concentration of unbound receptors is given by the difference between the total concentration of receptors, $[R]_T$ (assumed to be constant), and the concentration of receptors in the complex, $[R_T - C]$:

$$\frac{d[C]}{dt} = k_f[R_T - C][L]_0 - k_r[C] \tag{3.120}$$

The use of dimensionless groups will simplify the solution to Eq. (3.120). Let the concentration of the complex, $[C]$, be scaled to the total receptor concentration, such that $u = [C]/[R]_T$ and the dimensionless time, τ, is given by $\tau = k_r/t$. The ratio of the reverse to forward rate constants is simply the dissociation equilibrium constant (the inverse of our "normal" equilibrium constant), $K_D = k_r/k_f$, so that Eq. (3.120) now becomes

$$\frac{du}{d\tau} = \frac{(1-u)[L]_0}{K_D} - u \tag{3.121}$$

which we can solve using the boundary condition that $u = u_0$ at $\tau = 0$ to give

$$u(t) = u_0 \exp\left[-\left(1 + \frac{L_0}{K_D}\right)\tau\right] + \frac{L_0 K_D}{1 + \frac{L_0}{K_D}}\left\{1 - \exp\left[-\left(1 + \frac{L_0}{K_D}\right)\tau\right]\right\} \tag{3.122}$$

Though this equation seems a bit unwieldy, it is a rather straightforward description of the variation of the scaled complex concentration as a function of time for various L_0/K_D ratios. For example, if. $u_0 = 0$, as is the case for the beginning of the binding procedure where there is no complex yet formed, a series of plots results as shown in Figure 3.33 for various L_0/K_D ratios. Such a plot is general for different types of receptor–ligand pairs. Some typical values of the rate constants and receptor concentrations for investigated receptor–ligand pairs are shown in Table 3.11.

Figure 3.33 The change in dimensionless receptor–ligand binding complex concentration, u, with dimensionless time, τ, for various values of L_0/K_D. Reproduced, by permission, from Lauffenburger and Linderman [34]. Copyright © 1993 by Oxford University Press, Inc.

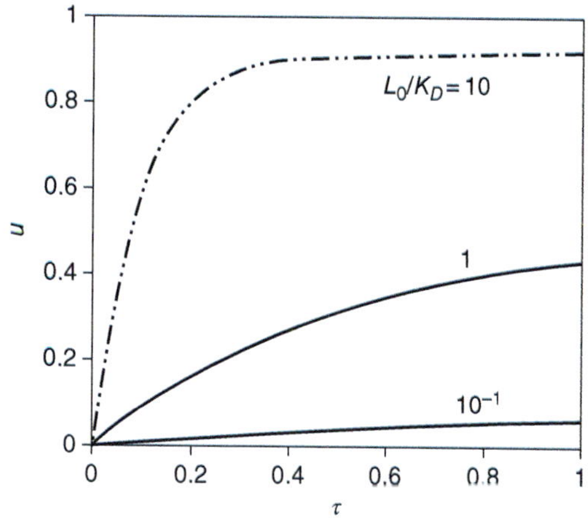

Table 3.11 Some Receptor–Ligand Binding Parameters

Receptor	Ligand	Cell Type	R_T (#/cell)	k_f (M^{-1} min^{-1})	k_r (min^{-1})	K_D (M)
Transferrin	Transferrin	HepG2	5×10^4	3×10^6	0.1	3.3×10^{-8}
FC_γ	2.4G2 Fab	Mouse macrophage	7.1×10^5	3×10^6	0.0023	7.7×10^{-10}
Chemotactic peptide	FNLLP	Rabbit neutrophil	5×10^4	2×10^7	0.4	2×10^{-8}
Interferon	Human interferon-α_2a	A549	900	2.2×10^8	0.072	3.3×10^{-10}
TNF	TNF	A549	6.6×10^3	9.6×10^8	0.14	1.5×10^{-10}
β-adrenergic	Hydroxybenzylpindolol	Turkey erythrocyte	–	8×10^8	0.08	1×10^{-10}
α_1-adrenergic	Prazosin	BC3H1	1.4×10^4	2.4×10^8	0.018	7.5×10^{-11}
Insulin	Insulin	Rat fat cells	1×10^5	9.6×10^6	0.2	2.1×10^{-8}
EGF	EGF	Fetal rat lung	2.5×10^4	1.8×10^8	0.12	6.7×10^{-10}
Fibronectin	Fibronectin	Fibroblasts	5×10^5	7×10^5	0.6	8.6×10^{-7}
Fc_g	IgE	Human basophils	–	3.1×10^6	0.0015	4.8×10^{-10}
IL-2 (heavy chain)	IL-2	T lymphocytes	2×10^3	2.3×10^7	0.015	6.5×10^{-10}
IL-2 (light chain)			1.1×10^4	8.4×10^8	24	2.9×10^{-8}
IL-2 (heterodimer)			2×10^3	1.9×10^9	0.014	7.4×10^{-12}

HepG2 = human hepatoma cell line; 2.4G2 Fab = Fab portion of 2.4G2 antibody against receptor; FNLLP = *N*-formylnorleucylleucylphenylalanine; A549 = human lung alveolar carcinoma; TNF = tumor necrosis factor; hydroxybenzylpindolol is an antagonist to the receptor; EGF = epidermal growth factor; IgE = immunoglobulin *E*; IL-2 = interleukin 2; prazosin is an antagonist to the receptor; BC3H1 = smooth muscle-like cell line; RBL = rat basophilic leukemia cell line.

Source: Lauffenburger and Linderman [39]. Copyright © 1993 by Oxford University Press, Inc.

References

1 Avrami, M., Kinetics of phase change I: general theory, *J. Chem. Phys.*, **7**, 1103 (1939).

2 Avrami, M., Kinetics of phase change II: transformation–time relations for random distribution of nuclei, *J. Chem. Phys.*, **8**, 212 (1940).

3 Avrami, M., Kinetics of phase change III: granulation, phase change, and microstructure, *J. Chem. Phys.*, **9**, 177 (1941).

4 Matusita, K., and S. Sakka, Kinetics study of the crystallization of glass by differential scanning calorimetry, *Phys. Chem. Glasses*, **20**, 81 (1979).

5 Kissinger, H. E., Reactive kinetics in differential thermal analysis, *Anal. Chem.*, **29**, 1702 (1957).

6 Baz, A., and T. Chen, Torsional stiffness of nitinol-reinforced composite drive shafts, *Composites Eng.*, **3**, 1119 (1993).

7 http://www.seas.upenn.edu/zshinar/researchb.html

8 http://aml.seas.ucla.edu/

9 http://impact.nace.org/economic-impact.aspx

10 JANAF, in *Thermochemical Tables*, D. R. Stull, and H. Prophet, eds., U.S. National Bureau of Standards, Washington, D.C., 1971.project directors

11 Parks, G. A., The isoelectric points of solid oxides, solid hydroxides, and aqueous hydroxo complex systems, *Chem. Rev.*, **65**, 177 (1965).

12 Hogenson, D. K., and J. H. Healy, Mathematical treatment of glass corrosion data, *J. Am. Cer. Soc.*, **45** (4), 178 (1962).

13 Vanderbilt, D. P., J. P. English, G. L. Fleming, F. W. McNeely, D. R. Cowsar, and R. L. Dunn, Biodegradable polyamides based on 4,4′-spirobibutyrolactone, in *Progress in Biomedical Polymers*, C. G. Gebelein, and R. L. Dunn, eds., Plenum Press, New York, 1990, p. 249.

14 Feijen, J., Biodegradable polymers for medical purposes, in *Polymeric Biomaterials*, E. Piskin, and A. S. Hoffman, eds., NATO ASI Series, Martinus Nijhoff Publishers, Dordrecht, 1986, p. 62.

15 Shih, C., A graphical method for the determination of the mode of hydrolysis of biodegradable polymers, *Pharmaceut. Res.*, **12**, 2036 (1995).

16 Ishii, H., and Y. Takahashi, *J. Electrochem. Soc.*, **135**, 1539 (1988).

17 Lackey, W. J., S. Vaidyaraman, and K. L. More, Laminated C—SiC matrix composites produced by CVI, *J. Am. Ceram. Soc.*, **80**, 113 (1997).

18 Gupte, S. M., and J. A. Tsamopoulos, Densification of porous materials by chemical vapor infiltration, *J. Electrochem. Soc.*, **136**, 2–555 (1989).

19 Ralls, K. M., T. H. Courtney, and J. Wulff, *Introduction to Materials Science and Engineering*, John Wiley & Sons, Inc, 1976.

20 Elliott, P., Overcome the challenge of challenge of corrosion, *Chem. Eng. Prog.*, **94**(5), 33.

21 Callister, W., *Materials Science and Engineering: An Introduction*, 5th ed., John Wiley & Sons, 2000, p. 574.

22 Callister, W., *Materials Science and Engineering: An Introduction*, 5th ed., John Wiley & Sons, Inc, 2000, p. 576.

23 Kingery, W. D., H. K. Bowen, and D. R. Uhlmann, *Introduction to Ceramics*, John Wiley & Sons, Inc, 1976.

24 Craig, B. D., *Fundamental Aspects of Corrosion Films in Corrosion Science*, Plenum Press, 1991, p. 4.

25 Gregg, R. A., and F. R. Mayo, Chain transfer in the polymerisation of styrene iii. The reactivities of hydrocarbons toward the styrene radical, *Discuss Faraday Soc.*, **2**, 328 (1947).

26 Billmeyer, F. W., *Textbook of Polymer Science*, 3rd ed., John Wiley & Sons, Inc, 1984.

27 Feijin, J., *Polymeric Biomaterials*, Springer Nature, 1986, p. 68.
28 Mark, J. Hampden-Smith, Toivo, T. Kodas, and Audunn, Ludvikssons, Chapter 5 "Chemical Vapor Deposition" eds., *Chemistry of Advanced Materials: An Overview*, John Wiley & Sons, Inc, 1998.
29 Galasso, F., *Chemical Vapor Deposited Materials*, CRC Press, 1991, p. 23.
30 Naslain, R., CVI composites, in *Ceramic–Matrix Composites*, R. Warren, ed., Chapman & Hall, London, 1992, pp. 199–244.
31 Theodore, M. Besmann, David, P. Stinton, Richard, A. Lowden, and Woo, Y. Lee, "Chemical Vapor Deposition (CVD) and Infiltration (CVI)," ed., *Carbide, Nitride, and Boride Materials Synthesis and Processing*, Chapman & Hall, London, UK, 1997, p. 563.
32 Theodore, M. Besmann, David, P. Stinton, Richard, A. Lowden, and Woo, Y. Lee, "Chemical Vapor Deposition (CVD) and Infiltration (CVI)," ed., Carbide, Nitride, and Boride Materials Synthesis and Processing, Chapman & Hall, London, 1997, p. 566.
33 Lackey, W. J., S. Vaidyaraman, and K. L. More, Laminated C-SiC matrix composites produced by CVI, *J. Am. Ceram. Soc.*, **80**, 113.
34 Lauffenburger, D. A., and J. J. Linderman, *Receptors*, Oxford University Press, Inc., 1993, p. 21.
35 Fontana, M. G., *Corrosion Engineering*, 3rd ed., McGraw-Hill Book Co, 1986, p. 457.
36 Young, L. J., Copolymerization reactivity ratios, in *Polymer Handbook*, J. Brandrup, and E. H. Immergut, eds., with the collaboration of W. McDowell, 2nd ed., Wiley-Interscience, New York, 1975, pp. II-105–II-386.
37 Billmeyer, F. W., *Textbook of Polymer Science*, Wiley, New York.
38 Amass, W., A. Amass, and B. Tighe, A review of biodegradable polymers: uses, current developments in the synthesis and characterization of biodegradable polyesters, blends of biodegradable polymers and recent advances in biodegradation studies, *Polym. Int.*, **47**, 89–144 (1998).
39 Lauffenburger, D. A., and J. J. Linderman, *Receptors*, Oxford University Press, Inc, 1993, p. 30.

Problems

Level I

3.I.1 Write the standard reaction potential for the half-reaction of chromium. What is the overall reaction when a chromium electrode and a hydrogen electrode are paired in an acid solution? Calculate the overall cell potential.

3.I.2 What happens when a piece of silver is dropped into a 1 M solution of hydrochloric acid? Is a gas produced? If so, what is it?

3.I.3 Calculate the free energies of a Ni–Cu and Zn–Ni electrochemical cell. Show your work.

3.I.4 The fraction crystallized of an amorphous $Mg_{72}Zn_{24}Ca_4$ alloy is 0.009 at 513 K and $t = 4.16$ minutes. Find the reaction order (a.k.a growth dimension or Avrami exponent) online for this alloy, or assume a value if you cannot find it, and calculate the crystallization rate.

3.I.5 Consider the condensation polyesterification reaction between ethylene glycol, HO–$(CH_2)_2$–OH, and terephthalic acid, HOOC–Ph–COOH, each of which has an initial concentration of 1.0 mol/l. Calculate the number average degree of polymerization at 1, 5, and 20 hours. The forward reaction rate constant for the polymerization reaction is 10.0 l/mol·h, and second-order, catalyzed kinetics can be assumed.

Level II

3.II.1 Calculate the standard reaction potential and free energy for an electrochemical cell comprising a Cd|0.05$M$$Cd^{2+}$ anode and 0.1$M$$Cl^-$|$Cl_2$ Pt cathode. You will need to look up the half-reaction potential for cadmium. Cite the source of your information and show your work.

3.II.2 The corrosion potential of iron immersed in deaerated acidic solution of pH = 3 is −0.720 V as measured at 25°C with respect to the normal calomel electrode. Calculate the corrosion rate of iron in millimeters per year, assuming the Tafel slope of the cathodic polarization curve, β_c, equals 0.1 V/decade, and the hydrogen ion exchange current, i_0, is 0.1×10^{-6} A/cm^2.

3.II.3 In an emulsion polymerization, all the ingredients are charged at time $t = 0$. The time to convert various amounts of monomer to polymer is indicated in the following table. Predict the time for 30% conversion. Polymerization is "normal"—that is, no water solubility or polymer precipitation.

Fraction of Original Monomer Charge Converted to Polymer	Time (h)
0.05	2.2
0.12	4.0
0.155	4.9

3.II.4 Construct a plot analogous to Figure 3.33 for the dissociation of the receptor–ligand complex; that is, for $u_0 = 1$. Make plots for $L_0/K_D = 10$, 1, and 0.1.

Level III

3.III.1 As a pure liquid is supercooled below the equilibrium freezing temperature (T_m), the free energy difference between the solid and liquid becomes increasingly negative. The variation of ΔG_v with temperature is given approximately by $\Delta G_v = (\Delta H_m/T_m)(T_m - T)$, where $\Delta H_m(<0)$ is the latent heat of solidification. Using this expression, show that exp.$(-N_0\Delta G*/RT)$ is a maximum when $T = T_m/3$. Why does the maximum nucleation rate occur at a temperature greater than $T_m/3$?

3.III.2 A review article on the CVD processes used to form SiC and Si_3N_4 by one of the pioneers in this area, Erich Fitzer [Fitzer, E., and D. Hegen, Chemical vapor deposition of silicon carbide and silicon nitride—Chemistry's contribution to modern silicon ceramics, *Angew. Chem. Int. Ed. Engl.*, **18**, 295 (1979)], describes the reaction kinetics of the gas-phase formation of these two technical ceramics in various reactor arrangements (hot wall, cold wall, and fluidized bed) for several precursor combinations, as indicated in the accompanying diagram. (a) Determine the activation energy(ies) for each ceramic/reactor/precursor type. (b) Comment on what causes the break in two of the curves at high temperatures.

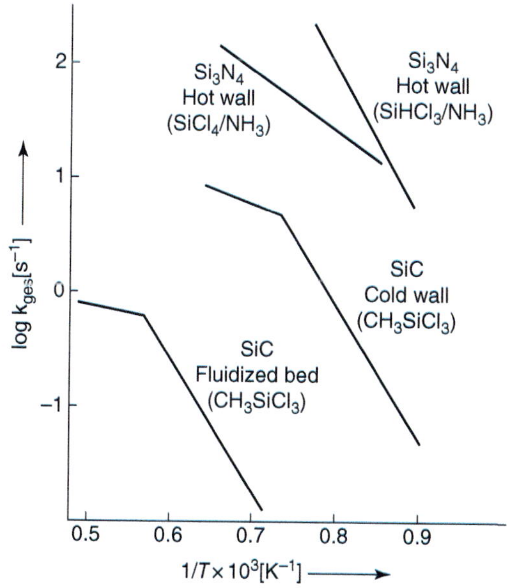

3.III.3 Consider three 10-m^3 reactors operating at 70°C. Each contains 1000 kg of styrene monomer. Additional ingredients initially are as follows:

	Reactor I	Reactor II	Reactor III
Inert liquid	8.00 m^3 benzene	8.00 m^3 water	8.00 m^3 water
Initiator	1.00 kg benzoyl peroxide	1.00 kg benzoyl peroxide	1.00 kg potassium persulfate
Polystyrene	10.0 kg polymer in solution	10.0 kg polymer dissolved in the monomer	10.0 kg polymer in stable latex form
Additive	–	1.00 kg poly(vinyl alcohol)	–

In each of the three reactors, does $-d[M]/dt$ (initial value) increase, decrease, or remain essentially the same when each of the following changes is made? (a) Amount of initiator is doubled; (b) amount of monomer is halved; (c) temperature is increased to 90°C; and (d) amount of initial polymer is doubled. Also, predict the effect of each change on the $\bar{x}_n$ of the polymer being formed initially.

4

Transport Properties of Materials

4.0 Introduction and Objectives

Perhaps the single largest subject area that differentiates engineering from the sciences is that of transport phenomena. For those of you who have had little or no background in transport processes, you may find some supplemental reading to be of great help in this chapter. There are many excellent textbooks and references on transport phenomena, including the landmark text by Bird, Stewart, and Lightfoot, which first brought the analogies between momentum, heat, and mass transfer to engineering students in a rigorous fashion. We will rely heavily on these analogies in this chapter; and without going into any great depth of coverage on any one transport topic, we will see how transport properties of liquid and solid materials go a long way toward describing many important phenomena in materials selection, design, and processing.

The analogy to which we refer is fundamentally quite simple. It states that the rate of flow per unit area, called a *flux*, of a fundamental quantity such as momentum, heat, or mass is proportional to a corresponding *driving force* that causes the flow to occur.

$$\text{Flux} = \text{"constant"} \times \text{Driving force} \tag{4.1}$$

In this chapter, we will focus mostly on the "constant" in Eq. (4.1), and not the flux or driving force, since these are specific to a process. The proportionality constant is a material property, however, which we can measure, correlate with process parameters, and hopefully predict. The term "constant" is in quotations because we will see that in some instances, it is really a function of the driving force. It is important to understand (a) the driving forces under which this proportionality constant operates and (b) the effect the constant has on transport phenomena.

The transport process about which most of us have an intuitive understanding is heat transfer so we will begin there. For heat to flow (from hot to cold), there must be a driving force, namely, a *temperature gradient*. The heat flow per unit area (Q/A) in one direction, say the y direction, is the heat flux, q_y. The temperature difference per unit length for an infinitesimally small unit is the temperature gradient, dT/dy. According to Eq. (4.1), there is then a proportionality constant that relates these two quantities, which we call the *thermal conductivity*, k. Do not confuse this quantity with rate constants of the previous chapter, or with the Boltzmann constant, k_B. The form of Eq. (4.1) for heat transfer, then, is called *Fourier's Law*:

$$q_y = -k\frac{dT}{dy} \tag{4.2}$$

The minus sign in Eq. (4.2) reflects the fact that for there to be heat flow in the $+y$ direction, the temperature gradient in that direction must be negative—that is, lower temperature in the direction

Materials Engineering and Science: Principles, Properties, and Processes, Second Edition. Brian S. Mitchell.

Companion website: www.wiley.com/go/materialsengineeringandscience2e

of heat flow. If the temperature gradient is expressed in units of K/m, and the heat flux is in $J/m^2 \cdot s$, then the thermal conductivity has units of J/K·m·s, or W/m·K. A related quantity is the *thermal diffusivity*, which is often represented by the lowercase Greek letter alpha, α. Thermal diffusivity is defined as $k/\rho C_p$, where k is the thermal conductivity, ρ is the density, and C_p is the *heat capacity* at constant pressure per unit mass. We will see in a moment why the term "diffusivity" is used to describe this parameter. We will generally confine our descriptions in this chapter to thermal conductivity.

The equations for one-dimensional momentum and mass flow are directly analogous to Fourier's Law. A *velocity gradient*, dv_x/dy, is the driving force for the bulk flow of momentum, or *momentum flux*, which we call the *shear stress* (shear force per unit area), τ_{yx}. This leads to *Newton's Law of Viscosity*:

$$\tau_{yx} = -\mu \frac{dv_x}{dy} \tag{4.3}$$

The proportionality constant, μ, is the *viscosity*. When the velocity gradient is cm/s·cm, or simply s^{-1}, we call it the *shear rate* because of the units of inverse time. Often, the lowercase Greek letter gamma with an overdot, $\dot{\gamma}$, is used to represent the shear rate, dv_x/dy. The shear stress of the x-directed moment in the y-direction, τ_{yx}, typically has units of $g/cm \cdot s^2$. Hence, viscosity has cgs units of g/cm·s, or *poise*. In the SI system, the unit of Pa·s is often used. Note that 1 Pa·s = 10 poise. In this chapter, we will have a lot to say about viscosity and how it varies for different materials in their molten or liquid states.

Finally, if we have a *concentration gradient* of one component (A) in a binary mixture ($A + B$), dc_A/dy, this leads to the bulk motion of molecules, known as the *molar flux* of component A in the y direction, J^*_{Ay}, which can be related to the concentration gradient using *Fick's Law*:

$$J^*_{Ay} = -D_{AB} \frac{dc_A}{dy} \tag{4.4}$$

The proportionality constant for mass transport is called the *molecular diffusivity* of A in B, and it has cgs units of cm^2/s when concentration is mol/cm^3 and distance is in centimeters. As a result, the molar flux has units of $mol/cm^2 \cdot s$. You should now recognize why the term "thermal diffusivity" is used in certain versions of Eq. (4.2).

Need More Help?
Try this video on analogies in transport processes: https://www.youtube.com/watch?v=qO5QUsCnMhw

You should be able to see the similarities in Eqs. (4.2) through (4.4). We will use these relationships a great deal in this chapter, so you should understand them fully and do some background reading if they are unfamiliar to you, but we will again concentrate on the proportionality constants: thermal conductivity, viscosity, and molecular diffusivity.

By the end of this chapter, you should be able to

- Locate or calculate thermal conductivities, heat capacities, and specific heats for any material.
- Locate or calculate viscosities for any liquid or molten material.
- Locate or calculate the binary-component diffusivity or self-diffusion coefficient for a gas, dilute solute, or melt.
- Estimate the diffusivity of a gas in a porous solid.

- Differentiate between Newtonian and non-Newtonian fluids, and know which behavior is appropriate for a certain type of material in the fluid state.
- Convert between thermal conductivity, thermal conductance, thermal resistivity, and thermal resistance for a material of specific geometry.
- Calculate permeabilities for substances, given diffusivity and solubility data.
- Calculate partition coefficients across membranes.

Finally, you will notice that there are a few things that are different in this chapter. It is primarily organized by the type of transport processes, then secondarily organized by material class—unlike the previous chapters, which were primarily organized by material type alone. This is done to add continuity to the transport subtopics while maintaining the analogies between them. Also, much of the material here is indicated as "optional"; that is, the information on momentum and mass transport properties typically are not taught in an introductory materials science course. This does not mean they are unimportant. Rather, it means that they are perhaps best read in conjunction with the chapter on materials processing (Chapter 7), since the driving forces arise due to the type of processing equipment being used—for example, casting, extrusion, or CVD processing. In any case, as engineers, we need to keep the similarities between the types of transport processes in mind as we study them, regardless of the relative extent to which each one is covered in a course.

4.1 Momentum Transport Properties of Materials*

In this section, we will examine the microscopic origins of viscosity for each material type, and we will see how viscosity is much more complex than simply serving as a proportionality constant in Eq. (4.3). Ultimately, we will find that viscosity is not a constant at all, but a complex function of temperature, shear rate, and composition, among other things.

4.1.1 Momentum Transport Properties of Metals and Alloys: Inviscid Systems

Most metallic systems above their melting point have very low viscosities, typically less than about 100 poise. Liquids with viscosities below this level are generally termed *inviscid*, that is, they are water-like in their fluidity. This is a relative statement, of course, since their viscosities are distinctly nonzero, but for all practical purposes the pressure required to cause these fluids to flow or overcome surface tension forces, for example, is vanishingly small. Nevertheless, the relative simplicity of metallic melts allows us to explore, at least briefly, the molecular origins of viscosity. We will then explore the temperature and composition dependences of metal and alloy viscosities.

The Molecular Origins of Viscosity

Viscosity—like all transport properties in this chapter—has its roots in the kinetic theory of gases. Though we concentrate on condensed phases in this book, it is important to look at the viscosity of gases, so that we may progress to the more complex topic of liquid viscosities, and so that we might see the analogies with the molecular origins of thermal conductivity and molecular diffusion in subsequent sections.

We already have much of the information we need to derive an expression for the viscosity of a gas. Recall from Chapter 1 that the atoms or molecules of a gas interact with one another. This interaction can be described in terms of repulsive and attractive energies, as given by Eq. (1.11). Recall also that the force, F, can be related to the energy, U, by Eq. (1.13). A potential energy function that has been extensively studied is the Lennard–Jones potential of Eq. (1.12). We can modify the form of Eq. (1.12) slightly to give the following potential energy function

$$U(r) = 4\varepsilon\left[\left(\frac{\sigma}{r}\right)^{12} - \left(\frac{\sigma}{r}\right)^{6}\right] \tag{4.5}$$

where σ is the *collision diameter* of the molecule and ε is the maximum energy of attraction between a pair of molecules. These two parameters are known for many substances and can be approximated for others using fluid critical properties, boiling point, or melting point. From this potential energy function, a relationship for viscosity (in poise) can be derived for gas at low density:

$$\mu = 2.6693 \times 10^{-5}\frac{\sqrt{MT}}{\sigma^2\Omega} \tag{4.6}$$

where T is the temperature of the gas in degrees Kelvin, M is the molecular weight of the gas, σ is in Å, and Ω is a dimensionless quantity that correlates with the dimensionless temperature, k_BT/ε. We will see that the dimensionless temperature plays a role in thermal conductivity and mass diffusivity as well. For rigid spheres (no attraction or repulsion), $\Omega = 1$, so that it represents a deviation of the real molecules from a rigid-sphere model. Note that the viscosity of gases at low density increases with temperature and that there is no dependence on pressure in this regime. So, we see that, in theory, there is indeed a link between the potential energy function of a gas and its viscosity.

For liquids—even simple monatomic liquids such as molten metals—this link between molecular interactions and a bulk property such as viscosity is still there, but it is difficult to derive a relationship from fundamental interaction energies that is useful for a wide variety of systems. Nonetheless, theoretical expressions for liquid viscosities do exist. Some are based on statistical mechanical arguments, but the model that is most consistent with the discussion so far is that of nonattracting hard spheres in a dense fluid. The viscosity (in Pa·s) of such a liquid is

$$\mu = 3.8 \times 10^{-8}\frac{\sqrt{MT}}{V^{2/3}}\left[\frac{PF^{4/3}\left(1 - \frac{PF}{2}\right)}{(1 - PF)^3}\right] \tag{4.7}$$

where M, T are the molecular weight and temperature as before, V is the molar volume of the liquid, and PF is the *packing fraction* or *packing density* of spheres in the liquid, which can be related to the collision diameter, σ, as

$$PF = \frac{\pi n_0 \sigma^3}{6} \tag{4.8}$$

where n_0 is the average number density (N/V). Some viscosities for elemental liquid metals as calculated from Eq. (4.7) are compared with experimental values in Table 4.1.

Notice the similarity between the relationship for liquid viscosity [Eq. (4.7)] and that for gaseous viscosity [Eq. (4.6)]. They both have a square root dependence on temperature and molecular weight and depend on the inverse square of the collision diameter [can you prove this for Eq. (4.7)?]. So, at least in principle, there is a fundamental relationship between the structure of a liquid and its viscosity.

Temperature Dependence of Pure Metal Viscosity

Practically speaking, empirical and semiempirical relationships do a much better job of correlating viscosity with useful parameters such as temperature than do equations like (Eq. 4.7). There are numerous models and their resulting equations that can be used for this purpose, and the interested

Table 4.1 Comparison of Calculated [Using Eq. (4.7) With PF = 0.45] and Experimental Values of Selected Liquid Metal Viscosities Near their Melting Points

Metal	μ_{cal} (mPa·s)	μ_{exp} (mPa·s)
Na	0.62	0.70
Mg	1.69	1.25
Al	1.79	1.2–4.2
K	0.50	0.54
Fe	4.55	6.92
Co	4.76	4.1–5.3
Ni	4.76	4.5–6.4
Cu	4.20	4.34
Zn	2.63	3.50
Ag	4.07	4.28
In	1.97	1.80
Sn	2.11	1.81
Au	5.80	5.38
Hg	2.06	2.04
Pb	2.78	2.61
Bi	2.54	1.63

Source: Lida and Guthrie [1]. Copyright © 1988 Oxford University Press.

student is referred to the many excellent references on this topic. A useful empirical relationship that we have already studied, and that is applicable to viscosity, is an Arrhenius-type relationship. For viscosity, this is

$$\mu = A\exp\left(\frac{H_\mu}{RT}\right) \tag{4.9}$$

where A and H_μ are constants. There are several methods for evaluating these two constants, but the simplest expressions relate them to the melting point of the metal, T_m. (We know that this makes some fundamental sense because the melting point is related to the intermolecular attraction between atoms.) The preexponential factor is given by

$$A(\text{cpoise}) = \frac{5.7\times 10^{-2}\sqrt{MT_m}}{V_m^{2/3}\exp(H_\mu/RT_m)} \tag{4.10}$$

The activation energy for viscous flow, H_μ, for "normal" metals is

$$H_\mu = 1.21T_m^{1.2} \tag{4.11}$$

and the activation energy for semi-metals is

$$H_\mu = 0.75T_m^{1.2} \tag{4.12}$$

The two relationships for activation energy as a function of melting point are shown graphically in Figure 4.1.

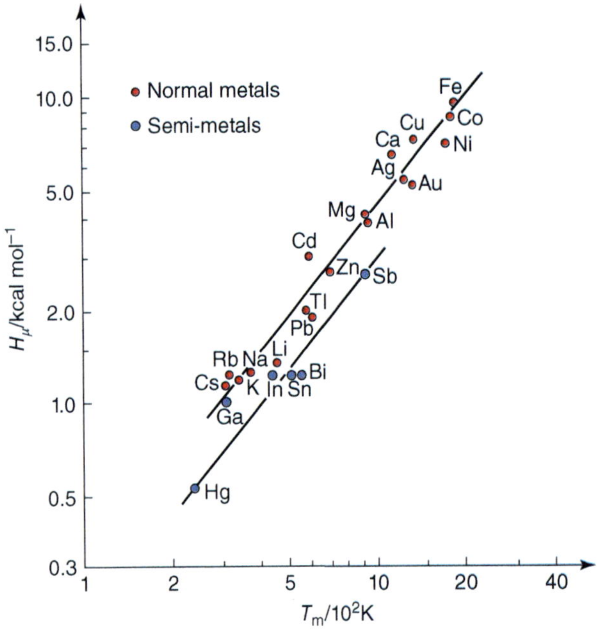

Figure 4.1 Dependence of activation energy for viscous flow on melting point for "normal" metals and semi-metals. Reproduced, by permission, from Iida and Guthrie [2]. Copyright © 1988 by Oxford University Press.

Composition Dependence of Alloy Viscosity

Attempts have been made to calculate the viscosity of a dilute liquid alloy from a theoretical standpoint but with little success. This is primarily because little is known about the interaction of dissimilar atoms in the liquid state. Empirical relationships for the viscosity of dilute liquid alloys have been developed, but these are generally limited to specific alloy systems—for example, mercury alloys with less than 1% impurities. The viscosities of binary liquid alloys have been empirically described using a quantity called the *excess viscosity*, μ^E (not to be confused with the excess chemical potential), which is defined as the difference between the viscosity of the binary mixture (alloy), μ_A, and the weighted contributions of each component, $x_1\mu_1$ and $x_2\mu_2$:

$$\mu^E = \mu_A - (x_1\mu_1 + x_2\mu_2) = -2(x_1\mu_1 + x_2\mu_2)\frac{H^E}{RT} \tag{4.13}$$

where x_1 and x_2 are the mole (atomic) fractions of each component, and H^E is the enthalpy of mixing. This expression indicates that the sign of the excess viscosity (either − or +) depends only on the sign of the enthalpy of mixing, which can also be either negative or positive. This relationship is only true some of the time, as borne out by experimental data [1]. The sign of excess viscosity can be a function of alloy composition, as well.

4.1.2 Momentum Transport Properties of Ceramics and Glasses

Newtonian Viscosity in Glasses

As we saw in Chapter 1, the structure of glasses is fundamentally different from metals. Unlike metals and alloys, which can be modeled as hard spheres, the structural unit in most oxide glasses is a polyhedron, often a tetrahedron. As a result, the response of a structural unit to a shear force is necessarily different in molten glasses than in molten metals. The response is also generally more complicated, such that theoretical descriptions of viscosity must give way completely to empirical expressions. Let us briefly explore how this is so.

Consider the melting of a common glass such as vitreous silica, SiO_2. Recall that the structural unit in the silicates is the $(SiO_4)^{4-}$ tetrahedron (see Figure 1.44). As heat is provided to the glass, Si—O bonds begin to rupture, and some structural units become mobile. Upon reaching the melting point, a certain portion of broken bonds is sufficient to cause flow in response to shear, but not all the bonds have been broken. For this reason, the SiO_2 melt has a high viscosity. As temperature is increased further above the melting point, more bonds are broken and the viscosity decreases. For silicate glasses, this change in viscosity can cover many order of magnitudes. Thus, viscosity–temperature plots for glasses are often given as semilogarithmic plots, as in Figure 4.2.

The large range of viscosities for glasses is related to its structure. For example, compared to a highly mobile liquid like water, the *relaxation time* is much longer in glasses. Relaxation time is the time needed for the structural components to adjust to the deformation forces resulting from viscous flow. In general, this is the same order of magnitude as the time required for a structural unit to move approximately one of its own diameters. The smaller the structural units and the less strongly they interact with each other, the shorter the relaxation time. For water, the relaxation time is of the order 10^{-12} s. For glasses near their softening points, this relaxation time could be around 10^2 s. It is interesting to note that almost all glasses have an inflection point in the log μ versus temperature curve at about 10^{15} Pa·s (10^{16} poise, just beyond where the data end in Figure 4.2), above which the viscosity begins to level off. This is the point at which brittleness begins to set in.

There are several named temperature regimes that have been assigned to glasses, within which certain characteristic melt properties, primarily related to formability, have been identified. From

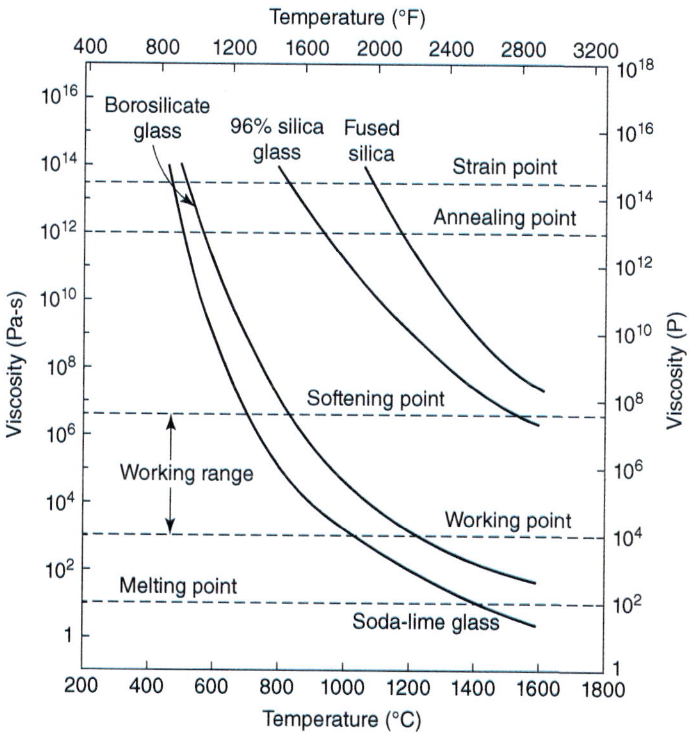

Figure 4.2 Dependence of viscosity on temperature for several silica-based glasses. Reproduced, by permission, from Callister [3]. Copyright © 2000 by John Wiley & Sons, Inc.

> **Need More Help?**
>
> Try this video on the viscosity of glasses:
>
> https://www.youtube.com/watch?v=z7eu5nBMKJc

highest to lowest temperature (as is found in a cooling operation), these temperatures include the *melting point* (~10 Pa·s), where the glass is fluid enough to be considered a liquid; the *working point* (~10^3 Pa·s), where the glass is easily deformed; the *softening point* (~10^6 Pa·s), where a glass piece can be handled without significant dimensional change; the *annealing point* (~10^{12} Pa·s), where the relaxation time is on the order 10^3 s; and the *strain point* (10^{13} Pa·s), where fracture occurs before plastic deformation in response to an applied force (see Chapter 5 for a comparison of fracture and plastic deformation). Other terminology used in the glass industry includes the *fictive temperature*, where the properties of the glass are the same as those of the melt in metastable equilibrium; the *devitrification temperature*, where the glass transforms from a noncrystalline state to one that is crystalline; and the *Littleton temperature*, where a glass rod heated at 5–10 K/m will deform at a rate of 1 mm/min under its own weight.

Most viscosity–temperature relationships for glasses take the form of an Arrhenius expression, as was the case for binary metal alloys. The Vogel–Fulcher–Tammann (VFT) equation is one such relationship.

$$\mu(T) = A\exp\left[\frac{E_\mu}{R(T - T_0)}\right] \tag{4.14}$$

where E_μ is the activation energy for viscosity, A is the preexponential factor, R is the gas constant, and T_0 is an additional adjustable parameter specific to a glass. Viscosity–temperature plots can be generated for glasses with three adjustable parameters using the VFT equation.

Cooperative Learning Exercise 4.1

Three-parameter models such as the VFT (Eq. 4.14) can be fit to data by first placing the equation in the following form:

$$\ln\mu = A + \frac{E_\mu}{R(T - T_0)}$$

then using three pairs of dependent/independent variable values: ($\ln\mu_1$, T_1); ($\ln\mu_2$, T_2), and ($\ln\mu_3$, T_3). With three such pairs of data, you can create three independent equations and solve them simultaneously using simple algebraic substitution or using a spreadsheet (see Problem 4.III.2 at the end of the chapter). Alternatively, the parameters can be determined sequentially using the following relations, which are represented for the VFT equation, where L_i is shorthand notation for $\ln\mu_i$:

$$A = \frac{(L_1 - L_2)(L_3T_3 - L_2T_2) - (L_2 - L_3)(L_2T_2 - L_1T_1)}{(L_1 - L_2)(T_3 - T_2) - (L_2 - L_3)(T_2 - T_1)}$$

$$T_0 = \frac{L_1T_1 - L_2T_2 - A(T_1 - T_2)}{L_1 - L_2} \qquad \frac{E_\mu}{R} = (T_i - T_0)(L_i - A)$$

Use the following viscosity–temperature data to complete the following exercise: $\ln\mu_1 = 13.3$, $T_g = 550$°C; $\ln\mu_2 = 7.6$, Littleton temperature = 715°C; $\ln\mu_3 = 4.0$, working point = 1000°C.

Person 1: Determine the VFT parameters A, E_μ/R, and T_0 from the data provided by solving the three independent equations created when substituting each data pair into the linearized form of the VFT equation.

Person 2: Determine the VFT parameters A, E_μ, and T_0 from the data provided by using the three relationships provided above.

Compare your answers!

Answer: $\ln \mu = -1.386 + 38030.3/(T - 288.8)$

Non-Newtonian Viscosities in Ceramic Slurries and Suspensions

As the melting point of inorganic materials such as oxides, borides, and carbides increases, melt processing becomes less practical due to increased energy requirements, difficulty in containment, or reduced formability. As a result, high-melting-point ceramics are often processed in the form of slurries, suspensions, or pastes. The distinction between a paste, slurry, or suspension depends upon several factors, such as (a) whether the ceramic particles, liquid phase, and trapped gases are continuous or discontinuous and (b) the value of particle parameters such as packing fraction (PF), degree of pore saturation (DPS), and mechanical compressibility (X), as summarized in Table 4.2. In a powder composed of spherical particles, the packing fraction (PF) is the fraction of the bulk volume occupied by spheres. As described in Section 1.1.1, the packing fraction for simple cubic is 0.52, and 0.74 for close-packed structures. The DPS is a measure of the pore saturation and is defined as the ratio of volume of the suspending liquid to pore volume (see Figure 4.3). The

Table 4.2 Characteristics of Ceramic/Liquid Mixture Consistency States[a]

Consistency	Particle Structure	Gas	Liquid	PF	DPS	X
Powder	C	C	D	L	<1	H
Granules	C	C	D	L-M	<1	M
Plastic	C	D	C	M-H	<1	L
Paste	C	D	C	M-L	= 1	0
Slurry	D	D	C	M-L	>1	0
Suspension	D	D	C	L	>1	0

Source: Reed [4]. Copyright © 1995 by John Wiley & Sons, Inc.
[a] C = continuous; D = discontinuous; H = high; M = moderate; L = low.

Figure 4.3 Liquid distribution in a powder leading to varying degree of pore saturation (DPS).

mechanical compressibility of a collection of particulates, X, is defined as $(\Delta V/V_0)/\Delta P$, where ΔV is the volume change resulting from an incremental pressure change ΔP, and V_0 is the initial volume. As more liquid is added to the particles, the mixture becomes essentially incompressible, and X approaches zero.

These distinctions are important from a processing standpoint, since the transfer of shear forces within the mixture changes significantly. These differences are shown schematically in Figure 4.4. A dry, bulk powder consists of discrete particles and random agglomerates produced by attractive van der Waals, electrostatic, or magnetic interparticle forces. Here, the flow resistance depends upon random particle attraction and agglomeration, and it tends to decrease with decreasing particle size. This occurs because the surface force relative to the gravitational force increases with decreasing particle size. When a wetting liquid of low viscosity is added to a dry powder, *agglomerates* form. Agglomerates that are purposefully made in this way are called *granules*, and the free-flowing material is referred to as being *granular*. The resistance to shearing flow is still low in these mixtures. On further increase of the liquid content, nearly all the powder will be transformed into granules with dry surfaces. At this point, only a small amount of liquid is required to cause nearly total agglomeration into a cohesive mass. The cohesive mass is very resistant to shearing, as indicated by the peak in Figure 4.4. The response of this cohesive mass to shear is plastic-like, so it is often called a *plastic body*. In the agglomerate, plastic, and paste states, shear stress may be transmitted by particles in contact with one another. From this point on, resistance to shear decreases as more liquid is added, and the liquid phase becomes continuous. The flow behavior of a paste depends on the content of colloidal particles and coagulation forces. A *paste* becomes a *slurry* upon further addition of liquid. Now the material may flow under its own weight. A dilute slurry is termed a *suspension*.

In addition to temperature, the viscosity of these mixtures can change dramatically over time, or even with applied shear. Liquids or solutions whose viscosity changes with time or shear rate are said to be *non-Newtonian*; that is, viscosity can no longer be considered a proportionality constant between the shear stress and the shear rate. In solutions containing large molecules and suspensions contain nonattracting anisometric particles, flow can orient the molecules or particles. This orientation reduces the resistance to shear, and the stress required to increase the shear rate

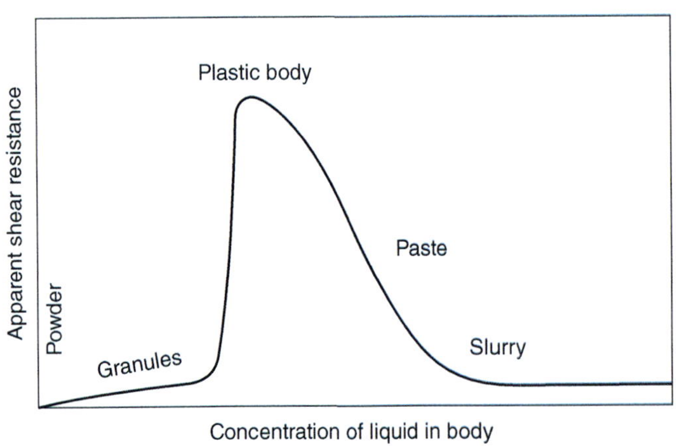

Figure 4.4 Processing consistency produced on mixing a powder with a wetting liquid. From Reed [4]. Copyright © 1995 by John Wiley & Sons, Inc. This material is used by permission of John Wiley & Sons, Inc.

diminishes with increasing shear rate. This behavior is often described by an empirical power law equation that is simply a variation of Eq. (4.3), and the fluid is said to be a *power law fluid*:

$$\tau = K\dot{\gamma}^n \tag{4.15}$$

Here, K is sometimes referred to as the *consistency index* and has units that depend on the value of the *power law index, n*—for example, N·s^n/m^2. The power law index is itself dimensionless. Typical values of K and n are listed in Table 4.3 In general, the power law index is independent of both temperature and concentration, although fluids tend to become more Newtonian (n approaches 1.0) as temperature increases and concentration decreases. The consistency factor, however, is more sensitive to temperature and concentration. To correct for temperature, the following relationship is often used:

$$K_\theta = K_0\mu_\theta/\mu_0 \tag{4.16}$$

where K_θ is the consistency index at temperature θ, K_0 is the consistency factor at the reference temperature, and μ_θ and μ_0 are the corresponding viscosities of the suspending fluid, which is usually, but not always, water.

For a power-law fluid, the viscosity is now a function of shear rate. Often, a non-Newtonian viscosity is designated with the lowercase Greek letter eta, η, to differentiate it from a Newtonian viscosity, μ:

$$\eta = K\dot{\gamma}^{n-1} \tag{4.17}$$

Table 4.3 Power-Law Parameters for Selected Fluids

Fluid	K (N·s^n/m^2)	n
23.3% Illinois yellow clay in water[a]	5.55	0.229
0.67% carboxymethylcellulose (CMC) in water[a]	0.304	0.716
1.5% CMC in water[a]	3.13	0.554
3.0% CMC in water[a]	9.29	0.566
33% lime in water[a]	7.18	0.171
10% napalm in kerosene[a]	4.28	0.520
4% paper pulp in water[a]	20.0	0.575
54.3% cement rock in water[a]	2.51	0.153
0.5% hydroxymethylcellulose (HMC) in water (293 K)[b]	0.84	0.509
2.0% HMC in water (293 K)[b]	93.5	0.189
1.0% polyethylene oxide in water (293 K)[b]	0.994	0.532
Human blood (293 K)[c]	0.00384	0.890
Ketchup[c]	33.2	0.242
Soybean oil[c]	0.04	1.0
Sewage sludge[c] (1.7–6.2 wt% sol.) w = wt% solids	$0.00113w^{3.4}$	0.817–0.843
Orange juice[c]	1.89	0.680
Corn syrup (48.4 total solids)[c]	0.0053	1.0

[a] Bird et al. [5].
[b] Bird et al. [6].
[c] Johnson [7].

In general, we will use μ to represent Newtonian viscosity, and η to represent non-Newtonian viscosity. Recognize, of course, that in certain limits and under certain conditions, non-Newtonian fluids can behave in a Newtonian manner. We will not attempt to switch between the two representations for viscosity under these circumstances.

The value of n in Eqs. (4.15) and (4.17) can be greater than or less than 1.0. For $n > 1.0$, the viscosity increases with shear rate, in what is a called a *shear-thickening* fluid. Moderately concentrated suspensions containing large agglomerates and concentrated, deflocculated slurries of ceramic powders behave in this manner. When $n < 1.0$, a *shear-thinning* fluid results, and less stress is required to incrementally increase the shear rate. Suspensions containing nonattracting, anisometric particles, along with liquids or solutions containing large molecules, are typically shear thinning, because laminar flow tends to orient the particles and reduces the resistance to shear. We will see in the next section that polymeric materials can be both shear-thinning and shear-thickening. These two types of power-law behavior are summarized graphically in the lower half of Figure 4.5. Notice that at low shear rates and over limited shear rate regimes, shear-thinning and shear-thickening fluids can behave in a pseudo-Newtonian manner.

The top three curves in Figure 4.5 represent the behavior of yet another class of non-Newtonian fluids that exhibit a *yield stress* (note how their curves begin further up on the shear stress axis than the previous group of curves). In ceramic systems, slurries that contain bonded molecules or strongly attracting particles require a finite stress called the yield stress, τ_y, to initiate flow. Once the yield stress has been overcome, the fluid can behave in either a Newtonian or non-Newtonian manner. Fluids that possess a yield stress, yet have a constant proportionality between stress and strain (shear rate) above the yield stress, are termed *Bingham fluids* (or Bingham plastics for organic macromolecules) and obey the general relationship:

$$\tau = \tau_y + \eta_p \dot{\gamma} \tag{4.18}$$

Bingham fluids that are either shear-thinning or shear-thickening above their yield stresses have corresponding power-law expressions incorporated into their viscosity models.

Further non-Newtonian types of behavior appear when time is introduced as a variable. Fluids whose viscosity decreases with time when sheared at a constant rate are called *thixotropic*, whereas fluids whose viscosity increases with time when sheared at a constant rate are termed *rheopectic*.

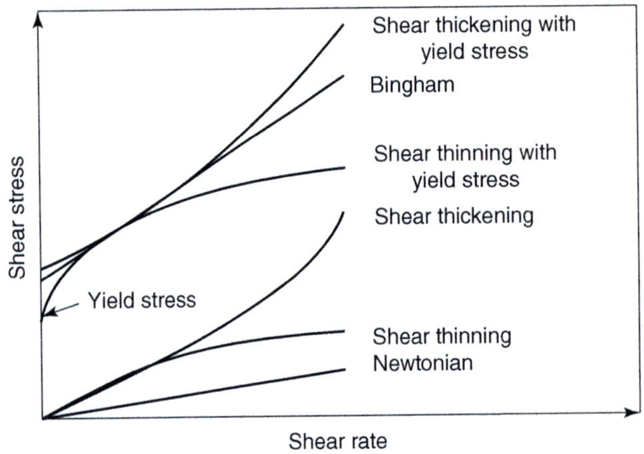

Figure 4.5 Variation of shear stress with shear rate for different models of flow behavior. From Reed [4]. Copyright © 1995 by John Wiley & Sons, Inc. This material is used by permission of John Wiley & Sons, Inc.

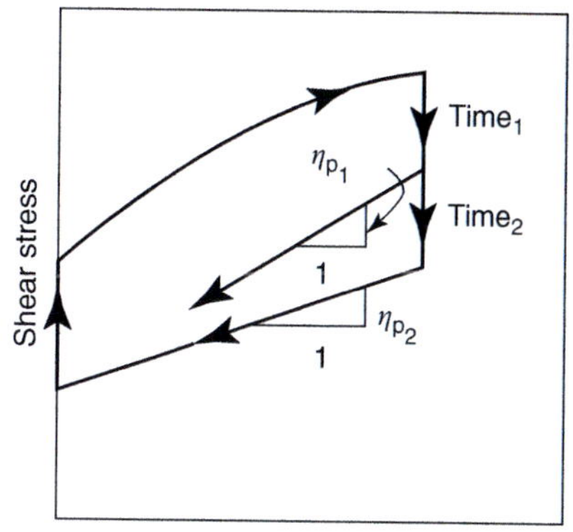

Figure 4.6 Thixotropic behavior, in which time at constant shear rate affects viscosity. From Reed [4]. Copyright © 1995 by John Wiley & Sons, Inc. This material is used by permission of John Wiley & Sons, Inc.

Thixotropic behavior is represented schematically in Figure 4.6. Note that the two viscosities (slope of shear stress versus shear rate) η_{P1} and η_{P2} are different, depending upon the time they were sheared at a given rate.

With this background of non-Newtonian behavior in hand, let us examine the viscous behavior of suspensions and slurries in ceramic systems. For dilute suspensions on noninteracting spheres in a Newtonian liquid, the viscosity of the suspension, η_S, is greater than the viscosity of the pure liquid medium, η_L. In such cases, a *relative viscosity*, η_r, is utilized, which is defined as η_S/η_L. For laminar flow, η_r is given by the *Einstein equation*

$$\eta_r = 1 + 2.5f_p^v \tag{4.19}$$

Cooperative Learning Exercise 4.2

Agglomeration in a slurry causes a change in the packing factor at which flow is blocked, f_{cr}^v, from 0.6 to 0.45 and causes a change in the hydrodynamic shape factor from 3.5 to 2.5. In both cases, the volume fraction of dispersed particles is 0.40.

Person 1: Calculate the relative viscosity for f_{cr}^v, = 0.6 and K_H = 3.5.

Person 2: Calculate the relative viscosity for f_{cr}^v, = 0.45 and K_H = 2.5.

Compare your answers. How would you expect these values to change if the shear rate were increased?

Answer: $\eta_r(0.6, 3.5) = 4.7$; $\eta_r(0.45, 2.5) = 9.0$

Increased shear rate would help prevent particle agglomeration, effectively increasing f_{cr}^v, and reducing the relative viscosity.

where f_p^v is the volume fraction of dispersed spheres. A more general relationship for the reduced viscosity allows for nonspherical particles, through the use of an apparent *hydrodynamic shape* factor, K_H:

$$\eta_r = 1 + K_H f_p^v \tag{4.20}$$

Need More Help?

Try this video on the viscosity particulate suspensions:

https://www.youtube.com/watch?v=34W5AQ5L7PU

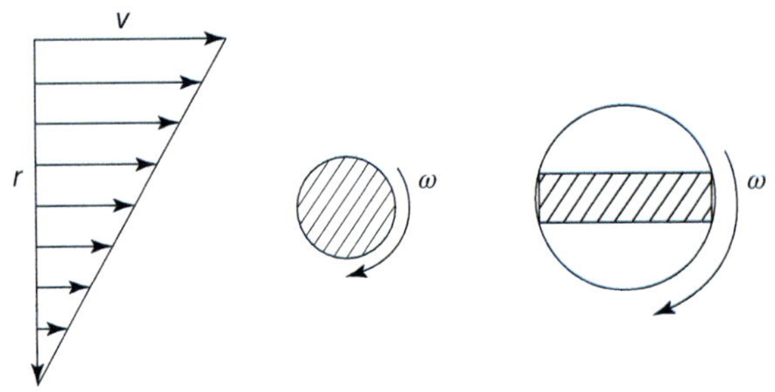

Figure 4.7 Effect of hydrodynamic shape factor on effective particle size in a flow field. From Reed [4]. Copyright © 1995 by John Wiley & Sons, Inc. This material is used by permission of John Wiley & Sons, Inc.

In suspensions of particles with an *aspect ratio* (length to diameter) greater than 1, particle rotation during flow results in a large effective hydrodynamic volume, and $K_H > 2.5$ (see Figure 4.7). At particle volume fractions above about 5–10%, interaction between particles during flow causes the viscosity relationship to deviate from the Einstein equation. In such instances, the reduced viscosity is better described by the following relationship:

$$\eta_r = \left[1 - f_p^v / f_{cr}^v\right]^{-K_H f_p^v} \tag{4.21}$$

Here, f_p^v is once again the volume fraction of dispersed spheres, and K_H is the apparent hydrodynamic shape factor, but a new parameter, f_{cr}^v, is introduced which accounts or the packing factor at which flow is blocked. This parameter can be determined graphically, as in Figure 4.8. The effect of volume fraction of dispersed particles on the relative viscosity for several different apparent hydrodynamic shape factors, as determined by Eq. (4.21), is shown in Figure 4.9. There are numerous other equations and correlations that attempt to predict the viscosity of suspensions and

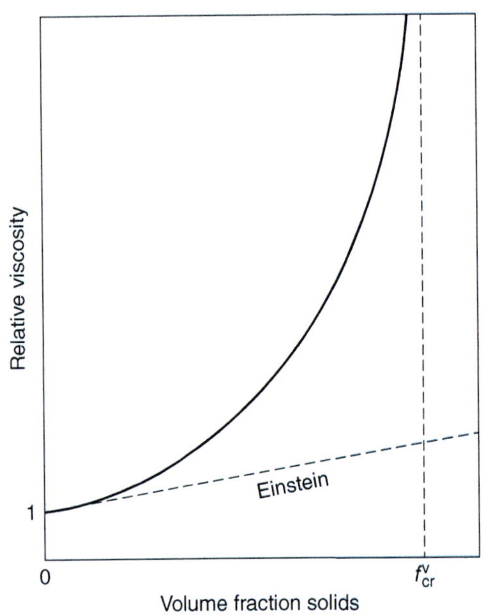

Figure 4.8 Effect of increasing the powder concentration in a slurry on relative viscosity. From Reed [4]. Copyright © 1995 by John Wiley & Sons, Inc. This material is used by permission of John Wiley & Sons, Inc.

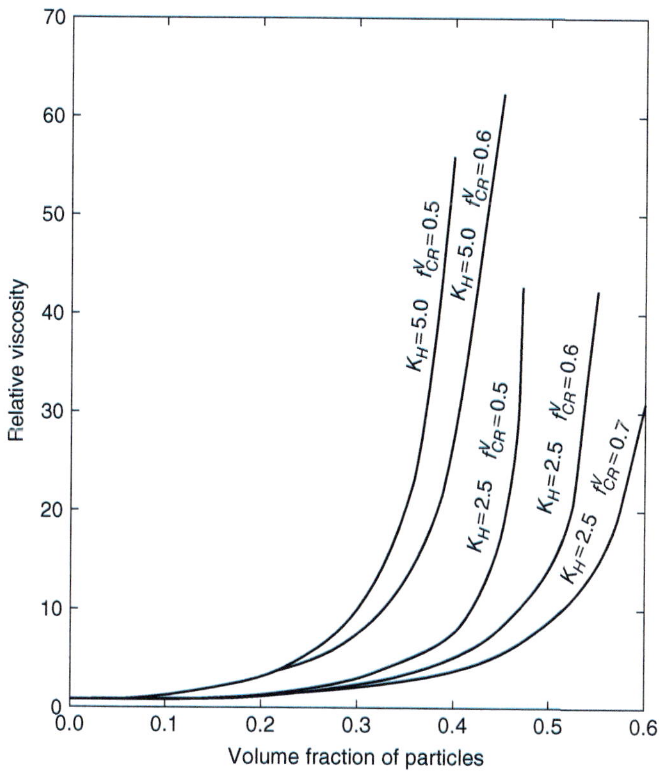

Figure 4.9 Dependence of relative viscosity on solids concentrations for several values of apparent hydrodynamic shape factors. From Reed [4]. Copyright © 1995 by John Wiley & Sons, Inc. This material is used by permission of John Wiley & Sons, Inc.

slurries. Some take into account adsorbed binder on the particles, and some attempt to predict such things as yield stress and the effects of time on rheological behavior. In most cases, these fluids behave in a highly non-Newtonian manner, and rheological properties can be predicted only for very narrow temperature and concentration ranges.

This description of non-Newtonian behavior in ceramic slurries and suspensions leads us directly into a description of viscosity in polymeric materials, which are oftentimes highly non-Newtonian in their behavior.

4.1.3 Momentum Transport Properties of Polymers

Unlike metals, in which momentum transport properties are generally limited to the molten state, and ceramics, in which momentum transport properties are primarily (but not always) described by the solution state, polymers can be found in either the solution or molten state. As a result, many of the principles that have been previously discussed apply to polymers, especially with regard to non-Newtonian behavior. There are, however, a few viscosity-related concepts, exclusive to polymers, that we will describe here.

Viscosity of Polymer Solutions

Polymer solutions are very common. Glues, pastes, and paints are just a few examples of commercially available aqueous suspensions of organic macromolecules. Also, recall from Chapter 3 that certain types of polymerization reactions are carried out in solution to assist in heat removal. The resulting polymers are also in solution, and their behavior must be fully understood in order to properly transport them and effect solvent removal, if necessary.

Unless they are formed directly in solution, polymers dissolve in two general steps. First, the polymer absorbs solvent to form a swollen *gel*. The second step involves the dissolution of the gel into the solution, sometimes called a *sol*. Networked or thermoset polymers may only undergo gel formation if they dissolve at all. Once the gel is in solution, the solution viscosity is generally described as it is for ceramic suspensions and slurries, with the relative viscosity

$$\eta_r = \frac{\eta}{\eta_s} \tag{4.22}$$

where η is the true solution viscosity, and η_s is the pure solvent viscosity. Keep in mind that not all polymer solutions are aqueous; that is, many polymers dissolve in nonpolar and/or organic solvents, depending on their chemical structure. Equation (4.22) shows us that the relative viscosity for pure solvents is exactly 1.0.

A Taylor series expansion in polymer concentration, c, is often used for the relative viscosity

$$\eta_r = 1 + [\eta]c + k'[\eta]^2c^2 + \cdots \tag{4.23}$$

where $[\eta]$ is called the *intrinsic viscosity* of the solution, and k' is called the *Huggins coefficient*. The expansion is often truncated at the second-order term. The intrinsic viscosity has units of reciprocal concentration. Both intrinsic viscosity and the Huggins coefficient are independent of concentration, although they can be functions of shear rate. You should be able to see from Eqs. (4.22) and (4.23) that the following relationship holds for the intrinsic viscosity:

$$[\eta] = \lim_{c\to 0}\left(\frac{\eta - \eta_s}{c\eta_s}\right) \tag{4.24}$$

Cooperative Learning Exercise 4.3

We are given the following measurements of relative viscosities of solutions of polyisobutylene $\left(\overline{M}_w = 2.5 \times 10^6\right)$ in decalin and in a 70% decalin and 30% cyclohexanol mixture at 298 K.

Person 1: Calculate the intrinsic viscosity, $[\eta]$, for the polyisobutylene sample in decalin.

Person 2: Calculate the intrinsic viscosity, $[\eta]$, for the polyisobutylene sample in decalincyclohexanol.

Compare your answers. Which is the "better" solvent?

Decalin		Decalin-Cyclohexanol	
c(g/cm^3)	η_r	c(g/cm^3)	η_r
0.000408	1.227	0.000409	1.155
0.000543	1.311	0.000546	1.210
0.000815	1.485	0.000818	1.326
0.001304	1.857	0.001169	1.488
0.001630	2.125	0.001637	1.740

Source: Adapted from Bird et al. [6].

Answers: Decalin 5.1 × 10^{2}cm^3/g; decalin-cyclohexanol 3.1 × 10^{2}cm^3/g.

Table 4.4 Summary of Viscometric Terms

Symbol	Name	Common Units
η	Solution viscosity	Poise or Pa·s
η_s	Solvent viscosity	Poise or Pa·s
$\eta_r = \eta/\eta_s$	Relative viscosity	Dimensionless
$\eta_{sp} = \eta_r - 1$	Specific viscosity	Dimensionless
$(\ln \eta_r)/c$	Inherent viscosity	dl/g
$\eta_{sp}/c = \eta_{sr}$	Reduced viscosity	dl/g

It may seem confusing to introduce yet another type of viscosity. In fact, there are many other types of viscosity (see Table 4.4), but we will concentrate on the intrinsic viscosity for polymer solutions. The intrinsic viscosity is important because it has been found to correlate well with many important parameters, including polymer molecular weight and type of solvent. Note that if Eq. (4.23) is truncated to the first two terms, the intrinsic viscosity is the same as the reduced specific viscosity, η_{sp}, of Table 4.4.

Polymer solvents are generally characterized as being either a *good solvent* or a *poor solvent*. A good solvent is defined as a solvent in which polymer–solvent interactions are more favorable than polymer–polymer interactions. Conversely, in a poor solvent, polymer–polymer interactions are more favorable than polymer–solvent interactions. So, in a good solvent, the polymer chains will tend to uncoil to interact more with the solvent, thereby dissolving more effectively. The Huggins coefficient has values in the range 0.3–0.4 for good solvents, with higher values for poor solvents.

Intrinsic viscosity has also been found to correlate with the molecular weight of the polymer (see Figure 4.10) according to the Mark–Houwink relationship:

$$[\eta] = K'\overline{M}_v^a \tag{4.25}$$

Need More Help?
Try this video on an introduction to polymer rheology: https://www.youtube.com/watch?v=ymxJ_i9bSNw

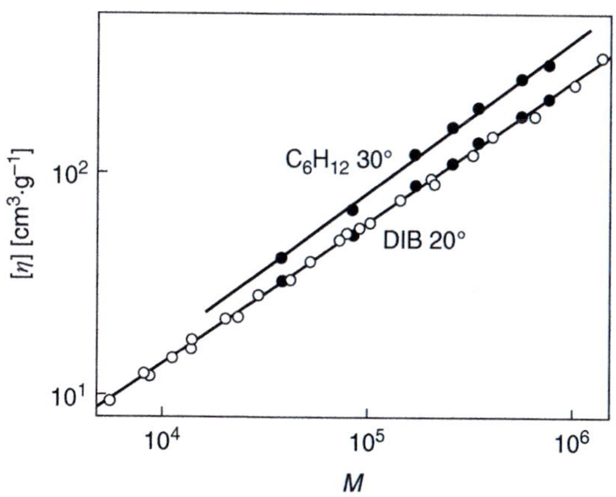

Figure 4.10 Intrinsic viscosity-molecular weight relationship for polyisobutylene in diisobutylene (DIB) and cyclohexane. Reproduced, by permission, from Strobl [8]. Copyright © 1997 by Springer-Verlag.

Table 4.5 Selected Values of Parameters for Mark–Houwink Equation

Polymer	Solvent	Temperature (°C)	$K' \times 10^5$	a
Cellulose triacetate	Acetone	25	8.97	0.90
SBR rubber	Benzene	25	54	0.66
Natural rubber	Benzene	30	18.5	0.74
	n-Propyl ketone	14.5	119	0.50
Polyacrylamide	Water	30	68	0.66
Polyacrylonitrile	Dimethyl formamide	25	23.3	0.75
Poly(dimethylsiloxane)	Toluene	20	20.0	0.66
Polyethylene	Decalin	135	62	0.70
Polyisobutylene	Benzene	24	107	0.50
	Benzene	40	43	0.60
	Cyclohexane	30	27.6	0.69
Poly(methyl methacrylate)	Toluene	25	7.1	0.73
Polystyrene				
Atactic	Toluene	30	11.0	0.725
Isotactic	Toluene	30	10.6	0.725
Poly(vinyl acetate)	Benzene	30	22	0.65
	Ethyl n-butyl ketone	29	92.9	0.50
Poly(vinyl chloride)	Tetrahydrofuran	20	3.63	0.92

Source: Rodriguez [9]. Copyright © 1982 by McGraw-Hill Book Company.

where K' and a are constants that depend on the particular polymer-solvent system, and $\bar{M}_v$ is the viscosity-average molecular weight (cf. Table 1.24). The constant a is often called the *Mark–Houwink exponent* and lies in the range of 0.5–0.8 (see Table 4.5), which leads to the result that the viscosity-average molecular weight lies between the number-average and weight-average molecular weights. When $a = 1.0$, $\bar{M}_v = \bar{M}_w$. Many other important characteristics of the polymer can be predicted using intrinsic viscosity, including chain stiffness and hydrodynamic coil radius, further emphasizing the importance of intrinsic viscosity.

Viscosity of Polymer Melts and Blends

Much regarding the shear-dependent viscosity of polymer melts has already been described in the context of ceramic slurries (cf. Section 4.1.2). There is an immeasurable amount of additional information on this topic, and much research has been conducted on predicting the rheological properties of polymer melts from their structure using what are termed *constitutive equations*. A review of these equations, even a perfunctory one, is beyond the scope of this text. However, we know enough about polymeric structure and fluid flow to describe two topics on polymer melts that have direct utility with regard to the processing of polymeric materials. The first deals with the effect of chain entanglements on polymer melt viscosity, and the second has to do with the viscosity of polymer melt blends. A third relevant topic—*viscoelasticity*—will be reserved until Chapter 5 when mechanical properties and elastic response are described.

The high viscosities of polymeric melts can be attributed to the high level of intermolecular bonding between polymer chains. In fact, the *interchain forces* can be stronger than primary bonds in the chain—for example, C—C backbone bonds, to the point that thermal degradation often occurs before vaporization in polymers. This phenomenon is chain-length-dependent; that is, the longer the chain, the more interatomic bonding that can occur. Polymer molecules with strong intermolecular forces such as hydrogen bonding or dipoles have stronger intermolecular bonds than primary bonds when the degree of polymerization, $\bar{x}_n$, exceeds about 200. Polymer molecules with weak intermolecular forces, such as dispersion forces, require longer chains—on the order of $x_n > 500$—for intermolecular forces to dominate.

Other than intermolecular forces, *chain entanglements* play an important role in determining polymer melt viscosity. As shown in Figure 4.11, there is a *critical chain length*, here represented by the weight-average degree of polymerization, $\bar{x}_w$, above which chain entanglements are significant enough to retard flow and increase the melt viscosity. Below this point, chain segments are too short to allow for significant entanglement effects. Typical values of the critical chain length are 610 for polyisobutylene, 730 for polystyrene, and 208 for poly(methyl methacrylate) [11]. Generally, the critical chain length is lower for polar polymers than for nonpolar polymers, again due to intermolecular forces.

Intermolecular forces also play an important role in determining the compatibility of two or more polymers in a *polymer blend* or *polymer alloy*. Although the distinction between a polymer blend and a polymer alloy is still the subject of some debate, we will use the convention that a polymer alloy is a single-phase, homogeneous material (much as for a metal), whereas a blend has two or more distinct phases because of polymer–polymer immiscibility (cf. Section 2.3.3). In general, polymers are incompatible with each other, so that most polymer–polymer mixtures are blends, in which the primary phase is continuous, and the secondary phase exists as droplets dispersed throughout the primary phase. As the droplet sizes become smaller, the two phases become

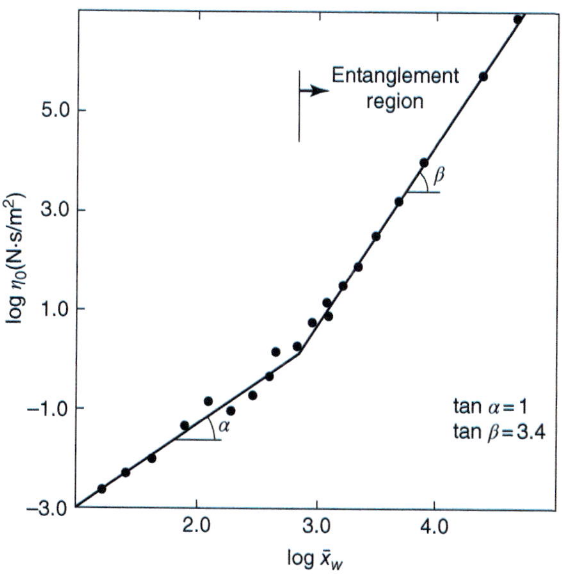

Figure 4.11 Effect of polymer chain entanglements on melt viscosity. From Tadmor and Gogos [10]. Copyright © 1979 by John Wiley & Sons, Inc. This material is used by permission of John Wiley & Sons, Inc.

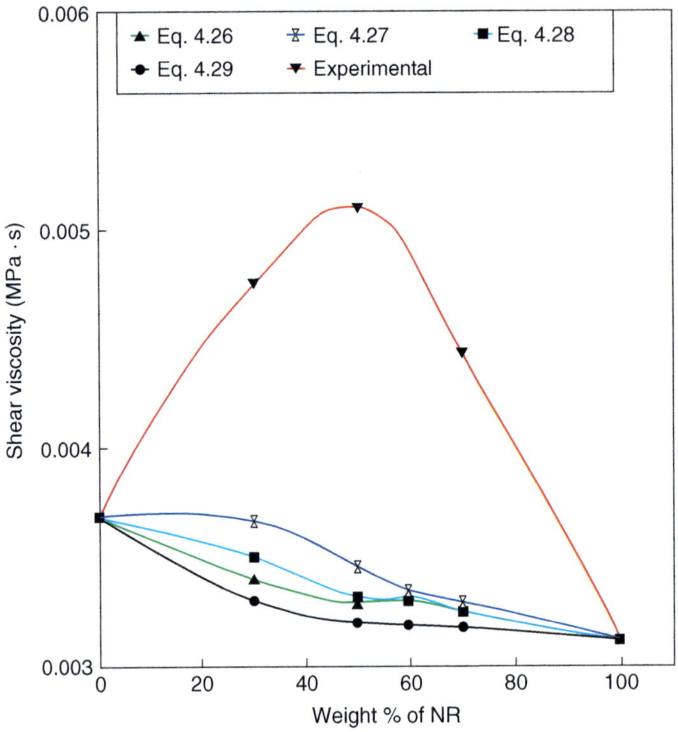

Figure 4.12 Viscosity of a natural rubber (NR)/poly(methyl methacrylate) (PMMA) polymer blend and predictions of Eqs. (4.26) through (4.29) at a shear rate of 333 s^{-1}. Adapted from Oommen et al. [12]. Copyright © 1997 by Elsevier.

indistinguishable, and an alloy begins to form. There are some immiscible polymer blends that form *bi-continuous structures* as well.

The rheological properties of polymer mixtures are complex and nearly impossible to predict. Figure 4.12 shows the viscosity of a natural rubber (NR)/poly(methyl methacrylate) (PMMA) blend (top curve) as a function of percentage NR [12]. For comparison, the predictions of four common equations are shown. The equations are as follows:

$$\eta_{\text{mix}} = \eta_1\phi_1 + \eta_2\phi_2 \tag{4.26}$$

$$\eta_{\text{mix}} = \eta_2 + \frac{\phi_1}{1/(\eta_1 - \eta_2) + \phi_2/2\eta_2} \tag{4.27}$$

$$\eta_{\text{mix}} = \eta_1 + \frac{\phi_2}{1/(\eta_2 - \eta_1) + \phi_1/2\eta_1} \tag{4.28}$$

$$\ln \eta_{\text{mix}} = \frac{\phi_1(\alpha - 1 - \gamma\phi_2)\ln \eta_1 + \alpha\phi_2(\alpha - 1 + \gamma\phi_1)\ln \eta_2}{\phi_1(\alpha - 1 - \gamma\phi_2) + \alpha\phi_2(\alpha - 1 + \gamma\phi_1)} \tag{4.29}$$

where η_1 and η_2 are the viscosities of components 1 and 2, and ϕ_1 and ϕ_2 are their corresponding volume fractions. In Eq. (4.29), α and γ are adjustable parameters [13].

Other models are based upon the immiscibility of polymer blends described above, and they model the system as Newtonian drops of the dispersed polymer with concentration ϕ_1 in a Newtonian medium of the second polymer with concentration $\phi_2 = 1 - \phi_1$. There exists some concentration, $\phi_{1I} = \phi_{2I} - 1$, at which phase inversion takes place; that is, at sufficiently high concentration, the droplet phase suddenly becomes continuous, and the second phase forms droplets. The phase inversion concentration has been shown to correlate with the viscosity ratio, $\lambda = \eta_1/\eta_2$, and the intrinsic viscosity for at least a dozen polymer alloys and blends:

$$\lambda = \left[\frac{\phi_m - \phi_{2I}}{\phi_m - \phi_{1I}}\right]^{[\eta]\phi_m} \tag{4.30}$$

where ϕ_m is the maximum packing volume fraction, which is about 0.84 for most polymer alloys and blends. Much work remains to be done in this important area of blend viscosity, which is a vital component of polymer processing and characterization.

4.1.4 Momentum Transport Properties of Composites

Given the vast number of possible matrix–reinforcement combinations in composites and the relative inability of current theories to describe the viscosity of even the most compositionally simple suspensions and solutions, it is fruitless to attempt to describe the momentum transport properties of composite precursors in a general manner. There are, however, two topics that can be addressed here in an introductory fashion: flow properties *of* matrix/reinforcement mixtures; and flow of matrix precursor materials *through* the reinforcement. In both cases, we will concentrate on the flow of molten polymeric materials or precursors, since most high-performance composites are polymer-based. Furthermore, the principles here are general, and they apply to the fluid-based processing of most metal–, ceramic–, and polymer–matrix composites.

Viscosity of Polymer–Fiber Melts

For most liquids with suspended particles that are processed into reinforced matrix composites, the viscosity can be described by the modified Einstein equation given in Eq. (4.20), which accounts for the variation in particle shape using the hydrodynamic shape factor, K_H. However, fiber-reinforced polymers are so ubiquitous and commercially important that some empirical studies have been performed that are worth describing here. Recent studies have included carbon [14], glass [15–17], and polyester [18] fibers in polyethylene [17], polypropylene [14, 15], polycarbonate [14, 15], polyamide [15, 16], and polyurethane [18] matrices. In addition to composition, reinforcing fibers can vary by length and diameter, but are normally classified as either long fibers (length more than about 5 mm) or short fibers (length less than about 5 mm, with a corresponding length/diameter ratio of 500–800). The effect of fiber fillers on polymer viscosity is generally the same, regardless of fiber length. As shown in Figure 4.13, the effect of the fibers is to cause more pronounced shear-thinning in the polymer (see 5 and 10 wt% curves) than is present without the fibers (0% curve), with plateaus at high and low shear rates. This is due to the tendency of fibers to align along the direction of flow, especially at high shear rates, thus counteracting any viscosity-building effect due to suspended particles that exist at low shear rates. The extent of shear thinning and the existence of a low-shear-rate plateau depends on the aspect ratio, fiber modulus, fiber concentration, and viscosity of the suspending fluid. The low shear rate plateau is usually not observed for suspensions of fibers with small aspect ratios at high concentrations in low-viscosity fluids.

Power-law expressions are still used to describe such polymer–fiber melts. Typical power-law parameters for selected fiber–polymer systems are shown in Table 4.6. Semiempirical expressions based on Eq. (4.23) have also been developed, as well as models based on energy dissipation. A complete review of these correlations is beyond the scope of this text, and the interested reader is referred to [19] for a more complete review of viscosity in fiber-reinforced polymer melts.

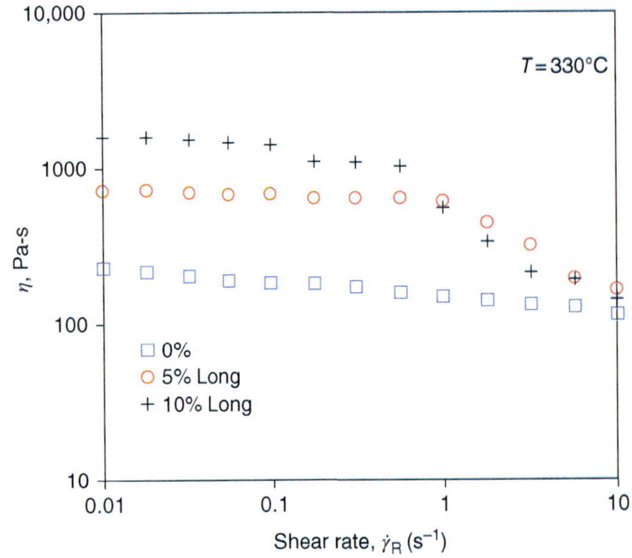

Figure 4.13 Effect of fiber content on shear viscosity for long glass fibers in polycarbonate. Reproduced, by permission, from Green and Wilkes [15]. Copyright © 1995 John Wiley & Sons.

Table 4.6 Power-Law Parameters for a Fiber-Filled Polyamide

Sample	$K(\text{N}{\cdot}\text{s}^n/\text{m}^2)$	n
Polyamide (PA) 6.6	213.1	0.81
PA 6.6 with 0.5 wt% blowing agent	109.2	0.80
30 vol% long glass-fiber-filled PA	1709.5	0.74
40 vol% long glass-fiber-filled PA	2400.6	0.73
30 vol% long glass-fiber-filled PA with 0.5 wt% blowing agent	1173.3	0.75

Adapted from Shuler et al. [16].

Viscosity of Polymers through Fiber Preforms

The flow of molten polymers, sometimes called resins if they are network polymers, through a fibrous preform is integral to a number of important composite processes, such as melt infiltration. In addition to nonreactive processes, there are a number of important reactive processing techniques such as *reaction injection molding* (RIM), *resin transfer molding* (RTM), and *reactive liquid composite molding* (RLCM) in which the polymer precursors flow and polymerize *in situ* as they flow through the fibrous preform. These techniques are described in more detail in Chapter 7, but it is worth describing the viscosity effects here first.

In the general case of a resin filling a fiber-packed mold, such as is found in melt infiltration (see Figure 4.14), fluid flow is equivalent to flow through porous media or packed columns, both of which have been studied for a long time. Typically, *Darcy's Law* is used for flow through porous media. This model describes the flow of Newtonian fluids in porous media, and it states that the volumetric flow rate, Q, through a constant-area packed channel is proportional to the

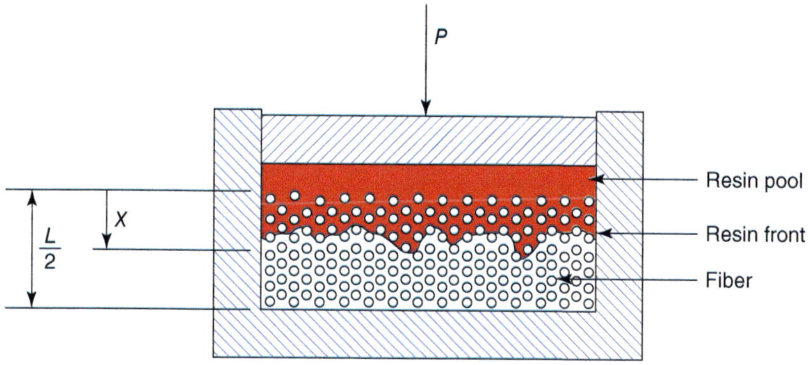

Figure 4.14 Schematic diagram of the melt infiltration process for flow past a unidirectional fiber assembly.

cross-sectional area, A, and the pressure drops across the channel, ΔP, and is inversely proportional to the length of the channel, L, and the viscosity of the fluid, μ:

$$Q = K_D \frac{A}{\mu} \left(\frac{\Delta P}{L} \right) \tag{4.31}$$

where K_D is called the *permeability* and has units of length squared. Not surprisingly, Eq. (4.31) has the same form as the Hagen–Poiseuille equation for volumetric flow in a cylindrical tube, with a correction factor for the permeability of the media (K_D), which in this case is fibers.

The permeability is given by the *Kozeny–Carman relationship*

$$K_D = \frac{\varepsilon^3}{(1-\varepsilon)^2} \left(\frac{1}{S^2 k} \right) \tag{4.32}$$

where ε is the *porosity*, S is the specific surface of the medium particle ($6/D$ for spheres, $4/D$ for cylinders), D is the diameter, and k is the so-called *Kozeny constant*. For the specific case of flow normal to the axial direction of a parallel array of fibers with diameter, D, as is found in the melt infiltration example, the permeability becomes $K_D = D^2\varepsilon^3/[80(1-\varepsilon)^2]$.

For the case of simultaneous reaction and flow, as is present in reactive molding, an expression is needed that explicitly relates the viscosity to both the temperature and the extent of reaction. The viscosity increases with time as the reaction proceeds (see Figure 4.15, top diagram). The temperature, however, has opposing effects; it will decrease the viscosity as the temperature is raised, but will also increase the reaction rate, leading to increased viscosity. To separate these effects, the polymerization reaction kinetics must be monitored independently. A viscosity–concentration correlation can be constructed by taking *isochrones* (t_1 and t_2 in Figure 4.15) of viscosity and extent of reaction and then replotting them on a viscosity–conversion diagram. The following relationship, proposed by Castro and Macosko [21], has been found useful in modeling the reactive molding process:

$$\mu = A_\mu \exp(E_\mu/RT) \left(\frac{C_g^*}{C_g^* - C^*} \right)^{A + BC^*} \tag{4.33}$$

Here A_μ and E_μ are the Arrhenius-type preexponential and activation energy terms, R is the gas constant, T is absolute temperature, C_g^* is the solidification (gel) composition, and C^* is the extent of reaction, $(C_0 - C)/C_0$.

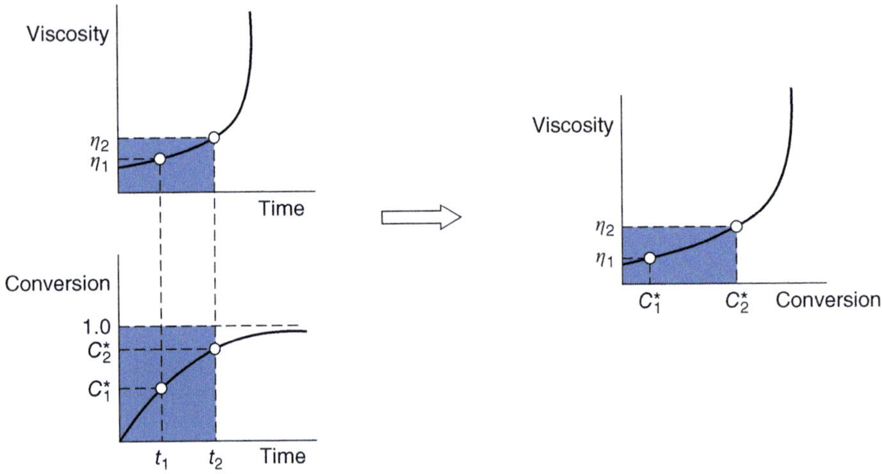

Figure 4.15 Mapping of viscosity rise versus time plot into viscosity–conversion plot. Reproduced, by permission, from Mallick [20].

Cooperative Learning Exercise 4.4

Darcy's Law can be utilized for non-Newtonian as well as Newtonian fluids, by using an effective viscosity in place of the Newtonian viscosity:

$$v_0 = \frac{Q}{A} = \frac{K_D \Delta P}{\eta_{\text{eff}} L}$$

where v_0 is the superficial velocity of the fluid flow front. In addition to the shear-rate-dependent non-Newtonian viscosities described earlier in this chapter, there are other models that are stress-dependent. One such model is the *Ellis Model*, in which the ratio of the zero-shear-rate viscosity, η_0, to the effective viscosity, η_{eff}, for the case of flow through porous media is given by

$$\frac{\eta_0}{\eta_{\text{eff}}} = 1 + \frac{4}{\alpha + 3}\left(\frac{\tau_{Rh}}{\tau_{1/2}}\right)^{\alpha - 1}$$

where $\tau_{1/2}$ is the shear stress at which $\eta_{\text{eff}} = \eta_0/2$, τ_{Rh} is the wall shear stress based on the hydrodynamic radius, and α is a constant that is the inverse of the power-law exponent, n.

Person 1: Calculate the permeability, K_D, for the case of a fiber preform made from unidirectional fibers of diameter 10 μm, and a porosity of 60%.

Person 2: Derive an expression for the effective viscosity for a 1.0% solution of poly(ethylene oxide) at 293 K, which has a $\eta_0 = 2.37\ \text{N·s/m}^2$, $\tau_{1/2} = 0.412\ \text{N/m}^2$. Consult Table 4.4 for the appropriate value of α.

Combine your answers to come up with an expression for the superficial velocity of a 1.0% solution of poly(ethylene oxide) within a mold consisting of 10-μm-diameter unidirectional fibers with 60% porosity.

Answers: $K_D = 1.7 \times 10^{-12}$; $\eta_{\text{eff}} = \dfrac{2.37}{1 + 1.8\,\tau_{Rh}^{0.88}}$; $v_0 = \left(7.12 \times 10^{-13} + 1.27 \times 10^{-12}\tau_{Rh}^{0.88}\right)\dfrac{\Delta P}{L}$

4.1.5 Momentum Transport Properties in Biologics

Many biological fluids are aqueous-based suspensions of particles. Such is the case with perhaps the most important of all biological fluids: blood. Blood is essentially a suspension of red blood cells in plasma. The concentration of red blood cells in blood is described as the *hematocrit*, or fraction of the blood volume represented by blood cells. The viscosity of human blood as a function of hematocrit is shown in Figure 4.16. Note that the viscosity of blood is substantially lower in small tubes than in large tubes, especially at high hematocrit values. This is due to the tendency of red blood cells to be carried in midstream (think "velocity profile" and "fully developed flow"!) such that in small blood vessels, the fluid at the wall is nearly completely plasma, which has a lower viscosity than that of whole blood (compare 0% to "normal value" hematocrit for large tubes in Figure 4.16). This has important implications for human anatomy. If the heart had to be large and powerful enough to propel blood throughout the body with "normal viscosities" such as are found in large tubes (e.g., arteries), then there would hardly be room in the chest cavity for lungs and other important organs.

Other biological fluids can be modeled as suspensions of particles in a solvent, such as was used for the description of suspensions and slurries in Section 4.1.2, namely, that the relative viscosity of the suspension is related to the hydrodynamic shape factor, K_H, and the volume fraction of particles, f_p^v, as in Eq. (4.20):

$$\eta_r = 1 + K_H f_p^v \tag{4.20}$$

Need More Help?

Try this video on blood flow and the Poiseuille equation: https://www.youtube.com/watch?v=Ok8AJAkkiw8

The biggest difference between biological particles and ceramic particles in the application of Eq. (4.20) is that while most ceramic particles are spherical ($K_H = 2.5$), most biological particles can be modeled as either *prolate ellipsoids* or *oblate spheroids* (or ellipsoids). Ellipsoids are characterized according to their shape factor, a/b, for which a and b are the dimensions of the semimajor and semiminor axes, respectively (see Figure 4.17). In a prolate ellipsoid, $a > b$, whereas in an oblate

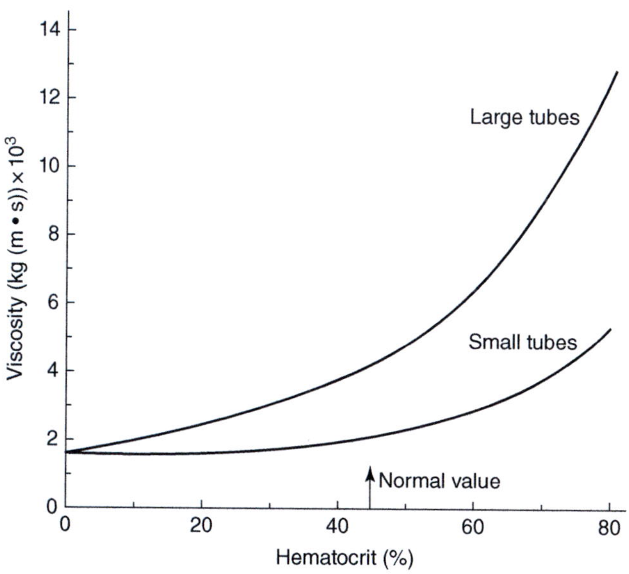

Figure 4.16 Viscosity of blood as a function of percentage blood volume represented by red blood cells (hematocrit). From Johnson [22]; Copyright © 1999 by John Wiley & Sons, Inc. This material is used by permission of John Wiley & Sons, Inc.

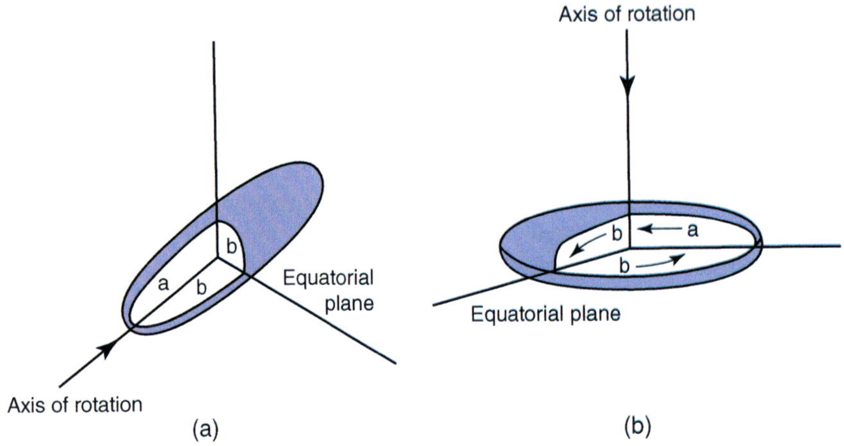

Figure 4.17 (a) Prolate and (b) oblate ellipsoids of revolution, showing the relationship between the semiaxes and the axis of revolution. Reprinted, by permission, from Hiemenz [23]. Copyright © 1986 by Taylor & Francis.

ellipsoid, $b > a$. In the extremes, $a \gg b$ approximates a cylinder, and $b \gg a$ approximates a disk, or platelet.

The hydrodynamic shape factor and axial ratio are related (see Figure 4.18) but are not generally used interchangeably in the literature. The axial ratio is used almost exclusively to characterize the shape of biological particles, so this is what we will utilize here. As the ellipsoidal particle becomes less and less spherical, the viscosity deviates further and further from the Einstein equation (see Figure 4.19). Note that in the limit of $a = b$, both the prolate and oblate ellipsoid give an intrinsic viscosity of 2.5, as predicted for spheres by the Einstein equation.

In practice, it is the viscosity that is experimentally determined, and the correlations are used to determine axial ratios and shape factors. The viscosity can be determined by any number of techniques, the most common of which is light scattering. In addition to ellipticity, *solvation* (particle swelling due to water absorption) can affect the viscosity of biological suspensions. An example of

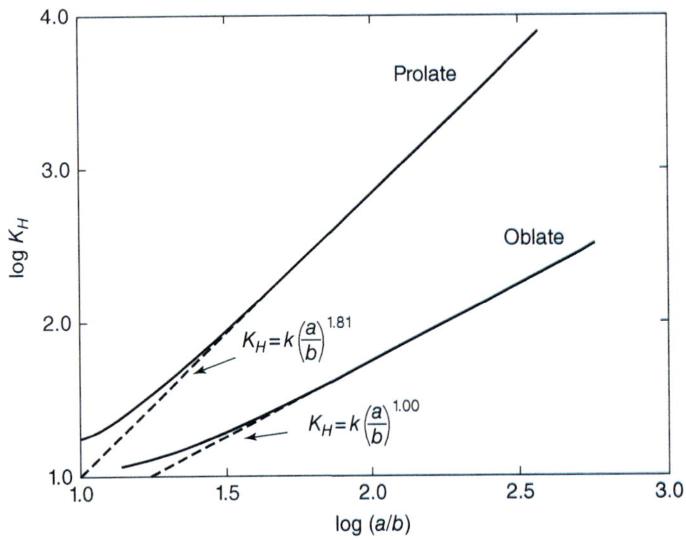

Figure 4.18 Theoretical values of the shape factor for ellipsoids. Adapted from Silver and Christiansen [24]. Copyright © 1999 by Springer-Verlag.

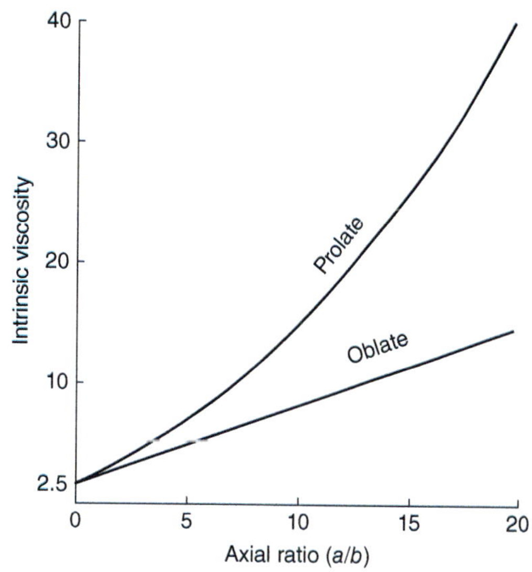

Figure 4.19 Intrinsic viscosity as a function of axial ratio for oblate and prolate ellipsoids. Intrinsic viscosity in ml/g. Reproduced, by permission, from Hiemenz [25]. Copyright © 1984 by John Wiley & Sons.

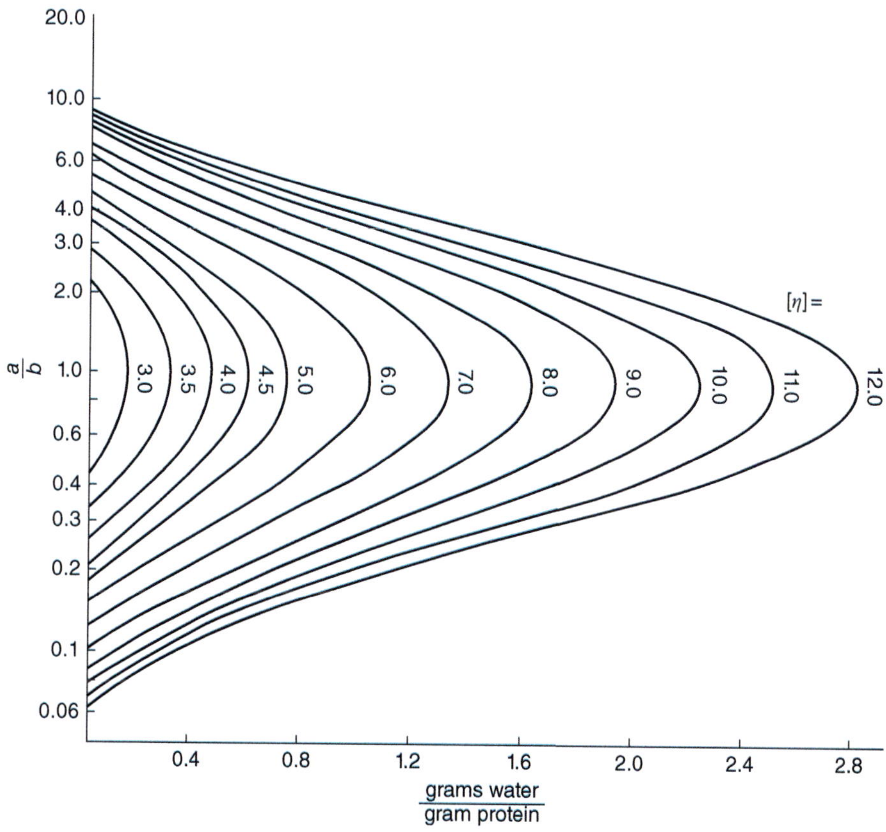

Figure 4.20 Variation of intrinsic viscosity of aqueous protein solutions with axial ratio and extent of solvation. Reproduced, by permission, from Hiemenz [26]. Copyright © 1984 by John Wiley & Sons.

how both solvation and particle ellipticity can affect solution viscosity is shown in Figure 4.20. Finally, for biological molecules that are macromolecules, such as most proteins, Eq. (4.23) can also be used to relate the relative viscosity to the intrinsic viscosity of the solution and the macromolecule concentration.

Cooperative Learning Exercise 4.5

Some plots of reduced specific viscosity (see Table 4.4) as a function of concentration for various samples of vitreous hyaluronan bovine protein are shown at right. Hyaluronan may be modeled as a prolate ellipsoid, but it is unknown whether it is a flexible or rigid chain.

Person 1: Pick one of the curves in the plot and determine the intrinsic viscosity by applying Eq. (4.23) and the definition of reduced specific viscosity. Use this value of intrinsic viscosity to estimate the axial ratio of the particle using Figure 4.19. Note that you may have to assume a linear approximation to prolate ellipsoid curve and extrapolate to obtain an estimate for a/b.

Person 2: Use the a/b ratio estimated by Person 1 to estimate the diffusivity of this protein. The diffusivity of prolate ellipsoids, D_{AB}, in units of cm^2/s, can be estimated using the following relationship

$$D_{AB} = \frac{(2.15 \times 10^{-13}\,\text{cm}^3/\text{s})[\ln(2a/b)]}{a}$$

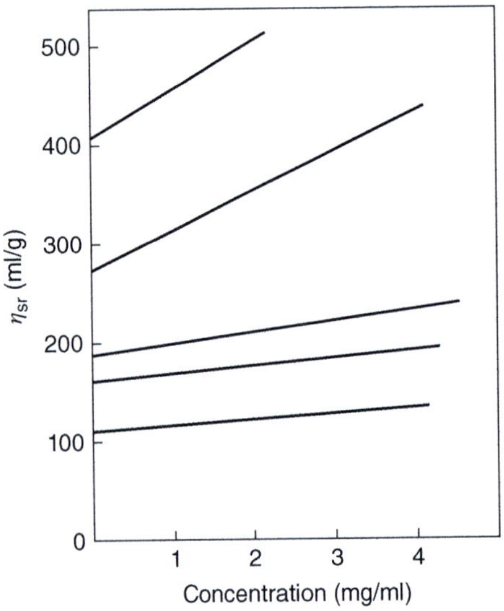

For ease of calculation, assume that the hyaluronan protein has a chain length, a = 215 nm. Compare the intrinsic viscosity and diffusivity results to come up with some conclusion regarding the shape and rigidity of this protein; *i.e.*, rigid chain or flexible chain.

Answers: The intrinsic viscosity of all the plots are determined by extrapolating back to zero concentration, and are found to range from about 110 to 410 ml/g. According to Figure 4.19, this corresponds to a/b values on the order of 100 or so. When this value is plugged into the equation for diffusivity, one obtains a value of about 10^{-12} m^2/s. As described in the text, the intrinsic viscosity is relatively high, and the diffusivity is relatively low, leading us to believe that the hyaluronan chain is somewhat rigid.

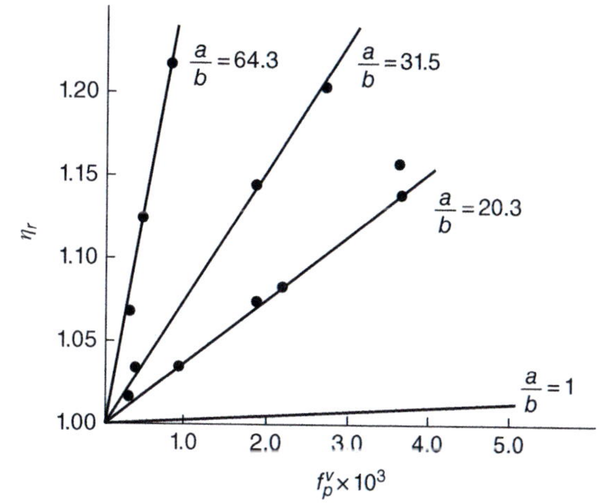

Figure 4.21 Effect of axial ratio and particle concentration on relative viscosity. Data are for tobacco mosaic virus particles. Adapted from Lauffer [28]. Copyright © 1944 by The American Chemical Society, Inc.

An example of how both particle concentration and axial ratio can be used to describe their effect on solution viscosity—in this case the reduced viscosity—is shown in Figure 4.21. Here, the deviations from the Einstein model ($a/b = 1$ at low concentrations) are high, since it is based upon a dilute solution of nonsolvated, spherical particles.

Later in the chapter, we will see that there is a relationship between particle shape and size, intrinsic viscosity, and the ability of the particle to diffuse through the solution, as characterized by the *diffusion coefficient*, or *diffusivity*. In general, molecules with high values of intrinsic viscosity (>100 ml/g) and relatively low diffusivities ($<10^{-9}$ m^2/s) are rigid or semiflexible rods. Molecules with both low diffusivities and intrinsic viscosities (~1 ml/g) are flexible rods, and molecules with high diffusivities ($>10^{-9}$ m^2/s) and low intrinsic viscosities are roughly spherical. Consult Section 4.3.5 for more information on diffusivities of biologics.

4.2 Heat Transport Properties of Materials

Of the three general categories of transport processes, heat transport gets the most attention for several reasons. First, unlike momentum transfer, it occurs in both the liquid and solid states of a material. Second, it is important not only in the processing and production of materials but in their application and use. Ultimately, the thermal properties of a material may be the most influential design parameters in selecting a material for a specific application. In the description of heat transport properties, let us limit ourselves to conduction as the primary means of transfer, while recognizing that for some processes, convection or radiation may play a more important role. Finally, we will limit the discussion here to theoretical and empirical correlations and trends in heat transport properties. Tabulated values of thermal conductivities for a variety of materials can be found in Appendix 4.

4.2.1 Heat Transport Properties of Metals and Alloys

Conduction, Convection, and Radiation

We begin this section with a brief review of the three primary modes of heat transfer: conduction, convection, and radiation.

In *conduction*, heat is conducted by the transfer of energy of motion between adjacent molecules in a liquid, gas, or solid. In a gas, atoms transfer energy to one another through molecular collisions. In metallic solids, the process of energy transfer via "free electrons" is also important. In *convection*, heat is transferred by bulk transport and mixing of macroscopic fluid elements. Recall that there can be *forced convection*, where the fluid is forced to flow via mechanical means, or *natural (free) convection*, where density differences cause fluid elements to flow. Since convection is found only in fluids, we will deal with it on only a limited basis. *Radiation* differs from conduction and convection in that no medium is needed for its propagation. As a result, the form of Eq. (4.1) is inappropriate for describing radiative heat transfer. Radiation is the transfer of energy through space by means of electromagnetic waves. The laws that govern the movement and absorption of electromagnetic waves such as light govern the transfer of heat by radiation. We will cover some of these topics later in Chapter 6.

The Molecular Origins of Thermal Conductivity

In an analogous fashion to the derivation from the potential energy function for the viscosity of a dilute gas at low density [see Eq. (4.6)], the thermal conductivity of a gas, k, based on the rigid sphere model, can be derived for a gas (in W/m·K):

$$k = 8.32 \times 10^{-2} \frac{\sqrt{T/M}}{\sigma^2 \Omega} \tag{4.34}$$

where T is the absolute temperature in degrees kelvin, M is the atomic weight of the monatomic gas, and σ and Ω are the same Lennard–Jones parameters used in Eq. (4.6). Notice the dependence on the same properties as viscosity—namely, atomic weight, molecular diameter, and temperature—although the thermal conductivity has an inverse dependence on the square root of molecular weight instead of the proportional dependence for viscosity. There is no variation of thermal conductivity with pressure for a rigid sphere, ideal, monatomic gas at low pressure, but the model does predict that thermal conductivity will be zero at absolute zero.

It can be shown that thermal conductivity and viscosity are related in this case as follows:

$$k = \frac{15R\mu}{4M} = \frac{5\hat{C}_v\mu}{2} \tag{4.35}$$

where M is again the atomic weight of the gas, R is the gas constant in appropriate units [1.9872 cal/mol·K for thermal conductivity (in cal/cm·K·s) and 831.4 J·cm/m·mol·K for thermal conductivity (in W/m·K)], and μ is the viscosity of the gas in poise (g/cm·s). The quantity, $\hat{C}_v$, is called the *heat capacity* per unit mass at constant volume and has the same units as R/M. In the final section of this chapter, we will see how both the viscosity and thermal conductivity (thermal diffusivity) are related to the molar diffusivity.

As with viscosity, the theoretical predictions become more complex as the atoms themselves become more complex and more dense. For a polyatomic gas, the thermal conductivity is given by extension of Eq. (4.35):

$$k = \frac{15R\mu}{4M}\left(\frac{4C_v}{15R} + \frac{3}{5}\right) \tag{4.36}$$

where R, μ, and M are as in Eq. (4.35) (with M the molecular weight of the polyatomic gas rather than the atomic weight). Here, C_v is the molar heat capacity, obtained by multiplying the per mass heat capacity by the molecular weight, $M\hat{C}_v$. Note that for an ideal monatomic gas, $C_v = 3R/2$, and Eq. (4.36) reduces to (4.35).

Equation (4.36) provides a simple method for estimating an important heat transfer dimensionless group called the *Prandtl number*. Recall from general chemistry and thermodynamics that there are two types of molar heat capacities, C_v, and the constant pressure heat capacity, C_p. For an ideal gas, $C_v = 3C_p/5$. The Prandtl number is defined as $\Pr = \hat{C}_p\mu/k$. Substitution of Eq. (4.36) into this relation, along with use of the relationship between C_v and C_p, gives the following for polyatomic gases:

$$\Pr = \frac{C_p}{C_p + 1.25R} \tag{4.37}$$

Cooperative Learning Exercise 4.6

The molecular weight of O_2 is 32.000; its molar heat capacity C_p at 300 K and its low pressure is 29.37 J/mol·K. The Lennard–Jones parameters for molecular oxygen are σ = 3.433 Å and Ω = 1.074.

Person 1: Estimate the viscosity of molecular oxygen at low pressure and 300 K. Then estimate the Prandtl number, Pr, using Eq. (4.37).

Person 2: Use Person 1's estimate of viscosity to estimate the thermal conductivity of molecular oxygen at low pressure and 300 K using Eq. (4.36). Note that you will need to modify Eq. (4.36) to include C_p instead of C_v. Beware of units and molar heat capacities!

Combine your answers to give an estimate for the Prandtl number, Pr, based on the definition $\Pr = \hat{C}_p\mu/k$. How does this compare with your estimation given by Eq. (4.37)?

Answers: Person 1: $\mu = 2.066 \times 10^{-4}$ g/cm · s; Pr = 0.74. Person 2: k = 0.023 W/m · K. Combined: Pr = 0.66.

The introduction of heat capacity into the relationships for thermal conductivity and the Prandtl number gives us an opportunity to make a clarification regarding these two quantities. Thermal conductivity is a true heat transport property; it describes the ability of a material to *transport* heat via conduction. Heat capacity, on the other hand, is a thermodynamic quantity and describes the ability of a material to *store* heat as energy. The latter, while not technically a transport property, will nonetheless be described in this chapter for the various material types, due in part to its theoretical relationship to thermal conductivity, as given by Eqs. (4.35) and (4.36), and, more practically, because it is often used in combination with thermal conductivity as a design parameter in materials selection.

Thermal Conductivities of Liquids

As was the case with viscosity, it is difficult to derive useful relationships that allow us to estimate thermal conductivities for liquids from molecular parameters. There is a theoretical development by Bridgman, the details of which are presented elsewhere [29], which assumes that the liquid molecules are arranged in a cubic lattice, in which energy is transferred from one lattice plane to the next at sonic velocity, v_s. This development is a reinterpretation of the kinetic theory model used in the last section, and with some minor modifications to improve the fit with experimental data, the following equation results:

$$k = 2.8(V/N)^{-2/3} k_B v_s \tag{4.38}$$

where V/N is the volume per molecule and k_B is Boltzmann's constant. This equation is limited to liquids well above their critical densities. The sonic velocity in liquids can be estimated by

$$v_s = \sqrt{\frac{C_p}{C_v}\left(\frac{\partial p}{\partial \rho}\right)_T} \tag{4.39}$$

where the quantity $(\partial p/\partial \rho)_T$ may be obtained from isothermal compressibility measurements or from an equation of state, and (C_p/C_v) is approximately unity for liquids away from their critical point.

More frequently, empirical relationships are used to estimate liquid thermal conductivities. A linear relationship with temperature is adequate, since liquid thermal conductivities do not vary considerably with temperature, and hardly at all with pressure.

Some experimental data for thermal conductivities of pure liquid metals are shown in Figure 4.22, and data for some binary metallic alloys are shown in Figure 4.23. These values generally will be lower than the thermal conductivities of the corresponding solid forms of the metals and alloys.

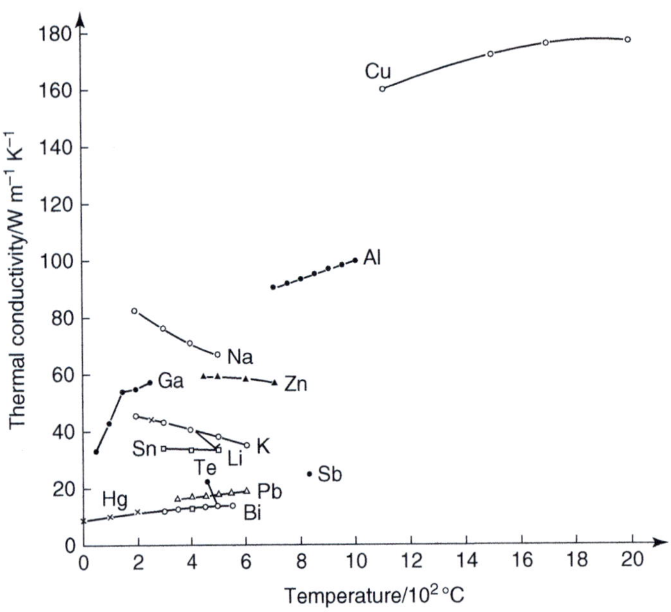

Figure 4.22 Thermal conductivity of some pure liquid metals as a function of temperature. Reproduced, by permission, from Iida and Guthrie [30]. Copyright © 1988 by Oxford University Press.

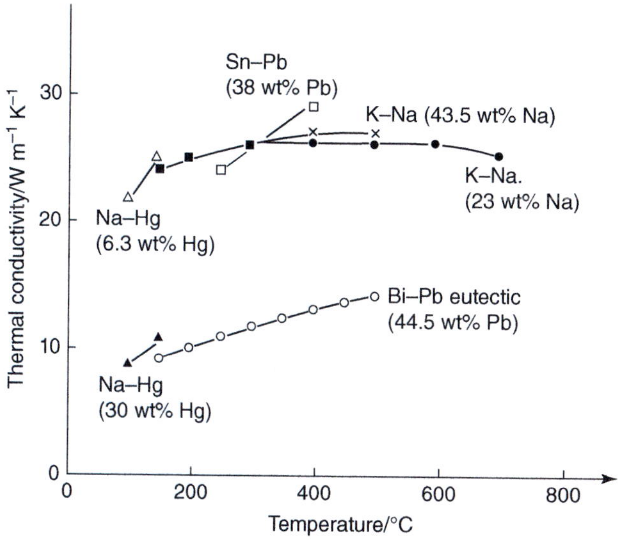

Figure 4.23 Thermal conductivity of some liquid alloys as a function of temperature. Reproduced, by permission, from Iida and Guthrie [30]. Copyright © 1988 by Oxford University Press.

Thermal Properties of Metallic Solids

In the preceding sections, we saw that thermal conductivities of gases, and to some extent liquids, could be related to viscosity and heat capacity. For a solid material such as an elemental metal, the link between thermal conductivity and viscosity loses its validity, since we do not normally think in terms of "solid viscosities." The connection with heat capacity is still there, however. In fact, a theoretical description of thermal conductivity in solids is derived directly from the kinetic gas theory used to develop earlier expressions. Equation (4.34), for example, can be derived from a simple expression for the thermal conductivity of a monatomic gas:

$$k = \frac{1}{3}\rho \hat{C}_v \bar{v} l \tag{4.40}$$

where ρ is the gas density, $\hat{C}_v$ is the constant volume heat capacity per unit mass (sometimes called the *specific heat*), $\bar{v}$ is the *mean molecular speed* (sometimes called the average particle velocity), and l is a quantity called the *mean free path*, which is a measure of the average distance traveled by the molecules between collisions.

We saw earlier that the molar heat capacity, C_v, equals $3R/2$ for an ideal monatomic gas. This is because almost all the energy of the gas is kinetic in nature; that is, an increase in temperature results in a direct increase in the translational kinetic energy of the atoms. For polyatomic gases, additional energy states are available, (e.g., vibrational and rotational, such that the gas can store energy in forms other than translational motion), and a higher heat capacity than that predicted by $3R/2$ results. In a solid, the atoms are similar to a vibrating diatomic molecule, where half of the thermal energy is the kinetic energy ($3R/2$) and half is the average potential energy ($3R/2$). In this simplistic approach, the heat capacity should then be roughly $3R$ for solids. This approximation is known as the *Law of Dulong–Petit*, and it is applicable to many nonmetallic solids at and above room temperature. It does not apply at low temperatures because the energy of atomic vibrations is quantized and the number of vibrational modes that are excited decreases as the temperature decreases below some value for each material. This temperature is known as the *Debye temperature*,

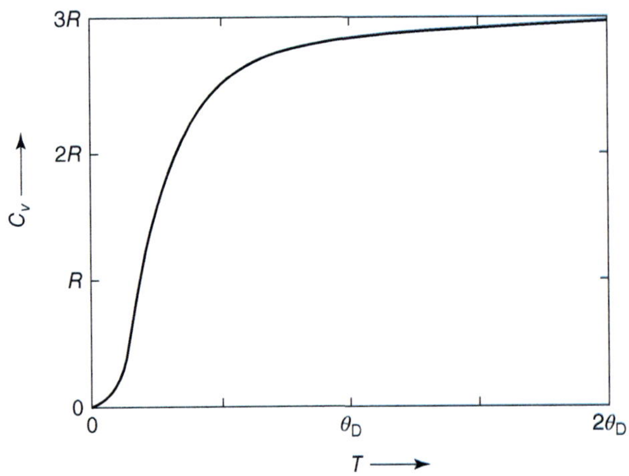

Figure 4.24 Molar heat capacity as a function of temperature, based on the Debye model. From Ralls et al. [31]. Copyright © 1976 by John Wiley & Sons, Inc. This material is used by permission John Wiley & Sons, Inc.

Table 4.7 Representative Values of Debye Temperature for Selected Elements

Element	θ_D(K)	Element	θ_D(K)	Element	θ_D(K)
Ag	225	Fe	467	Pt	229
Al	418	Ge	366	Si	658
Au	165	Hg	60–90	Sn (gray)	212
Be	1160	In	109	Sn (white)	189
Bi	117	La	132	Ta	231
C (diamond)	2000	Mg	406	Ti	278
Ca	219	Mo	425	Tl	89
Cd	300	Nb	252	V	273
Co	445	Ni	456	W	379
Cr	402	Pb	94.5	Zn	308
Cu	339	Pd	275	Zr	270

Source: Kittel [32]. Copyright © 1957 by John Wiley & Sons, Inc.

θ_D, above which the heat capacity approaches the value of Dulong and Petit ($3R$), and below which the heat capacity decreases monotonically with decreasing temperature (see Figure 4.24 and Table 4.7 for values of θ_D). In this region ($T \ll \theta_D$), the heat capacity is due to lattice vibrations (hence, a subscript l is added), and is given by:

$$C_{vl} = \frac{12\pi^4 R}{5}\left(\frac{T}{\theta_D}\right)^3 \tag{4.41}$$

Most nonmetals follow Eq. (4.41) at low temperatures, as illustrated in Figure 4.24. Most metals, however, show a linear decrease in heat capacity with temperature as absolute zero is approached. We require a slightly different model to describe the low-temperature behavior of C_v in metals.

> **Take a Peek!** Jump ahead to Section 6.0.1 to see how electronic structure contributes to the electrical conductivity in metals.

Recall from Figure 1.14 that metals have free electrons in what is called the valence band and have empty orbitals forming what is called the conduction band. In Chapter 6, we will see how this electronic structure contributes to the electrical conductivity of a metallic material. It turns out that these same electronic configurations can be responsible for thermal as well as electrical conduction. When electrons act as the thermal energy carriers, they contribute an electronic heat capacity, C_{ve}, that is proportional to both the number of valence electrons per unit volume, n, and the absolute temperature, T:

$$C_{ve} = \frac{\pi^2 n k_{\mathrm{B}}^2 T}{2E_F} \tag{4.42}$$

where k_{B} is Boltzmann's constant and E_F is the energy of the Fermi level, which is characteristic of an element, but in general is between 3 and 5 eV. The contribution of the electronic heat capacity to the overall heat capacity is usually quite small, but below about 4 K the electronic contribution becomes larger than the lattice contribution [Eq. (4.41)]. In this low-temperature regime, the total heat capacity is a sum of both the electronic and lattice contributions:

$$C_v = C_{vl} + C_{ve} = \alpha T^3 + \gamma T \tag{4.43}$$

where α is given by the coefficients in Eq. (4.41), and γ is given by the coefficients in Eq. (4.42).

Returning now to thermal conductivity, Eq. (4.40) tells us that any functional dependence of heat capacity on temperature should be implicit in the thermal conductivity, since thermal conductivity is proportional to heat capacity. For example, at low temperatures, we would expect thermal conductivity to follow Eq. (4.43). This is indeed the case, as illustrated in Figure 4.25. In copper, a pure metal, electrons are the primary heat carriers, and we would expect the electronic contribution to heat capacity to dominate the thermal conductivity. This is the case, with the thermal conductivity varying proportionally with temperature, as given by Eq. (4.42). For a semiconductor such as germanium, there are less free electrons to conduct heat, and lattice conduction dominates—hence, the T^3 dependence on thermal conductivity as suggested by Eq. (4.41).

This competition between electrons and the heat carriers in the lattice (phonons) is the key factor in determining not only whether a material is a good heat conductor or not, but also the temperature dependence of thermal conductivity. In fact, Eq. (4.40) can be written for either thermal conduction via electrons, k_e, or thermal conduction via phonons, k_p, where the mean free path corresponds to either electrons or phonons, respectively. For pure metals, $k_e/k_p \approx 30$, so that electronic conduction dominates. This is because the mean free path for electrons is 10 to 100 times higher than that of phonons, which more than compensates for the fact that C_{ve} is only 10% of the total heat capacity at normal temperatures. In disordered metallic mixtures, such as alloys, the disorder limits the mean free path of both the electrons and the phonons, such that the two modes of thermal conductivity are more similar, and $k_e/k_p \approx 3$. Similarly, in semiconductors, the density of free electrons is so low that heat transport by phonon conduction dominates.

So, in general, highly ordered materials such as pure metals, most pure ceramics, and crystalline polymers exhibit increasing thermal conductivity with decreasing temperature due to increases in both the electronic and phonon mean free paths at low temperatures. At sufficiently low temperatures, the mean free paths become constant, and the variation in heat capacity, as given by Eqs. (4.41) and (4.42), dominate the thermal conductivity. Disordered materials, on the other hand, such as solid–solution alloys, inorganic glasses, and glassy polymers, exhibit decreasing thermal

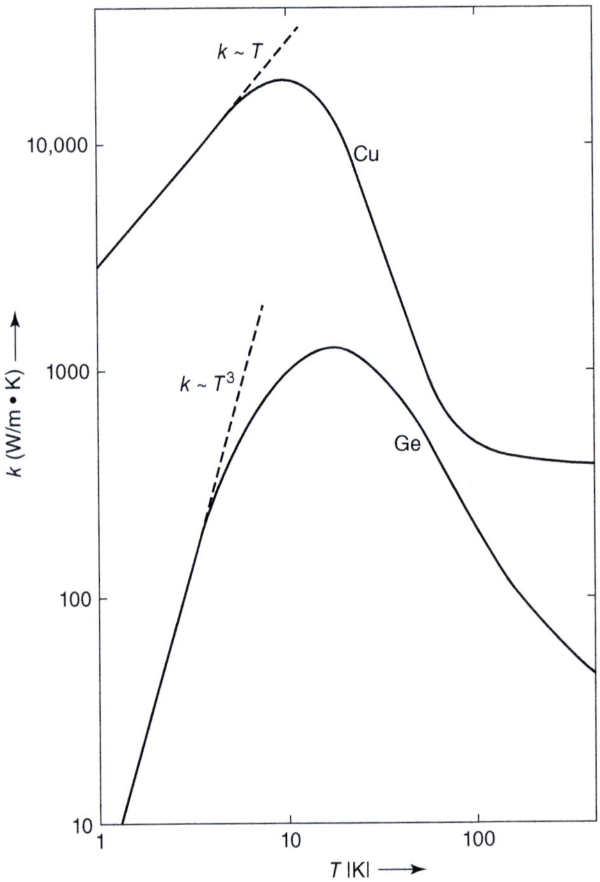

Figure 4.25 Temperature dependence of thermal conductivity for a pure metal (Cu) and a nonmetal (Ge), illustrating different temperature dependences at low temperature. From Ralls et al. [31]. Copyright © 1976 by John Wiley & Sons, Inc. This material is used by permission John Wiley & Sons, Inc.

conductivity with decreasing temperature. This is because the electronic and phonon mean free paths are relatively small at all temperatures, and temperature variation in heat capacity dominates the temperature dependence of thermal conductivity (see Figure 4.25). These trends are summarized in Figure 4.26. Note that highly ordered materials, such as copper, α-iron, polyethylene, and BeO, have significant regions in which thermal conductivity decreases with temperature. Disordered materials, such as PMMA, glass, and stainless steel, tend to increase thermal conductivity with temperature. The lowering of thermal conductivity with alloying is an important design consideration, as illustrated in Figure 4.27. Note that as more alloying elements are added to iron, the thermal conductivity drops at a given temperature. In the subsequent sections, we expand upon these generalizations for the remaining material classes.

4.2.2 Heat Transport Properties of Ceramics and Glasses

Heat Transport Properties of Ceramics

As a means of further illustrating the importance of the theoretical developments for heat capacity and thermal conductivity in the previous section, Figure 4.28 shows that the general trend of increasing heat capacity with temperature and the ultimate achievement of $C_v = 3R$ shown in

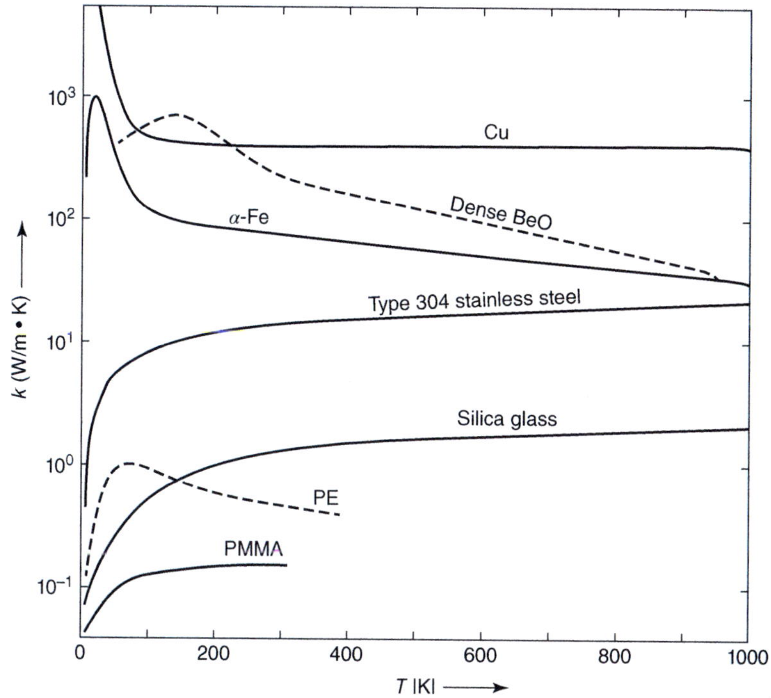

Figure 4.26 Variation of thermal conductivity with temperature for ordered (Cu, BeO, Fe, and PE) and disordered (SS, silica glass, and PMMA) with temperature. From Ralls et al. [31]. Copyright © 1976 by John Wiley & Sons, Inc. This material is used by permission John Wiley & Sons, Inc.

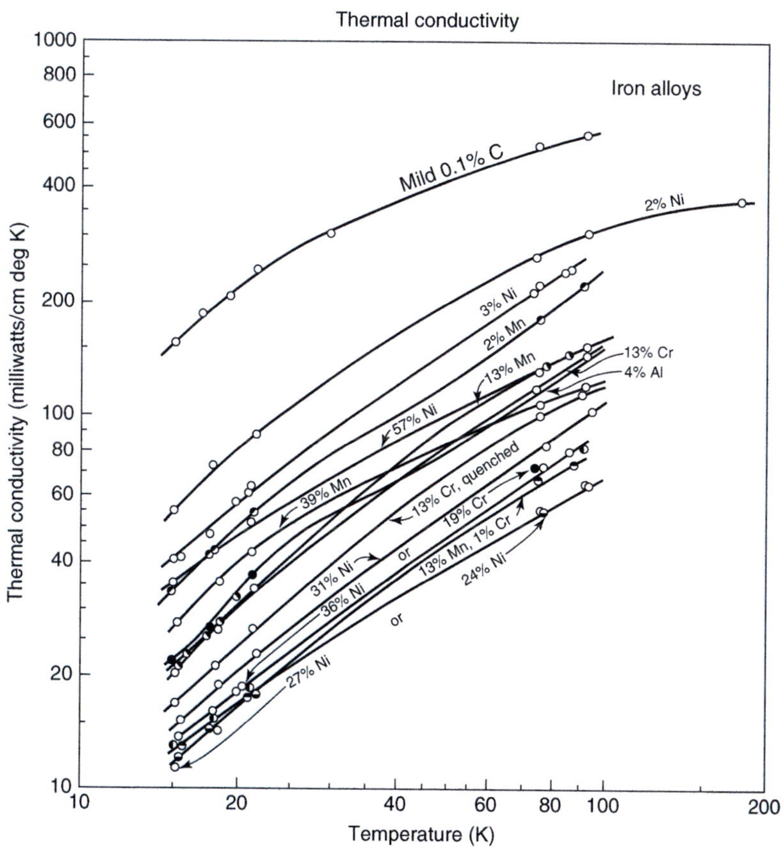

Figure 4.27 Thermal conductivity of some iron alloys. Reproduced, by permission, from Kittel [33]. Copyright © 1957 by John Wiley & Sons, Inc.

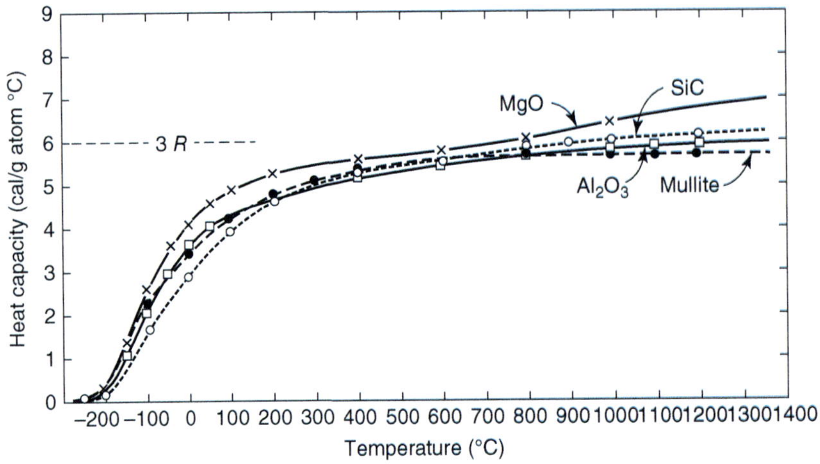

Figure 4.28 Heat capacity of some ceramic materials at different temperatures. From Kingery et al. [34]. Copyright © 1976 by John Wiley & Sons, Inc. This material is used by permission of John Wiley & Sons, Inc.

Figure 4.24 holds true for crystalline ceramic materials as well. The primary result of heat capacity theory as applied to ceramics is that the $3R$ value is achieved at about 1000°C for most oxides and carbides. Further increases in temperature do not strongly affect this value, and it does not depend strongly on the crystal structure. However, heat capacity does depend on porosity. Consequently, the heat energy required to raise the temperature of porous insulating fire brick is much lower than that required to raise the temperature of the dense fire brick material, even though its thermal conductivity is lower. This is one of the useful properties of insulating materials for furnaces, which can be rapidly heated and cooled due to their porosity. Other factors affecting heat capacity in ceramics are phase transformations, such as the α–β quartz transition, which results in step changes to the heat capacity, and the presence of point defects (cf. Section 1.1.3), which tend to increase the value of the heat capacity.

The discussion of the previous section would also lead us to believe that since most ceramics are poor electrical conductors (with a few notable exceptions) due to a lack of free electrons, electronic conduction would be negligible compared to lattice, or phonon, conduction. This is indeed the case, and we will see that structural effects, such as complexity, defects, and impurity atoms, have a profound effect on thermal conductivity due to phonon mean free path, even if heat capacity is relatively unchanged.

Materials with complex structures possess more thermal phonon scattering, which reduces the mean free path and lowers the thermal conductivity, as given by Eq. (4.40). For example, the magnesium aluminate spinel shown in Figure 1.42 has a lower thermal conductivity than either Al_2O_3 or MgO, although each has a similar heat capacity and structure. Similarly, the complex structure of mullite has a much lower thermal conductivity than magnesium aluminate spinel. A structure can also affect the temperature dependence of thermal conductivity, since in complex structures the phonon mean free path tends to approach lattice dimensions at high temperatures. Similarly, anisotropic crystal structures can have thermal conductivities that vary with direction, as illustrated for quartz in Table 4.8. This variation tends to decrease with increasing temperature, since anisotropic crystals become more symmetrical as the temperature is raised. Crystallite size also affects thermal conductivity, though its effect is evident only at higher temperatures due to electronic conductivity, as shown in Figure 4.29 for three common ceramic materials.

Table 4.8 Thermal Conductivity of Quartz

	Thermal Conductivity (W/m·K)		
Temperature (°C)	**Normal to c-Axis**	**Parallel to c-Axis**	**Ratio**
0	0.67	1.13	1.69
100	0.50	0.79	1.58
200	0.42	0.63	1.50
300	0.35	0.50	1.43
400	0.31	0.42	1.35

Source: Kingery et al. [34]. Copyright © 1976 by John Wiley & Sons, Inc.

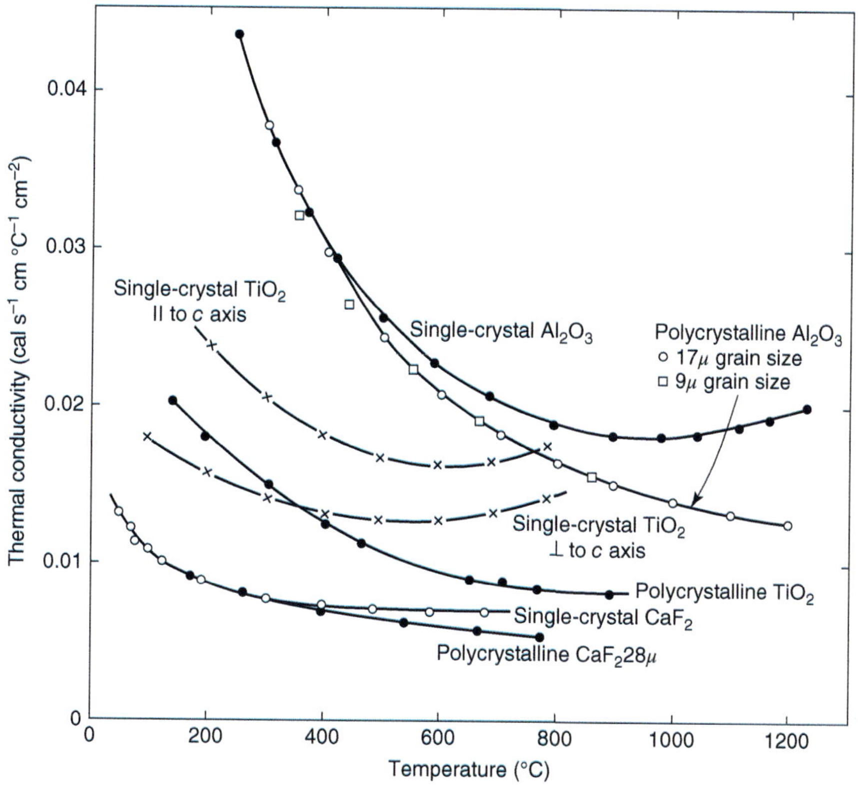

Figure 4.29 Thermal conductivity of single-crystal and polycrystalline Al_2O_3, TiO_2, and CaF. Multiply by 418.7 to obtain k in units of W/m·K. From Kingery et al. [34]. Copyright © 1976 by John Wiley & Sons, Inc. This material is used by permission of John Wiley & Sons, Inc.

Finally, in the same way that structural complexity lowers conductivity, the addition of impurities and variations in composition can affect thermal conductivity. As shown in Figure 4.30, the thermal conductivity is a maximum for simple elementary oxide and carbide structures, such as BeO. As the cation atomic number increases and the difference in atomic weight between the components becomes larger, the thermal conductivity decreases. In solid solutions, the additional

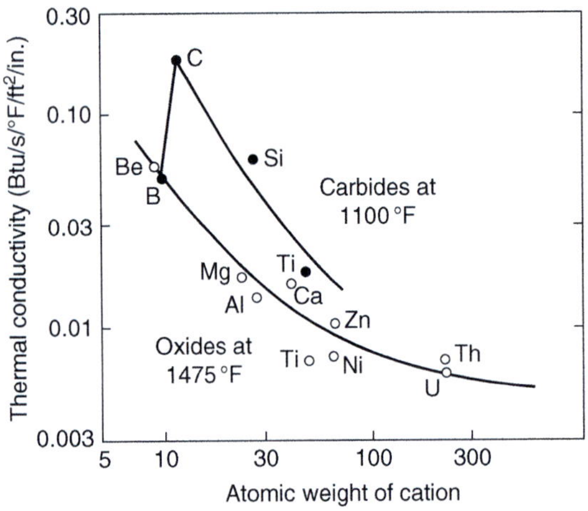

Figure 4.30 Effect of cation atomic number on thermal conductivity of some oxides and carbides. Multiply by 519.2 to obtain k in units of W/m·K. From Kingery et al. [34]. Copyright © 1976 by John Wiley & Sons, Inc. This material is used by permission of John Wiley & Sons, Inc.

phonon scattering for low concentrations of impurities is directly proportional to the volume fraction of impurity, and the impurity effect is relatively independent of temperature. The effect of solid–solution impurities over an extended range of compositions depends on mass difference, size difference, and binding energy difference of the constituents. In addition, temperature effects can be magnified in solid solutions, as illustrated in Figure 4.31. For the MgO—NiO solid solution shown, the effect of thermal conductivity lowering is even more pronounced at 200°C than at 1000°C.

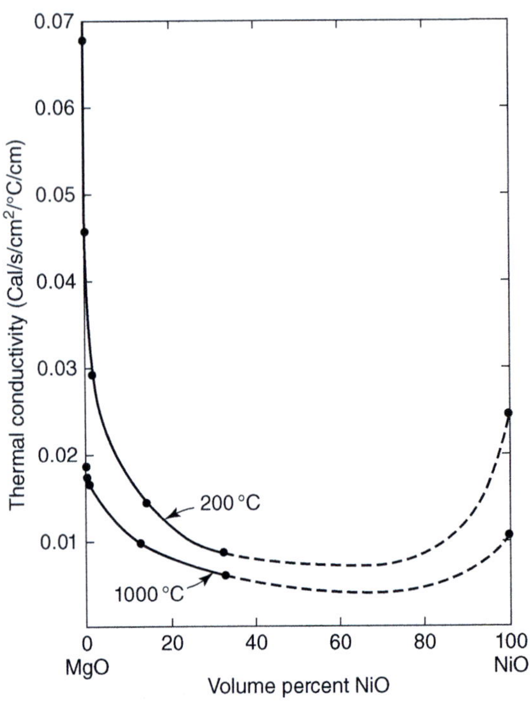

Figure 4.31 Thermal conductivity in the solid solution system MgO—NiO. Multiply by 418.7 to obtain k in units of W/m·K. From Kingery et al. [34]. Copyright © 1976 by John Wiley & Sons, Inc. This material is used by permission of John Wiley & Sons, Inc.

One final note is appropriate for this section. Because many oxide ceramics are used as insulating materials, the term *thermal resistivity* is often used instead of thermal conductivity. As will be the case with electrical properties in Chapter 6, resistivity and conductivity are merely inverses of one another, and the appropriateness of one or the other is determined by the context in which it is used. Similarly, *thermal conductance* is often used to describe the thermal conductivity of materials with standard thicknesses (e.g., building materials). Thermal conductance is the thermal conductivity divided by the thickness ($C = k/L$), and *thermal resistance* is the inverse of the product of thermal conductance and area ($R = 1/C{\cdot}A$).

Heat Transport Properties of Glasses

As was described in Chapter 1, glasses are structurally different from crystalline ceramics, and we would expect this to affect the heat capacity. This is true, and in general, the heat capacities of most oxide glasses approach 0.7 to 0.95 of the theoretical $3R$ value at the low-temperature end of the glass transition. On passing through the glass transition, the heat capacity generally increases by a factor of 1.3 to 3.0, because of increased configurational entropy. Eventually, the heat capacity of glasses reaches a value of $n3R$, where n is the number of atoms in the compound; thus $n = 3$ for SiO_2 and $C_p \approx C_v \approx 78$ J/mol·K above 1000°C. Several empirical correlations exist for estimating the heat capacity of glasses and related thermodynamic quantities such as enthalpy and thermal conductivity [35]. For example, the molar heat capacity of a multicomponent glass up to about 900°C can be estimated from the molar composition, x_i (entered in mol%), according to

$$C_p = \frac{1}{100}\left(\sum a_i x_i + T\sum b_i x_i + \frac{\sum c_i x_i}{T^2} + \frac{\sum d_i x_i}{T^{1/2}}\right) \tag{4.44}$$

where T is the temperature in K, and the factors a_i to d_i are presented in Table 4.9.

Based on the discussion of the previous section, we would also expect that structural complexity and disorder present in glasses would tend to decrease the thermal conductivity. This is true, especially when alkalis are added to sodium silicate glasses. Figure 4.32 shows the effect of replacing SiO_2 in Na_2O–SiO_2 glass with other oxides. These results reflect the fact that the stronger the bonds in the glass are, the higher the thermal conductivity. The dependence of thermal conductivity on

Table 4.9 Factors used to Determine Molar Heat Capacities of Glasses using Eq. (4.44)

Oxide	a_i	$10^3\ b_i$	$10^{-5}\ c_i$	d_i	Validity Region (K)
Na_2O	70.884	26.110	−3.5820	0	270–1170
K_2O	84.323	0.731	−8.2980	0	270–1190
MgO	46.704	11.220	−13.280	0	270–1080
CaO	39.159	18.650	−1.5230	0	270–1130
Al_2O_3	175.491	−5.839	−13.470	−1370.0	270–1190
SiO_2	127.200	−10.777	4.3127	−1463.9	270–1600
TiO_2	64.111	22.590	−23.020	0	300–800
FeO	31.770	38.515	−0.012	0	300–800
Fe_2O_3	135.250	12.311	−39.098	0	300–800

Source: Scholze [36]. Copyright © 1991 by Springer-Verlag.

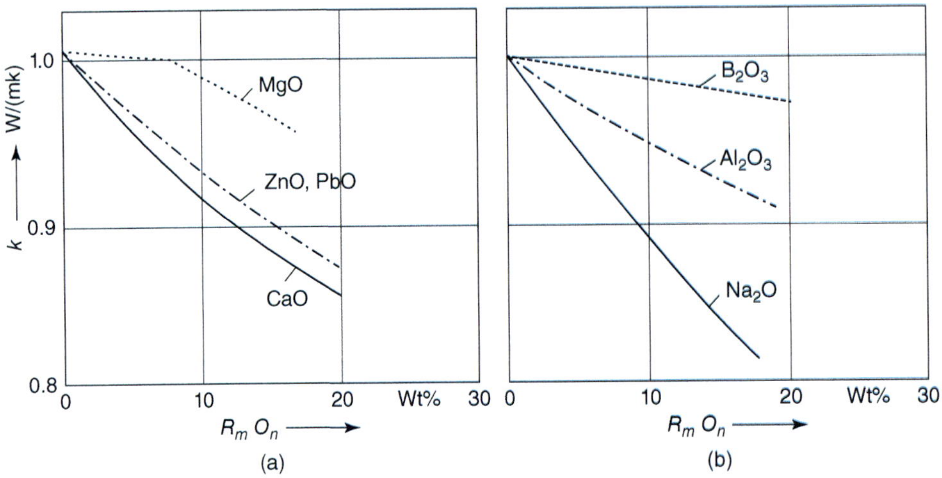

Figure 4.32 Change in thermal conductivity at 0°C of a Na_2O–SiO_2 glass (18–82 wt%) as SiO_2 is replaced proportionally by weight with other oxides. Reproduced, by permission, from Scholze [37]. Copyright © 1991 by Springer-Verlag.

bond strength allows an empirical expression to be developed in which the thermal conductivities of glass components are additive:

$$k = 10^{-1} \sum k_i x_i \tag{4.45}$$

where x_i is the weight fraction of component i, and k_i is its corresponding thermal conductivity factor, representative values for which can be found in Table 4.10. An alternative set of factors at higher temperatures has been developed by Primenko [38] and can be found elsewhere [39]. As

Table 4.10 Factors used to Determine Thermal Conductivity of Glasses Based on Composition using Eq. (4.45)

Oxide	k_i (W/m·K)[a]
Li_2O	−9.29
Na_2O	−4.79
K_2O	2.17
MgO	21.73
CaO	13.06
SrO	8.63
BaO	2.89
B_2O_3	8.216
Al_2O_3	13.61
SiO_2	13.33
TiO_2	−31.38
ZnO	8.00
PbO	2.68

Source: Scholze [36]. Copyright © 1991 by Springer-Verlag.
[a] Values from Ammar et al. [35].

a rough approximation, the thermal conductivity of a glass can be calculated from the density, ρ, by using

$$k = \frac{a}{\rho} + b \tag{4.46}$$

where the pair values for constants a and b of 2.09 and 0.17, or 2.30 and 0.21, are valid for temperatures between 0°C and 100°C.

4.2.3 Heat Transport Properties of Polymers

As with other disordered materials, the thermal conductivities of polymers are low due to phonon scattering. As a result, even though polymers tend to have heat capacities of the same order of magnitude as metals (1.5 to 3.5 J/g·K), their thermal conductivities (0.1 to 1.0 W/m·K) are 1000 times lower than metals. Polymers, therefore, are generally good insulators, as long as their use temperature is below their thermal stability temperature. Few correlations for heat capacity and thermal conductivity of polymers exist. As for all the materials in this chapter, representative values of thermal conductivity can be found in Appendix 4.

Cooperative Learning Exercise 4.7

Consider a glass with the following composition (in wt%): 72 SiO_2, 15 Na_2O, 11 CaO, and 2 Al_2O_3, and a density of 2.458 g/cm^3.

Person 1: Estimate the thermal conductivity at 30°C using Eq. (4.45).

Person 2: Estimate the thermal conductivity at 30°C using Eq. (4.46).

Compare your answers. What are some ways you could experimentally verify these estimates?

Answers: $k_1 = 1.06$; $k_2 = 1.02$

Thermal conductivities can be determined using any setup that would allow Fourier's Law of conduction to be applied. For example, a plate of glass could be placed on top of a heat sink (such as a large copper block), heated on one side, and the temperature difference at each face measured to determine the temperature gradient. Once the heat flow is measured, the proportionality constant, k, can be calculated.

The heat capacity of polymers at constant pressure increases with temperature up to the melting point. In practice, there is a sharp increase in heat capacity at the onset of melting, followed by a sharp drop at the end of the melting range (see Figure 4.33). There is then a slight increase in heat capacity with temperature in the melt.

The thermal conductivity of polymers is molecular-weight-dependent, and it increases with increasing molecular weight. Orientation also has an effect, and it can have a strong influence on thermal conductivity in highly crystalline polymers such as HDPE. As is evident from Figure 4.34, the thermal conductivity of most polymeric materials is, with a few notable exceptions for highly crystalline polymers such as polyethylene, less dependent upon temperature and, as a result, can only be influenced significantly through the incorporation of fillers. These effects will be discussed in the next section. Foaming with air or some other gas can also decrease thermal conductivity (see Figure 4.35). This has the same effect as the introduction of porosity in ceramic materials such as fire brick, due to the introduction of a lower thermal conductivity material (in this case, a gas) into the structure.

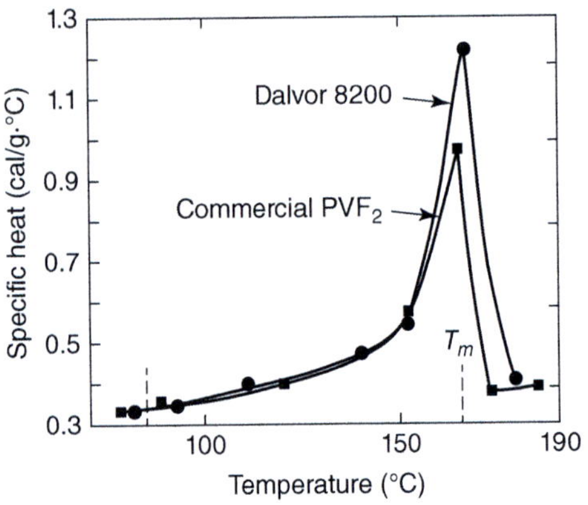

Figure 4.33 Specific heat of two commercial polyvinylidene fluoride samples with temperature. From Tadmor and Gogos [10]. Copyright © 1979 by John Wiley & Sons, Inc. This material is used by permission of John Wiley & Sons, Inc.

Cooperative Learning Exercise 4.8

Recall from the beginning of the chapter that a related quantity to the thermal conductivity is the *thermal diffusivity*, α, which is defined as $k/\rho C_p$, where k is the thermal conductivity, ρ is the density, and C_p is the *heat capacity* at constant pressure per unit mass, or specific heat. Below are the thermal properties of polycarbonate.

Person 1: Determine the density and thermal diffusivity for polycarbonate at 350 K from the plots.

Person 2: Determine the thermal conductivity and heat capacity for polycarbonate at 350 K from the plots.

Combine your information to calculate the thermal diffusivity at 350 K. How does this compare to the value obtained from the plot? Can you estimate the melting point of this polycarbonate from these plots?

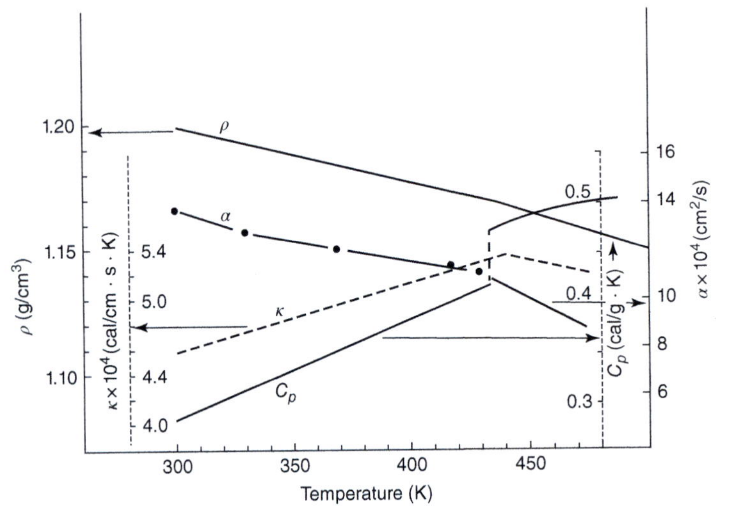

Answers: $\rho = 1.19$ g/cm^3; $\alpha = 12 \times 10^{-4}$cm^2/s; $k = 4.8 \times 10^{-4}$cal/m · s · K; $Cp = 0.33$ cal/g · K$\alpha_{\text{calc}} = k/\rho C_p = 4.8 \times 10^{-4}$ cal/m · s · K/(1.19 g/cm^3 × 0.33 cal/g · K) = 12×10^{-4}cm^2/s, same as on graph. $T_m \approx 434$ K = 162 ° C, as determined from graphs where discontinuity exists; thermal properties change!

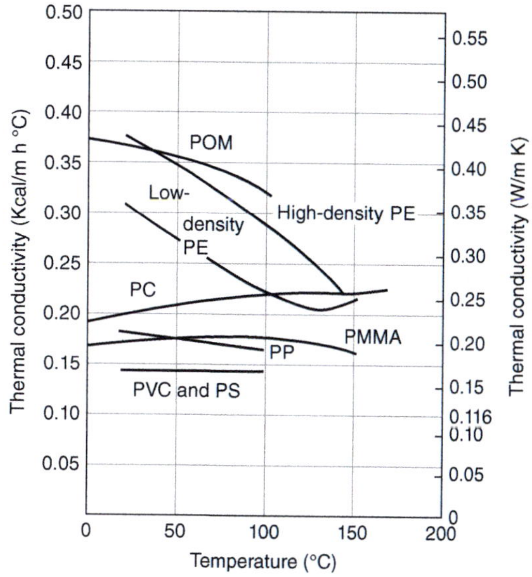

Figure 4.34 Temperature dependence of thermal conductivity for selected polymers. Reproduced, by permission, from Powell and Housz [40]. Copyright © 1998 by Taylor & Francis.

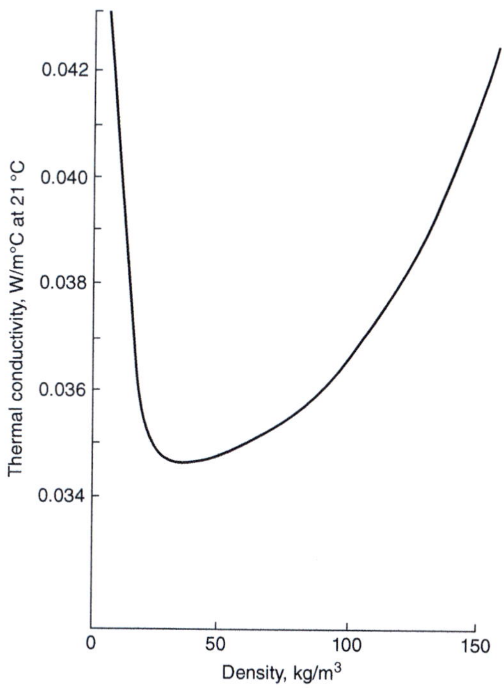

Figure 4.35 Thermal conductivity of expanded polystyrene beads as a function of density. Reproduced, by permission, from The Dow Chemical Company. Copyright © 1966.

4.2.4 Heat Transport Properties of Composites

As was the case with momentum transport, the heat transport properties of heterogeneous systems are difficult to correlate and virtually impossible to predict. There are two topics worthy of note, however, namely, the heat transport properties of filled composites, and the thermal conductivity of laminar composites.

Thermal Conductivity of Filled Composites

As we saw at the end of the previous section, the presence of gases in a material's structure can dramatically affect the heat transport properties. We expand upon this principle here, again concentrating on polymeric materials that have been reinforced with various fibers and fillers and that may also contain porous gas structures, as is found in foams.

To a first approximation, the heat capacity in these heterogeneous, multiphase systems, $\overline{C}_p$ can be approximated by a weighted average of the component heat capacities:

$$\overline{C}_p = \sum w_i C_{pi} \tag{4.47}$$

where w_i is the weight fraction of each component. Since the relative heat capacities (per gram) of gas, polymer, and metal are 2:4:1, the heat capacity of polymer tends to dominate in these types of multicomponent systems, acting as a heat sink during processing.

Due to the dependence on mean free path as described in Eq. (4.40), the thermal conductivity of heterogeneous systems is impossible to predict on heat capacity alone. As in previous sections, we do know that disorder tends to decrease thermal conductivity due to mean free path considerations, and this is indeed the case for fillers with high thermal conductivities, such as copper and aluminum in epoxy matrices (see Table 4.11). The thermal conductivity of the epoxy matrix increases only modestly due to the addition of even high percentages of thermally conductive fillers.

One way to deal with heterogeneous materials is to assume that the composite has an effective thermal conductivity, k_{eff}, which can then be used in heat transport expressions as previously performed for homogeneous materials. There are several expressions for k_{eff} that have been developed, depending upon the type and amount of reinforcing material. For spherical particles of small volume fractions, ϕ, we have

Table 4.11 Thermal Conductivity of Epoxy Resin Filled with Various Compounds

		Thermal Conductivity (W/m·K)		
Filler	**Vol% Filler**	**Filler**	**Epoxy Resin**	**Composite**
Aluminum, 30 mesh	63	208	0.20	2.5
Sand, coarse grain	64	1.17	0.20	1.0
Mica, 325 mesh	24	0.67	0.20	0.5
Alumina, tabular	53	30.3	0.20	1.0
Alumina, 325 mesh	53	30.3	0.23	1.4
Silica, 325 mesh	39	1.17	0.23	0.77
Copper powder	60	385	0.23	1.6
Alumina (tabular/325 mesh)	45/20	30.3	0.20	2.5

Source: Rodriguez [9]. Copyright © 1982 by McGraw-Hill Book Co.

$$\frac{k_{\text{eff}}}{k_0} = 1 + \frac{3\phi}{\left(\dfrac{k_1 + 2k_0}{k_1 - k_0}\right) - \phi} \tag{4.48}$$

where k_1 and k_0 are the thermal conductivities of the sphere and matrix phase materials, respectively. For large volume fractions, the thermal conductivity of the composite can be estimated with:

$$\frac{k_{\text{eff}}}{k_0} = 1 + \frac{3\phi}{\left(\dfrac{k_1 + 2k_0}{k_1 - k_0}\right) - \phi + 1.569\left(\dfrac{k_1 - k_0}{3k_1 - 4k_0}\right)\phi^{10/3} + \cdots} \tag{4.49}$$

Generally, the thermal interaction of spheres that can occur at high volume fractions is negligible, such that the simpler Eq. (4.48) is often used. For nonspherical reinforcement, such as aligned, continuous fibers, the thermal conductivity is anisotropic—that is, not the same in all directions. Rayleigh showed that the effective thermal conductivity along the direction of the fiber axis is

$$\frac{k_{\text{eff},z}}{k_0} = 1 + \left(\frac{k_1 - k_0}{k_0}\right)\phi \tag{4.50}$$

whereas the effective thermal conductivity in the perpendicular directions is identical:

$$\frac{k_{\text{eff},x}}{k_0} = \frac{k_{\text{eff},y}}{k_0} = 1 + \frac{2\phi}{\left(\dfrac{k_1 + k_0}{k_1 - k_0}\right) - \phi + \left(\dfrac{k_1 - k_0}{k_1 - k_0}\right)(0.30584\phi^4 + 0.013363\phi^8 \ldots)} \tag{4.51}$$

Additional expressions exist for nonspherical reinforcements and for solids containing gas pockets. The interested reader is referred to Ref. [42] for more information.

Thermal Conductivity of Laminar Composites

In the case of laminar composites or layered materials (cf. Figure 1.70), the thermal conductance can be modeled as heat flow through plane walls in a series, as shown in Figure 4.36. At steady state, the heat flux through each wall in the x direction must be the same, q_x, resulting in a different temperature gradient across each wall. Equation (4.2) then becomes

$$q_x = -k_A \frac{dT}{dx_A} = -k_B \frac{dT}{dx_B} = -k_C \frac{dT}{dx_C} \tag{4.52}$$

Figure 4.36 Heat flow through a multilayered wall.

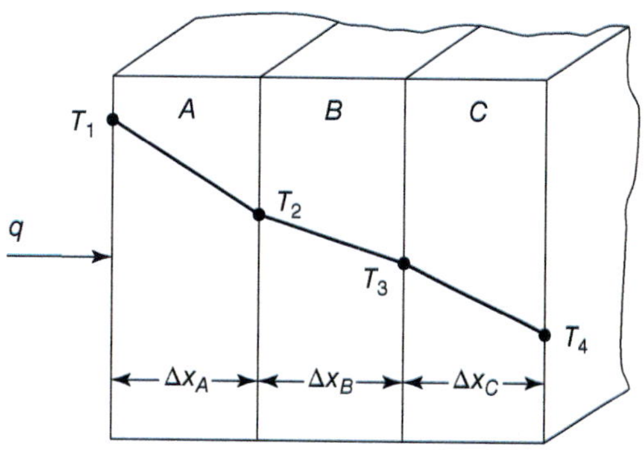

where the minus sign simply indicates the direction of heat flow, depending upon how the temperature gradient is defined. For the case of constant thermal conductivity with temperature, which is true for small temperature gradients, and a constant cross-sectional heat flow area, A, the heat flow in the x direction, $Q_x = q_x \cdot A$ becomes

$$Q_x = k_A A \frac{(T_1 - T_2)}{\Delta x_A} = k_B A \frac{(T_2 - T_3)}{\Delta x_B} = k_C A \frac{(T_3 - T_4)}{\Delta x_C} \tag{4.53}$$

Example Problem 4.1

Question: Derive an expression analogous to Eq. (4.54) for steady-state heat flow through concentric cylinders, as shown at the right. All cylinders are of the same length, L. Assume that thermal conductivity is constant with temperature, but that the material in each concentric ring is different.

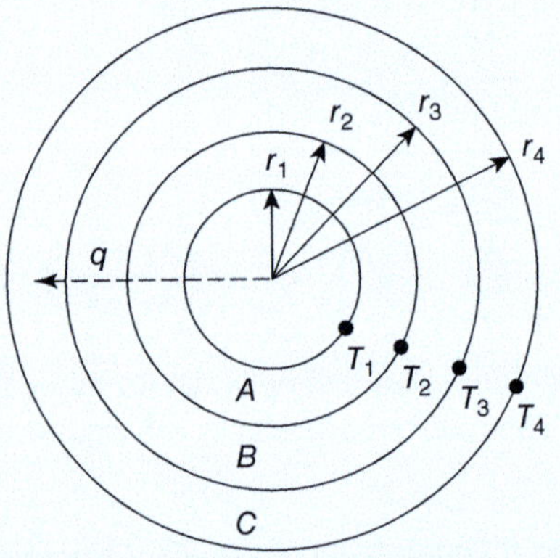

Answer: Fourier's Law in cylindrical coordinates can be written as

$$q = \frac{Q}{A} = -k\frac{dT}{dr}$$

The cross-sectional area normal to heat flow, A, is not constant, and varies with radius as $A = 2\pi rL$, where L is the length of the cylinders. For the inner cylinder, the substitution of the area expression into Fourier's Law and the separation of variables gives

$$\frac{Q_A}{2\pi L}\int_{r_1}^{r_2}\frac{dr}{r} = -k_A\int_{T_1}^{T_2} dT$$

and integration results in

$$Q_A = \frac{k_A 2\pi L}{\ln(r_1/r_2)}(T_1 - T_2)$$

Multiply the numerator and denominator by $(r_2 - r_1)$, and define a log-mean area, A_{lm} as follows

$$A_{A,lm} = \frac{(2\pi L r_2) - (2\pi L r_1)}{\ln(2\pi L r_2 / 2\pi L r_1)} = \frac{A_2 - A_1}{\ln(A_2/A_1)}$$

the heat flow through the inner cylinder becomes

$$Q = \frac{kA_{A,lm}(T_1 - T_2)}{(r_2 - r_1)} = \frac{(T_1 - T_2)}{(r_2 - r_1/kA_{A,lm})} = \frac{(T_1 - T_2)}{R_A}$$

where R_A is the resistance, which for the cylindrical geometry is defined as

$$R_A = \frac{(r_2 - r_1)}{k_A A_{A,lm}}$$

Finally, we recognize that at steady state, the heat flow through all the cylinders must be the same, $Q = Q_A = Q_B = Q_C$, so that

$$Q = \frac{(T_1 - T_2)}{R_A} = \frac{(T_2 - T_3)}{R_B} = \frac{(T_3 - T_4)}{R_C}$$

As was the case for the multilayer wall, the internal wall temperatures T_2 and T_3 can be eliminated through combination of the three equations to yield one equation

$$Q = \frac{(T_1 - T_4)}{R_A + R_B + R_C} = \frac{(T_1 - T_4)}{\sum R_i}$$

Note that this is the same result as Eq. (4.54), except that the resistance is defined differently.

Each of the three relations in Eq. (4.53) can be solved for the temperature difference, and the equations can be added to eliminate the internal surface temperatures, T_2 and T_3, such that only the external surface temperatures, T_1 and T_4 remain. The heat flow is then expressed as follows:

$$Q_x = \frac{(T_1 - T_4)}{\frac{\Delta x_A}{k_A A} + \frac{\Delta x_B}{k_B A} + \frac{\Delta x_C}{k_C A}} = \frac{(T_1 - T_4)}{R_A + R_B + R_C} \tag{4.54}$$

where $R_i = \Delta x_i/(k_i \cdot A)$ is called the *thermal resistance*, as introduced in Section 4.2.2. The total thermal resistance is simply the sum of the individual resistances, $R_{tot} = \sum R_i$. These "R values" are often reported in lieu of thermal conductivity values for common insulating materials, such as fiberglass and rock wool, for which thickness and sizes have been standardized, typically 1-ft^2 area and 1-in. thickness. Selected thermal conductivities for these and other insulating materials can be found in Appendix 4.

Cooperative Learning Exercise 4.9

Work in groups of three. Your plant is replacing asbestos used to insulate low-pressure steam pipes with a new multilayer composite material composed of a 2.5-cm-thick, polystyrene foam (k = 0.029 W/m·K) and a 2-mm-thick, outer, protective layer of polyethylene (k = 0.33 W/m·K). Assume that the pipe carries saturated steam at 130°C and that the outer surface of the insulation is 30°C. The piping is stainless steel (k = 14 W/m·K), with ID = 0.0254 m and OD = 0.0508 m. Use the equations developed in Example Problem 4.1 to solve the following problems, assuming a pipe length of 1 m.

Person 1: Calculate the log-mean heat transfer area for the stainless-steel pipe, then use this to calculate the thermal resistance through the pipe.

Person 2: Calculate the log-mean heat transfer area for the polystyrene foam only, then use this to calculate the thermal resistance through the foam.

Person 3: Calculate the log-mean heat transfer area for the outer polyethylene layer, then use this to calculate the thermal resistance through the polyethylene.

Combine your resistance to calculate the heat transfer rate through the 1-m section of pipe. Which of the resistances dominates the calculation?

Answers: $A_{ss,lm} = 0.115\ m^2$; $R_{ss} = 0.008$ K/W
$A_{PS,lm} = 0.229\ m^2$; $R_{PS} = 3.76$ K/W
$A_{PE,lm} = 0.323\ m^2$; $R_{PE} = 0.09$ K/W
$Q = 26.4$ W; polystyrene foam dominates

4.2.5 Heat Transport Properties of Biologics

Heat storage in biological systems is much the same as for the previous material classes. For example, the heat capacity of biological fluids, which are heterogeneous suspensions and mixtures, can be approximated by Eq. (4.47). In instances where the biological fluid or food substance is an aqueous-based suspension or solution of fat and solids, the specific heat (heat capacity per unit weight in J/g·K) can be approximated as

$$C_p = 4.2(x_{H_2O} + 0.5x_{fat} + 0.3x_{solids}) \tag{4.55}$$

where x_i is the mass fraction of water, fat, and solids, respectively.

Heat transfer in biological systems, on the other hand, is often a complex combination of conduction, convection, and radiation. For example, the insulating value of fur and feathers in animals is certainly measurable but is based on more than just the thermal conductivity of the biological materials constituting the fur or feathers, since gases become entrapped in these porous materials, and heat is being radiated to and from the surroundings. As a result, the description of heat transfer in biological systems is much more empirical, and it involves a separate set of terminologies from the traditional, engineering-oriented set of terms. Nonetheless, the principles are the same, and we relax our restriction on heat transfer via conduction only to better describe heat transfer in "biological units"—animals and plants.

Thermal Conductance in Biological Systems

As introduced in Section 4.2.2, thermal conductance is the thermal conductivity divided by the thickness ($C = k/L$), and *thermal resistance* is the inverse of the product of thermal conductance and area ($R = 1/C{\cdot}A$). These terms are widely used to describe the insulating value of skin coverings, such as fur and clothing, in biological systems. As mentioned in the introduction to this section, the insulating value of fur and clothing is derived from more than thermal conductivity considerations. It involves the reduction of permeability to water vapor transmission, the disruption of convective heat transport from the surface, and an increase in surface area. Nonetheless, the insulating value of clothing has been standardized in terms of thermal conductance and is expressed in units of *clo*. A clo is the amount of insulation that would allow 6.45 N·m/s of heat to transfer to the surroundings from a 1.0 m^2 area of skin across a 1°C temperature difference between the skin and the environment. The higher the clo value of clothing, the less heat will be transferred from the skin. We will continue to use common engineering units for our analysis.

Table 4.12 Conductance Values for Selected Pieces of Clothing

Clothing	C_{cl} (W/m^2·K)
Undergarments	34–130
Thermal undergarments	18
Shirts, vests, and blouses	22–46
Pants and Slacks	15–25
Sweaters and jackets	17–38
Sock	65–161
Stockings, pantyhose	640
Sandals	320
Boots	81

Source: Johnson [22]. Copyright © 1999 by John Wiley & Sons, Inc.

For clothing, Fourier's Law [Eq. (4.2)], can be expressed as

$$q_y = \frac{Q}{A} = -k_{cl}\frac{dT}{dy} \tag{4.2}$$

which at steady state and for constant thermal properties, can be expressed in terms of conductance of clothing, C_{cl}, as

$$Q = \frac{-k_{cl}A\Delta T}{\Delta y} = C_{cl}A(T_{sk} - T_{\infty}) \tag{4.56}$$

where T_{sk} is the temperature of the skin and T_{∞} is the temperature of the surrounding environment. Some representative values for the conductance of common items of clothing are given in Table 4.12. Clothing is often worn in layers, and the principles represented by Eq. (4.54)—that is, the additivity of resistances—apply here as well. Due to the added complexities of layered clothing in terms of simultaneous series and parallel heat conduction, the following empirical relationships have been developed for the thermal resistance of clothing, R_{cl} (in K/W), in men and women:

$$R_{cl,\text{men}} = \frac{0.494\sum(1/C_{cl}) + 0.0119}{A} \tag{4.57}$$

$$R_{cl,\text{women}} = \frac{0.524\sum(1/C_{cl}) + 0.0053}{A} \tag{4.58}$$

where A is the surface area in m^2. Clothing surface area is based on a 15–25% increase over the surface area of a nude body, A (in m^2), as determined by the formula of DuBois:

$$A = 0.07673W^{0.425}H_t^{0.725} = 0.2025m^{0.425}H_t^{0.725} \tag{4.59}$$

where W is the body weight in newtons, H_t is the height in meters, and m is the body mass in kilograms. Mean skin temperature, T_{sk}, is also affected by the presence of clothing. The following empirical relationship can be employed and is essentially independent of metabolic rate up to four or five times the rate at rest. For clothed humans,

$$T_{sk} = 25.8 + 0.267T_{\infty} \tag{4.60}$$

where T_{∞} is again the surrounding environmental temperature, in degrees Celsius.

Table 4.13 Selected Values of Thermal Conductance for Fur and Feathers[a]

Animal	Condition	*m* (kg)	Thickness (mm)	*C* ($W/m^2 \cdot K$)
Rabbit	Back	2.5	9	2.4
Horse	Flank	650	8	4.1
Pig	Flank	100	5	2.8
Chicken	Well-feathered	2	25	1.6
Cow	—	600	10	3.2
Dog	Winter	50	44	1.2
Polar bear	In air	400	63	1.2
Polar bear	In water	400	57	27
Beaver	In air	26	44	1.2
Beaver	In water	26	40	14
Squirrel	Summer	—	3	9.0
Monkey	Summer	—	8	3.2

Source: Johnson [22]. Copyright © 1999 by John Wiley & Sons, Inc.
[a] Includes skin and still air layer.

Fur on animals is treated much the same way as clothing on humans. Some typical values of conductance for furs and feathers are presented in Table 4.13. The total surface area of an animal, A (in m^2), can be estimated by

$$A = 0.09m^{2/3} \tag{4.61}$$

where m is the animal's body mass, in kilograms. The values in Table 4.13 include the effects of convection and conduction, although radiation was not a significant contributor in these instances. We will see in the next section how the effects of convection can be combined with thermal conductivity to describe heat transfer in these systems.

Convective Heat Transfer in Biological Systems*

Unlike the materials of the previous sections, for which processing temperatures, thermal conductivities, and processing environments vary greatly, even within a material subclass (e.g., the heat transport processes involved in creating SiC via chemical vapor deposition are much different from those involved in forming SiC by a solid-state reaction), biological materials generally function within a relatively narrow temperature range, have thermal conductivities within an order of magnitude of each other, and function primarily in only one gaseous environment—air. As a result, we are able to briefly discuss here the role of convection in heat transfer in biological systems. To be sure, convection plays an important role in the processing and application of all the other material classes as well, and as you probably are already aware, the dimensionless groups and principles discussed here are general to heat transport in all systems. We will revisit some of these topics in Chapter 7 when the processing of materials is described.

Convection involves the transfer of heat by means of a fluid, including gases and liquids. Typically, convection describes heat transfer from a solid surface to an adjacent fluid, but it can also describe the bulk movement of fluid and the associate transport of heat energy, as in the case of a hot, rising gas. Recall that there are two general types of convection: *forced convection* and *natural (free) convection*. In the former, fluid is forced past an object by mechanical means, such as a pump or a fan, whereas the latter describes the free motion of fluid elements due primarily to density differences. It is common for both types of convection to occur simultaneously in what is termed *mixed convection*. In such instances, a modified form of Fourier's Law is applied, called *Newton's Law of Cooling*, where the thermal conductivity is replaced with what is called the *heat transfer coefficient*, h_c:

$$q_y = \frac{Q_y}{A} = h_c dT \tag{4.62}$$

Unlike Fourier's Law, Eq. (4.62) is purely empirical—it is simply the definition for the heat transfer coefficient. Note that the units of h_c (W/m^2·K) are different from those for thermal conductivity. Under steady-state conditions and assuming that the heat transfer area is constant and h_c is not a function of temperature, the following form of Eq. (4.62) is often employed:

$$Q_y = h_c A \Delta T \tag{4.63}$$

where ΔT is generally taken as the difference between the solid surface temperature and the fluid temperature far from the surface (bulk fluid temperature).

Heat transfer coefficients generally cannot be calculated from first principles, and are normally derived from experiments with specific materials, fluids, geometries, and conditions. The use of dimensionless groups helps simplify this analysis. A complete review of dimensionless groups and their use in heat transfer analysis is not appropriate here. We will limit our discussion to three appropriate dimensionless groups: the Reynolds number, Re; Prandtl number, Pr; and Nusselt number, Nu. The *Reynolds number* is a ratio of inertial forces to viscous forces and is given by $\mathrm{Re} = Dv\rho/\mu$, where D is the characteristic diameter in which a fluid is flowing, v is the fluid velocity, μ is the fluid viscosity, and ρ is its density. The Prandtl number was introduced in Section 4.2.1 as a ratio of momentum diffusivity to thermal diffusivity and is defined as $\mathrm{Pr} = \hat{C}_p\mu/k$, where μ is again the fluid viscosity, k is its thermal conductivity, and $\hat{C}_p$ is the heat capacity. The Reynolds and Prandtl numbers provide a means for correlating measurable fluid properties (μ, ρ, k, and $\hat{C}_p$) and geometric considerations (D, v) with the heat transfer coefficient, h_c, in terms of the *Nusselt number*, Nu, which is defined as

$$\mathrm{Nu} = \frac{h_c D}{k} \tag{4.64}$$

where D is again a characteristic diameter and k is the thermal conductivity. Most correlations involve Pr and Re raised to an appropriate power:

$$\mathrm{Nu} = \text{constant}\ \mathrm{Re}^n \mathrm{Pr}^m \tag{4.65}$$

The values of m and n vary greatly depending on whether the flow is turbulent or laminar and depending on the specific geometry involved in heat transfer—for example, flow from smooth-walled tubes, or non-Newtonian fluid flow. We will not review all of these correlations here. We simply note that with proper descriptions of the surrounding fluid and the geometry of the biological system, values for heat transfer coefficients can be estimated and used in the analysis of heat transfer in biological materials. These correlations are useful in estimating physical properties such as thermal conductivity in biological systems where direct measurement may be difficult, as illustrated in Example Problem 4.2.

Example Problem 4.2 (Adapted from Johnson)

The metabolic rate of animals, M (in W), can be approximated by $M = 3.39\ m^{0.75}$, where m is the body mass of the animal in kilograms. About 15% of this heat is lost from the respiratory system, and another 50% is lost from the legs, tail, and ears. The remainder must be transported away from the body through the skin. Estimate the thermal conductivity of a 5-mm-thick hair coat of a 600-kg cow if the skin surface temperature is at 32°C and the ambient temperature is 15°C. Assume that the cow is standing still in a field, with a breeze of 0.5 m/s. For flow across cylinders where 1000 < Re < 50 000, the following correlation for Nusselt number may be used:

$$\text{Nu} = 0.26(\text{Re})^{0.6}(\text{Pr})^{0.3}$$

Answer: The heat generated by the cow is

$$M = 3.39(600\ \text{kg})^{0.75},\ \text{so}\ Q = 0.35(411) = 144\ N\cdot\text{m/s}$$

The total surface area of the cow, as given by Eq. (4.61) is

$$A = 0.09(600\ \text{kg})^{2/3} = 6.40\ m^2$$

We might estimate that about 20% of this surface area is in the legs and tail, which have already been accounted for in terms of heat loss, so that the remainder, about 5.12 m^2, is the surface area of the body. We can further model the cow's body as roughly cylindrical in shape and can estimate that its radius is 40 cm. This allows us to calculate the length of the cylinder (including the surfaces at both ends) as

$$A = 2\pi rL + 2\pi r^2 = 2\pi(0.4\ m)L + 2\pi(0.4\ m)^2 = 5.12\ m^2 \Rightarrow L = 1.64\ m$$

We will assume forced convection to calculate dimensionless groups. Physical properties of air must be estimated, and we will use the ambient air temperature to estimate them. The following physical properties can be found in tables or calculated:

$$\rho(15°\text{C}) = 1.226\ \text{kg/m}^3, \qquad \mu(15°\text{C}) = 1.8\times10^{-5}\ \text{kg/m}\cdot\text{s},$$
$$C_p = 1.0048\ \text{kJ/kg}\cdot\text{K}, \qquad k = 0.0253\ \text{W/m}\cdot\text{K}$$
$$\text{Pr} = (1004.8\ \text{J/kg}\cdot\text{K})(1.8\times10^{-5}\ \text{kg/m}\cdot\text{s})/(0.0253\ \text{J/m}\cdot\text{K}\cdot\text{s}) = 0.7$$
$$\text{Re} = (0.8\ \text{m})(0.5\ \text{m/s})(1.226\ \text{kg/m}^3)/(1.8\times10^{-5}\ \text{kg/m}\cdot\text{s}) = 27244$$
$$\text{Nu} = 0.26(27{,}244)^{0.6}(0.7)^{0.3} = 107 = h_cD/k$$

Solving for h_c from definition of Nu (see Eq. 4.64), we obtain

$$h_c = 107(0.0253\ \text{J/m}\cdot\text{K}\cdot\text{s})/(0.80\ \text{m}) = 2.17\ \text{W/K}$$

The total thermal resistance comes from a combination of forced convection from the body and thermal conduction through the skin:

$$R_{tot} = R_{conduction} + R_{convection} = L_{hair}/k_{hair}A + 1/h_cA = \Delta T/Q$$

Substituting values from above and solving for k_{hair} gives

$$(0.005\ \text{m})/k_{hair}(5.12\ \text{m}^2) + 1/(2.17\ \text{W/K})(5.12\ \text{m}^2) = (32-15\ \text{K})/(144\ \text{N}\cdot\text{m/s}),$$
$$k_{hair} = 0.035\ \text{W/m}\cdot\text{K}$$

4.3 Mass Transport Properties of Materials*

We bring the analogies of momentum, heat, and mass transfer in materials to a close with a description of mass transport, often called *mass diffusion*, or simply *diffusion*. Many of the phenomena we have already studied depend upon diffusion. For example, most phase transformations (excluding "diffusionless" transformations such as martensitic) rely on the movement of atoms relative to one another. Corrosion cannot occur unless there is a movement of atoms or ions to a surface that allows the appropriate chemical reactions to take place. And, as we will see in Chapter 7, some processing methods such as chemical vapor deposition are wholly dependent upon the transport of chemical reactants to the location where reaction conditions allow products to form. This is why the term "diffusion-controlled" is often used to describe the kinetics of a process—the reaction is rapid relative to the time it takes to transport reactants to and/or products from the reaction site. For these reasons, it is important to study diffusion in materials, even though the driving forces are complex and the mass transport properties of a material are highly dependent upon the diffusing species of interest.

Unlike momentum and heat transport, the entity being transported is not the same in all diffusion situations. We must describe not only the material in which the transport is taking place but also *what* is being transported (see Appendix 5, for example). The movement of hydrogen gas through metals, for example, is much different from the movement of liquid water through metals. As a result, the systems we will address will be highly specific and not at all general. Nonetheless, we can begin as in the previous sections of this chapter, with a description of mass transport properties from a fundamental viewpoint, and extend this to liquids and solids, with the recognition that as systems become more complex, theoretical predictions must give way to empiricisms and correlations.

4.3.1 Mass Transport Properties of Metals and Alloys

Recall from the beginning of this chapter that the transport of mass at steady state is governed by Fick's Law, given in one-dimension by Eq. (4.4):

$$J^*_{Ay} = -D_{AB}\frac{dc_A}{dy} \tag{4.4}$$

where J^*_{Ay} is the molar flux of component A in the y direction, dc_A/dy is the concentration gradient of A in the y direction, and D_{AB} is the molecular diffusivity, or *diffusion coefficient*, of A in component B. As in the previous sections, we will concentrate on the material-related properties of transport, which in this section is the molecular diffusivity. We begin with a description of the molecular origins of mass diffusivity, and we then proceed to correlations and predictive relationships for the specific material classes.

The Molecular Origins of Mass Diffusivity

In a manner directly analogous to the derivations of Eq. (4.6) for viscosity and Eq. (4.34) for thermal conductivity, the diffusion coefficient, or mass diffusivity, D, in units of m^2/s, can be derived from the kinetic theory of gases for rigid-sphere molecules. By means of summary, we present all three expressions for transport coefficients here to further illustrate their similarities.

$$\mu = 2.6693 \times 10^{-5}\frac{\sqrt{MT}}{\sigma^2\Omega} \tag{4.6}$$

$$k = 8.32 \times 10^{-2} \frac{\sqrt{T/M}}{\sigma^2 \Omega} \tag{4.34}$$

$$D = 2.6280 \times 10^{-7} \frac{\sqrt{T^3/M}}{P\sigma^2 \Omega} \tag{4.66}$$

As before, M is the molecular weight of the rigid sphere, T is the absolute temperature, and σ and Ω are the same Lennard–Jones parameters used in Eq. (4.6) (note that technically, $\Omega = 1$ for the rigid-sphere model). The pressure, P, in Eq. (4.66) is in units of atmospheres. Equation (4.66) gives the so-called *self-diffusion coefficient*, or diffusivity for a sphere diffusing through a gas of the same spheres (i.e., D_{AA}). Fick's Law [cf. Eq. (4.4)] related the molar flux to the concentration gradient in a mixture of gases, A and B. In this case, the diffusion coefficient for the rigid-sphere model is

$$D_{AB} = 2.6280 \times 10^{-7} \frac{\sqrt{T^3(M_A + M_B)/2M_A M_B}}{P\sigma_{AB}^2 \Omega_{AB}} \tag{4.67}$$

where M_A and M_B are the molecular weights of species A and B, respectively; $\sigma_{AB} = \frac{1}{2}(\sigma_A + \sigma_B)$, and Ω_{AB} is again approximately unity for rigid spheres. For rigid sphere gases, we note that the diffusivity of species B through A should be the same as the diffusivity for species A through B, that is, $D_{AB} = D_{BA}$.

Mass Diffusivity in Liquid Metals and Alloys

The hard-sphere model of gases works relatively well for self-diffusion in monatomic liquid metals. Several models based on hard-sphere theory exist for predicting the self-diffusivity in liquid metals. One such model utilizes the hard-sphere packing fraction, PF, to determine D (in cm^2/s):

$$D = \sigma C_{AW} \left(\frac{\pi RT}{M} \right)^{1/2} \frac{(1 - PF)^3}{8PF(2 - PF)} \tag{4.68}$$

where T and M are the absolute temperature and molecular weight, as before; R is the gas constant, σ is the collision diameter of the previous equations, and C_{AW}, the *Adler–Wainwright correction factor*, is a function of PF that can be determined from Figure 4.37. The packing fraction can be calculated using Eq. (4.8) as before. Recall from Table 4.1 that $PF = 0.45$ near the melting point. The values predicted by Eq. (4.68) for several liquid metals are compared with experimental values in Table 4.14.

Several empirical expressions also exist, including one that correlates self-diffusivity (in cm^2/s) with melting point, T_m:

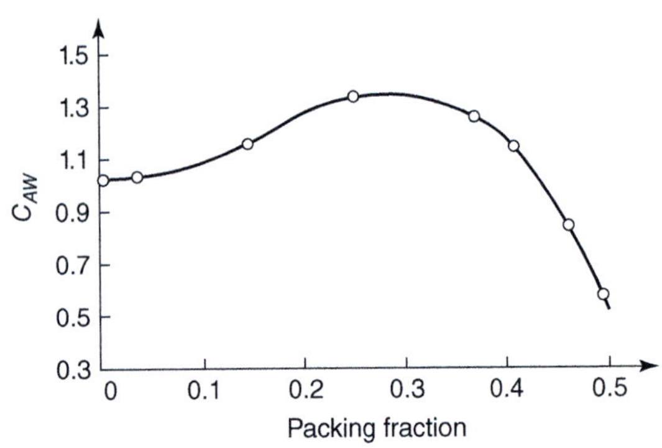

Figure 4.37 The Adler–Wainwright correction, C_{AW}, as a function of packing fraction, PF. Reproduced, by permission, from Iida and Guthrie [43]. Copyright © 1988 by Oxford University Press.

Table 4.14 Comparison of Experimental Melting Point Self-Diffusivities (in Units of 10^{-9} m^2/s) with Correlations for Selected Liquid Metals

Metal	*D*, Experimental	*D*, Eq. (4.68)	*D*, Eq. (4.69)
Li	5.76–6.80	7.01	6.72
Na	3.85–4.23	4.24	4.10
K	3.59–3.76	3.85	3.71
Cu	3.97	3.40	3.18
Zn	2.03–2.06	2.55	2.45
Ga	1.60–1.71	1.73	1.64
Rb	2.22–2.68	2.68	2.58
Ag	2.55–2.56	2.77	2.68
Cd	1.78	2.00	1.94
In	1.68–1.69	1.77	1.72
Sn	2.05–2.31	1.96	1.86
Hg	0.93–1.07	1.07	0.93
Pb	1.68–2.19	1.67	1.60
Al	—	—	4.87
Ni	—	—	3.90
Mg	—	—	5.63
Sb	—	—	2.66
Cs	—	—	2.31
Fe	—	—	4.16

Source: Iida and Guthrie [1]. Copyright © 1988 by Oxford University Press.

$$D = 3.5 \times 10^{-6} \left(\frac{T_m}{M}\right)^{1/2} V_m^{1/3} \tag{4.69}$$

where M is the molecular weight and V_m is the atomic volume at the melting point. The self-diffusivity values predicted by Eq. (4.69) for several liquid metals at their melting point are compared in Table 4.14 with values from the semiempirical hard sphere model of Eq. (4.68) and experimental values.

Notice the similarity in form of Eq. (4.66) with Eq. (4.7), which used the packing fraction to estimate the viscosity. In this case, however, the dependences are the inverse of packing fraction and molecular weight to those in Eq. (4.7). In other words, whatever factors that change viscosity will have the opposite effect on diffusion. As viscosity goes up, diffusivity goes down; the inverse is true as well. This relationship is stated explicitly in the *Stokes–Einstein equation*

$$D = \frac{k_B T}{4\pi \mu r} \tag{4.70}$$

where k_B is Boltzmann's constant, T is absolute temperature, μ is the viscosity of the "solvent" (in this case, the molten metal), and r is the radius of the diffusing hard sphere (in this case, a molten metal atom). Equation (4.70) is the Stokes–Einstein equation for self-diffusion. Another form of the

Stokes–Einstein equation can be developed for dilute solutions of large spheres in small, spherical molecules, which leads to a binary diffusion coefficient:

$$D_{AB} = \frac{k_B T}{6\pi \mu r} \quad (4.71)$$

Equation (4.71) is a starting point in the determination of diffusivities in liquid metal alloys, but in most real systems, experimental values are difficult to obtain to confirm theoretical expressions, and pair potentials and molecular interactions that exist in liquid alloys are not sufficiently quantified. Even semiempirical approaches do not fare well when applied to liquid alloy systems. There have been some attempts to correlate diffusivities with thermodynamic quantities such as partial molar enthalpy and free energy of solution, but their application has been limited to only a few systems.

Mass Diffusivity in Solids

Fick's Law [Eq. (4.4)] implies that a concentration gradient, dc/dy, is necessary for diffusion and molar flow to occur. In solid metals such as pure nickel, no concentration gradient exists, yet radioactive tracer experiments tell us that diffusion still takes place. As illustrated in Figure 4.38, a thin layer of radioactive nickel placed at one end of a pure nickel bar will slowly diffuse through the pure metal, a process that can be monitored easily by radiation detection. This tells us that even though no concentration gradient exists (the chemical potentials of radioisotopes are identical, hence there is no difference in concentration across the interface), there is still atomic movement. We know that above absolute zero, atomic vibrations occur in solids. It is these small amplitude vibrations that allow some atoms to overcome an activation energy barrier and move from one position to another in the lattice. This type of random walk process allows atomic interchange to take place—neighboring atoms simply switch places due to thermal vibrations.

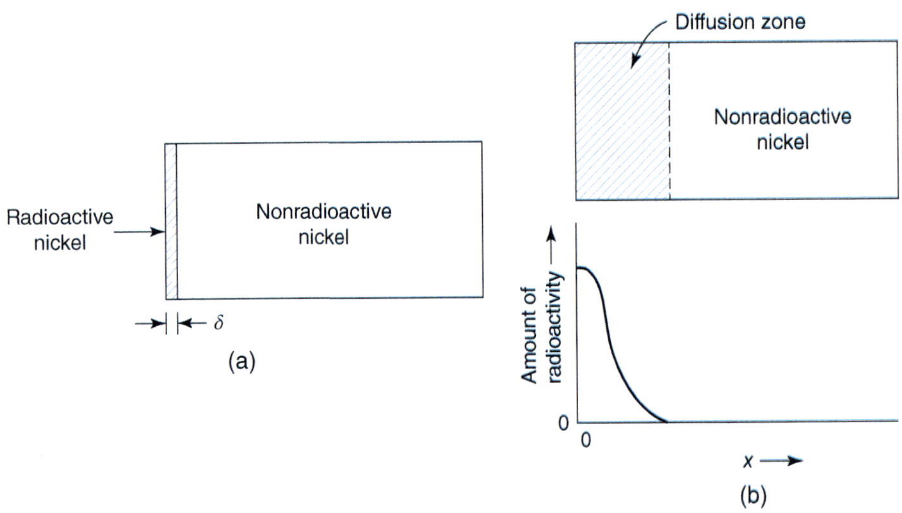

Figure 4.38 Schematic illustration of a self-diffusion experiment in which (a) a thin layer of radioactive nickel is deposited on one surface of a nonradioactive nickel specimen. After heating and time (b), the radioactive nickel has diffused into the sample, as monitored with a radioactive detector. From Ralls et al. [31]. Copyright © 1976 by John Wiley & Sons, Inc. This material is used by permission John Wiley & Sons, Inc.

Experiments such as the one illustrated in Figure 4.38 not only give us self-diffusion coefficients for certain substances but as the temperature of the experiment is varied, they give us the temperature dependence of the process and a measurement of the activation energy barrier to diffusion. Diffusion in solid systems, then, can be modeled as an activated process; that is, an Arrhenius-type relationship can be written in which activation energy, E_a, and temperature dependence are incorporated, along with a preexponential factor, D_0, sometimes called the *frequency factor*:

$$D = D_0 \exp(-E_a/RT) \tag{4.72}$$

Since this is an empirical expression, it can be applied to almost any diffusion process, but it does not tell us much about the diffusion mechanism. In general, three types of processes have been identified for diffusion in solids: an *interchange mechanism* (sometimes called an *exchange mechanism*), a *vacancy mechanism*, and an *interstitial mechanism*. These mechanisms are illustrated in Figure 4.39. A fourth mechanism, *ring interchange*, is possible, but rarely observed, and will not be considered here. In most metals, the vacancy mechanism (Figure 4.39a) dominates, because it requires a relatively small amount of energy, and there are sufficient vacancies in most metals. The interstitial mechanism (Figure 4.39b) is possible in metals, since a self-interstitial is a relatively mobile species requiring shorter distances of atomic movement, but the equilibrium number of self-interstitials is relatively small at almost all temperatures. Finally, the direct interchange mechanism (Figure 4.39c) is possible in metals but requires extremely high activation energy.

Activation energies for self-diffusion are expected to be in the range of 232–261 kJ/mol for the vacancy mechanism and 492–615 kJ/mol for the interstitial mechanism. Observed values are typically in the range of 164–203 kJ/mol. Some representative values for self-diffusion activation energies, as well as frequency factor in Eq. (4.72), D_0, are given in Table 4.15. Bear in mind that D_0

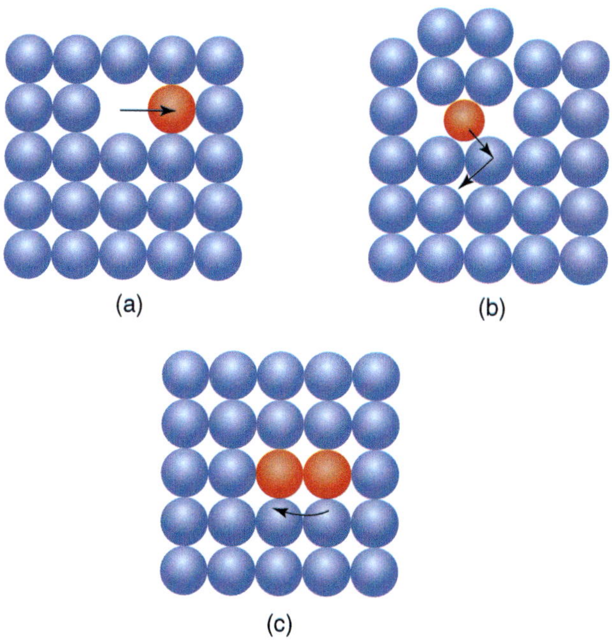

Figure 4.39 Illustration of (a) vacancy, (b) interstitial, and (c) interchange (exchange) mechanisms in atomic diffusion. From Ralls et al. [31]. Copyright © 1976 by John Wiley & Sons, Inc. This material is used by permission John Wiley & Sons, Inc.

Table 4.15 Frequency Factors, Activation Energies, and Melting Points for Self-Diffusion in Selected Metals

Species	D_0 (cm^2/s)	E_a (KJ/mol)	T_m (°C)
Pb	6.6	117	327
Zn (|| to *c* axis)	0.046	85	419
Zn (⊥ to *c* axis)	91	130	419
Ag	0.89	192	962
Au	0.16	222	1064
Cu	11	239	1085
Ni	—	293	1455
Co	0.31	280	1494
α-Fe	2300	306	1538
γ-Fe	5.8	310	—
Pt	0.048	233	1772

Source: Chalmers [44]. Copyright © 1959 by John Wiley & Sons, Inc.

contains the temperature-independent portions of diffusion, such as entropic effects, whereas E_a reflects all activation energy processes, including both atomic mobility and vacancy formation, in the case of diffusion via the vacancy mechanism. Notice in Table 4.15 that there is a general trend of increasing activation energy for diffusion with increasing melting point. Recall from Chapter 1 that intermolecular forces have a profound influence on a number of physical properties, such as melting point. The more strongly atoms are attracted to one another, the more energy it requires to pull them apart. This is reflected in both the melting point and the activation energy for diffusion, in which atoms must overcome attractive forces in order to move away from one another.

In alloys, attractive forces still play an important role, but the situation is slightly different insofar as solute atoms are sufficiently small relative to the host atom to allow the interstitial mechanism to dominate (see Figure 4.40). As a result, hydrogen, carbon, nitrogen, and oxygen diffuse interstitially

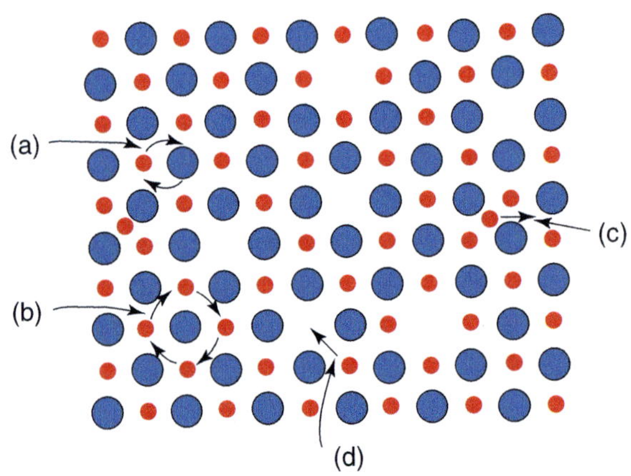

Figure 4.40 Illustration of diffusion mechanisms in alloys and ionic solids: (a) interchange (exchange); (b) ring rotation (rare); (c) interstitial migration; and (d) vacancy migration. From Kingery et al. [34]. Copyright © 1976 by John Wiley & Sons, Inc. This material is used by permission of John Wiley & Sons, Inc.

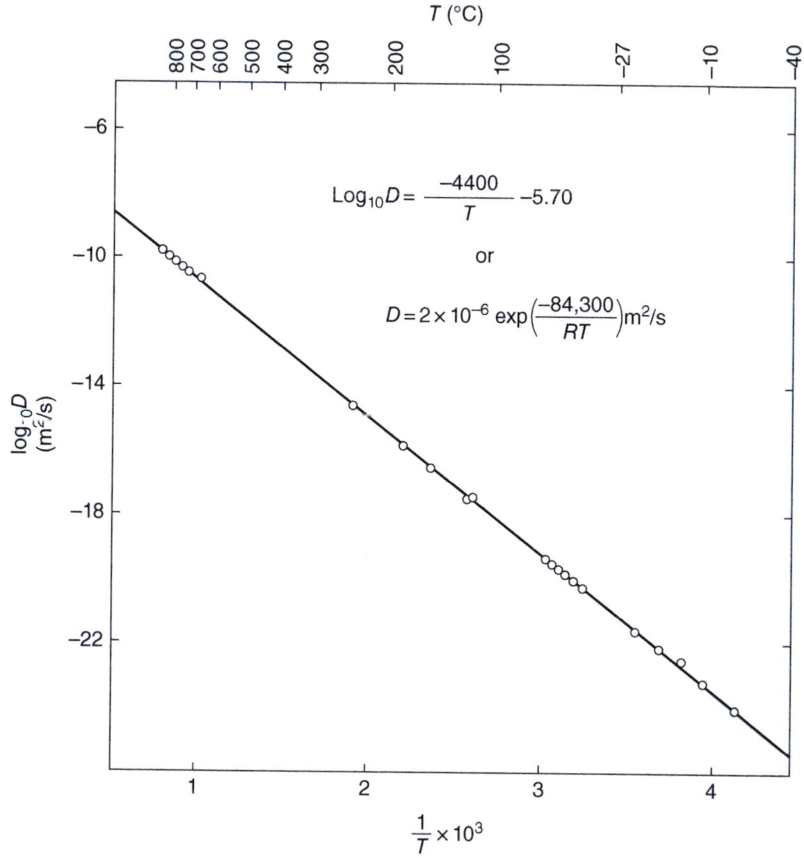

Figure 4.41 Variation with temperature of the diffusivity for carbon in BCC-Fe ($\alpha-$Fe). Activation energy is in units of J/mol. Reproduced, by permission, from Gaskell [45]. Copyright © 1992 by Macmillan Publishing Co.

in most metals with relatively low activation energy barriers. The diffusion of carbon in iron has been particularly well-studied in conjunction with the phase transformations in the C–Fe system (cf. Figure 2.8). An example of the excellent fit provided by Eq. (4.72) to experimental data in the C–Fe system is shown in Figure 4.41. Notice how the activation energy for this interstitial diffusion process is about 85 kJ/mol, which is much smaller than for vacancy-driven diffusion, due to the small size of the carbon atoms relative to iron. Not only do larger atoms, such as other metals, have difficulty diffusing interstitially, but also atomic interactions between dissimilar atoms greatly affect the activation energy for diffusion.

In addition to concentration and random walk diffusion, which are termed *ordinary diffusion*, there are other types of diffusion that can be important under certain circumstances. Atomic mobility can be brought about by thermal and pressure gradients. There are also such structurally driven diffusion processes as *dislocation diffusion*, *surface diffusion*, and *grain-boundary diffusion*. The latter is particularly important in polycrystalline materials where grain boundaries predominate. The grain boundaries provide a path of lower activation energy for movement than through the bulk solid. Moreover, grain boundary diffusion is less temperature-variant than bulk, or volume, diffusion in these materials, as illustrated for single crystalline and polycrystalline silver in Figure 4.42.

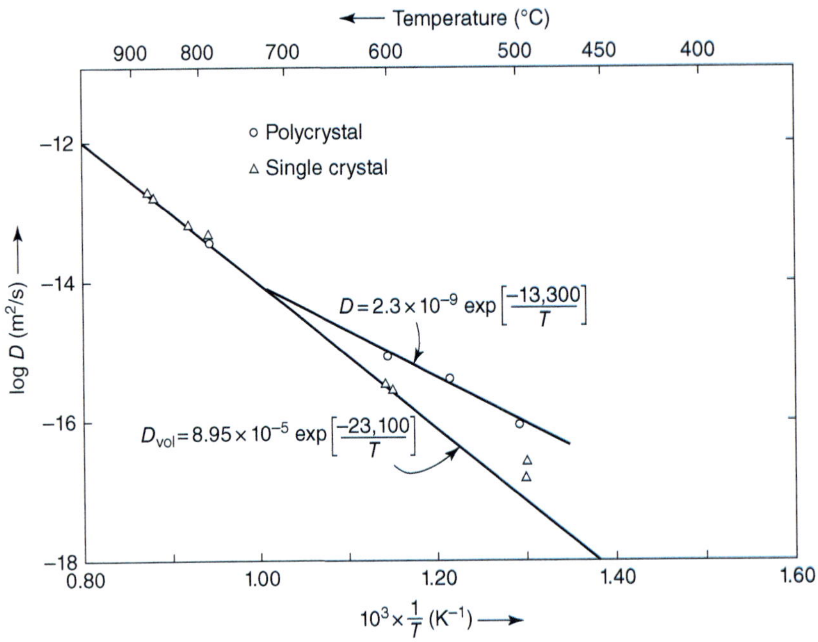

Figure 4.42 Self-diffusion coefficient in single and polycrystalline silver, illustrating the effect of grain boundary diffusion, especially at lower temperatures. From Ralls et al. [31]. Copyright © 1976 by John Wiley & Sons, Inc. This material is used by permission John Wiley & Sons, Inc.

Cooperative Learning Exercise 4.10

Consider the diffusivity data presented in Figure 4.42 for the self-diffusion of silver in polycrystalline and single-crystalline form. Use the correlations provided to calculate the following quantities.

Person 1: Calculate the self-diffusivity in polycrystalline silver (grain boundary diffusion), D_{GB}, at 500°C in m^2/s. What is the activation energy for this process in kJ/mol?

Person 2: Calculate the self-diffusivity in single-crystalline silver (volume diffusion), D_{vol}, at 500°C in m^2/s. What is the activation energy for this process in kJ/mol?

Combine your answer to get the ratio of grain boundary diffusion to volume diffusion for silver at 500°C. How would this change at higher temperatures? How would this change as the grain size decreased further?

Answers: $D_{GB}(500\,°\mathrm{C}) = 2.3 \times 10^{-9}\exp(-13,300/773) = 7.8 \times 10^{-17}\,\mathrm{m^2/s}$
$E_a = (13,300\ \mathrm{K})(8.314\ \mathrm{J/mol\cdot K}) = 111\ \mathrm{kJ/mol}$
$D_{vol}(500\,°\mathrm{C}) = 8.95 \times 10^{-5}\exp(-23,100/773) = 9.4 \times 10^{-18}\,\mathrm{m^2/s}$
$E_a = (23,100\ \mathrm{K})(8.314\ \mathrm{J/mol\cdot K}) = 192\ \mathrm{kJ/mol}$
$D_{GB}/D_{vol} = 7.8/0.94 = 8.3$. We would expect this ratio to decrease at higher temperature, eventually approaching 1.0 as volume diffusivity increases due to its stronger temperature dependence. This ratio should get larger with decreasing grain size as there are more grain boundaries per unit volume.

4.3.2 Mass Transport Properties of Ceramics and Glasses

As with thermal conductivity, we see in this section that disorder can greatly affect the mechanism of diffusion and the magnitude of diffusivities, so that crystalline ceramics and oxide glasses will be treated separately. Finally, we will briefly describe an important topic relevant to all material classes, but especially appropriate for ceramics such as catalyst supports—namely, diffusion in porous solids.

Diffusion in Crystalline Oxides

We wish to investigate the individual terms in Eq. (4.72) (i.e., D_0 and E_a), in more detail as they apply to crystalline oxide ceramics. To do so, we look first at a model ionic system, potassium chloride, KCl, realizing that the results will be general to all ceramics that have close-packed anion lattices and in which diffusion occurs by a vacancy mechanism.

Diffusion of K^+ ions in KCl occurs by interchange of the potassium ions with cation vacancies (see Figure 4.40d). It makes sense, then, that the diffusivity of potassium ions in KCl, $D_{K,Cl}$ is both a function of the potassium ion mobility, and the cation vacancy concentration, $[V'_K]$:

$$D_{K,Cl} = D_i [V'_K] \tag{4.73}$$

Let us first concentrate on the potassium ion mobility, D_i, in Eq. (4.73). As with any activated process, the potassium ions must overcome an activation energy barrier, which we will notate as a free energy $\Delta G^{\dagger}$, to move to the cation vacancy site, as illustrated in Figure 4.43. Recall from

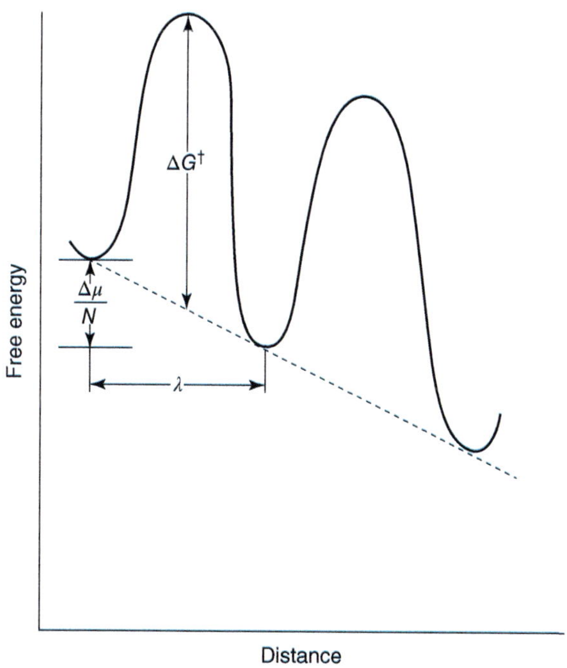

Figure 4.43 Diffusion in a potential gradient $\Delta\mu$, where $\Delta G^{\dagger}$ is the height of the activation energy barrier and λ is the jump distance. From Kingery et al. [34]. Copyright © 1976 by John Wiley & Sons, Inc. This material is used by permission of John Wiley & Sons, Inc.

Section 3.0.2 that the fraction of ions with the required energy to overcome the barrier is given by the Boltzmann distribution and depends on exp.$(-\Delta G^{\dagger}/k_{\mathrm{B}}T)$. It can be shown, then, that

$$D_i = \gamma\lambda^2 v \exp\left(-\Delta G^{\dagger}/k_{\mathrm{B}}T\right) \tag{4.74}$$

where γ is a nearest-neighbor factor on the order of unity, λ is the jump distance, and v is the frequency factor, which for solids has a value of about $10^{13}\ \mathrm{s}^{-1}$.

Turning now to the vacancy concentration in Eq. (4.73), recall from Chapter 1, and Eq. (1.36) in particular, that the concentration of vacancies is related to the free energy of formation of the vacancy, so that for a cation–anion vacancy pair, or Schottky defect, the concentration of vacancies, $[V'_K]$ is given by

$$n_V/N = [V'_k] = \exp(-\Delta G_s/2k_{\mathrm{B}}T) \tag{4.75}$$

where n_V is the number of defects (vacancies), N is the total number of lattice sights, T is the absolute temperature, k_{B} is Boltzmann's constant, and ΔG_s is the free energy change necessary to form a Schottky vacancy pair. We know that free energy can be broken down into entropic and enthalpic components [Eq. (2.12)], such that

$$[V'_k] = \exp(-\Delta G_s/2k_{\mathrm{B}}T) = \exp\left(\frac{-\Delta S_s T + \Delta H_s}{2k_{\mathrm{B}}T}\right) \tag{4.76}$$

If we perform a similar breakdown of $\Delta G^{\dagger}$ in Eq. (4.74) and substitute Eqs. (4.74) and (4.76) into (4.73), we obtain the following relationship for the diffusivity of potassium ions in KCl:

$$D_{\mathrm{K,Cl}} = \gamma\lambda^2 v \exp\left(\frac{\Delta S^{\dagger} + \Delta S_s/2}{k_{\mathrm{B}}}\right)\exp\left(\frac{-\Delta H^{\dagger} - \Delta H_s/2}{k_{\mathrm{B}}T}\right) \tag{4.77}$$

The important parts of Eq. (4.77) are the exponential terms. The first exponential—which contains the entropies associated with potassium ion movement and vacancy formation, respectively—comprises the temperature-independent contributions to D_0, as discussed in the previous section. The second exponential—which contains the enthalpies of the two processes and the temperature dependence—represents the activation energy, E_a, and temperature dependence of Eq. (4.72).

To complete our discussion on diffusion in KCl, then, Eq. (4.77) tells us that a plot of $D_{\mathrm{K,Cl}}$ versus $1/T$ should result in a straight line with slope of $(\Delta H^{\dagger}/k_{\mathrm{B}} + \Delta H_s/2k_{\mathrm{B}})$. As shown in Figure 4.44, this is the case only at high temperatures, where the *intrinsic properties* of the material dominate. In the lower-temperature region, impurities within the crystal fix the vacancy concentration. This is called the *extrinsic region*, where the diffusion coefficient is a function of divalent cation impurity concentration, such as calcium. The insert in Figure 4.44 shows that the cation vacancy concentration becomes fixed at low temperatures, and this is reflected in the fact that the slope of the diffusivity curve becomes $\Delta H^{\dagger}/k_{\mathrm{B}}$ only. The change in slope occurs at the point where the intrinsic defect concentration (attributable only to temperature) is comparable with the extrinsic defect concentration (attributable to impurities). When the Schottky formation enthalpy is in the range of 628 kJ/mol, as is typical for BeO, MgO, CaO, and Al_2O_3, the crystal must have an impurity concentration smaller than 10^{-5} before intrinsic diffusion can be observed at 2000°C. It is thus unlikely that intrinsic diffusion will be observed in these oxides because impurity levels of only parts per million are sufficient to control the vacancy concentrations.

The experimental diffusion coefficients for most oxide ceramics do not generally show this intrinsic/extrinsic behavior illustrated in KCl. This is probably because diffusion data are normally

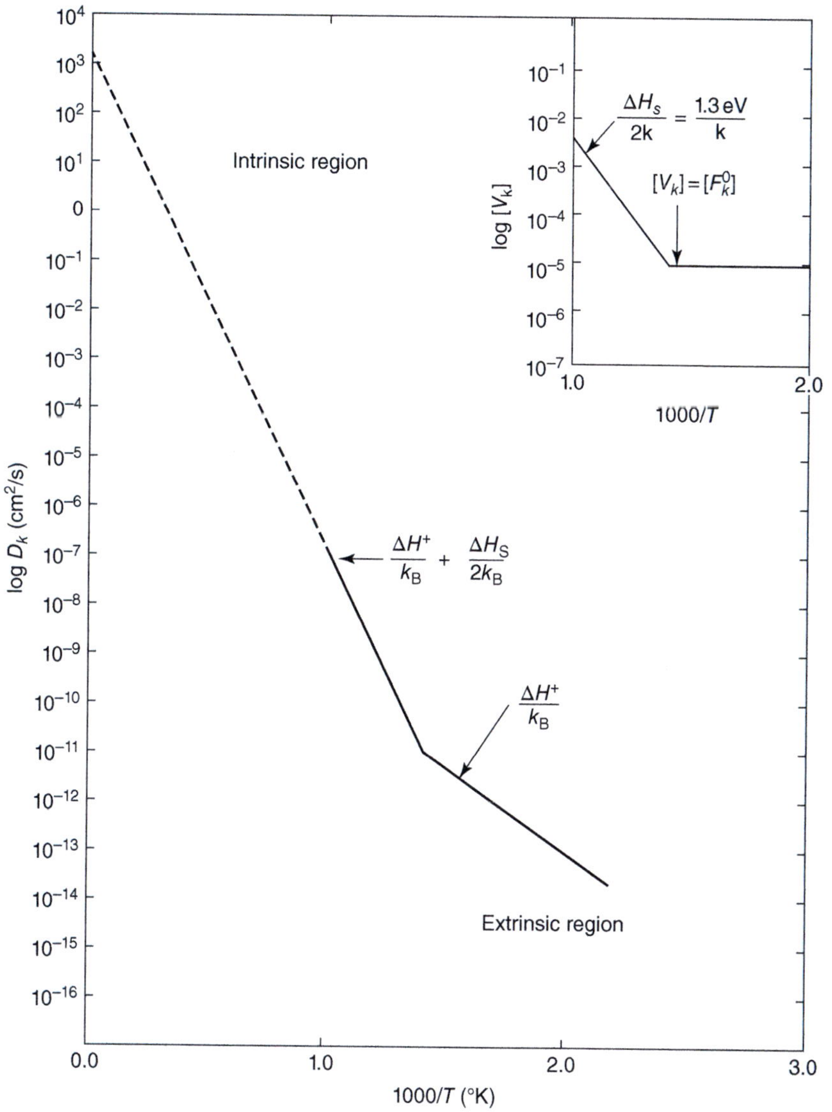

Figure 4.44 Diffusivity as a function of temperature for KCl with 10^{-5} atom fraction divalent cation impurities. Insert plot shows the variation of vacancies with temperature. Adapted from Kingery et al. [34]. Copyright © 1976 by John Wiley & Sons, Inc. This material is used by permission of John Wiley & Sons, Inc.

collected over a relatively small temperature range, as illustrated for a number of oxides in Figure 4.45. However, some diffusion in oxides is clearly extrinsic, such as for the diffusivity of oxygen in oxides containing the fluorite structure, such as UO_2, ThO_2, and ZrO_2. In these compounds, the addition of divalent or trivalent cation oxides, such as CaO, has been shown to fix the concentration of oxygen vacancies and make their concentration independent of temperature. Consequently, the oxygen ion diffusivity has a temperature dependence related only to the activation energy associated with oxygen ion mobility (121 kJ/mol). Cation diffusion coefficients, on the other hand, such as Mg diffusion in MgO, are more difficult to categorize as wholly intrinsic or extrinsic. The relatively high activation energies in these systems suggest that high-temperature measurements may be intrinsic and independent of minor impurities, but it is difficult to generalize.

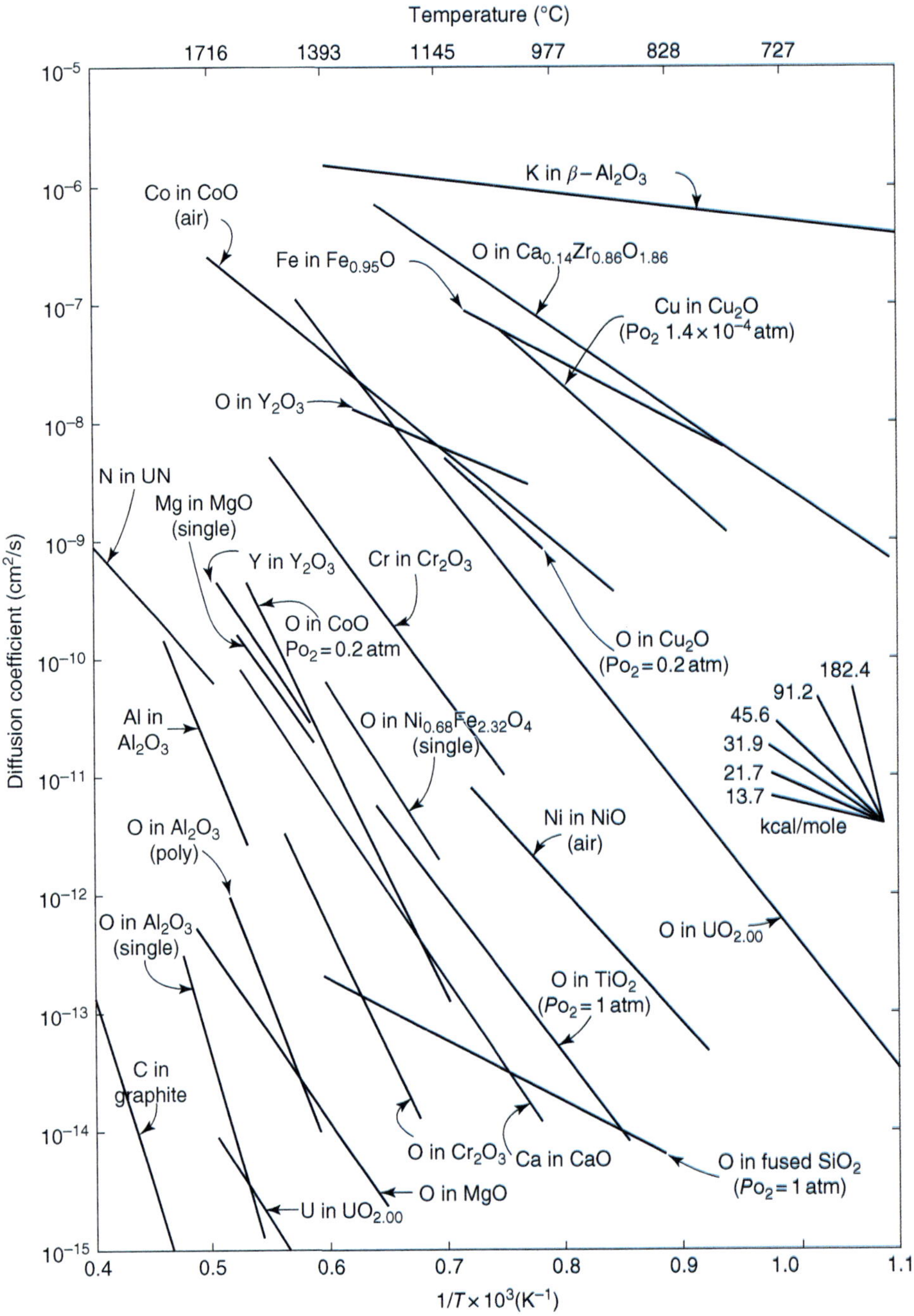

Figure 4.45 Diffusivities in some common ceramics as a function of temperature. The activation energy can be estimated from the slope of the insert. From Kingery et al. [34]. Copyright © 1976 by John Wiley & Sons, Inc. This material is used by permission of John Wiley & Sons, Inc.

Intrinsic diffusion behavior is dominant in nonstoichiometric oxides such as zinc oxide or cobaltous oxide in equilibrium with oxidizing or reducing atmospheres. In these systems, the interstitial and vacancy concentrations are a function of the partial pressure in the surrounding environment. The dependence on partial pressure is related to whether the nonstoichiometric oxide is metal-deficient or oxide-deficient. In metal-deficient oxides, such as FeO, NiO, and MnO, the concentration of vacant cation sites can be very large. For example, $Fe_{1-x}O$ contains 5–15% vacant iron sites. In this case, the solubility of oxygen in the metal oxide becomes an issue, and the metal ion diffusivity, D_M, can be a function of oxygen partial pressure, as illustrated in Figure 4.46a. The free energy of solution for oxygen, ΔG_O, must then be considered, which in turn leads to a functional dependence of D_M and vacancy concentration on the enthalpy of solution for oxygen, ΔH_O, as shown in Figures 4.46b,c, whereas the metal ion mobility, B_M, remains a function of the activation energy barrier only, as illustrated in Figure 4.46d.

A similar situation arises in oxygen-deficient oxides, such as CdO or Nb_2O_5, except that now three temperature regimes are possible, as shown in Figure 4.47. In the low-temperature regime, oxygen vacancy concentration is controlled by impurities, and extrinsic behavior is observed. At intermediate temperatures, the oxygen vacancy concentration changes due to a change in oxygen solubility with temperature. This is nonstoichiometric behavior. At high temperatures, thermal vacancies become dominant, and intrinsic behavior dominates.

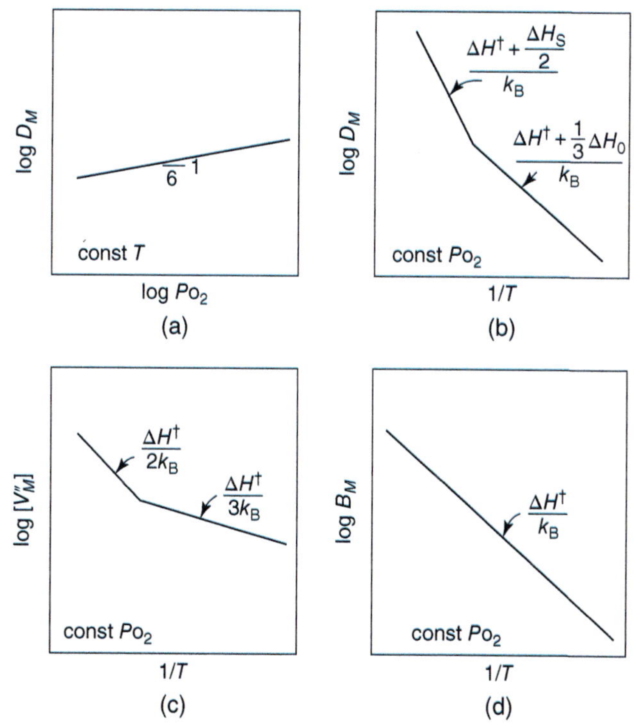

Figure 4.46 Schematic representation of the variation of metal ion diffusivity as a function of (a) oxygen partial pressure and (b) temperature, and (c) the metal ion vacancy concentration and (d) mobility with temperature. From Kingery et al. [34]. Copyright © 1976 by John Wiley & Sons, Inc. This material is used by permission of John Wiley & Sons, Inc.

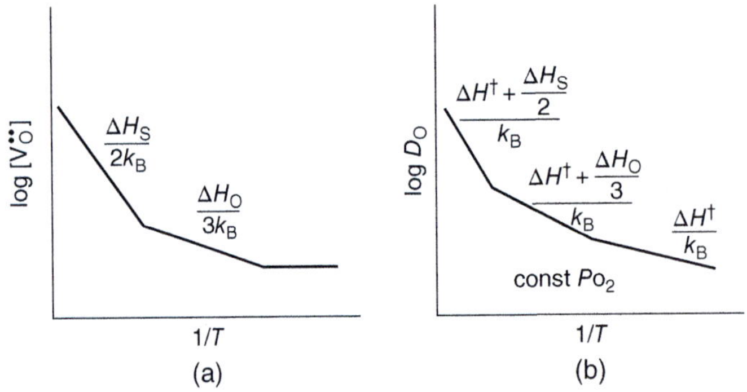

Figure 4.47 Schematic representation of temperature dependence of (a) oxygen vacancy and (b) oxygen diffusivity in oxygen-deficient oxide ceramics. From Kingery et al. [34]. Copyright © 1976 by John Wiley & Sons, Inc. This material is used by permission of John Wiley & Sons, Inc.

Chemical effects also occur in crystalline oxides; that is, impurity atoms diffuse at varying rates through oxides. In all cases, cation diffusion is much faster than oxygen diffusion, but similar cations, such as Ca and Mg, can behave very differently in different oxides. This is due to differences in chemical potential and activity, and mobility differences.

As in metals, dislocation, surface, and grain boundary diffusion can be important. Grain boundary diffusion, in particular, can lead to a diffusion coefficients that are 10^3 to 10^6 times greater than the bulk diffusion coefficient. However, the effect is not the same for both cations and anions. In some systems, only the ion that is expected to be present in excess at the grain boundary will exhibit enhanced grain boundary diffusion.

Diffusion in Glasses–Solubility and Permeability

The principles developed in previous sections apply to noncrystalline solids as well, but the terminology used is often different, owing to the applications for which diffusion in oxide glasses is important. There are two such situations that will be described here: (a) diffusion of gases through glass and (b) diffusion of cations through glass. The former is important in many applications, such as high-vacuum use, where glass is a common material of construction. The latter is important in glass containers, where cations in the glass can leach out over time depending on the solution it contains. Yet another application, the diffusion of gases through molten glasses, is important from an industrial standpoint in the fining of glasses but is beyond the scope of this text.

In both gas and cation diffusion, the movement of atoms through solid glasses is often described in terms of the *permeability*, P_M, which is directly proportional to the diffusivity, D:

$$P_M = SD \tag{4.78}$$

where S is the *solubility*. In this case, the permeability is defined as the volume of gas at standard temperature and pressure (STP) passing per second through a unit area of glass of thickness 1 cm, with a 1 atm pressure difference across the glass. The solubility is defined as the volume of gas at STP dissolved in a unit volume of glass per atmosphere of external gas pressure. The diffusivity is defined as before. It should be clear from Eq. (4.78) that determination of two of the parameters fixes the third. Solubility increases with temperature according to an Arrhenius-type relationship, where the heat of solution serves as the activation energy. Since diffusivity also follows an Arrhenius expression, as described in the previous sections, one would expect that the permeability would also show a similar temperature dependence, and this is indeed the case. An example of

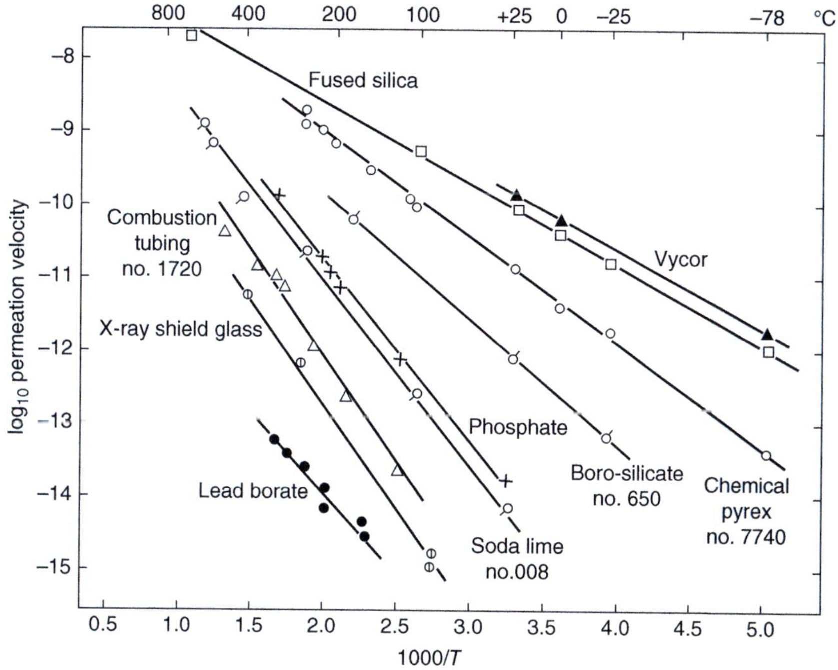

Figure 4.48 Permeability of helium through various glasses. From Kingery et al. [34]. Copyright © 1976 by John Wiley & Sons, Inc. This material is used by permission of John Wiley & Sons, Inc.

the logarithmic permeability dependence on inverse temperature is shown in Figure 4.48. The differences in permeabilities among the various glasses can be explained by considering the network-modifying cations as "hole blockers" in the glass network. Thus, the permeability would be expected to increase with an increase in the concentration of network-forming constituents such as SiO_2, B_2O_3, and P_2O_5 (cf. Section 1.2.4). Such an increase has been shown experimentally. As for the diffusivity of gases in glasses, the activation energy increases with increasing size of the gas molecule, as illustrated in Figure 4.49. For these gases, the frequency factor, D_0, is in the range of 10^{-4} to 10^{-3} cm^2/s.

In the case of oxygen transport, diffusion may take place either as molecular or network diffusion, although molecular diffusion apparently takes place some seven orders of magnitude faster than network diffusion. The movement of water through glasses can also be important, particularly in the dehydroxylation of hydrated glasses such as vitreous silica. The hydroxyl content has been shown to greatly affect the properties of glasses, including reductions in viscosity and glass transition temperatures, and increases in phase separation and crystallization rates with increasing hydroxyl content. Some data from a typical dehydroxylation experiment for vitreous silica are shown in Figure 4.50. Here, t is time and L is the sample thickness. The value of $t^{1/2}$ L at 50% hydroxyl removal is shown in the plot. This value is significant because the diffusivity can be estimated according to the following relationship:

$$D = \frac{0.049}{(t/L^2)_{50\%}} \tag{4.79}$$

where $(t/L^2)_{50\%}$ is the value of t/L^2 at removal of exactly 50% of the hydroxyl groups from the glass.

For the movement of cations in glasses, as the amount of modifying cations in silicate glasses is increased, the activation energy for their diffusive transport decreases, resulting in an increase in

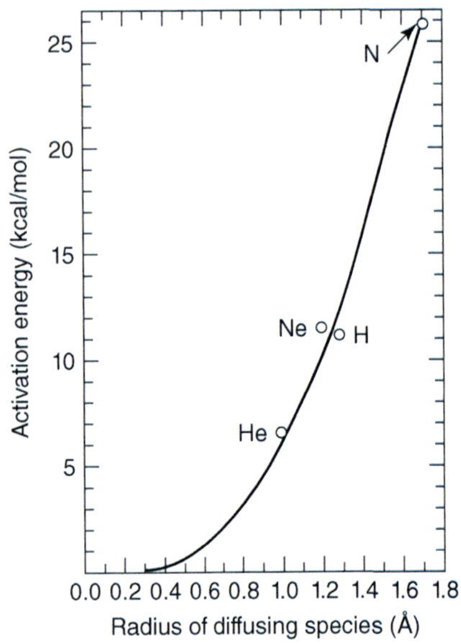

Figure 4.49 Activation energy for diffusion as a function of atomic radius of diffusing species in silica glass. From Kingery et al. [34]. Copyright © 1976 by John Wiley & Sons, Inc. This material is used by permission of John Wiley & Sons, Inc.

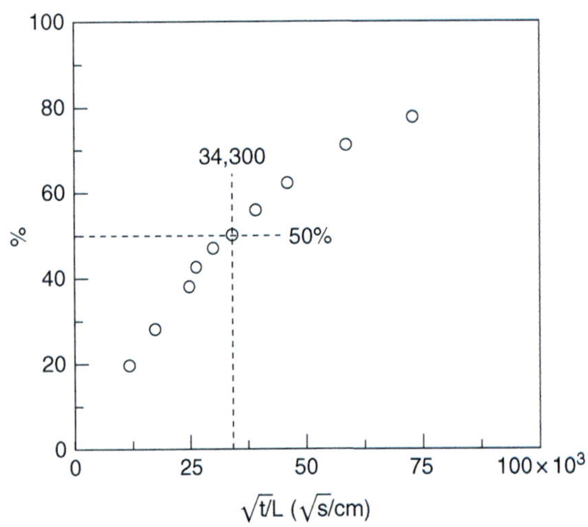

Figure 4.50 Experimental dehydroxylation curve for vitreous silica at constant temperature. Reproduced, by permission, from Shelby [46]. Copyright © 1993, The American Ceramic Society.

the diffusivity. The increase in cation concentration presumably leads to a breakup of the network and a decrease in the average interionic separation. Divalent cations diffuse much more slowly at a given temperature than monovalent cations, and their activation energies are much larger. The diffusivities of modifying cations in rapidly cooled glasses are generally higher than those in well-annealed glasses. These differences, often as large as an order of magnitude, reflect differences in specific volumes of the glasses, with the larger-volume, more open structure possessing the larger diffusivity.

The process of cation movement in glasses is important for a number of processes, such as ion exchange and dissolution. Many univalent cations in glasses can be substituted for one another, such as the complete substitution of Na ion by Ag ions, even at low temperatures. Ions such as Li^+ and K^+ can also be substituted for Na^+, although stresses can build up in the glass due to cation radius differences. This procedure can actually be utilized to strengthen glasses, as will be discussed in Chapter 5. In the dissolution of glasses, ions in solution attack the glass network through leaching of network modifiers, such as Na^+. This is particularly prevalent in acidic aqueous solutions, where the counter migration of H^+ ions into the network accelerates the leaching process, although an exact 1:1 exchange of Na^+:H^+ does not necessarily take place. Water is involved in the process as well. A theoretical description of this process, or even one in terms of diffusion theory, is difficult since the composition of the diffusing matrix is changing with time. A simple formula for determining the amount of leached-out alkalis, $[Q]$, is to assume a power-law dependence with time, $[Q] = kt^{\alpha}$, where α typically has values between 0.45 and 0.9. A square-root dependence on leaching time (i.e., $\alpha = 0.5$) is predicted by theory but is not necessarily experimentally observed. The properties and development of alkali- and hydroxyl-resistant glasses are important areas of investigation, and the interested reader is referred to the many fine texts on these subjects listed at the end of this chapter.

Diffusion in Porous Solids

To this point, we have primarily concentrated on the atomic-level movement of atoms involved in diffusion. There are macroscopic effects to consider as well. For example, liquids and gases moving through porous solids can often travel faster through the pores, which are also filled with either gas or liquid, than through the solid particles. If the pores are small relative to the mean free path of the diffusing species (Figure 4.51a), then the diffusion rate can be influenced strongly by collisions with the pore walls in a process known as *Knudsen diffusion*. Knudsen diffusion is rare for diffusing liquids, since their mean free paths are so small. It can be significant, however, for gases, especially at low pressures or high temperatures. Knudsen diffusion is characterized by the Knudsen number, $\text{Kn} = l/d$, where l is the mean free path of the diffusing species, and d is the pore diameter in the same units as l. For $\text{Kn} < 0.01$, Fickian diffusion occurs (Figure 4.51b). For $\text{Kn} > 100$, Knudsen diffusion dominates. In this case, a Knudsen mass diffusivity, D_{KA} (in m^2/s), is defined as follows:

$$D_{KA} = 48.5d\sqrt{T/M_A} \tag{4.80}$$

where d is again the pore diameter (in meters), T is the absolute temperature in degrees kelvin, and M_A is the molecular weight of the diffusing species, A, in g/mol. It is interesting to note that for a multicomponent gas, D_{KA} is completely independent of the diffusivity in the other gaseous species, since A collides with the walls of the pore and not with the other gases. Each gas will have its own Knudsen diffusion coefficient in this case. When $\text{Kn} \sim 1.0$, neither Fickian nor Knudsen diffusion dominates (Figure 4.51c), and *transition diffusion* occurs. The diffusivity in this case can be estimated in several ways, including a vectorial combination of the two diffusion coefficients.

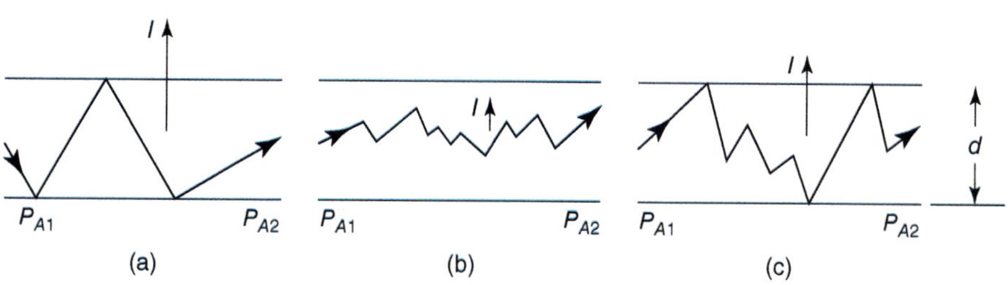

Figure 4.51 Diffusion in pores: (a) Knudsen diffusion, (b) Fickian diffusion, and (c) transitional diffusion.

In the specific instance of A diffusing through a catalytic pore and reacting at the end of the pore to form gaseous component B, equimolar counterdiffusion can be assumed, and an *effective transition region diffusivity*, D'_{NA}, is independent of concentration and can be calculated from the Knudsen and binary diffusion coefficients:

$$D'_{NA} = \frac{1}{1/D_{AB} + 1/D_{KA}} \tag{4.81}$$

This simplified diffusivity is sometimes used for diffusion in porous catalysts even when equimolar counterdiffusion is not occurring. This greatly simplifies the equations. When no reactions are occurring, the diffusivity is a function of concentration (in terms of mole fraction, x_A):

$$D_{NA} = \frac{1}{(1-\alpha x_A)/D_{AB} + 1/D_{KA}} \tag{4.82}$$

where α is the flux ratio factor, which for flow in an open system is related to the molecular weights of the two components:

$$\alpha = 1 - \sqrt{\frac{M_A}{M_B}} \tag{4.83}$$

Cooperative Learning Exercise 4.11

A gas mixture at a total pressure of 0.10 atm absolute and 298 K is composed of N_2 (component A) and He (component B). The mixture is diffusing through an open capillary having a diameter of 5×10^{-7} m. The mole fraction of N_2 at one end is $x_{A1} = 0.8$ and at the other end is $x_{A2} = 0.2$. The mean free paths for nitrogen and helium at 298 K are 654 and 1936 Å, respectively (1 Å = 10^{-10} m); and their collision diameters, σ, are 3.75 and 2.18 Å, respectively.

Person 1: Calculate the Knudsen number for N_2. Is the diffusion molecular, Knudsen, or transitional? Estimate the Knudsen diffusivity for nitrogen, D_{kA}, under these conditions.

Person 2: Estimate the hard-sphere binary diffusivity for nitrogen in helium, D_{AB}, under these conditions. You may assume that $\Omega_{AB} = 1.0$ in Eq. (4.67).

Combine your information to estimate the transitional diffusivity for nitrogen, D'_{NA}. You can use an average concentration for nitrogen in your calculation. How does it compare to the two diffusivities you calculated individually? How does it compare to the transitional diffusivity if there were equimolar counterdiffusion—that is reaction at the end of the pore?

Answers: Kn = 0.13 for N_2 $D_{KA} = 7.9 \times 10^{-5} m^2/s$ $D_{AB} = 5.8 \times 10^{-4} m^2/s$
$D_{NA} = 6.3 \times 10^{-5} m^2/s$ $D'_{NA} = 6.95 \times 10^{-5}\ m^2/s$; a difference of about 9%

When chemical reaction occurs, α must be defined in terms of molar fluxes. For the case of equimolar counterdiffusion, $\alpha = 1$, and Eq. (4.82) reduces to (4.81).

In real porous solids, the pores are not straight, and the pore radius can vary. Two parameters are used to describe the diffusion path through real porous solids: the *void fraction*, ε, defined as the ratio of pore area to total cross-sectional area, and the *tortuosity*, τ, which corrects for the fact that pores are not straight. The resulting effective diffusivity is then

$$D_{A,\text{eff}} = D_{AB}(\varepsilon/\tau) \tag{4.84}$$

Tortuosity values range from about 1.5 to over 10. A reasonable range of values for many commercial porous solids is about 2–6.

4.3.3 Mass Transport Properties of Polymers

Diffusion in Polymer Solutions

Take a Peek! Jump ahead to Section 5.1 to see an explanation of Hooke's Law and how a spring represents the recoverable elastic response of a material to applied stress.

A dilute polymer in a low-molecular-weight solvent can be modeled as a bead-spring chain (see Figure 4.52a). Each chain is a linear arrangement of N beads and $N - 1$ Hookean springs. The beads are characterized by a *friction coefficient*, represented by the lowercase Greek letter zeta, ζ, which describes the resistance to bead motion through the solvent. The model also allows for hydrodynamic interactions between the beads and the solvent molecules. The general result, which holds true for most solutions, relates the diffusivity to the friction coefficient:

$$D_{AB} = \frac{k_B T}{\zeta} \tag{4.85}$$

The form of Eq. (4.85) should look familiar. For large, spherical particles in a low-molecular-weight solvent, $\zeta = 6\pi \mu r$, where μ is the viscosity of the pure solvent and r is the large particle radius, and Eq. (4.85) becomes Eq. (4.71) a form of the Stokes–Einstein equation, which gives the binary diffusion coefficient, D_{AB},

$$D_{AB} = \frac{k_B T}{6\pi\mu r} \tag{4.71}$$

Eq. (4.70) for self-diffusion can be arrived at in a similar way through proper derivation of the friction coefficient. Equation (4.71) does not take into account hydrodynamic interactions. It is also necessary to come up with an equivalent radius, r, for a polymer chain, which can be difficult,

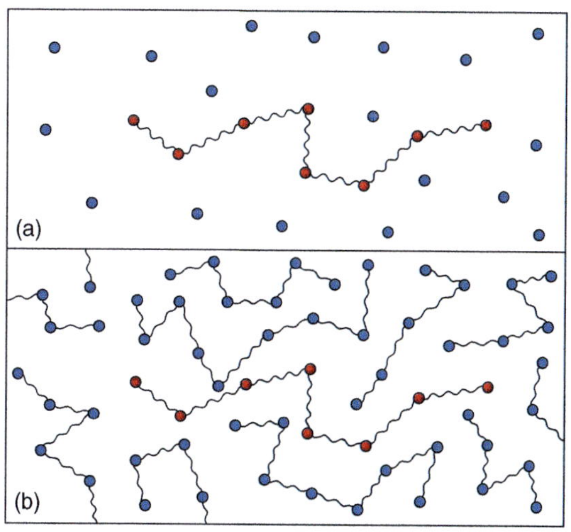

Figure 4.52 Single-molecule bead spring models for (a) dilute polymer solution and (b) polymer melt. From Bird et al. [47]. This material is used by permission of John Wiley & Sons, Inc.

especially when the conformation is such that the chain is extended and does not form a sphere at all. Nonetheless, a *radius of gyration*, r_g, is often used to characterize polymer chains in solution, and the resulting friction coefficient is $\zeta = 4\pi\,\mu r_g$.

More rigorous treatments include hydrodynamic interaction effects and relate the diffusivity directly to the molecular weight of the polymer chain. The general result is that the diffusivity of polymer chains in solution is related to the molecular weight of the polymer, M:

$$D_{AB} \propto M^{-\nu} \tag{4.86}$$

The value of ν can vary. Theory predicts that the diffusivity of polymer chains in solution should be proportional to the inverse square root of the number of beads, $N^{-1/2}$, and since N is proportional to M, we have $D_{AB} \propto M^{-1/2}$. In "good solvents," or solvents in which interactions between the polymer chain and solvent are favorable, and the chain takes on an extended structure, we have $\nu = 3/5$. If hydrodynamic interactions are ignored, we expect $D_{AB} \propto 1/M$, and $\nu = 1.0$.

Diffusion in Polymer Melts

We expect the diffusion situation to be more complex in polymer melts, since chains become entangled and it is difficult for them to move past one another. As a result, the bead-spring model of Figure 4.52b is inadequate. We further expect that chain entanglements should increase with the molecular weight of the polymer chain. A model has been developed that predicts these molecular weight effects in polymer melts. The *reptation model* assumes that an isolated polymer chain is contained in a hypothetical tube, as shown schematically in Figure 4.53. Under the constraints imposed by the tube, two types of chain motion are possible. First, there is a conformational change of the chain taking place within the tube, along with a "reptation" motion that causes the chain to translate through the tube, much like a snake wriggling through the grass. Eventually, this motion will carry the chain completely out of the hypothetical tube along what is termed the "*primitive path*." The time-dependent evolution of the primitive path leads to chain disentanglement, and it is the basis for determination of the diffusion coefficient. The diffusivity in polymer melts is predicted by this model to vary inversely with the square of molecular weight—that is, $\nu = 2$ in Eq. (4.86). This dependence has been observed experimentally, as shown in Figure 4.54 for a series

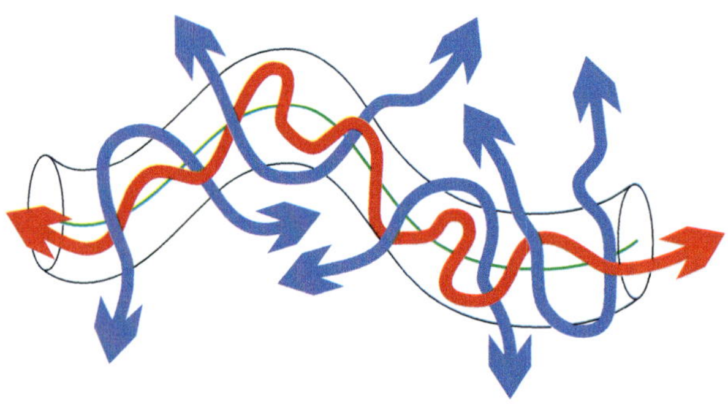

Figure 4.53 Schematic illustration of the reptation process in polymer melts, showing chain entanglements (blue arrows), the "wriggling motion" of the polymer chain (red arrow), and the primitive path of the polymer chain (green line). Reproduced, by permission, from Strobl [48]. Copyright © 1997 by Springer-Verlag.

of deuterated polyethylene (PE) chains in a PE melt. A non-entangled polymer melt is expected to follow the model of non-interacting spheres, and $\nu = 1.0$. There can be a transition from non-entangled to entangled melts, depending on time and conditions.

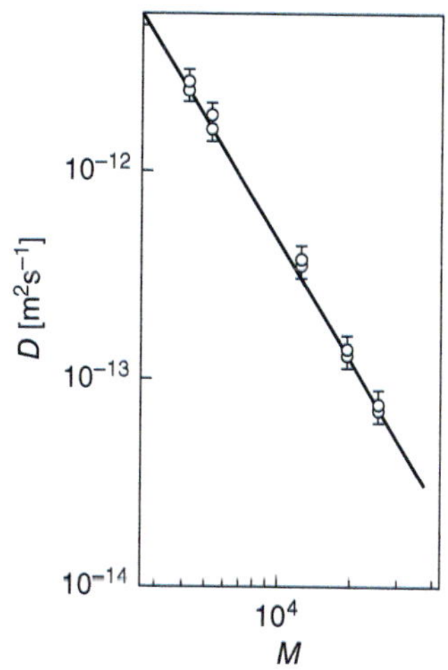

Figure 4.54 Diffusivity at 176°C for polyethylene of different molecular weights, M. The continuous line corresponds to $D_{AB} \propto M^{-2}$. Reproduced, by permission, from Strobl [49]. Copyright © 1997 by Springer-Verlag.

Permeability and Partition Coefficients

In the solid form, many polymers are used in packaging. The diffusion of gases through the packaging material can be important in a number of applications, especially where food and beverages are concerned. Polymeric films can also be used as *membranes*, such as in reverse osmosis applications, in fuel cells, or as replacements for biological materials in artificial kidneys, intestines, or other organs. Membranes are thin layers of materials used to control access of gases, liquids, or solids in solutions. We briefly describe here the transport of molecules through nonporous polymeric films, primarily for membrane applications. There are also porous membranes, which allow particles smaller than the pore diameter to pass through while retaining the larger particles, but the principles of diffusion in porous solids of the previous section are applicable there.

In nonporous membranes, diffusion occurs as it would in any other nonporous solid. However, the molecular species must first dissolve into the membrane material. This step can oftentimes be slower than the diffusion, such that it is the rate-limiting step in the process. As a result, membranes are not characterized solely in terms of diffusion coefficients, but in terms of how effective they are in promoting or limiting both solubilization and diffusion of certain molecular species or solutes. When the solute dissolves in the membrane material, there is usually a concentration discontinuity at the interface between the membrane and the surrounding medium (see Figure 4.55). The equilibrium ratio of the solute concentration in one medium, c_1, to the solute concentration in the surrounding medium, c_2, is called the *partition coefficient*, K_{12}, and can be expressed in terms of either side of the membrane. For the water–membrane–water example illustrated in Figure 4.55,

$$K_{12} = \frac{c_{m1}}{c_{w1}} = \frac{c_{m2}}{c_{w2}} \tag{4.87}$$

The partition coefficient is thus dimensionless. Values for partition coefficients vary greatly, depending on the solute and the membrane material.

The concept of permeability, P_M, described first in Section 4.3.2 also applies to membranes. Equation (4.78) relates the permeability to the diffusion coefficient and solubility. Some representative values of permeabilities for common gases in common polymer films are given in Table 4.16. The units of permeability in Table 4.16 are obtained when diffusivity is in units of m^2/s, and gas solubility is in units of m^3 gas·m^2/(m^3 solid·N). Note that carbon dioxide permeabilities are generally 3–4 times higher than for oxygen, due primarily to the higher solubility of carbon dioxide in most hydrocarbon materials.

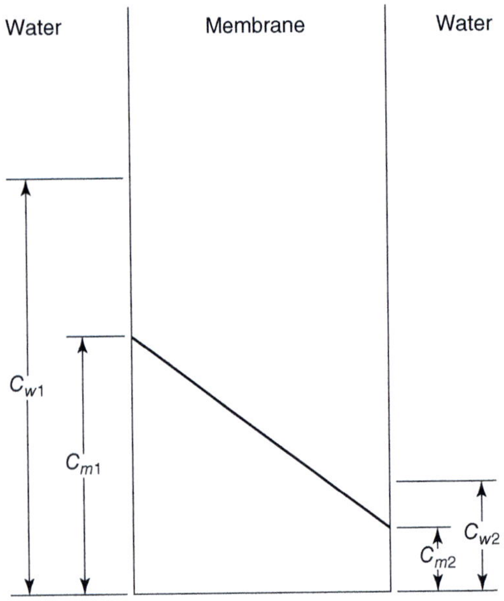

Figure 4.55 Schematic illustration of concentration drop at a membrane interface, and concentration gradient across the membrane. From Johnson [22]. Copyright © 1999 by John Wiley & Sons, Inc. This material is used by permission of John Wiley & Sons, Inc.

Table 4.16 Permeabilities of Selected Polymers

	Permeability $[10^{16}\ m^4/(s\ N)]$		
Film Type	**Oxygen**	**Carbon Dioxide**	**Water Vapor**
Low-density polyethylene	0.45–1.5	0.88–8.8	0.97–3.8
Linear low-density polyethylene	0.80–1.1	—	2.6–5.0
Medium-density polyethylene	0.30–0.95	0.88–4.4	1.3–2.4
High-density polyethylene	0.059–0.46	0.45–1.1	0.65–1.6
Polypropylene	0.15–0.73	0.88–2.4	0.65–1.7
Polyvinyl chloride	0.071–0.26	0.49–0.93	—
Polyvinyl chloride, plasticized	0.0088–0.086	0.088–6.3	>1.3
Polystyrene	0.023–0.088	1.1–3.0	18–25
Ethylene vinyl acetate copolymer (12%)	0.091–1.5	4.0–6.1	9.7
Ionomer	0.40–0.86	1.1–2.0	3.6–4.9
Rubber hydrochloride	0.015–0.15	0.059–0.59	>1.3
Polyvinylidene chloride	0.00091–0.0030	0.0067	0.24–0.81

Source: Johnson [22]. Copyright © 1999 By John Wiley & Sons, Inc.

To obtain the volume flow rate of gases through these films, a modified form of Fick's Law may be used

$$\dot{V} = \frac{P_M A \Delta P}{L} \tag{4.88}$$

Cooperative Learning Exercise 4.12

A 0.15-mm-thick film is required for use in packaging a pharmaceutical product at 30°C. The partial pressure of O_2 is 0.21 atm outside the package and is 0.01 atm inside the package. Use permeability data from Table 4.16 and assume that the resistance to diffusion outside and inside the membrane are negligible compared to the resistance of the film.

Person 1: Calculate the volume flow rate per unit area of oxygen through a linear low-density polyethylene film.

Person 2: Calculate the volume flow rate per unit area of oxygen through a low-density polystyrene film.

Compare your answers. Which is better for this type of application? Which do you think would be preferable under high relative humidity, assuming that water vapor could potentially damage the product?

Answers: (Average values of P_M used)
$\dot{V}/A = 1.3 \times 10^{-8}$ m^3/s·m^2 for polyethylene
$\dot{V}/A = 7.5 \times 10^{-10}$ m^3/s·m^2 for polystyrene
PS is better for this application (lower flow rate of oxygen across barrier), but its permeability for water is higher than PE.

where $\dot{V}$ is the volumetric flow rate of gas through the film in m^3/s, P_M is the permeability, A is the membrane cross-sectional area in m^2, L is the membrane thickness in meters, and ΔP is the partial pressure difference of the diffusing gas across the membrane in Pa.

The permeability is affected not only by temperature, but also by relative humidity. For example, cellophane is an excellent oxygen blocker when dry, but becomes poor when moist. The values for permeability in Table 4.16 are strictly for flat materials. Real permeabilities can vary greatly due to flexural cracks, overlapping, and delamination that are common in real materials. There have been attempts, however, to model the diffusion in strained materials [50].

4.3.4 Mass Transport Properties of Composites

Little is known about the mass transport properties of reinforced-composite materials. Certainly, there are no new relations or concepts that govern estimations of diffusivities that have not already been discussed. In most polymer–matrix composites, the transport properties of the polymer play an important role in diffusion through the composite. For example, hydrophilic polymers such as epoxy readily absorb water from the atmosphere. Thermoplastic polymers absorb relatively little moisture, since they are more hydrophobic but are more susceptible to uptake of organic solvents.

Structural effects can also be very important in diffusion through composites. For example, when pores are created either during formation or through de-bonding between phases, diffusion in porous solids as described in Section 4.3.2 can occur. More importantly, atomic mobility can occur along the matrix–reinforcement interphase and can lead to composite degradation. Not only is diffusion important in this case, but such transport processes as *wicking* and *capillary action* can occur, especially when liquids are involved. For this reason, *environmental barrier coatings* (EBCs) are often employed, which provide only an outer layer of protection, in this case a diffusion barrier, but contribute nothing from a structural standpoint. Recall from Chapter 1 that this is an example a complementary composite effect.

Environmental barrier coatings are a type of laminar composite. As with heat transfer, diffusion in laminar composites can be modeled as steady-state diffusion through a composite wall, as illustrated in Figure 4.56. Here, hydrogen gas is in contact with solid material A at pressure P_1 and in contact with solid B at pressure P_2. At steady state, the molar flux of hydrogen through both walls must be the same (i.e., $J^*_{H,Ax} = J^*_{H,Bx}$), and Fick's Law [Eq. (4.4)] in the x direction becomes

$$D_{HA}\frac{c_1 - c_1^*}{\Delta x_1} = D_{HB}\frac{c_2^* - c_2}{\Delta x_2} \quad (4.89)$$

The concentrations in the solid phases, c_1^* and c_s^*, are determined by the solubilities and diffusivities of hydrogen in A and B, and so they are not equal. The thermodynamic activity of hydrogen has a single value at the interface, however. (Refer to Section 3.0.1 for a description of thermodynamic activity.) Hence, the treatment of diffusion flux in a composite wall is simplified by considering activity gradients rather than concentration gradients. If the dissolution of hydrogen gas in the solid follows the reaction

$$\frac{1}{2}\mathrm{H_2(g)} \Leftrightarrow \mathrm{H\ (dissolved)} \quad (4.90)$$

then the Law of Mass Action [Eq. (3.4)] tells us that the equilibrium constant for the reaction, K, is given by

$$K = \frac{[H\ \mathrm{(dissolved)}]}{[\mathrm{H_2(g)}]^{1/2}} \quad (4.91)$$

Recall that the concentration (and activity) of molecular hydrogen in the gas phase can be represented by its partial pressure, $P_{\mathrm{H_2}}$, and the concentration of dissolved hydrogen is represented as c_H (either c_1 or c_2) in Figure 4.56, so that the equilibrium constant becomes

$$K = \frac{c_H}{P_{\mathrm{H_2}}^{1/2}} \quad (4.92)$$

There is an equilibrium constant for the dissolution of hydrogen in each solid phase, K_A and K_B, respectively. Similarly, the activity of hydrogen at the interface, a^*_H is related to the concentration at the interface and equilibrium constants, $c_1^* = K_A a^*_H$ and $c_2^* = K_B a^*_H$, so that Eq. (4.89) becomes

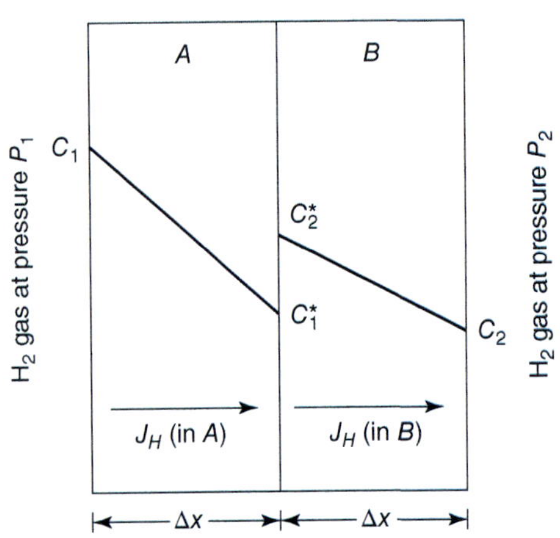

Figure 4.56 Schematic representation of concentration profiles of hydrogen diffusing through a composite wall. Reproduced, by permission, from Gaskell [51]. Copyright © 1992 by Macmillan Publishing.

$$D_{HA}K_A\frac{P_1^{1/2}-a_H^*}{\Delta x_1}=D_{HB}K_B\frac{a_H^*-P_2^{1/2}}{\Delta x_2} \tag{4.93}$$

and we can solve for the activity of the interface to obtain

$$a_{\mathrm{H}}^*=\frac{D_{HA}K_{\mathrm{A}}\Delta x_2P_1^{1/2}+D_{HB}K_{\mathrm{B}}\Delta x_1P_2^{1/2}}{D_{HA}K_{\mathrm{A}}\Delta x_2+D_{HB}K_{\mathrm{B}}\Delta x_1} \tag{4.94}$$

4.3.5 Mass Transport Properties of Biologics

As was the case for composites, there is little new in the way of fundamental concepts for mass transport in biologics that has not already been presented. However, it is possible to briefly describe extensions of some previously introduced topics that are of particular importance to biological materials—namely, diffusion of nonspherical molecules in solution, diffusion through biological membranes, and convective mass transfer in biological systems.

Diffusion in Biological Fluids

Recall from Section 4.3.3 that polymer molecules in dilute solutions can be characterized by a friction coefficient, ζ, which describes the resistance to motion through the solvent, and that the diffusivity of these molecules can be related to the friction coefficient by

$$D_{AB}=\frac{k_{\mathrm{B}}T}{\zeta} \tag{4.85}$$

Recall also from Section 4.1.5 that many biological molecules in solution are not spherical and can be modeled as either prolate or oblate ellipsoids (see Figure 4.17). A further factor affecting diffusion of these ellipsoids in dilute aqueous solutions is that the molecules can become solvated, resulting in an effective radius that is different than the unsolvated molecules. These concepts of asymmetry and solvation are illustrated in Figure 4.57. Accordingly, there are friction coefficients

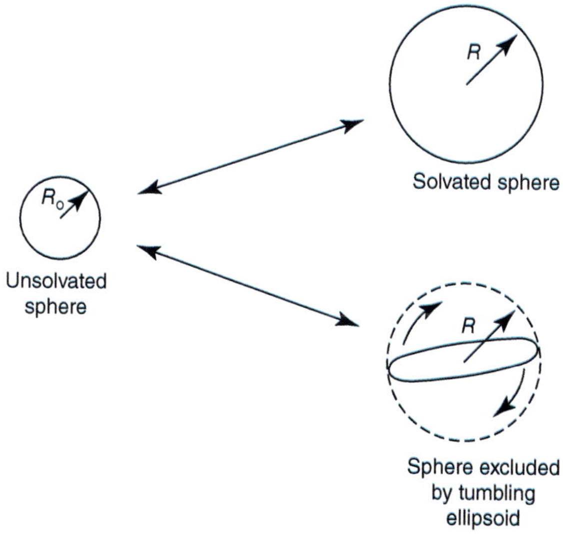

Figure 4.57 Schematic illustration of effective radius of solvated ellipsoids. Reproduced, by permission, from Hiemenz [52]. Copyright © 1984 by John Wiley & Sons.

for the unsolvated Stoke's sphere, ζ_0, the actual particle, ζ, and the equivalent solvated spheres, ζ^*. The ratio of the friction coefficients, ζ/ζ^*, then is a measure in the increase in ζ due to particle asymmetry, and the ratio ζ^*/ζ_0 is a measure of the increase in friction coefficient due to solvation. The first ratio can be related to the axial ratios (b/a) of the ellipsoids. For prolate ellipsoids $(b/a < 1)$

$$\frac{\zeta}{\zeta^*} = \frac{\left[1-(b/a)^2\right]^{1/2}}{(b/a)^{2/3}\ln\left\{\dfrac{1+\left[1-(b/a)^2\right]^{1/2}}{b/a}\right\}} \tag{4.95}$$

Cooperative Learning Exercise 4.13

Consider the diffusion of hydrogen through a composite wall comprising Ni and Pd of identical thicknesses at 400°C. The hydrogen gas pressure in contact with the Ni is 1.0 atm, and that in contact with the Pd is 0.1 atm. The following data on each metal are provided:

	Ni	Pd
Atomic weight, kg/kg-mol	58.71	106.4
Density at 400°C, kg/m^3	8770	11 850
Diffusivity of H at 400°C, m^2/s	4.9×10^{-10}	6.4×10^{-9}
Solubility of H at 400°C and 1 atm hydrogen pressure, atomic ppm	250	10 000

Person 1: Calculate the equilibrium constant for the dissolution of hydrogen in nickel, K_{Ni}.

Person 2: Calculate the equilibrium constant for the dissolution of hydrogen in palladium, K_{Pd}.

(*Hint*: Determine concentration in each metal by converting solubility to weight and volume.)

Combine your answers to calculate the activity of hydrogen at the interface.

Person 1: Use the activity of hydrogen at the interface to calculate the concentration of hydrogen in nickel at the interface.

Person 2: Use the activity of hydrogen at the interface to calculate the concentration of hydrogen in palladium at the interface.

Combine your information to calculate the flux of hydrogen (in kg/m^2·s) through the composite wall if both metals have a thickness of 1 mm.

Answers: $K_{Ni} = 0.0374/(1.0)^{0.5} = 0.0374$; $K_{Pd} = 1.11/(1.0)^{0.5} = 1.11$; $a^*_H = 0.318$
$c^*_1 = 0.0374(0.318) = 0.0119$ kg/m^3; $c^*_2 = 1.11(0.318) = 0.353$ kg/m^3
$J^*_{H-Ni} = 4.9 \times 10^{-10}(0.0374 - 0.0119)/0.001 = 1.25 \times 10^{-8}$ kg/m$^2\cdot$s
(Adapted from Gaskell)

and for oblate ellipsoids $(b/a > 1)$

$$\frac{\zeta}{\zeta^*} = \frac{\left[(b/a)^2 - 1\right]^{1/2}}{(b/a)^{2/3}\tan^{-1}\left[(b/a)^2 - 1\right]^{1/2}} \tag{4.96}$$

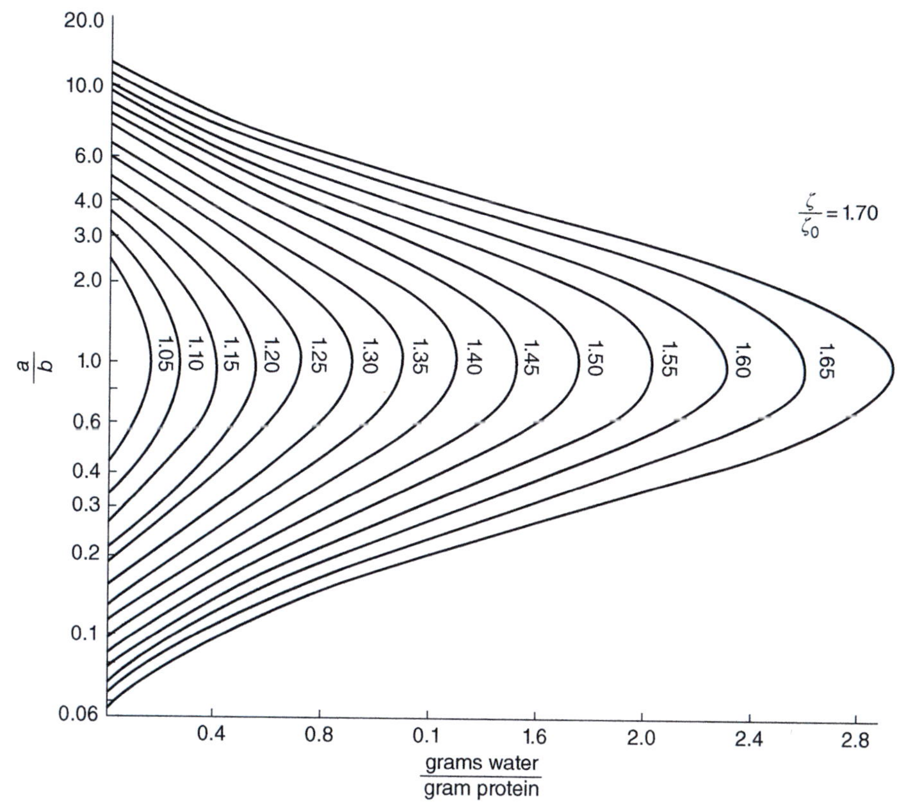

Figure 4.58 Effect of axial ratio, *a/b*, and concentration on friction coefficient ratio for aqueous protein dispersions. Reproduced, by permission, from Hiemenz [53]. Copyright © 1986 by Taylor & Francis.

The second ratio, ζ^*/ζ_0, is related to the density and mass of the solvated and unsolvated molecules:

$$\frac{\zeta^*}{\zeta_0} = \left[1 + \frac{m_{1b}\rho_2}{m_2\rho_1}\right]^{1/3} \tag{4.97}$$

where m_{1b} is the mass of bound solvent, m_2 is the mass of the solute molecule, and ρ_1 and ρ_2 are the density of the solvent and solute, respectively. Note that multiplication of the two ratios gives the third ratio, ζ/ζ_0. A graphical summary of the variation in this ratio with axial ratio and concentration for aqueous protein dispersions is given in Figure 4.58. With Eqs. (4.95)–(4.97) and Eq. (4.85), it is possible to estimate the diffusivity for solvated ellipsoids.

Diffusion Through Biological Membranes

Diffusion through polymeric membranes has already been described, and while there are many examples of membrane diffusion in biological systems, such as dialysis, reverse osmosis, and cell metabolism, the principles described previously are still applicable to biological materials. We take a moment here to elaborate on only one example topic, namely, the role of skin as a biological membrane. In a way, skin is a type of environmental barrier coating (cf. Section 4.3.4). Its purpose is to separate an organism from its environment and to provide a barrier to certain substances. Fortunately (and unfortunately), skin is not totally impervious to all substances, particularly air and water. Some animals can obtain a reasonably large fraction of their oxygen requirements through the skin. For example, humans obtain about 2% of the resting oxygen demand through the skin. Passive diffusion of water vapor through the skin is about 6% of the maximum sweat capacity in humans. Water can also be absorbed through the skin, where the mass diffusivity is about 2.5 to 25×10^{-12} m^2/s.

Cooperative Learning Exercise 4.14

The diffusivity of human hemoglobin at 20°C is 6.9×10^{-11} m^2/s. The viscosity of water at 20°C is approximately 0.01 poise, and the radius of a hemoglobin molecule in solution is approximately 2.64×10^{-9} m.

Person 1: Estimate the experimental friction coefficient, ζ, from the experimental diffusivity for human hemoglobin using Eq. (4.84).

Person 2: Estimate the unsolvated Stoke's sphere friction coefficient, ζ_0, using the Stokes–Einstein relationship for spheres, $\zeta_0 = 6\pi\,\mu r$.

Combine your answers to obtain the ratio of friction coefficients, ζ/ζ_0. If the axial ratio of hemoglobin is $a/b = 1.2$, use Figure 4.58 to estimate the extent of hydration for hemoglobin in solution.

Answer: $\zeta = (1.38 \times 10^{-23})(293)/6.9 \times 10^{-11}) = 5.86 \times 10^{-11}$kg/s;
$\zeta_0 = 6\pi(10^{-3})(2.64 \times 10^{-9}) = 4.98 \times 10^{-11}$kg/s
$\zeta/\zeta_0 = 5.86/4.98 = 1.18$; from Figure 4.58, fraction solvation is about 0.5 g water/g hemoglobin.

In a manner similar to a laminar composite material, skin is composed of two major layers: the nonvascular epidermis and the highly vascularized (blood-vessel-containing) dermis. The outer layer of the epidermis is the stratum corneum, which is about 10 μm thick and is composed of compacted dead skin cells. This is the greatest barrier to permeability of penetrating substances. The stratum corneum provides the greatest barrier against hydrophilic compounds, while the viable epidermis and dermis are most resistant to lipophilic compounds. Attempts have been made to increase the permeability of the skin, as in the case of transdermal drug delivery. For example, the diffusivity of the stratum corneum to certain compounds can increase as much as 10-fold when the layer is hydrated. Hydrating agents such as dimethyl sulfoxide (DMSO) and surfactants may be added to patches to enhance drug movement through the stratum corneum. Some values of permeability to various commercial solvents of female breast epidermis (300–600 μm thick) are given in Table 4.17. Note that these units of permeability are different than those in Table 4.16 for polymers. Units of m/s can be obtained by multiplying values in Table 4.17 by the appropriate solvent liquid density.

Convective Mass Transfer in Biological Systems*

In a manner analogous to convective heat transfer (cf. Section 4.2.5), mass transfer can occur not only by static diffusion but by convection as well. Convective mass transfer occurs when mass is moved in bulk, either as a component in a moving medium or by itself due to pressure or surface tension gradients. As in the case with convective heat transfer, we briefly describe here the dimensionless groups that are involved in convective mass transfer, not as a thorough analysis of the topic,

Table 4.17 Skin Permeability to Various Commercial Solvents

Solvent	**Permeability [10^{-7} kg/(m^2 s)]**
Dimethyl sulfoxide (DMSO)	489
N-Methyl-2-pyrrolidone	475
Dimethyl acetamide	297
Dimethyl formamide	272
Methyl ethyl ketone	147
Methylene chloride	66.7
Water	41.1
Ethanol	31.4
Butyl acetate	4.4
Gammabutyrolactone	3.1
Toluene	2.2
Propylene carbonate	1.9
Sulfolane	0.6

Source: Johnson [22]. Copyright © 1999 by John Wiley & Sons, Inc.

but rather to provide a means for estimating important physical parameters for the materials of interest—specifically, diffusivity. Although the focus here is on convective mass transfer in biologics, primarily due to some important processes such as drying and membrane separations that are important in biological processes, the discussion here is general to all materials classes.

The mass transfer analog to the Prandtl number, Pr, is the *Schmidt number*, Sc:

$$\mathrm{Sc} = \frac{\mu}{\rho D_{AB}} \tag{4.98}$$

where μ is the fluid viscosity, ρ is the fluid density, and D_{AB} the mass diffusivity of dilute solute through the fluid. The Schmidt number is the ratio of momentum diffusivity to mass diffusivity, just as the Prandtl number is the ratio of momentum diffusivity to thermal diffusivity. The Nusselt number [Eq. (4.64)] analog in convective mass transport is the *Sherwood number*, Sh:

$$\mathrm{Sh} = \frac{k_x l}{c D_{AB}} \tag{4.99}$$

where k_x (in molar units) is the mass transfer analog to the heat transfer coefficient, h_c, [see Eq. (4.94)], l is a characteristic length, c is the concentration (in molar units), and D_{AB} is again the diffusivity. The characteristic length, l, may be a distance along a straight path of mass transfer, or it may be a diameter or radius for mass transfer to and from circular or spherical geometries.

Just as there are correlations for the Nusselt number based on the Reynolds and Prandtl numbers [Eq. (4.65)], there are empirical correlations for the Sherwood number as a function of Reynolds and Schmidt numbers, which are of the general form

$$\mathrm{Sh} = \text{constant } \mathrm{Re}^n \mathrm{Sc}^m \tag{4.100}$$

The values of n and m vary depending on the system and the geometry, but typical values are $n = m = 0.5$ for creeping flow around a sphere in a gas–liquid system, and $n = m = 0.33$ for creeping flow around a sphere in liquid–solid systems.

References

1 Iida, T., and R. I. L. Guthrie, *The Physical Properties of Liquid Metals*, Oxford University Press, 1988.
2 Iida, T., and R. I. L. Guthrie, *The Physical Properties of Liquid Metals*, Oxford University Press, 1988, p. 187.
3 Callister, W., *Materials Science and Engineering: An Introduction*, 5th ed., John Wiley & Sons, Inc., 2000, p. 425.
4 Reed, J. S., *Principles of Ceramics Processing*, 2nd ed., John Wiley & Sons, Inc., 1995.
5 Bird, R. B., W. E. Stewart, and E. N. Lightfoot, *Transport Phenomena*, John Wiley & Sons, New York, 1960.
6 Bird, R. B., R. C. Armstrong, and O. Hassager, *Dynamics of Polymeric Liquids*, Vol. **1**, John Wiley & Sons, New York, 1977.
7 Johnson, A. T., *Biological Process Engineering*, Wiley-Interscience, New York, 1999.
8 Strobl, G., *The Physics of Polymers*, 2nd ed., Springer-Verlag, 1997, p. 295.
9 Rodriguez, F., *Principles of Polymer Systems*, 2nd ed., McGraw-Hill Book Company, 1982.
10 Tadmor, Z., and C. G. Gogos, *Principles of Polymer Processing*, John Wiley & Sons, Inc., 1979.
11 Cowie, J. M. G., *Polymers: Chemistry and Physics of Modern Materials*, 2nd ed., Chapman and Hall, New York, 1991, p. 252.
12 Oommen, Z., S. Thomas, C. K. Premalatha, and B. Kuriakose, Melt rheological behavior of natural rubber/poly(methyl methacrylate)/natural rubber-g-poly(methy methacrylate) blends, *Polymer*, **38**(22), 5611–5621 (1997).
13 Sood, R., M. Kulkami, A. Dutta, and R. A. Mashelkar, Viscosity of homogeneous polymer blends: An altered free-volume state model, *Polym. Eng. Sci.*, **28**, 20 (1988).
14 Caneiro, O. S., and J. M. Maia, Rheological behavior of (short) carbon fiber/thermoplastic composites. Part II: The influence of matrix type, *Polym. Compos.*, **21**(6), 970 (2000).
15 Green, J. P., and J. O., Wilkes steady state and dynamic properties of concentrated fiber-filled thermoplastics, *Polym. Eng. Sci.*, **32**(21), 1670 (1995).
16 Shuler, S. F., D. M. Binding, and R. B. Pipes, Rheological behavior of two- and three-phase fiber suspensions, *Polym. Compos.*, **15**(6), 427 (1994).
17 Jamil, F. A., M. S. Hameed, and F. A. Stephan, Rheological, mechanical, and thermal properties of glass-reinforced polyethylenes, *Polym.-Plast. Technol. Eng.*, **33**(6), 659 (1994).
18 Suhara, F., and S. K. N. Kutty, Rheological properties of short polyester fiber-polyurethane elastomer composite with different interfacial bonding agents, *Polym. Plast. Technol. Eng.*, **37**(1), 57 (1998).
19 Utracki, L. A., ed., *Two Phase Polymer Systems*, Hanser, Munich, 1991, p. 305.
20 Mallick, P. K., *Composites Engineering Handbook*, Taylor & Francis, 1997, p. 507.
21 Castro, J. M., and C. W. Macosko, Studies of mold filling and curing in the reaction injection molding process, *AIChE J.*, **28**, 250 (1982).
22 *Biological Process Engineering*, by A. T. Johnson; 1999, John Wiley & Sons, Inc.
23 Hiemenz, P., *Principles of Colloid and Surface Chemistry*, 2nd ed., Taylor & Francis, 1986, p. 26.
24 Silver, F. H., and D. L. Christiansen, *Biomaterials Science and Biocompatibility*, Springer-Verlag, 1999, p. 150.
25 Hiemenz, P., *Polymer Chemistry*, John Wiley & Sons, 1984, p. 596.
26 Hiemenz, P., *Polymer Chemistry*, John Wiley & Sons, 1984, p. 598.
27 Silver, F. H., and D. L. Christiansen, *Biomaterials Science and Biocompatibility*, Springer-Verlag, 1999, p. 151.
28 Lauffer, M. A., *J. Am. Chem. Soc.*, **66**, 1188 (1944).
29 Bird, R. B., W. E. Stewart, and E. N. Lightfoot, *Transport Phenomena*, 2nd ed., John Wiley & Sons, New York, 2001, p. 279.

30 Iida, T., and R. I. L. Guthrie, *The Physical Properties of Liquid Metals*, Oxford University Press, 1988, p. 241.
31 Ralls, K. M., T. H. Courtney, and J. Wulff, *Introduction to Materials Science and Engineering*, John Wiley & Sons, Inc., 1976.
32 Kittel, C., *Introduction to Solid State Physics*, 2nd ed., John Wiley & Sons, Inc., 1957, p. 132.
33 Kittel, C., *Introduction to Solid State Physics*, 2nd ed., John Wiley & Sons, Inc., 1957, p. 151.
34 Kingery, W. D., H. K. Bowen, and D. R. Uhlmann, *Introduction to Ceramics*, John Wiley & Sons, Inc., 1976.
35 Ammar, M. M. et al., Thermal conductivity of silicate and borate glasses, *J. Am. Ceram. Soc.*, **66**, C76–C77 (1983).
36 Scholze, H., *Glass*, Springer-Verlag, 1991.
37 Scholze, H., *Glass*, Springer-Verlag, 1991, p. 363.
38 Primenko, V. I., Theoretical method of determining the temperature dependence of the thermal conductivity of glasses, *Glass Ceram.*, **37**, 240–242 (1980).
39 Scholze, H., *Glass: Nature, Structure and Properties*, Springer-Verlag, New York, 1991, p. 364.
40 Powell, P. C., and A. J. I. Housz, *Engineering with Polymers*, Taylor & Francis, 1998, p. 288.
41 Tadmor, Z., and C. G. Gogos, *Principles of Polymer Processing*, John Wiley & Sons, Inc., 1979, p. 132.
42 Bird, R. B., W. E. Stewart, and E. N. Lightfoot, *Transport Phenomena*, 2nd ed., John Wiley & Sons, 2001, p. 282.
43 Iida, T., and R. I. L. Guthrie, *The Physical Properties of Liquid Metals*, Oxford University Press, 1988, p. 209.
44 Chalmers, B., *Physical Metallurgy*, John Wiley & Sons, Inc., 1959.
45 Gaskell, D. R., *An Introduction to Transport Phenomena in Materials Engineering*, Macmillan Publishing Co., 1992, p. 519.
46 Shelby, J. E., in *Experimental Techniques of Glass Science*, C. J. Simmons, and O. H. El-Bayoumi, eds., The American Ceramic Society, 1993, p. 374.
47 Bird, R. B., W. E. Stewart, and E. N. Lightfoot, *Transport Phenomena*, 2nd ed., John Wiley & Sons, Inc., 2002.
48 Strobl, G., *The Physics of Polymers*, 2nd ed., Springer-Verlag, 1997, p. 283.
49 Strobl, G., *The Physics of Polymers*, 2nd ed., Springer-Verlag, 1997, p. 286.
50 Hinestroza, J., D. De Kee, and P. Pintauro, Apparatus for studying the effect of mechanical deformation on the permeation of organics through polymer films, *Ind. Eng. Chem. Res.*, **40**, 2183 (2001).
51 Gaskell, D. R., *An Introduction to Transport Phenomena in Materials Engineering*, Macmillan Publishing, 1992, p. 498.
52 Hiemenz, P., *Polymer Chemistry: The Basic Concepts*, John Wiley & Sons, 1984, p. 626.
53 Hiemenz, P., *Principles of Colloid and Surface Chemistry*, 2nd ed., Taylor & Francis, 1986, p. 88.

Problems

Level I

4.I.1 Estimate the viscosity of liquid chromium at its melting temperature of 2171 K.

4.I.2 Estimate the viscosity of liquid iron at 1600°C. The density of liquid iron at 1600°C is 7160 kg/m^3.

4.I.3 Various molecular weight fractions of cellulose nitrate were dissolved in acetone and the intrinsic viscosity was measured at 25°C:

$M \times 10^{-3}$ (g/mol)	77	89	273	360	400	640	846	1550	2510	2640
$[\eta]$ (dl/g)	1.23	1.45	3.54	5.50	6.50	10.6	14.9	30.3	31.0	36.3

Use these data to evaluate the Mark–Houwink constants for this system.

4.I.4 Agglomeration in a slurry causes a change in f_{cr}^{v} from 0.6 to 0.45 and a change in K_H from 3.5 to 2.5. Calculate and compare the viscosity for each slurry when $f_p^v = 0.40$.

4.I.5 Calculate the thermal resistance of a 2-mm-thick, 1 cm^2 section of human skin. The thermal conductivity of human skin is 0.627 N/K·s, the specific heat is 3470 N·m/kg·K, and the density is 1100 kg/m^3.

Level II

4.II.1 The thermal conductivity of a material varies with temperature as

$$\ln k = 0.01T + 0.5$$

where k has units of W/m·K, and T has units of °C. Heat flows by conduction through a plane slab of this material of thickness 0.1 m, the left face of which is at 100°C and the right face of which is at 0°C. Calculate (a) the mean thermal conductivity of this material in the range 0–100°C, (b) the heat flux through the slab, and (c) the temperature at $x = 0.05$ m.

4.II.2 Hydrogen dissolved in iron obeys *Sievert's Law*; that is, at equilibrium we have

$$\frac{1}{2}H_2(g) \Leftrightarrow [H]_{\text{dissolved in iron}}$$

where [H] is the concentration of atomic hydrogen in the iron, in ppm by weight, in equilibrium with hydrogen gas at the pressure P_{H_2}. At 400°C and a hydrogen pressure of 1.013×10^5 Pa (1 atm), the solubility of hydrogen in iron is 3 ppm by weight. Calculate the flux of hydrogen through an iron tank of wall thickness 0.001 m at 400°C. The density of iron 400°C is 7730 kg/m^3, and the diffusivity of hydrogen is 10^{-8} m^2/s. Clearly state any assumptions you make.

4.II.3 A hemodialysis membrane with an effective area of 0.06 m^2, thickness of 50 μm, and permeability of 2.96×10^{-14} m^2 is used to filter urea and other impurities from the blood. The viscosity of blood plasma is 1.2×10^{-3} N·s/m^2. What is the expected filtration rate if the transmembrane pressure is 2.25 Pa?

4.II.4 Fick's Second Law of Diffusion relates the change in concentration of a diffusing species with time to the diffusion coefficient and the concentration gradient for non-steady-state diffusion:

$$\frac{\partial c_A}{\partial t} = D\frac{\partial^2 c_A}{\partial x^2}$$

The solution to this partial differential equation depends upon geometry, which imposes certain boundary conditions. Look up the solution to this equation for a semi-infinite solid in which the surface concentration is held constant, and the diffusion coefficient is assumed to be constant. The solution should contain the error function. Report the following: the boundary conditions, the resulting equation, and a table of the error function.

Level III

4.III.1 A relationship for the viscosity of a simple liquid metal near its melting point, T_m, was proposed by Andrade:

$$\mu = 1.68 \times 10^{-4} \frac{\sqrt{MT_m}}{V_m^{2/3}}$$

where μ is in mPa·s and V_m is the atomic volume at the melting point. Show that this relationship gives approximately the same value for viscosity (within a factor of about 0.3) as the hard-sphere model when a packing fraction of 0.45 is used. Use this formula to estimate the viscosity of molten silver near its melting point. How does this compare to an experimental value? Cite your references.

4.III.2 Use a spreadsheet to solve for the three constants in the Vogel–Fulcher–Tammann equation using the data in Cooperative Learning Exercise 4.1. Show your work.

4.III.3 Determine the VFT equation for a pure SiO_2 melt based on the following three viscosity–temperature points:

Glass transformation temperature = 1203°C	$\ln \mu = 13.3$
Littleton temperature = 1740°C	$\ln \mu = 7.6$
Working point = 2565°C	$\ln \mu = 4.0$

4.III.4 In Cooperative Learning Exercise 4.4, you derived an expression for the superficial velocity of a 1% solution of poly(ethylene oxide) through a bed of 10-μm-diameter fibers with 60% porosity. You were left with a parameter, τ_{Rh}, in the expression, which is the shear stress at the wall, based upon the hydrodynamic radius, R_h. The mean hydrodynamic radius is defined as the ratio of the particle cross section to the "wetted perimeter," which is conveniently related to the void fraction, or porosity, ε, and the wetted surface, a, as $R_h = \varepsilon/a$. The wetted surface, in turn, is related to the porosity and the "specific surface," S, as $a = S(1-\varepsilon)$. Finally, the specific surface is directly related to geometrical dimensions of the particle and is defined as the ratio of particle surface area to particle volume. Complete the derivation of an expression for the superficial velocity by calculating the specific surface for cylinders (this result is general for all cylindrical geometries) and calculating a value of τ_{Rh} for the case of 60% porosity. Your final expression for v_0 should have only the pressure drop across the fiber bed, ΔP, and the bed length, L.

4.III.5 Derive an expression for steady-state heat flow through three parallel planes as shown below. In this instance, the total heat flow, Q_{tot}, is a sum of the individual heat flows, Q_i. The cross-sectional areas of each plane, A_i, are not necessarily equivalent.

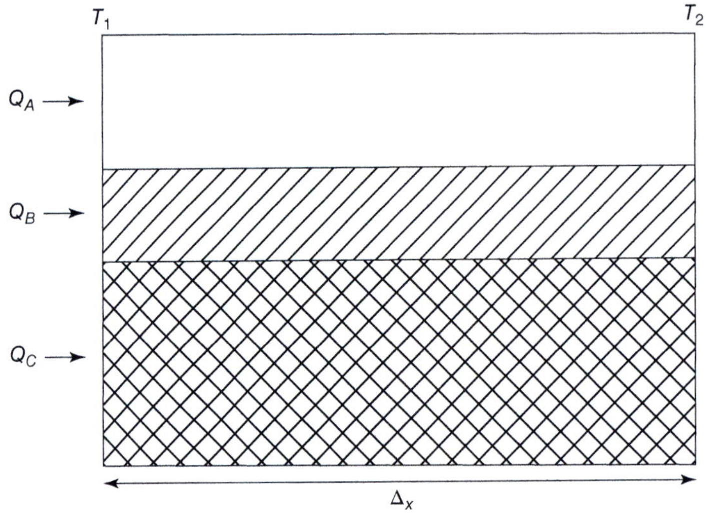

4.III.6 Composite membranes are used where the membrane of choice for a particular separation must be strengthened by addition of a second material. Assume a hollow fiber membrane with inside diameter of 1.2 mm and outside diameter of 1.3 mm. Mass diffusivity of the target species through this membrane is $3.0 \times 10^{-10}\ m^2/s$. A second material is layered on the outside of the first material to add strength. The mass diffusivity of the same target species through this material is $18 \times 10^{-10}\ m^2/s$. What thickness of the outside material can be added if the total diffusion does not exceed the resistance of the inner material by more than 25%?

5

Mechanics of Materials

5.0 Introduction and Objectives

The engineering principles of thermodynamics, kinetics, and transport phenomena, as well as the chemistry and physics of molecular structure, serve as the basis for the remaining topics in this book. In this chapter, we look at what for many applications is the primary materials selection criterion: mechanical properties. As in the previous chapter, we focus primarily on the properties of materials but briefly discuss the mechanics—both in the fluid state and solid state—that give rise to the properties. There is a great deal of new terminology in this chapter and it is cumulative; take time to understand all the definitions before proceeding to the next section.

We seek to understand the response of a material to an applied stress. In Chapter 4, we saw how a fluid responds to a shearing stress through the application of Newton's Law of Viscosity [Eq. (4.3)]. In this chapter, we examine other types of stresses, such as tensile and compressive, and describe the response of solids (primarily) to these stresses. That response usually takes on one of several forms: elastic, inelastic, viscoelastic, plastic (ductile), fracture, or time-dependent creep. We will see that Newton's Law will be useful in describing some of these responses and that the concepts of stress (applied force per unit area) and strain (change in dimensions) are universal to these topics.

By the end of this chapter, you should be able to:

- Determine strength and modulus from a stress–strain diagram.
- Qualitatively differentiate between stress–strain diagrams for different material classes.
- Describe the differences between the Kelvin and Maxwell models for viscoelasticity and identify their corresponding equations.
- Calculate Poisson's ratio.
- Relate flexural, bulk, and tensile moduli.
- Distinguish between ultimate strength, yield strength, and tensile strength.
- Distinguish between ductile and brittle materials.
- Understand the role of strain rate on mechanical properties.
- Identify the different types of mechanical creep and calculate creep rate.
- Apply time–temperature superposition principles to polymer moduli and calculate shift factors.
- Describe temperature, molecular weight, and strain-rate effects on the mechanical properties of polymers.
- Calculate axial and transverse moduli and strengths for unidirectional, continuous and discontinuous, fiber-reinforced composites, and laminate composites.

Materials Engineering and Science: Principles, Properties, and Processes, Second Edition. Brian S. Mitchell.

Companion website: www.wiley.com/go/materialsengineeringandscience2e

- Calculate elastic modulus of a nanocomposite using the Mooney equation.
- Describe the structure of soft and hard biologics that give rise to their unique mechanical properties.
- Use mechanical properties that help to select appropriate candidate materials for a specific application.

5.1 Mechanics of Metals and Alloys

In this section, we will describe the response of a material to an applied stress in terms of its atomic structure. Initially, we will focus on a single crystal of a metallic element, then work our way into topics related to polycrystalline materials.

5.1.1 Elasticity

The Molecular Origins of Elasticity

Recall from Section 1.0.4 that atoms are held together by interatomic bonds and that there are equations such as Eq. (1.13) that relate the interatomic force, F, to the potential energy function between the atoms, U, and the separation distance, r:

$$F = -\frac{dU}{dr} \tag{1.13}$$

It should make sense, then, that as we attempt to pull atoms apart or force them further together through applied stress, we can in principle relate how relatively difficult or easy this is to the potential energy function. We can develop a quantitative description of this process of pulling atoms apart—provided we do so over small deformations—in which the deformation is wholly recoverable; that is, the atoms can return to their original, undeformed positions, with no permanent displacement relative to one another. This is called an *elastic* response. The term "elastic" here does not imply anything specific to polymers in the same way that the more everyday use of the term does. It is used in the same sense that it is in physics and chemistry—a completely recoverable deformation.

A convenient way to visualize the elastic bond between two atoms is to place an imaginary spring between the atoms, as illustrated in Figure 5.1. For small atomic deformations, either away from each other with what is called a *tensile force* (shown in Figure 5.1) or toward each other with what is called a *compressive force* (not shown), the force stored in the spring causes the atoms to return to the undeformed, equilibrium separation distance, r_0, where the forces are zero (cf. Figure 1.3). Attractive forces between the atoms counteract the tensile and repulsive forces between the atoms counteract the compressive forces.

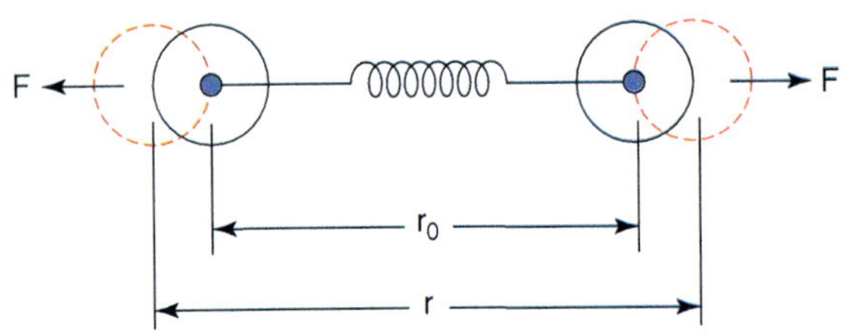

Figure 5.1 Schematic illustration of "bead-and-spring" model of atomic force between atoms. Ashby and Jones [1]/Reproduced with permission of Elsevier.

Example Problem 5.1

Derive an expression for the elastic modulus of an ionic solid in terms of the potential energy function parameters a, b, and n.

Solution: According to Eq. (5.5), the elastic modulus, E, is proportional to the stiffness of the theoretical springs that model the bonds between atoms in the solid, S_0. According to Eq. (5.1), S_0 is in turn proportional to the second derivative of the potential energy function, U, with respect to interatomic separation distance, r.

We begin with the potential energy function of Eq. (1.11) and recognize that for an ionic solid, $m = 1$, such that the potential energy function becomes $U = \frac{-a}{r} + \frac{b}{r^n}$. The strategy, then, is to differentiate U twice with respect to r, and evaluate at $r = r_0$.

The first derivative is $F = \frac{dU}{dr} = \frac{u}{r^2} - \frac{nb}{r^{n+1}}$, and the second derivative is $\frac{dF}{dr} = \frac{-2a}{r^3} + \frac{n(n+1)b}{r^{n+2}}$. All that is left now is to evaluate dF/dr at $r = r_0$. Equation (1.16) provides us with the equilibrium separation as a function of the potential energy parameters, which for the case of $m = 1$ becomes $r_0 = \left(\frac{nb}{a}\right)^{\frac{1}{n-1}}$. Substitution of r_0 into dF/dr with appropriate simplification gives the desired result:

$$E \propto S_0 = \left(\frac{dU}{dr}\right)_{r=r_0} = \frac{-2a}{\left(\frac{nb}{a}\right)^{\frac{3}{n-1}}} + \frac{n(n+1)b}{\left(\frac{nb}{a}\right)^{\frac{n+2}{n-1}}}$$

The stiffness of the spring, S_0, is a measure of how much force must be overcome to pull the atoms apart, and it is related to the potential energy function as:

$$S_0 = \left(\frac{d^2U}{dr^2}\right)_{r=r_0} \tag{5.1}$$

Notice that the stiffness is the slope (first derivative) of the force versus interatomic separation distance plot in the vicinity of r_0, as shown in Figure 1.3. So, for small deformations (i.e., $r \approx r_0$) we can relate the stiffness to the applied force, F:

$$S_0 = \frac{dF}{dr} \approx \frac{\Delta F}{\Delta r} = \frac{F - F_0}{r - r_0} = \frac{F}{r - r_0} \tag{5.2}$$

since $F = F_0 = 0$ at $r = r_0$ (see Figure 1.3).

Now consider a square array of N springs (bonds), as would be found in a single crystal, and illustrated in Figure 5.2. The total tensile force exerted across the unit area, F/A, is represented by the Greek lowercase sigma, σ_t, where the subscript "t" indicates a tensile force. It is the product of the stiffness of each spring, S_0, the number of springs per unit area, N/A, and the distance the springs are extended, $r - r_0$

$$\sigma_t = (N/A)S_0(r - r_0) \tag{5.3}$$

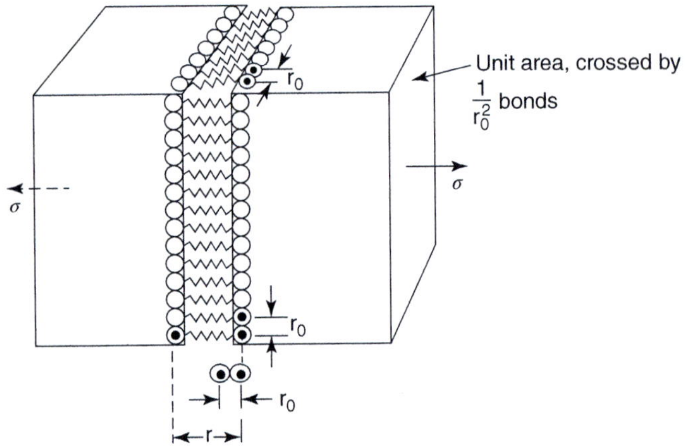

Figure 5.2 Illustration of multiple bead-and-springs in tensile separation of atomic planes. Ashby and Jones [2]/Reproduced with permission of Elsevier.

The number of springs per unit area is simply $1/r_0^2$, (see Figure 5.2), so that Eq. (5.3) becomes

$$\sigma_t = \left(\frac{S_0}{r_0}\right)\left(\frac{r - r_0}{r_0}\right) \tag{5.4}$$

We have written Eq. (5.4) with variables grouped as they are in order to define two very important quantities. The first quantity in parentheses is called the *modulus*—or in this case, the *tensile modulus, E*, since a tensile force is being applied. The tensile modulus is sometimes called *Young's modulus*, *elastic modulus*, or *modulus of elasticity*, since it describes the elastic, or recoverable, response to the applied force, as represented by the springs. The second set of parentheses in Eq. (5.4) represents the tensile *strain*, which is indicated by the Greek lowercase epsilon, ε_t. The strain is defined as the displacement, $r - r_0$, relative to the initial position, r_0, so that it is an indication of relative displacement and not absolute displacement. This allows comparisons to be made between tensile tests performed at a variety of length scales. Equation (5.4) thus becomes:

$$\sigma_t = E\varepsilon_t \tag{5.5}$$

Note that the strain is dimensionless but is often given in units of length/length (e.g., in./in. or cm/cm) to indicate the length scale of the displacement. Modulus then has the same units of stress, which are typically Pa (N/m^2) or psi (lb$_f$/in.^{2}). The stresses required to separate bonds are surprisingly high, and the resulting extensions are vanishingly small, such that the values of modulus are often reported in MPa (10^6 Pa) and GPa (10^9 Pa), or kpsi (10^3 psi, sometimes called "ksi") and Mpsi (10^6 psi).

Cooperative Learning Exercise 5.1

Work in groups of three. Look up the modulus of elasticity for each of the following substances in Appendix 6, and compute the tensile stress, σ_t, that arises when a cylindrical sample of each is strained to 0.1% of its original length. Assume that all deformations are perfectly elastic.

Person 1: Tungsten (W) *Person 2*: Tungsten carbide (WC) *Person 3*: Polycarbonate

Compare your answers. Assuming all samples have the same geometry (cross-sectional area), which of the three will require the greatest load to achieve this amount of strain?

Answers: σ_t(W) = (400 GPa)(0.001) = 400 MPa; σ_t(WC) = (686)(0.001) = 686 MPa; σ_t(Polycarbonate) = (2.4 GPa)(0.001) = 2.4 MPa. WC will require the greatest load (force).

Equation (5.5) is known as *Hooke's Law* and simply states that in the elastic region, the stress and strain are related through a proportionality constant, E. Note the similarity in form to Newton's Law of Viscosity [Eq. (4.3)], where the shear stress, τ, is proportional to the strain rate, $\dot{\gamma}$. The primary differences are that we are now describing a solid, not a fluid, the response is to a tensile force, not a shear force, and we do not (yet) consider time dependency in our tensile stress or strain.

The relationship between the potential energy and the modulus is an important one. If we know the potential energy function, even qualitatively or relatively to another material, we can make inferences as to the relative magnitude of the elastic modulus. Figure 5.3 shows the

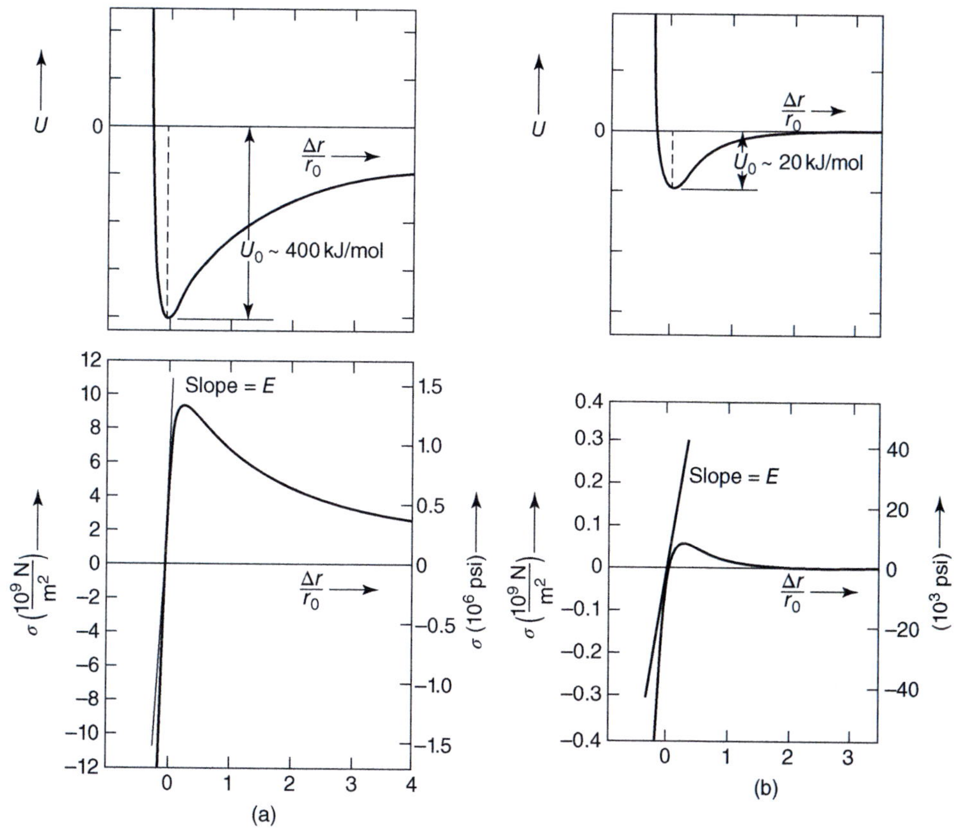

Figure 5.3 Schematic diagrams showing potential energy (*top*) and applied stress (*bottom*) versus linear strain for crystalline solids with (a) strong bonds and (b) weak bonds. From Ralls et al. [3]. Copyright © 1976 by John Wiley & Sons, Inc. This material is used by permission John Wiley & Sons, Inc.

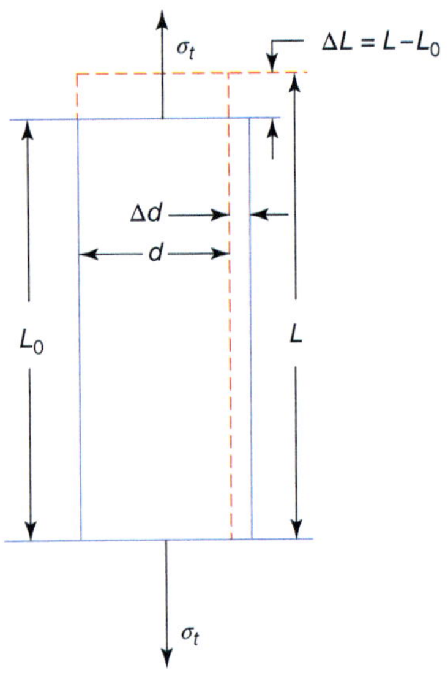

Figure 5.4 Schematic illustration of tensile strain and corresponding lateral strain. From Z. Jastrzebski [4]. Copyright © 1976 by John Wiley & Sons, Inc. This material is used by permission of John Wiley & Sons, Inc.

interrelationship between the potential energy (top curves), the force (represented per unit area as stress in the bottom curves), and tensile modulus (slope of force curves for two hypothetical materials). The deeper the potential energy well, the larger the modulus (compare Figure 5.3a and b).

Poisson's Ratio and Other Moduli

Let us examine the elastic deformation of a material on a more macroscopic level. Figure 5.4 shows a typical test material under tension on a scale somewhat larger than that represented in Figure 5.3. A stress, σ_t is once again applied, resulting in a strain ε_t, which we define as $\Delta L/L_0$. For small deformations, the stress and strain are related by Eq. (5.5). Notice, however, that for the material to (ideally) maintain a constant volume, the diameter, d, must decrease by an increment $\Delta d = d - d_0$. There is no requirement, of course, that volume be maintained, but in most real systems, there is a decrease in the cross section with elongation. This response is characteristic of a material and is given by *Poisson's ratio*, ν, which is defined as the ratio of the induced lateral strain to the axial strain:

$$\nu = -\frac{\Delta d/d_0}{\Delta L/L_0} \tag{5.6}$$

> **Need More Help?**
>
> Try these videos on the fundamentals of mechanical properties:
>
> Stress and Strain
> https://www.youtube.com/watch?v=0s5kBrk0d80
> https://www.youtube.com/watch?v=7Eiv7Qr4HQ8
>
> Hooke's Law, Modulus
> https://www.youtube.com/watch?v=DLE-ieOVFjl
>
> Poisson's Ratio
> https://www.youtube.com/watch?v=tuOlM3P7ygA

The minus sign in Eq. (5.6) is to account for the fact that Δd as defined above is usually negative. Thus, Poisson's ratio is normally a positive quantity, though there is nothing that prevents it from having a negative value. For constant volume deformations (such as in polymeric elastomers), $\nu = 0.5$, but for most metals, Poisson's ratio varies between 0.25 and 0.35. Values of Poisson's ratio for selected materials are presented in Appendix 6.

Poisson's ratio provides a basis for relating the lateral response of a material to the axial response. By definition, the forces that cause lateral deformation in a tensile experiment must be shear forces, since they are normal to the direction of the applied stress. In a generalized shear situation, as in Figure 5.5, the shear stress, τ, results in a shear strain, γ, which we must define slightly differently than in the tensile experiment. Here, the shear strain is the amount of shear, aa', divided by the distance of which the shear has occurred, ad, so $\gamma = aa'/ad$. There is still a proportionality between stress

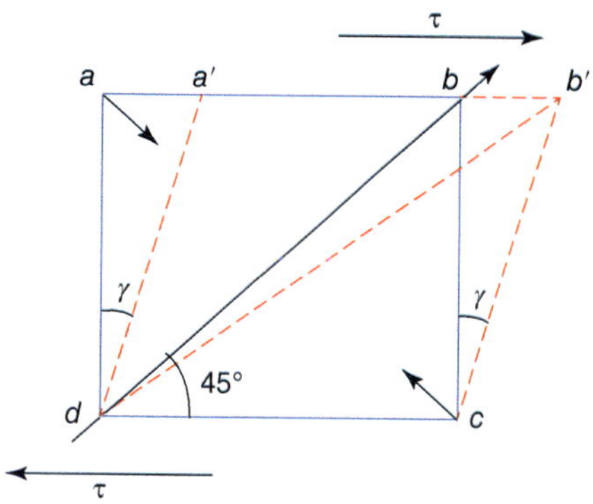

Figure 5.5 Schematic illustration of shear strain and stress. From Jastrzebski [4]. Copyright © 1976 by John Wiley & Sons, Inc. This material is used by permission of John Wiley & Sons, Inc.

and strain (for small deformations), but now we call the proportionality constant the *shear modulus*, G:

$$\tau = G\gamma \tag{5.7}$$

The shear modulus and elastic modulus are related through Poisson's ratio:

$$G = \frac{E}{2(1+\nu)} \tag{5.8}$$

In addition to the tensile and shear moduli, a *compressive modulus*, or *modulus of compressibility*, K, exists to describe the elastic response to compressive stresses (see Figure 5.6). The compressive modulus is also sometimes called the *bulk modulus*. It is the proportionality constant between the compressive stress, σ_c, and the bulk strain, represented by the relative change in bulk volume, $\Delta V/V_0$.

The compressive modulus and elastic modulus are also related through Poisson's ratio:

$$K = \frac{E}{3(1-2\nu)} \tag{5.9}$$

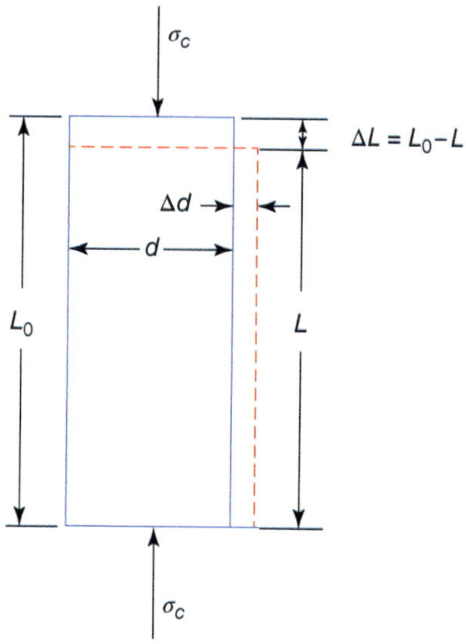

Figure 5.6 Schematic illustration of compressive strain and lateral strain. From Jastrzebski [4]. Copyright © 1976 by John Wiley & Sons, Inc. This material is used by permission of John Wiley & Sons, Inc.

From Eqs. (5.8) and (5.9), we can derive a relationship that relates all three moduli to one another:

$$E = \frac{9GK}{3K+G} \tag{5.10}$$

Equations (5.8)–(5.10) allow us to apply some convenient "rules of thumb." For most hard materials, $\nu \approx 1/3$, such that $K \approx E$ and $G \approx (3/8)E$. For elastomers, putties, gels, and colloidal systems, the compressive modulus is much higher than the other moduli, and the system is considered "incompressible." In this case, $E \approx 3G$.

Cooperative Learning Exercise 5.2

A tensile load of 5600 N is applied along the long axis (z direction) of cylindrical brass rod that has an initial diameter of 10 mm. The resulting deformation is entirely elastic, and it results in a diameter reduction (x direction) of 2.5×10^{-3} mm.

Person 1: Calculate the lateral strain, ε_x.

Person 2: Calculate the axial strain, ε_z. Assume that the modulus of elasticity for brass is 97 GPa.

Combine your information to calculate Poisson's ratio for this material.

Person 2: Use this value of Poisson's ratio to calculate the compressive modulus, K.

Person 1: Use the value of Poisson's ratio to calculate the shear modulus, G.

Use Eq. (5.10) to recalculate E from G and K to confirm your results.

Answers: $\varepsilon_x = -(2.5 \times 10^{-3}\text{mm})/(10\text{ mm}) = -2.5 \times 10^{-4}$; in order to calculate ε_z, we must first calculate the applied stress, $\sigma = (5600\text{ N})/[\pi(5 \times 10^{-3}\text{m})^2] = 71.3 \times 10^6\text{Pa}$, then use Hooke's Law and the elastic modulus to calculate the axial strain, $\varepsilon_z = \sigma/E = (71.3\text{ MPa})/(97000\text{ MPa}) = 7.35 \times 10^{-4}$; $\nu = -\varepsilon_x/\varepsilon_z = -(-2.5)/(7.35) = 0.34$. $K = 97\text{ GPa}/\{3[1-(2)(0.34)]\} = 101\text{ GPa}$; $G = 97\text{ GPa}/\{2[1+(0.34)]\} = 36.2\text{ GPa}$; $E = 9(36.2)(101)/[3(101)+36.2] = 97.0\text{ GPa}$. It checks out!

5.1.2 Ductility

To this point, we have limited the discussion to small strains—that is, small deviations from the equilibrium bond distance—such that all imposed deformations are completely recoverable. This is the elastic response region, one that virtually all materials possess (see Figure 5.7). What happens at larger deformations, however, is dependent to some extent on the type of material under consideration. Beyond the elastic region, we enter a realm of nonrecoverable mechanical response termed

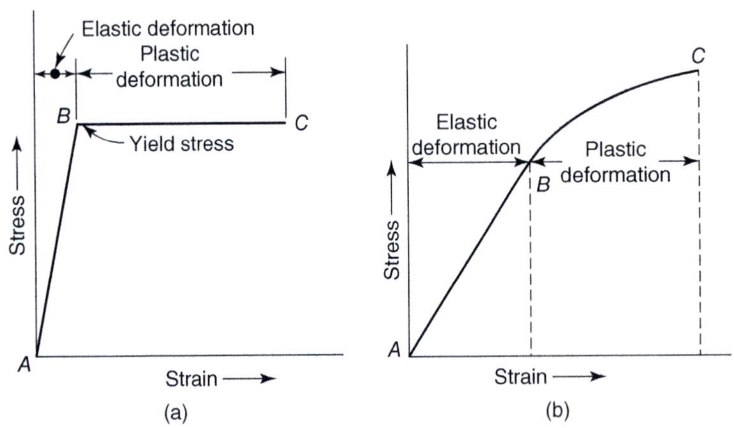

Figure 5.7 Illustration of (a) ideal elastic deformation followed by ideal plastic deformation and (b) typical elastic and plastic deformation in rigid bodies. From Jastrzebski [4]. Copyright © 1976 by John Wiley & Sons, Inc. This material is used by permission of John Wiley & Sons, Inc.

permanent deformation. There are two primary forms of permanent deformation: *viscous flow* and *plastic flow*. We will reserve the discussion of viscous flow to the appropriate section on mechanical properties of polymers since this is where viscous flow primarily occurs. Other types of nonelastic responses to applied forces, such as *fracture* and *creep*, will also be delayed until later sections. We concentrate now on a type of nonrecoverable deformation called plastic flow, or, more commonly in the terminology of metals and alloys, *ductility*. As illustrated in Figure 5.8, the plastic deformation region ideally occurs immediately following the elastic response region, and it results in elongation (strain) without additional stress requirement, hence, the flat region represented by segment BC in Figure 5.8a. More practically, there are microstructural phenomena that lead to the behavior shown in Figure 5.8b, namely, an increase in stress with % strain in the plastic deformation region. In the following section, we explore the molecular basis for these observations.

The Molecular Origins of Ductility—Slip in Single Crystals

Plastic deformation in single crystals occurs by one of two possible mechanisms: slip or twinning. *Slip* is the process by which bulk lattice displacement occurs by the motion of crystallographic planes relative to one another (see Figure 5.8a). The surface on which slip takes place is often a plane, known as the *slip plane*. The direction of motion is known as the *slip direction*. The visible intersection of a slip plane with the outer surface of the crystal is known as a *slip band*. *Twinning* is the process in which atoms subjected to stress rearrange themselves so that one part of the crystal becomes a mirror image of the other part. Every plane of atoms shifts in the same direction by an amount proportional to its distance from the twinning plane (see Figure 5.8b). Twinning usually requires a higher shear stress than slip does, and it occurs primarily in body-centered cubic (BCC) and hexagonal tightly packed (HCP) crystal structures. We will focus primarily on slip to describe the plastic deformation process.

Based on the concepts of intermolecular forces and shear modulus introduced in the previous section, it is relatively easy to estimate the theoretical stress required to cause slip in a single crystal. We call this the *critical shear stress*, σ_{cr}. Refer to Figure 5.9a, and consider the force required to shear two planes of atoms past each other. In the region of small elastic strains, the stress, τ, is related to the displacement, x, relative to the initial interplanar spacing, d, according to a modified form of Eq. (5.7) for the shear modulus, G:

$$\tau = G\frac{x}{d} \tag{5.11}$$

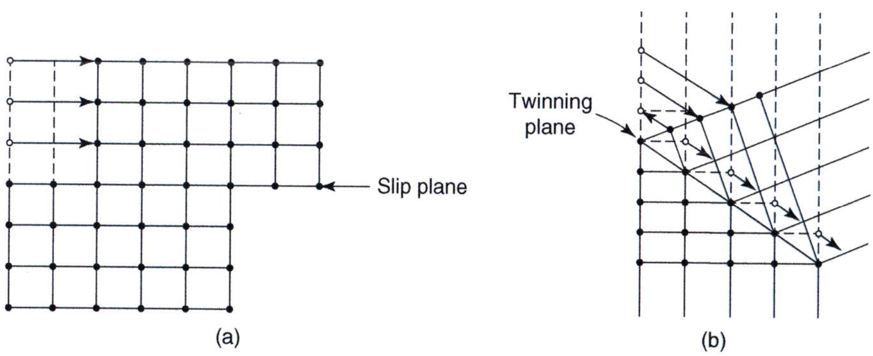

Figure 5.8 Schematic illustration of plastic deformation in single crystals by (a) slip and (b) twinning. From Jastrzebski [4]. Copyright © 1976 by John Wiley & Sons, Inc. This material is used by permission of John Wiley & Sons, Inc.

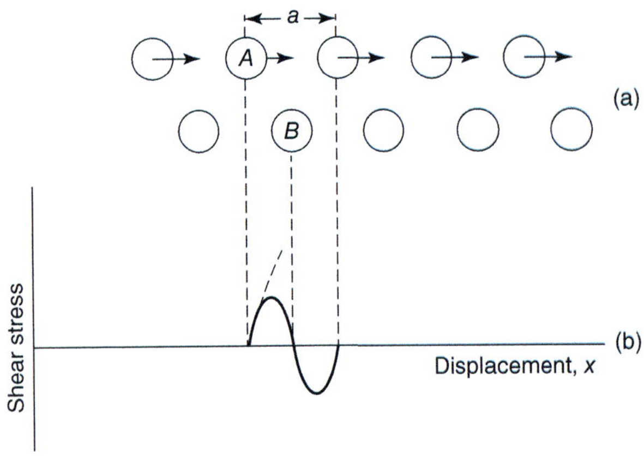

Figure 5.9 Schematic illustration of (a) relative shear of two planes of atoms in a strained material and (b) shear stress as a function of relative displacement of the planes from their equilibrium positions. Kittel [5]/Reproduced with permission of John Wiley & Sons.

As a first approximation, we may represent the stress–displacement relation by a sine function (Figure 5.9b), since when atom A in Figure 5.9a is displaced to the point where it is directly over atom B in the plane below, an unstable equilibrium exists and the stress is zero, so

$$\tau = \frac{Ga}{2\pi d}\sin(2\pi x/a) \tag{5.12}$$

where a is the interatomic spacing in the direction of shear. Note that the slope (represented by the dashed line) of the sine wave in Figure 5.10b is simply the shear modulus, G. For small displacements, x, Eq. (5.12) reduces to Eq. (5.11). The critical shear stress, τ_{cr}, at which the lattice becomes unstable and begins to plastically deform is given by the maximum value of τ in Eq. (5.12), or

$$\tau_{cr} = \frac{Ga}{2\pi d} \tag{5.13}$$

If we make the approximation that $a \approx d$, then

$$\tau_{cr} \approx \frac{G}{2\pi} \approx \frac{G}{6} \tag{5.14}$$

that is, the critical shear stress to plastic flow should be of the order of 1/6 the shear modulus. Really, the measured critical shear stress is much smaller than this for most single-crystal metals. Further theoretical, considerations reduced this value to $G/30$, but even this is much larger than the critical shear stress observed in most metals. Clearly, we cannot simply extend the theory of elasticity to large strains to determine shear strength—there are other considerations.

The primary consideration we are overlooking is that of crystal imperfections. Recall from Section 1.1.4 that virtually all crystals contain some concentration of defects. In particular, the presence of dislocations causes the actual critical shear stress to be much smaller than that predicted by Eq. (5.14). Recall also that there are three primary types of dislocations: edge, screw, and mixed. Although all three types of dislocations can propagate through a crystal and result in plastic deformation, we concentrate here on the most common and conceptually most simple of the dislocations, the edge dislocation.

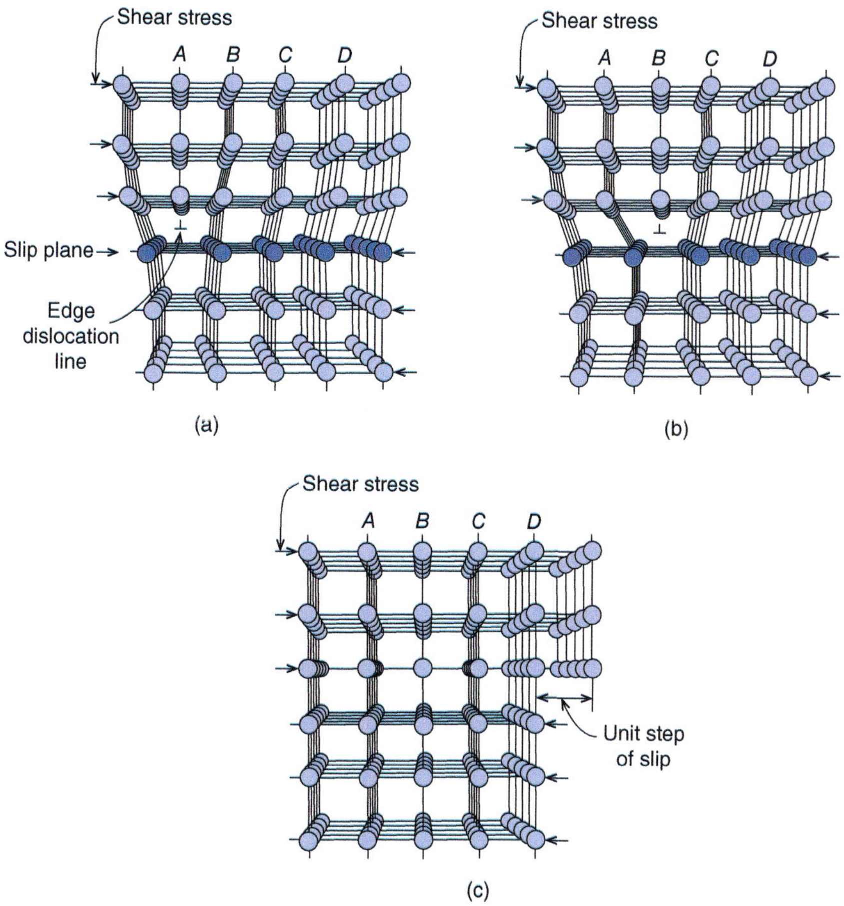

Figure 5.10 Schematic illustration of edge dislocation motion in response to an applied shear stress, where (a) the extra half-plane is labeled as *A* (cf. Figure 1.30), (b) the dislocation moves one atomic distance to the right, and (c) a step forms on the crystal surface as the extra half-plane exits the crystal. Callister [6]/ Reproduced with permission of John Wiley & Sons.

The movement of an edge dislocation in a single crystal in response to an applied shear stress is illustrated in Figure 5.10. An edge dislocation can move by slip only in its slip plane, which is the plane containing the line of the dislocation and the Burger's vector (cf. Figure 1.31). Dislocation densities range from 10^2 to 10^3 dislocations/cm^2 in the best silicon single crystals to as high as 10^{11} or 10^{12} dislocations/cm^2 in heavily deformed metal crystals. As we will see later on, heat treatment can reduce the dislocation density in metals by about five orders of magnitude.

Beside dislocation density, dislocation orientation is the primary factor in determining the critical shear stress required for plastic deformation. Dislocations do not move with the same degree of ease in all crystallographic directions or in all crystallographic planes. There is usually a preferred direction for slip dislocation movement. The combination of slip direction and slip plane is called the *slip system*, and it depends on the crystal structure of the metal. The slip plane is usually that plane having the most dense atomic packing (cf. Section 1.1.1). In face-centered cubic (FCC) structures, this plane is the (111) plane, and the slip direction is the [110] direction. Each slip plane may contain more than one possible slip direction, so several slip systems may exist for a

Table 5.1 Slip Systems in FCC, BCC, and HCP Metals

Metals	Slip Plane	Slip Direction	Number of Slip Systems
Face-Centered Cubic			
Cu, Al, Ni, Ag, Au	(111)	$[1\bar{1}0]$	12
Body-Centered Cubic			
α-Fe, W, Mo	(110)	$[\bar{1}11]$	12
α-Fe, W	(211)	$[\bar{1}11]$	12
α-Fe, K	(321)	$[\bar{1}11]$	24
Hexagonal Close-Packed			
Cd, Zn, Mg, Ti, Be	(0001)	$[11\bar{2}0]$	3
Ti, Mg, Zr	$(10\bar{1}0)$	$[11\bar{2}0]$	3
Ti, Mg	$(10\bar{1}1)$	$[11\bar{2}0]$	6

Table 5.2 Twinning Systems in FCC, BCC, and HCP Metals

Crystal Structure	Twinning Planes	Twinning Directions
fcc	(111)	$[11\bar{2}]$
bcc	(112)	$[11\bar{1}]$
hcp	$(10\bar{1}2)$	$[10\bar{1}1]$

particular crystal structure. For FCC, there are a total of 12 possible slip systems: four different (111) planes and three independent [110] directions for each plane. The possible slip systems for BCC and HCP metals are listed in Table 5.1. BCC structures have a high number of slip systems and, as a result, undergo extensive plastic deformation; that is, they are highly ductile. The opposite is true of HCP structures, which have relatively few slip systems. For reference, the corresponding twinning systems are listed in Table 5.2.

The final factor in determining the critical shear stress for plastic deformation is how the preferred slip planes are oriented relative to the applied stress. Figure 5.11 illustrates how a test sample may be oriented such that the preferred slip plane is not necessarily along the direction of the applied force, *F*. Shear forces will still exist, and their magnitude will depend not only on the applied stress but also on orientation. The angle ϕ in Figure 5.11 represents the angle between the normal to the slip plane and the applied stress direction. The angle λ represents the angle between the slip direction and the applied stress direction. Remember that the slip plane and the slip direction are not the same. It is also important to note that the three vectors in Figure 5.11 are not coplanar in general, that is, $\phi + \lambda \neq 90°$. The *resolved shear stress*, τ_r, then is the component of the applied stress, σ_t, acting along the particular slip plane of interest and can be determined as

$$\tau_r = \sigma_t \cos\phi \cos\lambda \tag{5.15}$$

As mentioned previously, several slip systems can operate simultaneously but there will be one system that has the orientation which affords it the largest resolved shear stress of all the slip systems in operation. This system will be the one that has the maximum geometric factor, $(\cos\phi\cos\lambda)_{max}$, since the applied stress is the same for all slip systems. So, the *maximum resolved shear stress*, $\tau_{r,max}$, is given by:

$$\tau_{r,\max} = \sigma_t(\cos\phi\cos\lambda)_{\max} \tag{5.16}$$

As the applied stress, σ_t, increases, the maximum resolved shear stress increases according to Eq. (5.16), finally reaching a critical value, called the *critical resolved shear stress*, τ_{cr}, at which slip along the preferred plane begins and plastic deformation commences. We refer to the applied stress at which plastic deformation commences as the *yield stress*, σ_y. The yield stress can be determined from Eq. (5.16) in the form of *Schmid's Law*, which states that slip takes place along a given slip plane and direction when the corresponding component of shear stress reaches a critical value:

$$\sigma_y = \frac{\tau_{cr}}{(\cos\phi\cos\lambda)_{\max}} \tag{5.17}$$

The minimum stress necessary to introduce yielding occurs when a single crystal is oriented such that $\phi = \lambda = 45°$, for which

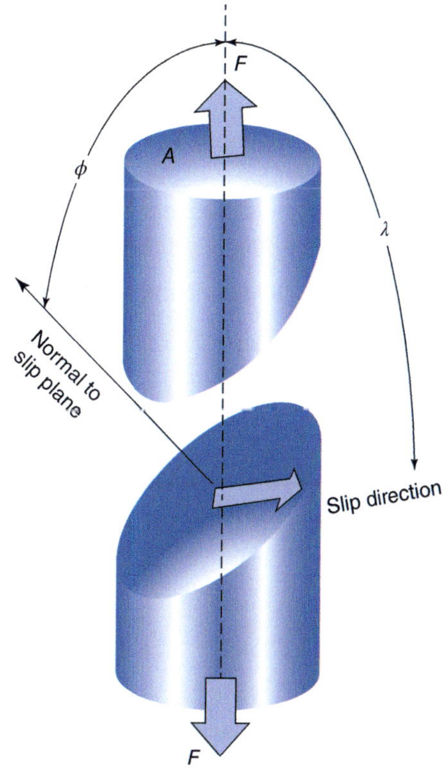

Figure 5.11 Schematic illustration of the relationship between tensile axis, slip plane, and slip direction used in calculating the resolved shear stress for a single crystal. Callister [7]/Reproduced with permission of John Wiley & Sons.

$$\sigma_y = 2\tau_{cr} \tag{5.18}$$

Equations (5.17) and (5.18) are valid for an applied tensile or compressive stress and can be used in the case of twinning as well. However, τ_{cr} for twinning is usually greater than τ_{cr} for shear. Some values of the critical resolved shear stress for slip in some common metals, and their temperature dependence, are shown in Figure 5.12. Note that the critical resolved shear stress for HCP and FCC (close-packed) structures rises only modestly at low temperatures, whereas that for the BCC and rock salt structures increases significantly as temperature decreases.

Cooperative Learning Exercise 5.3

Consider a single crystal of BCC Mo oriented such that a tensile stress of 52 MPa is applied along a [010] direction. From Table 5.1, we see that the slip plane for BCC Mo is (110) and the slip direction is [111].

Person 1: Determine the angle, ϕ, between the normal to the slip plane and the applied stress direction. You will find it helpful to draw a (110) plane in a BCC unit cell and draw the normal to this plane. Once you have determined this direction, you should be able to easily determine the angle it makes with the [010] direction by inspection.

Person 2: Determine the angle, λ, between the slip direction and the applied stress direction. You will find it helpful to draw a cubic cell with the [010] and $[\bar{1}11]$ directions as two sides of a triangle, the third side being a face diagonal. At this point, Table 1.8 may be helpful in establishing a relationship between the length of these sides in terms of the lattice parameter, a. You can then use some simple geometry to find λ.

Combine your information to calculate the resolved shear stress, τ_r, using Eq. (5.17).

Answers: (see figures) $\phi = 45°$; $\lambda = \tan^{-1}(2)^{0.5} = 54.7°$; $\tau_r = 52$ MPa(cos 45°)(cos 54.7°) = 21.3 MPa

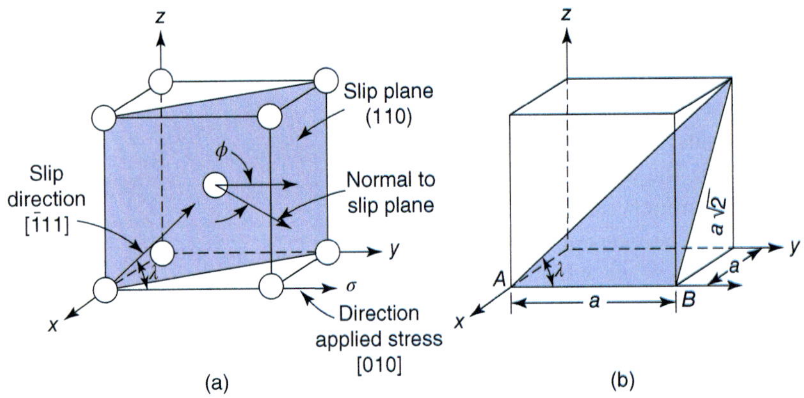

Callister [8]/Reproduced with permission of John Wiley & Sons.

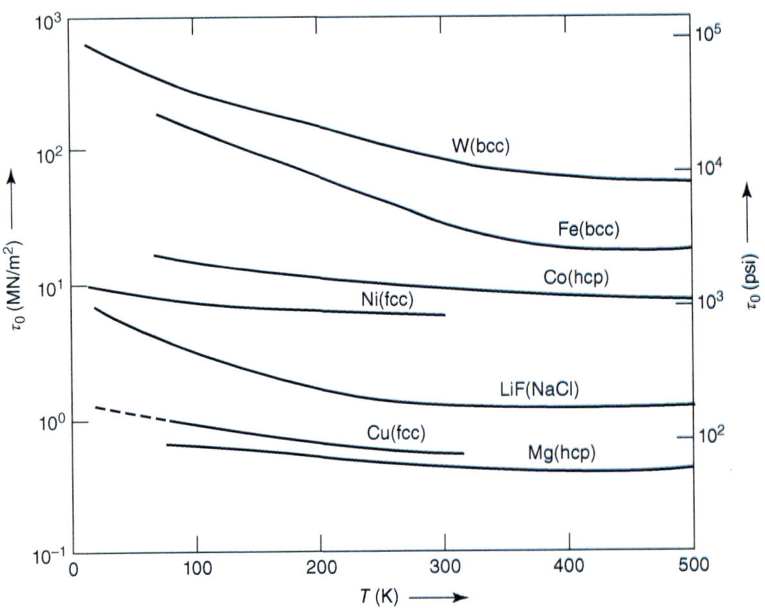

Figure 5.12 Temperature variation of critical resolved shear stress for single-crystal metals of different crystal structures. From Ralls et al. [3]. Copyright © 1976 by John Wiley & Sons, Inc. This material is used by permission John Wiley & Sons, Inc.

Slip and Recovery in Polycrystals

The yield stress, σ_y, sometimes called the *yield strength*, as introduced by Eqs. (5.19) and (5.20), is an important material property, particularly in ductile materials. It not only represents the change from a recoverable, elastic response to a plastic response, and the accompanying permanent deformation (see Figure 5.7) but also represents an important design parameter in materials selection. Obviously, when a material undergoes permanent deformation, it may not be useful for its specified purpose and may need to be replaced. As a result, the yield strength may be even more important mechanical property than the modulus for certain applications.

There are no theoretical expressions for the yield strength beyond what has already been described in terms of the critical resolved shear stress in single crystals. This problem is further complicated by the fact that most real materials are polycrystalline—that is, composed of many randomly oriented crystallites, or grains, with corresponding grain boundaries where the grains meet. As a result, only certain grains may be oriented favorably to allow slip to begin, but neighboring grains may not be so oriented, and the stress required to initiate plastic flow increases substantially in polycrystalline materials. It has been shown that at least five independent slip systems must be mutually operative for a polycrystalline solid to exhibit ductility. This means that some of the metals in Table 5.1, particularly the HCP metals such as zinc, do not have sufficient slip systems to undergo significant plastic deformation when in the polycrystalline structure.

Since the grain boundaries essentially act as barriers to slip in adjacent crystals (see Figure 5.13), it makes sense that the yield strength should depend on grain size. This is indeed the case, and the *Hall–Petch relationship* shows an inverse square-root dependence of yield strength on grain size, d:

$$\sigma_y = \sigma_0 + kd^{-1/2} \tag{5.19}$$

where σ_0 and k are empirically determined constants. Equation (5.19) has proved very useful for describing the variation in yield strength with grain size over small ranges of grain sizes, as illustrated for a copper–zinc–brass alloys in Figure 5.14. This relation is not valid, however, over large size ranges, nor for all grain sizes. The Hall–Petch relationship, and others like it, have found renewed interest in the area of nanostructured materials, since they predict enormously large yield strengths as grains approach the *nanocrystalline* range ($d \approx 10^{-9}$ m, or 1 nm).

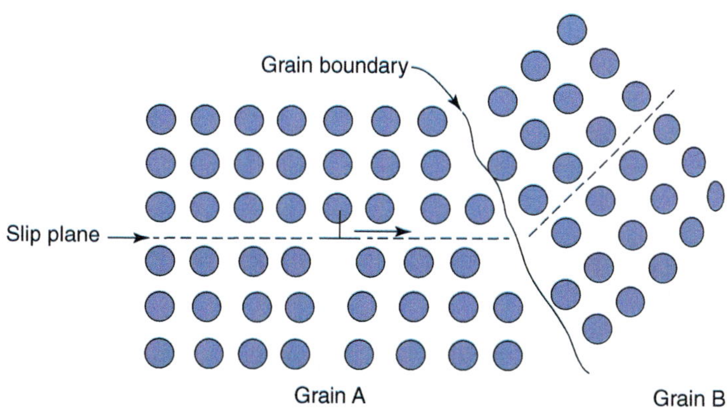

Figure 5.13 Illustration of grain boundaries acting as barriers to slip in polycrystalline materials. Callister [9]/ Reproduced with permission of John Wiley & Sons.

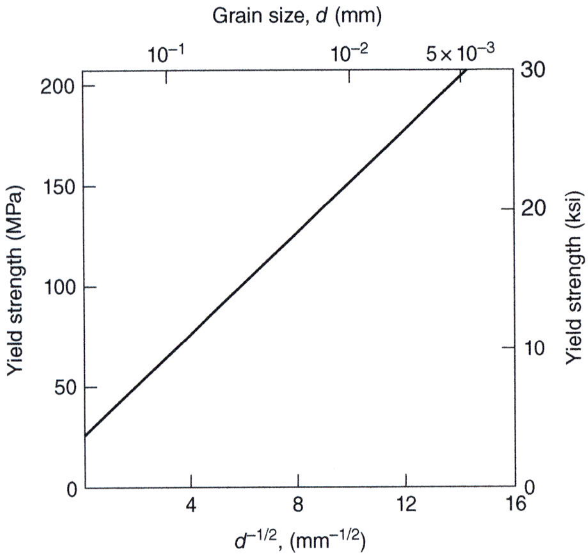

Figure 5.14 Influence of grain size on yield strength for a 70–30 Cu–Zn brass alloy. Callister [10]/Reproduced with permission of John Wiley & Sons.

Need More Help?
Try these videos on hardness and hardening in metals and alloys: Critical Resolve Shear Stress https://www.youtube.com/watch?v=79XTX6TtU7Q Strain Hardening https://www.youtube.com/watch?v=effCCaljeh8 Annealing and Recrystallization https://www.youtube.com/watch?v=xmuiMdSaGHk Hardness https://www.youtube.com/watch?v=cI5uXy4vmN8 https://www.youtube.com/watch?v=RJXJpeH78iU https://www.youtube.com/watch?v=G2JGNlIvNC4

Take a Peek! Jump ahead to Section 5.2.1 to read more about brittle fracture.

The presence of grain boundaries also affects slip in the material even after plastic deformation has commenced. A phenomenon known as *strain hardening*, or *work hardening* or *cold working*, is the result of constrained dislocation mobility and increased dislocation density as a material is strained. It is easy to observe this phenomenon; in fact, you have probably used it many times in your life without even knowing it. Simply take a small piece of metal such as a copper tube or an aluminum paper clip and bend it back and forth several times. In addition to becoming harder to bend and generating heat, the metal will eventually break. Why would a seemingly ductile metal suddenly break? This is strain hardening. As the metal is "worked" back and forth, the dislocations that allow slip and ductility to occur eventually increase in density and decrease in mobility such that the critical resolved shear stress necessary to continue plastic flow increases beyond the applied stress. At that point, either you cannot work the piece anymore or *brittle fracture* occurs. The critical resolved shear stress can be related to the *dislocation density*, ρ (not to be confused with the mass density):

$$\tau_{cr} = \tau_{cr,0} + A\rho^{1/2} \tag{5.20}$$

where A is an empirically determined constant that is approximately the product of the shear modulus, G, and the Burger's vector magnitude, b, and $\tau_{cr,0}$ is the critical resolved shear stress in the absence of interfering dislocations. This relationship has been borne out experimentally, as shown in Figure 5.15 for both single- and polycrystalline copper. The dislocation density in a metal increases with deformation or cold work, due to dislocation multiplication or the formation of new dislocations. Consequently, the average distance between dislocations decreases, and the motion of dislocations is limited by the other dislocations around it. Thus, as shown in Eq. (5.20), as the dislocation density increases, the stress necessary to deform the metal increases.

In addition to grain size control and strain hardening, a single-phase metal can be strengthened (yield strength increased) by various thermal and chemical methods, including solid solution formation and *dispersion strengthening*. Solute atoms introduced in the formation of solid solutions tend to segregate to dislocations because the total elastic strain energy associated with a dislocation and the solute atom is reduced. As a result, alloys are almost always stronger than pure metals because impurity atoms that go into solid solution ordinarily impose lattice strains (either compressive or tensile, depending on the relative size of the impurity atom) on the surrounding atoms. In this way, lattice strain field interactions between dislocations and these impurity atoms restrict dislocation movement. These same lattice interactions exist during plastic deformation, too, so a greater stress is necessary to first initiate and then continue plastic deformation in solid-solution-strengthened alloys. Besides

Take a Peek! Jump ahead to Section 5.4.1 to read more about dispersion-strengthened materials.

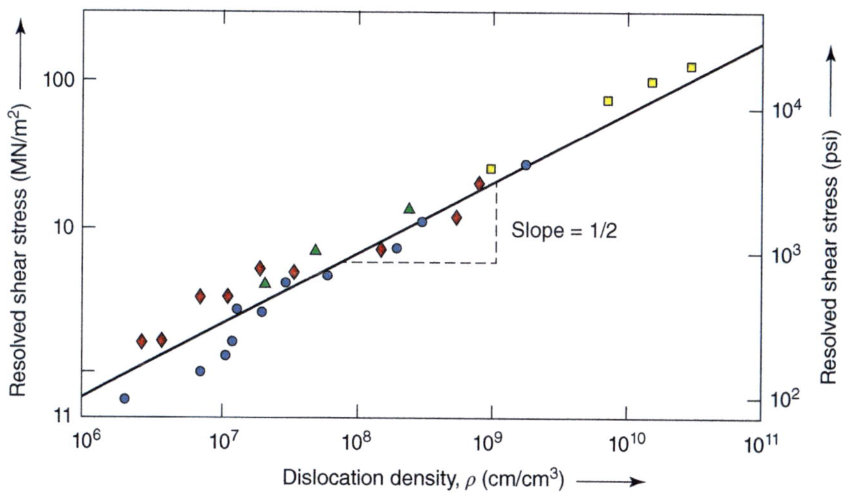

Figure 5.15 Resolved shear stress as a function of dislocation density for copper. Data are for □ polycrystalline copper; ○ single-crystal copper with one slip system operative; ◇ single-crystal copper with two slip systems operative; and Δ single-crystal copper with six slip systems operative. From Ralls et al. [3]. Copyright © 1976 by John Wiley & Sons, Inc. This material is used by permission John Wiley & Sons, Inc.

relative size of the solute atoms, the effective elastic modulus of a solute atom also determines its effectiveness as a solid-solution strengthener.

The same effect on dislocation movement during slip can be accomplished by two other techniques known as *precipitation hardening* and *dispersion strengthening*. In both cases, the principle is the same: fine dispersions of heterogeneities impede dislocation motion. In precipitation hardening, an alloy is subjected to a solution heat treatment (see Figure 5.16). First, an alloy of suitable composition (line *x–x* in Figure 5.16) is heated to a temperature above the solvus line but below the solidus line (point 1), and it is held at this temperature to allow complete homogenization of the solid solution. The alloy is quenched to form a supersaturated solution (point 2). The quenched alloy is then reheated and held at this temperature, called the *aging temperature* (point 3), for a specified time to allow the nonequilibrium second phase to precipitate out. Precipitation hardening is a very important method of strengthening many solid-solution alloys such as aluminum and magnesium alloys, copper–beryllium alloys, high nickel-based alloys, and some stainless steels. In dispersion-strengthened alloys, the atmosphere can be controlled during melt processing to form fine compound materials that do not dissolve readily in the base metal. These dispersed particles then act to restrict dislocation movement. A common example is *oxide-dispersion-strengthened* (ODS) *alloys* in which oxygen reacts with the parent metal to form higher-melting-point metal oxide particles.

Deformation in polycrystalline materials can lead to not only strain hardening and an increase in dislocation density, as described above, but also a change in grain shape. In some instances, all of these processes can be reversed through heat treatment in what is generically called *annealing*. Annealing refers to a heat treatment in which a material is exposed to an elevated temperature for an extended period of time and then slowly cooled. The annealing process usually consists of three stages: (1) heating to the desired temperature; (2) holding or "soaking" at that temperature; and (3) cooling at a slow rate, usually to room temperature. Time is an important parameter in each step. During annealing, several processes can occur that lead to what is termed "recovery," which is a restoration of properties and structures that are more similar to those found in the pre-cold-worked states. We concentrate here on those recovery processes during annealing that are specific

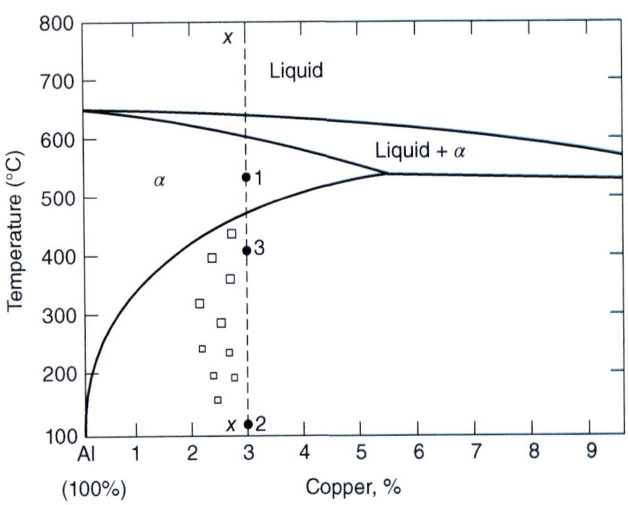

Figure 5.16 Partial Al–Cu equilibrium phase diagram illustrating precipitation hardening. From Jastrzebski [4]. Copyright © 1976 by John Wiley & Sons, Inc. This material is used by permission of John Wiley & Sons, Inc.

to metals and alloys, and we return in later sections to describe annealing in other materials, particularly inorganic glasses.

Annealing in metals can first lead to *stress relaxation* in which stored internal strain energy due to plastic deformation is relieved by thermally activated dislocation motion (see Figure 5.17). Because there is enhanced atomic mobility at elevated temperatures, dislocation density can decrease during the recovery process. At still higher temperatures, a process known as *recrystallization* is possible, in which a new set of strain-free and equiaxed (having equal dimensions in all directions) grains are formed. These recrystallized grains are similar to those in the pre-cold-worked metal and have relatively low dislocation densities. The recrystallization process is a diffusion-driven phenomenon similar to the crystallization processes described in Section 3.1.2. As shown in Figure 5.17, the tensile strength decreases and the ductility increases during recrystallization. It is worth reemphasizing that this is a time-dependent phenomenon, too. The degree of

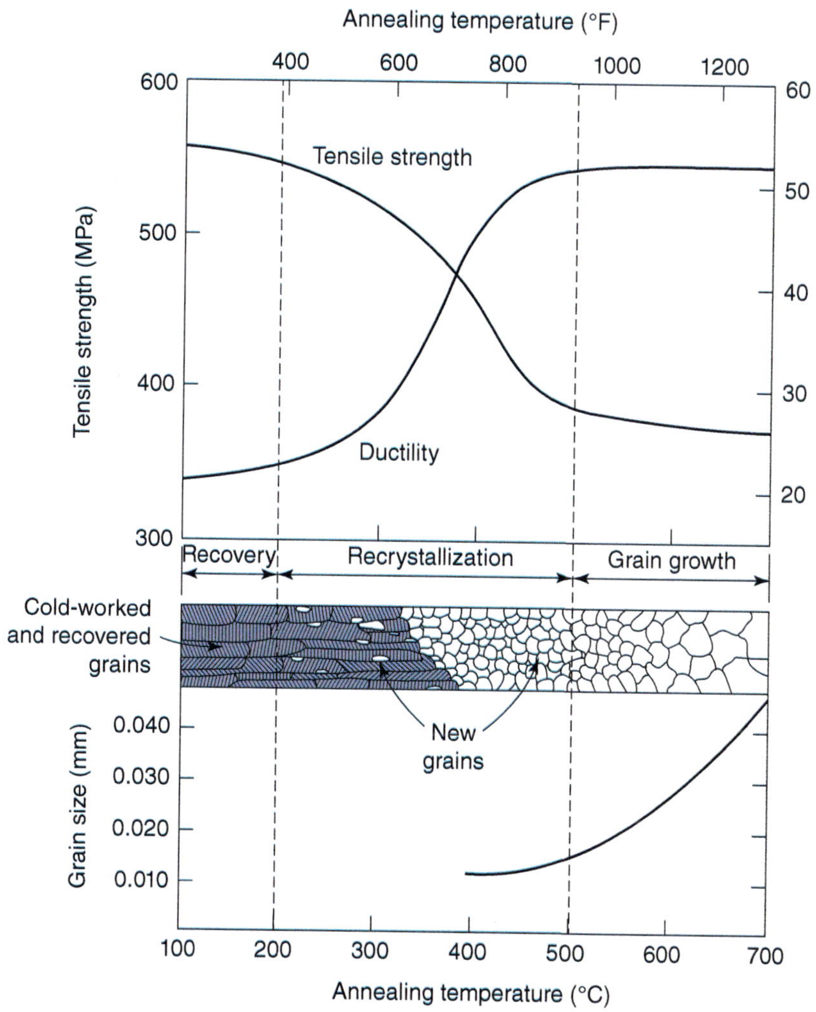

Figure 5.17 Influence of annealing temperature on tensile strength and ductility of a brass alloy. Grain structures during recovery, recrystallization, and grain growth are shown schematically. Callister [11]/ Reproduced with permission of John Wiley & Sons.

Table 5.3 Recrystallization and Melting Points of Some Common Metals and Alloys

	Recrystallization Temperature		Melting Temperature	
Metal	**°C**	**°F**	**°C**	**°F**
Lead	−4	25	327	620
Tin	−4	25	232	450
Zinc	10	50	420	788
Aluminum (99.999 wt%)	80	176	660	1220
Copper (99.999 wt%)	120	250	1085	1985
Brass (60Cu–40Zn)	475	887	900	1652
Nickel (99.99 wt%)	370	700	1455	2651
Iron	450	840	1538	2800
Tungsten	1200	2200	3410	6170

recrystallization and recovery depends upon how long the strained material is held at the specified temperature. Typically, the times necessary to obtain complete recrystallization are on the order of hours, and a *recrystallization temperature* is often defined as the temperature at which recrystallization reaches completion in one hour. The recrystallization temperature is about one-third to one-half of the absolute melting temperature, T_m. For pure metals, the recrystallization temperature is close to 0.3 T_m, but for commercial alloys, it may be as high as 0.7 T_m. Recrystallization temperatures for some common metals and alloys are listed in Table 5.3.

Strength and Hardness

One further property that is related to plastic deformation is *hardness*. Hardness is a measure of a material's resistance to localized plastic deformation—for example, from a scratch or small indentation. A description of hardness is included in this section because hardness is related to binding forces of molecules, just as strength and modulus are. We also used the term "hardening" to describe the increase in tensile strength as a metal is cold-worked. Hence, hardness is a recoverable quantity during annealing, just as yield strength and ductility are. The principles of strengthening through solid solutions and dispersants also apply here, such that alloys and dispersion-strengthened metals tend to be harder than the pure metals as well.

An unfortunate aspect of hardness is that it is difficult to quantify in an absolute manner. Most hardness scales are relative and mostly qualitative, and there is a proliferation of different hardness scales. Some of the more common scales are shown in Figure 5.18. In a typical hardness test, a hard indenter of a standard shape is pressed into the surface of a material under a specified load, causing first elastic and then plastic deformation. The resulting area of the indentation or depth of indentation is measured and assigned a numerical value. The value depends upon the apparatus and scale used. One of the simplest scales is the Mohs scale, in which a material is assigned a place on a scale depending only upon what it will scratch and what will scratch it, relative to the hardest (diamond) and softest (talc) materials known.

Since both hardness and strength are relative measures of resistance to plastic deformation, we would expect them to be proportional to each other. This is indeed the case, but since absolute values of hardness are hard to determine, it is difficult to assign a universal equation to this

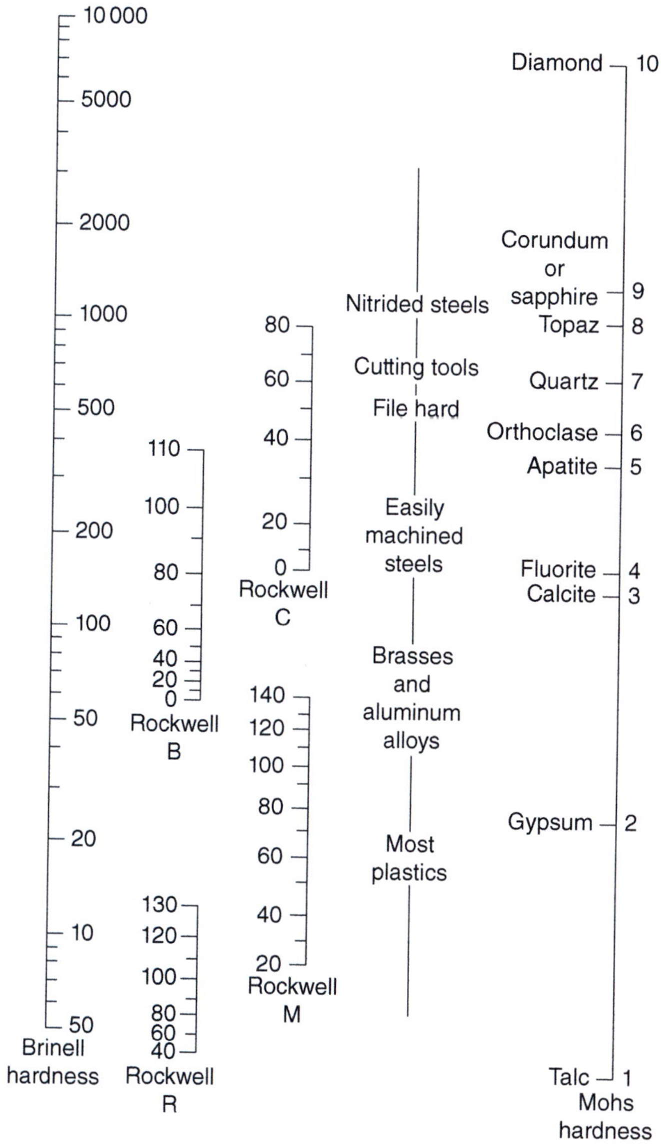

Figure 5.18 Comparison of some common hardness scales. From Jastrzebski [4]. Copyright © 1976 by John Wiley & Sons, Inc. This material is used by permission of John Wiley & Sons, Inc.

relationship. The relationship between tensile strength and Brinell hardness is shown in Figure 5.19 for steel, brass, and cast iron.

In the past two decades, more standardized measurements of hardness have been made using a technique called *nanoindentation*. This technique is essentially the same as that described above for traditional hardness testers, except that the indenters are much smaller (about 100 nm), as is the indentation area. Developments in instrumentation have made possible the measurements of loads in the nanonewtons range, and displacements of about 0.1 nm can be accurately measured [13]. The result of a nanoindentation experiment, shown schematically in Figure 5.20b, is a load versus

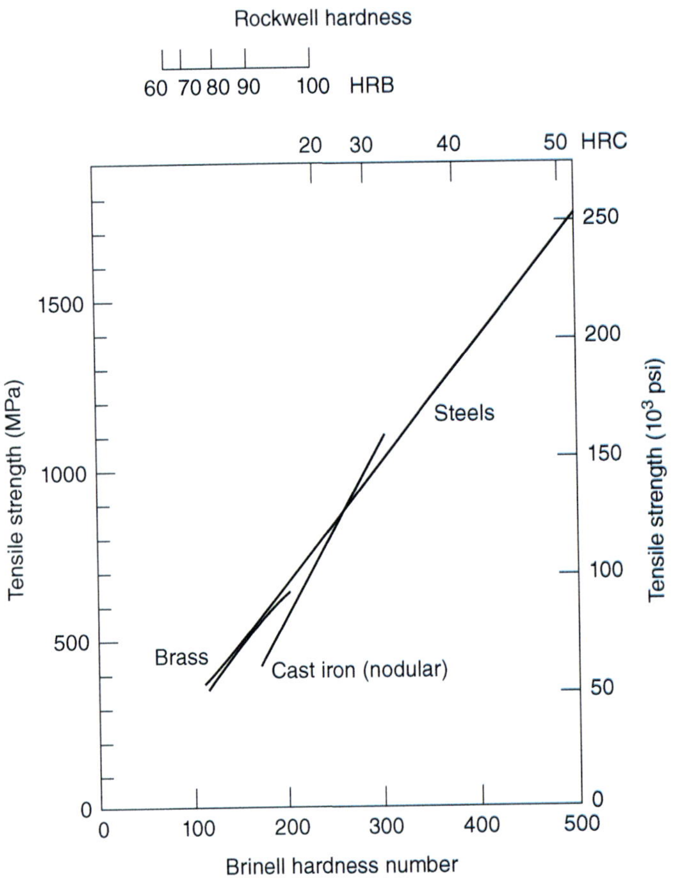

Figure 5.19 Relationship between hardness and tensile strength for steel, brass, and cast iron. Callister [12]/Reproduced with permission of John Wiley & Sons.

displacement curve, an example of which is shown in Figure 5.20a. In Figure 5.21a, h_{max} represents the displacement at the peak load, P_{max}; and h_c, the contact depth, is defined as the depth of the indenter in contact with the sample under load. The final displacement, h_f, is the displacement after complete unloading. The quantity, S, is called the initial unloading contact stiffness and is determined from the slope of the initial portion of the unloading curve, dP/dh, as indicated in Figure 5.20a.

Nanoindentation hardness, H, is defined as the indentation load divided by the projected contact area of the indentation. From the load–displacement curve, hardness at the peak load can be determined as

$$H = \frac{P_{max}}{A} \tag{5.21}$$

where A is the projected contact area. The hardness thus has units of N/m^2, or GPa. For an indenter with a known geometry, A is a function of the contact depth, h_c. For a perfect triangular-shaped, diamond-tipped indenter known as a *Berkovich indenter*, the projected contact area is

$$A = 24.56h_c^2 \tag{5.22}$$

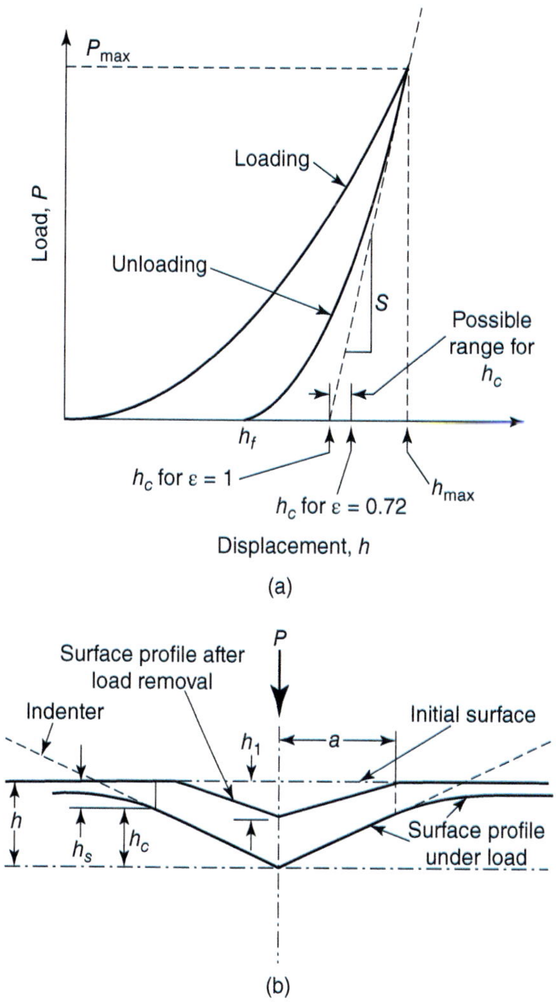

Figure 5.20 (a) A typical load–displacement curve and (b) the deformation pattern of an elastic–plastic sample during and after nanoindentation. Li and Bhushan [14]/Reproduced with permission of Elsevier.

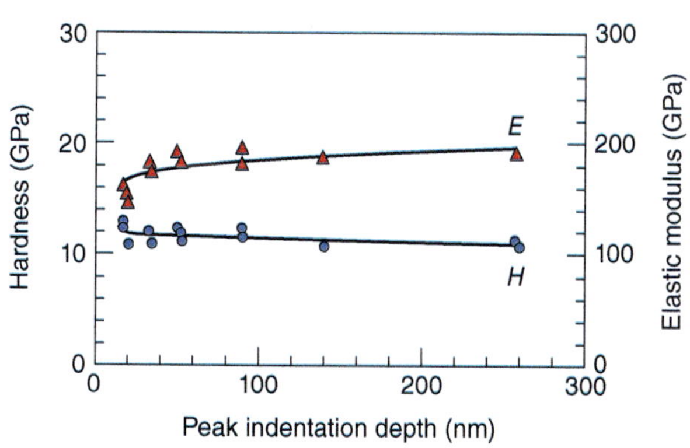

Figure 5.21 Hardness (H) and elastic modulus (E) as a function of indentation depth at various peak loads for Si(100). Li and Bhushan [15]/Reproduced with permission of Elsevier.

Often corrections must be made to A for imperfect tips and blunting effects.

The modulus from a nanoindentation test is often reported in terms of a *reduced elastic modulus*, E_r, to take into account the fact that at this size scale, the elastic response of the probe tip, E_i, as well as the modulus of the test material, E, must be considered:

$$E_r = \frac{1-\nu^2}{E} + \frac{1-\nu_i^2}{E_i} \tag{5.23}$$

where ν and ν_i are Poisson's ratio for the test material and the indenter material, respectively. For diamond, which is a common indenter material, $E_i = 1141$ GPa and $\nu_i = 0.07$. The reduced elastic modulus can be obtained from a load–displacement curve via the stiffness, S, according to:

$$S = 2\beta E_r \sqrt{A/\pi} \tag{5.24}$$

where A and E_r are as before, and β is a constant that depends on the geometry of the indenter. For a Berkovich-type indenter, $\beta = 1.034$.

There are many experimental details to consider in nanoindentation, but the end result is that both hardness and elastic modulus can be determined relatively routinely on very small samples, including thin films and nanocomposites. What is more, the hardness measurement is quantitative and reproducible. A typical hardness–modulus–indentation depth curve is shown in Figure 5.21 for a Si (100) plane.

5.1.3 Thermomechanical Effects

In addition to the temperature dependence of the properties such as strength and modulus, which we will discuss individually for each material class, there are two fundamental topics that are often described in the context of heat transfer properties or thermodynamics of materials—for example, thermal conductivity or specific heat—but are related more to mechanical properties because they involve dimensional changes. These two properties, *thermoelasticity* and *thermal expansion*, are closely related but will be described separately.

Thermal Expansion

Thermal expansion is a measure of the expansion or contraction of a solid when it is heated or cooled. In the most general case, the *coefficient of thermal expansion*, α, sometimes called the volume thermal expansion coefficient, is defined as the change in volume per unit volume, V, with temperature, T, at constant pressure, P:

$$\alpha_V = \frac{1}{V_0}\left(\frac{dV}{dT}\right)_P \tag{5.25}$$

where we use the subscript V to indicate volume thermal expansion. In many materials, the change of dimensions in any one direction may be different from that in the other two directions, depending on the crystallographic directions of the measurement. The change in linear dimensions with temperature along any one direction defines the *linear thermal expansion coefficient*, α_L:

$$\alpha_L = \frac{1}{L_0}\left(\frac{dL}{dT}\right)_P \tag{5.26}$$

For materials that are *isotropic*, that is, have the same properties in all directions, it can be shown that $\alpha_V = 3\alpha_L$. A material that has different properties in different directions is said to be *anisotropic*. Thus, a linear expansion coefficient, if no direction of measurement is explicitly stated, implies an

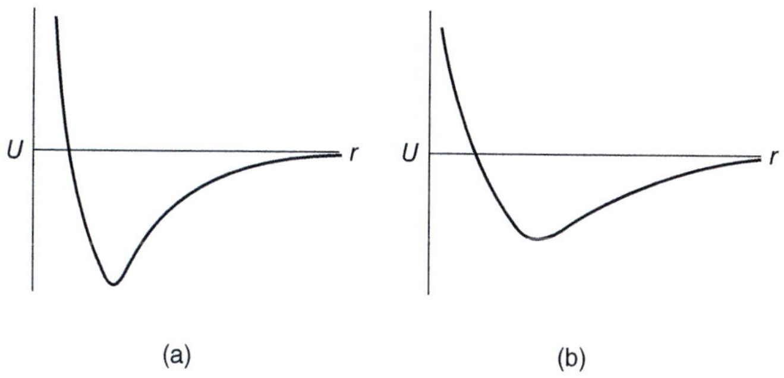

Figure 5.22 Potential energy diagrams for materials with (a) high melting point, high elastic modulus, and low thermal expansion coefficient and (b) low melting point, low elastic modulus, and high thermal expansion coefficient.

isotropic material. Conversely, a volume thermal expansion coefficient implies an anisotropic material, and one should exercise caution when deriving linear thermal expansion coefficients from volume-based measurements.

Like the melting point, specific heat, and elastic modulus, thermal expansion can be directly related to the binding force between molecules, and thus the potential energy function, as illustrated in Figure 5.22. Just as it affected the energy required to separate bonds during melting or separate bonds under an applied force, the bond strength between atoms affects the ability of atoms to deviate from their equilibrium positions due to increased thermal energy. Due to its origins in lattice vibration, the thermal expansion coefficient shows the same temperature dependence as heat capacity, and at lower temperatures it approaches zero as absolute zero is reached. It is, however, possible for a material to have a negative value of thermal expansion coefficient.

Thermoelastic Effect

A mechanical phenomenon that involves the thermal expansion coefficient is the thermoelastic effect, in which a material is heated or cooled due to mechanical deformation. The thermoelastic effect is represented by the following relation:

$$\left(\frac{dT}{d\varepsilon}\right)_S = \frac{-V_m \alpha E T}{C_V} \tag{5.27}$$

where $(dT/d\varepsilon)_S$ is the change in temperature with strain at constant entropy (adiabatic stretching or cooling), V_m is the molar volume, α is the thermal expansion coefficient, E is the tensile modulus, and C_v is the molar heat capacity at constant volume. Most materials have a positive value of α; that is, they expand upon heating, so that $(dT/d\varepsilon)_S$ is negative and the material cools upon straining. When α is negative, the material heats upon straining. This is the case for many polymer elastomers, such as a rubber band, which heats up upon being stretched.

5.1.4 Stress–Strain Diagrams

We turn our attention now to some practical aspects of mechanical property determinations. The important quantities such as modulus, strength, and ductility are typically summarized in graphical form on a *stress–strain diagram*. The details of how the experiment is performed and how the stress–strain diagram is generated are described for some common types of applied forces below.

The Tensile Test

A typical tensile test is performed in what is known as a universal testing apparatus, which is shown schematically in Figure 5.23. The specimen to be tested is typically in the shape of a "dog bone" such that the specimen is wider at the ends than in the center to facilitate mounting in the specimen grips. The sample cross section can be any geometry, as long as it is uniform along the distance between the grips, known as the *gauge length*. Circular and rectangular cross sections are the most common geometries. The gauge length can be any length, and is not standardized, though gauge lengths of 1 in. (25 mm) and 2 in. (50 mm) are common. One *grip*, typically the top grip, is attached to an immovable *load cell* that is mounted to the test frame. The load cell measures the force being imparted to the sample. The other grip is attached to a moveable portion of the frame called the *crosshead*. The crosshead moves downward at a specified rate, called the *crosshead speed*, which imparts a known *extension rate* to the sample. These rates are normally very slow, on the order of fractions of an inch per minute. The absolute extension in the sample could be estimated from the extension rate and time, but it is typically measured much more accurately with what is called an *extensometer* (eks′ ten **säm′** ə ter).

The tensile test is typically destructive; that is, the sample is extended until it plastically deforms or breaks, though this need not be the case if only elastic modulus determinations are desired. As described in the previous section, ductile materials past their yield point undergo plastic deformation and, in doing so, exhibit a reduction in the cross-sectional area in a phenomenon known as *necking*.

The "raw data" from a tensile test are the load versus elongation measurements made by the load cell and the extensometer, respectively. To eliminate sample geometry effects, the extension is divided by the initial length to obtain the dimensionless strain (which is occasionally multiplied

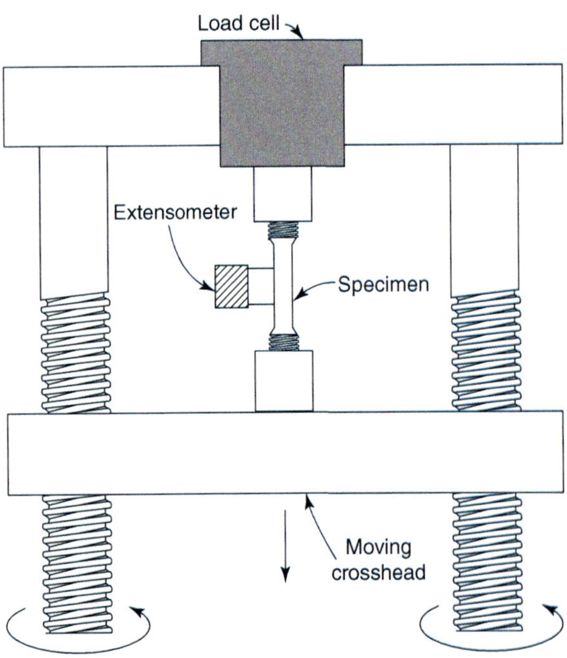

Figure 5.23 Schematic diagram of a tensile test. From Ralls et al. [3]. Copyright © 1976 by John Wiley & Sons, Inc. This material is used by permission John Wiley & Sons, Inc.

by 100 and reported as percentage elongation for samples that undergo large deformations), and the load is divided by the cross-sectional area to obtain the stress. Stress is plotted versus strain, and the result is a *stress–strain diagram*.

Cooperative Learning Exercise 5.4

Work with a neighbor. The tensile testing machine shown in Figure 5.23 can be used to measure the tensile properties of materials ranging from the most flexible and weak rubber to the strongest and stiffest steel. In order to measure the tensile properties of these vastly different materials, some things on the machine or the way a test is performed may need to be changed. What would you have to change in order to run these two very different types of materials? Name at least three things you can think of.

Answers: load cell; elongation rate; type of grip; type of extensometer.

(Courtesy of Brian P. Grady)

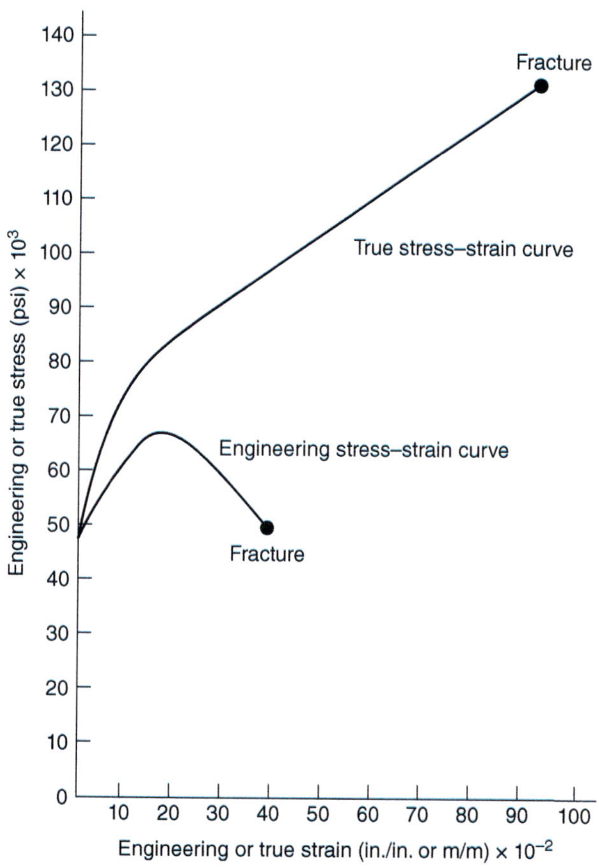

Figure 5.24 Comparison of engineering stress–engineering strain and true stress–true strain plots. Shackelford [16]/Reproduced with permission of Pearson.

The presence of necking presents a dilemma. The cross-sectional area is integral to the calculation of stress, yet it has a decreasing value as the experiment proceeds, especially where necking is prevalent. Two types of stress are usually encountered in this case. If the actual cross-sectional area corresponding to the load measurement is used, the *true stress* is obtained. This requires that the cross-sectional area be continuously monitored and matched to the corresponding load measurements. A more convenient method is to divide the measured load by the initial cross-sectional area, which is a constant, to obtain what is termed the *engineering stress*, $\sigma_{eng} = F/A_0$. There is a corresponding true strain, ε_{true}, which is the sum of all the instantaneous length changes divided by the instantaneous length (see Figure 5.4), which for small deformations is related to the instantaneous cross-sectional area, A:

$$\varepsilon_{\text{true}} = \ln\left(\frac{L_0 + \Delta L}{L_0}\right) \approx \ln\left(\frac{A_0}{A}\right) \tag{5.28}$$

Obviously, the engineering stress will be less than the true stress (since A_0 is always greater than or equal to A), and the engineering strain will be greater than the true strain at a given load for similar reasons. The result is that the true stress–true strain and engineering stress–engineering strain plots look different, as illustrated in Figure 5.24. It is customary to use engineering stress and strain for small deformations, and this is the convention we will use throughout the rest of the book, except where noted. Subscripts will be dropped from stress and strain, and the use of engineering values will be assumed.

Modulus and Strength Determinations

Once the stress–strain diagram has been generated, several important mechanical properties for a material can be obtained. Consider the stress–strain diagram for mild steel in Figure 5.25. Several points have been labeled for ease of identification. The elastic region of the diagram is easily identifiable. It begins at the origin and is the constant slope region of the graph. Here, the tensile

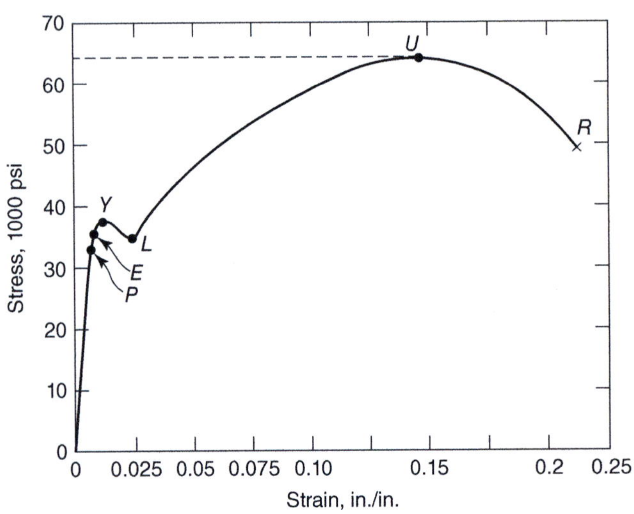

Figure 5.25 Stress–strain diagram for mild steel, illustrating different types of stress. From Jastrzebski [4]. Copyright © 1976 by John Wiley & Sons, Inc. This material is used by permission of John Wiley & Sons, Inc.

modulus, E, is given by Eq. (5.5), and it can be determined as the slope of the stress–strain curve in the elastic region. Where the elastic region ends is difficult to determine exactly. In Figure 5.25, two points have been labeled that can correspond to the end of the elastic region. The first, point P, is called the *proportional limit*, and it signals the end of the mathematical relationship between stress and strain that leads to a constant slope—that is, the proportional region. Within the region OP, Hooke's Law applies. The second point, point E, indicates the *elastic limit*. Up to point E, the strain is fully recoverable. The implication, however, is that while the response in the region OE may be elastic, there may not be a perfect Hookean response from point P to point E. In reality, it is difficult to differentiate between the proportional and elastic limits for most materials, and either one will produce an adequate determination of the tensile modulus.

Beyond point E, the material begins to plastically deform, and at point Y the yield point is achieved. The stress at the yield point corresponds to the yield strength, σ_y [see Eq. (5.17)]. Technically, point Y is called the *upper yield point*, and it corresponds to the stress necessary to free dislocations. The point at which the dislocations actually begin to move is point L, which is called the *lower yield point*. After point L, the material enters the ductile region, and in polycrystalline materials such as that of Figure 5.25, strain hardening occurs. There is a corresponding increase in the stress required to maintain plastic flow, and a maximum stress value is often achieved in this region, here at point U. The stress at point U is called the ultimate strength, σ_u, or the *ultimate tensile strength* (UTS). Sometimes the UTS is a more important design parameter than the yield strength because it indicates the maximum load that a material will bear, regardless of whether it deforms or not. It should be pointed out that not all materials strain harden, so the ultimate strength can be located at any point along the stress–strain curve; it is simply the maximum strength. We will see examples for which the maximum stress occurs at the yield point, in which case $\sigma_u = \sigma_y$. The final point in Figure 5.26 is point R, which represents the *breaking point*, or *rupture point* or *fracture point*, of the material. The stress at point R is called the *fracture strength*, σ_f. Again, we will see an example of stress–strain diagrams for which the fracture strength can vary position; it is simply the point at which the material physically separates into two or more pieces. If the material is not extended to the breaking point, then no fracture strength can be determined.

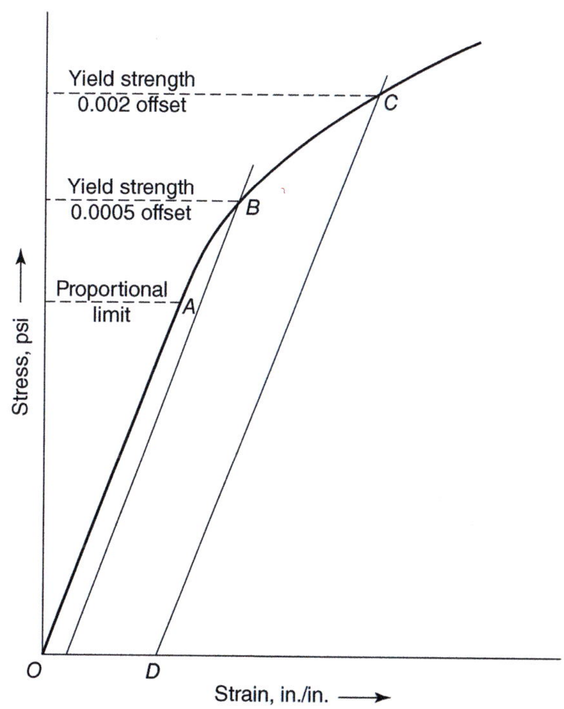

Figure 5.26 Stress–strain diagram indicating method for determining proof stress. From Jastrzebski [4]. Copyright © 1976 by John Wiley & Sons, Inc. This material is used by permission of John Wiley & Sons, Inc.

Yield Stress and Proof Stress

In Figure 5.25, we noted that there was a distinct point, point P, which indicated the end of the proportional region in the stress–strain diagram, up to which point Hooke's Law applied. Unfortunately, most ductile metals do not have a true proportional limit, or a true yield

Need More Help?

Try these videos on mechanical testing:

Proof Stress
https://www.youtube.com/watch?v=InEpKlYBkvk

Tensile Testing
https://www.youtube.com/watch?v=reS8tk4OtgM
https://www.youtube.com/watch?v=D8U4G5kcpcM

point. A common example of the stress–strain diagram in a tensile experiment is shown in Figure 5.26. Instead of having an upper and lower yield point, the modulus changes slowly and continuously until the ductile region is attained. This makes it difficult to determine the proportional limit and modulus, as well as the yield point and yield strength.

We must create an artificial convention for determining where the elastic region ends and the ductile region begins. This is achieved with what is known as the *proof stress*. The proof stress is defined as the stress required to produce a certain definite permanent strain. This is determined by the offset method. A line (DC in Figure 5.26) is drawn parallel to the straight-line portion of the stress–strain diagram (OA in Figure 5.26), starting at a predetermined strain for point D. Typically, this is a value of 0.2% strain (0.002), and the distance OD is known as the "0.2% offset." The yield point is then taken as the point where the line DC intersects the stress–strain diagram, or point C. The stress at point C is the proof stress, which is then taken as the yield stress and is referred to as the "yield stress at 0.2% offset." The modulus is still calculated in the straight-line portion of the curve, as indicated by line OA. It should be noted that other types of offsets can be used, such as the 0.05% offset indicated by the line ending at point B. The type of offset used should be clearly indicated in the yield strength determination for these materials.

Ductility, Toughness, and Resilience

Several additional useful quantities can be determined from a stress–strain diagram. These are illustrated in Figure 5.27. In addition to the tensile modulus, yield strength, and ultimate strength as described previously, the ductility, which we have previously used in a qualitative sense to describe a metal that undergoes plastic deformation, can be given a quantitative value by reporting the percent elongation at failure. This point is indicated by point 4 in Figure 5.27. Note that in most

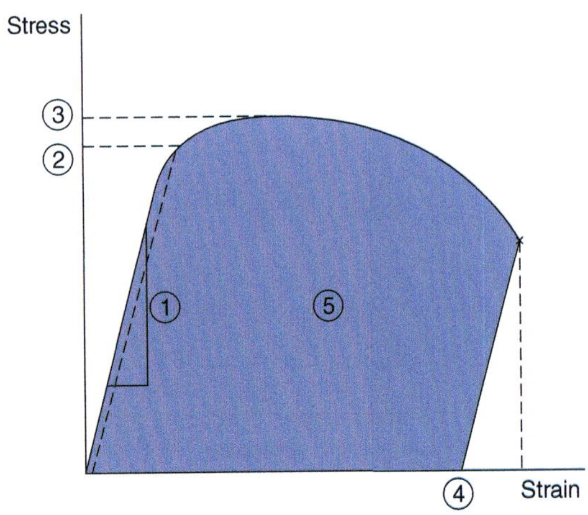

Figure 5.27 Stress–strain diagram showing (1) modulus, (2) yield strength, (3) ultimate tensile strength, (4) ductility, and (5) toughness. Note the use of proof stress in determination of yield stress. Shackelford [17]/Reproduced with permission of Pearson.

cases, there is an elastic recovery after fracture (the slope is equal to that given by point 1) that must be accounted for in determining the final elongation. Recall that elongation is:

$$\%\text{Elongation} = \left(\frac{L_f - L_0}{L_0}\right) \times 100 \tag{5.29}$$

The % elongation at fracture is thus dependent upon the initial gauge length, which should be cited when reporting values of ductility. Typically, a gauge length of 50 mm (2 in.) is used. Another quantity that is sometimes used to express ductility is the percentage *reduction in area* (%RA):

$$\%\text{RA} = \left(\frac{A_0 - A_f}{A_0}\right) \times 100 \tag{5.30}$$

Cooperative Learning Exercise 5.5

Work in groups of four. A tensile specimen of 0.505-in. diameter and 2-in. gauge length is subjected to a load of 10,000 lb_f that causes it to elastically deform at constant volume to a gauge length of 2.519 in.

Person 1: Calculate the engineering strain.

Person 2: Calculate the true strain

Person 3: Calculate the percentage elongation.

Person 4: Calculate the percentage reduction in area, %RA. (You may want to consult Person 2 on this.)

Compare your answers. How are they similar? How are they different?

Persons 1, 3: Determine the engineering stress. Use this value and the engineering strain to estimate the elastic modulus.

Person 2, 4: Determine the true stress. Use this value and the true strain to estimate the elastic modulus.

Compare your answers. Why are the moduli determined with these two methods the same or different?

Answers: ε = (2.519 − 2)/2 = 0.2595; ε_{true} = ln(2.519/2) = 0.2307; %elongation = (0.2595) × 100 = 25.95%; %RA = $(1 - A_f/A_0) \times 100$ = (1 − 0.79397) × 100 = 20.6. σ = 10, 000 $lb_f/[\pi(0.505\text{in}/2)^2]$ = 49.9 kpsi; E = 49.9 kpsi/0.2595 = 192.4 Mpsi σ_{true} = 10, 000(1.259)/$[\pi(0.505\text{in}/2)^2$ = 62.9 kpsi; E = 62.9/0.2307 = 272.5 Mpsi.

As illustrated in Figure 5.25, the true stress–strain and engineering stress–strain curves can be quite different, especially at high elongations such as this. In fact, relations such as (5.31) are probably not valid at these large elongations. Furthermore, the elastic limit has probably been exceeded at point, despite our assumptions.

Two remaining quantities, resilience, and toughness, are related to the ability of a material to absorb energy. *Toughness*, as indicated in Figure 5.28, is the area under the stress–strain curve up to the point of fracture. Thus, it is both a measure of a material's strength and ability to deform plastically. A material with low strength and high ductility may possess more toughness than one

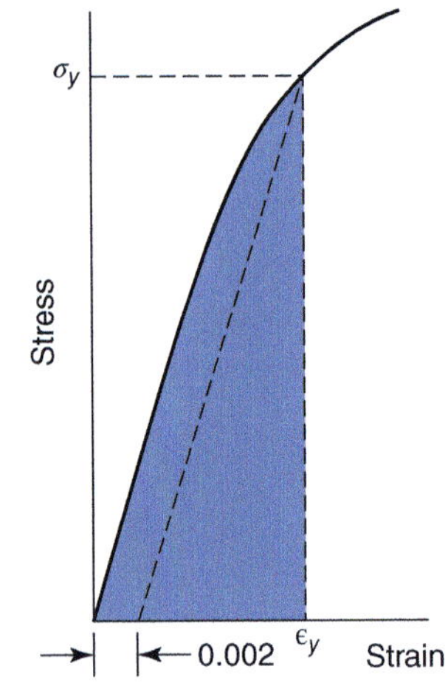

Figure 5.28 Schematic illustration of how modulus of resilience (shaded area) is determined. Callister [18]/ Reproduced with permission of John Wiley & Sons.

with high strength and low ductility. This is an important design consideration in many applications. The continued use of metals in automobile manufacturing (for frames and other structural components) is the result, in part, of safety considerations. There are certainly stronger materials than aluminum, less dense materials than steel, and less costly materials than either, but the combination of properties, including the ability to absorb energy during impact, is what keeps metals attractive for these applications.

A quantity related to toughness is the *resilience*, which is the ability of a material to absorb energy when it is deformed elastically, and then, upon unloading, to recover the energy. Resilience is quantified with the *modulus of resilience*, U_r, which is the strain energy per unit volume required to stress a material from an unloaded state up to the point of yielding. As shown in Figure 5.28, this is simply the area under the curve to the yield point. Again, note how the yield point is determined from the 0.2% offset in metals. Assuming a linear elastic region up to the yield point (which we know is an approximation in most situations), the modulus of resilience can be found from the yield strength, σ_y, and modulus, E:

$$U_r = \frac{\sigma_y^2}{2E} \tag{5.31}$$

5.1.5 Mechanical Properties of Metals and Alloys

After a somewhat lengthy introduction into the theoretical and experimental aspects of mechanical properties, which we will see apply to all material classes, we come finally to the specifics of mechanical properties of metals and alloys.

Temperature Dependence of Strength

The modulus and strength of selected metals and alloys are listed in Appendix 6. However, there are some generalizations that can be made and trends to be observed in the mechanical properties of metals and alloys.

At elevated temperatures, the thermal recovery processes described in Section 5.1.2 can occur concurrently with deformation, and both strength and strain hardening are consequently reduced. The latter effect results in decreasing the difference between yield and tensile strengths until at sufficiently high temperatures, they are essentially equal. At lower temperatures, temperature has a marked influence on deformation in crystalline materials. Temperature can affect the number of active slip systems in some metals, particularly HCP metals (see Section 5.1.2), and this determines the ductility of polycrystalline metals such as Zn and Mg. Furthermore, dislocations move less readily as temperature is decreased. As a result, yield and tensile strengths of polycrystalline materials increase and their ductility, measured in %RA [see Eq. (5.30)], decreases with decreasing temperature, as illustrated in Figure 5.29. Below a certain temperature known as the *brittle-to-ductile transition temperature*, metals can behave in a brittle manner; that is, they fracture before they undergo extensive plastic deformation. This is particularly true for BCC metals below the brittle-to-ductile transition temperature. However, FCC metals seldom show such a transition, and their ductility

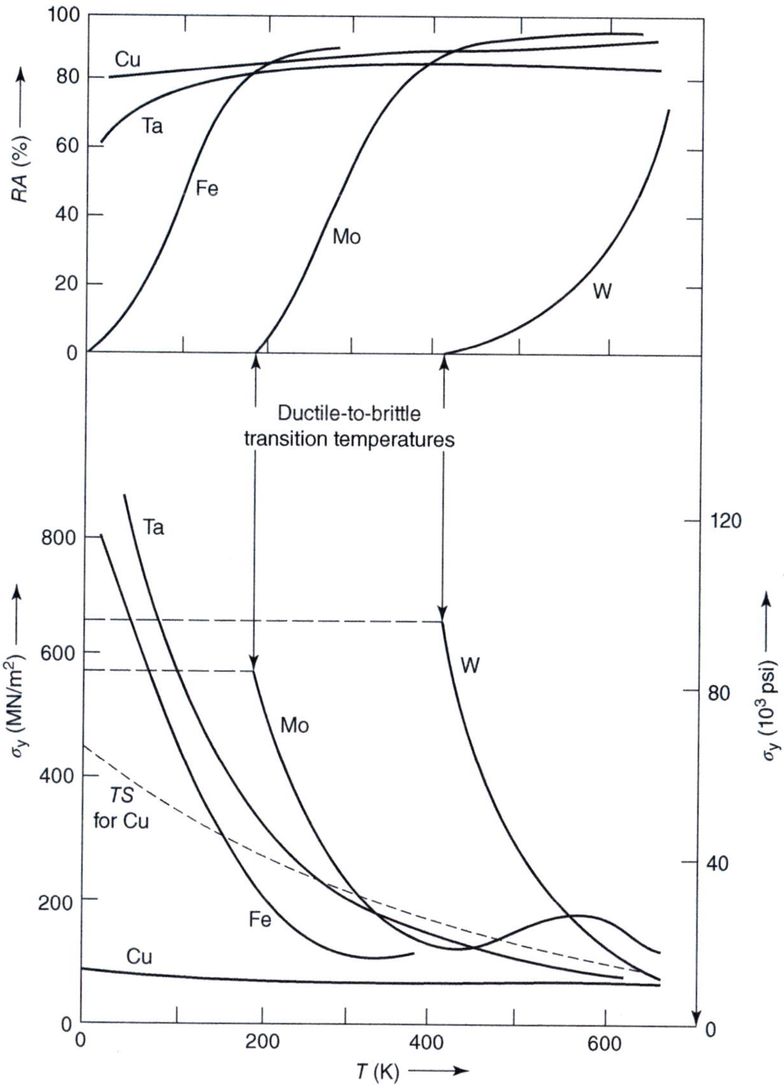

Figure 5.29 Variation of percentage reduction in area (%RA, top graph) and yield strength (bottom graph) with temperature for selected metals. From Ralls et al. [3]. Copyright © 1976 by John Wiley & Sons, Inc. This material is used by permission John Wiley & Sons, Inc.

decreases only slightly with decreasing temperature. This phenomenon is related to the temperature dependence of the critical resolved shear stress previously shown in Figure 5.12.

Strain-Rate Dependence of Strength

The plastic behavior of crystalline materials also depends on tensile strain rate, $\dot{\varepsilon}$. To this point, we have basically assumed that tests are performed at a constant strain rate that does not vary much from experiment to experiment. Strain rate is an important parameter, however, as illustrated in Figure 5.30. Here, the tensile strength (*TS*, or σ_u) for polycrystalline copper is shown both as a function of temperature and strain rate. The temperature dependence is as discussed in the previous section; that is, *TS* increases with decreasing temperature. The strain rate dependence on tensile

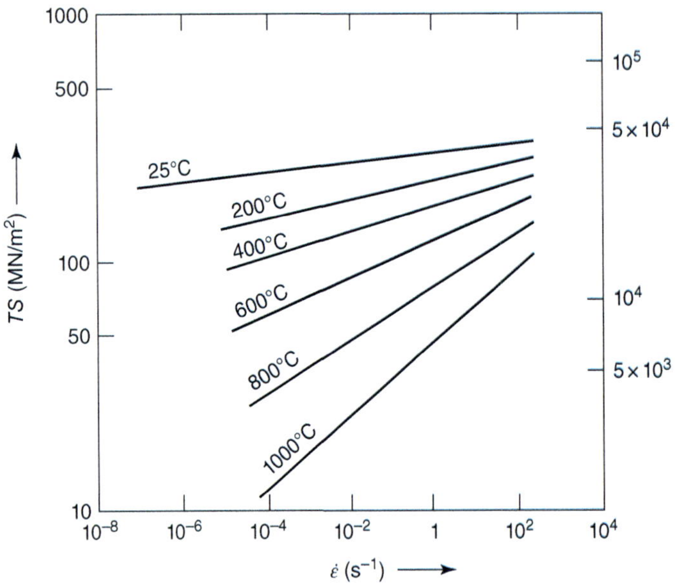

Figure 5.30 Dependence of tensile strength on strain rate and temperature for polycrystalline copper. From Ralls et al. [3]. Copyright © 1976 by John Wiley & Sons, Inc. This material is used by permission John Wiley & Sons, Inc.

strength is more pronounced at elevated temperatures and, as illustrated by the logarithmic coordinates in Figure 5.29, can be described by the relationship:

$$\sigma_u = K\dot{\varepsilon}^m \tag{5.32}$$

where K is a proportionality constant and m is a temperature-dependent measure of the strain-rate sensitivity of a material. A similar relationship holds for the yield stress, σ_y. The value of m is usually small at low temperatures but can be appreciable at high temperatures. For FCC metals at low temperatures, m is of the order 0.01. At low temperatures, BCC metals are more strain rate sensitive, but m is still only about 0.1. In the next section, we describe materials that have large strain-rate exponents on the order of 0.5 and that can undergo great extensions.

Cooperative Learning Exercise 5.6

Using the data in Figure 5.30, select two strain rates for which data are provided.

Person 1: Determine the value of m in Eq. (5.36) for copper at 1000°C.

Person 2: Determine the value of m in Eq. (5.36) for copper at 25°C.

Compare your results. Do they support the claim that strain-rate dependence is more pronounced at elevated temperatures? What would the value of m be if there were no strain-rate dependence?

Answer: In both cases, Eq. (5.32) can be used to obtain $\ln(\sigma_2/\sigma_1) = m \ln(\varepsilon_2/\varepsilon_1)$, from which m (1000°) ≈ 0.1; m(25°) ≈ 0.02. There is indeed more dependence at higher temperatures. The value of m would be zero for no dependence.

Superelasticity and Superplasticity

Superplastic materials are polycrystalline solids that are able to undergo large uniform strains prior to failure. For deformation in uniaxial tension, elongations to failure in excess of 200% are usually indicative of superplasticity, although several materials can attain extensions greater than 1000%. There are two main types of superplastic behavior: (a) *micrograin* or *microstructural superplasticity* and (b) *transformation* or *environmental superplasticity*. Micrograin superplasticity is exhibited by materials with a fine grain size, usually less than 10 μm, when they are deformed within the strain rate range 10^{-5} to 10^{-1}/s at temperatures greater than 0.5 T_m, where T_m is the melting point in degrees Kelvin. Superplastic deformation is characterized by low flow stresses; this, combined with the high uniformity of plastic flow, has led to considerable commercial interest in the superplastic forming of components. The most important mechanical characteristic of a superplastic material is its high strain-rate sensitivity of flow stress, as given previously in Eq. (5.29). For superplastic behavior, m would be greater than or equal to 0.5; for most superplastic materials, m lies in the range 0.4–0.8. The presence of a neck in a material subject to tensile straining leads to a locally high strain rate and, for a high value of m, to a sharp increase in the flow stress within the necked region. Hence, the neck undergoes strain rate hardening that inhibits its further development. Thus, a high strain-rate sensitivity results in a high resistance to neck development and the high tensile elongation characteristic of superplastic materials.

Historical Highlight

Superplasticity was first described a long time ago. Find out more: https://link.springer.com/article/10.1007/s10853-009-3780-5#Sec4

There are two main types of superplastic alloys: *pseudosingle-phase* and *microduplex*. In the former class of material, a combination of hot and cold working and heat treatment is employed to develop a fine-scale distribution of dispersoids so that on recrystallization the alloy will have a grain size less than 5 μm. Ideally, the dispersion of particles will also prevent any further grain growth during superplastic deformation; however, most of the second-phase particles have usually redissolved into solution at the superplastic forming temperatures. The precipitation-strengthened aluminum alloys based on the Al–Cu, Al–Mg, Al–Zn–Mg, Al–Li, and Al–Cu–Li alloys can be included in this group. Some typical properties of the Al-based superplastic alloys are presented in Table 5.4. Other materials include dispersion-strengthened copper alloys where silicide or aluminide particles are used to limit grain growth. The aluminum alloys, which from a commercial viewpoint are the most important of the pseudosingle-phase materials, can be further subdivided into those which are recrystallized prior to superplastic forming and those which acquire their fine grain structure only after a limited amount of deformation at the forming temperature. The microduplex materials are thermomechanically processed to give a fine grain or phase size. Grain growth is limited by having a microstructure that consists of roughly equal proportions of two or more chemically and structurally different phases. This latter group of materials includes α/β titanium alloys, α/γ stainless steels, α/β and $\alpha/\beta/\kappa$ copper alloys, and eutectics. Some properties of selected titanium-based superplastic alloys are presented in Table 5.4.

A property related to superplasticity is *superelasticity* or pseudoelasticity. Many of the same types of alloys that possess superplasticity also possess superelasticity, particularly titanium-based alloys, and these are the same alloys that were described in Section 3.1.2 as shape memory alloys. Recall that these materials have a transformation temperature (cf. Figure 3.5) below which they are in an austenitic–martensitic state and above which they are in the austenitic state. If the material is tested

Table 5.4 Properties of Selected Alloys Exhibiting Superplasticity

Alloy Composition	Temperature of Applicability (°C)	Superplasticity (% Elongation)	*m*
Al-Based Alloys (Balance Al)			
6%Cu–0.4%Zr–0.3%Mg	400–480	1800	0.45–0.7
5.5%Zn–2.0%Mg–1.5% Cu–0.2%Cr	510–530	1400	0.5–0.8
2.7%Cu–2.2%Li–0.7% Mg–0.12%Zr	510–530	800	0.4–0.6
4.8%Cu–1.3%Li–0.4% Mg–0.4%Ag–0.14%Zr	470–530	1000	0.45
5%Ca–5%Zn	450–550	600	0.4–0.5
4.7%Mg–0.7%Mn–0.15%Cr	480–550	670	0.4–0.65
2.5%Li–1.2%Cu–0.6% Mg–0.1%Zr	500–540	1000	0.4–0.6
Ti-Based Alloys (Balance Ti)			
6%Al–4%V	790–940	1400	0.6–0.8
5.8%Al–4%Sn–3.5%Zr–0.5% Mo–0.3%Si–0.05%C	950–990	400	0.35–0.65
4%Al–4%Mo–2%Sn–0.5%Si	810–930	1600	0.48–0.65
4.5%Al–3%V–2%Fe–2%Mo	750–830	700	0.5–0.55
14%Al–20%Nb–3%V –2%Mo	940–980	1350	0.4–0.6
6%Al–2%Sn–4%Zr–2%Mo	880–970	900	0.5–0.7

Source: J. Pilling [19].

just above its transformation temperature to austenite, the applied stress transforms the austenite to martensite and the material exhibits increasing strain at constant applied stress; that is, considerable deformation occurs for a relatively small applied stress. When the stress is removed, the martensite reverts to austenite and the material recovers its original shape. This effect makes the alloy appear extremely elastic; in some cases, strains of up to about 10% can be fully recovered (see Figure 5.31). Superelastic shape memory alloys are used in the nearly indestructible frames of eyeglasses, and they make excellent springs. Many biomedical applications use superelastic wires and tubes, including catheters and guide wires for steering catheters and orthodontics.

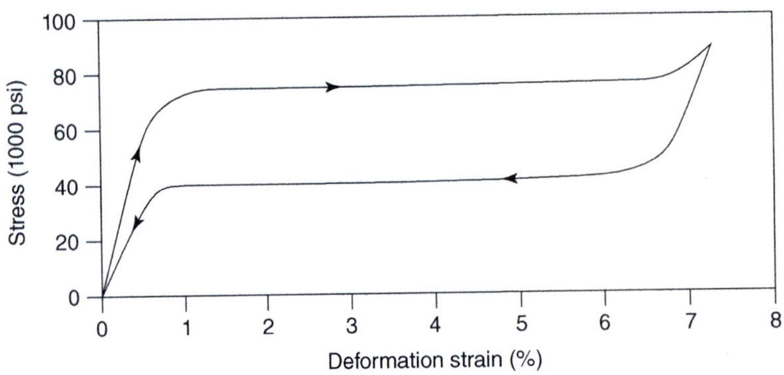

Figure 5.31 Example of superelasticity in a Ni–Ti alloy. A deformation of 7% is fully recovered. (ShapeMemory Applications, Inc.).

5.2 Mechanics of Ceramics and Glasses

In the previous sections, we saw how materials can behave in elastic and ductile manners and how plastic flow past the yield point could be considered a form of destructive failure due to permanent deformation. In this section, we will see that there are additional types of destructive failure. In particular, we will concentrate on two types of failure known as brittle fracture and creep.

These types of failure are common in, but not limited to, brittle materials such as ceramics and glasses. Brittle materials are characterized by their lack of plastic deformation, as illustrated in Figure 5.32. The ductile materials such as metals have a distinct yield point, after which plastic flow begins, and the process of cross-sectional area reduction known as necking occurs (cf. Section 5.1.4). The necking phenomenon is illustrated in Figure 5.32a and b, after which point failure can occur by either highly ductile fracture (known as rupture) or moderately ductile fracture. In brittle materials, little or no plastic deformation precedes the fracture, as in Figure 5.32c.

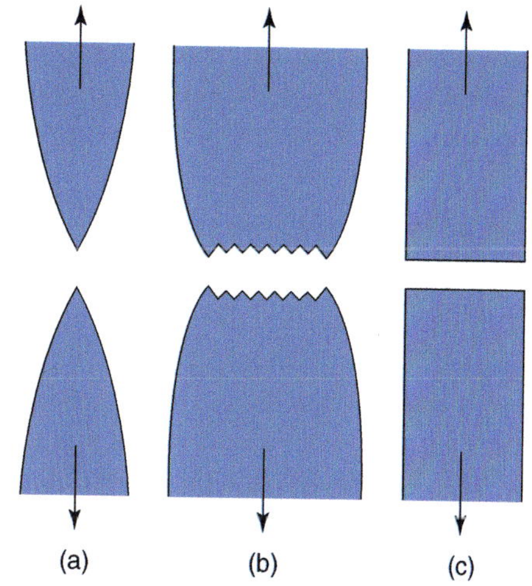

Figure 5.32 Schematic illustration of (a) rupture, (b) ductile fracture, and (c) brittle fracture. Callister [20]/Reproduced with permission of John Wiley & Sons.

In terms of the stress–strain diagram, brittle materials exhibit primarily an elastic response, with little or no strain beyond the elastic limit prior to fracture (Figure 5.33). The modulus may be higher

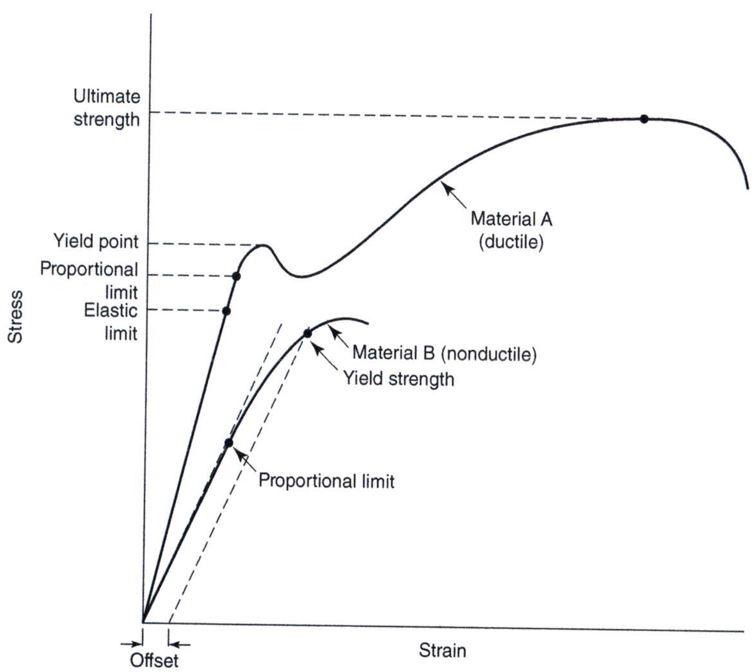

Figure 5.33 Comparison of typical stress–strain diagrams for ductile (top curve) and brittle (bottom curve) materials. Somayaji [21]/Reproduced with permission of Pearson.

or lower than a ductile material. As illustrated earlier by point R in Figure 5.25, the stress at fracture is known as the fracture strength, σ_f. We will devote the next section to the development of theoretical expressions for fracture strength. A description of creep will then be given, followed by an overview of the mechanical properties of glasses and ceramics.

5.2.1 Fracture Mechanics

The Molecular Origins of Fracture

Let us first calculate the theoretical fracture strength of a material—that is, the energy required to separate the atomic planes of a solid. Consider pulling a bar of unit cross section, much as we did schematically in Figure 5.9, except that instead of a shear stress, τ, as was used there, we now are applying a tensile stress, σ, in an attempt to pull the atomic planes apart. Cohesive stresses between atomic planes, σ, vary with the change in interplanar separation, x, where $x = a - a_0$. Here, a_0 is the original (equilibrium) interplanar spacing, for which the corresponding stress is zero, and a is the interplanar spacing at any applied stress.

The variation of the tensile stress with interplanar separation can be approximated by a sine curve of wavelength λ (see Figure 5.34), much like we did in Figure 5.10 for an applied shear stress [cf. Eq. (5.12)]:

$$\sigma = \sigma_c \sin\left(\frac{2\pi x}{\lambda}\right) \tag{5.33}$$

The quantity σ_c is called the *theoretical cleavage strength* (we will use the terms fracture and cleavage interchangeably here) and it is the maximum stress required to separate, or cleave, the planes. Since cleavage occurs at an interplanar separation of $x = \lambda/2$, the work per unit area required to separate planes, also known as the *strain energy*, is the area under the curve in Figure 5.37 up to that point,

$$W_\varepsilon = \int_0^{\lambda/2} \sigma \, dx \tag{5.34}$$

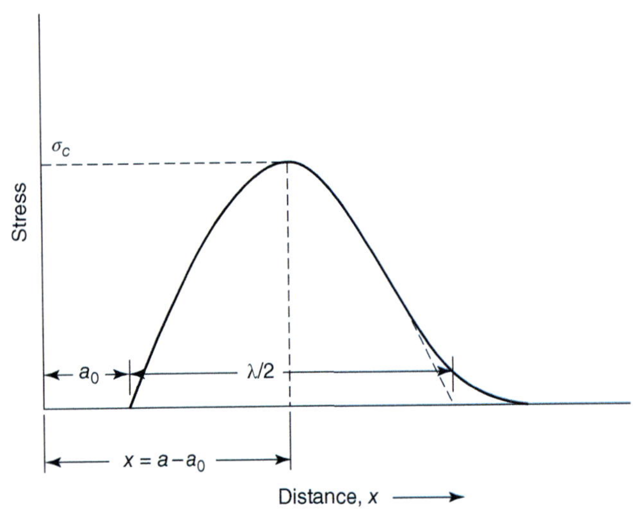

Figure 5.34 Sine-wave representation of stress variation with interatomic separation distance for two atomic planes. From Jastrzebski [4]. Copyright © 1976 by John Wiley & Sons, Inc. This material is used by permission of John Wiley & Sons, Inc.

or, if we substitute the expression for stress from Eq. (5.33) into Eq. (5.34),

$$W_\varepsilon = \int_0^{\lambda/2} \sigma_c \sin\left(\frac{2\pi x}{\lambda}\right) dx = \frac{\lambda \sigma_c}{\pi} \tag{5.35}$$

Recall from Section 2.4.1 that there is work associated with creating any new surface. The fracture of the solid at $\lambda/2$ creates two new surfaces where none existed before (see Figure 5.35c), so that the work of cohesion between the two surfaces, W_ε, is given by Eq. (2.69), $W_\varepsilon = 2\gamma$, where γ is the surface energy of the solid. We can equate this work with the work calculated in Eq. (5.35) to give:

$$\frac{\sigma_c \lambda}{\pi} = 2\gamma \tag{5.36}$$

We also know that to a first approximation, Hooke's Law applies to the elastic region, which for brittle materials is effectively up to the point of cleavage, so that the stress at cleavage, σ_c, is given by Hooke's Law as:

$$\sigma_c = E\varepsilon_c \tag{5.37}$$

where E is the elastic modulus, and ε_c is the strain at cleavage, $(a - a_0)/a_0$. The interplanar separation $x = a - a_0$ can be estimated as $\lambda/2$ at cleavage, so that Eq. (5.37) becomes:

$$\sigma_c = \frac{E\lambda}{2a_0} \tag{5.38}$$

Solving Eq. (5.36) for λ and substituting into Eq. (5.38) gives:

$$\sigma_c^2 = \frac{E\gamma\pi}{a_0} \tag{5.39}$$

or, since $\sqrt{\pi}$ is a numerical factor close to 1 (we are looking for only order of magnitude effects here), we have:

$$\sigma_c \approx \sqrt{\frac{E\gamma}{a_0}} \tag{5.40}$$

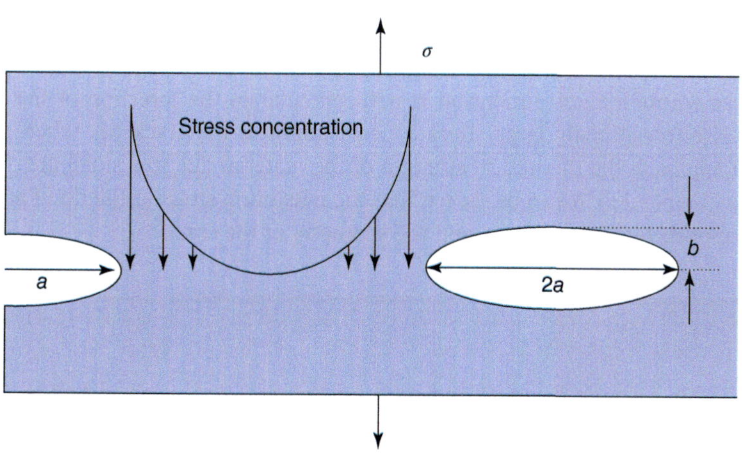

Figure 5.35 The Griffith model for microcrack-induced brittle failure. Cracks at the surface have a length of a, whereas internal cracks have a length of $2a$.

Cooperative Learning Exercise 5.7

Use values of modulus from Appendix 6, along with surface energy from Appendix 3, to estimate the following quantities. In both cases, take an interplanar distance of 10^{-8} cm as an order-of-magnitude estimate.

Person 1: Calculate the σ_c and the ratio σ_c/E for MgO.

Person 2: Calculate the σ_c and the ratio σ_c/E for Au.

Compare your answers. Based on these results, what would you expect the theoretical ratio of σ_c/E to be?

Answers: σ_c(MgO) = $[(210 \times 10^9 \text{N/m}^2)(1.0\ \text{J/m}^2)/(10^{-10}\text{m})]^{1/2}$ = 56.5 GPa $\approx E/4$; $\sigma_c \approx E/6$ σ_c(Au) = $[(77 \times 10^9 \text{N/m}^2)(1.4\ \text{J/m}^2)/(10^{-10}\text{m})]^{1/2}$ = 32.8 GPa; $\sigma_c \approx E/2.5$; generally $E > \sigma_c > E/10$.

The functional dependences of Eq. (5.40) are important: the cleavage strength is proportional to the square root of both modulus and surface energy and is inversely proportional to the square root of the equilibrium interplanar distance. These relationships return in subsequent sections. For now, work through the Cooperative Learning Exercise 5.7 to see some important generalizations about the theoretical cleavage strength and modulus.

Fracture in Real Brittle Materials—Griffith Theory

In the Cooperative Learning Exercise 5.7, we estimated that the theoretical cleavage strength might lie somewhere between the value for the modulus and $E/10$. The fracture strength of most ceramic materials is more like $E/1000$. What is wrong with Eq. (5.40)? We took experimental values for modulus and surface energy, which, granted, may not be very accurate, but are not off by a combined four orders of magnitude necessary to cause these errors. The problem is with the denominator. Just as dislocations caused the theoretical shear strength to differ from the yield strength, there are structural imperfections that cause the experimental fracture strength to differ from the theoretical cleavage strength.

It was proposed by Griffith in 1920 that microscopic cracks exist both within and on the surface of all real materials, which are deleterious to the strength of any material that does not possess ductility. The presence of cracks whose longest dimension is perpendicular to the direction of the applied tensile stress gives rise to especially large stress concentrations at the crack tip, where the real localized stress can approach the theoretical strength of the material due to a small area over which it is applied (see Figure 5.35). It can be shown that the maximum stress at the tip of a crack, σ_m, is related to the applied tensile stress, σ_0, and the geometry of the crack

$$\sigma_m = 2\sigma_0 \left(\frac{a}{\rho}\right)^{1/2} \tag{5.41}$$

where ρ is the radius of curvature of the crack tip, b^2/a, where b is the height of the crack and a is its length as defined in Figure 5.35, and a is the length of a surface crack or *half* the length of an interior crack (see Figure 5.35). Do not confuse the length of the crack with the interplanar spacing; the same terminology is used for a reason, as will be shown momentarily, but the values are certainly not the same.

For a crack to propagate, $\sigma_m > \sigma_c$; that is, the maximum stress at the crack tip must exceed the theoretical cleavage strength. Since the radius of curvature cannot be less than the order of an atomic diameter, σ_m must always be finite but can attain σ_c even though the bulk of the material is under fairly low applied stress. The stress required to cause crack propagation is then the stress that leads to fracture. Using surface energy and work arguments like those in the development of the theoretical cleavage strength from last section, it can be shown that the fracture strength according to this crack propagation model of Griffith is:

$$\sigma_f = \left(\frac{2\gamma E}{\pi a}\right)^{1/2} \tag{5.42}$$

where γ is again the surface energy of the material, E is its elastic modulus, and a is now one-half the length of an internal crack or the length of a surface crack. Notice the similarity in form between Eqs. (5.42) and (5.40). As we surmised, it was not our values of modulus or surface energy that were erroneous in the application of Eq. (5.40), but rather the size parameter. Instead of interplanar separations on the order of angstroms (Å, 10^{-10} m) as in Eq. (5.40), it is actually cracks, on the order of microns (10^{-6} m), that lead to fracture as in Eq. (5.42) and account for the nearly 100-fold (square root of 10^4) difference in the observed and calculated values of the fracture strength estimated from the elastic modulus referred to at the beginning of this section.

Equation (5.42) is an important result. It shows clearly that fracture strength decreases as the size of the crack increases, and it also shows that fracture strength can be related directly to surface energy and elastic modulus. The Griffith equation applies only to completely brittle materials. For materials that exhibit some plastic deformation prior to fracture, the surface energy term may be replaced by a term $\gamma + \gamma_p$, where γ_p is the plastic deformation energy associated with crack extension. Values of γ_p can be up to 1000 times greater than γ, such that the latter can often be disregarded for materials that undergo some plastic deformation prior to fracture. Finally, the Griffith equation tells us that if we can reduce crack sizes, or eliminate them altogether, the fracture strength of a material can be manipulated. We will see that this has important applications in glass science.

Failure Modes and Fracture Toughness

Once the critical fracture stress has been achieved and the crack begins to propagate, it can do so in one of three primary modes (see Figure 5.36). Mode I is a tensile, or opening, mode. Modes II and III are sliding and tearing modes, respectively. We will limit our present description to the opening mode, though you should be aware that corresponding values for the quantities developed in this section apply to the other two modes as well.

The parameter K_1, known as the mode I *stress intensity factor*, is used to characterize the distribution of stresses at the crack tip and is a function of the applied stress, the crack size (length), and the sample geometry. Stress intensity factors are conveniently defined in terms of the stress components on the crack plane. In most cases, fracture is dominated by tensile loading, so that the important stress intensity factor is K_1. The value of K_1 at which fracture occurs is the critical value of the stress intensity factor at which propagation initiates, K_{1c}, also known as the *fracture toughness*. Do not confuse the fracture toughness with the toughness defined in Section 5.1.4, which was the area under the stress–strain curve. The fracture toughness is specific to fracture in brittle

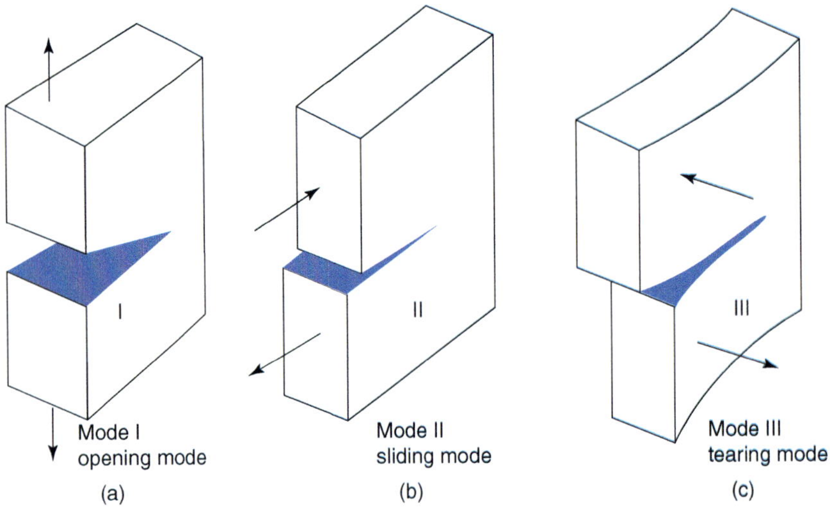

Figure 5.36 The three modes of crack surface displacement: (a) opening mode; (b) sliding mode; and (c) tearing mode.

materials, and it indicates the conditions of flaw size and stress necessary for brittle fracture. It is defined as:

$$K_{1c} = Y\sigma_c\sqrt{\pi a} \tag{5.43}$$

where σ_c is the theoretical cleavage strength, a is crack length as defined in Figure 5.35 (length of a surface crack, or half the length of an internal crack), and Y is a geometric factor that depends on the sample geometry and the manner of load application. The units of fracture toughness are thus $\text{MPa·m}^{1/2}$, or $\text{psi·in}^{1/2}$. The magnitude of K_{1c} diminishes with increasing strain rate, decreasing temperature, and increasing grain size.

Cooperative Learning Exercise 5.8

The crack detection limit for a device that inspects steel structural beams is 3 mm. A structural steel with a fracture toughness of 60 $\text{MPa·m}^{0.5}$ has no detectable surface cracks. Assume a geometric factor of unity to determine:

Person 1: The load at which this structural steel may undergo fast fracture if a surface crack just at the detection limit goes undetected.

Person 2: The % increase in crack size required for this structural steel to undergo fast fracture at an applied load of 500 MPa.

Answers: σ_c = 618 MPa; a = 4.58 mm (52.7% increase).

Fatigue

A failure phenomenon that is closely related to brittle fracture is caused by repeated exposure to cyclic stresses. This type of failure, called *fatigue*, results in fracture below the ultimate strength of ductile materials, σ_u, or below the fracture strength, σ_f, of brittle materials. Even in ductile materials, the result of fatigue is brittle-like failure, insofar as there is little or no plastic deformation prior to failure. This process, like fracture, is caused by the initiation and propagation of cracks.

Fatigue is estimated to comprise approximately 90% of all metal failure, though polymers and ceramics are also susceptible to it. Fatigue is particularly important because it can cause catastrophic failure in even the largest of structures due to small, virtually unnoticeable stress fluctuations over long periods of time.

Though stresses, especially environmental stresses, are generally not uniform, there are four categories of cyclical stresses to consider. The first is called a *reversed stress cycle* (Figure 5.37a), in which the stress alternates equally between compressive and tensile loads. A *repeated stress cycle* occurs when the maximum and minimum stress of a reversed stress cycle are asymmetric relative to the zero stress level, as in Figure 5.37b. Other types of cyclical stresses include a *fluctuating stress cycle*, which has a maximum and minimum tensile *or* compressive load (but not both), and a *random stress cycle*, in which the stresses are neither periodic nor uniform in the magnitudes. In most cases, the stress cycle is characterized by the *mean stress*, $\bar{\sigma} = (\sigma/min_{max})$, the *range of stress*, $\sigma_r = \sigma_{max} - \sigma_{min}$, and the *stress amplitude*, $S = (\sigma_{max} - \sigma_{min})/2$. Note that for a reversed stress cycle, the mean stress is zero, so that the stress amplitude is usually a more meaningful quantity.

Fatigue tests are normally performed by applying one of the stress cycles described above until the test specimen fractures. The *number of cycles to failure*, N_f, at a specified stress amplitude, S, sometimes called the *fatigue life*, then characterizes the material. For some materials, a *fatigue limit* is reached (see Figure 5.38), which is defined as the value of S below which no failure will occur,

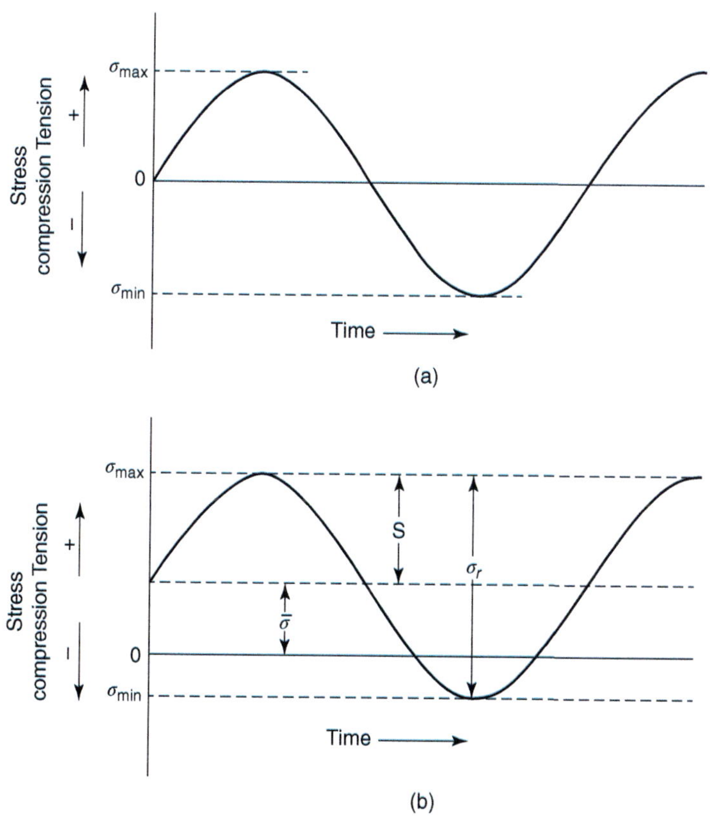

Figure 5.37 Variation of stress with time that accounts for fatigue failure by (a) a reversed stress cycle and (b) a repeated stress cycle. Callister [22]/Reproduced with permission of John Wiley & Sons.

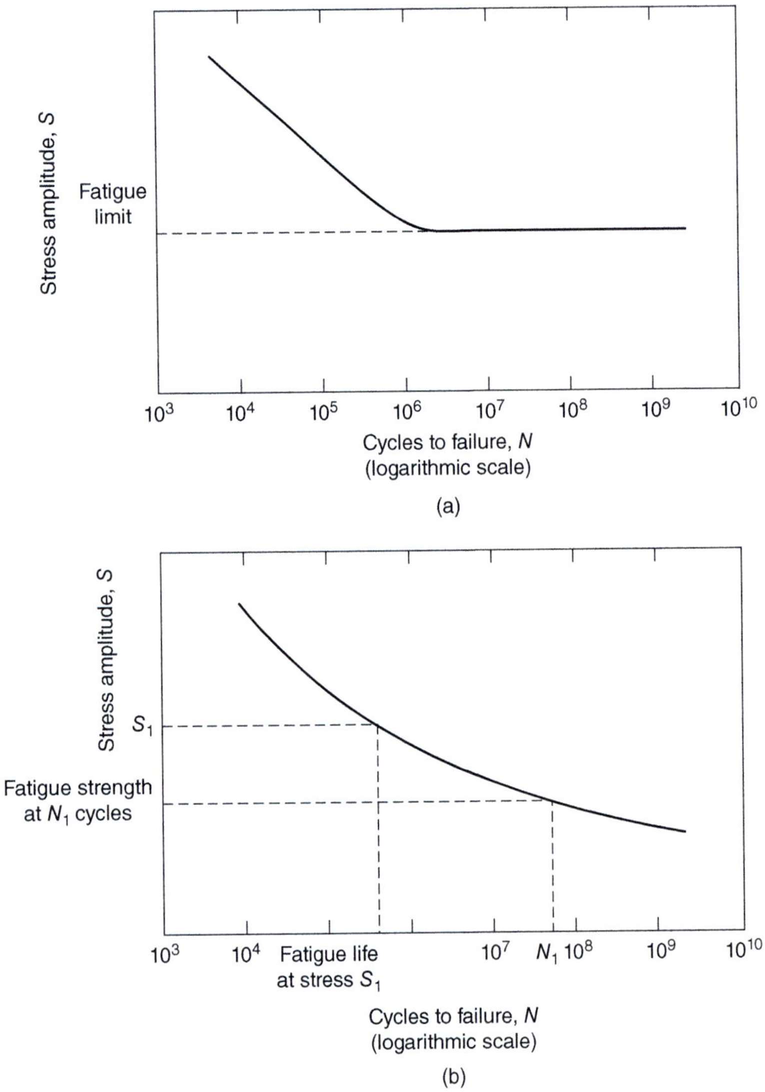

Figure 5.38 Stress amplitude versus cycles to failure illustrating (a) fatigue limit and (b) fatigue life. Callister [23]/Reproduced with permission of John Wiley & Sons.

regardless of the number of cycles, N. The *fatigue strength*, sometimes called the *endurance limit*, is defined as the stress level at which failure will occur for some specified number value of N_f—for example, 10^7 cycles.

As mentioned earlier, fatigue is the result of crack initiation and propagation. Crack growth under fatigue is estimated by the *Paris equation*:

$$\frac{da}{dN} = A(\Delta K)^m \tag{5.44}$$

where a is once again the crack length as defined earlier (length of a surface crack or half the length of an internal crack), N is the number of cycles (not necessarily to failure), A and m are

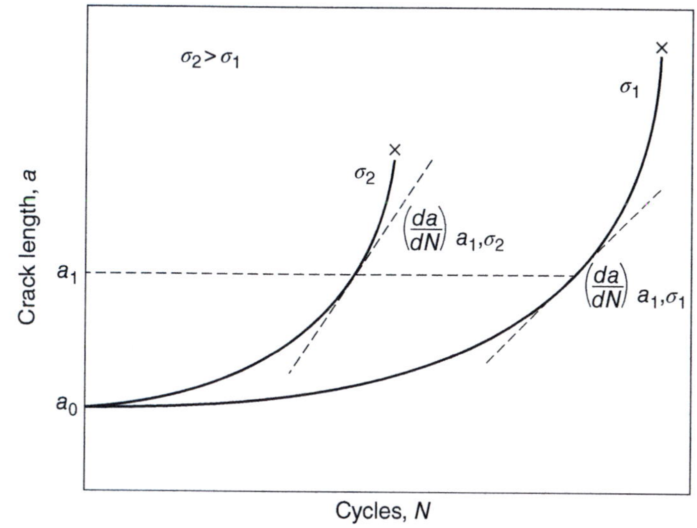

Figure 5.39 Determination of crack growth rate, *da/dN* from crack length versus number of cycles data. Callister [24]/Reproduced with permission of John Wiley & Sons.

material-dependent constants, and ΔK is the range of stress intensity factor at the crack tip, $K_{max} - K_{min}$. Utilization of Eq. (5.43) shows that:

$$\Delta K = Y(\sigma_{max} - \sigma_{min})\sqrt{\pi a} \tag{5.45}$$

The values of A and m are determined using the Paris equation by plotting da/dN at various values of ΔK on logarithmic coordinates. The values of da/dN are determined from plots of crack length versus number of cycles, as shown in Figure 5.39. The corresponding crack lengths and stress ranges can then be used to determine ΔK using Eq. (5.45). Typical values of m range between 1 and 6. It should be noted that the Paris equation applies only to a specific region of crack formation. As shown in Figure 5.39, the crack growth rate is initially small, and fatigue cracks can be nonpropagating. At high crack growth rates, crack growth is unstable. Only at intermediate values of crack growth rates does the Paris equation apply.

Need More Help?

Try these videos on brittle fracture and fatigue:

Fracture
https://www.youtube.com/watch?v=HyzUBphwXRI
https://www.youtube.com/watch?v=jJMSvgcZaGA

Griffith Theory
https://www.youtube.com/watch?v=Uvfb8ZRuLKE
https://www.youtube.com/watch?v=HyzUBphwXRI

Ductile-Brittle Transition
https://www.youtube.com/watch?v=uh5a2d7Abec
https://www.youtube.com/watch?v=pD_LhWFc4Oc

Embrittlement
https://www.youtube.com/watch?v=zH01GdfGlew

Fatigue
https://www.youtube.com/watch?v=o-6V_JoRX1g

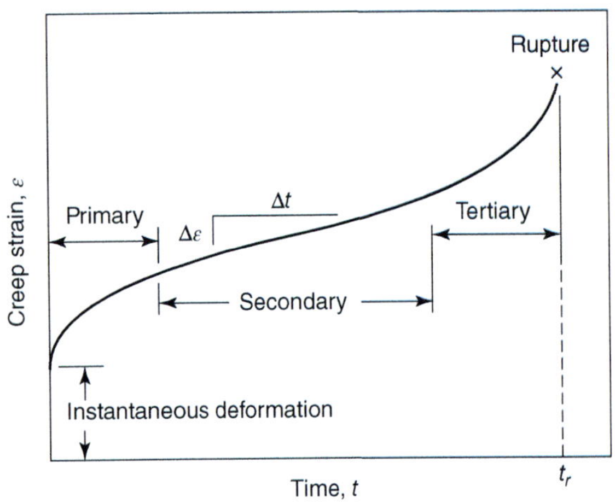

Figure 5.40 Representative creep curve illustrating primary, secondary, and tertiary creep. Callister [25]/Reproduced with permission of John Wiley & Sons.

5.2.2 Creep

Fatigue is an example of the influence of time on the mechanical properties of a material. Another example of a time-dependent mechanical property is *creep*. Creep, sometimes called *viscoplasticity*, is defined as time-dependent deformation under constant stress, usually at elevated temperatures. Elevated temperatures are necessary because creep is typically important only above $T_{mp}/2$, where T_{mp} is the absolute melting point of the material.

A typical creep experiment involves measuring the extent of deformation, called the *creep strain*, ε, over extended periods of time, on the order of thousands of hours, under constant tensile loads and temperature. The resulting plot of creep strain versus time (Figure 5.40) shows the resulting *creep rate*, $\dot{\varepsilon} = d\varepsilon/dt$, which is the slope of the curve at a specified time. A typical creep curve exhibits three distinct regions, with different functional dependences of strain on time for each. The first region is called *primary creep*, or *transient creep*, in which the primary creep rate, $\dot{\varepsilon}_p$, continuously decreases with time. A power–law relationship can be applied in this region:

$$\dot{\varepsilon}_p \propto t^n \tag{5.46}$$

where t is time and n is a negative number. This region represents the strain-hardening region of creep. A similar power law expression can be used in the *tertiary creep* regime to describe the increase in tertiary creep rate, $\dot{\varepsilon}_t$, with time, except that n is now (usually) a positive number. Values of n range from −2 to 20 for tertiary creep, with values of $n = 4$ being typical. The creep rate increases during tertiary creep until failure, termed *rupture*, occurs. The stage of creep that usually occupies the longest time is *secondary creep*, or *steady-state creep*, which is characterized by a constant creep rate, $\dot{\varepsilon}_s$. In this area, strain hardening is balanced by thermal recovery.

Need More Help?
Try these videos on creep: https://www.youtube.com/watch?v=viaVPY-XrwQ https://www.youtube.com/watch?v=rz7qUVdVAWs

The variation of creep with time as a function of both load and temperature is illustrated in Figure 5.41. Arrhenius-type relationships have been developed for steady-state creep as a function of both variables such as:

$$\dot{\varepsilon}_s = K\sigma^m \exp\left(\frac{-E_c}{RT}\right) \tag{5.47}$$

where σ is the applied stress, R is the gas constant, T is absolute temperature, K and m are material constants, and E_c is called the activation energy for creep.

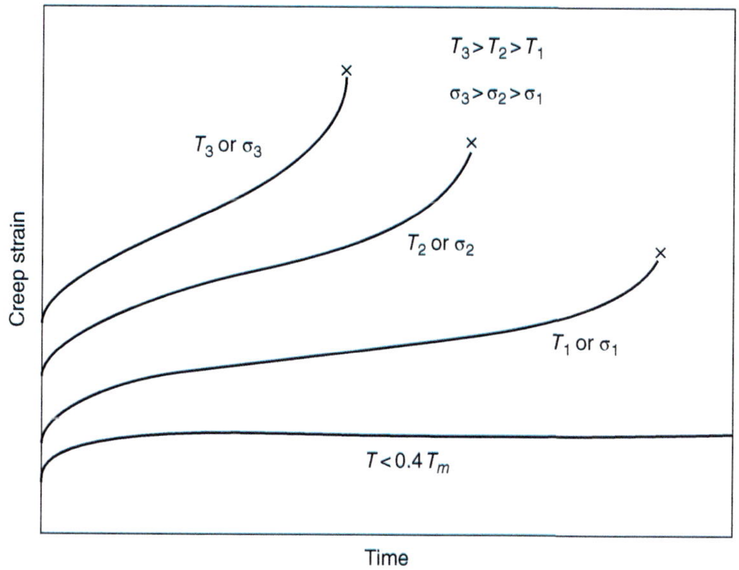

Figure 5.41 Influence of applied stress and temperature on creep behavior. Callister [26]/Reproduced with permission of John Wiley & Sons.

Cooperative Learning Exercise 5.9

A proposed jet engine turbine blade material exhibits a steady-state creep rate of 0.0001 in/in-sec at 500°C under a stress of 20,000 psi. The activation energy for creep is 40 kcal/mol.

Person 1: At what temperature will this creep rate be doubled at the same stress level?

Person 2: At what temperature will this creep rate be tripled at the same stress level?

Compare your answers. Agree upon a temperature at which you both estimate these blades will operate and calculate the corresponding creep rate.

Answer: In both instances, use a ratio of Eq. (5.47) at the two different temperatures to solve for T(0.0002) = 521°C; T(0.0003) = 534°C. In military jet engines, the inlet gas temperatures can exceed 1400°C. At the sated loads, the creep rate is over 10^6 times that at 500°C!

5.2.3 Mechanical Properties of Ceramics and Glasses

In this section, we describe the elastic, ductile, fracture, fatigue, and creep properties that are specific to certain classes of ceramic and glass materials, with primary emphasis on those mechanical properties that make the material favorable for certain applications from a design point of view.

Crystalline Ceramics

The moduli of ceramic materials are generally higher than those for metals, being in the range of 70–500 GPa. For example, the elastic modulus as a function of temperature for a common engineering ceramic, polycrystalline Al_2O_3, is shown in comparison to moduli for three common metals in

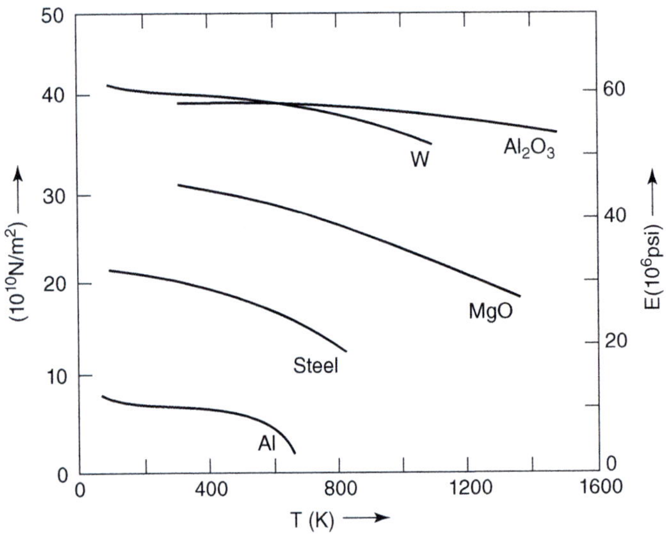

Figure 5.42 Comparison of elastic modulus between polycrystalline Al_2O_3 and some common metals as a function of temperature. From Ralls et al. [3]. Copyright © 1976 by John Wiley & Sons, Inc. This material is used by permission John Wiley & Sons, Inc.

Figure 5.42. Note that the modulus is relatively independent of temperature, which is common in polycrystalline ceramics.

The moduli for single crystals can be even higher than for the polycrystalline analogs, depending upon the direction of orientation of the crystal relative to the applied stress. For example, the maximum modulus for MgO lies along the [111] direction and has a value of $E_{[111]} = 336$ GPa, whereas the minimum value is along the [100] direction, for which $E_{[100]} = 245$ GPa. As is the case for most polycrystalline materials, the modulus of a polycrystalline sample of MgO lies somewhere in between these two values, around 310 GPa.

One may not normally think of ceramics as having significant ductile responses, since they are mostly brittle materials that normally fracture around their yield point, but slip is a definite possibility, especially in single-crystal ceramics. As was the case for metals, plastic deformation in crystalline ceramics occurs by a slip dislocation mechanism (cf. Section 5.1.2). The slip systems in some common ceramic crystals are shown in Table 5.5. Note that the number of independent slip systems is low compared to the metallic structures of Table 5.3, especially the BCC metals.

The yield strength and ductility of single crystals are highly dependent upon temperature and strain rate. These effects are illustrated in Figure 5.43 for single crystal Al_2O_3. Note the presence of both an upper yield stress and lower yield stress, much as there was for metals, and that they both decrease exponentially with increasing temperature and decreasing strain rate. This drop in yield point values can be explained in terms of dislocation multiplication, rather than dislocation movement. Notice again how the modulus is relatively unaffected by either temperature or strain rate. At low temperatures and high strain rates, brittle behavior is observed. Finally, the alloying of single-crystal Al_2O_3 with different cations such as Fe, Ni, Cr, Ti, and Mg leads to a strengthening effect, much as precipitation hardening in metals.

As was the case for metals, polycrystalline ceramics show a dependence of mechanical properties on grain size, as given earlier by the Hall–Petch relation in Eq. (5.19), which in this case applies to fracture strength, σ_f.

Table 5.5 Slip Systems in Some Ceramic Crystals

Crystal	Slip System	Number of Independent Systems	Comments
C (diamond), Si, Ge	$(111)[1\bar{1}0]$	5	At $T > 0.5T_m$
NaCl, LiF, MgO, NaF	$(110)[1\bar{1}0]$	2	At low temperatures
NaCl, LiF, MgO, NaF	$(110)[1\bar{1}0]$		At high temperatures
	$(001)[1\bar{1}0]$	5	
	$(111)[1\bar{1}0]$		
TiC, UC	$(111)[1\bar{1}0]$	5	At high temperatures
PbS, PbTe	$(001)[1\bar{1}0]$	3	
	(110) [001]		
CaF_2, UO_2	$(001)[1\bar{1}0]$	3	
CaF_2, UO_2	$(001)[1\bar{1}0]$		At high temperatures
	(110)	5	
	(111)		
C (graphite), Al_2O_3, BeO	$(0001)[11\bar{2}0]$	2	
TiO_2	$(101)[10\bar{1}]$	4	
	(110) [001]		
$MgAl_2O_4$	$(111)[1\bar{1}0]$	5	
	(110)		

Source: Kingery et al. [27]. Copyright © 1976 by John Wiley & Sons, Inc.

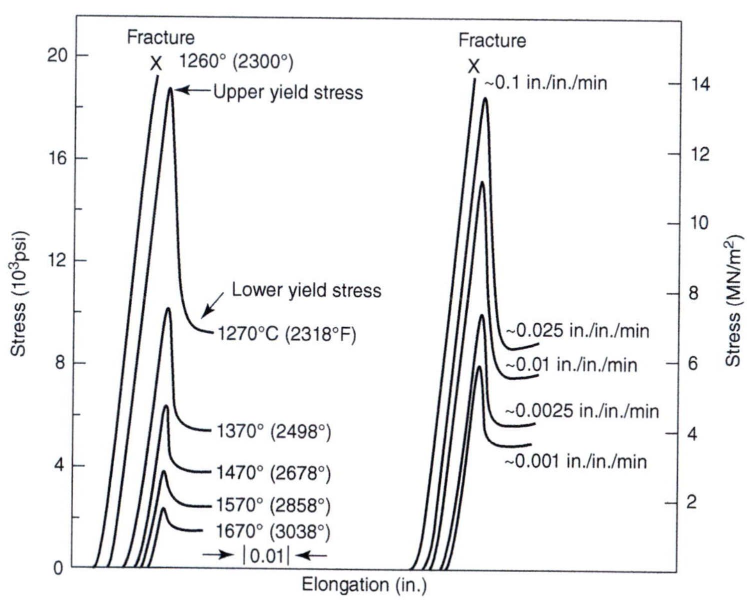

Figure 5.43 Effect of temperature and strain rate on deformation behavior in single-crystal Al_2O_3. From Kingery et al. [27]. Copyright © 1976 by John Wiley & Sons, Inc. This material is used by permission of John Wiley & Sons, Inc.

$$\sigma_f = \sigma_0 + kd^{-1/2} \tag{5.19}$$

Remember that in brittle materials, fracture occurs as a result of the most severe flaw (crack) in the sample. Hence, the correlations of strength with average grain size must reflect processing techniques that produce samples in which the size of the most severe flaw scales with the average grain size. In cases where the sizes of the initial flaws are limited by the grains and scale with the grain size, the strengths vary with a modified form of Eq. (5.19):

$$\sigma_f = kd^{-1/2} \tag{5.48}$$

This behavior is reflected at larger grain sizes in Figure 5.44, whereas Eq. (5.19) is followed at smaller grain sizes.

Typically, the most severe flaws are associated with porosity, surface damage, abnormal grains, or foreign inclusions. The effect of porosity on both modulus and strength is illustrated in Figures 5.45 and 5.46, respectively. The solid curve in Figure 5.45 represents the following equation, which holds for some ceramics:

$$E = E_0\left(1 - 1.9P - 0.9P^2\right) \tag{5.49}$$

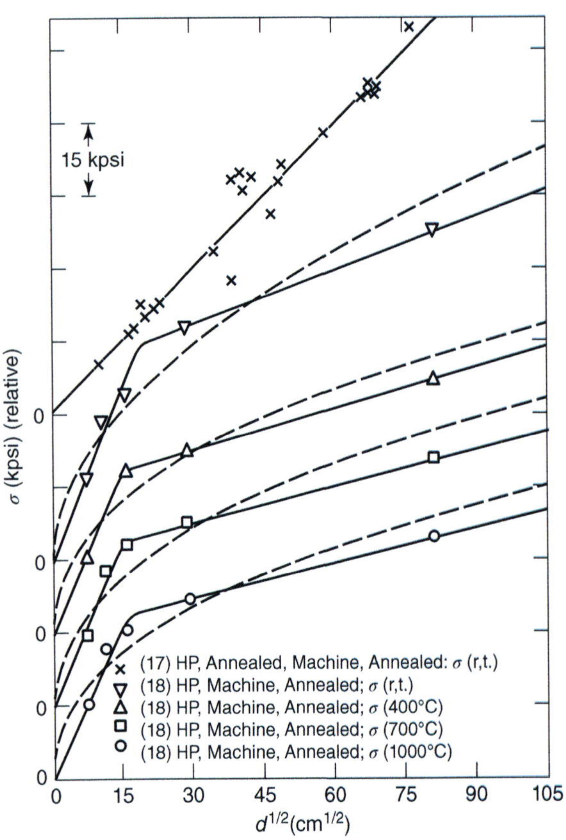

Figure 5.44 Effect of grain size on strength in polycrystalline Al_2O_3 for various processing methods (HP, hot pressed, rt, room temperature). From Kingery et al. [27]. Copyright © 1976 by John Wiley & Sons, Inc. This material is used by permission of John Wiley & Sons, Inc.

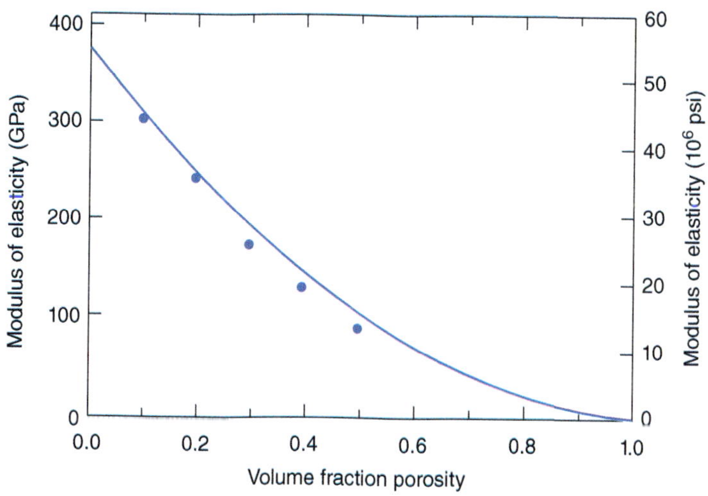

Figure 5.45 Effect of porosity on the elastic modulus of Al_2O_3 at room temperature. Callister [28]/ Reproduced with permission of John Wiley & Sons.

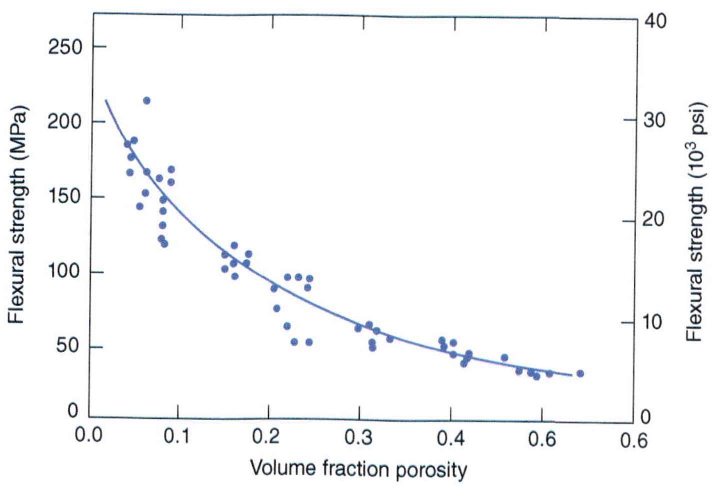

Figure 5.46 Effect of porosity on the flexural strength of Al_2O_3 at room temperature. Callister [28]/ Reproduced with permission of John Wiley & Sons.

where E_0 is the modulus of the nonporous material, and P is the volume fraction porosity. A similar relationship can be used to estimate the flexural strength of a porous ceramic, σ_{fs}, as a function of porosity:

$$\sigma_{fs} = \sigma_0 \exp(-nP) \tag{5.50}$$

where σ_0 and n are experimentally determined constants. This relationship is represented by the solid line in Figure 5.46.

The creep properties of ceramics are of particular importance, especially at high temperatures. In general, the principles of Section 5.2.2 still apply, but the ceramic creep mechanisms can be very complex, and factors such as diffusional creep, grain boundary creep, and substructure formation

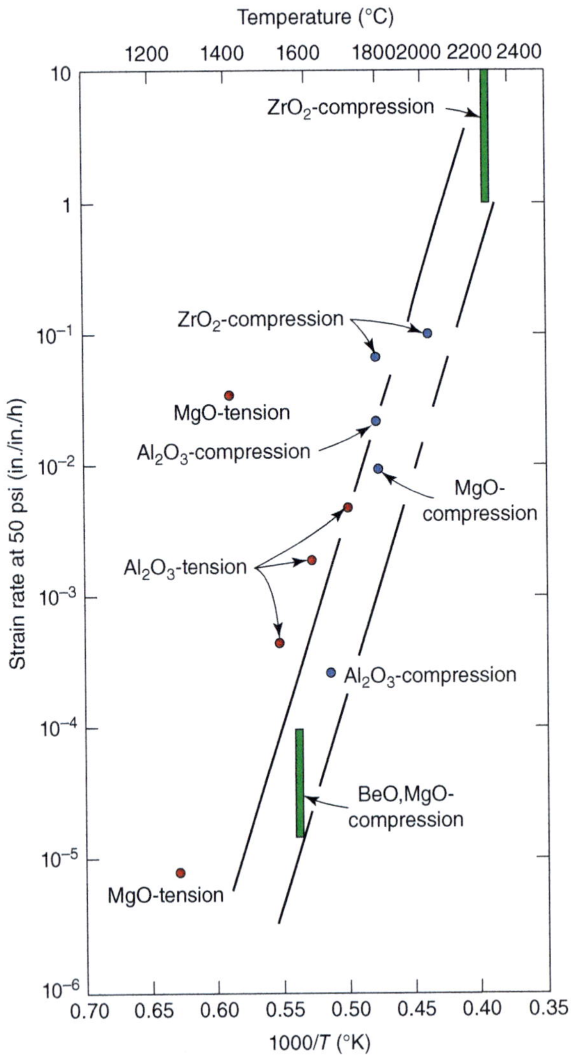

Figure 5.47 Low-stress creep rates of several polycrystalline oxides. From Kingery et al. [27]. Copyright © 1976 by John Wiley & Sons, Inc. This material is used by permission of John Wiley & Sons, Inc.

must be considered in a complete description of creep in ceramics. Porosity and grain size are also factors in creep, as they were in modulus and strength determinations. A summary of these effects may be seen graphically in Figure 5.47, in which the creep rates of several polycrystalline oxide ceramics are presented as a function of temperature. All of the values fall within a band, represented by the dashed lines, and variations in creep rate values for a given material (e.g., Al_2O_3 in tension or compression) can be attributed to differences in microstructure (pores, grain size, etc.).

Glasses

Due to the disordered structure, the modulus and strength of most inorganic glasses are less than those in most crystalline ceramics (see Figure 5.48). However, due to other favorable properties,

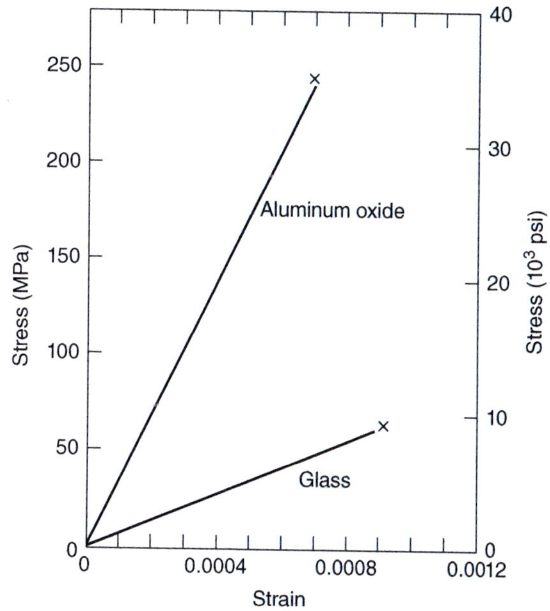

Figure 5.48 Typical stress–strain diagrams for aluminum oxide and glass. Callister [29]/ Reproduced with permission of John Wiley & Sons.

such as optical transparency and ease of formability, glasses find wide industrial use and, as a result, have been the subject of much investigation.

The mechanical properties of silicate-based glasses have been extensively studied due to their widespread use as containers, reinforcing fibers, and optical fibers. Consequently, there is a wealth of information on the influence of many variables on the mechanical properties of these glasses. We shall concentrate on (a) composition and temperature effects on modulus and strength and (b) an important method of strength enhancement in glass fibers.

Several empirical expressions exist for estimating the modulus of multicomponent glasses based on composition. One particularly useful one that provides an absolute value for the modulus, E, is based upon the density of the glass, ρ, and the packing density, V_t, dissociation energy, U_i, mol%, p_i, and molecular weight, M_i, of the components:

$$E = \frac{0.2\rho \sum V_i p_i \sum U_i p_i}{\sum M_i p_i} \tag{5.51}$$

Values for V_i and U_i for selected metal oxide glass components are listed in Table 5.6. In a similar manner, the compressive modulus, K, and Poisson's ratio, ν, can be determined using these parameters:

$$K = V_t^2 \sum U_i p_i \tag{5.52}$$

and

$$\nu = 0.5 - \frac{0.139}{V_t} \tag{5.53}$$

respectively, where

$$V_t = \frac{\rho \sum V_i p_i}{\sum M_i p_i} \tag{5.54}$$

Table 5.6 Factors Used to Determine Elastic Modulus in Multicomponent Glasses Using Eq. (5.51)

Oxide	V_i (mol%)	U_i (kJ/cm^3)
Li_2O	8.0	80.4
Na_2O	11.2	37.3
K_2O	18.8	23.4
BeO	7.0	125.6
MgO	7.6	83.7
CaO	9.4	64.9
SrO	10.5	48.6
BaO	13.1	40.6
B_2O_3	20.8	77.8
Al_2O_3	21.4	134.0
SiO_2	14.0	64.5
TiO_2	14.6	86.7
ZrO_2	15.1	97.1
P_2O_5	34.8	62.8
As_2O_5	36.2	54.8
ZnO	7.9	41.5
CdO	9.2	31.8
PbO	11.7	17.6

Source: Scholze [30]/Springer Nature.

Strength in glasses is also composition-dependent, as shown in Figure 5.49. Correlations similar to Eqs. (5.51)–(5.54) for the variation of strength with composition do exist; however, it is the effect of surface cracks that more profoundly influences strength in real glasses. There are two important techniques that can be used to improve the strength of glasses: *thermal tempering* and *chemical tempering*. Tempering utilizes the fact that glasses, and crystalline ceramics for that matter, can be strengthened by developing a state of compression on their surfaces.

Tempered glasses are useful because failure normally occurs under an applied tensile stress, and failure in ceramics and glasses is almost always initiated at the surface. When a permanent compressive stress, called a *residual compressive stress*, is placed on a surface, either through thermal or chemical means, the applied stress must first overcome this residual compression before the surface is brought into tension under which failure can occur (see Figure 5.50). Notice that the residual stress is compressive in nature at the surface of the plate and is tensile in the center (shaded areas). When a transverse load (dash line) is placed on the plate (recall that a bending-type load creates a tensile force at the bottom and a compressive force at the top), the resultant stress (solid line) creates a large compressive force at the top of the plate, which it can accommodate, but the tensile force at the bottom of the plate is counteracted by the residual compressive stress, such that the resultant stress is only mildly tensile in nature, and well below the fracture stress. A similar situation occurs for glass fibers (cylinders) in tension, which possess compressive residual stresses on their surfaces.

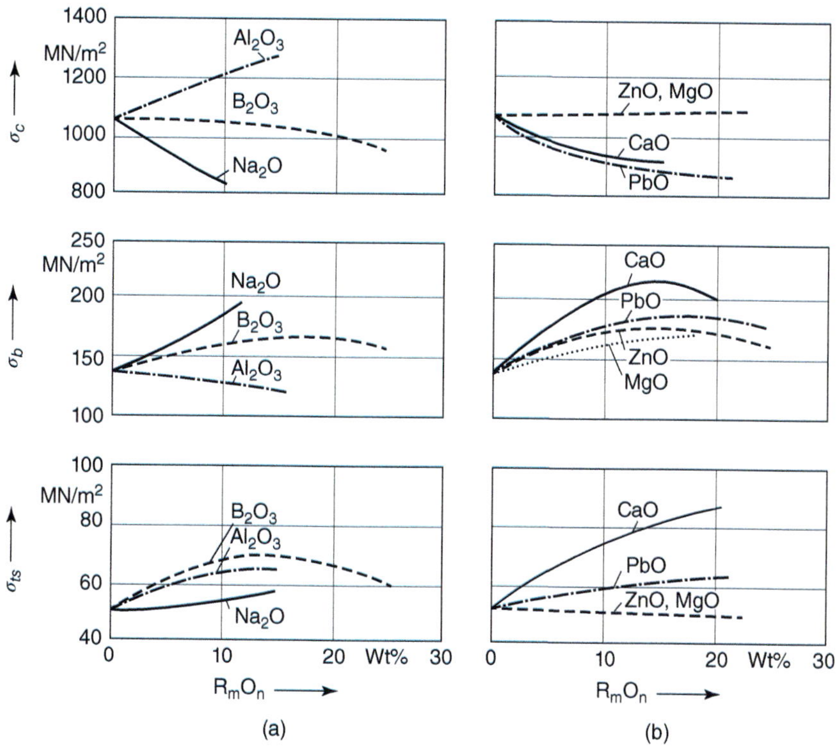

Figure 5.49 Change in compressive strength, σ_c, bending strength, σ_b, and tensile strength, σ_t, of a Na_2O–SiO_2 glass (20–80%) as SiO_2 is replaced proportionately by weight with other oxides. Scholze [31]/ Reproduced with permission of Springer Nature.

In thermal tempering, the residual compressive stresses are introduced into the surface of the glass through thermal treatment. A glass initially at a uniform temperature above the glass transition temperature, T_g, is rapidly cooled, or *quenched*, to a new temperature by the use of cold air or oil baths. The outside of the glass, which initially cools more rapidly than the interior, becomes rigid while the interior is still molten. The temperature difference between the surface and the midplane of the glass rapidly reaches a maximum upon quenching, and the interior contracts at a greater rate than the rigid outside due to thermal expansion differences. The stresses induced in the later stages of cooling remain. The resulting stress profile is nearly parabolic in shape (see Figure 5.50), with the magnitude of the compressive stresses at the surface being approximately twice the maximum tension in the interior of the glass. Obviously, heat transfer rate and the heat transfer coefficient at the surface play important roles in determining the residual stress distribution.

Another way to introduce compressive stresses on the surface of glasses is to change the chemical composition of the surface through the selective replacement of surface atoms in a process called chemical tempering. This process increases the molar volume of the surface relative to the interior. This is normally done through the substitution of large ions for small ones. Diffusivity is an important consideration in this process, since the ion being inserted into the surface of the glass must have sufficient mobility to displace the existing ion. The most common example of chemical tempering is the replacement of Na^+ ions in sodium–silicate glasses with K^+ ions.

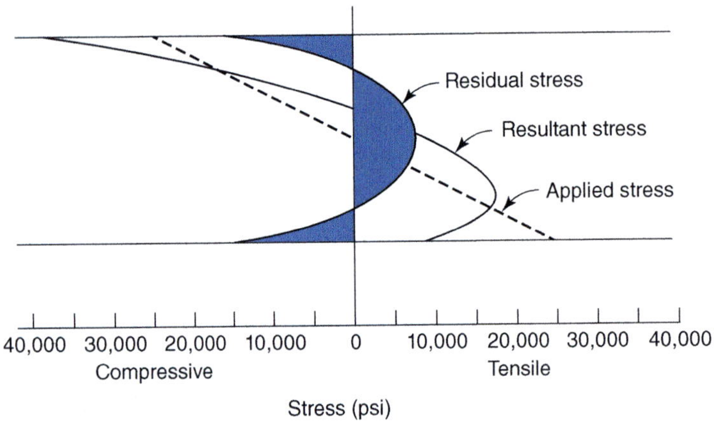

Figure 5.50 Residual stress, applied stress, and resultant stress distribution for transverse loading of a tempered glass plate. From Kingery et al. [27]. Copyright © 1976 by John Wiley & Sons, Inc. This material is used by permission of John Wiley & Sons, Inc.

The compressive stress on the surface associated with the fractional change in molar volume, $\Delta V/V$, can be approximated by:

$$\sigma = \frac{E(\Delta V/V)}{3(1-2\nu)} \tag{5.55}$$

where E is the tensile modulus and ν is Poisson's ratio. The compression depth of chemically strengthened glass is controllable, but in practice is limited to a few hundred microns.

In a related glass processing technique called *etching*, the surface of the processed glass is treated with an acid such as hydrofluoric acid, HF, which removes a thin, outer layer of the glass. With this procedure, cracks can be eliminated, or at least reduced in length. As the crack length decreases, the strength of the glass increases in accordance with Eq. (5.42).

In some instances, residual stresses are not desirable, for example, when the glass is to be used as an optical component. Residual stresses can be removed through a process called *annealing*, in which the glass is heated to a uniform temperature near the transformation temperature and held for a specific amount of time. The stresses are resolved due to increased atomic mobility at elevated temperatures. The glass is then slowly cooled through the critical viscosity range at a rate sufficiently slow to keep residual stresses from developing. The glass is then rapidly cooled.

5.2.4 Cement and Concrete*

A great irony of the materials, engineering, and science disciplines is that the single largest non-natural material of construction used in terms of volume—namely, *cement*—finds little or no technical coverage in an introductory text. To assist in remedying this situation, we present here some introductory information on cements and concretes, with an emphasis on their mechanical properties, so that proper consideration can be given to them during materials selection and design. First, we must provide some background. The term "cement" refers to a substance that can be used to hold together aggregates such as sand or stone. A cement can, in principle, be any kind of adhesive; it need not be inorganic, and it may be a single compound or a mixture. In this section, we will use the term to refer to a specific inorganic mixture of calcium, aluminum, and silicon oxides known as *hydraulic cements*. Hydraulic cements are set through reaction with water, as opposed to *air-set cements* that are principally organic and set through a loss of water.

Historical Highlight

The Romans pioneered the use of water-cured, or hydraulic, cement. Learn more of the history of this revolutionary material here:
https://www.worldcementassociation.org/about-cement/our-history

Hydraulic cements are so named because they *set* (initial stiffening) and *harden* (develop strength) through a series of *hydration reactions*. The starting materials of limestone or chalk ($CaCO_3$), gypsum ($CaSO_4$, kaolin (Al_2O_3), shale or sand (SiO_2), and slags containing various metal oxides such as MgO and FeO are mixed and burned to form a fused mass called a *clinker*, which is then ground to form cement powder. During the burning process, decomposition of the carbonates occurs:

$$CaCO_3 \rightarrow CaO + CO_2 \tag{5.56}$$

A shorthand notation is often used to designate the oxides (e.g., C for CaO and A for Al_2O_3; see Table 5.7), and it is also used to designate the compounds formed between the components during heating, such as calcium aluminate (CA) and tricalcium silicate (C_3S):

$$CaO + Al_2O_3 \rightarrow CaAl_2O_4 \equiv CA \tag{5.57}$$

$$3CaO + SiO_2 \rightarrow Ca_3SiO_5 \equiv C_3S \tag{5.58}$$

The primary components of Portland cement prior to the hydrolysis reactions are C_3S (3CaO–SiO_2, tricalcium silicate); C_2S (2CaO–SiO_2, dicalcium silicate); C_3A (3CaO–Al_2O_3, tricalcium aluminate); and C_4AF (4CaO–Al_2O_3–Fe_2O_3, tetracalcium aluminoferrite).

Upon addition of water, the hydration reactions initiate, and the hydraulic cement begins to gain strength. This process is very complex, but the strengthening effect is due basically to the formation of three types of hydration products: *colloidal products* such as $C_2S{\cdot}xH_2O$, which have a size of less than 0.1 μm; *submicrocrystalline products* such as $Ca(OH)_2$, Al^{3+}, Fe^{3+}, and SO_4^{2-} phases with sizes from 0.1 to 1 μm; and *microcrystalline products*, primarily of $Ca(OH)_2$, with particle sizes greater than 1 μm. The most common type of hydraulic cement, *Portland cement*, usually contains mostly colloidal products.

The setting and hardening processes can take many days, and the ultimate strength of the cement may not be realized for years in some instances. This is illustrated in Figure 5.51 for several types of Portland cement, where the compressive strength is plotted as a function of time. Recall that most applications of cement such as buildings and bridges call for compressive loads, such that compressive modulus and strength are the operative mechanical properties.

Table 5.7 Shorthand Notation of Important Oxides in Cement Compositions

C	CaO	F	Fe_2O_3	N	Na_2O
A	Al_2O_3	M	MgO	K	K_2O
L	Li_2O	T	TiO_2	P	P_2O_5
S	SiO_2	H	H_2O	f	FeO
C	CO_2	S	SO_3		

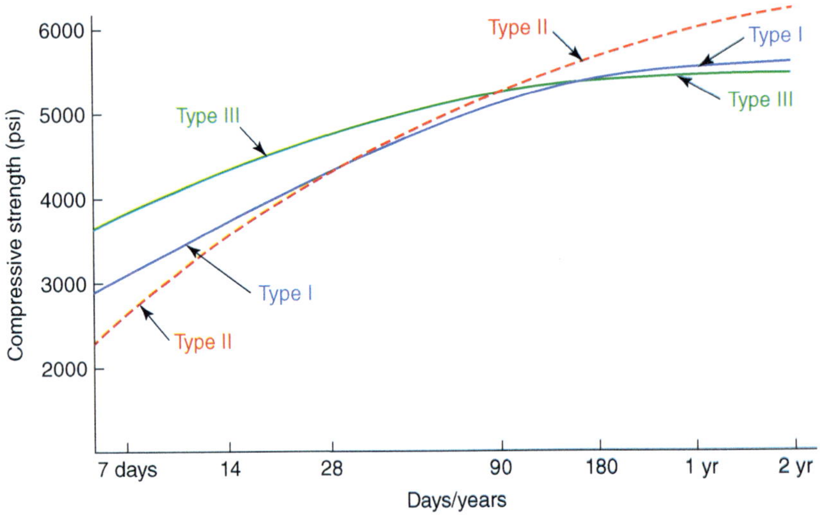

Figure 5.51 Compressive strength of concrete as it develops with time. Somayaji [33]/Reproduced with permission of Pearson.

There are different types of Portland cement, depending upon the intended application: Type I is general purpose; Type II is moderate sulfate-resistant; Type III is high early strength for applications where the cement must set rapidly; Type IV is low heat of hydration, which is required for applications such as dams since the hydration reactions are highly exothermic; and Type V is sulfate-resistant. Sulfate resistance is important when using cement that may be exposed to some sulfur-containing groundwaters and seawaters since sulfur tends to expand and disintegrate cement.

The material of construction known as *concrete* is obtained by combining cement with *aggregates* and *admixtures* in addition to the water required for hydration. The aggregates take up about 60–75% of the volume of the concrete. Aggregates are classified according to size as either fine (about the size of sand) or coarse (about the size of gravel). The aggregate size and quantity are selected to control the consistency, workability, and ultimate mechanical properties of the concrete. The effect of aggregate size on the compressive strength of concrete is shown in Figure 5.52. The increase in strength that comes from coarser gradation is due solely to a reduction in water requirements for the reduced amount of cement. Other important effects on the ultimate mechanical properties of concrete are cure temperature, air content, water-to-cement ratio, and, of course, the use of reinforcements.

Specific Properties

One aspect of ceramics that makes them particularly attractive for structural applications is their low density relative to most structural metals, especially for high-temperature applications. One way to illustrate this is to report mechanical properties such as modulus and strength on a per unit weight basis, resulting in quantities known as *specific properties*. Hence, the tensile modulus divided by the density is known as the *specific modulus*, and the ultimate strength divided by the density is known as the *specific strength*. Table 1.32 shows the specific moduli and strengths of some common ceramics in fiber form, as well as values for some common metals for comparison. This effect is most clearly seen by utilizing the well-known design charts of Ashby [34]. The first is a plot of elastic modulus (Young's modulus) versus strength (yield strength, compressive strength, tensile strength, or tear strength, as indicated) in Figure 5.53. Materials generally fall into groups (e.g., engineering ceramics, engineering alloys, etc.) with individual materials listed where possible.

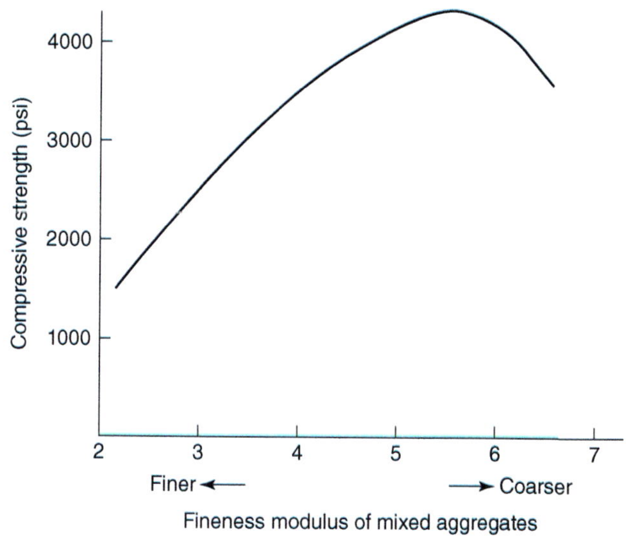

Figure 5.52 Effect of aggregate size on compressive strength of concrete. Somayaji [33]/Reproduced with permission of Pearson.

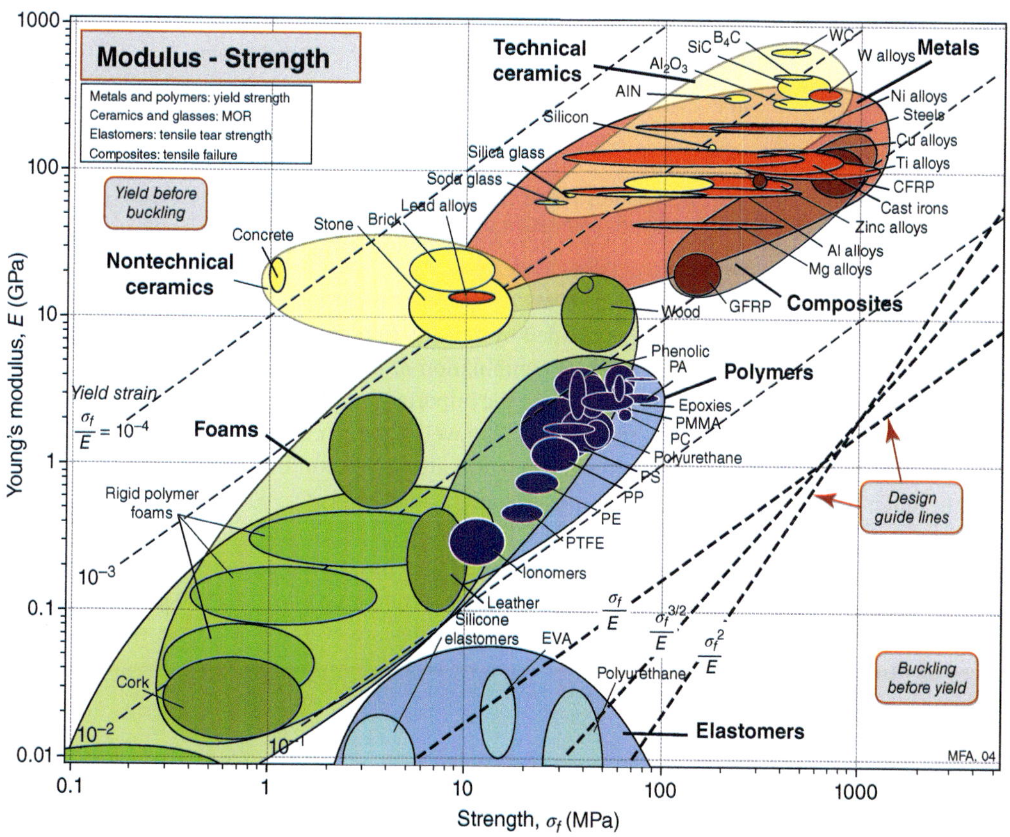

Figure 5.53 Design chart of modulus versus strength (see inset for type of strength determination). Ashby [32]/Reproduced with permission of Elsevier.

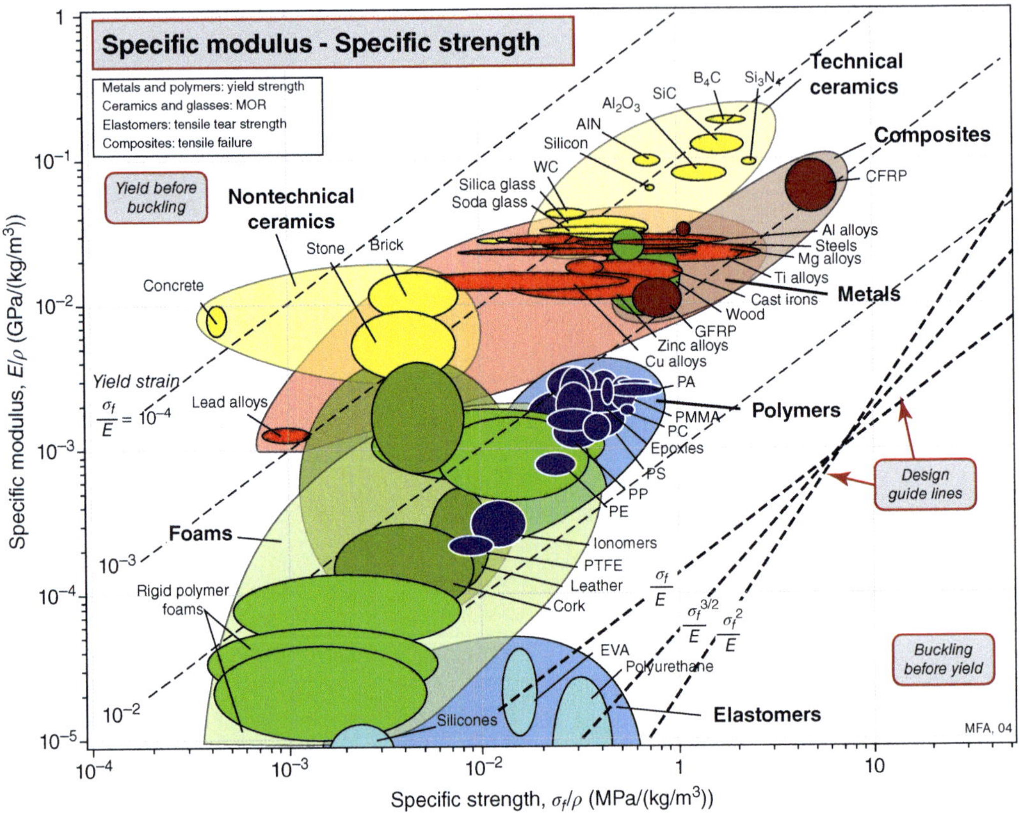

Figure 5.54 Design chart of specific modulus versus specific strength (see inset for type of strength determination). Ashby [35]/Reproduced with permission of Elsevier.

Figure 5.54 is the corresponding plot in terms of specific properties; that is, the modulus and strength values of Figure 5.54 have been divided by density. In both charts, materials at the upper right-hand side of the diagram have the best combination of modulus and strength.

We select here an extreme example to illustrate the importance of specific properties. Suppose we are looking for a refractory material. Tungsten, W, is a well-known refractory metal with a melting point of 3387°C. Its region of modulus and strength values are found in Figure 5.53. A similar ceramic material in terms of modulus and strength would be Al_2O_3, found adjacent to W in Figure 5.53, with a melting point of around 2050°C. Now locate these materials in Figure 5.54 (W is the unlabeled region above Cu alloys). Notice how W has moved downward and to the left relative to Al_2O_3, due solely to its high density (19.3 g/cm^3) compared to Al_2O_3 (3.98 g/cm^3). Similar arguments apply to the use of polymers, for which densities are even lower. These materials are described in the next section.

5.3 Mechanics of Polymers

In this section, we describe the mechanical properties of a class of materials that continues to grow in terms of use in structural applications. As issues related to energy consumption and climate change continue to increase demands for lightweight, recyclable materials, the development of

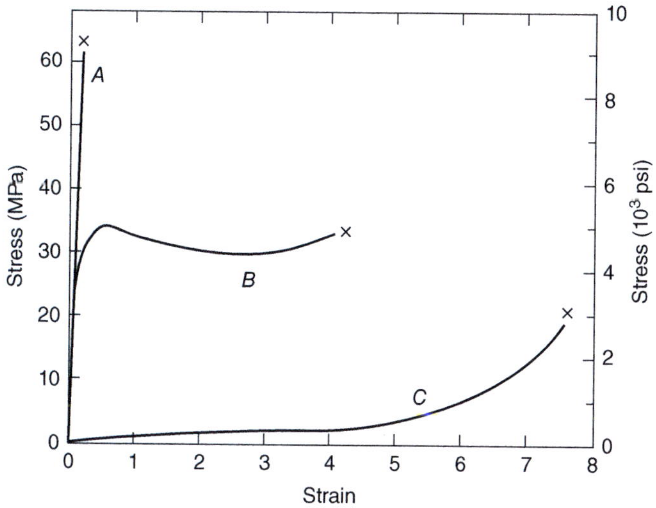

Figure 5.55 The stress–strain behavior of brittle polymer (curve A), ductile polymer (curve B), and highly elastic polymer (curve C). Callister [36]/Reproduced with permission of John Wiley & Sons.

new polymers and the characterization of recycled polymers will continue to dominate research and development efforts in this area.

Stress–strain diagrams for polymers can exhibit many of the same characteristics as both brittle materials (Figure 5.55, curve A) and ductile materials (Figure 5.55, curve B). In general, highly crystalline polymers (curve A) behave in a brittle manner, whereas amorphous polymers can exhibit plastic deformation, as in curve B. We will see that these phenomena are highly temperature-dependent, even more so with polymers than they are with metals or ceramics. We will also see that due to the unique structures of some crosslinked polymers, recoverable deformations at high strains up to the point of rupture can be observed in materials known as elastomers (Figure 5.55, curve C).

Despite the similarities in brittle and ductile behavior to ceramics and metals, respectively, the elastic and permanent deformation mechanisms in polymers are quite different, owing to the difference in structure and size scale of the entities undergoing movement. Whereas plastic deformation (or lack thereof) could be described in terms of dislocations and slip planes in metals and ceramics, the polymer chains that must be deformed are of a much larger size scale. Before discussing polymer mechanical properties in this context, however, we must first describe a phenomenon that is somewhat unique to polymers—one that imparts some astounding properties to these materials. That property is *viscoelasticity*, and it can be described in terms of fundamental processes that we have already introduced.

5.3.1 Viscoelasticity

As the term implies, viscoelasticity is the response of a material to an applied stress that has both a *viscous* and an *elastic* component. In addition to a recoverable elastic response to an applied force, polymers can undergo permanent deformation at high strains, just as was the case for metals and some glasses, as described previously. The mechanism of permanent deformation is different in polymers, however, and can resemble liquid-like, or viscous flow, just like we described in Chapter 4. Let us first develop two important theoretical models to describe viscoelasticity, then describe how certain polymers exhibit this important property.

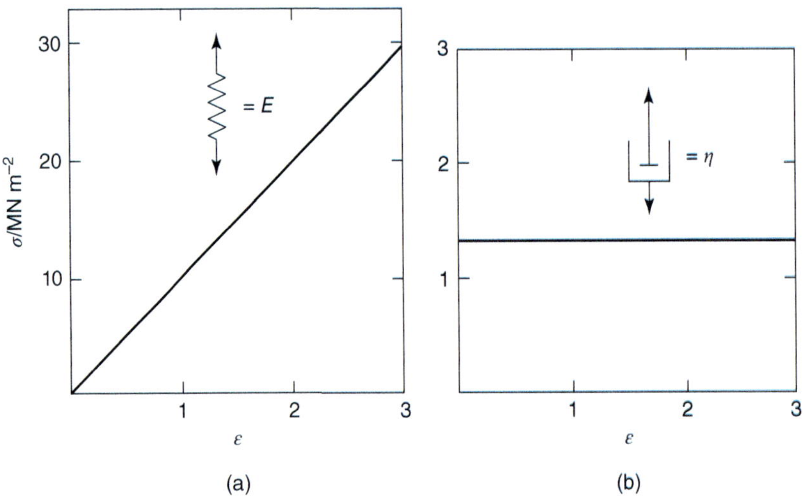

Figure 5.56 Stress–strain behavior of (a) spring of modulus E and (b) dashpot of viscosity η. Cowie [37]/ Reproduced with permission of Taylor & Francis.

In both models of viscoelasticity, the elastic response is represented by a Hookean *spring* of elastic modulus E (Figure 5.56a), and the viscous flow is represented by a piston-like device known as a *dashpot* of viscosity η (Figure 5.56b). Note that we use the non-Newtonian viscosity, η, rather than the Newtonian viscosity, μ, because we make no assumptions about the viscous behavior of the polymer. We can also generalize these models for both tension and shear by using either the elastic modulus, E, or the shear modulus, G. In fact, we will use these moduli more interchangeably than we have previously. Keep in mind that both the spring and dashpot are completely imaginary devices—much as the "frictionless pulley" of introductory physics—and are utilized only to assist in the development of the two models of viscoelasticity.

The Maxwell Model

The first model of viscoelasticity was proposed by Maxwell in 1867, and it assumes that the viscous and elastic components occur in series, as in Figure 5.57a. We will develop the model for the case of shear, but the results are equally general for the case of tension. The mathematical development of the *Maxwell model* is fairly straightforward when we consider that the applied shear stress, τ, is the same on both the elastic, τ_e, and viscous, τ_v, elements,

$$\tau = \tau_e = \tau_v \tag{5.59}$$

since any applied stress is transmitted completely through the spring, once it is extended, to the dashpot below. The total shear strain, γ, is the sum of the elastic strain, γ_e, and the viscous strain, γ_v,

$$\gamma = \gamma_e + \gamma_v \tag{5.60}$$

since the total displacement is the sum of the displacements in the spring and the dashpot, which act independently. We can take the time derivative of Eq. (5.60) to obtain the total strain rate, $\dot{\gamma} = d\gamma/dt$:

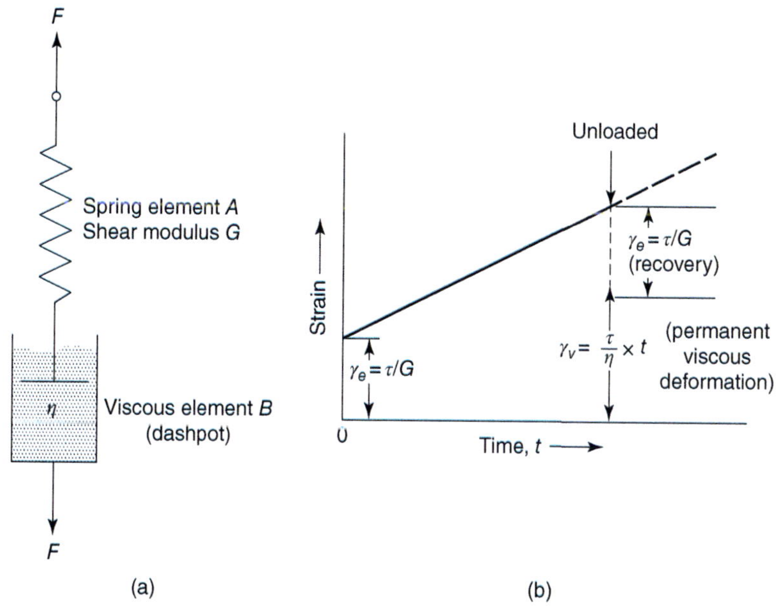

Figure 5.57 (a) Maxwell spring and dashpot in series model of viscoelasticity and (b) constant stress conditions resulting in time-dependent strain. From Jastrzebski [4]. Copyright © 1976 by John Wiley & Sons, Inc. This material is used by permission of John Wiley & Sons, Inc.

$$\frac{d\gamma}{dt} = \frac{\gamma_e}{dt} + \frac{\gamma_v}{dt} \tag{5.61}$$

Recall from Eq. (5.7) that the shear strain can be related to the shear stress through the shear modulus, G, according to Hooke's Law, where we now add subscripts to differentiate the elastic quantities from the viscous quantities:

$$\tau_e = G\gamma_e \tag{5.7}$$

This relation can be differentiated with respect to time and solved for the strain rate:

$$\frac{d\gamma_e}{dt} = \frac{d\tau_e/dt}{G} \tag{5.62}$$

Recall also from Section 4.0 that the viscous shear rate, $\dot{\gamma}_v$, can be related to the viscous shear stress through the viscosity, η, according to Newton's Law of Viscosity, Eq. (4.3):

$$\tau_v = \eta \frac{d\gamma_v}{dt} \tag{4.3}$$

where the subscript "v" has once again been added to differentiate the various shear stresses and strains. [In this version of Eq. (4.3), the minus sign has been incorporated into the direction of the shear stress through a change in coordinate system for ease of mathematical manipulation.] Equation (4.3) can be solved for the shear rate and substituted into Eq. (5.61), along with Eq. (5.62), to yield:

$$\frac{d\gamma}{dt} = \frac{1}{G}\frac{d\tau_e}{dt} + \frac{1}{\eta}\tau_v \tag{5.63}$$

The utility of the Maxwell model is that it predicts permanent deformation due to flow at any applied stress. As shown in Figure 5.57b, at zero time a stress is applied and the spring deforms according to Hooke's Law. Immediately following zero time, the dashpot begins to flow under the restoring forces of the spring, which loses its stored energy to the dashpot for dissipation. If the stress is kept constant, then $d\tau/dt = 0$, and Eq. (5.63) becomes Newton's Law of Viscosity, and the Maxwell body is indistinguishable from a liquid. Once the stress is removed, the spring recovers, again according to Hooke's Law, and the permanent deformation that remains is due solely to the viscous flow.

Under conditions of constant shear, $d\gamma/dt = 0$, Eq. (5.63) becomes an ordinary differential equation, which can be solved by separation of variables and integration using the boundary conditions $\tau = \tau_0$ at $t = 0$ and $\tau = \tau$ at $t = t$ to give the following relation for the shear stress, τ, as a function of time

$$\tau = \tau_0 \exp\left(\frac{-Gt}{\eta}\right) \tag{5.64}$$

Under constant shear conditions, Eq. (5.64) predicts that the stress will exponentially disappear, so that after a time $t = \eta/G$, the stress will have decayed to $1/e$ of its original value. This time is referred to as the *relaxation time*, t_{rel}, for the material, so that Eq. (5.64) becomes:

$$\tau = \tau_0 \exp\left(\frac{-t}{t_{rel}}\right) \tag{5.65}$$

This *stress relaxation* process is important, and we shall revisit it after developing the second model of viscoelasticity.

The Kelvin–Voigt Model

A similar development can be followed for the case of a spring and dashpot in parallel, as shown schematically in Figure 5.58a. In this model, referred to as the *Kelvin–Voigt model* of viscoelasticity, the stresses are additive

$$\tau = \tau_e + \tau_v \tag{5.66}$$

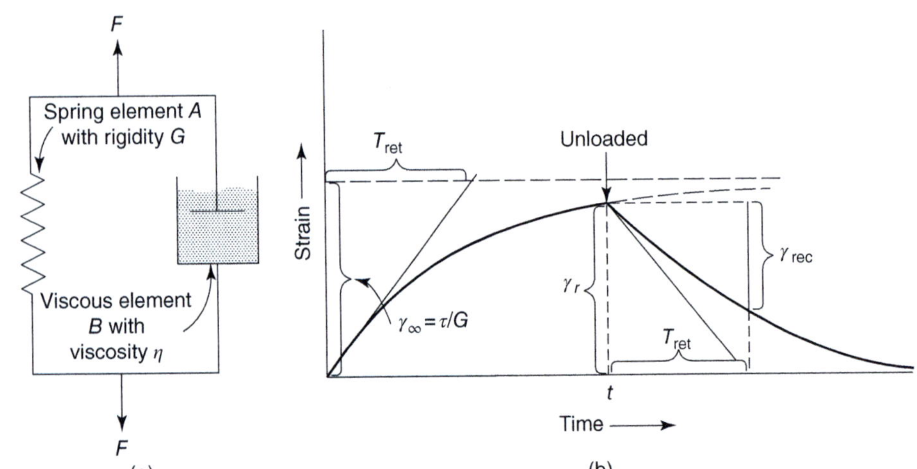

Figure 5.58 (a) Kelvin–Voigt spring and dashpot in parallel model of viscoelasticity and (b) resulting time-dependent strain. From Jastrzebski [4]. Copyright © 1976 by John Wiley & Sons, Inc. This material is used by permission of John Wiley & Sons, Inc.

and the strain is distributed

$$\gamma = \gamma_e = \gamma_v \tag{5.67}$$

Hooke's Law and Newton's Law can once again be substituted, this time into Eq. (5.66), to give:

$$\tau = G\gamma + \eta\frac{d\gamma}{dt} \tag{5.68}$$

Under conditions of constant stress, $d\tau/dt = 0$, Eq. (5.68) can be solved to give a relation for the total strain as a function of time:

$$\gamma = \frac{\tau}{G}\left[1 - \exp\left(\frac{-Gt}{\eta}\right)\right] \tag{5.69}$$

Need More Help?

Try these videos on viscoelasticity:

https://www.youtube.com/watch?v=csy3Qwlg7p4
https://www.youtube.com/watch?v=9gPc_yPFngk

When the Kelvin–Voigt element is stressed under these constant stress conditions, part of the energy is stored in the spring, while the remainder of the energy is dissipated in the dashpot. Eq. (5.69) indicates that when the sample is subjected to the stress, τ, its elastic response is retarded by the viscous element and the strain does not appear instantaneously but increases gradually, asymptotically approaching its final value of γ_∞, the strain at infinite time, as shown in Figure 5.58b. Upon releasing the stress, recovery will occur because of the elasticity of the spring, but it is again retarded by the dashpot. The strain vanishes gradually and the sample slowly returns to its original shape, theoretically at infinite time. This time-dependent relaxation process is an example of *anelasticity*, which is time-dependent elastic behavior. Although all materials exhibit some anelasticity, it is generally neglected in everything but polymers, where it can be significant. In fact, the entire time-dependent elastic behavior described in this section as viscoelasticity is a form of anelasticity.

Practically, the time needed to achieve the equilibrium after removal of the stress is called the *retardation time*, t_{ret} which is equal to the ratio η/G, as in the Maxwell model. Equation (5.69) can then be written

$$\gamma_r = \frac{\tau}{G}\left[1 - \exp\left(\frac{-t}{t_{ret}}\right)\right] \tag{5.70}$$

where γ_r is the strain at time t. If the retardation time is small, the recovery will take place at finite time. The Kelvin–Voigt model illustrates the behavior of materials exhibiting a retarded elastic deformation, as occurs in such phenomena as the *elastic aftereffect*, *creep recovery*, and *elastic memory*.

When a material is subjected to a tensile or compressive stress, Eqs. (5.59) through (5.70) should be developed with the shear modulus, G, replaced by the elastic modulus, E, the viscosity, η, replaced by a quantity known as *Trouton's coefficient* of viscous traction, λ, and shear stress, τ, replaced by the tensile or compressive stress, σ. It can be shown that for incompressible materials, $\lambda = 3\eta$, because the flow under tensile or compressive stress occurs in the direction of stress as well as in the two other directions perpendicular to the axis of stress. Recall from Section 5.1.1 that for incompressible solids, $E = 3G$; therefore, the relaxation or retardation times are λ/E.

Example Problem 5.2

In Cooperative Learning Exercise 5.9, you studied the effect of temperature on the creep rate of an engine turbine blade. Calculate the viscosity in poise that can be assigned to this material at 500°C under a stress of 20,000 psi.

Solution: For the case of continuous deformation under a constant tensile stress, we can write, by analogy to Newton's Law of Viscosity [Eq. (4.3)], a relationship between tensile strain rate, $\dot{\varepsilon}$, tensile stress, σ, and Trouton's coefficient, λ. As

$$\sigma = \lambda\dot{\varepsilon}$$

For incompressible materials, Trouton's coefficient can be related to $\lambda = 3\eta$, so that

$$\eta = \frac{\sigma}{3\dot{\varepsilon}}$$

From CLE 5.9, $\dot{\varepsilon} = 0.0001$ in./in.-s at 500°C, and $\sigma = 20{,}000$ psi, such that

$$\eta = \frac{(20{,}222\ \text{psi})(6.895 \times 10^4\ \text{dyn/cm}^2 \cdot \text{psi})}{3(0.0001 s^{-1})}$$

$$\eta = 4.6 \times 10^{12}\ \text{poise}$$

Time–Temperature Superposition—The WLF Equation

Recall that the stress relaxation, t_{rel} for the Maxwell model is simply the ratio of the viscosity to the modulus. These values are easily measured in an oscillating-load apparatus, much like that described for fatigue testing in Section 5.2.1. The time-dependent tensile modulus for a typical amorphous polymer, poly(methyl methacrylate) (PMMA), measured by stress relaxation at various temperatures is shown in Figure 5.59. We could take a cross section of the modulus–time curve at some fixed relaxation time, say 0.001 hours, 1 hour, and 100 hours, and replot the data as log E versus temperature, as in Figure 5.60. In doing so, we notice two reasonably well-defined plateaus at all relaxation times. The first, at $E = 10^{10}$ Pa, is the region of glassy polymer behavior. The second plateau, at $E = 10^6$ Pa, is the region of rubbery polymer behavior. Between the two plateaus is the glass transition, which occurs at $T = T_g$. Notice that T_g is dependent on relaxation rate. At still higher temperatures, rubbery flow and then liquid-like flow occur.

Another method for obtaining a composite curve is to shift the curves in Figure 5.59 relative to some reference temperature, T_R, as shown in Figure 5.61. For example, the log E versus time curve at 115°C in Figure 5.59 could be selected as the reference curve; that is, T_R = 115°C. Curves at 110°C, 100°C, and 40°C could be shifted to the left, for example, and the curve at 135°C could be shifted to the right, as illustrated in Figure 5.61. The result is a modulus–time master curve, and the process of shifting the curves relative to a reference point is called the *time–temperature superposition principle*. This is not simply a graphical exercise. The process, which has been borne out experimentally for several amorphous polymers, is carried out by calculating a *shift factor*, a_T, according to the *Williams–Lendel–Ferry (WLF) equation*:

$$\log a_T = \frac{-17.44(T - T_g)}{51.6 + T - T_g} \tag{5.71}$$

The WLF equation holds over the temperature range from T_g to about T_g + 100 K. The constants in Eq. (5.71) are related to the free volume. This is a procedure analogous to the one we used to generate time–temperature–transformation (TTT) diagrams for metallic phase transformations in Section 3.1.2.

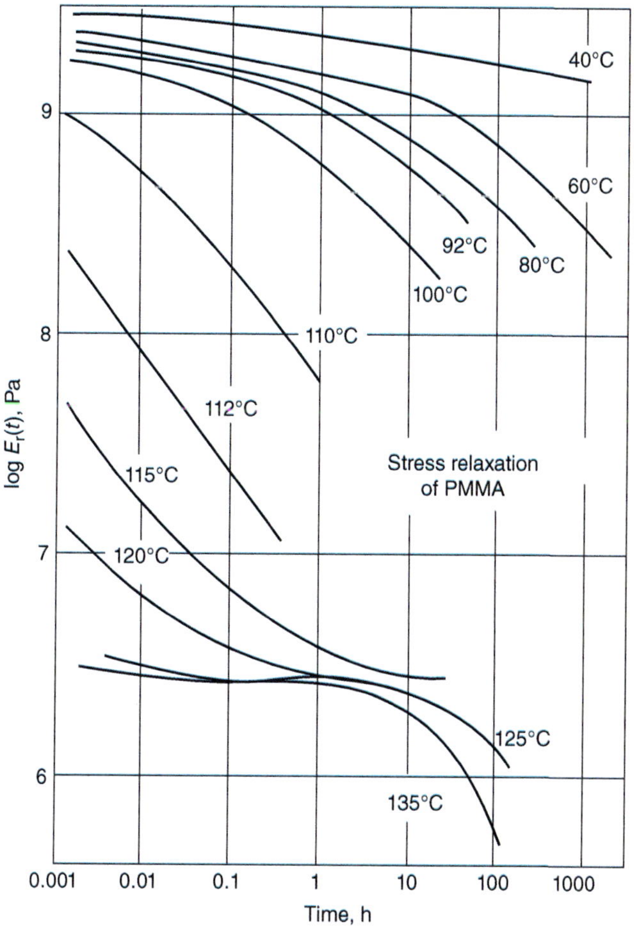

Figure 5.59 Time-dependent modulus for unfractionated poly(methyl methacrylate). Rodriguez [38]/ Reproduced with permission of Hemisphere Publishing Corporation.

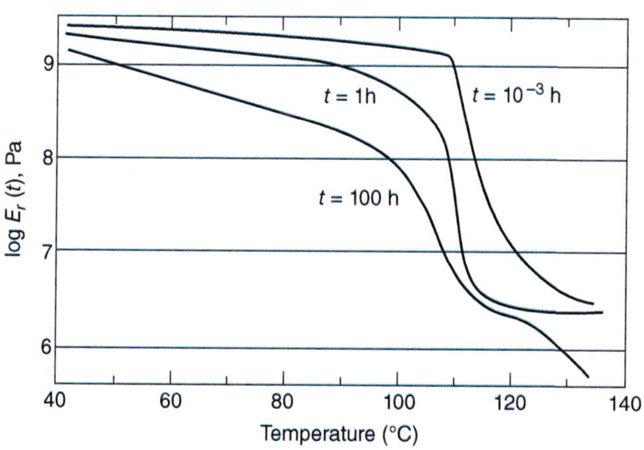

Figure 5.60 Modulus–temperature master curve based on cross sections of Figure 5.59. Rodriguez [38]/Reproduced with permission of Hemisphere Publishing Corporation.

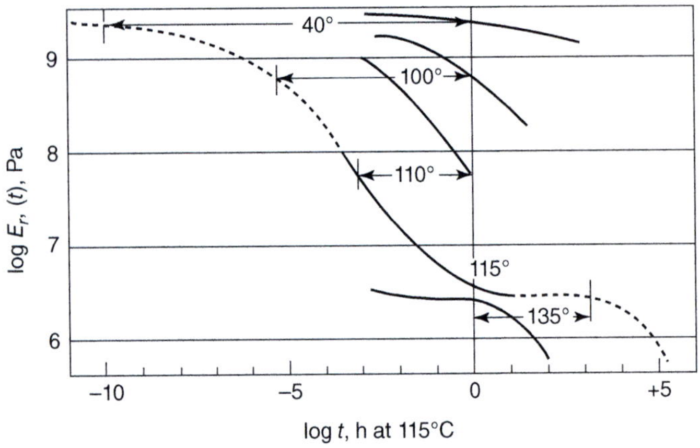

Figure 5.61 Modulus–time master curve based on WLF-shift factors using data from Figure 5.60 with a reference temperature of 114°C. Rodriguez [38]/Reproduced with permission of Hemisphere Publishing Corporation.

Cooperative Learning Exercise 5.10

Work in groups of three. The shift factor, a_T, in the WLF Equation [Eq. (5.71)], is also a ratio of stress relaxation times, t_{rel}, in the polymer at an elevated temperature, T, relative to some reference temperature, T_0, and can be related via an Arrhenius-type expression to the activation energy for relaxation, E_{rel} as

$$a_T = exp\left[\frac{E_{rel}}{R}\left(\frac{1}{T} - \frac{1}{T_0}\right)\right] \quad \text{(CLE 1.1)}$$

You have developed a new semicrystalline polymer, which has a typical activation energy for relaxation of E_{rel} = 120 kJ/mol. You wish to know the creep compliance for 10 years at 27°C. You know that, in principle, you can obtain the same information in a much shorter period of time by conducting your compliance tests at a temperature above 27°C.

Person 1: Calculate the shift factor and the required test duration if you performed the tests at 32°C.

Person 2: Calculate the shift factor and the required test duration if you performed the tests at 77°C.

Person 3: Calculate the shift factor and the required test duration if you performed the tests at 180°C.

Compare your answers. Which is the most reasonable temperature for running the compliance tests? What are some things that you should check before deciding on a final temperature?

Answers: $a_T(305) = 0.45$, 4.5 year; $a_T(350) = 1.30 \times 10^{-3}$, 91 hours; $a_T(453) = 0.9 \times 10^{-7}$, 27 seconds; 77°C gives the most reasonable test duration, 32°C is too long and 180°C may be too short to collect sufficient data. You should determine T_g and/or T_m to make sure that the is neither exactly on the glass transition temperature, nor above the crystalline melting point.

The time–temperature superposition principle has practical applications. Stress relaxation experiments are practical on a time scale of 10^1 to 10^6 seconds (10^{-2} to 10^3 hours), but stress relaxation data over much larger time periods, including fractions of a second for impacts and decades for creep, are necessary. Temperature is easily varied in stress relaxation experiments and, when used to shift experimental data over shorter time intervals, can provide a master curve over relatively large time intervals, as shown in Figure 5.61. The master curves for several crystalline and amorphous polymers are shown in Figure 5.62.

5.3.2 Mechanical Properties of Polymers

To a much greater extent than either metals or ceramics, the mechanical properties of polymers show a marked dependence on a number of parameters, including temperature, strain rate, and morphology. In addition, factors such as molecular weight and temperature relative to the glass transition play important roles that are not present in other types of materials. It is impossible to cover, even briefly, all of these effects. We concentrate here on the most important effects that can affect selection of polymers from a mechanical design point of view.

Temperature and Strain Rate Effects

Polymers such as polystyrene (PS) and PMMA, with a high tensile modulus at ambient temperatures fall into the category of brittle materials, since they tend to fracture prior to yielding. "Tough" polymers, or those that undergo some deformation prior to failure, can be typified by cellulose acetate. These two categories of behavior are illustrated in Figure 5.63, along with their respective temperature dependences.

It can be seen from Figure 5.63 that the effect of temperature on the characteristic shape of the curves is significant. As the temperature increases, both the rigidity and the yield strength decrease,

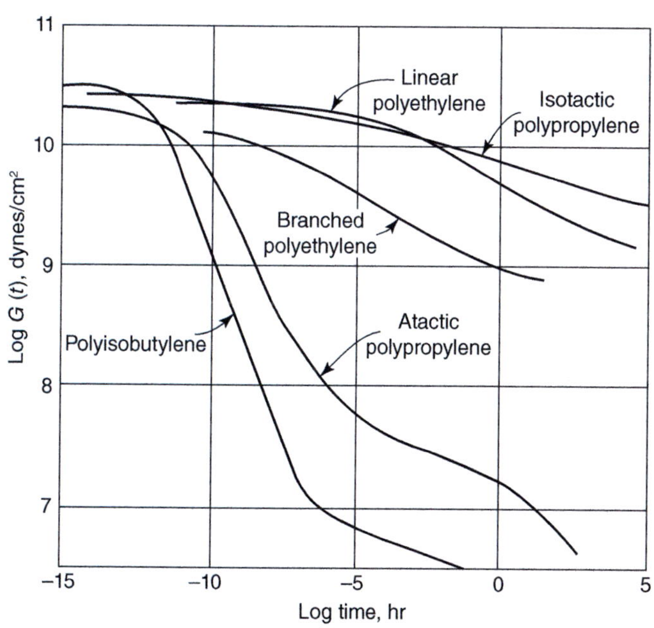

Figure 5.62 Stress–relaxation master curve for several polymers. From Billmeyer [39]. Copyright © 1984 by John Wiley & Sons, Inc. This material is used by permission of John Wiley & Sons, Inc.

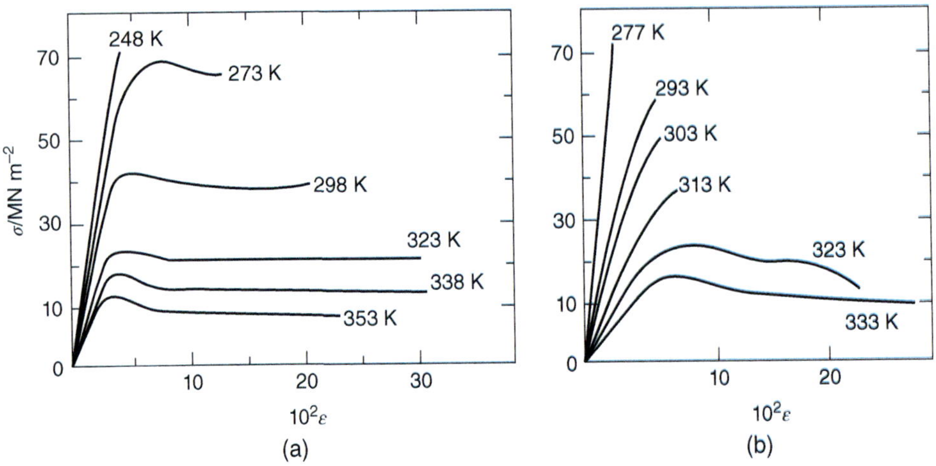

Figure 5.63 Influence of temperature on the stress–strain response of (a) cellulose acetate and (b) poly(methyl methacrylate). Cowie [40]/Reproduced with permission of Taylor & Francis.

while the elongation generally increases. For cellulose acetate, there is a ductile–brittle transition at around 273 K, above which the polymer is softer and tougher than the hard, brittle polymer below 273 K. For PMMA, the hard, brittle characteristics are retained at a much higher temperature, but it eventually reaches a soft, tough state at about 320 K. Thus, if the requirements of high rigidity and toughness are to be met, the temperature is important.

As was illustrated in Figure 5.43 for a ceramic material, temperature and strain rate have similar, if not opposite, effects on the stress–train relationship in polymers. Figure 5.64 shows the effect of increasing strain rate on the stress–strain diagram for polyethylene. Compare this effect with decreasing temperature for the polymers in Figure 5.63, where similar effects are observed—that is, increasing yield strength and modulus. Despite the similarities in yield strength behavior with temperature and strain rate between polymers, metals, and ceramics, the mechanisms

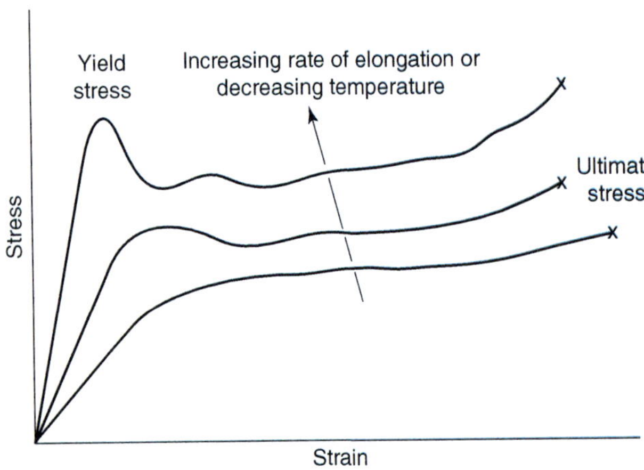

Figure 5.64 Effect of increasing strain rate and decreasing temperature on stress–strain curves for polyethylene. Rodriguez [41]/Reproduced with permission of Hemisphere Publishing Corporation.

are quite different, as alluded to earlier. Specifically, the necking of polymers is affected by two physical factors that are not normally significant in metals. The first is the dissipation of mechanical energy as heat, which can raise the temperature in the neck, causing significant softening. The magnitude of this effect increases with strain rate. The second is deformation resistance of the neck, which has a higher strain rate than the surrounding polymer, and results in the strain-rate dependence of yield strength, as shown in Figure 5.64. The relative importance of these two opposing effects depends on the material, the length and thickness of the specimen, and the test conditions, especially the strain rate, which can vary greatly in different forming operations.

In semicrystalline polymers such as polyethylene, yielding involves significant disruption of the crystal structure. Slip occurs between the crystal lamellae, which slide by each other, and within the individual lamellae by a process comparable to glide in metallic crystals. The slip within the individual lamellae is the dominant process and leads to molecular orientation since the slip direction within the crystal is along the axis of the polymer molecule. As plastic flow continues, the slip direction rotates toward the tensile axis. Ultimately, the slip direction (molecular axis) coincides with the tensile axis, and the polymer is then oriented and resists further flow. The two slip processes continue to occur during plastic flow, but the lamellae and spherulites increasingly lose their identity, and a new fibrillar structure is formed (see Figure 5.65).

A model has been developed by Eyring to describe the effects of temperature and strain rate on flow stress in polymers. The fundamental idea is that a segment of a polymer must pass over an energy barrier in moving from one position to another in the solid. This is an activated process, similar to the many activated processes that we have already described and that were first illustrated in Figure 3.1. There is a forward rate and a reverse rate across the barrier, as for all activated processes. The significant point introduced by Eyring is that the application of a shear stress modifies the barrier height, so that in the direction of stress, the rate of segment jumping, formerly slow, now becomes fast enough to give rise to measurable strain change. The final form of the Eyring equation rests on the reasonable assumptions that the imposed strain rate is proportional to the

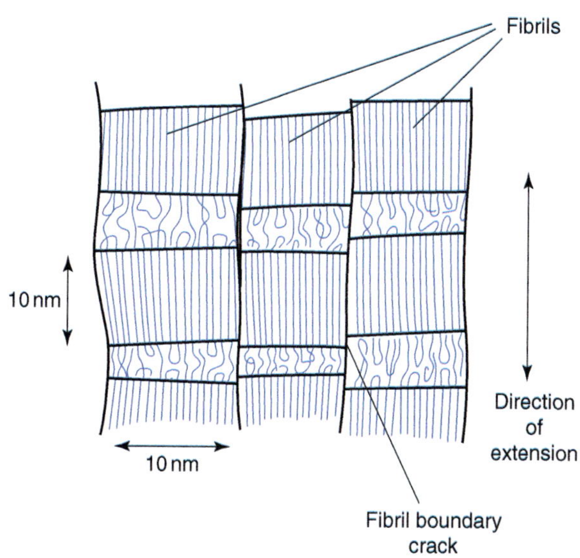

Figure 5.65 Fibrillar structure of an oriented polymer crystal as the result of an applied tensile force. McCrum et al. [42]/Reproduced with permission of Oxford University Press.

net rate at which segments jump preferentially in the direction of the shear stress and that the dominant shear stress in a tensile test is the maximum shear stress, which at yield is $\sigma_y/2$. It should be of no surprise that the final Eyring result contains an Arrhenius-type expression based on the activation energy barrier for segment jumping, ΔH:

$$\dot{\varepsilon}_y = \dot{\varepsilon}_0 \exp\left(\frac{-\Delta H}{RT}\right) \exp\left(\frac{\sigma_y V^*}{2RT}\right) \tag{5.72}$$

where V^* is the activation volume, R is the gas constant, T is absolute temperature, $\dot{\varepsilon}_y$ is the strain rate at yield, and $\dot{\varepsilon}_0$ is a constant. It is usually desirable to investigate the effect of changing strain rate on yield stress, so that Eq. (5.72) is rewritten in the form:

$$\frac{\sigma_y}{T} = \left(\frac{2}{V^*}\right)\left[\frac{\Delta H}{T} + 2.303R\ log\left(\frac{\dot{\varepsilon}_y}{\dot{\varepsilon}_0}\right)\right] \tag{5.73}$$

Figure 5.66 shows plots of σ_y/T against log $\dot{\varepsilon}_y$ for polycarbonate for a series of temperatures between 21.5°C and 140°C, where the T_g of polycarbonate is 160°C. Note that the predictions of the Eyring model are obeyed, namely, that σ_y/T increases linearly with log $\dot{\varepsilon}_y$, and at constant log $\dot{\varepsilon}_y$, σ_y/T increases with decreasing temperature. Thus, the yield strength is dependent on both strain rate and temperature.

The effect of temperature relative to the glass transition, T_g, is illustrated further with the shear modulus, G. The variation of shear modulus with temperature for three common engineering polymers is shown in Figure 5.67. In general, the temperature dependence is characterized by a shallow decline of log G with temperature with one or two abrupt drops. The shallow decline is due to thermal expansion effects. As the temperature increases, the molecules in the solid move further apart. This effect occurs in all solids. The abrupt drops in modulus are attributable to viscoelastic

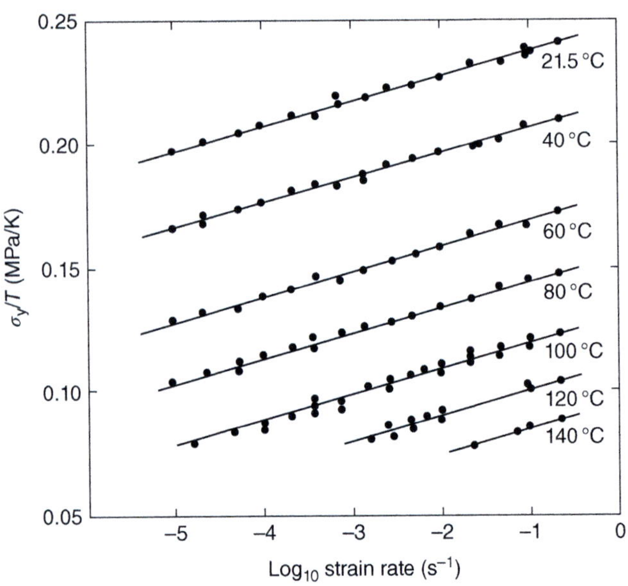

Figure 5.66 Eyring plot for polycarbonate. McCrum et al. [43]/Reproduced with permission of Oxford University Press.

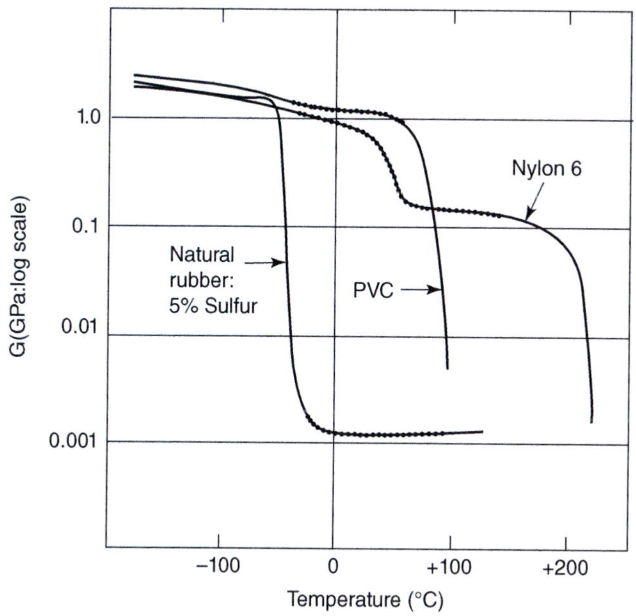

Figure 5.67 Dependence of shear modulus on temperature for three common engineering polymers: crosslinked natural rubber, amorphous polyvinyl chloride (PVC), and crystalline Nylon 6. The typical use temperatures are indicated by dotted lines. McCrum et al. [44]/Reproduced with permission of Oxford University Press.

relaxation processes due to some type of molecular movement—for example, a side group rotation or backbone movement. Note that the modulus of all three polymers at −200°C is in the range of 1–5 GPa. For natural rubber, which is slightly crosslinked, the modulus falls abruptly on heating from −50°C to −25°C. This is the *glass-to-rubber transition*. The modulus then levels off at a value that depends upon the degree of crosslinking. We will describe the mechanical properties of elastomers in more detail later. For polyvinyl chloride (PVC), a small viscoelastic relaxation occurs at −40°C, followed by a major drop at 70°C attributable to the glass-to-rubber transition. The Nylon 6 (polycaprolactam) sample has a crystallinity of about 50% and shows a series of relaxations when heated. The major relaxations at 50°C and 220°C are attributable to the glass-to-rubber and crystalline melting transitions, respectively. This behavior is typical of crystalline polymers, which normally possess several relaxations, whereas amorphous polymers such as PVC typically have only one transition.

Molecular Weight and Composition Effects

As we saw in the previous section, degree of crystallinity affects the mechanical behavior and characteristics of polymers. A related factor is molecular weight. To the extent that molecular weight affects the glass transition (see Figure 1.68) and the use temperature relative to T_g, in turn, affects the mechanical properties such as shear modulus (see Figure 5.68), molecular weight affects mechanical properties. The relationship between T_g and the number-average molecular weight can be approximated by

$$T_g = T_g^{\infty} - \frac{k}{\overline{M}_n} \tag{5.74}$$

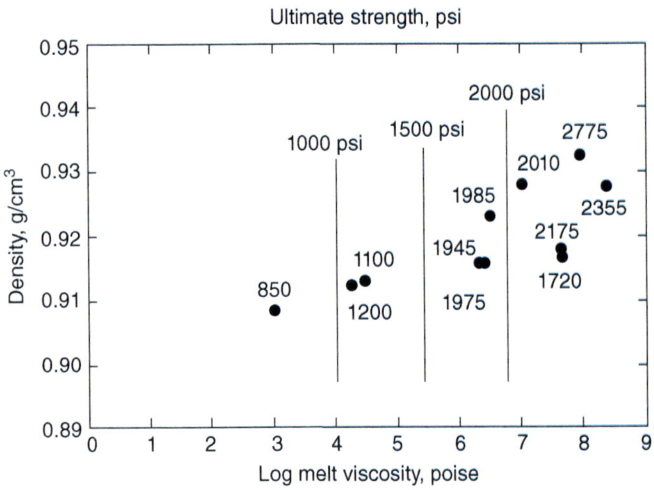

Figure 5.68 Dependence of ultimate tensile strength of polyethylene on molecular weight (melt viscosity) and crystallinity (density). From Billmeyer [39]. Copyright © 1984 by John Wiley & Sons, Inc. This material is used by permission of John Wiley & Sons, Inc.

where T_g^∞ is the glass transition temperature at infinite molecular weight, and k is about 2×10^5 for PS and 3.5×10^5 for atactic poly(α-methyl styrene). The form of this empirical equation is useful for relating several polymer properties, including tensile strength, to the number-average molecular weight

$$\text{Property} = a - \frac{b}{\overline{M}_n} \tag{5.75}$$

where a and b are constants, although the actual variation with molecular weight sometimes depends on an average between the number- and weight-average molecular weights. If a polymer exhibits a yield point and then undergoes extensive elongation before tensile failure, its UTS increases with increasing molecular weight. This effect is illustrated in Figure 5.68 for polyethylene, where molecular weight is represented as melt viscosity, since there is a direct correlation between the two (see Figure 4.11). A similar plot can be made for tensile modulus.

The effects of crystallinity on tensile strength are also included in Figure 5.68 in terms of the polymer density. As the crystalline content of the polymer rises, so does its density, and there is a trend toward higher tensile strength for more highly crystalline polymers. Taken together, crystallinity and molecular weight influence several mechanical properties, including hardness, flexural fatigue resistance, elongation at tensile break, and even impact strength. The chance of brittle failure is reduced by raising molecular weight, which increases brittle strength, and by reducing crystallinity. For amorphous polymers, impact strength is found to depend on the weight-average molecular weight. As the degree of crystallinity decreases with temperature during the approach to the crystalline melting point, T_m, stiffness, and yield strength decrease correspondingly. These factors often set limits on the temperature at which a polymer is useful for mechanical purposes.

Equation (5.75) implies that the tensile strength of a mixture of polymer components (blend or alloy) can be related to the weight average of the component tensile strengths, $\sigma_{t,i}$:

$$\overline{\sigma}_t = \sum w_i \sigma_{t,i} \tag{5.76}$$

where w_i is the weight fraction of component i. In most real polymer blends, the relationship between composition and mechanical properties is much more complicated than Eq. (5.76) would suggest. As described in Section 2.3.3, the ultimate structure of a blend—that is, one phase or two phase, and the resulting microstructure—depends on how the polymers interact. For example, some mechanical properties of a polycarbonate/PMMA blend are shown in Figure 5.69. These blends are invariably two-phase structures. At both 70% and 90% PC, the blends exhibit improved tensile modulus, tensile strength, and yield strength, relative to pure PC, without sacrificing elongation at break. Hence, the area under the stress–strain curve, or toughness, is greater for these two blends than for pure PC. The toughness enhancement can be attributed to the brittle–ductile transition of the brittle phase, but the diameter of the dispersed phase must be smaller than 1 μm to observe this effect. As the amount of PMMA is further increased (% PC decreases), a significant drop-off in elongation at break and toughness is observed, even though the yield and tensile strengths do not change significantly. This observation reflects the brittle behavior of PMMA, as first illustrated in Figure 5.63b. This example should serve as an illustration only. The properties of specific polymer–polymer blends are highly variable, and experimentation should give way to generalizations in these cases.

As first introduced in Section 4.2.3, polymer foams are formed when the second phase is a gaseous one such as air or CO_2. The primary benefit of structural foams in relation to solid polymers is the increased specific modulus. This benefit can only be realized when the foam has the proper skin-core structure, as illustrated in Figure 5.70. As a result of manipulations during processing, an

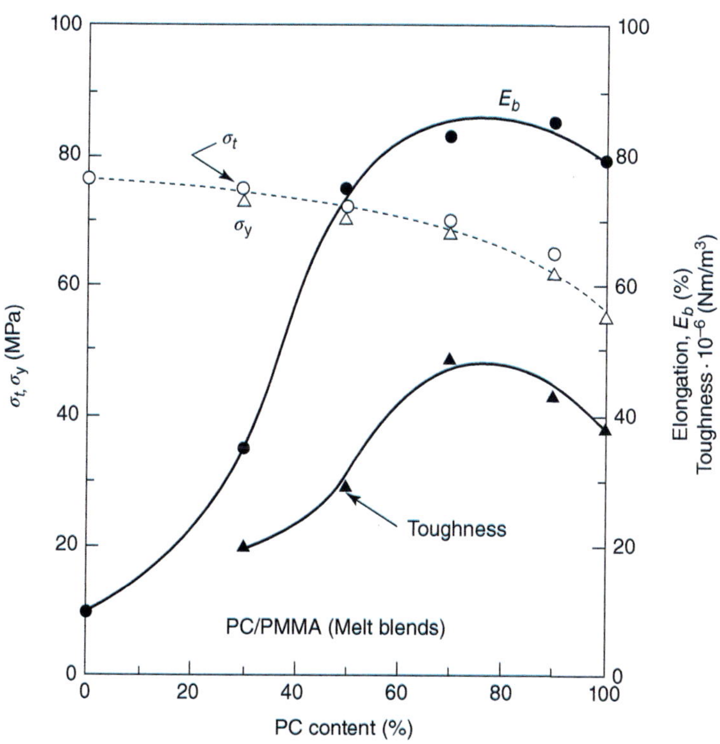

Figure 5.69 Toughness, elongation at break, E_b, yield strength, σ_y, and tensile strength σ_t as a function of composition for a PC/PMMA blend. Kyu et al. [45]/Reproduced with permission of Carl Hanser Verlag GmbH & Co.

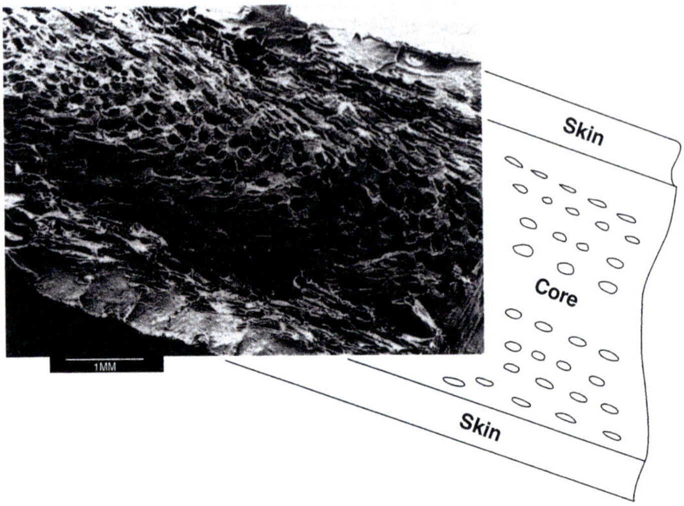

Figure 5.70 Skin-core morphology of an injection-molded polypropylene structural foam. Reproduced with permission from Hornsby [46]. Copyright © 1991, Hanser Gardner Publications.

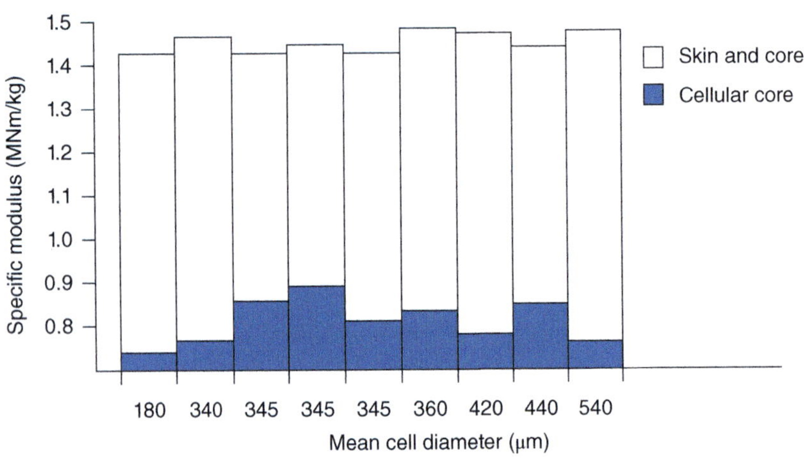

Figure 5.71 Effect of gas cell size on the specific flexural modulus of 6-mm-thick polypropylene structural foam. Samples were tested at 23°C with and without their outer skins. Hornsby [47]/Reproduced with permission of Carl Hanser Verlag GmbH & Co.

injected molded polypropylene foam, for example, will have a nonuniform density through its cross section, with larger, nearly spherical gas cells in the center and small, elongated cells nearer the surface. This form of cell morphology reflects the changing shear and thermal history experienced by the foaming polymer during processing. The outer layer, or skin, contains very few pores. Without the proper skin formation, this functionally graded material will not exhibit adequate mechanical response, since it is the outer skin that bears the initial load, particularly in bending tests. This is illustrated in Figure 5.71, where the effect of gas cell size on specific modulus is less of an effect than is the removal of the outer skin.

In terms of copolymerization, the addition of a comonomer to a crystalline polymer usually causes a marked loss in crystallinity, unless the second monomer crystallizes isomorphously with the first. Crystallinity typically decreases very rapidly, accompanied by reductions in stiffness, hardness, and softening point, as relatively small amounts (10–20 mol%) of the second monomer are added. In many cases, a rigid, fiber-forming polymer is converted to a highly elastic, rubbery product by such minor modifications. The dependence of mechanical properties on copolymer composition in systems that do not crystallize results primarily from changes in intermolecular forces as measure by cohesive energy (cf. Section 2.3.2). Higher cohesive energy results in higher stiffness and hardness and generally improves mechanical properties. The effect of temperature on the mechanical properties of amorphous copolymers depends on structure, as shown in Figure 5.72. Polymer A and polymer B are two amorphous homopolymers with mechanical properties as indicated. A random copolymer of the two has a comparable modulus–temperature spectrum to the two homopolymers, exhibiting a nearly three-decade drop in shear modulus, with differences only in the relaxation temperature. The relaxation temperature is high for polymer B, low for polymer A, and intermediate to the two for the copolymer. However, the block and graft versions of the copolymer exhibit two relaxation temperatures, corresponding to the two phases of A and B that form in these copolymers. One relaxation process is typical of the PS phase, while the other is typical of the polybutadiene phase. By varying the proportions of comonomer, it is possible to obtain a low styrene content, rubbery solid with glassy inclusions, or a high styrene content, glassy solid with rubbery inclusions.

Finally, the use of low-molecular-weight species to improve flow properties called *plasticizers* normally reduces stiffness, hardness, and brittleness. Plasticization is usually restricted to amorphous polymers or polymers with a low degree of crystallinity because of the limited compatibility of plasticizers with highly crystalline polymers. Other additives, such as antioxidants, do not affect the mechanical properties significantly by themselves, but can substantially improve property retention over long periods of time, particularly where the polymer is subject to environmental degradation.

Elastomers

In addition to the numerous ways we have grouped polymers (cf. Section 1.3.2), polymers can also be grouped according to mechanical behavior into the categories of plastics, fibers, and elastomers. Though we will not elaborate upon these rather nonspecific distinctions here, we simply note that elastomers exhibit some unique mechanical behavior when compared to conventional plastics and metals, as in Figure 5.73. The most notable characteristics are the low modulus and high deformations. Elastomers readily undergo deformation and exhibit large, reversible elongations under small, applied stresses.

Although many synthetic elastomers are now available, natural rubber must still be regarded as the standard elastomer because of the excellently balanced combination of desirable qualities. The most important synthetic elastomer is styrene–butadiene rubber (SBR), which is used predominantly for tires when reinforced with carbon black. Nitrile rubber (NR) is a random copolymer of acrylonitrile and butadiene and is used when an elastomer is required that is resistant to swelling in organic solvents.

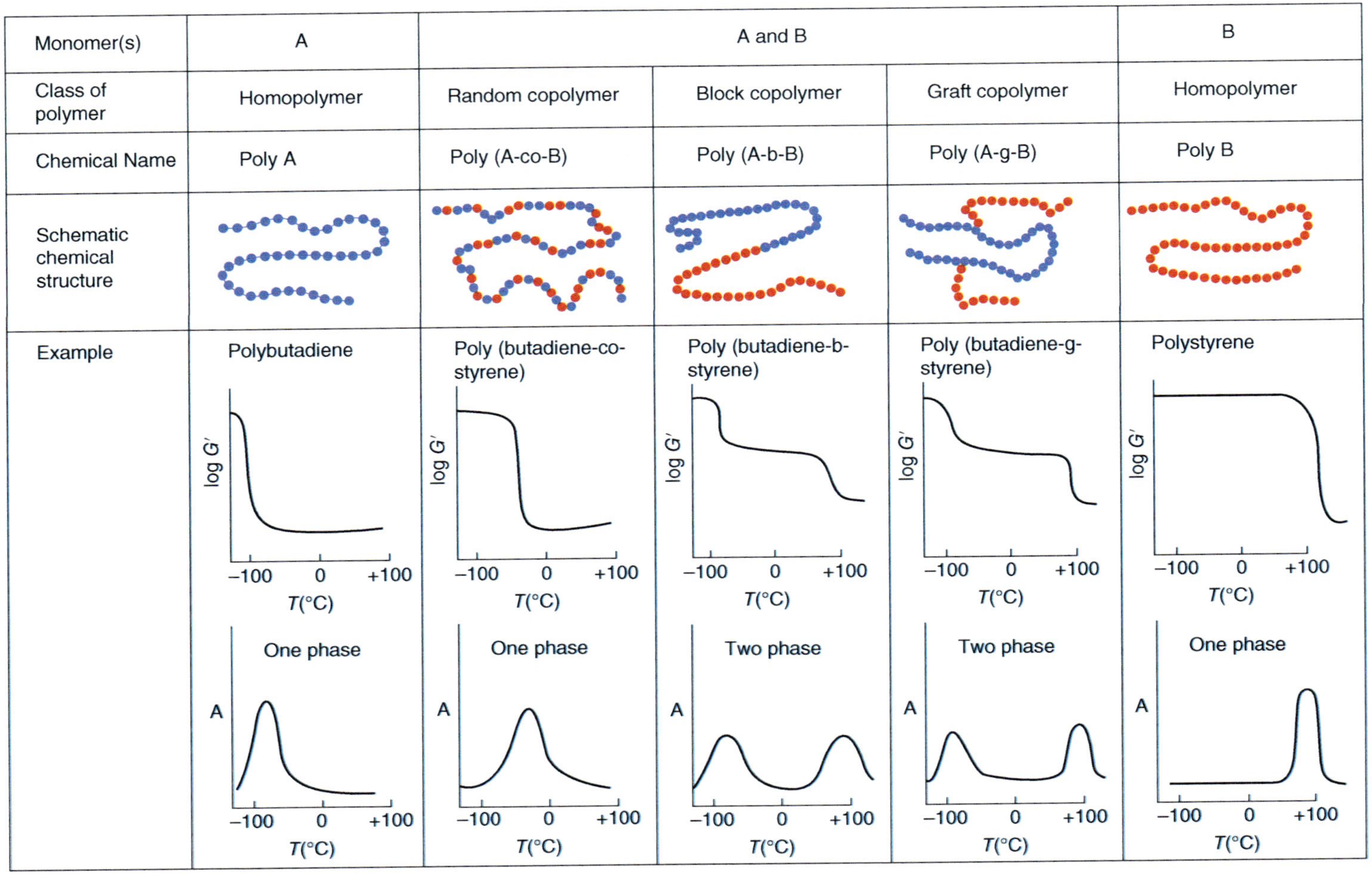

Figure 5.72 Effect of temperature on shear modulus for random, block, and graft copolymers. Bottom curves are the derivative of the log *G* curves. McCrum et al. [48]/Reproduced with permission of Oxford University Press.

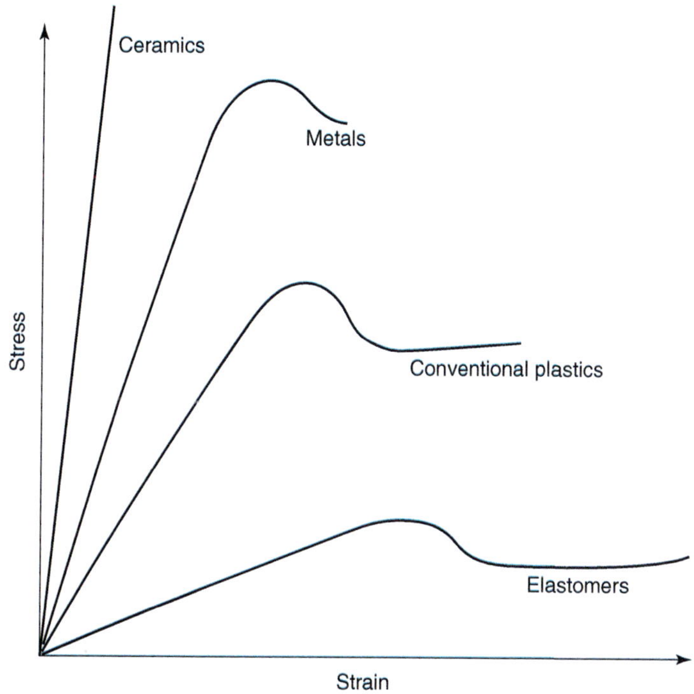

Figure 5.73 Comparison of idealized stress–strain diagrams for metals, amorphous polymers, and elastomers.

> **Historical Highlight**
>
> It's no coincidence that the rise of polymers and polymer science coincided with the Second World Word. One example is the development of synthetic rubber. Read more about the Synthetic Rubber Program and Rubber Reserve Company here:
> https://www.loc.gov/rr/scitech/trs/trschemical_rubber.html

Elastomers exhibit this behavior due to their unique, crosslinked structure (cf. Section 1.3.2). It has been found that as the temperature of an elastomer increases, so does the elastic modulus. The elastic modulus is simply a measure of the resistance to the uncoiling of randomly oriented chains in an elastomer sample under stress. Application of a stress eventually tends to untangle the chains and align them in the direction of the stress, but an increase in temperature will increase the thermal motion of the chains and make it harder to induce orientation. This leads to a higher elastic modulus. Under a constant force, some chain orientation will take place, but an increase in temperature will stimulate a reversion to a randomly coiled conformation and the elastomer will contract.

Unlike noncrosslinked polymers—for which mechanical properties vary little with molecular weight alone, as long as the molecular weight exceeds that of a polymer segment—elastomers, and network polymers in general, exhibit an increase in elastic modulus with crosslink density, as shown in Figure 5.74. Recall that a network polymer technically has an infinite molecular weight. For a finite molecular weight, the crosslinks are temporary ones and are due to entanglements.

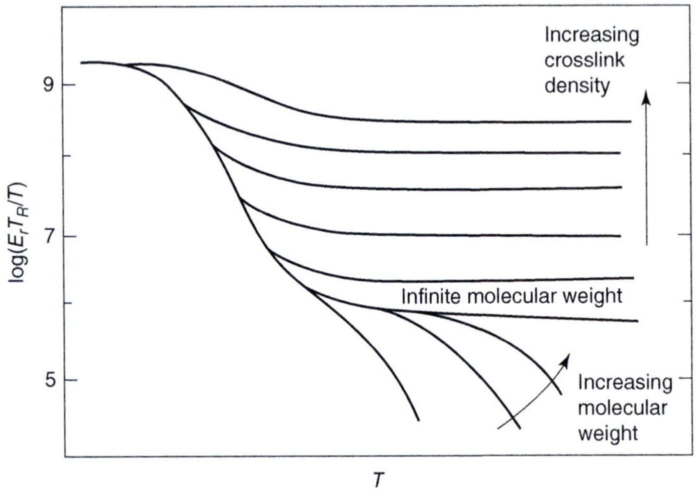

Figure 5.74 Qualitative effects of increasing molecular weight and crosslink density on modulus where T_R is any reference temperature. Rodriguez [49]/Reproduced with permission of Hemisphere Publishing Corporation.

5.4 Mechanics of Composites

The effect of dispersoids on the mechanical properties of metals has already been described in Section 5.1.2. In effect, these materials are composites, since the dispersoids are a second phase relative to the primary, metallic matrix. There are, however, many other types of composite materials, as outlined in Section 1.4, including laminates, random-fiber composites, and oriented fiber composites. Since the chemical nature of the matrix and reinforcement phases, as well as the way in which the two are brought together (e.g., random versus oriented), vary tremendously, we shall deal with specific types of composites separately. We will not attempt to deal with all possible matrix-reinforcement combinations, but rather focus on the most common and industrially important composites from a mechanical design point of view.

The starting point for these descriptions will be the law of mixtures, which was first introduced in Eq. (1.49) and which can be used to describe, to a first approximation, the composite property, P, that results from a combination of the reinforcement and matrix properties, P_r and P_m, respectively:

$$P_c = P_r V_r + P_m V_m \tag{5.77}$$

where V_r and V_m are the volume fractions of the reinforcement and matrix components, respectively. This relationship will not apply in all instances but will serve as a starting point for descriptions of specific composite systems.

5.4.1 Mechanical Properties of Particle-Reinforced Composites

Dispersion Strengthened Metals and Alloys

In this section, we elaborate upon the dispersion-strengthened metals first described in Section 5.1.2. The hard particle constituent is a small component in comparison to the softer matrix phase, seldom exceeding 3% by volume, and the particles are only a few micrometers in size. Dispersion-hardened alloys differ from precipitation-hardened alloys in that the particle is added to the matrix, usually by nonchemical means. As a result, the characteristics of the particles largely control the properties of the composite, particularly strength. The finer the interparticle spacing for a given alloy series, the better the ductility. Cold working is generally required to achieve high strength levels.

The production method for dispersion-strengthened metals has a considerable effect on the properties of the final product. The particular production method depends on the materials involved and includes such techniques as surface oxidation of ultrafine powders, internal oxidation of dilute solid–solution alloys, mechanical mixing of fine metal powders, direct production from liquid metals, and colloidal processing. The surface oxidation of ultrafine powders method is used to form sintered aluminum powder (SAP) alloys, in which each aluminum particle is surrounded by a thin oxide layer. The powder is then compacted, sintered, and cold-worked to form a strong, dispersion-strengthened alloy. The unique characteristics of SAP alloys are their high-temperature strength retention. Instead of softening at a temperature about half the melting point, recrystallization and grain growth are avoided and an appreciable fraction of the material's low-temperature strength persists up to 80% of its melting temperature (see Figure 5.75). The SAP technique has also been employed in the production of lead, tin, and zinc oxide-dispersion-strengthened alloys.

In the colloidal technique, the size and distribution of a dispersed thoria (ThO_2) phase are controlled to produce dispersion-strengthened alloys, primarily with nickel as the metallic phase. The so-called TD (thoria-dispersed) nickel has modest strength at room temperature but retains this strength nearly to its melting point. TD nickel is 3–4 times stronger than pure nickel in the 870–1315°C range, and oxidation resistance of the alloy is better than that of nickel at 1100°C.

Other dispersion-strengthened systems include Al_2O_3-reinforced copper, lead, and iron and TiO_2 and MgO-reinforced copper. In general, the strength of dispersion-strengthened materials is not particularly impressive at low temperatures relative to the metal phase melting point. The primary benefit, as discussed previously, is in the high-temperature strength retention. This effect is summarized in Figure 5.76 for several dispersion-strengthened materials in comparison to pure aluminum and Inconel metals.

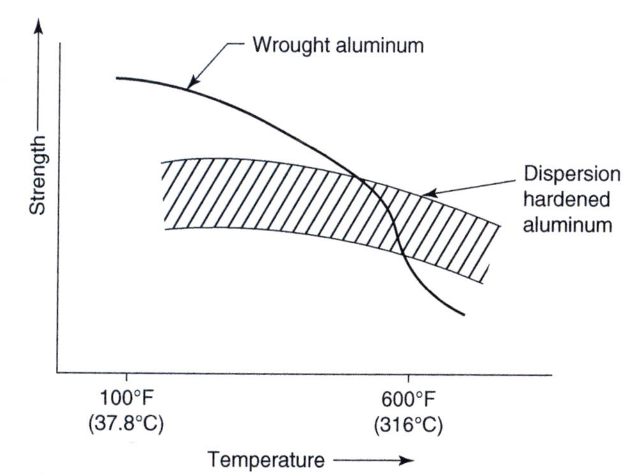

Figure 5.75 Strength of dispersion-strengthened aluminum compared with wrought aluminum. Schwartz [50]/Reproduced with permission of McGraw-Hill.

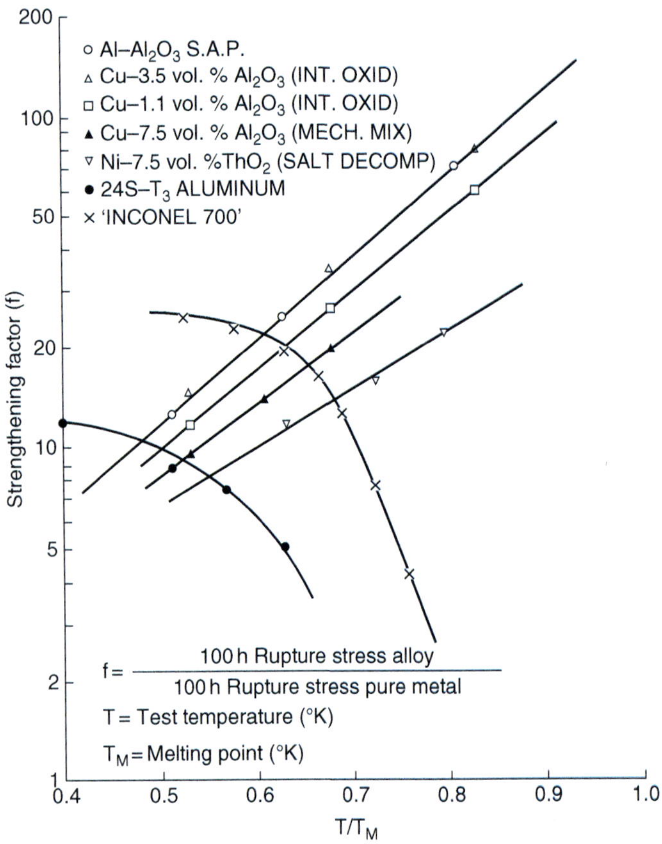

Figure 5.76 The strength of different dispersion-strengthened alloys relative to the pure metal strength as a function of relative temperature. Preparation techniques include sintered aluminum powder (SAP), internal oxidation, and salt decomposition. Kelly [51]/Reproduced with permission of Elsevier.

Need More Help?

Try these videos on dispersion strengthening:

https://www.youtube.com/watch?v=isXN5wqmk5E

and cermets:

https://www.youtube.com/watch?v=YZcuWzy_BCY

Cermets

Recall from Section 1.4.5 that a cermet is a particular composite that is similar to a dispersion-strengthened metal, except that the grains are generally larger (cf. Figure 1.73). The metal matrix usually accounts for up to 30% of the total volume, though a wide a variety of compositions are possible. In fact, in oxide-based cermets, either the oxide or the metal may be the matrix phase. For example, the thermal shock resistance of a 28% Al_2O_3–72% Cr is good but can be greatly enhanced by reversing the proportions of the constituents. In the same system, composition reversal reduces the elastic modulus at room temperature by a factor of 10, from 361 to 32.4 GPa. Other oxide-based cermets include Al_2O_3 in stainless steel and in a Cr–Mo alloy. The oxide-based cermets have been extensively applied as a tool material in high-speed cutting of difficult-to-machine materials.

Carbide-based cermets fall into three general categories: tungsten carbide-based, chromium carbide-based, and titanium carbide-based. Tungsten carbide is widely used as a cutting-tool material and is high in rigidity, compressive strength, hardness, and abrasion resistance. Cobalt is added up to 35% by volume to assist in bonding, and the properties of the cermet vary accordingly. As with the oxide-based cermets, increasing binder volume produces such property improvements as improved ductility and toughness. Compressive strength, hardness, and tensile modulus are reduced as binder content increases. Combined with superior abrasion resistance, the higher impact strength cermets result in die life improvements of up to 7000%. Chromium carbide-based cermets offer phenomenal resistance to oxidation, excellent corrosion resistance, relatively high thermal expansion, relatively low density, and the lowest melting point of the stable carbides. Also important is the similarity of thermal expansion to that of steel. Titanium carbide, used principally for high-temperature applications, has good oxidation and thermal-shock resistance, good retention of strength at elevated temperatures, and a high elastic modulus. Because of the relatively poor oxidation resistance of cobalt at elevated temperatures, nickel is more commonly used as the matrix phase. Finally, refractory cermets such as tungsten-thoria and nickel-barium carbonate are used in high-temperature applications.

Rubber-Reinforced Polymer Composites

One of the major developments leading to plastics with outstanding toughness has been the production of rubber-modified glassy polymers such as the acrylonitrile butadiene–styrene (ABS) resins made by polymerizing a continuous glassy matrix in the presence of small rubber particles. Only a small amount (5–15%) of rubber is needed to make this improvement, and the optimum size of the rubber particles is 1–10 μm. The mechanism by which toughness is developed in these materials can be explained in terms of the crack propagation mechanism first described in Section 5.2.1. Most glassy thermoplastics undergo a specific type of crack formation and propagation called *crazing*, in which a network of fibrils connect the two bulk surfaces in a polymeric crack, as illustrated in Figure 5.77. The incorporation of rubber particles into the glassy matrix serves to bridge the microvoids and arrest crazing and crack propagation, resulting in increased toughness. This effect is illustrated in Figure 5.78, in which the ductility is improved in PS through the addition of small rubber particles to produce high-impact polystyrene (HIPS).

5.4.2 Mechanical Properties of Fiber-Reinforced Composites

As first described in Section 1.4.2, there are several ways of further classifying fiber–matrix composites, such as according to the fiber and matrix type—for example, glass–fiber-reinforced polymer

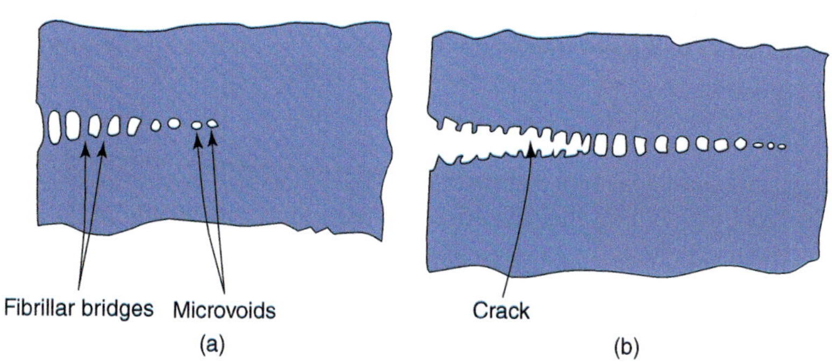

Figure 5.77 Schematic illustrations of (a) a craze showing microvoids and fibrillar bridges and (b) a craze followed by a crack. Callister [52]/Reproduced with permission of John Wiley & Sons.

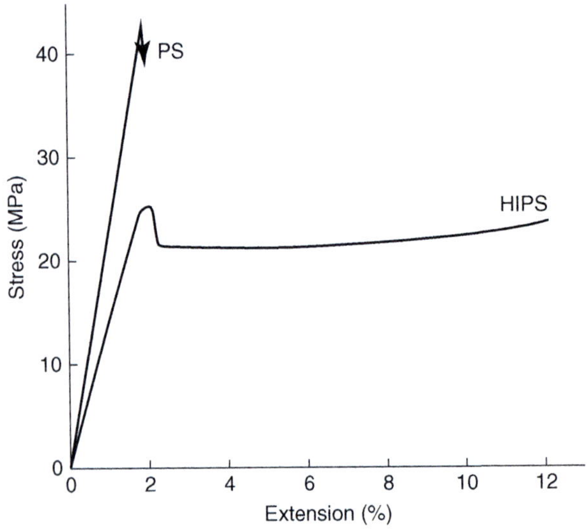

Figure 5.78 Stress–strain curves for polystyrene (PS) and high-impact polystyrene (HIPS). McCrum et al. [53]/ Reproduced with permission of Oxford University Press.

composites (GFRP) or by fiber orientation. In this section, we utilize all of these combinations to describe the mechanical properties of some important fiber-reinforced composites. Again, not all possible combinations are covered but the principles involved are applicable to most fiber-reinforced composites. We begin with some theoretical aspects of strength and modulus in composites.

Recall from Section 1.4, and from Figure 1.71 in particular, that the fiber reinforcement can be classified as either continuous or discontinuous. In the next two sections we describe the mechanical properties of these two classes of fiber reinforcement in composites, with the emphasis on reinforcement of polymer–matrix composites. Generally, the highest strength and stiffness are obtained with continuous reinforcement. Discontinuous fibers are used only when manufacturing economics dictate the use of a process where the fibers must be in this form—for example, injection molding. We begin our description of discontinuous fiber-reinforced composites with some concepts related to the distribution of stresses that are independent of fiber composition. These developments will be limited to unidirectionally aligned fibers (cf. Figure 1.72)—that is, fibers with their axes aligned. First, we will concentrate on continuous, unidirectional fibers, and then the strength and modulus of discontinuous, unidirectional fiber-reinforced composites will be analyzed.

Fiber–Matrix Coupling

Consider the case of fibers that are so long that the effects of their ends can be ignored. Furthermore, consider the case where these continuous fibers are all aligned along a preferred direction, as illustrated in Figure 5.79. Three orthogonal axes, 1, 2, and 3, are defined within the composite. When loads are applied to the composite block, we can assume that the block deforms as if the stress and strain are uniform within each component. For example, if a stress σ_1 is applied parallel to axis 1, as in Figure 5.80a, the fibers and matrix are approximately coupled together in parallel; that is, the fibers and matrix elongate equally in the direction of σ_1. The fiber

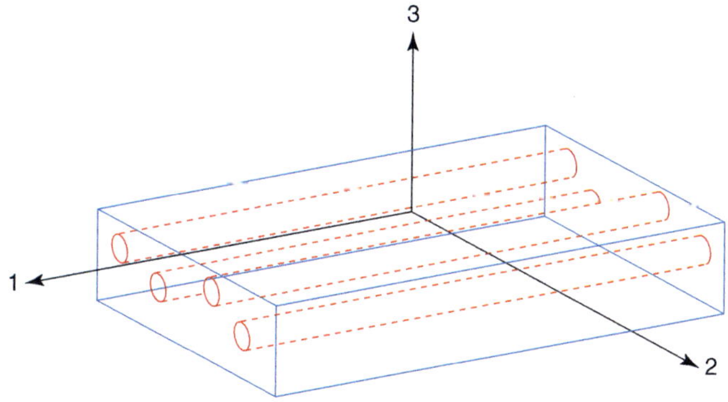

Figure 5.79 Definition of axes in a continuous, unidirectional fiber-reinforced composite. McCrum et al. [54]/ Reproduced with permission of Oxford University Press.

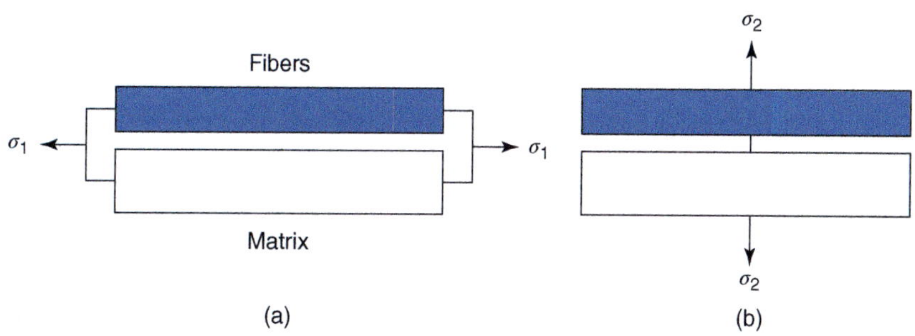

Figure 5.80 Fiber–matrix coupling in continuous, unidirectional fiber-reinforced composites: (a) Coupling for tensile stress parallel to fiber axis. (b) Coupling for stress perpendicular to fiber axis. McCrum et al. [54]/ Reproduced with permission of Oxford University Press.

and matrix axial strains along axis 1—ε_{f1} and ε_{m1}, respectively—equal the total strain in the composite:

$$\varepsilon_{f1} = \varepsilon_{m1} = \varepsilon_1 \tag{5.78}$$

Fibers extend the entire length of the composite, so that at any section the area fractions occupied by fibers and matrix equal their respective volume fractions, V_f and $V_m = 1 - V_f$. The total stress, σ_1, must then equal the weighted sum of stresses in fibers and matrix, σ_{f1}, and σ_{m1}, respectively:

$$\sigma_1 = V_f\sigma_{f1} + \left(1 - V_f\right)\sigma_{m1} \tag{5.79}$$

This is the rule of mixtures for stress, as given in Eq. (5.77). Fibers and matrix are assumed to carry pure axial tension, with no stress in the 2–3 plane; that is, $\sigma_{f2} = \sigma_{f3} = \sigma_{m2} = \sigma_{m3} = 0$. We treat the fibers and matrix as acting in a purely elastic manner and neglect the viscoelastic effects of the polymer matrix. Hooke's Law then applies to both the fiber and the matrix

$$\sigma_{f1} = E_f\varepsilon_{f1} \tag{5.80}$$

$$\sigma_{m1} = E_m\varepsilon_{m1} \tag{5.81}$$

as well as to the composite, where E_1 is the composite modulus

$$\sigma_1 = E_1 \varepsilon_1 \tag{5.82}$$

Substitution of the stresses from Eqs. (5.81) and (5.82) into Eq. (5.82) and division by ε_1 yields a prediction for the axial composite tensile modulus:

$$E_1 = V_f E_f + (1 - V_f) E_m \tag{5.83}$$

This, too, is the simple rule of mixtures of Eq. (5.77) applied to modulus. Care must be exercised to use the correct fiber modulus, E_1 in the case of anisotropic fibers such as carbon and Kevlar®. In Eq. (5.83), it is the axial tensile modulus that must be used. The second term in Eq. (5.83) usually makes only a small contribution, since $E_m \ll E_f$. Except at very low fiber volume fractions, E_1 is then given to a good approximation by

$$E_1 \approx V_f E_f \tag{5.84}$$

In contrast, when a stress acts in the 2–3 plane, as in Figure 5.80b, the matrix plays a crucial load-bearing role. Fibers and matrix now couple approximately in series, and the whole tensile force is assumed to be carried fully by both the fibers and matrix. The tensile forces in the fiber and matrix, σ_{f2} and σ_{m2}, are therefore equal to each other and to the overall stress in the composite, σ_2:

$$\sigma_{f2} = \sigma_{m2} = \sigma_2 \tag{5.85}$$

Displacement in the fibers and matrix in the direction of σ_2 can be approximated as being additive. The total strain is therefore the weighted sum of strains in the fibers and matrix:

$$\varepsilon_2 = V_f \varepsilon_{f2} + (1 - V_f) \varepsilon_{m2} \tag{5.86}$$

Again, the stress in each component is approximated as being pure uniaxial tension; that is, there is no stress in the 1–3 plane, and $\sigma_{f1} = \sigma_{f3} = \sigma_{m1} = \sigma_{m3} = 0$. Once again, Hooke's Law applies to the fiber, matrix, and composite, and when the strains from these relationships are substituted into Eq. (5.86), an expression for the transverse tensile modulus of the composite, E_2 is obtained:

$$E_2 = \frac{E_f E_m}{(1 - V_f) E_f + V_f E_m} \tag{5.87}$$

The axial and transverse tensile moduli for a continuous, unidirectional glass–fiber-reinforced epoxy matrix composite as predicted by Eqs. (5.83) and (5.87) are given as a function of volume fraction fiber, V_f, in Figure 5.81. Since $E_f \gg E_m$, Eq. (5.87) reduces to the approximate expression:

$$E_2 \approx \frac{E_m}{1 - V_f} \tag{5.88}$$

which indicates that fibers now play the subordinate role for transverse tensile loads in continuous, unidirectional fiber-reinforced composites.

Similar arguments can be applied to Poisson's ratio, and the free contraction parallel to axis 2 when a tensile stress is applied parallel to axis 1, ν_{12}, is found to obey the rule of mixtures

$$\nu_{12} = V_f \nu_f + (1 - V_f) \nu_m \tag{5.89}$$

The other Poisson's ratio, ν_{21} applying to the 1–2 plane does not equal ν_{12}, but can be found from ν_{12}, E_1, and E_2.

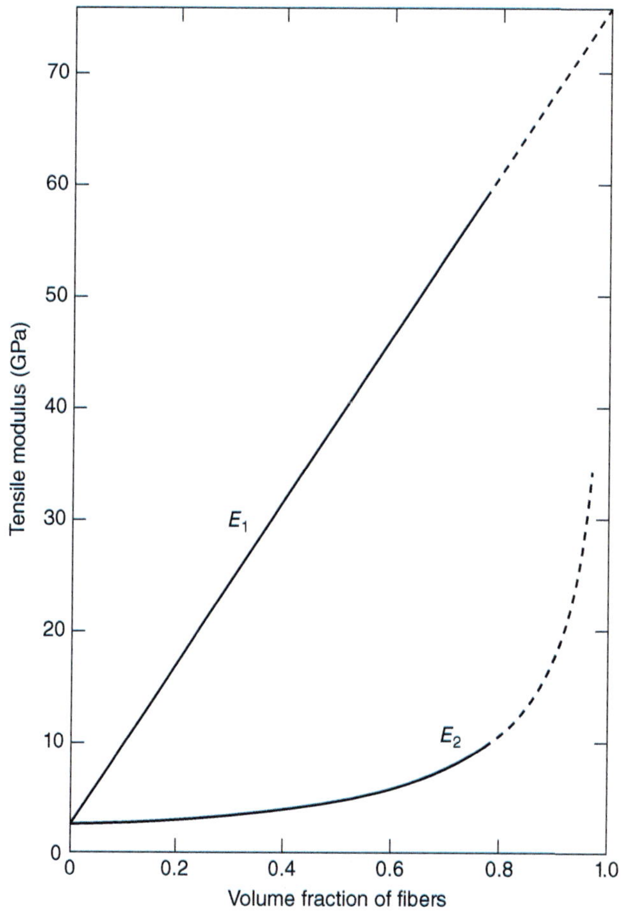

Figure 5.81 Predicted tensile moduli for continuous, unidirectional glass–fiber-reinforced epoxy matrix composite. McCrum et al. [55]/Reproduced with permission of Oxford University Press.

When a shear stress acts parallel to the fibers, τ_{12}, the composite deforms as if the fibers and matrix were coupled in series. Hence, the shear strain, γ_{12}, can be found from a relation similar to Eq. (5.86), and the corresponding shear modulus for the composite, G_{12}, is

$$G_{12} = \frac{G_f G_m}{(1 - V_f)G_f + V_f G_m} \tag{5.90}$$

The same models of mechanical coupling can be used to predict the coefficient of linear thermal expansion in the composite, α_1, based on the moduli, and thermal expansion coefficients of the fiber and matrix, α_f and α_m, respectively:

$$\alpha_1 = \frac{V_f E_f \alpha_f + (1 - V_f)E_m \alpha_m}{V_f E_f + (1 - V_f)E_m} \tag{5.91}$$

Here, α_f is assumed to be independent of direction within the fiber, but for highly anisotropic fibers such as carbon and Kevlar, allowances must be made for their variation in thermal expansion coefficient with direction. The thermal expansion coefficient perpendicular to the fiber axis, α_2, can also be derived:

$$\alpha_2 = V_f \alpha_f (1 + \nu_f) + (1 - V_f) \alpha_m (1 + \nu_m) - \alpha_1 \nu_{12} \tag{5.92}$$

It is usually the case that $\alpha_f \ll \alpha_m$ and $E_f \gg E_m$, such that Eqs. (5.91) and (5.92) predict widely differing coefficients of thermal expansion parallel and perpendicular to the fibers. This effect is illustrated in Figure 5.82, where the thermal expansion coefficients in the axial and transverse direction for a continuous, unidirectional glass–fiber-reinforced epoxy matrix composite are plotted as a function of fiber volume fraction. Clearly, the fibers are highly effective at reducing thermal expansion in the aligned direction, but the thermal expansion at right angles to the fibers is not only higher than in the axial direction but may actually exceed either of the component expansion coefficients. A further difference between α_f and α_m is that microscopic thermal stresses are generated, which may significantly affect the strength of the composite.

Finally, we turn our attention to the prediction of strength in continuous, unidirectional fiber composites. Consider again the case in which a tensile stress acts parallel to the fibers (cf. Figure 5.80a).

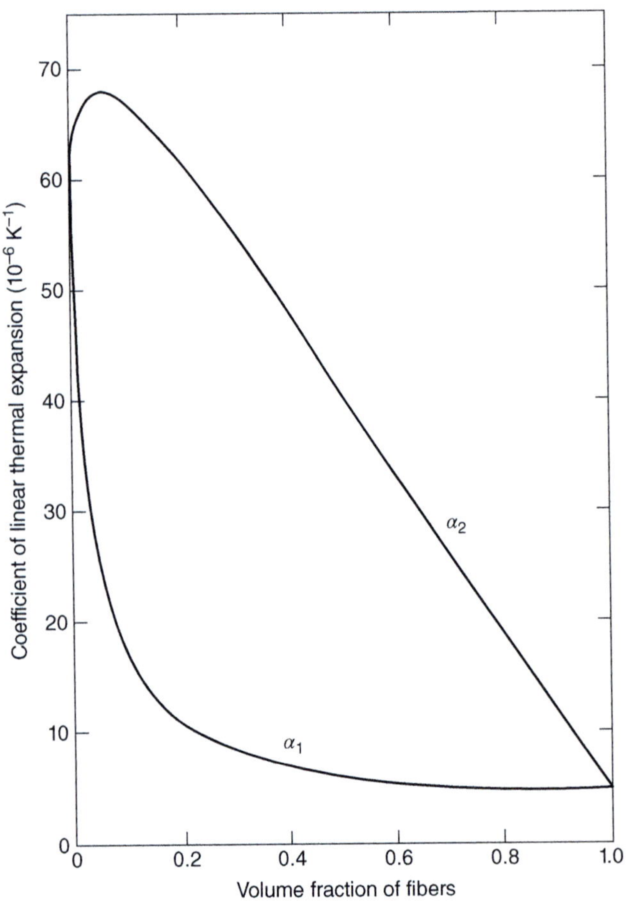

Figure 5.82 Predicted thermal expansion coefficients in the axial and transverse direction for continuous, unidirectional glass–fiber-reinforced epoxy matrix composite. McCrum et al. [56]/Reproduced with permission of Oxford University Press.

The sequence of events varies, depending on which of the two components is more brittle—that is, which extension at fracture is smaller, ε_f^* or ε_m^*, where the asterisk indicates fracture (failure). Let us consider each case independently.

In the case where $\varepsilon_f^* < \varepsilon_m^*$, as in the case of carbon-reinforced epoxy, we employ the simple assumption of parallel coupling between the fibers and the matrix, for stress parallel to axis 1. We further assume that the fibers and the matrix fail independently, at the same stresses and strains as when the tensile tests are carried out on the pure materials. Our analysis begins by superimposing the fiber and matrix stress–strain curves, multiplied by their respective volume fractions, as in Figure 5.83. The total stress, σ_1, at strain, ε_1, is given by Eq. (5.79) as the sum of the two components (dashed lines), which in Figure 5.83 have already been multiplied by their respective volume fractions. The result is the solid line in Figure 5.83a and b. As strain is increased, a point is reached around ε_f^* where the fibers begin to break, and their contribution to the total stress falls to zero. What happens next depends on the volume fraction of fibers, V_f. At low V_f, (Figure 5.83a), there is sufficient matrix to carry the tensile load even after all the fibers have failed. The total stress at strain ε_1 is, from this point on, just the contribution from the matrix alone, $(1 - V_f)\sigma_m$. On further straining, the composite fails when the matrix, which now contains broken fibers, finally fails at $\varepsilon_1 = \varepsilon_m^*$. On the other hand, at high V_f, when the fibers break, there is insufficient matrix to maintain the stress. The point of failure of the fibers then marks the failure of the composite (Figure 5.89b). In this case, the stress just prior to failure is the sum of $V_f\sigma_f^*$ and $(1-V_f)\sigma_m'$, where $\sigma_m' = \sigma_m\varepsilon_f^*$, the stress carried by the matrix at the failure strain of the fibers. So, there are two criteria for the strength at failure. At low V_f

$$\sigma_1^* = (1 - V_f)\sigma_m^* \tag{5.93}$$

and at high V_f

$$\sigma_1^* = V_f\sigma_f^* + (1 - V_f)\sigma_m' \tag{5.94}$$

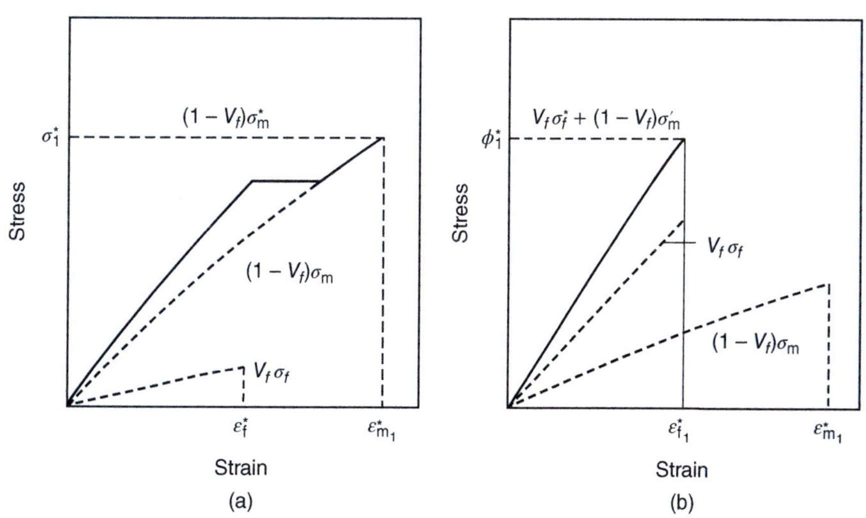

Figure 5.83 Schematic illustration of stress–strain curves for continuous, unidirectional fiber-reinforced composites containing brittle fibers in a ductile matrix. Contributions from fibers and matrix are shown as dashed lines at (a) low fiber volume fractions and (b) high fiber volume fractions. McCrum et al. [57]/Oxford University Press.

Cooperative Learning Exercise 5.11

Recall from Section 5.1.3 that thermal expansion is the change in dimensions of a solid due to heating or cooling. As a result, a thermally induced strain can result when the material is heated, as dictated by Eq. (5.26). For small deformations at constant pressure, we can thus approximate the strain as

$$\varepsilon = \frac{\Delta L}{L} = \alpha_T \Delta T$$

where ΔT is the temperature change due to heating or cooling. For unidirectionally aligned fiber composites, then there will be a different strain in the axial and transverse directions for a given temperature change since the coefficients of thermal expansion are different in the two directions.

Consider, then, a composite material that consists of 60% by volume continuous, uniaxially aligned, glass fibers in an epoxy matrix. Take the tensile modulus and Poisson's ratio of glass to be 76 GPa and 0.22, respectively, and of the epoxy to be 2.4 GPa and 0.34, respectively. Take the coefficients of thermal expansion to be 5×10^{-6} K^{-1} and 60×10^{-6} K^{-1} for glass and epoxy, respectively.

Person 1: Calculate the thermal expansion coefficient in the axial direction, α_1 for this composite.

Person 2: Calculate Poisson's ratio for the composite, ν_{12}.

Combine your information to calculate the thermal expansion coefficient in the transverse direction, α_2 for this composite.

The composite is now subjected to a temperature rise of 100 K.

Person 1: Calculate the strain in axial direction as a result of the temperature rise.

Person 2: Calculate the strain in the transverse direction as a result of the temperature rise.

Compare your results. In which direction will the greatest percentage change in dimensions occur?

Answers: $\alpha_1 = 6.13 \times 10^{-6}$K^{-1}; $\nu = 0.268$; $\alpha_2 = 34.2 \times 10^{-6}$K^{-1}. $\varepsilon_1 = 6.13 \times 10^{-4}$; $\varepsilon_2 = 3.42 \times 10^{-3}$. The greatest percentage change occurs in the transverse direction, although the absolute change will depend upon the original dimensions.

These two estimates for the fracture strength are shown schematically in Figure 5.84 as two intersecting straight lines. The greater of the two represents the actual strength of the composite and is indicated by a solid line. Figure 5.84 shows an important fact: a minimum fiber fraction, $V_{f,\min}$, is required before the fibers exert any strengthening effect. The minimum fiber fraction can be found from Eq. (5.94):

$$V_{f,\min} = \frac{\sigma_m^* - \sigma_m'}{\sigma_f^* - \sigma_m'} \tag{5.95}$$

For example, the minimum fiber fraction for a carbon-fiber-reinforced epoxy matrix composite is 0.03.

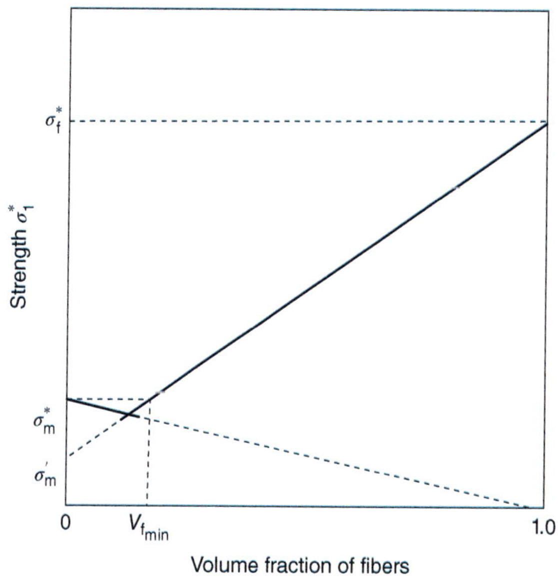

Figure 5.84 Schematic illustration of tensile strength as a function of fiber volume fraction for a continuous, unidirectional fiber-reinforced, ductile matrix composite. McCrum et al. [59]/Oxford University Press.

In the second case where $\varepsilon_f^* > \varepsilon_m^*$, two criteria for the failure of the composite are obtained. The analysis is similar to the previous case but is omitted here. At low V_f

$$\sigma_1^* = V_f \sigma_f' + (1 - V_f)\sigma_m^* \tag{5.96}$$

and at high V_f

$$\sigma_1^* = V_f \sigma_f^* \tag{5.97}$$

The plots of tensile strength versus volume fraction fibers corresponding to those in Figure 5.84 for $\varepsilon_f^* > \varepsilon_m^*$ are shown in Figure 5.85 for $\varepsilon_f^* > \varepsilon_m^*$. Again, the solid line represents the behavior of the composite. Equations (5.93)–(5.94) and (5.96)–(5.97) have been found to predict well the experimental tensile strengths of brittle fiber/ductile matrix and ductile fiber/brittle matrix composites at a variety of fiber volume fractions in unidirectional, continuous fiber-reinforced composites.

5.4.3 Mechanical Properties of Laminate Composites

In this section, we examine the mechanics of laminate composites, first through a generalized description of their mechanics, then with some specific stress–strain behavior.

In the case of unidirectionally reinforced layers, or *laminae* (not to be confused with the similar-looking *lamellae*, cf. Section 1.3.6), a composite is formed by laminating the layers together. If all the layers were to have the same fiber orientation (see Figure 5.86a), the material would still be weak in the transverse direction. Therefore, angle-ply laminates are preferred for most applications. Typically, the laminae are stacked so that the fiber directions are symmetrical about the mid-plane of the composite, as in Figure 5.86b. A *symmetrical* stacking sequence ensures that in-plane stresses do not produce out-of-plane twisting or bending, as in Figure 5.87. The stacking sequence can be either 0°/90°, as in the bottom two plies of Figure 5.86b, termed *cross-ply*, −45°/+45° as in the third

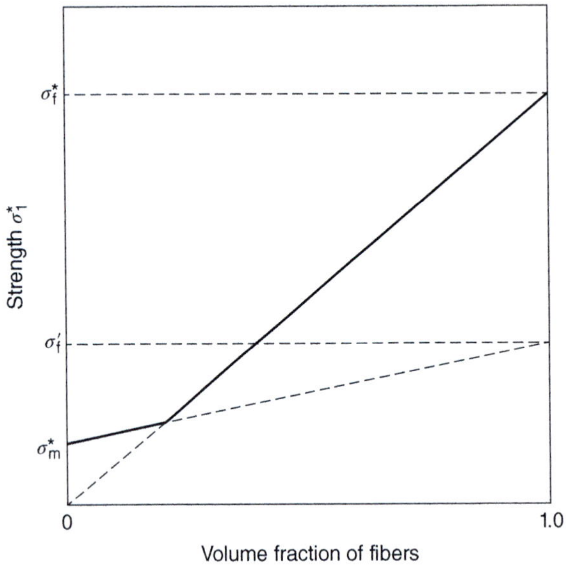

Figure 5.85 Schematic illustration of tensile strength versus fiber volume fraction for continuous, unidirectional ductile fibers in a brittle matrix. McCrum et al. [60]/Oxford University Press.

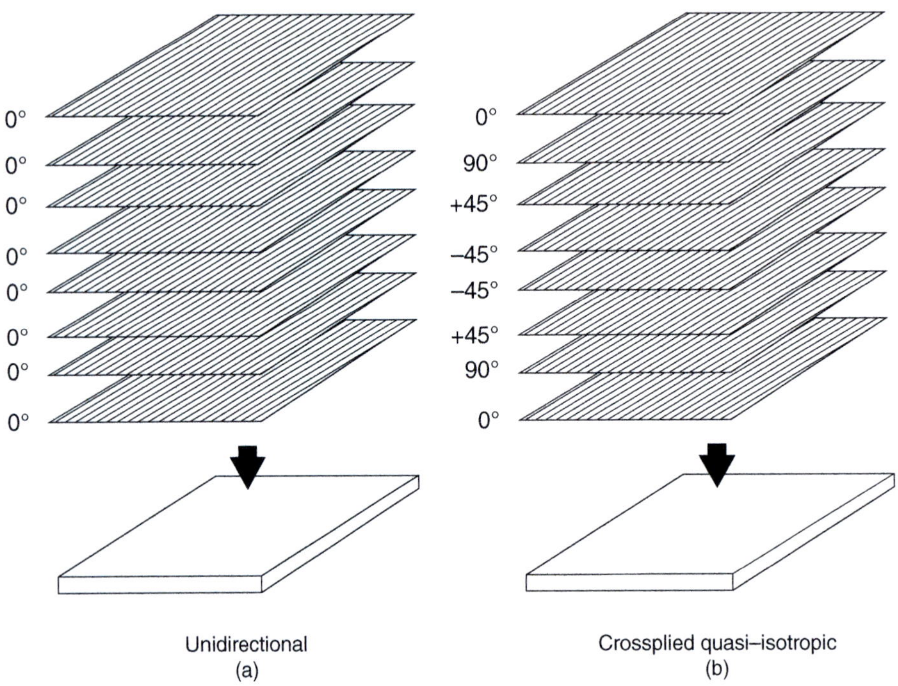

Figure 5.86 Schematic illustration of unidirectional (a) and crossplied (b) laminar composites. Schwartz [61]/ Reproduced with permission of McGraw-Hill.

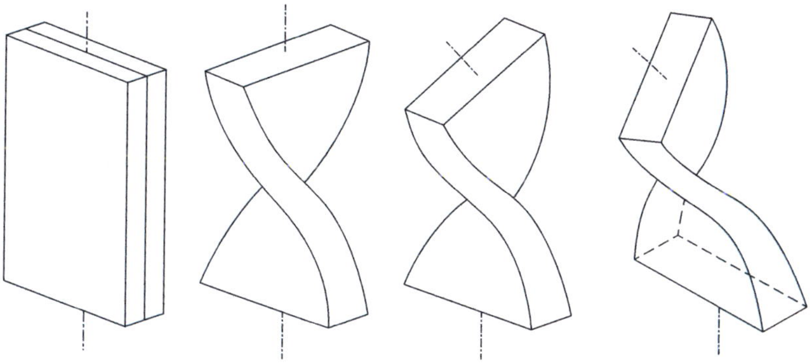

Figure 5.87 Twisting, bending, and shear of an unsymmetrical laminate under direct stress. Powell and Housz [62]/Reproduced with permission of Stanley Thorned Publishers.

and fourth ply of Figure 5.86b, termed *angle-ply*, or a combination of the two. The layers can be of the same thickness, or different thickness, and multiple layers may be oriented in the same direction before changing the orientation of the next multiple layers—for example, three layers at 0°, three layers at +45°, three layers at −45°, and so on. The outer layers of the composite may be selected to provide a protective coating, as described previously.

Need More Help?
Try these videos on mechanical properties composites: https://www.youtube.com/watch?v=gnnq0B2FXuU https://www.youtube.com/watch?v=-Jz0BFpXJss https://www.youtube.com/watch?v=KyyERdNuKtw

Strength of Laminate Composites

The strength of angle-ply laminates can be calculated using the same type of elastic analysis as for modulus in the previous section. The strains produced in the laminate by a given set of applied stresses are first calculated, using the computed laminate moduli. Stresses corresponding to these strains are then calculated for each layer of the laminate. These stresses are then expressed in terms of stresses parallel and normal to the fibers, and the combination of stresses is compared with strength criteria for the unidirectional material. Results for a three-direction carbon-fiber-reinforced epoxy composite (0, $\pm\theta$) under tensile stress in the $\theta = 0°$ direction are shown in Figure 5.88. The composite contains 60 vol% fibers. Here, R is the fraction of plies which are angled at $\pm\theta$, and 1-R is the fraction at $\theta = 0$.

In cross-ply laminates, the stress–strain behavior is slightly nonlinear, as illustrated in Figure 5.89. The stress–strain behavior of a unidirectional lamina along the fiber axis is shown in the top curve, while the stress–strain behavior for transverse loading is illustrated in the bottom curve. The stress–strain curve of the cross-ply composite, in the middle, exhibits a knee, indicated by strength σ_k, which corresponds to the rupture of the fibers in the 90° ply. The 0° ply then bears the load, until it too ruptures at a composite fracture strength of σ_f.

One additional problem area for fiber-reinforced thermosetting laminates is *interlaminar splitting*, which occurs particularly easily when the void content is high. Measurements of *interlaminar shear strength* (ISS) are made by subjecting short bar specimens to three-point bending. Deformation of laminated composites can also occur because of changes in temperature and the absorption of moisture. This is known as the *hygrothermal effect*, and it causes the polymer matrix to undergo both dimensional and property changes with temperature and humidity.

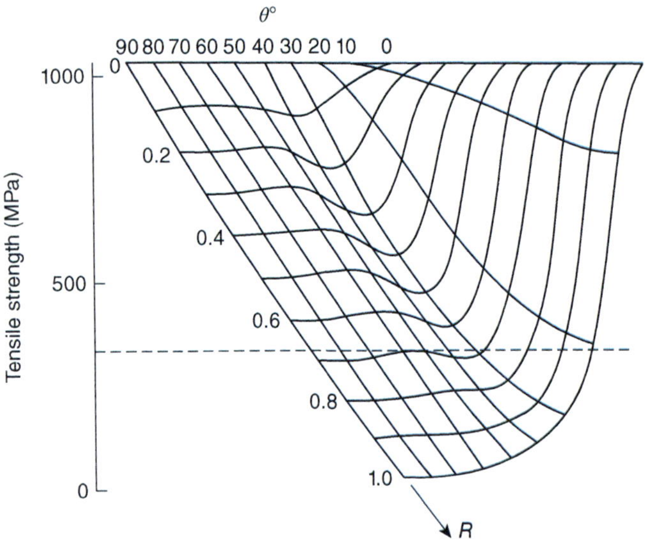

Figure 5.88 Strength chart for a (0, ±θ) unidirectional carbon-fiber-reinforced epoxy laminate composite. McCrum et al. [63]/Reproduced with permission of Oxford University Press.

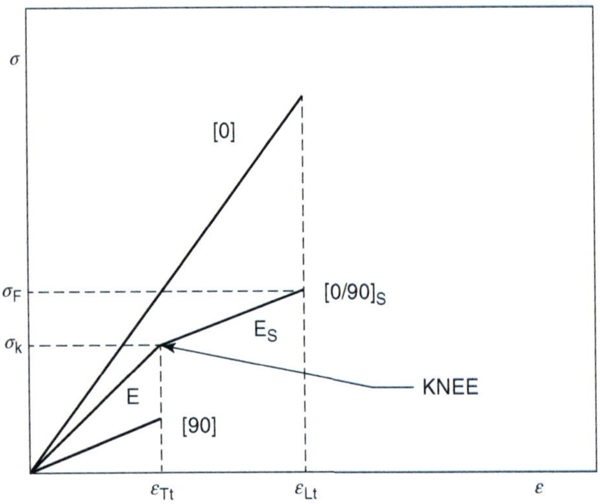

Figure 5.89 Schematic tensile stress–strain diagram for a symmetric (0, ±90°) cross-plied laminate tested at 2 = 0°. Mallick [58]/Reproduced with permission of Taylor & Francis.

5.4.4 Mechanical Properties of Nanocomposites

We saw in Section 1.4.7 that nanocomposites are an increasingly important subset of composite materials and reviewed the structure of two examples in exfoliated clay nanocomposites and carbon nanotube-reinforced composites. We can even classify some previously described materials like dispersion-strengthened alloys as nanocomposites depending on the size of the dispersed phase. Here, we take a closer look at the mechanical properties of some nanocomposites and how they provide technological benefits beyond simple combination or summation effects.

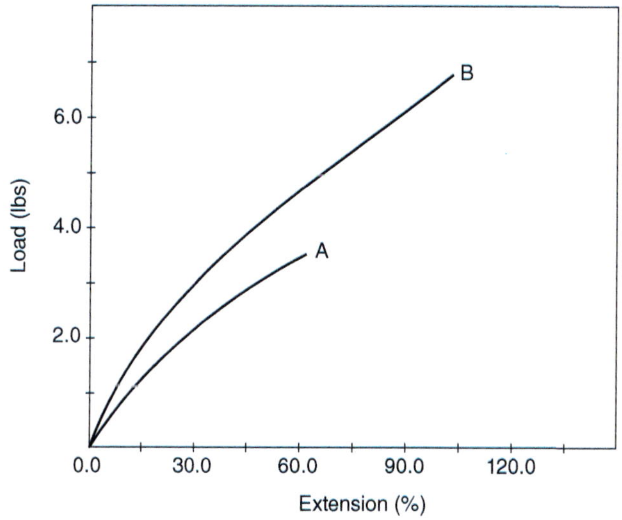

Figure 5.90 Stress–strain curves for (A) a pristine polyurethane elastomer and (B) a polyurethane–clay nanocomposite. Wang and Pinnavaia [64]/Reproduced with permission of American Chemical Society.

Take, for example, the polymer–clay nanocomposites of Figure 1.77. In the case of montmorillonite (MMT) clay intercalated with a polyurethane elastomer matrix, not only is the elastic modulus improved over the pure elastomer as illustrated in Figure 5.90 but also the elongation at break is over 50% higher in the nanocomposite compared to the pure elastomer. This marked increase in toughness is characteristic of these types of nanocomposites and is due in large part to the nature of the matrix-filler interphase. Not only is the surface area of the nanofiller much larger than in traditional composites (see Section 1.1.6) but also the strong chemical interactions between the polymer and the exfoliated clay surface ensure that there is little slippage of the nanofiller during mechanical loading. Under this interfacial condition, the modulus of the nanocomposite, E, relative to the matrix modulus, E_1, is given by a modified *Mooney equation*:

$$ln\frac{E}{E_1} = \frac{k_E \emptyset_2}{\left[1 - \frac{\emptyset_2}{\emptyset_m}\right]} \tag{5.98}$$

where ϕ_2 is the volume fraction of the filler and ϕ_m is the maximum value of filler volume fraction. Here, k_E is called the *Einstein coefficient* and is related to the aspect ratio (length to thickness, L/t) of the nanofiller under the aforementioned nonslip condition:

$$k_E = 2.5(L/t)^{0.645} \tag{5.99}$$

This strong interaction between the polymer matrix and the nanoscale clay filler also affects the matrix properties. For example, the glass transition of PMMA is markedly different in a MMT clay nanocomposite than by itself (see Table 5.8). There is even a difference in T_g depending on whether the nanoclay is intercalated or exfoliated (controlled by how the nanocomposite is processed), further illustrating the importance of the matrix-nanofiller interaction.

The improvement in mechanical properties goes beyond modulus, strength, and toughness. Abrasion and wear resistance are improved in nanocomposites, and in unexpected ways. Fillers generally increase the wear resistance of softer materials like polymers, but only if the filler particles

Table 5.8 Effect of Nanoclay Filler on the Glass Transition Temperature of PMMA

Type of Nanocomposite	Nanostructure	T_g, °C
Pure PMMA	—	104.6
PMMA/10 wt% MMT	Intercalated	112.0
PMMA/10 wt% MMT (emulsion)	Partially exfoliated	122.6

Source: Adapted from Meneghetti and Qutubuddin [65]. T_g as determined by differential scanning calorimetry (DSC).

are larger than the abrading particles. When the filler is comparable in size to the abrading particle it tends to become dislodged from the surface of the composite and wear resistance is compromised. This is not the case for nanocomposites. The incorporation of nanoscale $CaCO_3$ in a PMMA matrix resulted in a decrease of material loss due to abrasion by a factor of two for only 3 wt% loading of the nanofiller [66]. Other nonmechanical properties such as gas and liquid permeability, thermal stability, flame resistance, as well as electrical and optical properties for certain types of nanoscale fillers and matrices are all improved in the nanocomposite.

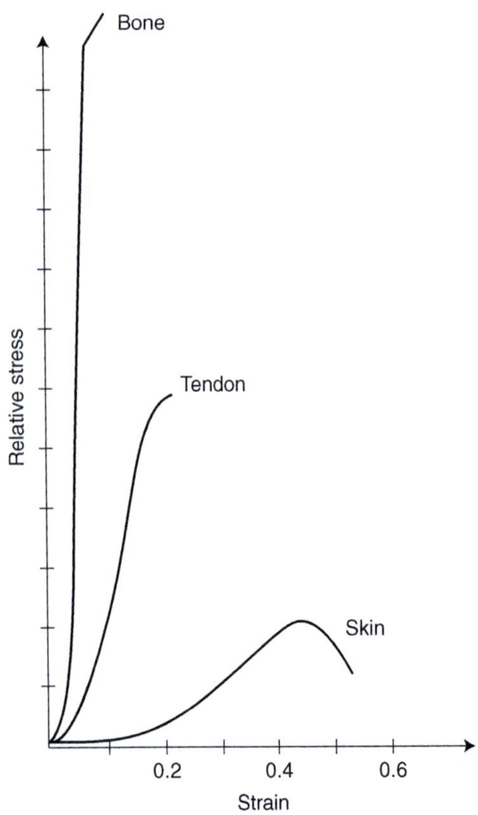

Figure 5.91 Schematic comparison of stress–strain curves for some soft and hard biological materials. Silver and Christiansen [67]/Reproduced with permission of Springer Nature.

5.5 Mechanics of Biologics

Most of the terms, concepts, and equations for studying the mechanical properties of biological materials have already been developed in the previous sections of this chapter. What remains is to relate the unique structural characteristics of biologics to their mechanical properties. As in Chapter 1, we will divide the biologics into soft and hard classifications, since their structure and mechanical responses are uniquely different. These differences are summarized in Figure 5.91, in which the stress–strain curves of some soft and hard biologics are compared. The types of materials used as substitutes for these two classifications are also different. In each case, we will begin by examining the structure and mechanical properties of true biological materials, then briefly describe materials that can be used to replace them.

5.5.1 Mechanical Properties of Soft Biologics

Because most biological structures are composed of α helices, β sheets (cf. Section 1.5.1), and collagen triple helixes, it is important to understand the mechanical properties of these basic structures.

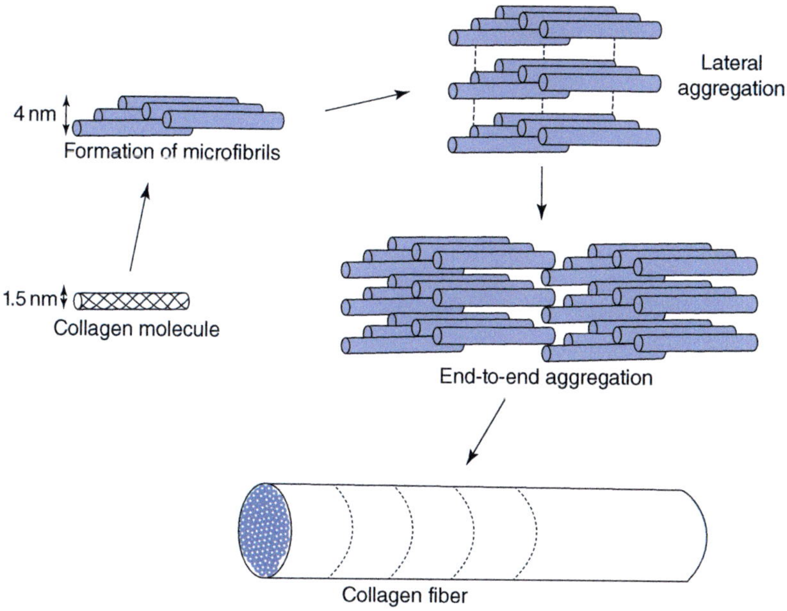

Figure 5.92 Structure of collagen fibers based on collagen molecule, microfibrils, and fibrillar aggregation. Yannas [68]/Reproduced with permission of Elsevier.

For example, the mechanical properties of tendon collagen fibers are believed to reflect the mechanical properties of the collagen triple helix which serves as the building block for the collagen protein fiber (see Figure 5.92, cf. Figure 1.89). UTS values as high as 40 MPa, strains up to 10%, and high strain moduli exceeding 500 MPa have been reported for isolated collagen fibers with diameters from 50 to 100 μm. These properties are related to the stiffness of the triple helix, which has been estimated to have an elastic modulus as high as 4 GPa.

Although the mechanical properties of connective tissues such as tendons and ligaments are dependent upon the properties of the collagen fibrils, their diameters, and the content of the collagen in the tissue, they are also dependent upon several other factors, including the orientation of the collagen fibers and the extent of tissue hydration. The latter is believed to occur due to the interaction of water with the hydrophobic regions of the collagen protein.

In tissues such as tendons that function to transmit loads along a single direction, the collagen fibers are oriented in only one direction. UTS s as high as 120 MPa, along with strains of about 10%, have been reported for tendons. For ligaments, these values are 147 MPa and 70%, respectively (see Table 5.9). Other tissues, such as skin, bear loads within a plane and need to have collagen fibers

Table 5.9 Selected Mechanical Properties of Some Soft Biologics

Tissue	Maximum Strength (MPa)	Maximum Strain (%)	Elastic Modulus (MPa)
Arterial wall	0.24–1.72	40–53	1.0
Tendons/ligaments	50–150	5–50	100–2500
Skin	2.5–15	50–200	10–40
Hyaline cartilage	1–18	10–120	0.4–19

Source: Silver and Christiansen [69]/Springer Nature.

oriented in the plane. In the next section, we look more closely at the structure–mechanical property relationships in skin.

Skin

Tissue such as tendons and ligaments are termed *oriented* networks because the collagen fibers are oriented in a preferred direction. Other collagen-containing tissues such as skin are termed *orientable*. Because skin normally bears loads within the plane of the skin, the collagen fibers become oriented with the direction of loading and reorient during mechanical deformation. Thus, skin can bear loads in any direction within its plane. For this reason, the collagen network in skin must be loose enough to be able to rearrange. Other types of soft tissue such as aorta and cartilage contain some orientable collagen networks.

The stress–strain curve for skin is shown in Figure 5.93. For strains up to about 30%, the collagen network offers little resistance to deformation, which is like tendon in this region. Collagen fibers are initially wavy and become straight upon loading. Between 30% and 60% strains, the collagen fibers become aligned and resist further extension. Over 60% elongation, the fibers begin to yield and fail. At high strains, the modulus is about 22.4 MPa, and the UTS is 14 MPa, based on the true cross-sectional area. The *elastic fraction* of skin is defined as the ratio of the total force divided by the force at equilibrium and is a measure of the viscoelastic nature of a material. For completely elastic materials, the elastic fraction is 1.0, whereas for viscoelastic materials, the value lies between 0 and 1.0. In skin, the elastic fraction increases from about 0.50 to 0.73 with increasing strain, as illustrated in Figure 5.94. This contrasts with other types of collagen-containing soft tissue such as aorta.

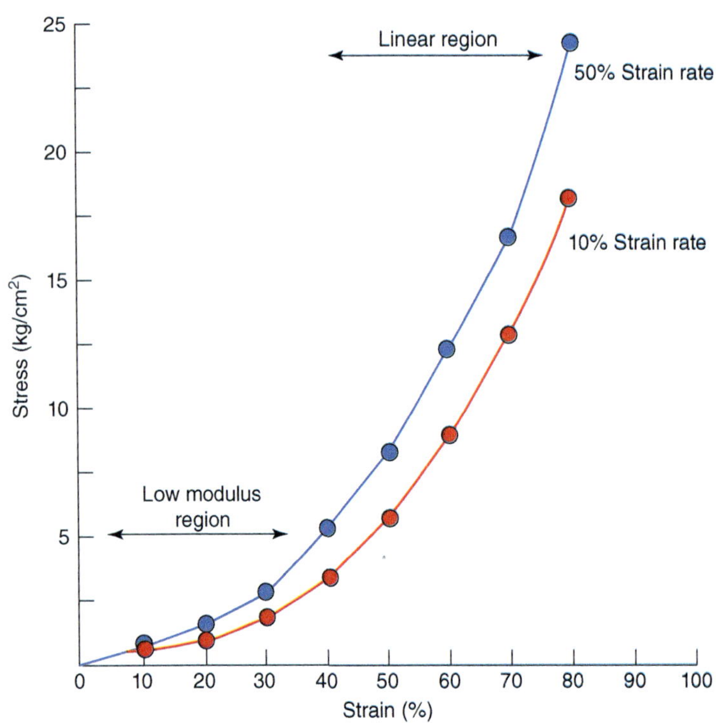

Figure 5.93 Stress–strain curves for wet back skin at various strain rates. Silver and Christiansen [70]/ Reproduced with permission of Springer Nature.

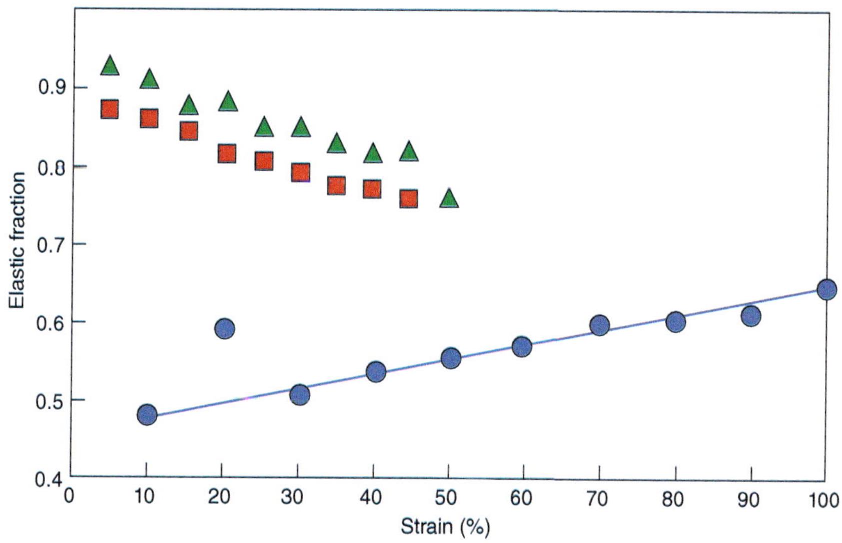

Figure 5.94 Effect of strain on elastic fraction of aorta in the circumferential (green triangles) and axial (red squares) directions, as well as on skin (blue circles). Silver and Christiansen [71]/Reproduced with permission of Springer Nature.

Historical Highlight

The directional dependence of the mechanical properties of skin was first observed by ***Karl Langer***. Read more about him and the eponymous Langer lines used in surgery and forensic science: https://jamanetwork.com/journals/jamadermatology/article-abstract/1912372

The mechanical behavior of skin is time-dependent and strain-rate-dependent, and for this reason, the elastic fraction is less than 1.0. The time dependence of the mechanical properties and the strain dependence of the elastic fraction imply an interesting structural transition that occurs in skin with strain. This behavior is attributable to the collagen network. At low strains, the elastic fraction is low and appears to involve the viscous realignment of the collagen fibers with the applied stress. At higher strains, the elastic fraction approaches 0.75, which is near the value for tendon material. In fact, several types of connective tissue exhibit an elastic fraction approaching 0.75 at high strains. This suggests that the collagen network limits the high-strain deformation in all types of connective tissues.

Although the proteins in skin are also composed of about 5% elastic fibers, they do not appear to affect the mechanical properties of the tissue. The elastic fibers are believed to contribute to the recoil of the skin, which gives it the ability to be wrinkle-free when external loads are removed. As humans age, the elastic fiber network of the skin is lost, and wrinkles begin to appear. The mechanical role of the elastic fibers is very different in vascular tissue, however.

Fatigue testing of skin in uniaxial tension shifts the stress–strain curve to the right within the linear region, and the elastic fraction approaches 0.85. Therefore, mechanical tests must be carried out carefully to eliminate any preconditioning effects. Further complicating testing is the fact that the mechanical properties of skin depend on the location from which the specimen is obtained.

The contribution of the epithelial layer to the mechanical properties depends on its thickness. For most regions, the stratum corneum is thin and does not contribute significantly to the mechanical properties of skin; however, in places where it is thick, such as the palms of the hand, it needs to be considered.

Sutures

A *suture* is a biomedical product used to hold tissue together following separation by surgery or trauma. Sutures are broadly categorized according to the type of material (natural and synthetic) from which they are made, the permanence of the material (absorbable or nonabsorbable), and the construction process (braided and monofilament). As shown in Table 5.10, the most popular natural materials used for sutures are *silk* and *catgut* (animal intestine). The synthetic materials are exclusively polymeric, except for fine-sized stainless steel sutures. Approximately half of today's sutures are nonabsorbable and remain indefinitely intact when placed in the body. Polymers such as polypropylene, nylon, poly(ethylene terephthalate), and polyethylene, as well as their copolymers, are used as sutures. *Absorbable sutures* were commercially introduced with poly(glycolic acid) (PGA), and they were followed by copolymers of glycolide and lactide. A PGA suture will lose strength over a 28-day period. More recently, novel absorbable polymers of polydioxanone and poly(glycolide-*co*-trimethylene carbonate) have been developed for surgical use.

Regardless of whether a suture is natural or synthetic, or if it is absorbable or permanent, it must meet the strength requirements necessary to close a wound under a given clinical circumstance. Suture strength is generally measured and reported in both *straight pull* and *knotted pull* tests.

Table 5.10 Representative Mechanical Properties of Some Commercial Sutures

Suture Type	St. Pull (MPa)	Kt. Pull (MPa)	Elongation to Break (%)	Subjective Flexibility
Natural materials				
Catgut	370	160	25	Stiff
Silk	470	265	21	Very supple
Synthetic absorbable				
Poly(glycolic acid)	840	480	22	Supple
Poly(glycolide-*co*-lactide)	740	350	22	Supple
Poly(*p*-dioxanone)	505	290	34	Moderately stiff
Poly(glycolide-*co*-trimethylene carbonate)	575	380	32	Moderately stiff
Synthetic nonabsorbable				
Poly(butylene terephthalate)	520	340	20	Supple
Poly(ethylene terephthalate)	735	345	25	Supple
Poly[p(tetramethylene ether)	515	330	34	Supple
Terephthalate-*co*-tetramethylene terephthalate]				
Polypropylene	435	300	43	Stiff
Nylon 66	585	315	41	Stiff
Steel	660	565	45	Rigid

Source: Goupil [72]/With permission of Elsevier.

The former is essentially a traditional tensile test; the second is obtained after the suture has been knotted. As illustrated in Table 5.10, knotted (Kt.) pull strengths are approximately one-half to two-thirds of the straight (St.) pull strength values. There are other characteristics a suture must have in addition to strength, including knot security, strength retention, resistance to infection, flexibility, and ease of tying.

Artificial Soft Biologics

In addition to sutures, polymers are used for a number of biomedical applications, as illustrated in Figure 5.95. Polymers used for hard structural applications such as dentures and bones are presented in this figure but will be described in the next section. In this section, we will concentrate on polymers for soft biological material applications and will limit the description to mechanical properties as much as possible.

Artificial biologics, whether soft or hard, can be categorized as either temporary (short-term) or permanent (long-term) in their intended application. Most, but certainly not all, polymers for biomedical applications are of the short-term type and include sutures, drug delivery devices, temporary vascular grafts and stents, tissue scaffolds, and orthopedic fixation devices. Short-term devices

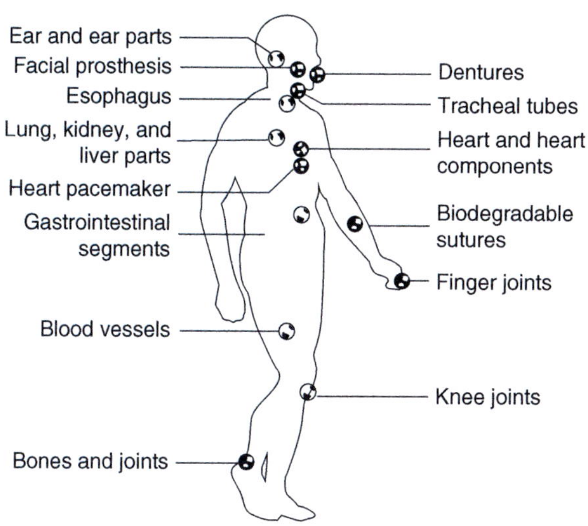

Ear and ear part: acrylic, polyethylene, silicone, poly(vinyl chloride) (PVC)
Dentures: acrylic, ultrahigh molecular weight polyethylene (UHMWPE), epoxy
Facial prosthesis: acrylic, PVC, polyurethane (PUR)
Tracheal tubes: acrylic, silicone, nylon
Heart and heart components: polyester, silicone, PVC
Heart pacemaker: polyethylene, acetal
Lung, kidney, and liver parts: polyester, polyaldehyde, PVC
Esophagus segments: polyester, polyaldehyde (PP), PVC
Blood vessels: PVC, polyester
Biodegradable sutures: PUR
Gastrointestinal segments: silicones, PVC, nylon
Finger joints: silicones, UHMWPE
Bones and joints: acrylic, nylon, silicone, PUR, PP, UHMWPE
Knee joints: polyethylene

Figure 5.95 Common applications of polymers as soft biologics. Visser et al. [73]/Reproduced with permission of Elsevier.

can either be removable or degradable. We will concentrate for the moment on biodegradable polymers and will make no distinction between often-confused terms such as biodegradable, bioresorbable, bioerodable, and bioabsorbable. In all cases, we describe a polymer that is converted under physiological conditions into water-soluble materials.

Some of the common biodegradable polymers used in short-term biomedical applications, and their corresponding mechanical properties, are listed in Table 5.11. Polyhydroxybutyrate (PHB), polyhydroxyvalerate (PHV), and their copolymers are derived from microorganisms and are used in such applications as controlled drug release, sutures, and artificial skin. Polycaprolactone is a semicrystalline polymer that is used in some applications as a degradable staple. Polyanhydrides are very reactive and hydrolytically unstable, which is both an advantage and a limitation to their use. A wide variety of drugs and proteins, including insulin, bovine growth factors, heparin, cortisone, and enzymes, have been incorporated into polyanhydride matrices. Poly(ortho esters) have also been used for controlled drug delivery. Poly(amino acids) are of interest due to the presence of

Table 5.11 Selected Mechanical Properties of Some Biodegradable Polymers

Polymer	Glass Transition (°C)	Melting Temperature (°C)	Tensile Strength (MPa)	Tensile Modulus (MPa)	Flexural Modulus (MPa)	Elongation Yield (%)	Elongation Break (%)
Poly(glycolic acid) (MW: 50,000)	35	210	n/a	n/a	n/a	n/a	n/a
Poly(lactic adds)							
L-PLA (MW: 50,000)	54	170	28	1200	1400	3.7	6.0
L-PLA (MW: 100,000)	58	159	50	2700	3000	2.6	3.3
L-PLA (MW: 300,000)	59	178	48	3000	3250	1.8	2.0
D, L-PLA (MW: 20,000)	50	—	n/a	n/a	n/a	n/a	n/a
D, L-PLA (MW: 107,000)	51	—	29	1900	1950	4.0	6.0
D, L-PLA (MW: 550,000)	53	—	35	2400	2350	3.5	5.0
Poly(β-hydroxybutyrate) (MW: 422,000)	1	171	36	2500	2850	2.2	2.5
Poly(ε-caprolactone) (MW: 44,000)	−62	57	16	400	500	7.0	80
Polyanhydrides							
Poly(SA-HDA anhydride) (MW: 142,000)	n/a	49	4	45	n/a	14	85
Poly(ortho esters)							
DETOSU: t-CDM: 1,6-HD (MW: 99,700)	55	—	20	820	950	4.1	220
Polyiminocarbonates							
Poly(BPA iminocarbonate) (MW: 105,000)	69	—	50	2150	2400	3.5	4.0
Poly(DTH iminocarbonate) (MW: 103,000)	55	—	40	1630	n/a	3.5	7.0

Source: Kohn and Langer [74]/With permission of Elsevier.

amino acid groups that can potentially be modified, but have found few biomedical applications because they are highly insoluble and difficult to process. Many of these biodegradable polymers, as well as nonbiodegradable fibers, are fabricated into fabrics for a variety of biomedical applications.

An example of how these materials can be assembled into a useful artificial soft biologic is *artificial skin*. The use of artificial skin allows for prompt closure of the largest of burn wounds, thereby limiting wound contracture and providing permanent coverage with a readily available, easily stored skin replacement material. Key characteristics of artificial skin are biocompatibility and resistance to infection.

Need More Help?

Try these videos on mechanical properties of soft biologics:

https://www.youtube.com/watch?v=Y7mNFwcjGnc

Several types of synthetic skin grafts exist, but are all essentially *bilaminate membranes*, as shown in Figure 5.96. A small pore diameter is required in the outer layer to prevent the wound from drying out. However, a large pore diameter is required in the lower layer to promote the ingrowth of fibroblasts and long-term adhesion to the wound. One such bilaminate artificial skin consists of an inner layer of collagen-chondroitin-6-sulfate and an outer layer of silastic. The collagen binds to the wound bed and is invaded by neovasculature and fibroblasts. The collagen chondroitin 6-sulfate very significantly inhibits wound contracture, and with time the collagen is resorbed and is completely replaced by remodeled dermis. The pore size of the collagen layer is typically 50 μm. The silicone layer is 0.1 mm thick and provides a bacterial barrier in addition to the required water flux rate of 1–10 mg/cm^2 hours. The silicone layer also provides the mechanical rigidity needed to suture the graft in place, thus preventing movement of the material during wound healing. The collagen layer adheres to the wound rapidly, as evidenced by the peel strength of the graft, which is 9 N/m at 24 hours and 45 N/m at 10 days. Histological cross sections of wounds closed with this artificial skin show complete replacement of the bovine collagen at seven weeks with remodeled human dermis.

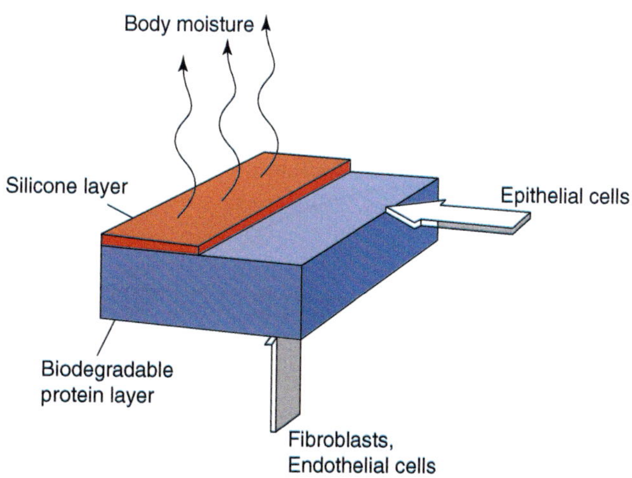

Figure 5.96 Schematic diagram of bilaminar artificial skin. Kane et al. [75]/Reproduced with permission of Elsevier.

5.5.2 Mechanical Properties of Hard Biologics

In contrast to soft biologics, whose mechanical properties primarily depend upon the orientation of collagen fibers, the mechanical properties of mineralized tissues, or hard biologics, are more complicated. Factors such as density, mineral content, fat content, water content, and sample preservation and preparation play important roles in mechanical property determination. Specimen orientation also plays a key role, since most hard biologics such as bone are composite structures. For the most part, we will concentrate on the average properties of these materials and will relate these values to those of important, man-made replacement materials.

Bone

The structure of bone was described in Section 1.5.2. Recall that bone is a composite material, composed primarily of a calcium phosphate form called hydroxyapatite (HA). The major support bones consist of an outer load-bearing shell of *cortical* (or compact) bone with a medullary cavity containing *cancellous* (or spongy) bone toward the ends.

The stress–strain curves for cortical bones at various strain rates are shown in Figure 5.97. The mechanical behavior is as expected from a composite of linear elastic ceramic reinforcement (HA) and a compliant, ductile polymer matrix (collagen). In fact, the tensile modulus values for bone can be modeled to within a factor of two by a rule-of-mixtures calculation on the basis of a 0.5 volume fraction HA-reinforced collagen composite. The UTS for cortical bone varies from 100 to 300 MPa, the modulus varies up to 30 GPa, and the strain at failure is typically only 1% or 2%. Bone has either isotropic symmetry, requiring five stiffness coefficients, or orthotropic symmetry, which requires

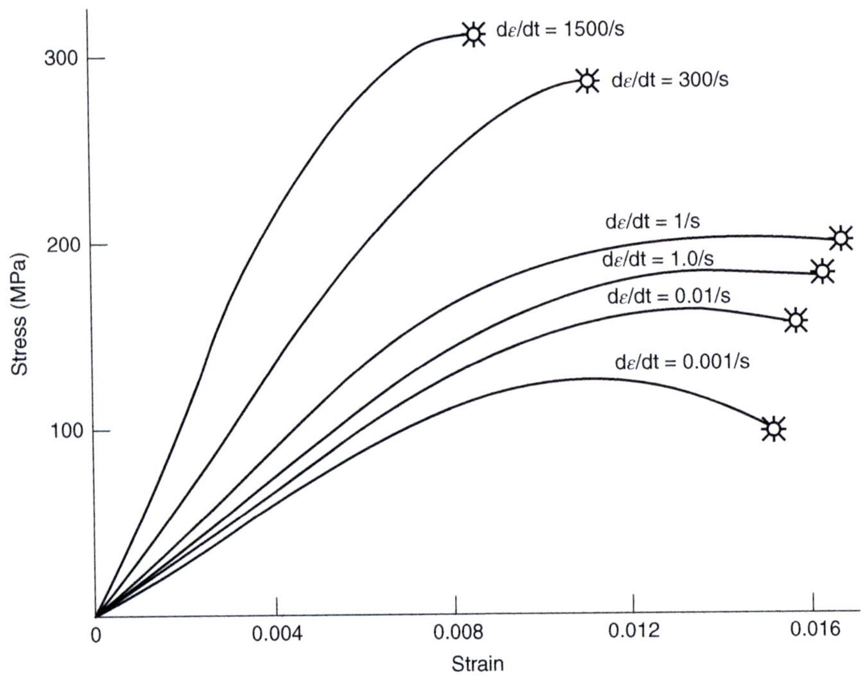

Figure 5.97 Stress–strain curves at various strain rates for cortical bone. Silver and Christiansen [76]/ Reproduced with permission of Springer Nature.

nine such coefficients. These coefficients are particularly important for the modeling of bone resorption and implant performance.

Figure 5.97 shows that at higher strain rates the modulus increases and the bone behaves in a more brittle manner. Values for the critical stress intensity factor, K_{1c}, of 2–12 MN/m$^{3/2}$ and critical strain energy of 500–6000 J/m^2 have been determined. Conversely, bone is much more viscoelastic in its response at very low strain rates. Recent experiments have shown that bone is a complex viscoelastic material; that is, time–temperature superposition cannot be used to obtain its properties. However, the nonlinearity in its mechanical behavior tends to occur outside the range of physiological interest, such that linear models are often useful for creep and stress relaxation behavior. Mineralization tends to increase modulus and UTS, as well as reducing the strain at failure. Mineralization does not affect the viscoelastic behavior of bone, however, and the increasing elasticity with decreasing strain rate observed in Figure 5.97 is followed, even with increased mineralization.

The mechanical properties of cancellous bone are dependent upon the bone density and porosity, and the strength and modulus are therefore much lower than those for cortical bone. The axial and compressive strengths are proportional to the square of the bone density, and moduli can range from 1 to 3 GPa.

Teeth

The structure of a human tooth was given in Figure 1.87. Recall that the two most prominent components of teeth are enamel and dentin and that the primary components of each are HA and HA/collagen, respectively. The mechanical properties of dentin and enamel—and, where available, their demineralized forms—are presented in Table 5.12. Notice the severe decrease in modulus and tensile strength as dentin loses its mineral content. A knowledge and understanding of these properties are important for developing suitably compatible dental amalgams, adhesives, and prosthetics.

Artificial Hard Biologics

A summary of polymeric materials used in biomedical applications was given in Figure 5.95. Recall that many of these applications were for replacement of soft tissues, such as skin and tendons. In this section, we concentrate not only on those polymeric materials used for hard biologics but also on all classes of materials that find use in the bonding, replacing, and growth-scaffolding of hard

Table 5.12 Selected Mechanical Properties of Human Dentin and Enamel

	Dentin		
Property	**Mineralized**	**Demineralized**	**Enamel**
Density (g/cm^3)	2.14–2.18	—	2.95–2.97
Elastic modulus (GPa)	14.7	0.26	84–130
Tensile strength (MPa)	105.5	29.6	10.3
Compressive strength (MPa)	297	—	384
Shear strength (MPa)	138	—	90.2
Knoop hardness (kg/mm^2)	68	—	355–431
Poisson's ratio	0.31	—	0.33

Source: W. J. O'Brien [77].

biologics. A schematic summary of some of these applications that involve metallic, ceramic, polymeric, and composite components is given in Figure 5.98. We will approach each of these material classes in order and will describe only a few examples of the biomedical applications for each. In most cases, these applications will be as structural replacements, and in all cases, we will concentrate on the specific mechanical properties that make each material suitable for the particular

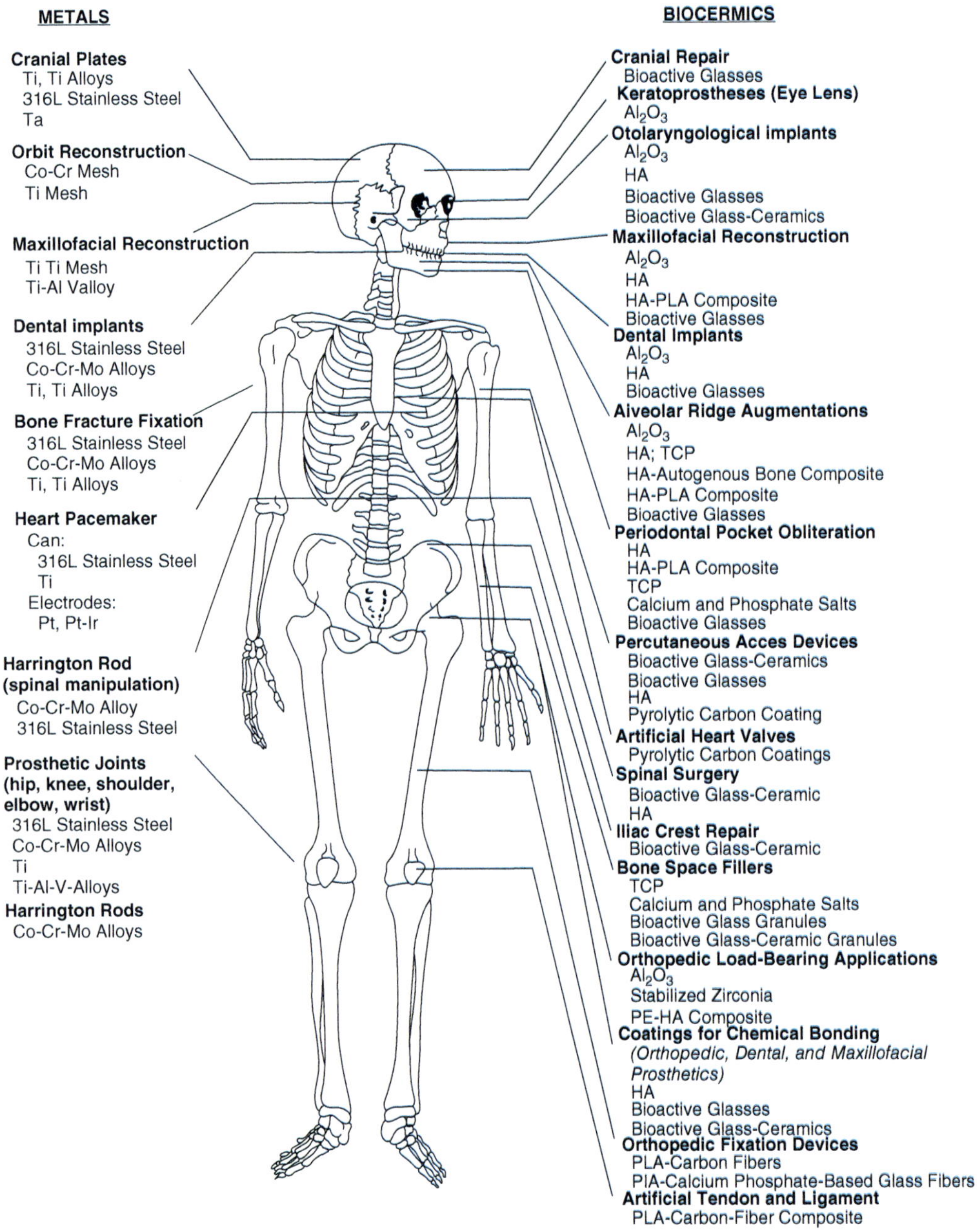

Figure 5.98 Clinical uses of inorganic biomaterials. Hench [78]/Reproduced with permission of American Chemical Society.

Table 5.13 Comparison of Some Mechanical Properties of Current Implant Materials with Those of Cortical Bone

Materials	Young's Modulus (GPa)	Ultimate Tensile Strength (MPa)	Critical Stress Intensity Factor K_{1C} (MN m$^{-3/2}$)	Critical Strain Energy Release Rate, G_{1C} (J m^{-2})
Alumina	365	6–55	3	40
Cobalt–chromium alloys	230	900–1540	100	50,000
Austenitic stainless steel	200	540–1000	100	50,000
Ti–6wt% Al–4wt% V	106	900	80	10,000
Cortical bone	7–30	50–150	2–12	600–5000
PMMA bone cement	3.5	70	1.5	400
Polyethylene	1	30	—	8000

Source: Kelly [79]. Copyright © 1994 by Elsevier Science Ltd.

application. Topics such as biocompatibility, longevity, and infection resistance are beyond the scope of this text.

Metals were some of the first materials to be used for the replacement of hard biologics. This was a reasonable selection since the principal function of the long bones of the lower body is to act as load-bearing members. The *hip prosthesis* has been the most active area of joint replacement research. Both stainless steel, such as 316L, and Co–Cr alloys became the early materials of choice, since their stiffness, rigidity, and strength considerably exceed those of natural bone (see Table 5.13). Titanium and the so-called *aviation alloy* Ti–6Al–4V are now also used for the femoral portion of hip prostheses. In addition to strength and modulus, corrosion resistance and fatigue life are of prime importance. Also, the metals must be nontoxic, mutagenic, or carcinogenic within the body, either of themselves or due to the release of chemical components as a result of interactions with various body fluids. In fact, these are requirements for all classes of materials in these types of biomedical applications. In most prosthetic applications, the metal or alloy is coated with, and held in place by, polymers.

Need More Help?

Try these videos on mechanical properties of hard biologics:

https://www.youtube.com/watch?v=Js4QY9ENYGM

Polymers have found two specific applications in orthopedics. First, they are used for one of the articulating surface components in joint prostheses. Thus, they must have a low coefficient of friction and low wear rate when they are in contact with the opposing surface, which is usually made of metal or ceramic. Initially, poly(tetrafluoroethylene) (PTFE) was used in this type of application, but its accelerated creep rate and poor stress corrosion caused it to fail *in vivo*. Ultimately, this material was replaced with ultrahigh-molecular-weight polyethylene (UHMWPE). The second-place polymers find biomedical application as a structural interface between the implant and bone tissue. In this case, the appropriate mechanical properties of the polymer are of major importance. The first type of use was with PMMA as a grouting material to fix both the stem of the femoral component and the acetabular component in place and to distribute the loads more uniformly from the implants to the bone. Since high interfacial stresses result from the accommodation of a high-modulus prosthesis within the much lower-modulus bone, the use of a lower-modulus interposition material has been introduced as an alternative to PMMA fixation. Thus, polymers such as polysulfone have been used as porous coatings on the implant's metallic core to permit mechanical interlocking through bone and/or soft tissue ingrowth into the pores. This requires that the polymers have surfaces that resist creep under the stresses found in clinical situations and have high enough yield strength to minimize plastic deformation. As indicated earlier, the mechanical properties of concern in polymer applications are yield stress, creep resistance, and wear rate. These factors are controlled by such polymer parameters as molecular chain structure, molecular weight, and degree of branching.

Ceramics and glasses have played an increasingly important role in implants in recent years. This has occurred because of two quite disparate uses. First, there is the use associated with improved properties such as resistance to further oxidation, high stiffness, and low friction and wear in articulating surfaces. This requires the use of full-density, controlled, small, uniform-grain-size (usually less than 5 μm) materials. The small grain size and full density are important since these are the two principal bulk parameters controlling the ceramic's mechanical properties. Clearly, any void within the ceramic's body will increase stress, thereby degrading the mechanical properties. Thus, full-density, small-grain-size aluminum oxide, Al_2O_3, has proved quite successful as the material in matched pairs of femoral head and acetabular components in total hip arthroplasty. Other ceramics for this use include TiO_2 and Si_3N_4. The second application takes advantage of the *osteophilic surface* of certain ceramics and glass ceramics. These materials provide an interface of such biological compatibility with bone-forming cells called *osteoblasts* that these cells lay down bone in direct apposition to the material with a direct chemicophysical bond. Special compositions of glass ceramics, termed *bioglasses*, were originally used for orthopedic implants. The bioglasses have nominal compositions in the ranges 40–50% SiO_2, 5–31% Na_2O, 12–35% CaO, and 0–15% P_2O_5, plus in some cases MgO, B_2O_3, or CaF_2. The model proposed for the "chemical bond" formed between the glass and bone is that the former undergoes a controlled surface degradation, producing a SiO_2-rich layer and a Ca, P-rich layer at the interface. This latter layer eventually crystallizes as a mixture of hydroxy-carbonate apatite structurally integrated with collagen, which permits subsequent bonding by newly formed mineralized tissues.

There is an entirely different series of ceramic-based compounds that have been shown to be osteophilic. These include HA, which you will recall is the form of the naturally occurring inorganic component of calcified tissues, and calcite, $CaCO_3$, and its Mg analog, dolomite, among others. The most extensive applications in both orthopedics and dentistry have involved HA. The elastic properties of HA and related compounds such as fluorapatite (FAp), chlorapatite (ClAp), and cobalt apatite (Co_3Ap) are compared with those of bone, dentin, and enamel in Table 5.14. The use of both HA and glass ceramics as claddings on the metallic stems of prostheses is still another method of providing fixation instead of using PMMA. The fracture toughness properties of these ceramic-based systems are compared to natural bone in Figure 5.99.

Composites provide an attractive alternative to the various metal-, polymer- and ceramic-based biomaterials, which all have some mismatch with natural bone properties. A comparison of modulus and fracture toughness values for natural bone provides a basis for the approximate mechanical compatibility required for artificial bone in an exact structural replacement, or to stabilize a bone–implant interface. A precise matching requires a comparison of all the elastic stiffness coefficients. From Table 5.12, we see that a possible approach to the development of a mechanically compatible artificial bone material is to stiffen one of the polymers already used as an implant with a suitable second phase. The general effect has been well demonstrated for polymers in that a progressive increase in the volume fraction of the second phase will correspondingly increase the elastic modulus but decrease the fracture toughness. Hence, starting with a ductile polymer, an

Table 5.14 Selected Mechanical Properties of Various Calcium-Bearing Minerals Compared to Bone and Teeth

Material	Bulk Modulus (10^{10} N m^{-2})	Shear Modulus (10^{10} N m^{-2})	Elastic Modulus (10^{10} N m^{-2})	Poisson's Ratio
FAp (mineral)	9.40	4.64	12.0	0.26
HAp (mineral)	8.90	4.45	11.4	0.27
HAp (synthetic)	8.80	4.55	11.7	0.28
ClAp (synthetic	6.85	3.71	9.43	0.27
Co_3Ap (synthetic, type B)	8.17	4.26	10.9	0.28
CaF_2	7.74	4.70	11.8	0.26
Dicalcium phosphate dihydrate	5.50	2.40	6.33	0.31
Bone (human fervour, powdered)	2.01	0.800	2.11	0.33
Dentine (bovine, powdered)	3.22	1.12	3.02	0.35
Enamel (bovine, powdered)	6.31	2.93	7.69	0.32

Source: Katz [80]/With permission of Elsevier.

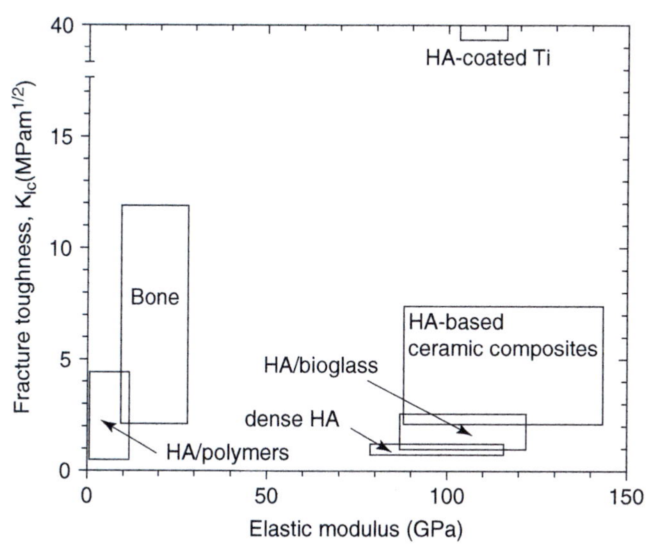

Figure 5.99 Fracture toughness and modulus of some biomaterials for bone replacement. Suchanek and Yoshimura [81]/Reproduced with permission of Cambridge University Press.

eventual transition into brittle behavior is observed. Because PMMA is brittle, whereas PE is ductile, with an elongation to fracture of about 100%, PE provides the more suitable starting materials and has comparable mechanical behavior to collagen. Various materials could be considered as additions to PE that could stiffen the materials at the expense of fracture toughness, but the selection of HA has proved particularly appropriate.

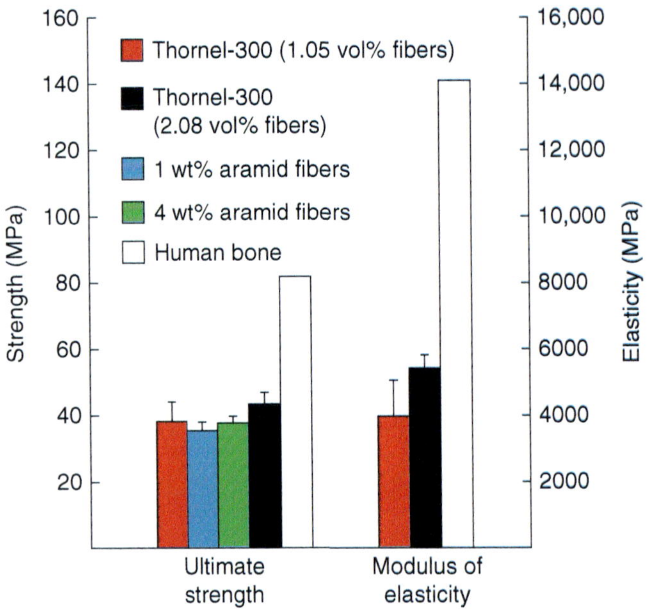

Figure 5.100 Comparison of tensile properties of fiber-reinforced bone cement (PMMA) and human compact bone. Reprinted, by permission, from Kelly [82]. Copyright © 1994 by Elsevier Science Publishers, Ltd.

Starting with polyethylene granules of average molecular weight 400,000, along with HA particles of 0.5–20 μm in size, composites of HA-reinforced PE are prepared to various volume fractions by a compounding and molding technique that minimizes polymer degradation. The composites are pore-free, with a homogeneous distribution of HA particles of volume fractions from 0.1 to 0.5. The rule of mixtures applies to the prediction of tensile modulus in these composites, and values increase with increasing HA content up to about 9 GPa at the 50% volume level. The increase in modulus is accompanied by a decrease in the elongation-to-fracture. However, the composite remains unusually ductile and maintains its fracture toughness, up to about 40% volume HA. From 40% to 50% volume fraction, the elongation to fracture decreases rapidly, with brittle fracture observed at $V_f = 0.45$. At these volume fractions, a fracture toughness of 3–8 $MN/m^{3/2}$ is observed (cf. Figure 5.99). Hence, in terms of cortical bone replacement, an HA–PE composite with 50% volume HA has a tensile modulus within the lower band of the values for bone and a comparable fracture toughness.

Another example of composites in biomedical applications is graphite-fiber-reinforced bone cement. Self-curing PMMA is used extensively as a bone cement in orthopedic surgery for fixation of endoprostheses and the repair of bone defects. However, such bone cement is weak in tension and is also significantly weaker than compact bone. The reinforcement of bone cement with graphite fibers has been shown to increase its tensile, compressive, and shear strengths, as well as increase fatigue life and reduce creep. The tensile modulus of fiber-reinforced PMMA also increases with increasing fiber content when the fiber volume percentage is kept small, as illustrated in Figure 5.100, where Thornel is an ultrahigh-strength graphite fiber, and the aramid fiber in use is Kevlar-29. Incorporation of 2% Thornel-300 fibers improves the tensile strength of PMMA to 50%, and its modulus to about 40%, of that of compact human bone. The energy absorption capacity of the composite is increased due to a phenomenon known as *fiber pullout* wherein the process of destroying the fiber-matrix interphase and pulling the fiber from the matrix takes additional energy.

References

1 Ashby, M. F., and D. R. H. Jones, *Engineering Materials 1*, 2nd ed., Elsevier, 1996, p. 44.
2 Ashby, M. R., and D. R. H. Jones, *Engineering Materials 1*, 2nd ed., Elsevier, 1996, p. 59.
3 Ralls, K. M., T. H. Courtney, and J. Wulff, *Introduction to Materials Science and Engineering*, John Wiley & Sons, Inc., 1976.
4 Jastrzebski, Z., *The Nature and Properties of Engineering Materials*, 2nd ed., John Wiley & Sons, Inc., 1976.
5 Kittel, C., *Introduction to Solid State Physics*, 2nd ed., John Wiley & Sons, 1957, p. 517.
6 Callister, W., *Materials Science and Engineering: An Introduction*, 5th ed., John Wiley & Sons, Inc., 2000, p. 155.
7 Callister, W., *Materials Science and Engineering: An Introduction*, 5th ed., John Wiley & Sons, Inc., 2000, p. 160.
8 Callister, W., *Materials Science and Engineering: An Introduction*, 5th ed., John Wiley & Sons, Inc., 2000, p. 162.
9 Callister, W., *Materials Science and Engineering: An Introduction*, 5th ed., John Wiley & Sons, Inc., 2000, p. 166.
10 Callister, W., *Materials Science and Engineering: An Introduction*, 5th ed., John Wiley & Sons, Inc., 2000, p. 167.
11 Callister, W., *Materials Science and Engineering: An Introduction*, 5th ed., John Wiley & Sons, Inc., 2000, p. 175.
12 Callister, W., *Materials Science and Engineering: An Introduction*, 5th ed., John Wiley & Sons, Inc., 2000, p. 140.
13 Li, X., and B. Bhushan, A review of nanoindentation continuous stiffness measurement technique and its applications, *Mater. Charact.*, **48**, 11–36 (2002).
14 Li, X., and B. Bhushan, *Materials Characterization*, Vol. **48**, Elsevier, 2002, p. 13.
15 Li, X., and B. Bhushan, *Materials Characterization*, Vol. **48**, Elsevier, 2002, p. 14.
16 Shackelford, J. F., *Introduction to Materials Science for Engineers*, 5th ed., Pearson, 2000, p. 192.
17 Shackelford, J. F., *Introduction to Materials Science for Engineers*, 5th ed., Pearson, 2000, p. 190.
18 Callister, W., *Materials Science and Engineering: An Introduction*, 5th ed., John Wiley & Sons, Inc., 2000, p. 130.
19 J. Pilling, Superplasticity on the Web, http://callisto.my.mtu.edu:591/FMPro?-db=sp&-format=sp.html&-view.
20 Callister, W., *Materials Science and Engineering: An Introduction*, 5th ed., John Wiley & Sons, Inc., 2000, p. 186.
21 Somayaji, S., *Civil Engineering Materials*, 2nd ed., Pearson, 2001, p. 24.
22 Callister, W., *Materials Science and Engineering: An Introduction*, 5th ed., John Wiley & Sons, Inc., 2000, p. 210.

23 Callister, W., *Materials Science and Engineering: An Introduction*, 5th ed., John Wiley & Sons, Inc., 2000, p. 212.

24 Callister, W., *Materials Science and Engineering: An Introduction*, 5th ed., John Wiley & Sons, Inc., 2000, p. 217.

25 Callister, W., *Materials Science and Engineering: An Introduction*, 5th ed., John Wiley & Sons, Inc., 2000, p. 226.

26 Callister, W., *Materials Science and Engineering: An Introduction*, 5th ed., John Wiley & Sons, Inc., 2000, p. 227.

27 Kingery, W. D., H. K. Bowen, and D. R. Uhlmann, *Introduction to Ceramics*, John Wiley & Sons, Inc., 1976.

28 Callister, W., *Materials Science and Engineering: An Introduction*, 5th ed., John Wiley & Sons, Inc., 2000, p. 413.

29 Callister, W., *Materials Science and Engineering: An Introduction*, 5th ed., John Wiley & Sons, Inc., 2000, p. 410.

30 Scholze, H., *Glass*, Springer Nature, 1991.

31 Scholze, H., *Glass*, Springer Nature, 1991, p. 273.

32 Ashby, M. F., *Materials Selection in Mechanical Design*, 2nd ed., Elsevier, 1999, p. 42.

33 Somayaji, S., *Civil Engineering Materials*, 2nd ed., Pearson, 2001, p. 127.

34 Ashby, M. F., *Materials Selection in Mechanical Design*, 2nd ed., Butterworth Heinemann, Oxford, 1992.

35 Ashby, M. F., *Materials Selection in Mechanical Design*, 2nd ed., Elsevier, 1999, p. 44.

36 Callister, W., *Materials Science and Engineering: An Introduction*, 5th ed., John Wiley & Sons, Inc., 2000, p. 475.

37 Cowie, J. M. G., *Polymers: Chemistry & Physics of Modern Materials*, 2nd ed., Taylor & Francis, 1991, p. 277.

38 Rodriguez, F., *Principles of Polymer Systems*, 2nd ed., Hemisphere Publishing Corporation, 1982, p. 217.

39 Billmeyer, F. W., *Textbook of Polymer Science*, 3rd ed., John Wiley & Sons, Inc., 1984.

40 Cowie, J. M. G., *Polymers: Chemistry & Physics of Modern Materials*, 2nd ed., Taylor & Francis, 1991, p. 283.

41 Rodriguez, F., *Principles of Polymer Systems*, 2nd ed., Hemisphere Publishing Corporation, p. 249.

42 McCrum, N. G., C. P. Buckley, and C. B. Bucknall, *Principles of Polymer Engineering*, 2nd ed., Oxford University Press, 1997, p. 72.

43 McCrum, N. G., C. P. Buckley, and C. B. Bucknall, *Principles of Polymer Engineering*, 2nd ed., Oxford University Press, 1997, p. 191.

44 McCrum, N. G., C. P. Buckley, and C. B. Bucknall, *Principles of Polymer Engineering*, 2nd ed., Oxford University Press, 1997, p. 154.

45 Kyu, T., J. M. Saldanha, and M. J. Kiesel, Toughness Enhancement in Polycarbonate/Polymethylmethacrylate Blends via Phase Separation, in *Two-Phase Polymer Systems*, L. A. Utracki, ed., Carl Hanser Verlag GmbH & Co., 1991, p. 271.

46 Hornsby, P. R., Structure-Property Relationships in Plastics Structural Foams, in *Two-Phase Polymer Systems*, L. A. Utracki, ed., Hanser Gardner Publications, 1991, p. 102.

47 Hornsby, P. R., Structure-Property Relationships in Plastics Structural Foams, in *Two-Phase Polymer Systems*, L. A. Utracki, ed., Carl Hanser Verlag GmbH & Co., 1991, p. 115.

48 McCrum, N. G., C. P. Buckley, and C. B. Bucknall, *Principles of Polymer Engineering*, 2nd ed., Oxford University Press, 1997, p. 173.

49 Rodriguez, F., *Principles of Polymer Systems*, 2nd ed., Hemisphere Publishing Corporation, 1982, p. 221.

50 Schwartz, M., *Composite Materials Handbook*, 2nd ed., McGraw-Hill, 1992, p. 1.29.

51 Kelly, A., *Composite Materials*, Elsevier, 1966, p. 62.

52 Callister, W., *Materials Science and Engineering: An Introduction*, 5th ed., John Wiley & Sons, 2000, p. 493.

53 McCrum, N. G., C. P. Buckley, and C. B. Bucknall, *Principles of Polymer Engineering*, 2nd ed., Oxford University Press, 1997, p. 200.

54 McCrum, N. G., C. P. Buckley, and C. B. Bucknall, *Principles of Polymer Engineering*, 2nd ed., Oxford University Press, 1997, p. 258.

55 McCrum, N. G., C. P. Buckley, and C. B. Bucknall, *Principles of Polymer Engineering*, 2nd ed., Oxford University Press, 1997, p. 259.

56 McCrum, N. G., C. P. Buckley, and C. B. Bucknall, *Principles of Polymer Engineering*, 2nd ed., Oxford University Press, 1997, p. 265.

57 McCrum, N. G., C. P. Buckley, and C. B. Bucknall, *Principles of Polymer Engineering*, 2nd ed., Oxford University Press, 1997, p. 267.

58 Mallick, P. K., *Fiber-Reinforced Composites*, Taylor & Francis, 1988, p. 187.

59 McCrum, N. G., C. P. Buckley, and C. B. Bucknall, *Principles of Polymer Engineering*, 2nd ed., Oxford University Press, 1997, p. 268.

60 McCrum, N. G., C. P. Buckley, and C. B. Bucknall, *Principles of Polymer Engineering*, 2nd ed., Oxford University Press, 1997, p. 269.

61 Schwartz, M., *Composite Materials Handbook*, 2nd ed., McGraw-Hill, 1992, p. 3.71.

62 Powell, P. C., and A. J. I. Housz, *Engineering with Polymers*, Stanley Thorned Publishers, 1998, p. 201.

63 McCrum, N. G., C. P. Buckley, and C. B. Bucknall, *Principles of Polymer Engineering*, 2nd ed., Oxford University Press, 1997, p. 408.

64 Z. Wang and T.J. Pinnavaia, Nanolayer Reinforcement of Elastomeric Polyurethane, *Chem. Mater.* 1998 **10** (12), 3769–3771, DOI: https://doi.org/10.1021/cm980448n.

65 Meneghetti, P., and S. Qutubuddin, Synthesis, thermal properties and applications of polymer-clay nanocomposites, *Thermochim. Acta*, **442**(1–2), 74–77 (2006).

66 Avella, M., M. E. Errico, and E. Maruscelli, Novel PMMA/$CaCO_3$ nanocomposites abrasion resistant prepared by an in situ polymerization process, *Nano Lett.*, **1**(4), 213–217 (2001).

67 Silver, F. H., and D. L. Christiansen, *Biomaterials Science and Biocompatibility*, Springer Nature, 1999, p. 18.

68 Yannas, I. V., CHAPTER 2 - Classes of Materials Used in Medicine, in *Biomaterials Science*, B. D. Ratner, A. S. Hoffman, F. J. Schoen, and J. E. Lemons, eds., Elsevier, 1996, p. 86.

69 Silver, F. H., and D. L. Christiansen, *Biomaterials Science and Biocompatibility*, Springer Nature, 1999.

70 Silver, F. H., and D. L. Christiansen, *Biomaterials Science and Biocompatibility*, Springer Nature, 1999, p. 203.

71 Silver, F. H., and D. L. Christiansen, *Biomaterials Science and Biocompatibility*, Springer Nature, 1999, p. 202.

72 Goupil, D., Chapter 7, Application of Materials in Medicine and Dentistry, in *Biomaterials Science*, B. D. Ratner, A. S. Hoffman, F. J. Schoen, and J. E. Lemons, eds., Elsevier, 1996, p. 356.

73 Visser, S. A., R. W. Hergenrother, and S. L. Cooper, CHAPTER 2 - Classes of Materials Used in Medicine, in *Biomaterials Science*, B. D. Ratner, A. S. Hoffman, F. J. Schoen, and J. E. Lemons, eds., Elsevier, 1996, p. 57.

74 Kohn, J., and R. Langer, CHAPTER 2 - Classes of Materials Used in Medicine, in *Biomaterials Science*, B. D. Ratner, A. S. Hoffman, F. J. Schoen, and J. E. Lemons, eds., Elsevier, 1996, p. 64.
75 Kane, J. B., R. G. Tompkins, M. L. Yarmush, and J. F. Burke, Chapter 7, Application of Materials in Medicine and Dentistry, in *Biomaterials Science*, B. D. Ratner, A. S. Hoffman, F. J. Schoen, and J. E. Lemons, eds., Elsevier, 1996, p. 367.
76 Silver, F. H., and D. L. Christiansen, *Biomaterials Science and Biocompatibility*, Springer Nature, 1999, p. 208.
77 W. J. O'Brien. Biomaterials Properties Database, https://www.zubnistranky.cz/bio.htm.
78 Hench, L., in *Materials Chemistry*, L. V. Interrante, L. A. Casper, and A. B. Ellis, eds., American Chemical Society, 1995, p. 525.
79 Kelly, A., ed., revised edition, *Concise Encyclopedia of Composite Materials*, Elsevier Science Ltd, 1994, Artificial Bone, p. 16.
80 Katz, L. J., in *Biomaterials Science*, B. D. Ratner, A. S. Hoffman, F. J. Schoen, and J. E. Lemons, eds., Chapter 7, Application of Materials in Medicine and Dentistry, Elsevier, 1996, p. 335.
81 Suchanek, W., and M. Yoshimura, *J. Mater. Res.*, **13**(1), 94 (1998).
82 Kelly, A., ed., *Concise Encyclopedia of Composite Materials*, Elsevier Science Publishers, Ltd., 1994, p. 270, revised edition.

Problems

Level I

5.I.1 The following three stress–strain data points are provided for a titanium alloy: $\varepsilon = 0.002778$ at or $\sigma = 300$ MPa; $\varepsilon = 0.005556$ at $\sigma = 600$ MPa; $\varepsilon = 0.009897$ at $\sigma = 900$ MPa. Calculate the modulus of elasticity for this alloy.

5.I.2 Search and download a data file that contains the stress–strain data for a material of your choice. Import the data into a spreadsheet and plot the data appropriately to generate a stress–strain curve. Calculate the tensile modulus and ultimate strength of your material based on your plot. Be sure to include the source of your data and provide a narrative around how the plot was generated and the required quantities calculated.

5.I.3 Calculate the composite modulus for epoxy reinforced with 70 vol% boron filaments under isotropic strain conditions.

5.I.4 Prove that the true strain is related to the engineering strain by the factor one plus Poisson's ratio. Show your work and clearly state your assumptions.

5.I.5 According to the Kelvin (Voigt) model of viscoelasticity, what is the viscosity (in Pa·s) of a material that exhibits a shear stress of 9.32×10^9 Pa at a shear strain of 0.5 cm/cm over a duration of 100 seconds? The shear modulus of this material is 5×10^9 Pa.

5.I.6 A high-strength steel has a yield strength of 1460 MPa and a K_{1c} of 98 MPa·$m^{\frac{1}{2}}$. Calculate the size of a surface crack that will lead to catastrophic failure at an applied stress of $\frac{1}{2}\sigma_y$.

5.I.7 An alloy is evaluated for potential creep deformation in a short-term laboratory experiment. The creep rate is found to be 1% per hour at 800°C and 5.5×10^{-2}% per hour at 700°C. (a) Calculate the activation energy for creep in this temperature range. (b) Estimate the creep rate to be expected at a service temperature of 500°C. (c) What important assumption underlies the validity of your answer to part b?

5.I.8 A glass plate contains an atomic-scale surface crack. Assume that the crack tip radius is approximately equal to the diameter of an O^{2-} ion. Given that the crack is 1 μm long and the theoretical strength of the defect-free glass is 7.0 GPa, calculate the breaking strength of the glass plate.

Level II

5.II.1 In normal motion, the load exerted on the hip joint is 2.5 times body weight. (a) Estimate the corresponding stress (in MPa) on an artificial hip implant with a cross-sectional area of 5.64 cm^2 in an average-sized patient, and (b) calculate the corresponding strain if the implant is made of Ti–6Al–4V alloy. Clearly state your assumptions and sources of data.

5.II.2 In the absence of stress, the center-to-center atomic separation distance of two Fe atoms is 0.2480 nm along the [111] direction. Under a tensile stress of 1000 MPa along this direction, the atomic separation distance increases to 0.2489 nm. (a) Calculate the modulus of elasticity along the [111] direction. (b) How does this compare with the average modulus for polycrystalline iron with random grain orientation? (c) Calculate the center-to-center separation distance of two Fe atoms along the [100] direction in unstressed α-iron. (d) Calculate the separation distance along the [100] direction under a tensile stress of 1000 MPa, if the modulus in this direction is 125 GPa.

5.II.3 Look up the elastic modulus of a typical commercial-grade glass fiber and that of a typical epoxy resin. Calculate the upper limit elastic modulus of a unidirectionally reinforced composite from these two materials with 65% glass fiber by volume. Cite the source of your information and show your work.

Level III

5.III.1 In a recent article [Shinzato et al., A new bioactive bone cement: Effect of glass bead filler content on mechanical and biological properties, *J. Biomed. Mater. Res.*, **54**(4), 491 (2000)], the mechanical properties of glass-bead-filled PMMA bone cement are described. Four plots from this article are reproduced below. "GBC" refers to the glass bead composite, and the number refers to the weight percent of glass beads; for example, GBC50 has 50 wt% glass beads in the PMMA matrix. The data from all tests were obtained from cylindrical samples with a gauge length of 20 mm. (a) Calculate the radius (in mm) of the cylindrical samples used for the compressive tests. Note that the compressive strengths of samples GBC30, GBC40, and GBC50 shown in Figure (a) were taken as the inflection point in the stress–strain diagrams shown in Figure (d), rather than the break point, as in samples GBC60 and GBC70. (b) Estimate Poisson's ratio for this material. (c) What is the force (in terms of weight, kg) required to break the GBC samples in a three-point bending test? (d) Sketch the stress–strain diagrams that result from the three-point bend test for the following three materials: glass beads (assume they are a generic silicate glass), PMMA, and GBC50. You should have three separate curves, each representing the relative behaviors of the three materials.

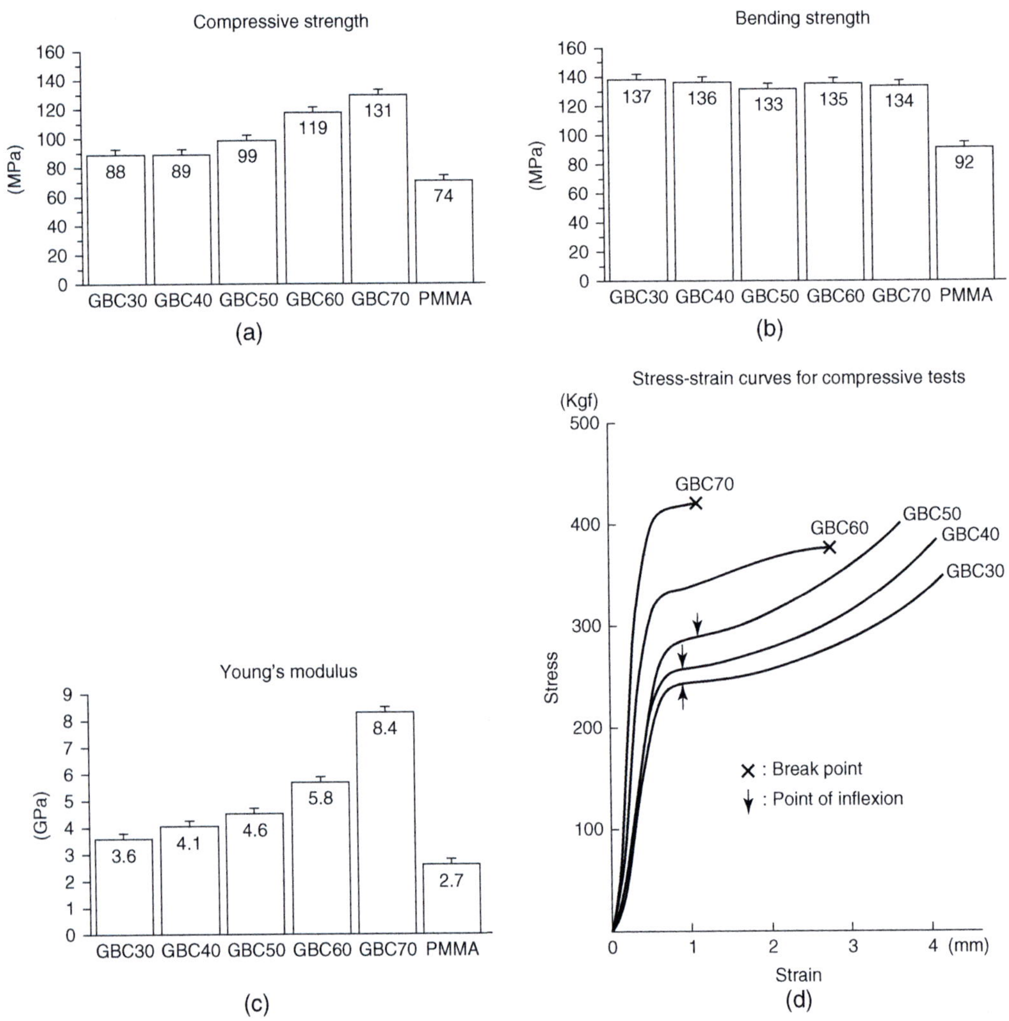

5.III.2 In another article on bone cements [Vallo, C, Influence of filler content on static properties of glass-reinforced bone cement, *J. Biomed. Mater. Res.*, **53**(6), 717 (2000)], the following data for the critical intensity factor in these same glass-bead-filled PMMA composites as was used in Problem 5.III.1 are given:

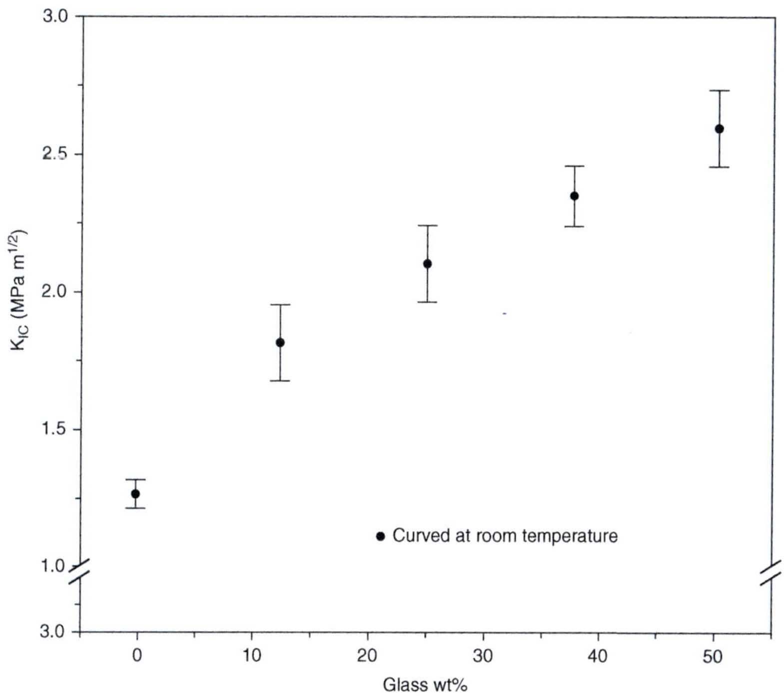

Using the values of the compressive strength from Problem 5.III.1 as representative of the stress at the fracture instability (critical stress for crack propagation), estimates the crack length at the fracture instability, a (in mm), for the GBC material. Clearly indicate which composition you are using in your calculation and any assumptions you must make.

6

Electrical, Magnetic, and Optical Properties of Materials

This chapter is organized much like Chapter 4. We will describe three related properties of materials and subdivide each section into the five materials classes. Just as in Chapter 4, there are some fundamental equations that relate important fluxes to driving forces and in doing so create proportionality constants that are material properties. Again, we will be more concerned with the constants—material properties—than with the driving forces. In fact, we will be even less concerned with the driving forces and fluxes of electricity and magnetism in this chapter than we were with momentum, heat, and mass transport topics since the latter have more to do with materials processing—which we will touch upon in Chapter 7—whereas the former are more the realm of solid-state physics. Both the origins of the driving forces and their resultant fluxes are important to the engineer, however. These are the design parameters that must often be met. The point is that the derivations of many of these fundamental equations will be dispensed with in favor of their applications. The third topic of this chapter is related to the first two, but no fundamental equation of the driving-force form exists. It has to do with how electromagnetic radiation interacts with materials and is a logical extension of the description of electrical and magnetic phenomena.

The first topic is *electrical conductivity*. The driving force for the flow of electrons that leads to electrical conduction is the *electric field*, $\mathcal{E}$, and the resulting flow of electrons is characterized by the *current density*, **J**. The flow and driving force are related through *Ohm's Law:*

$$\mathbf{J} = \sigma \mathcal{E} \tag{6.1}$$

where **J** is the current density in units of amps/m^2, $\mathcal{E}$ is the electric field strength in units of volts/m, and σ is the *electrical conductivity* in units of ohms^{-1}/^{m} or $(\Omega \cdot \text{m})^{-1}$. It is indeed unfortunate that the symbol for conductivity is also the symbol we utilized for the strength of a material, but it is rare that the two are used in the same context, so that confusion should be easily avoided.

The second section will address the topic of *magnetism*. The magnetic driving force is the *magnetic field*, **H**, and is related to a quantity known as the magnetic flux density, or *magnetic induction*, **B**, through the magnetic analog to Ohms Law:

$$\mathbf{B} = \mu \mathbf{H} \tag{6.2}$$

where **B** has units of tesla (Webers/m^2), and **H** has units of amps/m. The proportionality constant, μ, is called the *magnetic permeability*. We will have much more to say about the units of permeability in Section 6.1, since there is some variability in its definition. It is once again unfortunate that the symbol for magnetic permeability is identical to that of another important engineering parameter, Newtonian viscosity, but there is little opportunity for confusion in real applications.

Materials Engineering and Science: Principles, Properties, and Processes, Second Edition. Brian S. Mitchell.

Companion website: www.wiley.com/go/materialsengineeringandscience2e

The third topic has to do with how materials interact with radiation that has both an electrical component and a magnetic component—that is, electromagnetic radiation. This topic is often generically referred as "optical properties" of materials, even though the full range of electromagnetic radiation, of which the optical portion is only a small segment, can be considered. There is no unifying, driving-force-containing equation to describe the optical properties of materials. However, there is one useful relationship that we will use to build most of the descriptions of Section 6.2 upon which relates the intensity of incident electromagnetic radiation, I_0, to the transmitted, absorbed, and reflected intensities, I_T, I_A, and I_R, respectively:

$$I_0 = I_T + I_A + I_R \tag{6.3}$$

By the end of this chapter, you should be able to:

- Relate conductivity, resistivity, resistance, and conductance.
- Calculate conductivity from charge carrier concentration, charge, and mobility.
- Differentiate between a conductor, insulator, semiconductor, and superconductor.
- Differentiate between an intrinsic and an extrinsic semiconductor.
- Differentiate between a *p*- and *n*-type semiconductor.
- Identify different types of electrical polarizability, and determine if they are relevant to a given chemical structure.
- Describe how a *p*–*n* semiconductor junction works.
- Differentiate between diamagnetism, paramagnetism, ferromagnetism, ferrimagnetism, and antiferromagnetism.
- Describe the temperature dependence of magnetic susceptibility.
- Define the Curie temperature for a substance.
- Define the Neél temperature for a substance.
- Determine coercivity and remnant induction from a hysteresis loop.
- Relate transmissivity, absorptivity, and reflectivity.
- Define refractive index.
- Calculate reflectivity from refractive index differences at an interface.
- Describe the interactions of light with materials that result in color.

Need More Help?
Try this video on Ohm's Law: https://www.youtube.com/watch?v=Cztil0re5Eo

6.0 Electrical Properties of Materials

Let us examine Eq. (6.1) more closely to see where the driving force, electrical current, and electrical conductivity come from. The electric field, $\mathcal{E}$, is a vector quantity, as indicated by the boldfaced type. For the time-being, we will limit the description to an electric field in one direction only and will consider only the magnitude of the electric field, $\mathcal{E}$. An electric field is generated by applying a voltage, V, across a distance, l, as illustrated in Figure 6.1. The electric field strength is then given by

$$\mathcal{E} = V/l \tag{6.4}$$

When a conductive material is placed within the electric field, current begins to flow, as characterized by the current density, $\mathbf{J}$, of Eq. (6.1). The current density is also a vector quantity, but since our field is in only one dimension, current will similarly flow only in one direction, so that we will use only the scalar quantity from here on, J. The current density is simply the *current*, I, per unit area in the specimen, A:

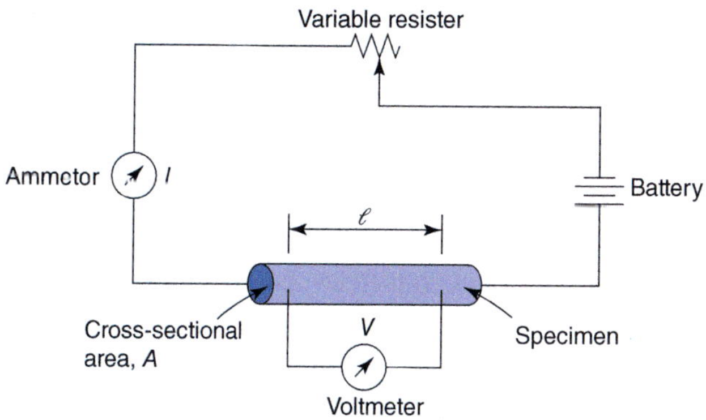

Figure 6.1 Schematic diagram of conductivity measurement in a material. Callister [13]. Copyright © 2000 by John Wiley & Sons, Inc.

$$J = \frac{I}{A} \tag{6.5}$$

where I is in units of ampere (amps, A, for short) and A is in units of m^2. The current, in turn, is defined as the rate of charge passage per unit time:

$$I = \frac{dq}{dt} \tag{6.6}$$

where q is the charge in units of coulombs, so that an ampere is equal to a coulomb/s. Substitution of Eqs. (6.4) and (6.5) into Eq. (6.1) and isolation of V gives

$$V = I\left(\frac{l}{\sigma A}\right) \tag{6.7}$$

We define the quantity in parentheses as the *resistance*, R, to obtain the more familiar form of Ohm's Law:

$$V = I \times R \tag{6.8}$$

Cooperative Learning Exercise 6.1

Consider a wire, 3 mm in diameter and 2 m long. Use the data in Appendix 7 to answer the following questions.

Person 1: What is the resistance in the wire if it is made of copper?

Person 2: What is the resistance in the wire if it is made of stainless steel?

Exchange the results of your calculations, check them, and use them to make the following calculations, assuming that a potential difference of 0.05 V is placed across the ends of the wire.

Person 1: If the wire is made of stainless steel, calculate the current flow, current density, and magnitude of the electric field.

Person 2: If the wire is made of copper, calculate the current flow, current density, and magnitude of the electric field.

Answers: $R_{Cu} = 4.84 \times 10^{-3}\ \Omega$; $R_{SS} = 0.202\ \Omega$. $I_{SS} = 0.25$ A, $J_{SS} = 35{,}000$ A/m^2, $\mathcal{E}_{SS} = 0.025$ V/m; $I_{Cu} = 10.3$ A; $J_{Cu} = 1.5 \times 10^6$ A/m^2, $\mathcal{E}_{Cu} = 0.025$ V/m.

The relationship between conductivity and resistance is an important one, and it can be further generalized through the introduction of two additional quantities called the *conductance*, G, and the *resistivity*, ρ. The resistivity is simply the resistance per unit length times the cross-sectional area, $\rho = RA/l$, and as such is a geometry-independent material property, as is the conductivity, σ. Similarly, the conductance is the conductivity per unit area times length, $G = \sigma l/A$. It is important to remember that the geometry of the sample must be known to calculate its resistance and/or conductance, whereas the resistivity and conductivity are inherent material properties that can vary with temperature and composition but are generally independent of sample geometry. Moreover, it is useful to keep in mind that conductivity and resistivity are inverses of one another, and if we know one, we can find the other.

Conductivity requires a *charge carrier*. There are two types of charge carriers: electrons and ions. The structural descriptions of Chapter 1 will be helpful in determining the primary type of charge carrier within a material, if any. In subsequent sections, we explore the molecular origins of each type of conductivity, investigate the important parameters that cause conductivity to vary in materials, and describe additional electrical conduction phenomena that have revolutionized our lives.

6.0.1 Electrical Properties of Metals and Alloys

Electrical conduction in metals and alloys occurs by the motion of electrons. It can be shown that the conductivity is proportional to the number of electrons per unit volume, n_e, the charge per electron, q_e, and the *electron mobility*, μ_e:

$$\sigma = n_e q_e \mu_e \tag{6.9}$$

where the units of n_e are m^{-3}, q_e is the magnitude of charge on an electron (1.602×10^{-19} coulombs), and μ_e is in $m^2/V \cdot s$. Any of these three quantities can and do affect the overall conductivity of a material, leading to an astounding 27 orders of magnitude in the variation of electrical conductivity values. This immense variation in a physical property causes us to further classify materials according to their ability to conduct electricity. Those materials for which the value of σ is low, ranging from 10^{-20} to 10^{-10} $(\Omega \cdot m)^{-1}$, are termed *insulators*, while those with intermediate values on the order of 10^{-6} to 10^{4} $(\Omega \cdot m)^{-1}$ are called *semiconductors*. Materials with high electrical conductivities above 10^{7} $(\Omega \cdot m)^{-1}$, such as most metals and alloys, are called *conductors*. Notice that the range of conductivities for these classes is still very large and that there are significant gaps between them for which absolute classification is difficult. In fact, a given material can cross over from one classification to another, depending on such factors as temperature. As a result, it is important to understand the origins of electrical conduction in these materials.

Need More Help?
Try these videos on band gap theory: https://www.youtube.com/watch?v=ots5zxbrlUk https://www.youtube.com/watch?v=iBAzZWx7xO8

The Molecular Origins of Electrical Conduction

Recall from Section 1.0.4 that the bonds in metals tend to form bands of electrons. The electrons involved in bonding are found in the valence band, inner electrons not involved in bonding are found in the core band, and the remaining unfilled orbitals of the outer bands form the conduction band. It is the conduction band that gives metals and alloys the ability to freely conduct electrons. As shown in Figure 6.2, the "distance" between the conduction and valence bands in a solid,

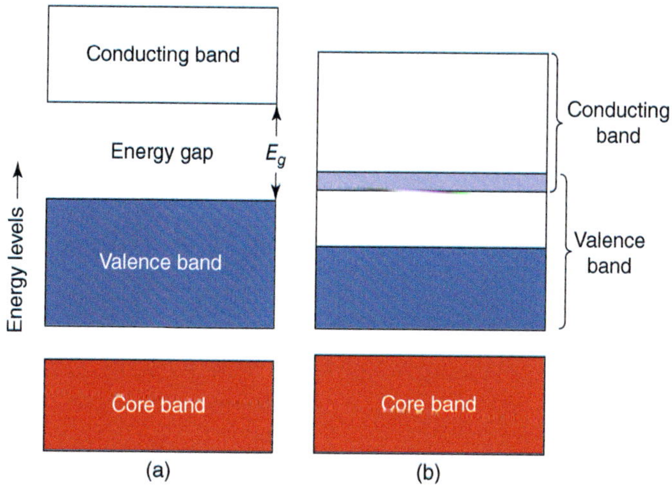

Figure 6.2 Schematic illustration of band gaps in (a) an insulator and (b) a conductor. From Jastrzebski [14]. Copyright © 1976 by John Wiley & Sons, Inc. This material is used by permission of John Wiley & Sons, Inc.

called the *band gap*, E_g, can be used to characterize the electrical behavior of a material. Insulators have large band gaps of several electron volts (eV), whereas conductors have small band gaps, or even overlapping valence and conduction bands, as illustrated in Figure 6.2b. Metallic solids are good electrical conductors because their valence band is not completely filled or because it may overlap with a second empty valence band, thereby causing the total energy band to be only partially filled.

According to the band theory, the motion of electrons under an external electric field is possible only for electrons in partially filled energy bands. The *probability function*, $f(E)$, that describes the probability that an energy level E is occupied by an electron is given by Fermi as

$$f(E) = \frac{1}{\exp\left[(E - E_f)/k_B T\right] + 1} \tag{6.10}$$

where E_f is called the Fermi energy, k_B is Boltzmann's constant, and T is the absolute temperature. The Fermi energy in a metal represents the kinetic energy of the electrons having the maximum energy level that some electrons can attain. At absolute zero, the Fermi–Dirac distribution law requires that the Fermi function, $f(E)$, must have a value of 1 or 0 (see Figure 6.3a). Thus, at $T = 0$ K we have $f(E) = 1$ and $E < E_f$. At temperatures greater than 0 K, $f(E)$ changes from 1 to 0 over an energy range of about one unit of $k_B T$. At $E = E_f$ we obtain $f(E) = \frac{1}{2}$ (see Figure 6.3b). If $E - E_f \gg k_B T$, the exponential term in the denominator of Eq. (6.10) becomes very large relative to unity, and $f(E)$ becomes

$$f(E) = \exp\left[\frac{-(E - E_f)}{k_B T}\right] \tag{6.11}$$

For electron energies, E, below the Fermi level, $E_f - E \gg k_B T$, and Eq. (6.11) can be approximated as

$$f(E) \approx 1 - \exp\left[\frac{E - E_f}{k_B T}\right] \tag{6.12}$$

indicating that the probability of occupation of the energy band is nearly unity.

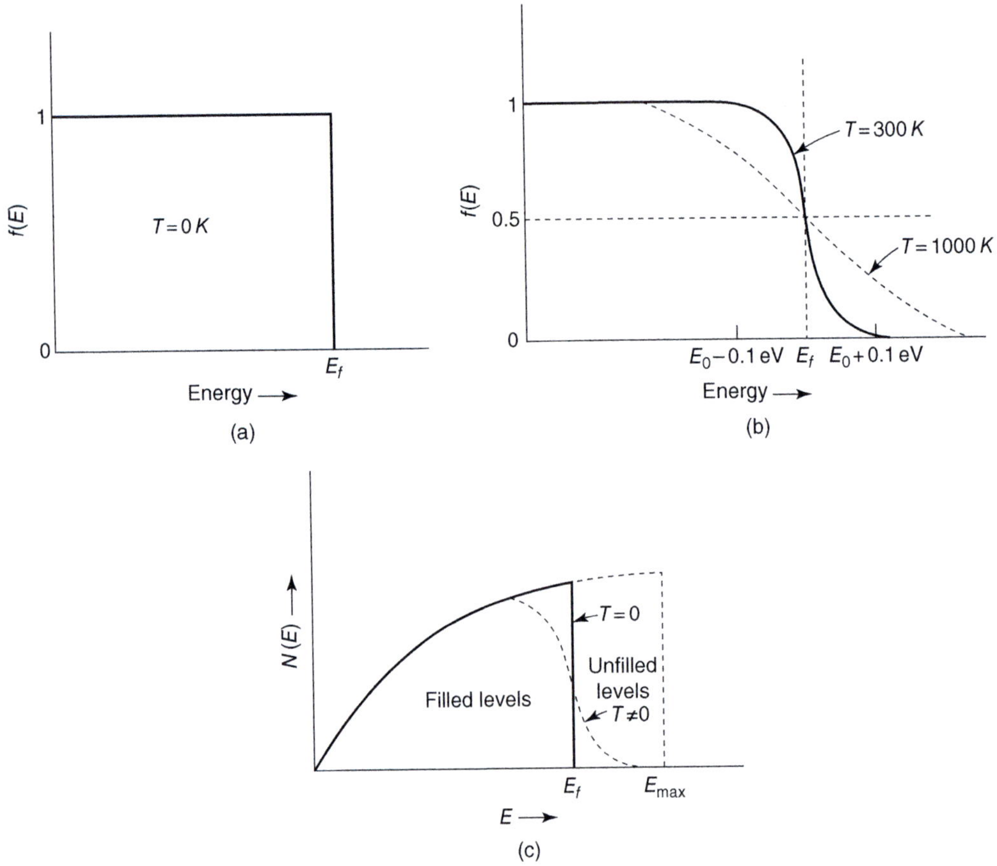

Figure 6.3 The Fermi distribution function (a) at absolute zero and (b) at a finite temperature. (c) The population density of electrons in a metal as a function of energy. From Jastrzebski [14]. Copyright © 1976 by John Wiley & Sons, Inc. This material is used by permission of John Wiley & Sons, Inc.

The state densities are not uniform across the energy band, and their *population density*, $N(E)$, is the greatest in the center of the band. The number of electrons in the band, n_e, can then be evaluated and used in Eq. (6.9) by integrating the product of the density of state $N(E)$ and the probability of their occupation, $f(E)$, over the band energy range (see Figure 6.3c). Thus, for metals,

$$n_e = \int_0^{E_f} N(E) \cdot f(E)\, dE \tag{6.13}$$

The energy distribution in the conduction zone extends over several electron volts and is about 5 eV for some metals at absolute zero. Thus, only a small fraction of the electrons in the energy band can be excited above the Fermi level, and only those within an energy range of the order $k_B T$ can be excited thermally.

In addition to the number of electrons, the other factor in Eq. (6.9) that affects conductivity is electron mobility, μ_e. The mobility is the average charge carrier velocity, or *drift velocity*, $\bar{v}$, divided by the electric field strength, $\mathcal{E}$:

$$\mu_e = \bar{v}/\mathcal{E} \tag{6.14}$$

In a typical metal, $\mu_e = 5 \times 10^{-3}$ m^2/V · s, which gives a drift velocity of 5×10^{-3} m/s for an electric field of 1 V/m. The average electron velocity, in turn, is related to an important structural parameter, the mean free path, l, which we utilized in Section 4.2.1 to describe thermal conductivity:

$$l = \tau \bar{v} \tag{6.15}$$

where τ is a quantity known as the *relaxation time* and has units of s^{-1}. The relaxation time is closely related to the mean time of flight between electron collisions and to the mean free path of the conduction electrons, as indicated above.

As the mobilities are likely to depend on temperature only as a simple power law over an appropriate region, the temperature dependence on conductivity will be dominated by the exponential dependence of the carrier concentration. We will have more to say about carrier mobility in the section on semiconductors.

In summary, metals are good electrical conductors because thermal energy is sufficient to promote electrons above the Fermi level to unoccupied energy levels. At these levels ($E > E_f$), the accessibility of unoccupied levels in adjacent atoms yields high mobility of conduction electrons known as *free electrons* through the solid.

Resistivity in Metals and Alloys

Resistivity is the reciprocal of conductivity. We will find it useful to momentarily concentrate on resistivity instead of conductivity to explain some important electrical phenomena. If the metal crystal lattice were perfect and there were no lattice vibrations, the electrons would pass through the lattice unscattered, encountering no resistance. But, as we saw in the description of thermal conductivity (cf. Section 4.2), lattice vibrations and phonon scattering play a role in disrupting the mean free path of electrons. Working backward through the previous equations, then, shows that a reduction in mean free path, l [Eq. (6.15)], due to phonon scattering and lattice vibrations decreases the electron mobility [Eq. (6.14)], which in turn can reduce electrical conductivity [Eq. (6.9)] or increase resistivity. This effect leads to the temperature-dependent portion of resistivity, ρ_t. Since the thermal lattice vibrations increase directly with increasing temperature, the electrical resistivity of metals will increase with temperature in a linear manner. At high temperatures approaching the melting point, the resistivity usually rises somewhat more rapidly than linearly with temperature. Only at very low temperatures can lattice defects and impurity atoms affect conductivity and resistivity, rather than thermal lattice vibrations. This type of resistivity is known as the *residual resistivity*, ρ_r, and is relatively temperature-independent. The two types of resistivity are additive, such that the total resistivity is given by

$$\rho = \rho_t + \rho_r \tag{6.16}$$

This relationship is shown in Figure 6.4 for a series of copper alloys. The variation of resistivity with temperature is approximately linear over a wide temperature range, and near absolute zero the alloys have a relatively temperature-independent residual resistivity. Notice that the level of nickel impurities increases, so do both the residual and thermal vibration resistivities. This is directly related to a reduction in the electron mean free path.

Often, the variation of resistivity with temperature in the linear region is approximated with an empirical expression of the form

$$\rho_t = \rho_0 + aT \tag{6.17}$$

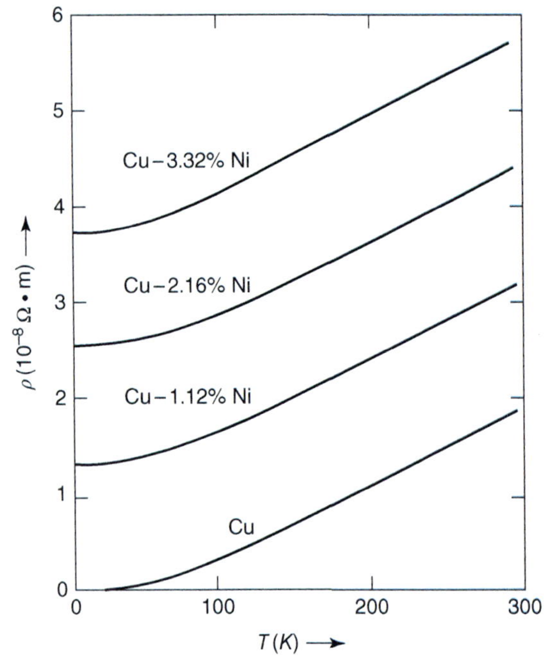

Figure 6.4 Variation of resistivity with temperature for several copper alloys. From Ralls et al. [15]. Copyright © 1976 by John Wiley & Sons, Inc. This material is used by permission John Wiley & Sons, Inc.

Need More Help?
Try this video on resistivity: https://www.youtube.com/watch?v=k01ycafePJg

where ρ_0 and a are experimentally determined constants for a particular metal. Do not confuse the empirical parameter ρ_0 with the residual resistivity, ρ_r—the former is an equation-fitting parameter only for the linear region of the resistivity-temperature plot, whereas the latter is a fundamental property of a material. The resistivities of some metals and alloys are presented in Appendix 7.

The variation of resistivity with composition is also expressed in an empirical fashion. For two-phase alloys consisting of phases α and β, the rule of mixtures can be used to approximate the alloy resistivity from the individual metal resistivities:

$$\rho = V_\alpha \rho_\alpha + (1 - V_\alpha)\rho_\beta \tag{6.18}$$

where V_α is the volume fraction of phase α.

Other factors such as cold deformation and processing conditions can affect resistivity. The cold deformation effect is not as pronounced as the addition of impurity or alloying elements, but the effect of processing on resistivity can be large, as illustrated in Figure 6.5 for a Cu_3Au compound. The effect on resistivity is again related to structure. The rapidly quenched compound maintains its disordered structure, which causes an increase in resistivity relative to the slowly cooled, ordered structure. In the ordered structure, gold atoms occupy the corner positions and copper atoms occupy the face-centered positions of the face-centered cubic (FCC) unit cell, which results in less electron scattering than the disordered structure in which copper and gold atoms occupy FCC atomic sites at random within each unit cell. Only at the order–disorder transition temperature

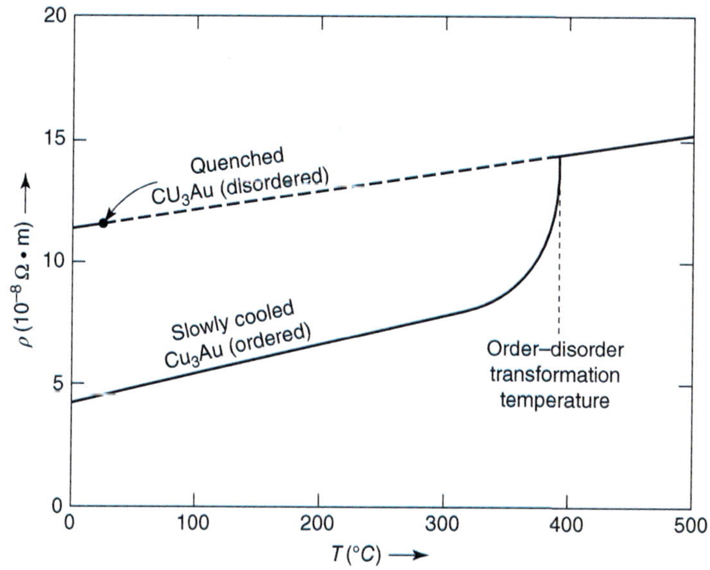

Figure 6.5 Temperature variation of the resistivity for Cu_3Au under two different processing conditions. From Ralls et al. [15]. Copyright © 1976 by John Wiley & Sons, Inc. This material is used by permission John Wiley & Sons, Inc.

do the two resistivity curves become coincident, corresponding to a transformation from the ordered to the disordered FCC structure as a result of increased thermal motion.

For metals in general, any mechanical or chemical action that alters the crystalline perfection will raise the residual resistivity and, therefore, the total resistivity, according to Eq. (6.16). Thus, vacancies in metals, in contrast to those in ionic solids, increase the resistivity. The reason for this lies in the inherent differences between conduction mechanisms in these two classes of materials. The difference between ionic and electronic conduction will be elaborated upon in Section 6.0.2.

Historical Highlight

Read [1], [2] (and hear!) more about the history of superconductors from the American Institute of Physics: https://history.aip.org/exhibits/mod/superconductivity/01.html

Superconduction

You may have noticed in Figure 6.4 that the resistivity of pure copper approaches zero at absolute zero temperature; that is, the residual resistivity is zero. An expanded scale shows that this is not really the case. Figure 6.6 shows that the residual resistivity in pure copper is about 10^{-10} Ω · m. This is the "normal" behavior of many metals. Some metals, however, such as Pb, lose all resistivity abruptly and completely at some low temperature. This phenomenon is called *superconductivity*, and the materials that exhibit it are called *superconductors*. The temperature at which the resistivity vanishes is called the *critical transition temperature*, T_c. For example, lead has a T_c = 7.19 K, and it has a resistivity that is less than 10^{-25} Ω · m at 4.2 K. Many elemental metals (see Figure 6.7), solid–solution alloys, and intermetallic compounds exhibit superconductivity. Often, such materials have relatively high electrical resistivities above T_c, compared to other nonsuperconducting metals such

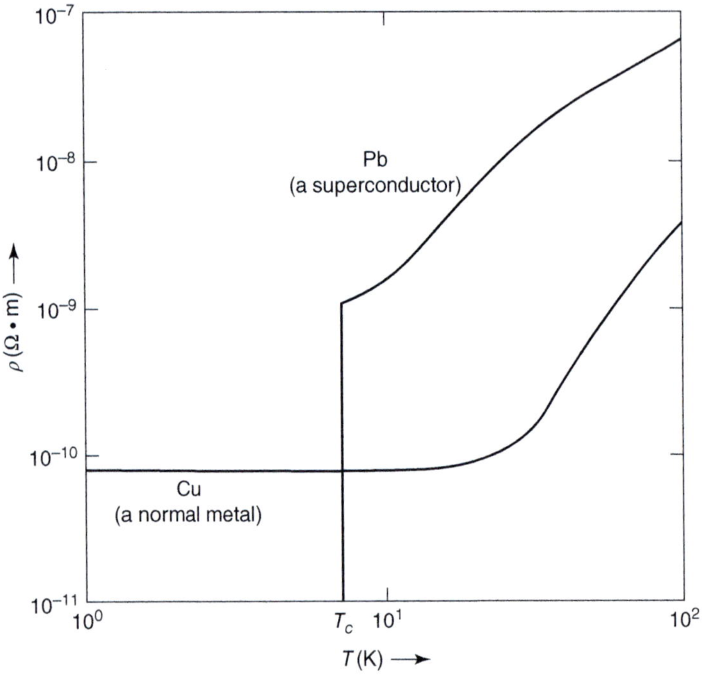

Figure 6.6 Variation of resistivity at low temperatures for a nonsuperconductor and a superconductor. From Ralls et al. [15]. Copyright © 1976 by John Wiley & Sons, Inc. This material is used by permission John Wiley & Sons, Inc.

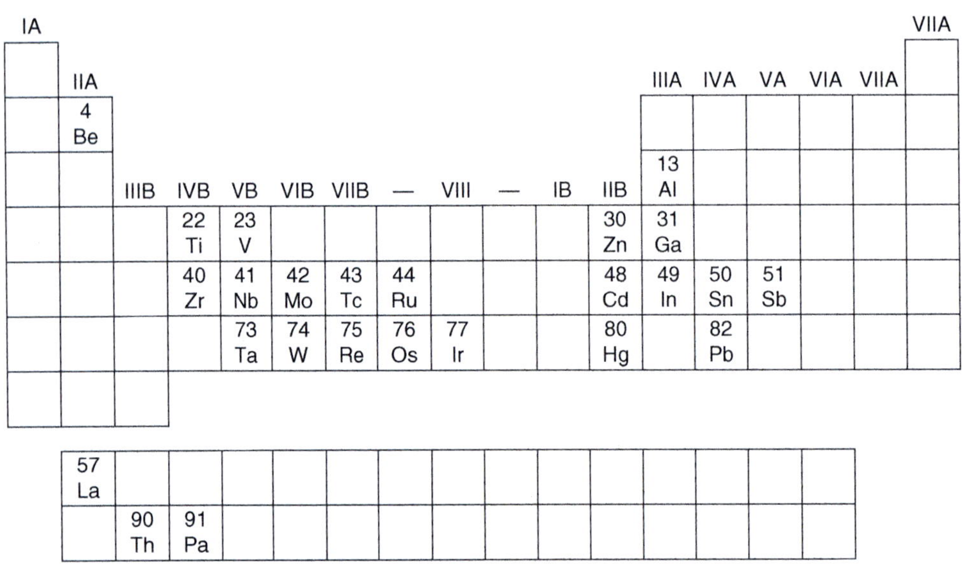

Figure 6.7 The superconducting elements. Amato/Reproduced with permission of Ivan Amato.

as Ag and Cu. Observed T_c values range from a few millidegrees Kelvin to over 21 K for intermetallic compounds such as Nb_3Ge.

Need More Help?

Try this video on superconductivity:

https://www.youtube.com/watch?v=CEUo9VswoVI

In 1957 the research team of Bardeen, Cooper, and Schrieffer produced a theory, now known as the *BCS theory*, that managed to explain all the major properties of superconductivity. The essence of the theory is that superconductivity is caused by electron–lattice interaction and that the superconducting electrons consist of paired ordinary electrons called a *Cooper pair*. We give here a qualitative description of how Cooper pairs form and how they lead to superconductivity.

Figure 6.8a shows the energy–momentum curve of an ordinary conductor with seven electrons in their discrete energy levels. In the absence of an electric field, the current from electrons moving to the right is exactly balanced by the electrons moving to the left, and the net current is zero. When an electric field is applied, all the electrons acquire some extra momentum, and this is equivalent to shifting the whole distribution in the direction of the electric field, as shown in Figure 6.8b. When the electric field is once again removed, the faster electrons will be scattered into lower energy states due to collisions with the vibrating lattice or impurity atoms, and the original distribution in Figure 6.8a is reestablished. For the model of Figure 6.8b, it means that the electron is scattered from energy level a to energy level b in order to reestablish the original distribution.

In the case of a superconductor, it becomes energetically more favorable for the electrons to pair up. Those of opposite momenta pair up to form a new particle called a *superconducting electron* or Cooper pair. This link between two electrons is shown in Figure 6.9a by an imaginary spring. The velocity of the entire center of mass is zero. According to the *de Broglie relationship* ($\lambda = h/p$, where λ is the wavelength, h is Planck's constant, and p is the momentum), this means that the wavelength associated with the new particle, λ, is infinitely long. This is valid for all the Cooper pairs. So, there is a large number of identical particles all with infinite wavelength—that is, a quantum phenomenon on a macroscopic scale. An applied electric field will displace all the particles again as shown in Figure 6.9b, but this time, when the electric field disappears, there is no return to the previous state.

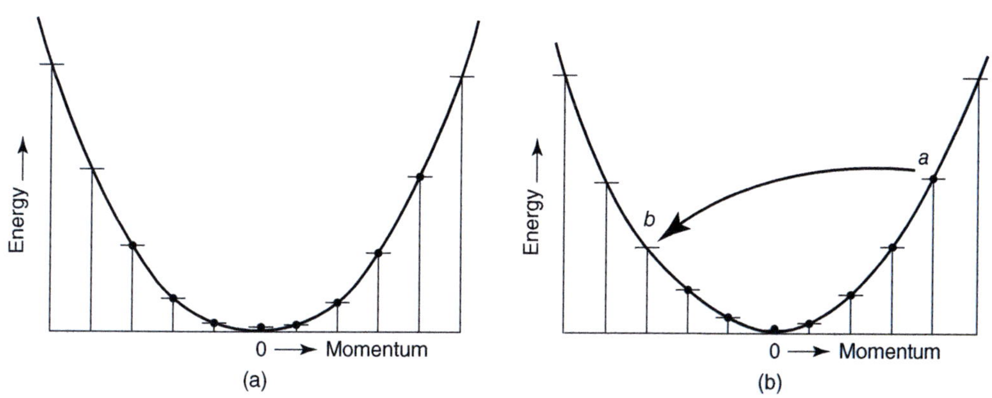

Figure 6.8 One-dimensional schematic representation of the energy–momentum curve for seven electrons in a conductor with (a) no applied electric field and (b) an applied electric field. Solymar and Walsh [16]/ Reproduced with permission of Oxford University Press.

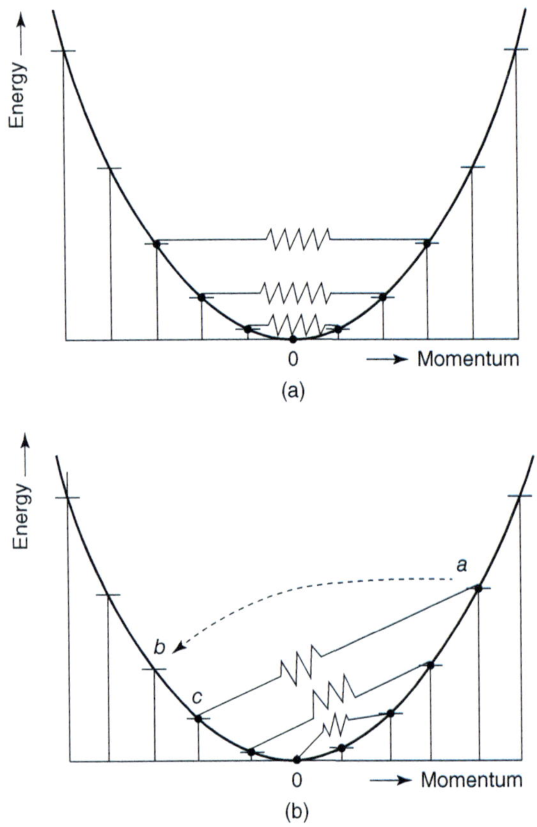

Figure 6.9 The energy–momentum curve for seven electrons in a superconductor with (a) no applied electric field and (b) an applied electric field. Solymar and Walsh [17]/Reproduced with permission of Oxford University Press.

Scattering from energy level *a* to energy level *b* is no longer possible because then the electrons at both *b* and *c* would become pairless, which is energetically unfavorable. So, the asymmetrical distribution remains and there will be more electrons going to the right than to the left, and the current persists, in theory, forever. As the temperature rises, thermal energy eventually overcomes the pairing interaction, and the superconductivity effect vanishes.

One technologically significant application for superconductivity is in the use of electromagnets utilizing a phenomenon known as the *Meissner Effect*. It would be beneficial to have high magnetic fields without any power dissipation. Early experimenters, however, found that above certain magnetic fields, the superconductor became normal, regardless of the temperature. Thus, in order to have zero resistance, a superconductor must be held not only below the critical transition temperature, T_c, but also below a *critical magnetic field*, H_c. The interrelation between H_c and T_c is well described by the formula

$$H_c = H_0\left[1 - \left(\frac{T}{T_c}\right)^2\right] \tag{6.19}$$

where H_0 is the critical magnetic field at absolute zero. This relationship is graphically illustrated in Figure 6.10, where the material is a "normal" conductor above the curve and a superconductor below

Take a Peek! Jump to §6.1.2 to learn more about ***Oxide Superconductors*** and then on to ***The Meissner Effect*** in §6.2.2 to complete the discussion on superconductors.

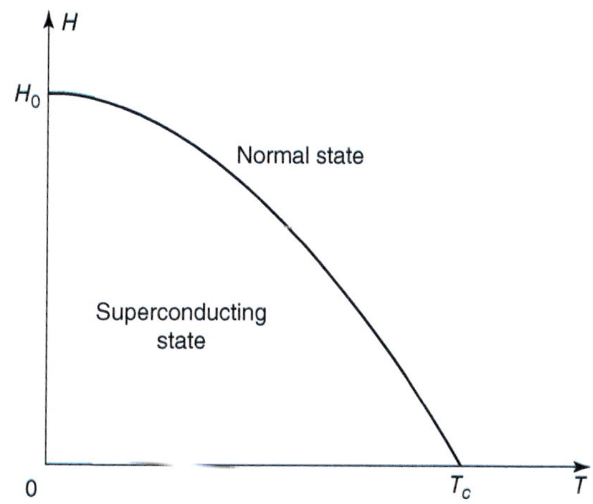

Figure 6.10 Temperature dependence of critical magnetic field for a normal superconductor. Solymar and Walsh [18]/Reproduced with permission of Oxford University Press.

Table 6.1 Critical Temperature and Critical Magnetic Field of Selected Superconducting Elements

Element	T_c (K)	$H_0 \times 10^{-4}$ (A/m)	Element	T_c (K)	$H_0 \times 10^{-4}$ (A/m)
Al	1.175	0.8347	Pb	7.196	6.39
Ga α	1.083	0.464	Sn β	3.722	2.43
Hg α	4.154	3.27	Ta	4.47	6.597
Hg β	3.949	2.70	Th	1.38	1.273
In	3.409	2.24	Ti α	0.40	0.446
Mo	0.915	0.771	Zn	0.857	0.431

Source: Ref. [58].

the curve. The values of H_0 and T_c for a number of elemental superconductors are given in Table 6.1. We will have more to say about superconductors in the next section.

Need More Help?

Try these videos on semiconduction:

https://www.youtube.com/watch?v=60Qz051rD_w
https://www.youtube.com/watch?v=s6rQI7t9XM4

Intrinsic Semiconduction

We continue our description of band gaps in order to elaborate upon those materials with conductivities in the range 10^{-6} to $10^{4}\,(\Omega \cdot \text{m})^{-1}$. At first, it may not seem important to study materials that neither readily conduct electricity nor protect against it, as do insulators. However, the development of semiconducting materials continues to revolutionize our lives.

Recall from Figure 6.2 that the gap between the valence and conduction bands called the band gap, E_g, can be used to classify materials as conductors, insulators, or semiconductors; also recall that for semiconductors the value of E_g is typically on the order of 1–2 eV. The magnitude of the band gap is characteristic of the lattice alone and varies widely for different crystals. In semiconductors, the valence and conduction bands do not overlap as in metals, but there are enough electrons in the valence band that can be "promoted" to the conduction band at a certain temperature to

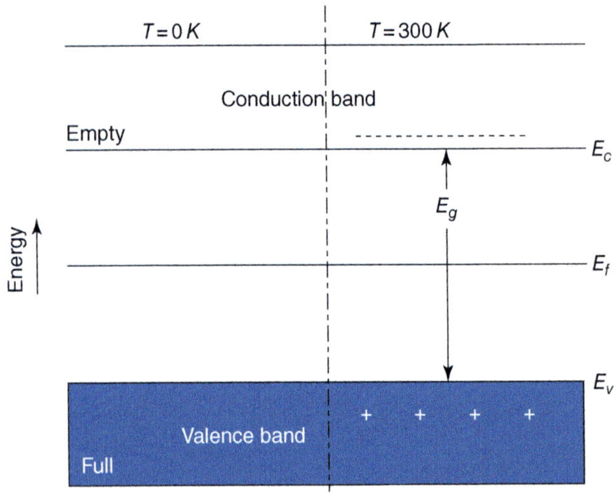

Figure 6.11 Energy bands of an intrinsic semiconductor: E_f is the Fermi energy level; E_c is the lower edge of the conduction band; E_v is the upper edge of the valence band; and E_g is the band gap. From Jastrzebski [14]. Copyright © 1976 by John Wiley & Sons, Inc. This material is used by permission of John Wiley & Sons, Inc.

allow for limited electrical conduction. For example, in silicon, the energies of the valence electrons that bind the crystal together lie in the valence band. All four valence electrons for each atom are tied in place, forming covalent bonds. There are no electrons in the conduction band at absolute zero, and the material is effectively an insulator (see Figure 6.11). At an elevated temperature, enough electrons have been thermally promoted to the conduction band to permit limited conduction. Promotion of electrons leaves behind positively charged "holes" in the valence band which maintain charge neutrality. The term *hole* denotes a mobile vacancy in the electronic valence structure of a semiconductor that is produced by removing one electron from a valence band. Holes, in almost all respects, can be regarded as moving positive charges through the crystal with a charge magnitude and mass the same as that of an electron.

Conduction that arises from thermally or optically excited electrons is called *intrinsic semiconduction*. The conduction of intrinsic semiconductors usually takes place at elevated temperatures, since sufficient thermal agitation is necessary to transfer a reasonable number of electrons from the valence band to the conduction band. The elements that are capable of intrinsic semiconduction are relatively limited and are shown in Figure 6.12. The most important of these are silicon and germanium.

In intrinsic semiconductors, the number of holes equals the number of mobile electrons. The resulting electrical conductivity is the sum of the conductivities of the valence band and conduction band charge carriers, which are holes and electrons, respectively. In this case, the conductivity can be expressed by modifying Eq. (6.9) to account for both charge carriers:

$$\sigma = n_e q_e \mu_e + n_h q_h \mu_h \tag{6.20}$$

where n_h, q_h, and μ_h are now the number, charge, and mobility of the holes, respectively, and contribute to the overall conductivity in addition to the corresponding quantities for the electrons, indicated by the subscript e. The charge of the holes has the same numerical value as that of the electron, and since one hole is created for every electron that is promoted to the conduction band, $n_e = n_h$, such that Eq. (6.20) simplifies to

$$\sigma = n_e q_e (\mu_e + \mu_h) \tag{6.21}$$

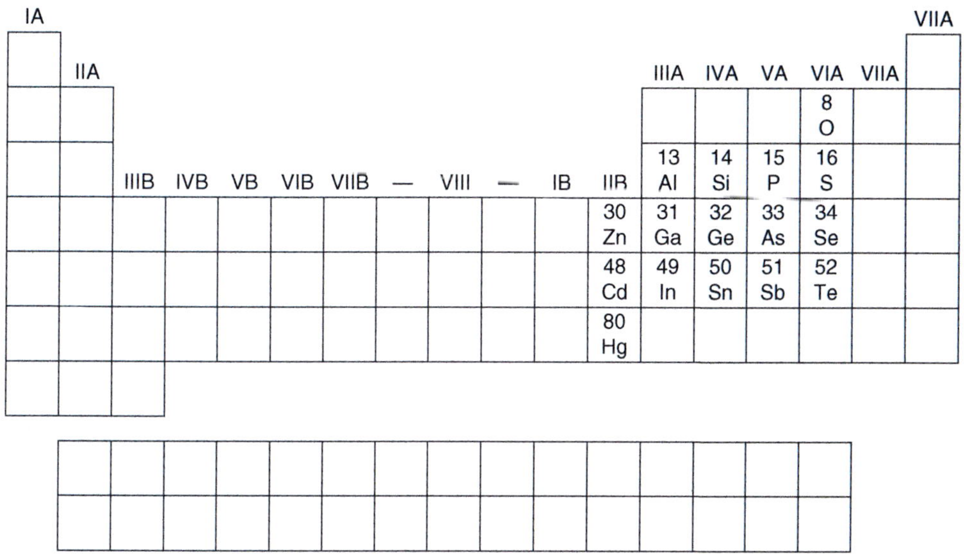

Figure 6.12 The intrinsic semiconducting elements. I. Amato/Reproduced with permission of Ivan Amato.

Table 6.2 Selected Room-Temperature Electronic Properties of Group IV Elements

Element	E_g (eV)	σ (ohm$^{-1}\cdot$m^{-1})	μ_e (m^2/V·s)	μ_h (m^2/V·s)
C (diamond)	7	10^{-14}	0.18	0.14
Silicon	1.21	4.3×10^{-4}	0.14	0.05
Germanium	0.785	2.2	0.39	0.19
Gray tin	0.09	3×10^{5}	0.25	0.24
White tin	0	10^{7}	—	—
Lead	0	5×10^{6}	—	—

Source: Ralls et al. [15]. Copyright © 1976 by John Wiley & Sons, Inc.

The mobilities of holes are always less than those of electrons; that is $\mu_h < \mu_e$. In silicon and germanium, the ratio μ_e/μ_h is approximately three and two, respectively (see Table 6.2). Since the mobilities change only slightly as compared to the change of the charge carrier densities with temperature, the temperature variation of conductivity for an intrinsic semiconductor is similar to that of charge carrier density.

It can be shown from quantum mechanics that the effective density of states for the conduction and valence bands, N_c and N_v, respectively, are

$$N_c = 2\left(\frac{2\pi m_e^* k_B T}{h^2}\right)^{3/2} \tag{6.22}$$

and

$$N_v = 2\left(\frac{2\pi m_h^* k_B T}{h^2}\right)^{3/2} \tag{6.23}$$

where m_e^* and m_h^* are the effective mass of electrons and holes, respectively, k_B is Boltzmann's constant, T is absolute temperature, and h is Planck's constant. If we multiply these densities of state by the distribution of Eq. (6.11), in which we replace the energy, E, with the respective energy of the conduction and valence bands (E_v is lower in energy than the Fermi energy, E_f, and E_c is higher), we can obtain the number of mobile charge carriers in each band:

$$n_e = N_c \exp\left[\frac{-(E_c - E_f)}{k_B T}\right] \tag{6.24}$$

and

$$n_h = N_v \exp\left[\frac{-(E_f - E_v)}{k_B T}\right] \tag{6.25}$$

The product of the mobile positive and negative charge carriers is then obtained by multiplication of Eqs. (6.24) and (6.25), and recalling that $n_e = n_h$ we obtain

$$n_h \cdot n_e = n_e^2 = N_c N_v \exp\left[\frac{-(E_c - E_v)}{k_B T}\right] \tag{6.26}$$

and since $E_c - E_v = E_g$, the band gap energy is

$$n_e = (N_v N_c)^{1/2} \exp\left[\frac{-E_g}{2k_B T}\right] \tag{6.27}$$

Equation (6.27) illustrates that the product of charge carriers, $n_e n_h$, is independent of the position of the Fermi level and depends on the energy gap. It is a further result of this relationship that

$$E_f = \frac{E_c + E_v}{2} \tag{6.28}$$

Thus, the Fermi level is halfway between the conduction and valence bands.

The term $(N_v N_c)^{1/2}$ in Eq. (6.27) depends on the band structure of the semiconductor and is usually a constant for a specific material, outside of its temperature dependence. Thus, we can remove the temperature dependence of N_v and N_c from Eqs. (6.22) and (6.23) and simplify Eq. (6.27) to

$$n_e = CT^{3/2} \exp\left[\frac{-E_g}{2k_B T}\right] \tag{6.29}$$

where C is a constant. Substitution of Eq. (6.29) into Eq. (6.21) gives

$$\sigma = CT^{3/2} q_e(\mu_e + \mu_h) \exp\left[\frac{-E_g}{2k_B T}\right] \tag{6.30}$$

Neglecting the variation of the $T^{3/2}$ term, which is negligible compared to the variation with temperature in the exponential term, and recalling that the mobilities are less sensitive to temperature than are the charge carrier densities, Eq. (6.30) can be rewritten as

$$\sigma = \sigma_0 \exp\left(\frac{-E_g}{2k_B T}\right) \tag{6.31}$$

where σ_0 is the overall constant.

Equation (6.31) indicates that the conductivity of intrinsic semiconductors drops nearly exponentially with increasing temperature. At still higher temperatures, the concentration of thermally excited electrons in the conduction band may become so high that the semiconductor behaves more like a metal.

Quantum Confinement

The fixed band gap model of Figure 6.11 applies to a large assembly of atoms or what is referred to as a *bulk mode material*. We saw in Chapter 1, how the surface to volume ratio of a spherical assembly of atoms increases exponentially as the size of the assembly decreases. This increase in surface atoms relative to internal or bulk atoms also affects the electronic states of the assembly. In short, the electrons and holes become more physically confined as the assembly decreases in size, eventually approaching a quantum mechanical limit called the *exciton Bohr radius*, a_B. An exciton is an electron–hole bound state that results from Coulombic attraction as described in Chapter 1 (Figure 6.13a). The exciton Bohr radius is related to the dielectric constant or relative permittivity of the semiconductor, ε_r, which we will learn about in Section 6.0.2:

$$a_B = \frac{4\pi\varepsilon_r\hbar^2}{m_{eh}^* e^2} \tag{6.32}$$

where $\hbar$ is the reduced Planck's constant (h divided by 2π), m_{eh}^* is the reduced mass of the exciton $(1/m_e^* + 1/m_h^*)^{-1}$, and e is the elementary charge.

The quantum mechanical details of quantum confinement are beyond the scope of this text, but in qualitative terms the band gap becomes discrete below the exciton Bohr radius, meaning the band gap energy is assembly size dependent (Figure 6.13b). The exciton Bohr radius is a property of an atom, so the assembly size below which the band gap becomes variable depends on the element or compound. In practical terms, it means that small assemblies of atoms that are confined in three dimensions—known as *quantum dots* because of the resulting zero-dimensional geometry—have properties that can be very different from their properties in the bulk state. One direct artifact of the discrete band gaps in quantum dots is that the emission of a photon upon relaxation of the electron from the excited state can now take on multiple values depending upon which levels it transitions from and to. This is called a *tunable bandgap* and can be manipulated to produce multiple emission colors from a single quantum dot depending upon the energy input to the system. We will see in Section 6.2.3, how controlled emissions in the visible spectrum from a variety of materials

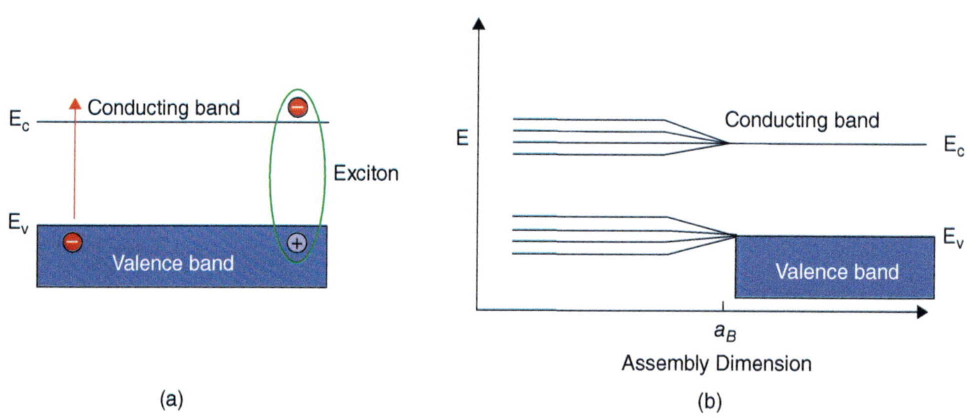

Figure 6.13 Schematic illustration of (a) an exciton (right) formed from the promotion of an electron from the valence to the conducting band (left) and (b) discrete band gaps transition as assembly size falls below the exciton Bohr radius.

Need More Help?
Try these videos on quantum dots: https://www.youtube.com/watch?v=Sma3v_1xv-k https://www.youtube.com/watch?v=fqSUDLTjaCA

can be used for televisions, monitors, and displays. Finally, in addition to quantum confinement in a zero dimension quantum dot, quantum confinement can occur in one and two-dimensional structures as well, leading to *quantum wires* and *quantum tubes*, respectively. Quantum confinement occurs not only in elements but in compounds. In fact, many of the technologically useful quantum confinement materials are made from compound semiconductors which are described in Section 6.0.2.

Extrinsic Semiconduction

The charge carrier density that leads to intrinsic semiconduction can also be increased through impurities of either higher or lower valence. For example, if pentavalent substitutional atoms such as P, As, or Sb are placed into a covalently bonded tetravalent material such as Si or Ge, in a process known as *doping*, only four of their five valence electrons are required to participate in covalent bonding. Since the fifth electron remains weakly bound to the impurity or *donor* atom, it is not entirely free in the crystal. Nevertheless, the binding energy, which is of the order 0.01 eV, is much less than that of a covalently bonded electron. This "extra" electron can be easily detached from the impurity or donor atom. The energy state of this electron is indicated by E_d, since it is the donor level, and the energy required to excite it to the conduction band can be represented by the band model shown in Figure 6.14. Notice that the energy gap between the donor and conducting bands is much smaller than that between the valence and conduction bands, or the normal band gap, E_g. Thermal agitation, even at room temperature, is sufficient to transfer this electron to the conduction band, leaving behind a positively charged hole in the donor band. The conductivity in this case is due to motion of electrons in the conduction band, and this type of material is called an *n-type semiconductor*. This is one type of *extrinsic semiconductor*, or *impurity semiconductor*, in which an insulating or intrinsic semiconducting material is made to increase its semiconduction

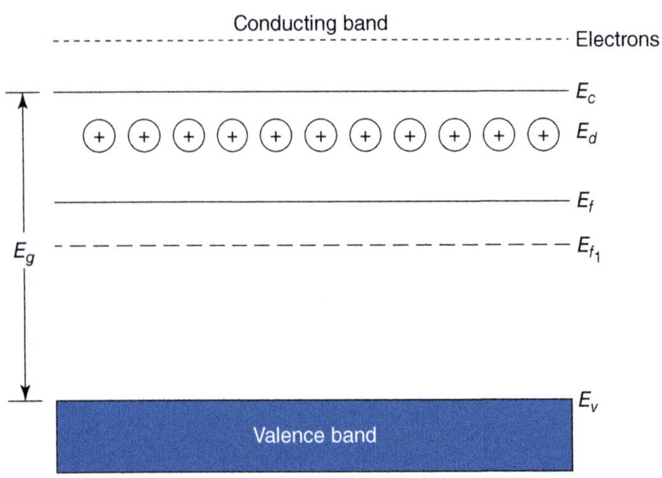

Figure 6.14 Schematic illustration of energy bands in an *n*-type, extrinsic semiconductor. From Jastrzebski [14]. Copyright © 1976 by John Wiley & Sons, Inc. This material is used by permission of John Wiley & Sons, Inc.

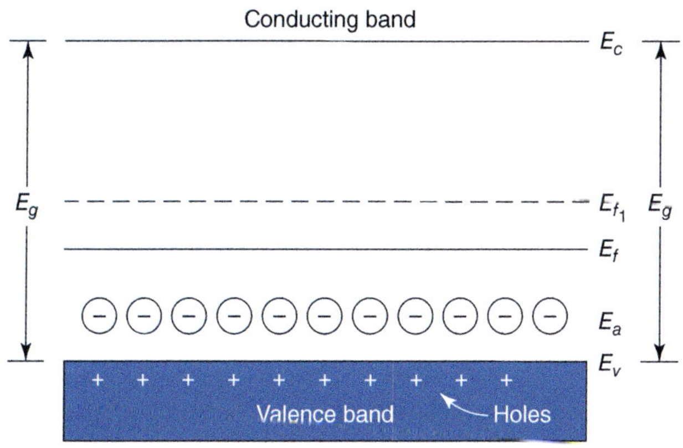

Figure 6.15 Schematic illustration of energy bands in a *p*-type, extrinsic semiconductor. From Jastrzebski [14]. Copyright © 1976 by John Wiley & Sons, Inc. This material is used by permission of John Wiley & Sons, Inc.

capabilities through the addition of impurity atoms. Notice also, in Figure 6.14, that the Fermi level of the *n*-type extrinsic semiconductor, E_f, is higher than that of the undoped, intrinsic semiconductor, E_{f1}.

The other possibility is to substitute with an impurity of lower valence, such as a trivalent element in a tetravalent lattice. Examples include the addition of boron, aluminum, gallium, indium, and thallium to silicon or germanium. In this case, a *p-type*, extrinsic semiconductor is created. Since the substitutional atoms are deficient in bonding electrons, one of their bonding orbitals will contain a hole that is capable of "accepting" an electron from elsewhere in the crystal. As shown in Figure 6.15, the binding energy is small, and the promotion of an electron from the valence band to the acceptor band, E_a, leaves a hole in the valence band that can act as a charge carrier. In summary, then, for an *n*-type semiconductor, electrons in the conduction band are charge carriers and holes participate in bonding, whereas for a *p*-type semiconductor, the electrons participate in bonding, and the holes are the charge carriers.

In both types of extrinsic semiconductors, the doping elements are chosen so that both acceptor and donor levels are located closer to the corresponding energy bands, and only a small energy gap is involved when exciting electrons. Intrinsic semiconductors are prepared routinely with initial total impurity contents less than one part per million (ppm), along with unwanted donor or acceptor atoms at less than one part in 10^7. To such highly purified materials, controlled amounts of substitutional donor or acceptor impurities are intentionally added on the order of 1–1000 ppm to produce extrinsic semiconductors having specific room temperature conductivities.

Unlike intrinsic semiconductors, in which the conductivity is dominated by the exponential temperature and band-gap expression of Eq. (6.31), the conductivity of extrinsic semiconductors is governed by competing forces: charge carrier density and charge carrier mobility. At low temperatures, the number of charge carriers initially increases with temperature, as shown in Figure 6.16 for an *n*-type semiconductor, because thermal activation promotes donor electrons to the conduction band. At intermediate temperatures, most of the donor electrons have been promoted and the number of charge carriers is nearly independent of temperature in what is known as the *exhaustion range*. At higher temperatures, the number of valence (bonding) electrons excited to the conduction band greatly exceeds the total number of electrons from substitutional donor atoms, and the extrinsic semiconductor behaves like an intrinsic semiconductor (dashed line in Figure 6.16).

Cooperative Learning Exercise 6.2

The following electrical conductivity characteristics for both the intrinsic and extrinsic forms of a semiconductor have been determined at room temperature. Recall that $q_e = q_h = 1.602 \times 10^{-19}$ coulombs.

	σ $(\Omega \cdot \mathrm{m})^{-1}$	n_e (m^{-3})	n_h (m^{-3})
Intrinsic	2.5×10^{-6}	3.0×10^{13}	3.0×10^{13}
Extrinsic	3.6×10^{-5}	4.5×10^{14}	2.0×10^{12}

Person 1: Use the data to write a relationship for the intrinsic conductivity of this substance as a function of the electron and hole mobilities.

Person 2: Use the data to write a relationship for the extrinsic conductivity of this substance as a function of the electron and hole mobilities.

Assuming that the mobilities are the same in both the intrinsic and extrinsic states, combine your information to solve for the electron and hole mobilities of this substance. Can you tell whether the substance is an n-type of p-type extrinsic semiconductor by looking at the data?

Adapted from Callister, p. 654, Problem 19.33.

Answers: Equation (6.20) can be used for both; $0.52 = \mu_e + \mu_h$; $112.4 = 225\mu_e + \mu_h$. Solve simultaneously to get $\mu_e = 0.50\ \mathrm{m^2/V\text{-}s}$; $\mu_h = 0.02\ \mathrm{m^2/V\text{-}s}$. This is an n-type semiconductor, since $n_e >> n_h$.

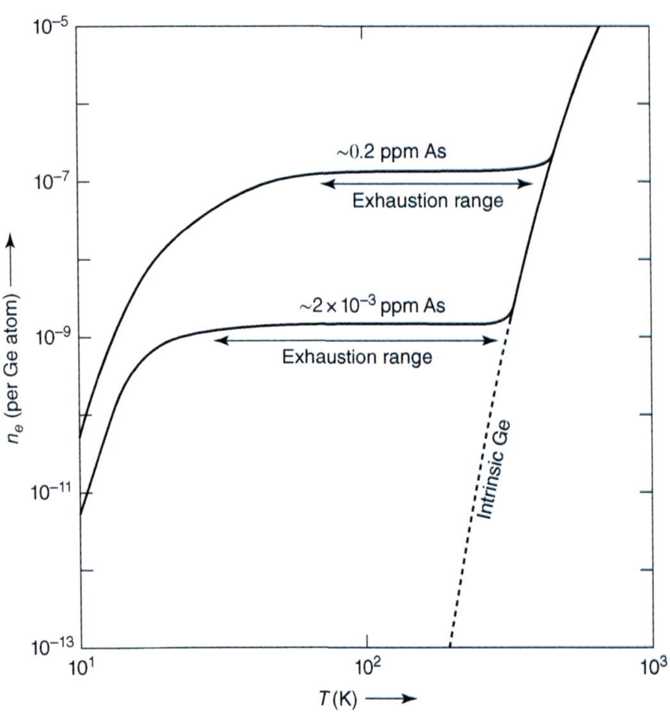

Figure 6.16 Temperature dependence of charge carrier concentration of n-type extrinsic germanium with two different As impurity levels. From Ralls et al. [15]. Copyright © 1976 by John Wiley & Sons, Inc. This material is used by permission John Wiley & Sons, Inc.

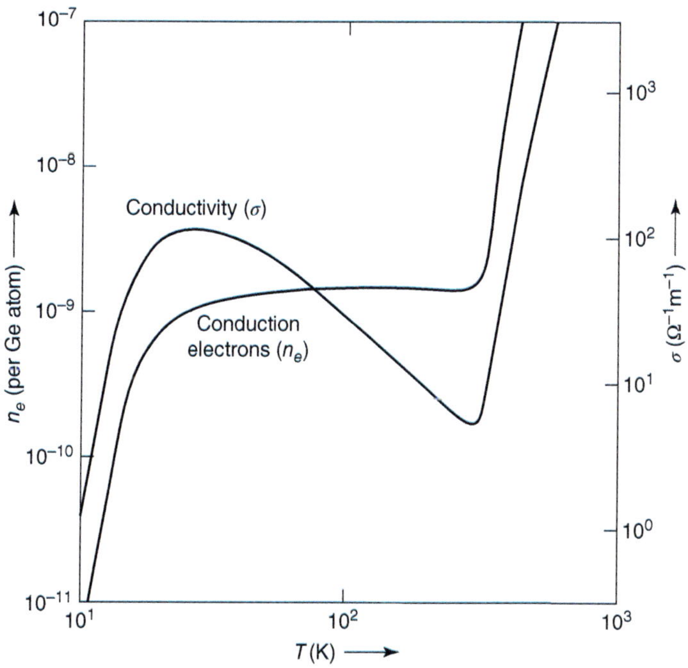

Figure 6.17 Temperature dependence of charge carrier density and conductivity of extrinsic semiconductor Ge doped with 2 ppb As. From Ralls et al. [15]. Copyright © 1976 by John Wiley & Sons, Inc. This material is used by permission John Wiley & Sons, Inc.

Based on charge carrier density (conduction electrons) alone, then, the conductivity of an extrinsic semiconductor should vary with temperature as shown in Figure 6.17. However, charge carrier mobility has a temperature-dependent effect. As mentioned in the previous section, the temperature dependence of carrier mobility has little effect on intrinsic conductivity. However, when extrinsic conductivity predominates and donor levels are close to the conduction band, the carrier mobility can determine the temperature dependence of conductivity. This occurs in the exhaustion range where mobility typically decreases with increasing temperature because of increased phonon scattering. Thus, in this region, characterized by an approximately constant number of charge carriers and a decreasing carrier mobility with temperature, the temperature dependence of conductivity of an extrinsic semiconductor is similar to that of a pure metal (cf. resistivity plot for pure copper in Figure 6.4). In summary, at low temperatures, the conductivity varies with temperature as the charge carrier concentration, at higher temperatures, the charge carrier mobility dominates and the conductivity decreases with temperature, and at still higher temperatures, the extrinsic semiconductor behaves as an intrinsic semiconductor. This behavior is summarized in Figure 6.17.

Semiconductor Junctions and Devices*

Semiconductor crystals can be made in such a way that an n-type region can be made adjacent to a p-type region. This is normally accomplished by selective doping during the growth of a single crystal. The boundary between such regions within a single crystal is called the *p–n junction*, and it has important applications in the development of transistors, rectifiers, and solar energy cells. Let us examine some of these devices.

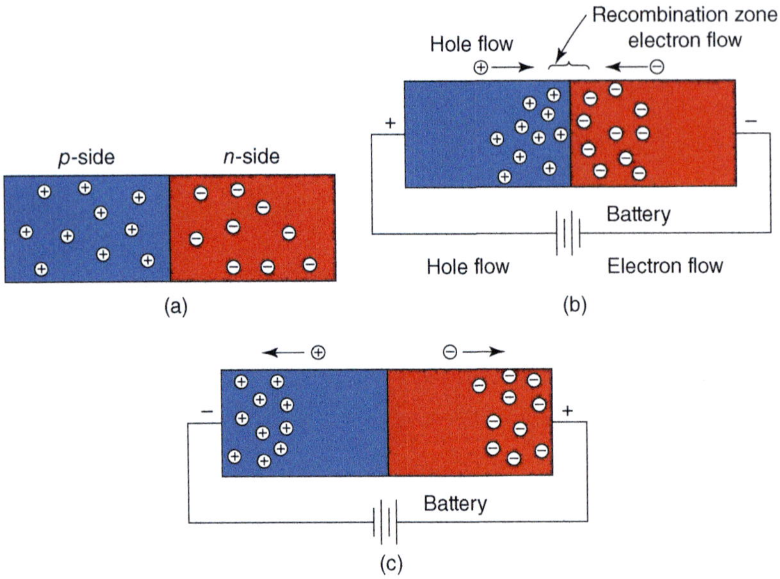

Figure 6.18 Schematic illustration of electron and hole distribution for a *p*–*n*-rectifying junction for (a) no applied electrical potential, (b) forward bias, and (c) reverse bias. Callister [19]. Copyright © 2000 by John Wiley & Sons, Inc.

The *rectifier*, or *diode*, is an electronic device that allows current to flow in only one direction. There is low resistance to current flow in one direction, called the *forward bias*, and a high resistance to current flow in the opposite direction, known as the *reverse bias*. The operation of a *p*–*n* rectifying junction is shown in Figure 6.18. If initially there is no electric field across the junction, no net current flows across the junction under thermal equilibrium conditions (Figure 6.18a). Holes are the dominant carriers on the *p*-side, and electrons predominate on the *n*-side. This is a dynamic equilibrium: holes and conduction electrons are being formed due to thermal agitation. When a hole and an electron meet at the interface, they recombine with the simultaneous emission of radiation photons. This causes a small flow of holes from the *p*-region to the *n*-region and a flow of electrons from the *n*-region to the *p*-region. New holes and electrons continue to form to balance this flow, and an equilibrium is established with no external current flowing through the system.

If the *n*-region is connected to the negative terminal and the *p*-region to the positive terminal of a battery (Figure 6.18b), holes will move from left to right while electrons will move from right to left. These meet each other and recombine as before in the *recombination zone*. This produces an abundant number of current carriers flowing across the junction; hence, the resistance in the forward direction is low. Only a small voltage is necessary to maintain a large current flow.

For reverse bias (Figure 6.18c), holes are attracted to the negative terminal, and electrons are attracted to the positive terminal. Thus, electrons and holes move away from each other and the junction region will be relatively free from charge carriers, in which case it is called the *barrier region*. Recombination will not occur to any appreciable extent, so the junction is now highly insulative. There may be, however, a small flow of current as the result of the thermal generation of holes and electrons in the semiconductor, termed *leakage current*. The current–voltage behavior of the *p*–*n* junction under both forward and reverse biases is shown in Figure 6.19. If the voltage is allowed to vary sinusoidally (Figure 6.20a), as in an alternating current, the current also varies

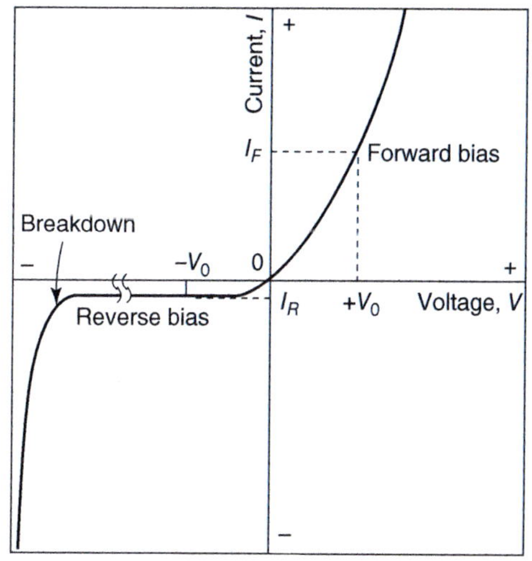

Figure 6.19 The current–voltage characteristics of a *p–n* junction. For forward and reverse biases. Callister [20]/Reproduced with permission of John Wiley & Sons.

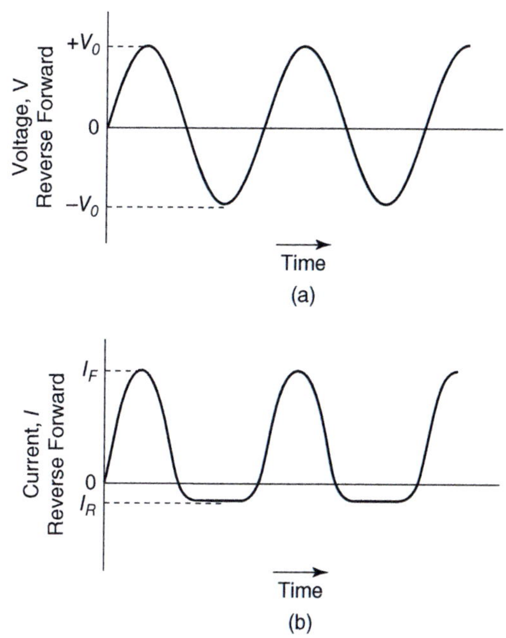

Figure 6.20 (a) Voltage–time and (b) current–time for a *p–n*-rectifying junction. Callister [20]/ Reproduced with permission of John Wiley & Sons.

(Figure 6.20b), but the maximum current for the reverse bias, I_R, is extremely small compared to that for the forward bias, I_F.

The *p–n* junction can be fabricated so that the leakage current in the reverse bias is extremely small, and potentials of up to several hundred volts can be applied before *breakdown* will occur. Breakdown occurs at high reverse voltages (over several hundred volts), when large numbers of charge carriers are generated, giving rise to an abrupt increase in current. Atoms may actually be ionized during breakdown due to the acceleration of electrons and holes in a phenomenon known as the *avalanche effect*. Some devices, called *Zener diodes*, are designed to utilize breakdown behavior as voltage regulators. The voltage breakdown can be varied by controlling the dopant level; generally, the higher the impurity concentration, the lower the breakdown voltage. In silicon

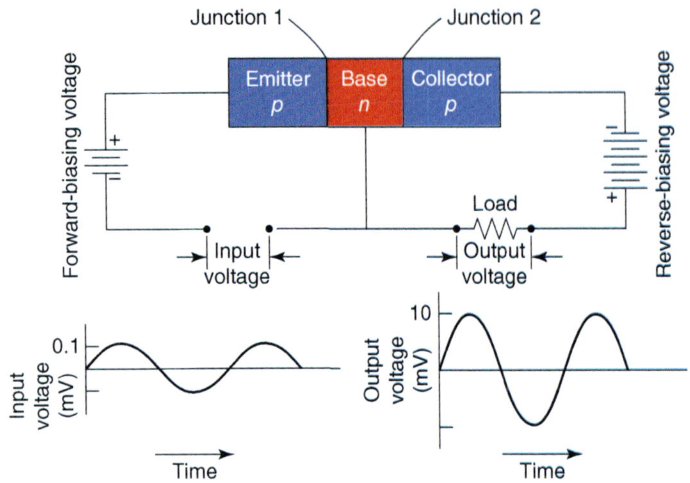

Figure 6.21 Schematic diagram of a *p*–*n*–*p* junction transistor and associated input and output voltage characteristics. Callister [21]/Reproduced with permission of John Wiley & Sons.

semiconductors, for example, the breakdown voltage can be varied from about 10 to about 1000 V by regulating the impurity concentration.

Another type of semiconductor junction device is a transistor, which is capable of two primary types of functions. The first is to amplify an electrical signal. The second is to serve as a switching device for the processing and storage of information. We describe the first type here and reserve the description of the second type—metal-oxide-semiconductor field-effect transistors (MOSFETs)—to the section on ceramics since these are the primary materials of construction.

The junction transistor is composed of two *p*–*n* junctions arranged back-to-back in either the *p*–*n*–*p* or *n*–*p*–*n* configurations. The former configuration is shown in Figure 6.21. It is composed of a thin *n*-type *base* region, sandwiched between a *p*-type *emitter* and a *p*-type *collector*. The first *p*–*n* junction is forward-biased, and the second *n*–*p* junction is reverse-biased. Since the emitter is *p*-type and Junction 1 is forward-biased, a large number of holes enter the *n*-type base region. These holes combine with electrons in the base. However, in a properly prepared junction transistor, the base is thin enough to allow most of the holes to pass through the base without recombination and into the *p*-type collector. A small increase in input voltage within the emitter–base circuit produces a large increase in current across Junction 2 due to the passage of holes. The large increase in collector current also results in a large increase in voltage across the load resistor. Thus, a voltage signal that passes through a junction transistor experiences *amplification*, as illustrated by the input and output voltage plots in Figure 6.21. A similar effect is observed in an *n*–*p*–*n* junction transistor, except that electrons instead of holes are injected across the base to the collector.

In addition to their ability to amplify electrical signals, transistors, and diodes may act as switching and information storage devices. Computer information is stored in terms of binary code—for example, zeros and ones. Transistors and diodes can operate as switches that have two states: on and off, or conducting and nonconducting. Thus, a conducting circuit can represent a zero while a nonconducting circuit can represent a one. In this way, information and logical operations can be stored and retrieved from a series of semiconducting circuits known as *integrated circuits*. Inexpensive integrated circuits are mass produced by forming layers of high-purity silicon in a precisely detailed pattern and introducing specific dopants into specific areas by diffusion or ion implantation. The chemical vapor deposition (CVD) processes described in Section 3.4.1 are applied in *semiconductor device fabrication*, which is described in more detail in ***Advanced Topics***.

6.0.2 Electrical Properties of Ceramics and Glasses

Many of the fundamental relationships and concepts governing the electrical properties of materials have been introduced in the previous section. In this section, we elaborate upon those topics that are more prevalent or technologically relevant in ceramics and glasses than in metals, such as electrical insulation and superconductivity, and introduce some topics that were omitted in Section 6.0.1, such as dielectric properties.

Electrical Resistance

At room temperature, many ceramics (and polymers, for that matter) have electrical resistivities that are approximately 20 orders of magnitude higher than those of metals (see Figure 6.22). As a result, they are often used as electrical insulators. With rising temperatures, however, insulating materials experience an increase in electrical conductivity, which may ultimately be greater than that for semiconductors. In fact, some oxide ceramics such as ReO_3 can achieve metallic levels of electrical conductivity at high temperatures. In all cases, the electrical conductivity can be interpreted in terms of carrier concentrations and carrier mobilities as previously described, though we shall see that the carriers may be different for ceramics and glasses than for metals.

Despite the potential for high electrical conductivities, we will concentrate on the more common high-resistivity aspects of ceramics in this section and will describe some of the factors that affect the use of ceramics and glasses as electrical insulators. For ceramics and glasses having electrical resistivities above about $10^5\ \Omega \cdot \text{cm}$, the concentration–mobility product of the charge carriers is small; as a result, minor variations in composition, impurities, heat treatment, stoichiometry, porosity, oxygen partial pressure, and other variables can have a significant effect on the resistivity. The effect of porosity is like that described for thermal conductivity in Section 4.2. As porosity increases, the electrical resistivity increases proportionally for small values of porosity. At higher porosities, the effect is more substantial. The effect of grain boundaries in polycrystalline materials is related to the mean free path of the ions. Except for very thin films or extremely fine-grained materials (less than 0.1 μm), the effects of grain boundary scattering are small compared to lattice

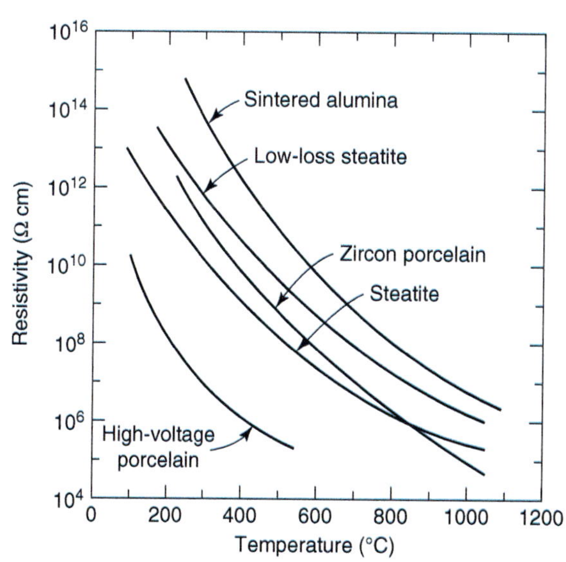

Figure 6.22 Temperature dependence of resistivity for some typical oxide ceramics. From Kingery et al. [22]. Copyright © 1976 by John Wiley & Sons, Inc. This material is used by permission of John Wiley & Sons, Inc.

scattering. Consequently, the grain size in uniform-composition materials has little effect on resistivity. However, substantial effects can result from impurity concentrations at grain boundaries. Particularly for oxide materials, there is a tendency to form a glassy silicate phase at the boundary between particles. Ceramic compositions with the highest glassy phase have lower resistivities at moderate temperatures. In glasses, the alkali content should be kept to a minimum when high resistivities are required. The reasons for this are related to the electrical conduction via ions. For now, we maintain our emphasis on the electrical insulating properties of ceramics and glasses in order to describe a unique application known as dielectrics.

Dielectrics, Ferroelectricity, and Piezoelectricity

A *dielectric* is a material separating two charged bodies, as illustrated in Figure 6.23b. For a substance to be a good dielectric, it must be an electrical insulator. As a result, any electrical insulator is also called a dielectric.

The interaction of a dielectric with an electric field is conveniently described in terms of a parallel plate capacitor. When a voltage is applied between two parallel plates in a vacuum (Figure 6.23a), one plate becomes positively charged and the other becomes negatively charged. If an electrical conductor were to be placed in contact with the plates, current would flow. In a vacuum, however,

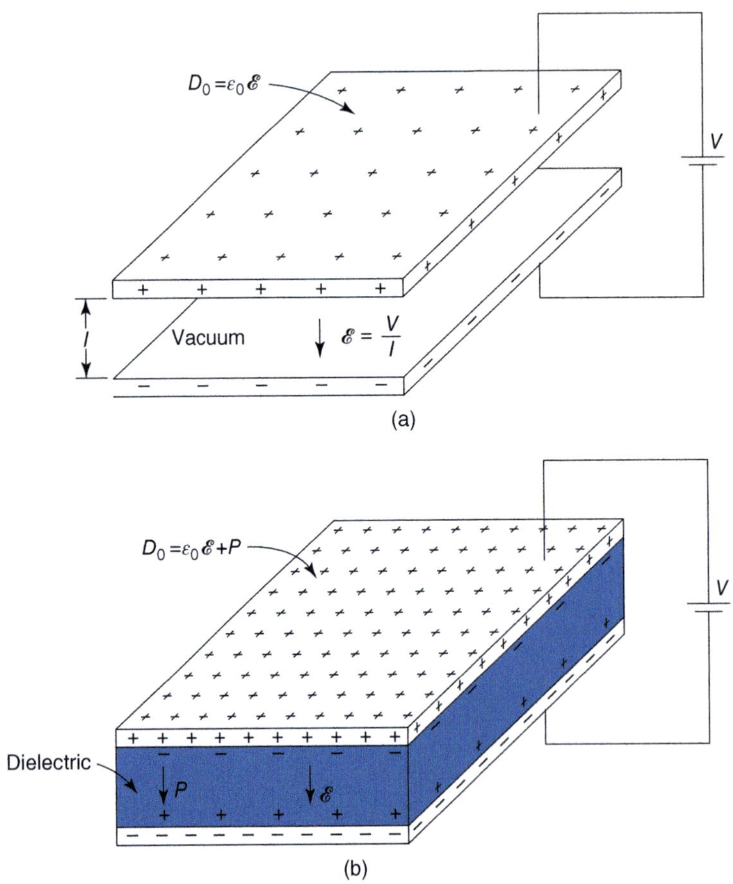

Figure 6.23 A parallel-plate capacitor (a) in a vacuum and (b) with a dielectric material between the plates. Callister [23]/Reproduced with permission of John Wiley & Sons.

no current flows, and the electric field strength is a function of the applied voltage, V, and the separation distance between the plates, l, as given earlier by Eq. (6.4):

$$\mathcal{E} = V/l \tag{6.4}$$

The magnitude of the charge per unit area on either plate called the *electric displacement or flux density*, $\mathbf{D}_0$, is directly proportional to the field:

$$\mathbf{D}_0 = \varepsilon_0 \mathcal{E} \tag{6.33}$$

where ε_0 is a constant called *electric permittivity* and has a value of 8.854×10^{-12} C/V · m (or farad/m, F/m, where a farad equals a coulomb/volt). The electric displacement, therefore, has units of C/m^2. Notice the similarity in form of Eq. (6.33) to Eq. (6.2) for the magnetic flux in a magnetic field. As before, both $\mathbf{D}_0$ and $\mathcal{E}$ are vector quantities, but for our purposes a description of their magnitudes is sufficient, so the boldface will be dropped.

Since D_0 is the surface charge density on the plate in vacuum, it can be related to the capacitance of a parallel plate capacitor in vacuum, C_0, which is defined as

$$C_0 = q/V \tag{6.34}$$

where q is the magnitude of charge on each plate and V is the applied voltage. With A the plate area and l the separation between plates, along with $q = D_0A$, Eq. (6.34) becomes

$$C_0 = \varepsilon_0 \frac{A}{l} \tag{6.35}$$

when a vacuum exists between the plates.

If we now place a dielectric (insulator) material between the plates, as in Figure 6.23b, a displacement of charge within the material is created through a progressive orientation of permanent or induced dipoles. Recall that a dipole is simply any type of charge separation, as first described in Section 1.0.3 and illustrated in Figure 6.24, and that it may be permanent or temporarily induced. The magnitude of the electric dipole, p_e, is simply proportional to the charge, q, and the separation distance, a, as $p_e = qa$.

Need More Help?

Try these videos on dielectrics and dielectric constant:

https://www.youtube.com/watch?v=NTASxHFYSP8
https://www.youtube.com/watch?v=7Fmqo90tts8

The interaction between permanent or induced electric dipoles with an applied electric field is called *polarization*, which is the induced dipole moment per unit volume. Polarization causes positive charge to accumulate on the bottom surface next to the negatively charged plate and causes negative charge to accumulate toward the positively charged plate on the top. This tends to decrease the effective surface charge density on either plate. The expected decrease in effective charge corresponds to the polarization, P, of the material, which equals the induced dipole moment per unit volume of polarizable material. The electric field is now found in Eq. (6.33) by accounting for the reduction in displacement due to polarization:

$$\mathcal{E} = (D - P)/\varepsilon_0 \tag{6.36}$$

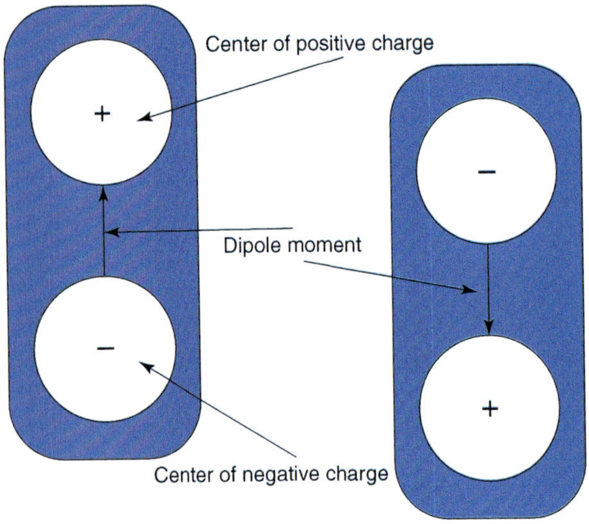

Figure 6.24 Schematic illustration of an electric dipole.

The displacement can now be found by rearranging Eq. (6.36) to

$$D = \varepsilon_0 \mathcal{E} + P \tag{6.37}$$

Equation (6.37) is often written

$$D = \varepsilon \mathcal{E} \tag{6.38}$$

where ε is called the *permittivity*.

The ratio of permittivity with the dielectric to the permittivity in vacuum, $\varepsilon/\varepsilon_0$, is called the *relative permittivity*, ε_r, or *dielectric constant*. The dielectric constant is a material property. Some values of dielectric constants for common ceramic and glass insulators are given in Table 6.3. Since a polarizable material causes an increase in charge per unit area on the plates of a capacitor, the capacitance also increases, and it can be shown that the dielectric constant is related to the capacitance and displacement in vacuum and with the dielectric material as follows:

$$\varepsilon_r = \frac{\varepsilon}{\varepsilon_0} = \frac{D}{D_0} = \frac{C}{C_0} \tag{6.39}$$

Cooperative Learning Exercise 6.3

A voltage of 10 V is applied across a parallel-plate capacitor with a plate separation of 2×10^{-3} m and plate area of 6.45×10^{-4} m^2, both in vacuum and then with a dielectric material placed between the plates. The polarization, P, due to the presence of the dielectric material is 2.22×10^{-7} C/m^2.

Person 1: Calculate the displacement in vacuum, D_0, and the capacitance in vacuum, C_0.

Person 2: Calculate the displacement, D, due to the presence of the dielectric.

Combine your information to calculate the dielectric constant and the capacitance in the presence of the dielectric.

Answers: Equation (6.33) $D_0 = 8.854 \times 10^{-12}$ (10/0.002) $= 4.427 \times 10^{-8}$ C/m^2, Eq. (6.35) $C_0 = 8.854 \times 10^{-12}$ $(6.45 \times 10^{-4}/0.002) = 2.85 \times 10^{-12}$ F; Eq. (6.37) $D = 8.854 \times 10^{-12}$ (10/0.002) $+ 2.22 \times 10^{-7} = 2.66 \times 10^{-7}$ C/m^2. $\varepsilon_r = D/D_0 = 6.0$; $C = \varepsilon_r C_0 = 1.17 \times 10^{-11}$ F.

Table 6.3 Dielectric Constants of Selected Ceramics and Glasses at 25°C and 10 Hz.

Glasses	ε_r	Inorganic Crystalline Materials	ε_r
Silica glass	3.8	Mica	3.6
Vycor glasses	3.8–3.9	Potassium chloride	4.75
Pyrex glasses	4.0–6.0	Diamond	5.5
Soda–lime–silica glass	6.9	Mullite ($3Al_2O_3 \cdot 2SiO_2$)	6.6
High-lead glass	19.0	Magnesium oxide	9.65

Source: Ralls et al. [15]. Copyright © 1976 by John Wiley & Sons, Inc.

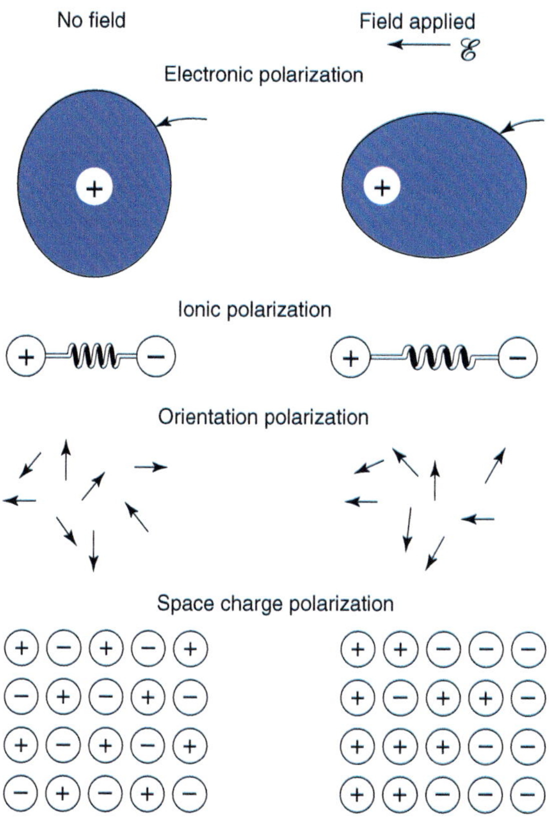

Figure 6.25 Schematic illustration of different types of polarization. From Kingery et al. [22]. Copyright © 1976 by John Wiley & Sons, Inc. This material is used by permission of John Wiley & Sons, Inc.

We can further describe the polarization, P, according to the different types of dipoles that either already exist or are induced in the dielectric material. The polarization of dielectric material may be caused by four major types of polarization: electronic polarization, ionic (atomic) polarization, orientation polarization, and space-charge (interfacial) polarization. Each type of polarization is shown schematically in Figure 6.25 and will be described in succession. In these descriptions, it will be useful to introduce a new term called *polarizability*, α, which is simply a measure of the ability of a material to undergo the specific type of polarization.

Electronic polarization arises because the center of the electron cloud around a nucleus is displaced under an applied electric field, as shown in Figure 6.25. The resulting polarization, or dipole per unit volume, is

$$P_e = N\alpha_e\mathcal{E}_{\text{loc}} \tag{6.40}$$

where N is the number of atoms or molecules, α_e is the electronic polarizability, and $\mathcal{E}_{\text{loc}}$ is the local electric field that an atom or molecule experiences. The local electric field is greater than the average field, $\mathcal{E}$, because of the polarization of other surrounding atoms.

Ionic polarization occurs in ionic materials because an applied field acts to displace cations in the direction of the applied field while displacing anions in a direction opposite to the applied field. This gives rise to a net dipole moment per formula unit. For an ionic solid, the atomic polarization is given by

$$P_i = N\alpha_i\mathcal{E}_{\text{loc}} \tag{6.41}$$

where N is the number of formula units per unit volume, and α_i is the ionic polarizability. Both ionic and electronic polarizations contribute to the formation of induced dipoles.

Orientation polarization can occur in materials composed of molecules that have permanent electric dipole moments. The permanent dipoles tend to become aligned with the applied electric field, but entropy and thermal effects tend to counter this alignment. Thus, orientation polarization is highly temperature-dependent, unlike the forms of induced polarization which are nearly temperature-independent. In electric fields of moderate intensity, the orientation polarization is proportional to the local electric field, as for the other forms of polarization:

$$P_o = N\alpha_o\mathcal{E}_{\text{loc}} \tag{6.42}$$

but in strong electric fields, this proportionality is not maintained since saturation must occur when all the permanent dipoles are aligned. Orientation polarization occurs mostly in gases and liquids where molecules are free to rotate. In some solids, the rotation of polar molecules may be restricted by lattice forces, which can reduce orientation polarization. For example, the dielectric constant of water is about 82, while that of ice is about 10. The orientation polarizability, α_o, can be further elaborated upon, since the permanent electric dipole moment per molecule, p_e, is a measurable quantity. The two quantities are related by a relationship similar to the Curie Law, which we shall elaborate upon in Section 6.1:

$$\alpha_o = \frac{p_e^{\,2}}{3k_BT} \tag{6.43}$$

where k_B is Boltzmann's constant and T is absolute temperature. A list of permanent dipole moments for selected polar molecules is given in Table 6.4.

The final type of polarization is *space–charge polarization*, sometimes called *interfacial polarization*, and results from the accumulation of charge at structural interfaces in heterogeneous materials. Such polarization occurs when one of the phases has a much higher resistivity than the other, and it is found in a variety of ceramic materials, especially at elevated temperatures. The space–charge polarization, P_{sc}, has a corresponding space–charge polarizability, α_{sc}. The two are related via a relationship of the form of the other types of polarization.

The total polarization, then, is the sum of all the contributions from the different types of polarization:

$$P = P_e + P_i + P_o + P_{sc} \tag{6.44}$$

Table 6.4 Permanent Dipole Moments of Selected Polar Molecules in the Gaseous Phase

Molecule	Dipole Moment (Debye)
AgCl	6.08
LiCl	7.12887
NaCl	9.00117
KCl	10.269
CsCl	10.387
LiF	6.3274
NaF	8.156
KF	8.585
RbF	8.5465
CsF	7.884
HF	1.826178
HCl	1.1086
HBr	0.8272
HI	0.448
HNO_3	2.17
H_2O	1.8546
MgO	6.2
NH_3	1.4718
CH_3Cl	1.8963
CH_2Cl_2	1.60
$CHCl_3$	1.04
C_2H_5OH	1.69
$C_6H_5NO_2$	4.22

Source: Ref. [59].

With appropriate substitution of the expressions for the various types of polarization, it is easy to show that the total polarizability, α, is also the sum of the various types of polarizability:

$$\alpha = \alpha_e + \alpha_i + \alpha_o + \alpha_{sc} \tag{6.45}$$

The total polarization, P, and hence the dielectric constant, ε_r, subjected to an alternating electric field depend upon the ease with which the permanent or induced dipoles can reverse their alignment with each reversal of the applied field. The time required for dipole reversal is called the *relaxation time*, and its reciprocal is called the *relaxation frequency*. When the time duration per cycle of the applied field is much less than the relaxation time of a particular polarization process, the dipoles cannot change their orientation rapidly enough to remain oriented in the applied field, and they "freeze" their reorientation process. This particular polarization process will not contribute to the total polarization. The order in which the polarization processes cease to contribute to the total polarization is related to their relative size scales. Obviously, electronic polarization is the smallest since it deals with electron density. Ionic polarization is next in size scale, since it deals

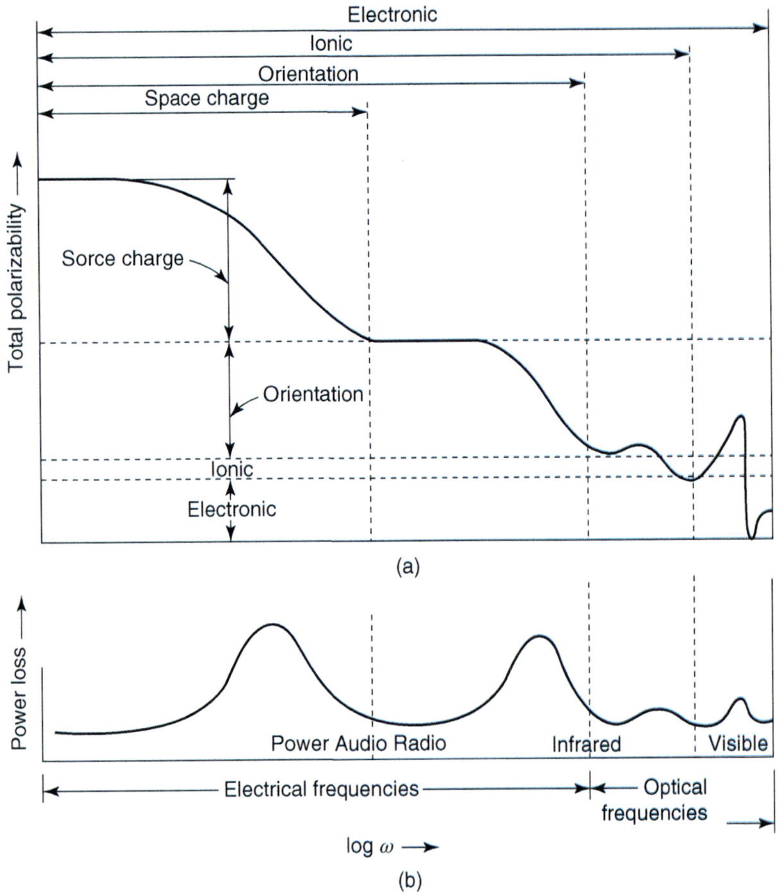

Figure 6.26 Frequency dependence of (a) total polarizability and (b) power loss. From Ralls et al. [15]. Copyright © 1976 by John Wiley & Sons, Inc. This material is used by permission John Wiley & Sons, Inc.

with the separation of ions, followed by orientation polarization, which is on the order of molecular dimensions and can sometimes be large. Finally, space–charge polarization, if it exists at all, is on the scale of interfacial dimensions. The larger the size scale, the more difficult it is for that type of polarization to respond to an alternating electric field. Thus, we would expect space charge to "freeze" up first with increasing electric field, followed by orientation, ionic, and electronic polarization, respectively. This process is shown in a plot of polarizability (which recall is proportional to polarization) versus the logarithm of the alternating electric field frequency in Figure 6.26a.

As the frequency of the applied field approaches the relaxation frequency, the polarization response increasingly lags the applied field. The reorientation of each type of dipole is opposed by internal friction, which leads to heating in the sample and power loss. The *power loss* depends on the degree to which the polarization lags the electric field. This lag can be quantified by measurement of the displacement–field relationship in a sinusoidally varying field. It is found that if the alternating field is given by

$$\mathcal{E} = \mathcal{E}_0 \sin(\omega t) \tag{6.46}$$

then the polarizability, P, is given by

$$P = P_0 \sin(\omega t - \delta) \tag{6.47}$$

where δ is the phase angle that the polarization lags the field, and ω is the angular frequency of the field. It can be shown that the power loss is related to the phase angle and the applied field:

$$\text{Power loss} = 2\pi\mathcal{E}^2\omega\varepsilon_0 \tan\delta \tag{6.48}$$

where $\tan\delta$ is called the *loss tangent*. For small phase angles, $\tan\delta \approx \delta$. As shown in Figure 6.26b, the power loss for a particular polarization type increases as that process becomes slower to respond to frequency changes, reaches a maximum as the electric field frequency reaches the relaxation frequency, and decreases after that polarization type has reached its relaxation frequency.

The loss tangent is an important quantity. The product of the loss tangent and the dielectric constant, $\varepsilon_r \tan\delta$, is called the *loss factor* and is the primary criterion for judging the usefulness of a dielectric as an insulator material. For this purpose, it is desirable to have a low dielectric constant and a small loss angle. For applications in which it is desirable to obtain a high capacitance in the smallest physical space, a high dielectric constant must be used, and a low value of the loss angle is needed.

At high enough frequencies, the dielectric will experience electrical breakdown and the insulator will become a conductor. As in a semiconductor, breakdown is initiated by the field-induced excitation of a number of electrons to the conduction band with sufficient kinetic energy to knock other electrons into the conduction band. The result is the avalanche effect described earlier. The magnitude of the electric field required to cause dielectric breakdown is called the *dielectric strength*, or *breakdown strength*. The dielectric strength for oxide ceramic insulators lies in the range of 1–20 kV/mm, with the value of 9 kV/mm for Al_2O_3 being typical.

Need More Help?

Try this video on dielectric strength:

https://www.youtube.com/watch?v=XEWfw7CHST8

The temperature dependence of the dielectric constant is highly dependent upon molecular structure, as well as the predominant polarization mechanism. For large molecules, orientation polarization is difficult below the material melting point, whereas orientation polarization can persist below the melting point in smaller molecules. Some substances undergo structural transformations in the solid state from higher- to lower-ordered structures which can restrict molecular rotation. The effects of temperature and frequency on the dielectric constant are not entirely independent. For example, the frequency effect on electronic and ionic polarization is negligible at frequencies up to about 10^{10} Hz, which is the limit of normal use. Similarly, the effect of temperature on electronic and ionic polarization is small. At higher temperatures, however, there is an increasing contribution resulting from ion mobility and crystal imperfection mobility. The combined effect is to give a sharp rise in the apparent dielectric constant at low frequencies with increasing temperature, as illustrated for alumina porcelain, whose primary component is Al_2O_3, in Figure 6.27. The loss tangent is also shown in Figure 6.27 for reference. Alumina is an important component of most ceramic dielectrics. The other base components are silica, SiO_2, and magnesia, MgO.

Two phenomena related to the electric dipoles in a material are *ferroelectricity* and *piezoelectricity*. Ferroelectricity is defined as the spontaneous alignment of electric dipoles by their mutual interaction in the absence of an applied electric field. The source of ferroelectricity arises from

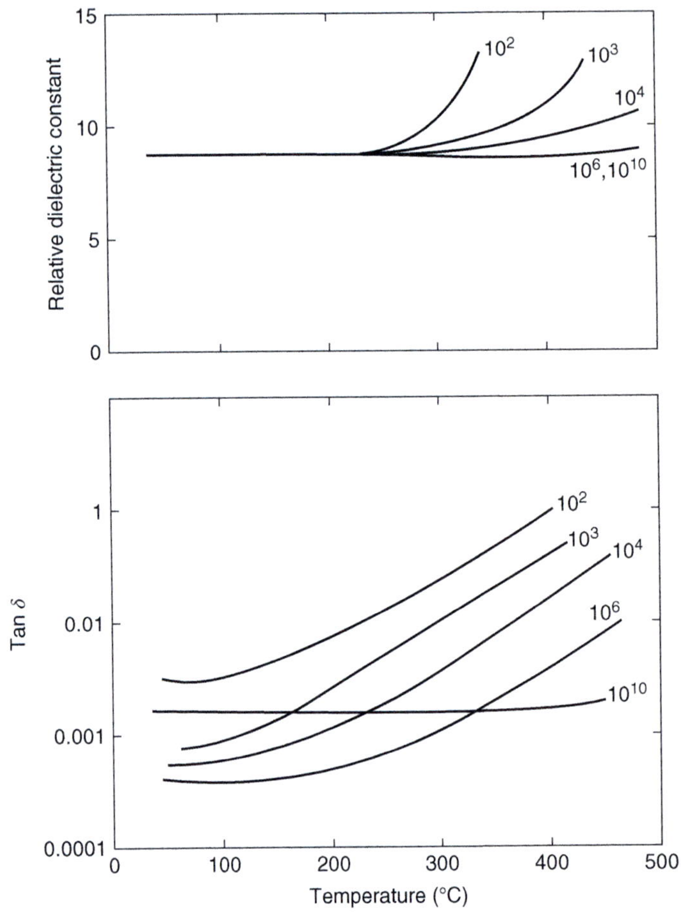

Figure 6.27 Frequency and temperature dependence of dielectric constant and tan δ for alumina porcelain. From Kingery et al. [22]. Copyright © 1976 by John Wiley & Sons, Inc. This material is used by permission of John Wiley & Sons, Inc.

the fact that the local field increases in proportion to the polarization. Thus, ferroelectric materials must possess permanent dipoles. One of the most important ferroelectric materials is barium titanate, $BaTiO_3$. The spontaneous polarization of $BaTiO_3$ is related directly to the positions of the ions within the tetragonal unit cell, as shown in Figure 6.28a. Divalent barium ions occupy each corner position, with the tetravalent titanium ion near, but just above, the center of the cell. The oxygen ions are located below the center of the (001) plane and just below the centers of the (100) and (010) planes, as illustrated in Figure 6.28b. Consequently, per unit cell, the centers of positive and negative charge do not coincide, and there is spontaneous polarization. At a certain temperature called the *ferroelectric Curie temperature* (120°C for barium titanate), barium titanate becomes cubic with Ti^{4+} ions located at the center of the unit cell and O^{2-} ions occupying face-centered positions, and it loses its spontaneous polarization. Ferroelectric materials have extremely high dielectric constants at relatively low applied field frequencies. Consequently, capacitors made from these materials can be significantly smaller than capacitors made from other dielectric materials.

Closely related to ferroelectricity is *piezoelectricity* in which polarization is induced and an electric field is established across a specimen by the application of external force (see Figure 6.29a,b). Reversing the direction of the external force, from tension to compression, reverses the direction of the

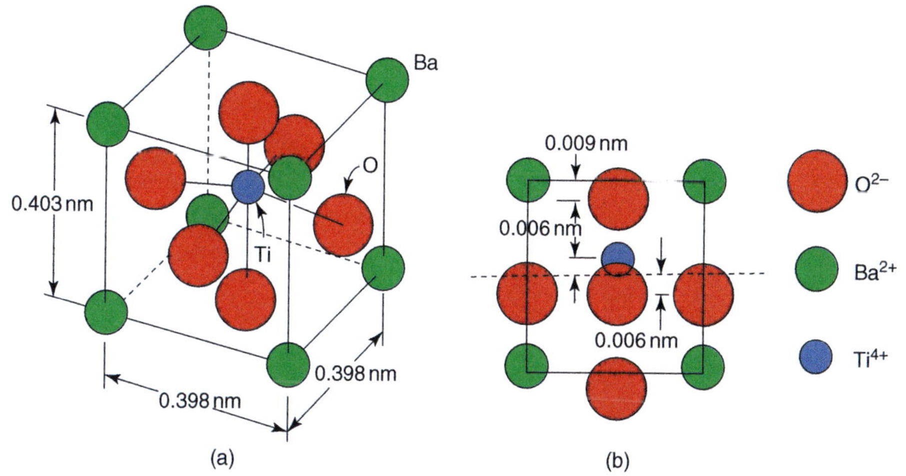

Figure 6.28 View of (a) the tetragonal unit cell of $BaTiO_3$ and (b) the ionic positions projected in a [100] direction onto a (100) face. From Ralls et al. [15]. Copyright © 1976 by John Wiley & Sons, Inc. This material is used by permission John Wiley & Sons, Inc.

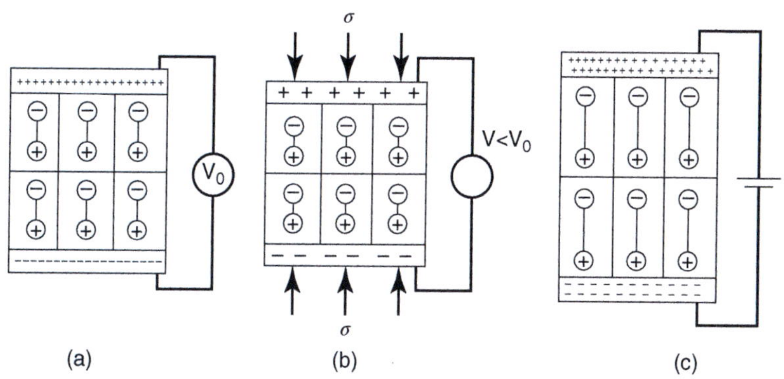

Figure 6.29 Schematic illustration of the piezoelectric effect that occurs in (a) an unstressed and (b) stressed piezoelectric material. Mechanical deformation can also occur when (c) a voltage is applied to a piezoelectric material. From Ralls et al. [15]. Copyright © 1976 by John Wiley & Sons, Inc. This material is used by permission John Wiley & Sons, Inc.

field. Alternatively, the application of an external electric field alters the net dipole length and causes a dimensional change, as in Figure 6.29c. Piezoelectric materials can be used as transducers—devices that convert mechanical stress into electrical energy and vice versa. Such devices are used in microphones and mechanical strain gauges. Barium titanate is also a piezoelectric material.

> **Need More Help?**
>
> Try these videos on polarization:
>
> https://www.youtube.com/watch?v=5O0yWvQhWkU
>
> piezoelectric effect:
>
> https://www.youtube.com/watch?v=wcJXA8IqYI8
>
> and ferroelectricity:
>
> https://www.youtube.com/watch?v=tdnJv0RAlYQ

Before proceeding, it may be useful to summarize some of the myriad of terms and symbols that have been used to this point. Such a summary is presented in Table 6.5. With the exception of a few

Table 6.5 Electrical Symbols, Terms, and SI Units

Quantity	Symbol	SI Units	
		Primary	Derived
Electric charge	q	Coulomb (C)	(None)
Electric potential	V	$kg \cdot m^2/s^2 \cdot C$	volt, V
Electric current	I	C/s	ampere, amp, A
Electric field strength	$\mathcal{E}$	$kg \cdot m/s^2 \cdot C$	V/m
Resistance	R	$kg \cdot m^2/s \cdot C^2$	ohm, Ω
Resistivity	ρ	$\mathbf{kg \cdot m^3/s \cdot C^2}$	$\mathbf{\Omega \cdot m}$
Conductance	G	$s \cdot C^2/kg \cdot m^2$	Ω^{-1}
Conductivity	σ	$\mathbf{s \cdot C^2/kg \cdot m^3}$	$\mathbf{(\Omega \cdot m)^{-1}}$
Electrical capacitance	C	$s^2 \cdot C^2/kg \cdot m^2$	farad, F
Permittivity	ε	$s^2 \cdot C^2/kg \cdot m^3$	F/m
Dielectric Constant (Relative Permittivity)	ε_r	**(dimensionless)**	**(dimensionless)**
Dielectric displacement	D	C/m^2	$F \cdot V/m^2$
Electric polarization	P	C/m^2	$F \cdot V/m^2$
Polarizability	α	$\mathbf{s^2 \cdot C^2/kg}$	$\mathbf{F \cdot m^2}$
Permanent Dipole Moment	p	$\mathbf{C \cdot m}$	**Debye, D**
Dielectric loss angle	δ	(none)	(none)
Charge carrier mobility	μ	$s \cdot C/kg$	$m^2/V \cdot s$
Charge carrier density	n	m^{-3}	(none)
Magnetic field strength	H	$C/m \cdot s$	A/m

Items in boldface are material properties.

more terms dealing with superconductivity, we have introduced all the variables we will need to describe the electrical properties of materials.

Ion-Conducting Ceramics and Glasses

In addition to the conduction of charge via electrons, charge can be conducted via ions. Ions are present in most crystalline ceramic materials such as oxides and halides. This process is termed *ionic conduction* and may occur either in conjunction with or separately from electronic conduction. As a result, we must expand our definition of conductivity to include both types of conduction:

$$\sigma = \sigma_{\text{electronic}} + \sigma_{\text{ionic}} \tag{6.49}$$

In a manner analogous to conduction via electrons and holes, the ionic conductivity can be related to the charge carrier density, n_i, charge carrier mobility, μ_i, and carrier charge, q_e, as originally given in Eq. (6.9) for electronic conduction, except that the charge carriers are now ions, which have multiple charges, or valences, z_i, so that Eq. (6.9) must be modified slightly:

$$\sigma_{\text{ionic}} = f n_i (z_i q_e) \mu_i \tag{6.50}$$

An additional factor, f, has been included to account for the number of equivalent sites to which an ion may transfer in a particular crystal structure; for example, $f = 4$ for an ion vacancy in the cubic crystal structure.

For an ion to move through the lattice under an electric field driving force, it must have sufficient thermal energy to pass over an energy barrier, which in this case is an intermediate position between lattice sites. This activation energy barrier to motion has already been described in Section 4.3.2 for diffusion in crystalline solids, so it will not be described here. It makes sense, though, that the ion mobility, μ_i, should be related to the ion diffusivity (diffusion coefficient), D_i, and can be derived from Eq. (6.14) presented earlier:

$$\mu_i = \frac{z_i q_e D_i}{k_B T} \tag{6.51}$$

where z_i is the valence of the ion, q_e is the electronic charge, and k_B and T are Boltzmann's constant and absolute temperature, as before. Substitution of Eq. (6.51) into Eq. (6.50) gives a relationship for the ionic conductivity:

$$\sigma_{\text{ionic}} = \frac{f n_i (z_i q_e)^2 D_i}{k_B T} \tag{6.52}$$

Typically, there is a predominate charge carrier in ionic solids. For example, in sodium chloride, the mobility of sodium ions is much larger than the mobility of chloride ions. This phenomenon can be temperature-dependent, however. As shown in Table 6.6, the fraction of total conductivity attributable to positive ions (K^+) in KCl drops from 0.96 at 435°C to 0.88 at 600°C. Similarly, the conductivity of CuCl is wholly attributable to electronic conduction at room temperature, but completely due to ionic conduction at 366°C. This temperature dependence is illustrated further for NaCl in Figure 6.30. At high temperatures (to the left in Figure 6.30), the number of sodium ion vacancies [n_i in Eq. (6.52)] is a thermodynamic property, and the conductivity varies with temperature as the product of the vacancy concentration and the diffusion coefficient, each of which has an exponential temperature dependence. This is the intrinsic range. At lower temperatures (to the right in Figure 6.30), the concentration of sodium ions is not in thermal equilibrium but is determined by minor solutes. As a result, this is the extrinsic region, and the temperature dependence of conductivity depends only on the diffusion coefficient. The activation energy for mobility can be determined from the data in Figure 6.30 for this region, but the slope of the curve in the intrinsic region gives the sum of the activation energy for mobility and lattice-defect formation.

Several types of compounds show exceptionally high ionic conductivity. Such phases fall into three broad categories: halides and *chalcogenides* of silver and copper, in which the metal atom is disordered over several alternative sites; oxides with the β-alumina structure, in which a monovalent cation is mobile; and oxides of fluorite structure, with large concentrations of defects caused either by a variable valence cation or solid solution with a second cation of lower valence. Chalcogenides are chiefly based on the elements of group VI—that is, S, Se, and Te. The silver and copper halides and chalcogenides often have simple arrays of anions. The cations occur in disorder in the interstices among the anions. The number of available sites is larger than the number of cations. In the highly conductive phases, the energy barrier between neighboring sites is very small, and channels are provided along which cations are free to move. The β-aluminas are hexagonal structures with approximate composition $AM_{11}O_7$. The mobile ion A is a monovalent species such as Na, K, Rb, Ag, Te, or Li, and M is a trivalent ion, Al, Fe, or Ga. The conductivities for several β-aluminas are plotted as a function of temperature in Figure 6.31. The crystal structure consists of planes of atoms parallel to the basal plane. Four planes of oxygens in a cubic close-packed sequence comprise a slab within which aluminum atoms occupy octahedral and tetrahedral sites as in spinel. The spinel blocks are bound together by a rather open layer of the monovalent ion and oxygen. This loosely bound layer is thought to be disordered and provides a two-dimensional channel for atom motion. Finally, the high dopant levels in the fluorite-type solid solutions lead to large defect concentrations

Table 6.6 Ionic and Electronic Contributions to Total Electrical Conductivity for Some Selected Compounds and Glasses

Compound	Temperature (°C)	Cationic Conductivity ($\sigma^{+}_{ionic}/\sigma_{total}$)	Anionic Conductivity ($\sigma^{-}_{ionic}/\sigma_{total}$)	Electronic Conductivity ($\sigma_{e,h}/\sigma_{total}$)
NaCl	400	1.00	—	—
	600	0.95	0.05	—
KCl	435	0.96	0.04	—
	600	0.88	0.12	—
KCl + 0.02% $CaCl_2$	430	0.99	0.01	—
	600	0.99	0.01	—
AgCl	20–350	1.00	—	—
AgBr	20–300	1.00	—	—
BaF_2	500	—	1.00	—
PbF_2	200	—	1.00	—
CuCl	20	—	—	1.00
	366	1.00	—	—
ZrO_2 + 7% CaO	>700	—	1.00	10^{-4}
$Na_2O \cdot 11Al_2O_3$	<800	1.00(Na^+)	—	$<10^{-6}$
FeO	800	—	—	1.00
ZrO_2 + 18% CeO_2	1500	—	0.52	0.48
ZrO_2 + 50% CeO_2	1500	—	0.15	0.85
$Na_2O \cdot CaO \cdot SiO_2$ glass	—	1.00(Na^+)	—	—
15% ($FeO \cdot Fe_2O_3$) $\cdot$ $CaO \cdot SiO_2 \cdot Al_2O_3$ glass	1500	0.1(Ca^+)	—	0.9

Source: Kingery et al. [22]. Copyright © 1976 by John Wiley & Sons, Inc.

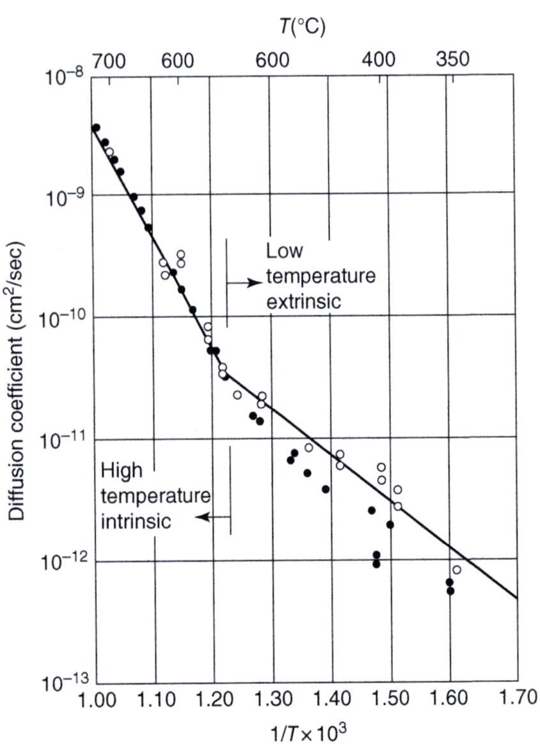

Figure 6.30 Diffusion coefficient measured directly (open circles) and calculated from electrical conductivity data (closed circles) for Na^+ in sodium chloride. From Kingery et al. [22]. Copyright © 1976 by John Wiley & Sons, Inc. This material is used by permission of John Wiley & Sons, Inc.

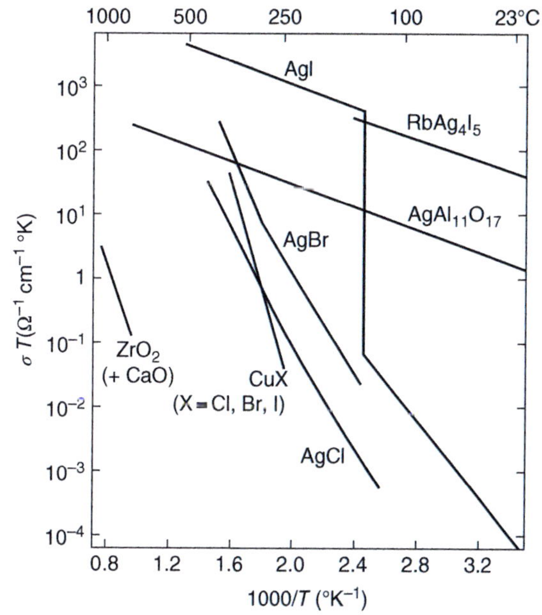

Figure 6.31 Conductivity of some highly conducting solid electrolytes. From Kingery et al. [22]. Copyright © 1976 by John Wiley & Sons, Inc. This material is used by permission of John Wiley & Sons, Inc.

and vacancy ordering. In such materials, rapid oxygen migration occurs. This is believed to be due to the high concentration of vacancies. The effect of structure on the ionic conductivity of ceramics leads us naturally into one of the most exciting technological developments of recent decades: high-temperature superconductors (HTSC).

Oxide Superconductors

In Section 6.0.1, we learned that some metals lose their residual resistivity near absolute zero and become what are called superconductors. We also learned that for many years, there was believed to be an upper-temperature limit of 30 K for the critical transition temperature, T_c, below which a substance begins to superconduct. In 1986, however, the first indication of superconductivity above 30 K was found in barium-doped La_2CuO_4. Then researchers found that high pressure could increase T_c to 50 K. Work began to find combinations of atoms of different sizes and valences that could simulate the pressure effect through chemical substitution. One of the variations was to substitute yttrium into the perovskite structure of $BaCuO_3$. Of these so-called yttrium–barium–copper (YBC) compounds, $YBa_2Cu_3O_7$ (termed 1–2–3 superconductor after the chemical formula) was found to superconduct at 92 K. Why all the fuss over a few tens of degrees? The answer is stated simply with the boiling points of three common gases: T_b (helium) = 4.2 K; T_b (nitrogen) = 77 K; T_b (oxygen) = 90 K. In order to cool the superconductors below T_c, a liquefied gas must be used. It is far less costly to liquefy nitrogen and oxygen (the major components of air) than helium. For any application in which liquid nitrogen can replace liquid helium, the refrigeration cost is about 1000 times less. In this way, HTSC, with $T_c > 77$ K, are of great technological interest, especially when utilizing the Meissner Effect described earlier.

Take a Peek! Jump to ***The Meissner Effect*** to learn more about the technological benefits of oxide superconductors.

Another striking difference the HTSC compounds exhibited was the anomalous behavior when placed in a magnetic field. Recall that there is also a critical magnetic field, H_c, above which the superconductor loses its superconductivity, and that this critical field is a function of temperature.

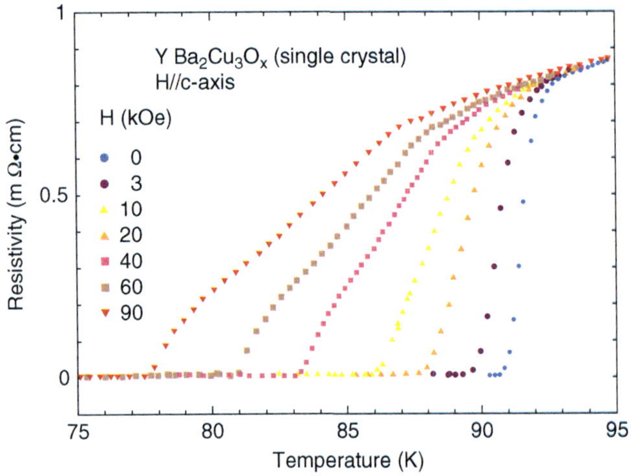

Figure 6.32 Variation of superconducting transition range in a magnetic field. Sheahan [24]/Reproduced with permission of Springer Nature.

In the metal and compound-based superconductors of earlier, the transition from superconducting to nonsuperconducting at a specified temperature is a relatively sharp one as the critical field is reached. This relationship was illustrated in Figure 6.10 and Eq. (6.19). When compounds belonging to the new class of HTSC are placed in a magnetic field, however, the temperature region for the superconducting–nonsuperconducting transition broadens as the field is increased, as illustrated for $Yba_2Cu_3O_7$ in Figure 6.32. At low fields, the resistivity drops sharply over a very narrow temperature range, whereas at high fields the transition range is quite large—nearly 15 K. Compare this resistivity–temperature behavior with that of Pb in Figure 6.6. The strong interdependence of temperature and magnetic field in superconductors suggests that there is not only a range of temperatures over which superconductivity develops in some materials but also a range of magnetic fields. Thus, two classes of superconductors are defined: Type I, also called "hard" superconductors, which exhibit a sharp transition (both in temperature and magnetic field); and Type II, also known as "soft" superconductors, which have a range of transition values for both temperature and magnetic field. In Type II superconductors, the magnetic field starts penetrating the material at a *lower critical field*, H_{c1}. Penetration increases until at the *upper critical field*, H_{c2}, the material is fully penetrated and the normal state is restored.

In addition to a critical temperature and critical field, all superconductors have a *critical current density*, J_c, above which they will no longer superconduct. This limitation has important consequences. A logical application of superconductors is as current-carrying media. However, there is a limit, often a low one, to how much current they can carry before losing their superconducting capabilities. The relationship between J_c, H_c, and T_c for a Type II superconductor is shown in Figure 6.33. Notice that the $H_c - T_c$ portion of this plot has already been presented in Figure 6.10 for a Type I superconductor.

Most of the oxide Type II HTSC are based upon the copper oxide lattice. Some examples and their transition temperatures are given in Table 6.7. In all cases, the structure is essentially a laminar one, with planes of copper oxide in the center (see Figure 6.34), through which the superconducting current flows. These materials are thus highly anisotropic—very little current can flow perpendicular to the copper oxide planes. As a result, great pains are taken to produce these materials in a highly oriented manner and to use them such that current flows parallel to the copper oxide planes. The role of elements other than copper and oxygen is secondary. In yttrium–barium–copper oxide

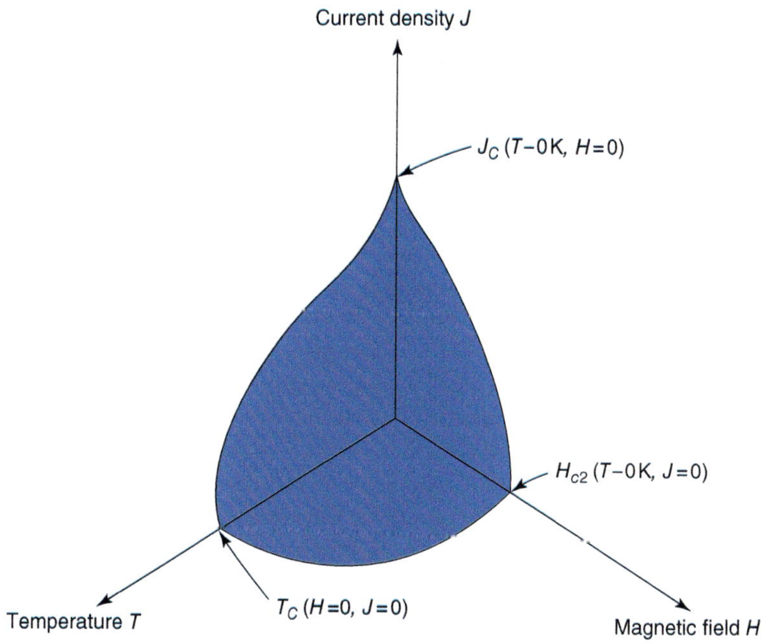

Figure 6.33 The relationship between temperature, magnetic field, and current density in a Type II superconductor. Callister [25]/Reproduced with permission of John Wiley & Sons.

Table 6.7 Some HTSC Materials Based on the Copper Oxide Lattice

Compound	T_c (K)
$YBa_2Cu_3O_7$	92
$(BiPb)_2Sr_2\ Ca_2Cu_3O_x$	105
$TlBa_2Ca_2Cu_3O_y$	115
$HgBa_2Ca_2Cu_3O_y$	135

Figure 6.34 Unit cell structure of $YBa_2Cu_3O_7$ superconductor. Sheahan [26]/Reproduced with permission of Springer Nature.

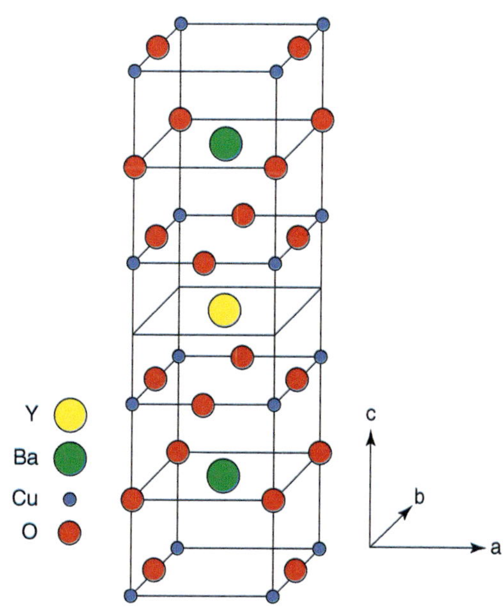

(YBCO), yttrium is only a spacer and a contributor of charge carriers. Nearly any of the rare earth elements (holmium, erbium, dysprosium, etc.) can be substituted for yttrium without significantly changing T_c.

It is beyond the scope of this text to describe the mechanism for charge transfer in HTSC. While there is still a great deal of discussion on the applicability of BCS theory (cf. Section 6.0.1) to HTSC materials, it is safe to say that many of the principles still apply—for example, density of states at the Fermi level. Finally, it would be misleading to suggest that these materials are the technological revolution they were first imagined to be. Like most other oxide ceramics, these materials are very brittle and attempts to fabricate them in useful forms, such as wire, have proved problematic. The additional difficulty is with current density, as described earlier. For large-scale applications such as particle accelerator magnets, current densities on the order of 10^{10} A/m^2 are required. A high critical magnetic field would also be beneficial for some applications, where magnetic fields of 30 tesla (T) (30,000 gauss) are required. Oxide superconductors meet neither of these requirements. As a result, metallic superconductors are still the materials of choice for these high-end applications, since they are more easily formed into wires, and have higher J_c and H_c values, even though their T_c is low. Metallic superconductors of the niobium family are particularly useful in these regards and include NbTi, Nb_3Sn, NbZr, Nb_3Al, and Nb_3Ge.

Compound Semiconductors

The niobium-based superconducting compounds lead us naturally into another use for intermetallics—namely, semiconductors. This topic, too, was introduced earlier in this chapter and we shall build upon those principles here to describe the semiconducting properties of compounds, ceramics, and glasses. The classification of intermetallics as ceramics may be confusing at this point, but we do so due to the strong covalent nature of their bonds that are more like oxide ceramics than to the metallic bond of elemental semiconductors. It is this bond structure that influences the electrical properties; hence, we discuss them together in this section.

Recall that elemental semiconductors such as silicon, germanium, tin, and diamond belong to group IV in the periodic table. They are characterized by four valency electrons, resulting in a tetrahedral arrangement of atoms in a diamond-like structure (cf. Figure 1.40). A large number of compounds possess structural and electrical characteristics that are similar to those of elemental semiconductors, many of which are compounds formed between elements of groups III and V of the periodic table, termed III–V compounds. In these cases, slight deviations from stoichiometry can lead to extrinsic conduction. One of the more successful examples is gallium arsenide, GaAs. Gallium is a group III element and arsenic is a group V element, and the resulting average valency in a 50–50 compound is four, just as for a group IV element. However, since the elemental substances have different electronegativities (cf. Table 1.4), the bond will be partly ionic in nature and the band gap will be higher than in the corresponding elemental semiconductor, Ge (see Table 6.8; some values may differ slightly from Table 6.2). These larger band gaps extend the useful range of conduction since intrinsic conduction becomes important only at correspondingly higher temperatures. The general decrease in band gap with increasing atomic number of the II–V compounds is a result of an increasing tendency toward metallic bonding.

Some semiconducting compounds can be of the II–VI types, which also has an average valence of four, but these have much more ionic character than III–V compounds. Their band gaps are thus larger, and in some cases, they may even be viewed as insulators. For example, ZnS, with a band-gap energy of 3.6 eV, is an insulator, whereas ZnSe has a band gap of 2.8 eV, which is closer to a semiconductor. A wide variety of compounds between transition metals and nonmetals also exhibit semiconductivity, but these materials are used infrequently because impurity control is much more difficult to achieve.

Table 6.8 Electronic Properties of Some Elemental and Compound Semiconductors at Room Temperature

		Mobility, μ ($m^2/V \cdot s$)		Effective Mass, m^*/m_0		
Material	**Energy Gap (eV)**	**Electrons**	**Holes**	**Electrons**	**Holes**	**Dielectric Constant**
Silicon (Si)	1.1	0.135	0.048	0.97	0.5	11.7
Ge	0.67	0.390	0.190	1.60	0.3	16.3
Sn (gray)	0.08	0.200	0.100			
GaAs	1.40	0.800	0.025	0.072	0.5	12.5
GaSb	0.77	0.400	0.140	0.047	0.5	15.0
GaP	2.24	0.050	0.002			10.0
InSb	0.16	7.800	0.075	0.013	0.60	17.0
InAs	0.33	3.300	0.040	0.02	0.41	14.5
InP	1.29	0.460	0.015	0.07	0.40	14.0
PbS	0.40	0.060	0.020			

Source: Jastrzebski [14]. Copyright © 1976 by John Wiley & Sons, Inc.

For nonstoichiometric compounds, the general rule is that when there is an excess of cations or a deficiency of anions, the compound is an *n*-type semiconductor. Conversely, an excess of anions or deficiency of cations creates a *p*-type semiconductor. There are some compounds that may exhibit either *p*- or *n*-type behavior, depending on what kind of ions are in excess. Lead sulfide, PbS, is an example of this duality. An excess of Pb^{2+} ions creates an *n*-type semiconductor, whereas an excess of S^{2-} ion creates a *p*-type semiconductor. Similarly, many binary oxide ceramics owe their electronic conductivity to deviations from stoichiometric compositions. For example, Cu_2O is a well-known *p*-type semiconductor, whereas ZnO with an excess of cations as interstitial atoms is an *n*-type semiconductor. A partial list of some impurity-controlled compound semiconductors is given in Table 6.9.

The conductivity of these materials can be controlled by the number of defects. In a *p*-type semiconductor such as Cu_2O, in which vacancies are formed in the cation lattice when the oxygen

Table 6.9 Some Impurity Compound Semiconductors

		n-Type			
TiO_2	Nb_2O_5	CdS	Cs_2 Se	$BaTiO_3$	Hg_2S
V_2O_5	MoO_2	CdSe	BaO	$PbCrO_4$	ZnF_2
U_3O_8	CdO	SnO_2	Ta_2O_5	Fe_3O_4	
ZnO	Ag_2S	Cs_2S	WO_3		
		p-Type			
Ag_2O	CoO	Cu_2O	SnS	Bi_2Te_3	MoO_2
Cr_2O_3	SnO	Cu_2S	Sb_2S_3		Hg_2O
MnO	NiO	Pr_2O_3	CuI		
		Amphoteric			
Al_2O_3	SiC	PbTe		Ti_2S	
Mn_3O_4	PbS	UO_2			
CO_3O_4	PbSe	IrO_2			

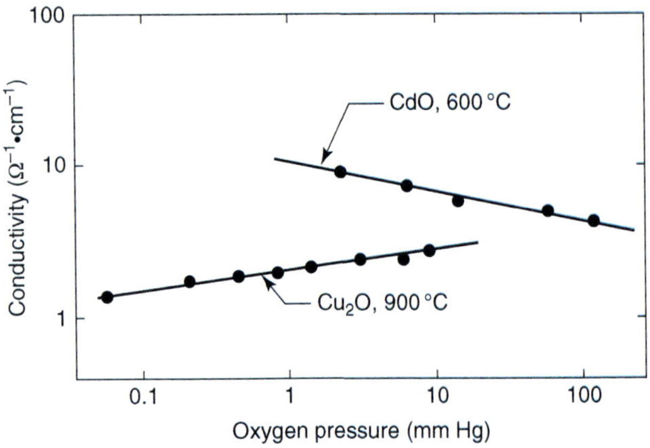

Figure 6.35 Electrical conductivity of two oxide semiconductors as a function of oxygen partial pressure. From Kingery et al. [22]. Copyright © 1976 by John Wiley & Sons, Inc. This material is used by permission of John Wiley & Sons, Inc.

partial pressure is increased, we can develop relationships between conductivity and oxygen partial pressure.

$$\sigma \propto K(T)^{1/4} P_{O_2}^{1/8} \tag{6.53}$$

Experimentally, as illustrated in Figure 6.35, the electrical conductivity of Cu_2O is found to be proportional to $P_{O_2}^{1/7}$, which is in reasonable agreement with the prediction of Eq. (6.53). The variation of another oxide semiconductor, CdO, with temperature, is also shown in Figure 6.35 for comparison.

As a final example of oxide-based semiconductors, recall that some chalcogenides exhibit high ionic conductivities. Chalcogenides also play a role in the high conductivities found in some glasses (see Figure 6.36). In addition to the group VI elements, chalcogenide glasses sometimes contain elements of group V, such as P, As, Sb, and Bi, as well as other ingredients like Ge or the halides. These conductivities are electronic, and not ionic, in nature. Their melting points are low, sometimes below 100°C, and they are quite sensitive to corrosion. The conductivity in these glasses can be explained in terms of a special type of glass structure defects around which pairs of positive and negative centers are present, and with which electrons couple. In addition to their semiconductivity, these glasses have a unique *switching behavior*. If a voltage is applied at the normal high-resistivity state, called the "off state," a critical current–voltage pair of values is eventually achieved at which point the glass suddenly switches into a low-resistivity state, called the "on state." Each state is stable, and the process is reversible. This is a type of storage, or memory, effect.

MOSFETs

A type of semiconductor device that utilizes oxide ceramics is a *metal-oxide-semiconductor field-effect transistor*, abbreviated as MOSFET. Just like semiconductor junction devices, the MOSFET is composed of n- and p-type semiconductor regions within a single device, as illustrated in Figure 6.37. One such device consists of two small islands of p-type semiconductor with and n-type silicon substrate. The islands are joined by a narrow p-type channel. The oxide portion of the MOSFET is an insulating layer of silicon dioxide that is formed by surface oxidation of the silicon. Gate, drain, and source connectors are attached. The MOSFET differs from the junction transistor in that a single type of charge carrier, either an electron or a hole, is utilized, instead of both. The conductivity of the channel is varied by the presence of an electric field imposed on the gate. For example, a positive

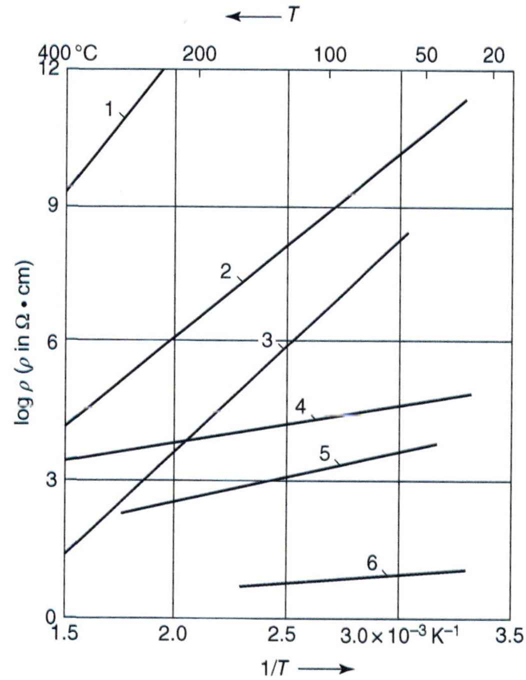

Figure 6.36 Temperature dependence of resistivity for some semiconducting glasses: (1) vitreous silica; (2) soda–lime silicate glass; (3) AsSeTe chalcogenide glass; (4) silicate glass with 18 mol% ($Fe_3O_4 + MnO$); (5) vanadium phosphate glass; and (6) borosilicate glass with 45 mol% ($Fe_3O_4 + MnO$). Scholze [27]/Reproduced with permission of Springer Nature.

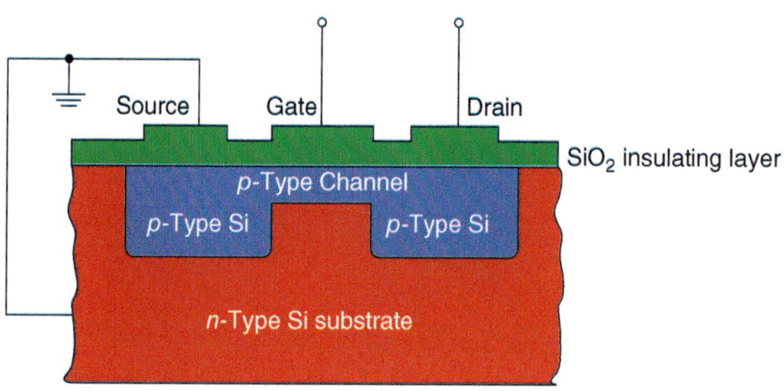

Figure 6.37 Schematic cross-sectional view of a MOSFET transistor. Callister [28]/Reproduced with permission of John Wiley & Sons.

field on the gate will drive holes out of the channel, thereby reducing the electrical conductivity. Thus, a small alteration in the field at the gate will produce a relatively large variation in current between the source and the drain. In this respect, the MOSFET is similar to the junction transistor, insofar as both create a signal amplification. The primary difference is that the gate current is exceedingly small in comparison to the base current of the junction transistor. As a result, MOSFETs are used where signal sources to be amplified cannot sustain an appreciable current.

Need More Help?
Try this video on MOSFETs: https://www.youtube.com/watch?v=spiw1hQrW-w

MOSFETs are now the basis for most digital storage media such as flash drives and solid-state drives (SSDs). Digital information in the form of 1 and 0 s is stored in the MOSFET in the form of charge in a *floating gate*. The presence of charge in the floating gate corresponds to a 1 while the absence of charge represents a zero.

6.0.3 Electrical Properties of Polymers

In this section, we consider the insulating properties of polymers as dielectrics, and we also describe their electrical- and ion-conducting capabilities. Given the highly insulative nature of most polymers, we begin with dielectric properties and then describe special types of polymers that conduct either electrons or ions.

Dielectric Polymers

Nonpolar polymers generally have dielectric constants in the 2–3 range while polar polymers can have values up to 7. Typical dielectric strengths of polymers are in the range of 20–50 kV/mm, which are approximately 2–10 times higher than ceramics and glasses and are hundreds of times higher than conducting metals and alloys.

Data from mechanical and dielectric measurements can be related in a qualitative if not quantitative, manner. Formally, the dielectric constant can be regarded as the equivalent of the mechanical compliance. This highlights the fact that the dynamic mechanical techniques of Section 5.3 measure the ability of a polymer to resist movement, whereas the dielectric technique measures the ability of the system to move, at least the dipolar portions. Interestingly, the dielectric loss appears to match the loss modulus more closely than the loss compliance when the data are compared for the same system. Both techniques respond in a similar fashion to a change in the frequency of the measurement. When the frequency is increased, the transitions and relaxations that are observed in a sample appear at higher temperatures. We expect maximum losses at transition temperatures for both electrical and mechanical deformations, as illustrated for poly(chlorotrifluoroethylene) in Figure 6.38. Just as mechanical fatigue after many alternating cycles can cause a polymer to fail at a lower stress, so dielectric insulation can fatigue under electrical stresses.

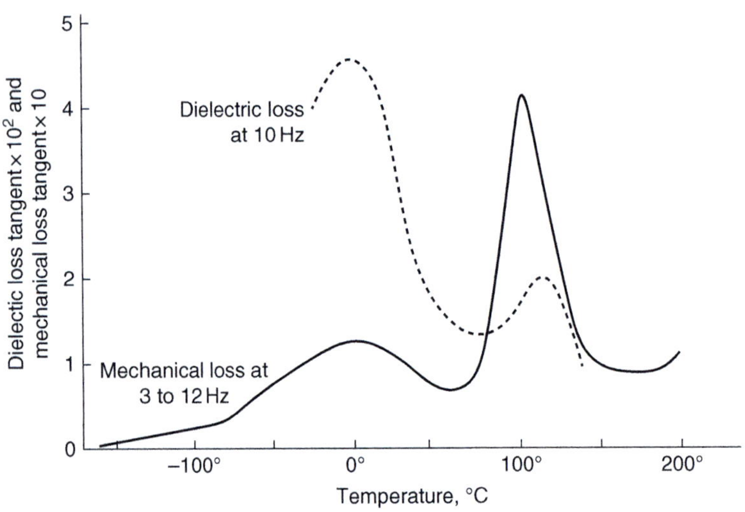

Figure 6.38 Mechanical and dielectric loss tangents for poly(chlorotrifluoroethylene). Rodriguez [29]/ Reproduced with permission of Hemisphere Publishing Corporation.

Electrically Conducting Polymers

High values of resistivity are common for organic polymers, with 10^{12} to 10^{18} $\Omega \cdot cm$ being typical. Electrical resistance can be lowered markedly by the addition of conductive fillers, such as carbon black or metal fillers, but the conductivity is due to the filler in this case. When we consider electronic conduction solely in homogeneous polymers, band theory is not totally suitable because the atoms are covalently bonded to one another, forming polymeric chains that experience weak intermolecular interactions. Thus, macroscopic conduction will require electron movement, not only along chains but also from one chain to another.

Polymers that display electronic conductivity are usually insulators in the pure state but, when reacted with an oxidizing or reducing agent, can be converted into polymer salts with electrical conductivities comparable to metals. Some of these polymers are listed in Figure 6.39, along with the conductivities of metals and ceramics for comparison. The preparation and mechanism of conductivity in a few of these polymers will be elaborated upon.

Need More Help?

Try this video on conductive polymers:
https://www.youtube.com/watch?v=u3O2d2KFUto

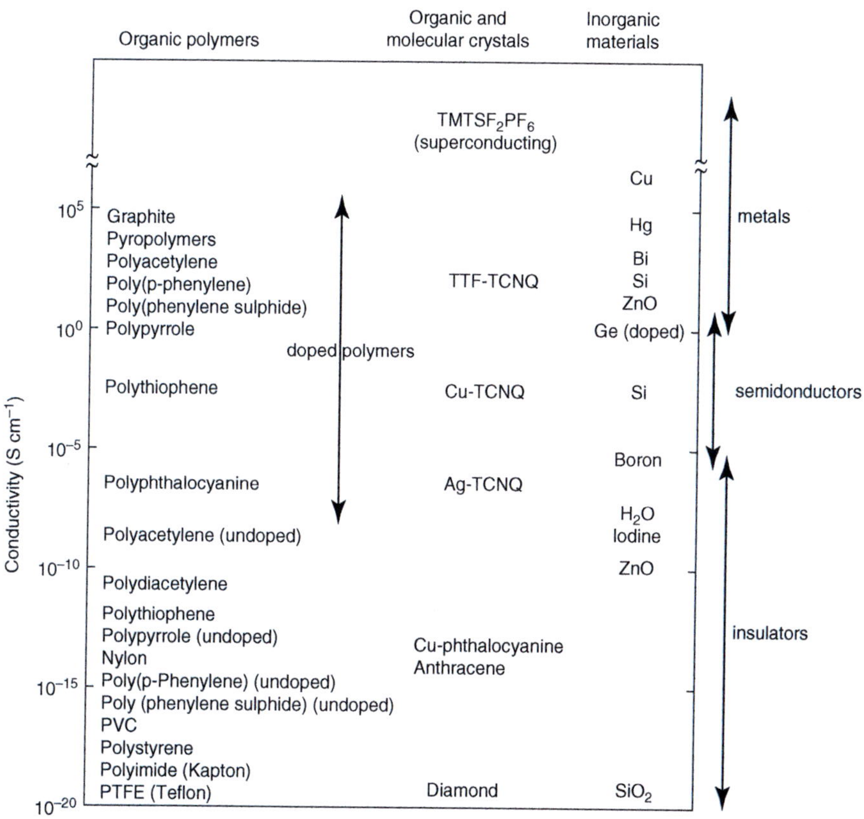

Figure 6.39 Conductivity ranges for doped and undoped polymers, inorganic materials, and molecular crystals. One Siemen (S) = 1 Ω^{-1}. Cowie [30]/Reproduced with permission of Taylor & Francis.

Table 6.10 Structures and Conductivities of Doped Conjugated Polymers

Polymer	Structure	Typical Methods of Doping	Typical Conductivity $(S\ cm)^{-1}$
Polyacetylene	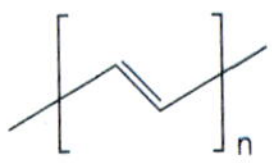	Electrochemical, chemical (AsF_5, I_2, Li, K)	$500–1.5 \times 10^5$
Polyphenylene	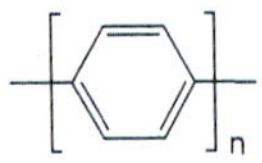	Chemical (AsF_5, Li, K)	500
Poly(phenylene sulfide)	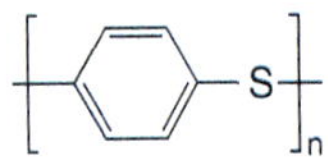	Chemical (AsF_5)	1
Polypyrrole	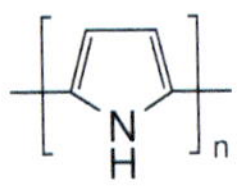	Electrochemical	600
Polythiophene	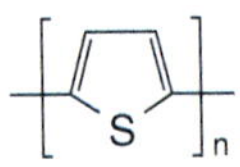	Electrochemical	100
Poly(phenyl-quinoline)	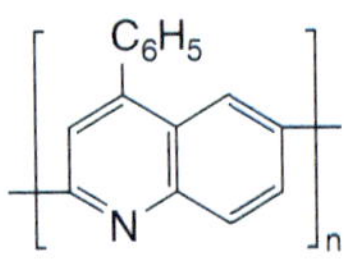	Electrochemical, chemical (sodium naphthalide)	50

One Siemen equals $1\ \Omega^{-1}$.
Source: Cowie 31]/Taylor & Francis.

One of the first polymers to exhibit electrical conductivity was polyacetylene. Normally, a very poor conductor in the pure state, it was found that a highly conductive polymer could be formed by reacting it with I_2. The result is a dramatic increase of over 10^{10} in conductivity. Polyacetylene is a *polyconjugated polymer*, the structure of which is shown in Table 6.10. Polyconjugation refers to the existence of multiple C—C bonds in the polymer, typically along the backbone. In a polyconjugated system, the π orbitals are assumed to overlap, and they form a valance and a conduction band as predicted by band theory. If all the bond lengths were equal—that is, delocalization that led to each bond having equal double bond character—then the bands would overlap and the polymer would behave as a quasi-one-dimensional metal having good conductivity. Experimental evidence has shown that this is an unstable system that will undergo lattice distortion. The alternative is to have single and double bonds alternate along the backbone, which leads to an energy gap between the valence and conduction bands. The *trans* structure of polyacetylene has such an alternating single-bond–double-bond backbone. The *trans* structure of polyacetylene is also unique because it has a twofold degenerate ground state in which sections A and B in Figure 6.40 are mirror images, and

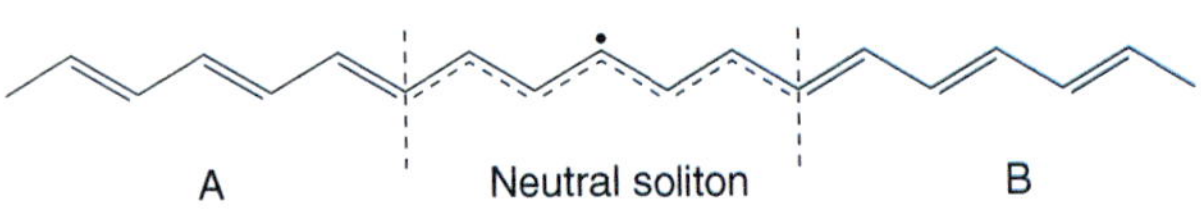

Figure 6.40 Isomerization of *cis* (A) and *trans* (B) sequences in polyacetylene that meet to form a soliton. Cowie [31]/Reproduced with permission of Taylor & Francis.

the single and double bonds can be interchanged without changing the energy. Thus, if the *cis* configuration begins to isomerize to the *trans* structure from different locations in a single chain, an A sequence may form and eventually meet a B sequence, as shown, but in doing so, a free radical called a *soliton*, is produced. The soliton is a relatively stable electron with an unpaired spin and is located in a nonbonding state in the energy gap, midway between the conduction and valence bands. It is the presence of these neutral solitons which gives *trans*-polyacetylene the characteristics of an intrinsic semiconductor with conductivities of 10^{-7} to 10^{-8} $(\Omega \cdot \text{cm})^{-1}$.

The conductivity of polyacetylene can be magnified by doping. Exposure of a polyacetylene film to dry ammonia gas leads to a dramatic increase in conductivity of 10^3 $(\Omega \cdot \text{cm})^{-1}$. Controlled addition of an acceptor, or *p*-doping, agent such as AsF_5, I_2, or $HClO_4$ removes an electron and creates a positive soliton (or a neutral one if the electron removed is not the free electron). Similarly, a negative soliton can be formed by treating the polymer with a donor, or *n*-doping, agent that adds an electron to the mid-gap energy level. This can be done by an electrochemical method. At high doping levels, the soliton regions tend to overlap and create new mid-gap energy bands that may merge with the valence and conduction bands, allowing freedom for extensive electron flow. Thus, in polyacetylene, the charged solitons are responsible for making the polymer a conductor.

Poly(*p*-phenylene) has all the structural characteristics required of a potential polymer conductor. It is an insulator in the pure state but can be doped using methods similar to those of polyacetylene to form both *n*- and *p*-type semiconductors. However, because poly(*p*-phenylene) has a higher ionization potential, it is more stable to oxidation and requires strong *p*-dopants. Examination of the structure shows that the soliton defect cannot be supported in poly(*p*-phenylene) because there is no degenerate ground state. Instead, it is assumed that conduction occurs because the mean free path of charge carriers extends over a large number of lattice sites and the residence time on any one site is small compared with the time it would take for a carrier to become localized. If, however, a carrier is trapped, it tends to polarize the local environment, which relaxes into a new equilibrium position. This deformed section of the lattice and the charge carrier then form a species called a *polaron*. Unlike the soliton, the polaron cannot move without first overcoming an energy barrier so movement is by a hopping motion. In poly(*p*-phenylene) the solitons are trapped by the changes in polymer structure because of the differences in energy and so a polaron is created which is an isolated charge carrier. A pair of these charges is called a *bipolaron*. In poly(*p*-phenylene) and most other polyconjugated conducting polymers, the conduction occurs via the polaron or bipolaron (see Figure 6.41).

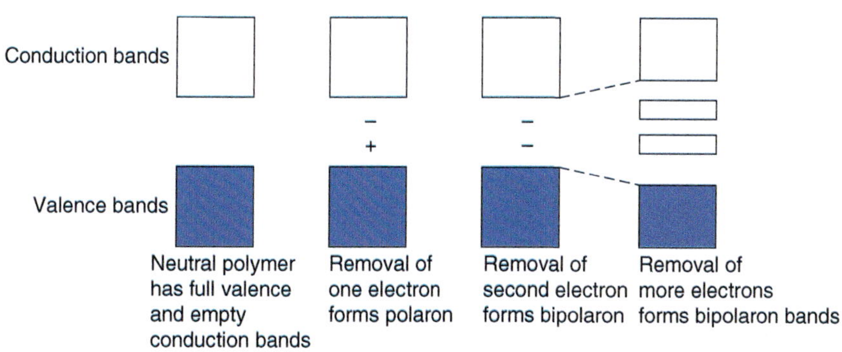

Figure 6.41 Proposed band structure for an oxidized *p*-type semiconducting polymer. Cowie [32]/Reproduced with permission of Taylor & Francis.

There are other types of semiconducting polymers as well, some of the more important of which are listed in Table 6.10. The conduction mechanism in most of these polymers is the polaron model described above. Applications for these polymers are growing and include batteries, electromagnetic screening materials, and electronic devices.

Ion-Conducting Polymers

Recall that electrical conductivity can be the result of electronic conductivity, in which electrons are the charge carriers, ionic conductivity, in which ions are the charge carriers, or a combination of both. In the previous section, we saw how some conjugated polymers can conduct electrons. In this section, we concentrate on ion-conducting polymers. There are generally two types of polymers in this category: those that conduct primarily hydrogen ions and those that conduct primarily larger anions. In both cases, the term *polymer electrolyte* is used to describe the type of ion-transporting polymeric material, although it is more commonly used to describe the latter. This should not be confused with the term *ionomer*, which describes polymers that contain ionic groups primarily involved in creating ionic bonds between chains. This is not to say that ionomers are incapable of ionic conductivity—they may well be. It is rather a matter of terminology usage. Ionomers typically have applications other than as ion conductors.

We begin by examining the structure and properties of polymer electrolytes. Initial investigations of ionically conductive polymers were principally focused on poly(ethylene oxide), PEO, when it was discovered that PEO can solvate a wide variety of salts, even at very high concentrations. The solvation occurs through the association of the metallic cations with the oxygen atoms in the polyether backbone (see Figure 6.42). The solvated ions can be quite mobile, and thus give rise to significant bulk ionic conductivities. Pure PEO is a semicrystalline polymer possessing both an amorphous phase and a crystalline phase at room temperature. Significant ionic transport occurs within the amorphous phase. This explains the dramatic decrease in ionic conductivity seen in many PEO-based systems for temperatures below the melting point of pure crystalline PEO

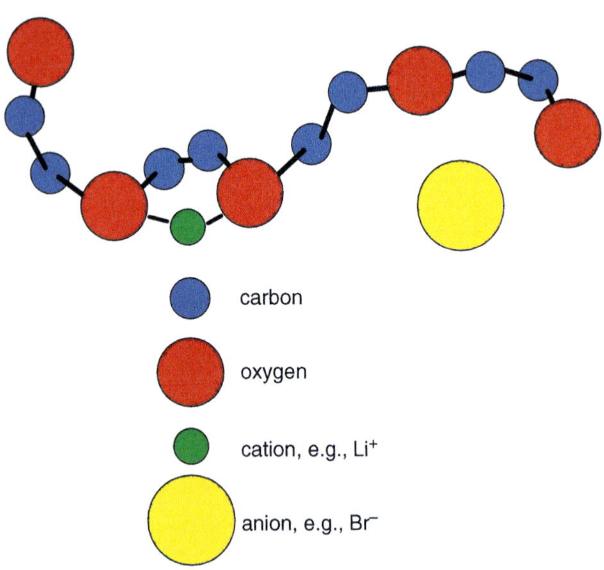

Figure 6.42 Schematic illustration of salt solvation by a PEO chain. Interrante et al. [33]/Reproduced with permission of American Chemical Society.

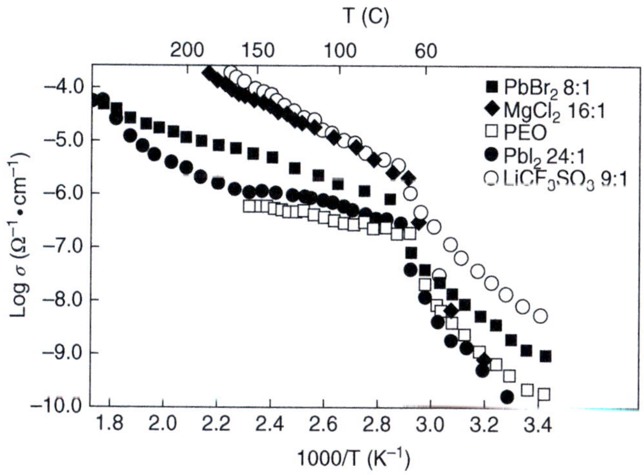

Figure 6.43 Temperature dependence of conductivity for various PEO-based polymer electrolytes containing divalent cations. Interrante et al. [34]/Reproduced with permission of American Chemical Society.

($T_m = 66°C$), as shown in Figure 6.43. The crystalline PEO regions are nonconductive and serve to hinder bulk ionic transport.

The mobility of the ions in polymer electrolytes is linked to the local segmental mobility of the polymer chains. Significant ionic conductivity in these systems will occur only above the glass transition temperature of the amorphous phase, T_g. Therefore, one of the requirements for the polymeric solvent is a low glass-transition temperature; for example, $T_g = -67°C$ for PEO.

One problem associated with polymer electrolytes arises from the opposing effects of increasing the salt concentration. Higher salt concentrations generally imply a higher density of charge carriers, which increases conductivity. However, through their interactions with the polyether chains, inorganic salts increase the T_g as their concentrations increase. This increase in T_g tends to lower the mobility of the ions and thus decreases ionic conductivity. The balance of these two effects leads to a maximum in the conductivity at a specific salt concentration, which varies from system to system.

One method of reducing crystallinity in PEO-based systems is to synthesize polymers in which the lengths of the oxyethylene sequences are relatively short, such as through copolymerization. The most notable linear copolymer of this type is oxymethylene-linked poly(oxyethylene), commonly called amorphous PEO, or aPEO for short. Other notable polymer electrolytes are based upon polysiloxanes and polyphosphazenes. Polymer blends have also been used for these applications, such as PEO and poly(methyl methacrylate), PMMA. The general performance characteristics of the polymer electrolytes are to have ionic conductivities in the range of 10^{-3} $(\Omega \cdot cm)^{-1}$ or (S/cm).

Still higher total conductivities are achievable in a special class of hydrogen-ion (proton)-conducting polymers. These polymers are particularly useful for fuel-cell and chlor-alkali processing applications in which the efficient transfer of protons is critical. These polymers are produced in membrane form and are therefore referred to as *proton-exchange membranes* (PEM).

The most common polymeric PEM is Nafion®, which is manufactured in various forms by Du Pont. The generalized chemical structure of Nafion is shown in Figure 6.44. It is a perfluorinated polymer, similar to Teflon®, with small amounts of sulfonic (SO_3^-) or carboxylic functional groups. The microstructure of Nafion® is actually quite complex. In general, the models, supported

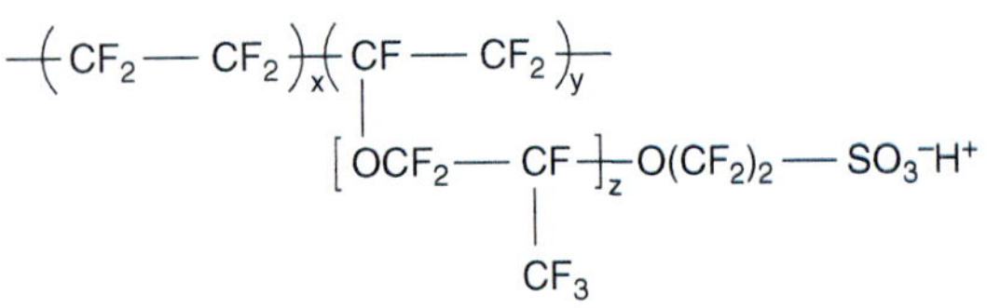

Figure 6.44 The generalized chemical structure of Nafion.

Table 6.11 Proton Conductivities of Selected Polymers

Polymer	Conductivity $(\Omega \cdot cm)^{-1}$
Nafion®	2.2×10^{-1}
Polybenzimidazole (PBI)	2×10^{-4}–8×10^{-4}
Poly(vinylidene fluoride)	10^{-2}

primarily by small-angle X-ray scattering (SAXS) results, predict that the ionic portions of Nafion tend to aggregate due to electrostatic interactions to form ionic clusters. It has been determined that the clusters are on the order of 30–50 Å in size, resulting in (1) a phase-separated polymer with discrete hydrophobic regions formed from the polymer backbone and (2) hydrophilic regions formed from the ionic clusters. These distinct regions are functionally important. Nafion is believed to conduct protons due to extensive hydration that takes place in the ionic clusters. Waters of hydration are necessary, since according to one theory the protons are transported through the polymer in the form of hydronium ions, H_3O^+ [3].

The proton conductivity of Nafion is orders of magnitudes higher than most other proton-conducting polymers (see Table 6.11). The conductivity, however, is highly dependent upon (a) the relative humidity in the membrane and (b) the ability of the membrane to stay hydrated during operation. One drawback of tetrafluoroethylene-based polymers is the inherent difficulty in processing due to limited solubility in most solvents. As a result, only certain types of geometries, such as thick films, are commercially available.

6.0.4 Electrical Properties of Composites

Because electronic and ionic conductions are so structure-sensitive, the simple rule-of-mixtures approach to estimating the conductivity and resistivity of composites is not normally of use. As a result, the conductivity of specific composites for specific applications must be experimentally determined. In the next two sections, we examine two examples of how composites can be used in electrical applications, and we describe the influence of each component on the electrical properties. The first example involves the electrical insulating properties of polymers, and the second one involves enhancing the electrically conducting properties of polymers.

Dielectric Properties of Glass-Fiber-Reinforced Polymers

As described in several of the previous sections on composites, the most common type of glass fiber used for reinforcement purposes is E-glass, which was originally developed for electrical insulation purposes, hence its name ("E" stands for "electrical grade"). It is not surprising, then, that when combined with organic resins such as epoxy, a class of materials results, which can be readily molded into complex shapes that possess excellent insulting properties and high dielectric strength.

Table 6.12 Some Electrical Properties of Selected Reinforcing Glass Fibers

Property	E-Glass	S-Glass	D-Glass
Dielectric constant, 10^6 Hz	5.80	4.53	3.56
Dielectric constant, 10^{10} Hz	6.13	5.21	4.00
Loss tangent, 10^6 Hz	0.001	0.002	0.0005
Loss tangent, 10^{10} Hz	0.0039	0.0068	0.0026

Source: Lubin [60]/John Wiley & Sons.

Table 6.13 Some Electrical Properties of Selected Resins

Property	Phenolic	Epoxy	Polyester	Silicone
Resistivity, Ω · cm	10^{12}–10^{13}	10^{16}–10^{17}	10^{14}	10^{11}–10^{13}
Dielectric strength, V/μm	14–16	16–20	15–20	7.3
Dielectric constant, 60 Hz	6.5–1.5	3.8	3.0–4.4	4.0–5.0

Source: Lubin [60]/John Wiley & Sons.

Table 6.14 Some Electrical Properties of Selected Glass-Fiber-Reinforced Resins

Property	GFR Nylon	GFR Polypropylene	GFR Phenolic	GFR (Laminate)	GFR Polyester	GFR Silicone (Laminate)
Resistivity, Ω · cm	2.6–5.5×10^{15}	1.7×10^{16}	10^{10}–10^{11}	6.6×10^7–10^9	10^{12}–10^{13}	2–5×10^{14}
Dielectric strength, V/μm	16–18	12–19	15–17	26–30	8–16	28.5
Dielectric constant, 60 Hz	4.6–5.6	2.3–2.5	7.1–7.2	4.4	4.6–5.2	3.9–4.2

Source: Shackelford and Alexander [61]/Taylor & Francis.

Some electrical properties of reinforcing fibers, composite resins, and the resulting composites are given in Tables 6.12–6.14, respectively. These values should be taken as approximate only, especially for the composites, since fiber orientation, content, and field strength have an enormous impact on the dielectric properties of these materials. Some of the most widespread electrical applications for glass-fiber-reinforced epoxy systems are in printed circuit boards and electrical housing such as junction boxes.

One interesting application of the dielectric properties of the resins in these composite systems is that it provides an opportunity to monitor the cure characteristics of the resin in the composite. In-process cure monitoring ensures that each molded part has properly cured before the mold is opened. The dielectric loss [see Eq. (6.48)] has been shown to increase rapidly at the beginning of the cure cycle (see Figure 6.45), attain a peak, then reduce to a constant value as the curing reactions are completed. At the beginning of the cure cycle, the resin viscosity in the uncured prepreg is relatively high, so that the dipole and ion mobilities are restricted. This results in low loss factors

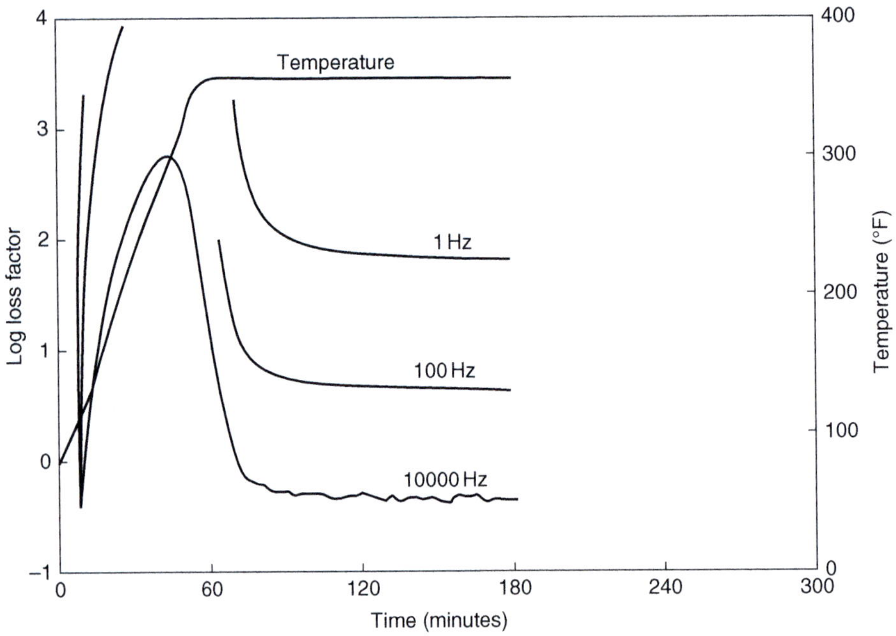

Figure 6.45 Dielectric loss factor as a function of cure time and frequency of the oscillating electric field in a fiber-reinforced polymer. Mallick [35]/Reproduced with permission of Taylor & Francis.

just after the uncured prepreg is placed in the mold. As the temperature of the prepreg increases in the mold, the resin viscosity is reduced and the loss factor increases owing to greater dipole and ion mobilities (see Figure 6.26). As soon as the gel point is reached, the resin viscosity increases rapidly and the loss factor decreases. At a full degree of cure, the loss factor levels off to a constant value.

Cooperative Learning Exercise 6.4

Assume that the conductivity of a unidirectional, continuous fiber-reinforced composite is a summation effect just like elastic modulus and tensile strength; that is, an equation analogous to Eq. (5.83) can be used to describe the conductivity in the axial direction, and one analogous to (5.87) can be used for the transverse direction, where the modulus is replaced with the corresponding conductivity of the fiber and matrix phase. Perform the following calculations for an aluminum matrix composite reinforced with 40 vol% continuous, unidirectional Al_2O_3 fibers. Use average conductivity values from Appendix 7.

Person 1: Calculate the conductivity of the composite in the axial direction.

Person 2: Calculate the conductivity of the composite in the transverse direction.

Compare your answers. How many orders of magnitude difference is there in the two conductivities?

Answers: $\sigma_{axial} = (0.4)(10^{-11}) + (0.6)(3.45 \times 10^7) = 20.7 \times 10^6$ $(\Omega\text{-m})^{-1}$; $\sigma_{trans} = (3.45 \times 10^7)(10^{-11})/[(0.4)(3.45 \times 10^7) + (0.6)(10^{-11})] = 2.5 \times 10^{-11}$ $(\Omega\text{-m})^{-1}$. 17 orders.

The technique for monitoring the dielectric loss factor is relatively simple. Two metal electrodes are placed opposite each other at critical locations on opposite sides of the mold. When the sheet molding compound (SMC), is placed between the electrodes, a capacitor is formed. The dielectric power loss is monitored continually throughout the molding cycle, as outlined earlier.

Need More Help?

Try this video on thermoelectric and Seebeck effects:

https://www.youtube.com/watch?v=5iTehoVGYeE

Metal Oxide-Polymer Thermistors

The variation of electrical properties with temperature heretofore described can be used to tremendous advantage. These so-called *thermoelectric effects* are commonly used in the operation of electronic temperature-measuring devices such as *thermocouples, thermistors*, and resistance–temperature detectors (RTDs). A thermocouple consists of two dissimilar metals joined at one end. As one end of the thermocouple is heated or cooled, electrons diffuse toward the cold end, and a *Seebeck voltage* is produced in each wire. Owing to the intentional dissimilarity between the Seebeck voltage in the two wires that compose the thermocouple, a voltage difference is established between the two wires, which can be measured and calibrated to temperature. A thermistor, on the other hand, directly correlates the change in resistivity of a substance to temperature. Although thermocouples are more widely utilized due to their inherent stability and larger operating temperature range, the thermistor concepts allow us to illustrate a thermoelectric effect that utilizes composite materials.

Thermistors can exhibit two primary resistance effects (see Figure 6.46): the *negative temperature coefficient* (NTC) effect, and the *positive temperature coefficient* (PTC) effect. In composites, these transitions arise from an individual component, such as a semiconductor–metal transition exhibited by V_2O_3 metallic filler, or result from a product property of the composite. Traditionally, PTC thermistors have been prepared from doped-$BaTiO_3$, such that the grains are semiconducting and the grain boundaries are insulating. The PTC effect occurs during heating through the Curie point, at which point the crystal structure changes from tetragonal to cubic, accompanied by a ferroelectric transition. In the ferroelectric state, the spontaneous polarization compensates for the insulating grain boundaries, resulting in a low resistivity. Although $BaTiO_3$ thermistors exhibit relatively large (six orders of magnitude) changes in resistivity with temperature, they have high room-temperature resistivities and are expensive to produce.

Composite-based PTC thermistors are potentially more economical. These devices are based on a combination of a conductor in a semicrystalline polymer—for example, carbon black in

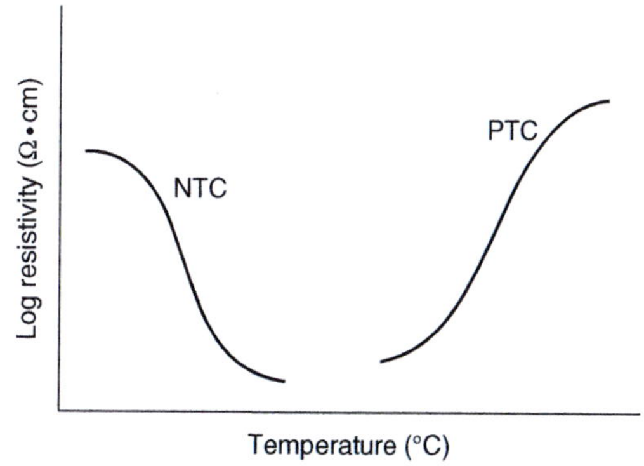

Figure 6.46 Schematic illustration of negative temperature coefficient (NTC) and positive temperature coefficient (PTC) effects. Moffatt et al. [36]/Reproduced with permission of John Wiley & Sons.

polyethylene. Other fillers include copper, iron, and silver. Important filler parameters in addition to conductivity include particle size, distribution, morphology, surface energy, oxidation state, and thermal expansion coefficient. Important polymer matrix characteristics in addition to conductivity include the glass transition temperature, T_g, and thermal expansion coefficient. Interfacial effects are extremely important in these materials and can influence the ultimate electrical properties of the composite.

There is a relatively sharp increase in resistivity near the melting temperature of the polymer, T_m, at which point a discontinuous change in specific volume occurs. The PTC phenomenon in these composites is believed to be the result of a separation of the conducting particles due to the thermal expansion of the polymer. Because the transition occurs at the melting point, one may control the PTC transition by utilizing a different polymer matrix. Composites prepared with crystalline polymers can also exhibit dramatic NTC effects just above the melting point, due to the relaxation of the polymer structure. This effect may be eliminated by crosslinking the polymer or adding a third filler to stabilize the matrix.

Although a majority of these composite thermistors are based upon carbon black as the conductive filler, it is difficult to control in terms of particle size, distribution, and morphology. One alternative is to use transition metal oxides such as TiO, VO_2, and V_2O_3 as the filler. An advantage of using a ceramic material is that it is possible to easily control critical parameters such as particle size and shape. Typical polymer matrix materials include PMMA, epoxy, silicone elastomer, polyurethane, polycarbonate, and polystyrene.

The resistivity–temperature behavior for a V_2O_3–PMMA composite is shown in Figure 6.47. When the NTC transition in the V_2O_3 is combined with the PTC effect in the composite, a resistivity–temperature curve with a square-well appearance results. As the amount of V_2O_3 is increased, the PTC effect begins to appear near 50 vol% V_2O_3. The optimum composition is that which yields the lowest room-temperature resistivity and the largest PTC effect. As the amount of conducting filler is increased, the temperature at which the PTC effect occurs moves to higher temperatures. The PTC exhibited in the composite is related to a phenomenon called *percolation*, which deals with the number and properties of neighboring entities and the existence of a sharp

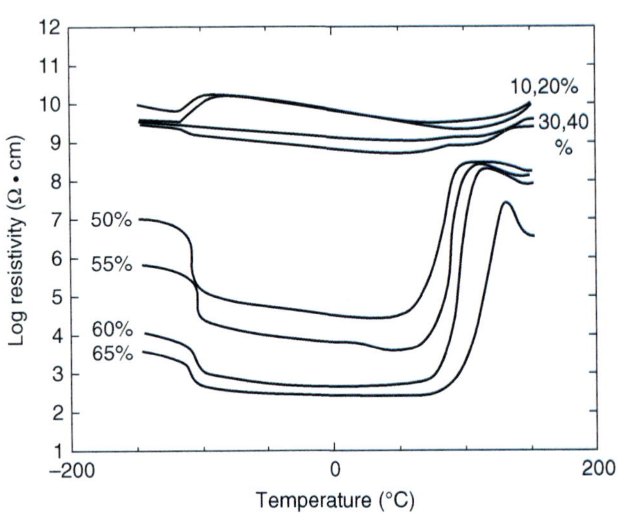

Figure 6.47 Temperature dependence of resistivity for a V_2O_3–PMMA composite of various volume fractions V_2O_3 at 1 kHz. Moffatt et al. [37]/Reproduced with permission of John Wiley & Sons.

transition at which long-range connectivity in the system appears or disappears. At low filler concentrations, a high resistivity is due to the insulating phase. As the volume fraction is increased, conducting paths start to form, resulting in a decrease in resistivity. This region is known as the *percolation threshold*. At higher filler concentrations, a saturation region is reached wherein the resistivity is relatively constant due to extensive particle contacts and eventually approaches the value of the pure conductor.

6.0.5 Electrical Properties of Biologics

The principles introduced in the previous sections have important applications in the description of many biological processes. Both ionic and electron charge transfers play important roles in bioprocesses such as nerve and muscle cell function. While an understanding of these processes would be a worthwhile academic exercise, there is little new to be learned about the materials involved—that is, structure–property relationships for the biological materials. What would be useful to us is a description of the electrical properties of the biological materials—for example, the electrical conductivity of various tissues, or the ion-conducting capacity of membranes. Unfortunately, while some of this information may indeed exist, it is difficult to find all the necessary information, and it is even more difficult to make generalizations. In other words, the human body, and other biological units like it, are complex chemoelectric devices that are not yet fully understood from a materials standpoint. There are any number of ongoing scientific studies aimed at better understanding these biological processes and utilizing them for the development of biomaterials, such as the stimulation of nerve cell growth using electrically conducting polymers [4], the electrical properties of amphiphilic lipid membranes [5], and the electrical properties of hydroxyapatite thin films [6]. These topics, while fascinating, are much too specific for this introductory text. Instead, let us concentrate on two, more general, classes of artificial biomedical devices that may give us some insight into the electrical processes in biologics: biosensors and bioelectrodes.

Biosensors

The term *bioelectrode* is broadly used to denote a class of devices that transmit information into or out of the body as an electrical signal. Those bioelectrodes that transmit information out of the body are generally called *biosensors*. Any type of sensor falls into one of two general categories: physical or chemical. Physical parameters of biological interest include pressure, volume, flow, electrical potential, and temperature. Chemical sensing involves the determination of the concentration of chemical species in a volume of gas, liquid, or tissue. The species can vary in size from a hydrogen ion to a live pathogen. When the species is complex, such as a parasite or a tumor, an interaction with another biological entity is required to recognize it. In this section, we give two examples of chemical biosensors: ion sensors and glucose sensors.

Many simple ions such as K^+, Na^+, Cl^-, and Ca^{2+} are normally kept within a narrow range of concentrations in the body, and they must be monitored during critical care. Potentiometric sensors for ion, also called *ion-selective electrodes* or ISEs, utilize a membrane that is primarily semipermeable to one ionic species. The ionic species is used to generate a voltage that generally obeys the Nernst equation [cf. Section 3.1.3 and Eq. (3.24)]:

$$E = C + \frac{RT}{zF}\, ln\left[a_i + k_{ij}\left(a_j\right)^{z/y}\right] \tag{6.54}$$

where E is the potential in response to an ion, i, of activity a_i and charge z; C is a constant; k_{ij} is the selectivity coefficient; and j is any interfering ion of charge y and activity a_j.

Glasses exist that function as selective electrodes for many different monovalent and some divalent cations. Alternatively, a hydrophobic membrane can be made semipermeable if a hydrophobic molecule called an *ionophore* that selectively binds an ion is dissolved in it. The selectivity of the membrane is determined by the structure of the ionophore. Some ionophores are natural products, such as gramicidin, which is highly specific for K^+, whereas others such as crown ethers and cryptands are synthetic. Ions such as S^{2-}, I^-, Br^-, and NO_3^- can be detected using quaternary ammonium cationic surfactants as a lipid-soluble counterion. ISEs are generally sensitive in the 10^{-1} to 10^{-5} M range but are not perfectly selective. The most typical membrane material used in ISEs is polyvinyl chloride (PVC) plasticized with dialkyl sebacate or other hydrophobic chemicals.

Biosensors are also available for glucose, lactate, alcohol, sucrose, galactose, uric acid, alpha amylase, choline, and L-lysine. All are amperometric sensors based on O_2 consumption or H_2O_2 production in conjunction with the turnover of an enzyme in the presence of substrate. In the case of glucose oxidase reaction, the normal biological reaction is:

$$\text{Glucose} + O_2 + H_2O \rightleftarrows \text{Gluconic acid} + H_2O_2 \tag{6.55}$$

Under many circumstances, the concentration of oxygen is rate-limiting, so the sensor often measures not glucose, but the rate at which oxygen diffuses to the enzyme to reoxidize its cofactor. There are two electrochemical ways to couple the reaction to electrodes: (a) monitor depletion of oxygen by reducing what is left at an electrode or (b) monitoring buildup of H_2O_2 by oxidizing it to O_2 and protons:

$$H_2O_2 \rightleftarrows O_2 + 2H^+ + 2e^- \tag{6.56}$$

An alternative method is to use electrochemical mediators that are at a higher concentration than O_2 and can therefore be shuttled back and forth between the protein and the electrode faster than the enzyme is reduced, so that the arrival of the glucose is always rate-limiting. A typical chemical that works in this way is ferrocene, which is an iron cation between two cyclopentadienyl anions, as shown in Figure 6.48. It exists in neutral and +1 oxidation state that are readily interconvertible at metal or carbon electrodes.

Most of the sensors are macroscopic and are employed in the controlled environment of a clinical chemistry analyzer. However, complete glucose sensors are available that contain disposable glucose oxidase-based electrodes, power supply, electronics, and readout in a housing the size of a ballpoint pen. The ultimate goal is an implantable glucose sensor, toward which much progress is being made.

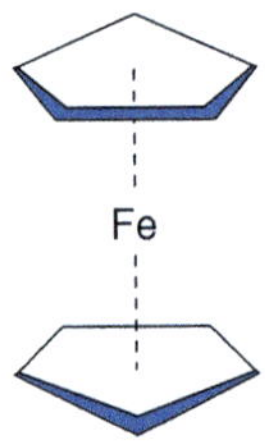

Figure 6.48 The structure of ferrocene.

Electrical Stimulation Devices

Bioelectrodes that transmit electrical signals into the body are generally known as *electrical stimulation* devices, examples of which include cardiac pacemakers, transcutaneous electronic nerve stimulators (TENs) for pain suppression, and neural prostheses such as auditory stimulation systems for the deaf and phrenic nerve stimulators for artificial respiratory control. In these and other similar devices, electrodes transmit current to appropriate areas of the body for direct control of, or indirect influence over, target cells.

Need More Help?
Try this video on bioelectrodes: https://www.youtube.com/watch?v=HGtgpkloAAA

Electrically excitable cells, such as those of the nervous system or heart, possess a potential difference across their membranes of approximately 60–90 mV. The inside of the cell is negative with respect to the outside. Such can transmit electrical signals, called *action potentials*, along their lengths. During an action potential, the cellular membrane (cf. Figure 1.85) changes polarity for a duration of about 1 ms, so that the inside of the cell becomes positive with respect to the outside. Excitable cells can be artificially stimulated by electrodes that introduce a transient electric field of proper magnitude and distribution. In general, the field generated near a cathodically driven electrode can be used to depolarize most efficiently an adjacent excitable cell's axonal process above a threshold value at which an artificially generated action potential results. Cathodes will tend to draw current outward through nearby cell membranes. In their passive state, such membranes can be electrically modeled as parallel arrangements of resistors and capacitors. Outward current therefore elicits depolarizing resistive potential drops and capacitive charging. Depending upon the specific application and electrode properties, a single action potential in a nerve cell might be elicited using currents on the order of a few microamps to a few milliamps, for durations of microseconds to milliseconds.

The materials used in bioelectrodes for electrical stimulation vary, depending upon the specific application. For example, intramuscular stimulation requires an electrode with excellent long-term strength and flexibility, but not very high stimulation charge densities. Stimulation of the visual cortex through placement of electrodes on the cortical surface, however, requires little mechanical flexibility in an electrode material with relatively moderate charge needs. Intracortical stimulation electrodes must be fabricated on a very small scale and must be capable of injecting very high charge densities. In all instances, the device must be biocompatible. In most neural control devices, the surface area of the exposed material is relatively small (<1 cm^2), so problems associated with passive biocompatibility are minor. The noble metals such as platinum, iridium, rhodium, gold, and palladium are generally preferred for the conductor material due to their low chemical reactivity and high resistance to corrosion. The insulating materials around the conductive electrodes, however, have a much higher volume and surface area, and they represent the largest problems associated with biocompatibility. Thin layers of medical-grade silicone rubber, poly (tetrafluoroethylene), polyimide, or epoxy are typically used for this purpose. Other devices are fabricated using lithography and thin-film technology that are typically insulated with ceramics such as SiO_2 and Si_3N_4. These insulting materials must be carefully selected to provide pinhole-free coatings, good adhesion to the conductors, and biocompatibility with the tissue.

Once inserted, charge transfer between the metal electrode and biological tissue requires a change in charge carriers from electrons in the metal to ions in the tissue fluid. This change in charge carriers occurs through one of two mechanisms. The first is a capacitive mechanism involving only the alignment of charged species at the electrode–tissue interface, in which charging and discharging occur at the electrode double layer. This mechanism is preferred because no chemical change occurs in either the tissue or the electrode. The amount of charge that can be transferred solely by capacitive charging is only about 20 μC/cm^2 of "real" electrode surface area, which is the actual area of the electrode–tissue interface. The second mechanism involves the exchange of electrons across the electrode–tissue interface and therefore requires that some chemical species be oxidized or reduced. Metal electrodes almost always inject charge by this so-called *faradaic charge transfer process* because the amount of charge required greatly exceeds that available from the capacitive mechanism alone. The types of reactions that occur at the electrode–tissue interface include monolayer oxide formation and reduction, the electrolysis of water, or the oxidation of chloride ions.

Capacitor electrodes are considered the ideal type of stimulation electrode because the introduction of a dielectric at the electrode–tissue interface allows charge flow completely by charging and discharging the dielectric film. The oxide film that is produced on tantalum by anodic polarization withstands substantial voltage without significant current leakage and permits charge flow without the risk of faradaic charge transfer reactions. Anodized titanium has higher dielectric strength than anodized tantalum, but its current leakage is too high for electrode applications. Capacitor electrodes based on tantalum–tantalum pentoxide have a charge storage capacity of about 100–150 μ C/cm^2. Their use is limited to neural prosthesis applications having electrodes about 0.05 cm^2 or larger in surface area, such as peripheral nerve stimulators. It is not possible to obtain adequate charge storage capacity in electrodes smaller then 10^{-3} cm^2.

6.1 Magnetic Properties of Materials

Just as materials have a response when placed in an electric field, they can have a response when placed in a magnetic field. We will see in this section that many of the concepts of permanent dipoles and dipole alignment in response to an applied field that was described in the context of electrical fields apply to magnetic fields as well. There are a few differences, however, and we will also see that there are fewer materials with specialized magnetic properties than there were with specialized electrical and electronic properties. The magnetic properties of materials are nonetheless important, and they are applied in a number of technologically important areas.

6.1.1 Magnetic Properties of Metals and Alloys

The Molecular Origins of Magnetism

Recall from Section 1.0.3 that charge separation in a molecule leads to an electric dipole, and that this dipole can be either permanent (intrinsic) or induced. And just as the magnitude of the electric dipole moment, p_e, was proportional to the charge and separation distance, so is the magnitude of the *magnetic dipole moment*, p_m, proportional to the pole strength, m, and the separation distance, a:

$$p_m = m \cdot a \tag{6.57}$$

Intrinsic magnetic dipole moments arise from two sources: (a) the orbital motion of electrons about the nucleus and (b) the net spin of unpaired electrons. In the first case, the intrinsic magnetic dipole is not observed until acted on by an external field. In the presence of an external magnetic field, the component of the magnetic moment that is parallel to the applied field is given by $m_l(eh/4\pi m_e)$, where m_l is the magnetic quantum number, e and m_e are the magnitude of the electronic charge and mass, respectively, and h is Planck's constant. Any filled shell or subshell contributes nothing to the orbital magnetic moment of an atom, but occupied outer electronic orbitals contribute to a net atomic magnetic moment (parallel to the applied field) if the summation of magnetic quantum numbers is not equal to zero. The quantity $eh/4\pi m_e$ is a fundamental quantity in magnetism called the *Bohr magneton* and is assigned the symbol μ_B. The Bohr magneton has a value of 9.27×10^{-24} A · m^2. The quantity $h/2\pi$ is the angular momentum of the electron. The second type of intrinsic magnetic dipole moment is associated with the spin of an electron. When the spin quantum number, m_s, is +½ (spin up), the magnetic moment is $+\mu_B$ (parallel to the magnetic field), and when $m_s = -½$ (spin down), the magnetic moment is $-\mu_B$ (antiparallel to the magnetic field). For free atoms like Mg, no permanent magnetic moment exists because there are no unpaired electrons, and the individual orbital momentums sum to zero. Free atoms like Na and molecules like O_2, each of which has unpaired electrons (see Figure 1.6), possess a permanent magnetic dipole moment. When both spin and orbital components of the magnetic dipole moment exist, they are additive, though not in a simple arithmetic manner.

Recall from Section 1.0.4 that the pairing of electrons in molecular orbitals of molecules results in two types of magnetism. Diamagnetism is the result of all paired electrons in the molecular orbitals, whereas paramagnetism results from unpaired electrons in the molecular orbitals. Let us now examine how these two types of magnetism behave in the presence of an imposed magnetic field. This can be done by comparing the *magnetic induction, B*, or internal magnetic field strength, within a substance in the presence of an external magnetic field, H, to the induction produced by the field alone in a vacuum, B_0.

First, consider the case of a field in a vacuum. The field can be created by passing current, I, through a coil of N turns (called a *solenoid*) of total length, l, as shown in Figure 6.49. The field is simply given by:

$$H = \frac{N \cdot I}{l} \tag{6.58}$$

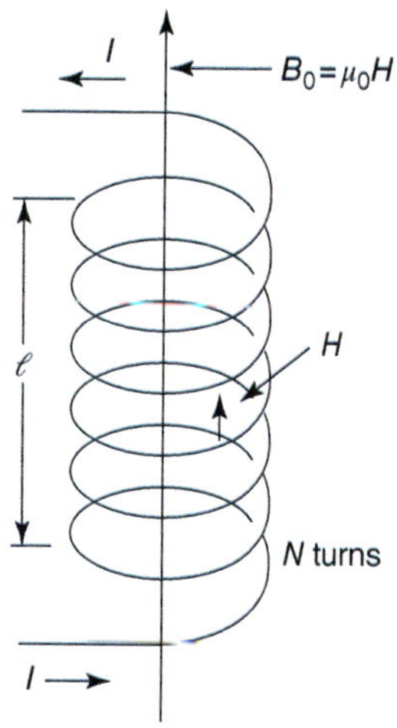

Figure 6.49 Schematic diagram of a solenoid in a vacuum. Ralls et al. [15]. Copyright © 1976 by John Wiley & Sons, Inc. This material is used by permission John Wiley & Sons, Inc.

Recall from Eq. (6.2) that the induction and field are related as follows:

$$B_0 = \mu_0 H \tag{6.59}$$

where the subscript "0" has been added to indicate a vacuum, and boldface has been dropped to indicate that the inductance and field—normally vectors—are treated as scalar quantities only. The constant μ_0 is the *magnetic permeability in vacuum* and has a value of $4\pi \times 10^{-7}$ Wb/A · m.

If a substance is now placed inside the solenoid, as shown in Figure 6.50, the magnetic field remains the same according to Eq. (6.58) as long as the current remains the same, but due to interactions between the field and the substance, the inductance is different (lower or higher), and Eq. (6.59) becomes

$$B = \mu_0 H + \mu_0 M \tag{6.60}$$

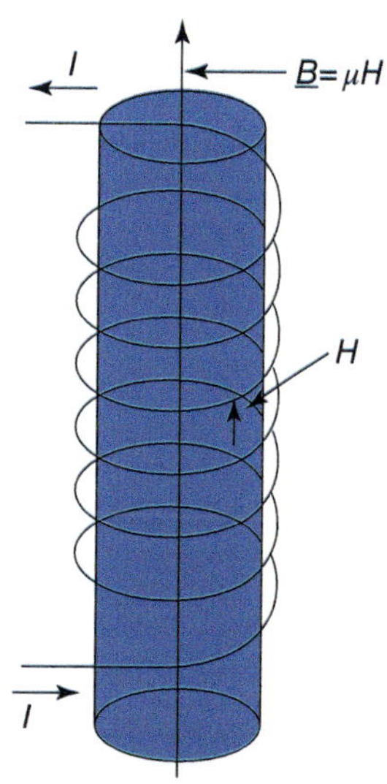

Figure 6.50 Schematic diagram of a substance within a solenoid. Ralls et al. [15]. Copyright © 1976 by John Wiley & Sons, Inc. This material is used by permission John Wiley & Sons, Inc.

where the second term contains the *magnetization, M*, of the substance. Magnetization represents the net magnetic dipole moment per unit volume that is aligned parallel to the external field. For many substances, a direct proportionality exists between the magnetization and the magnetic field:

$$M = \chi H \tag{6.61}$$

with the dimensionless proportionality constant, χ, called the *magnetic susceptibility*. Do not confuse the magnetic susceptibility with the electronegativity of an element, which is represented by the same variable. The benefit of introducing magnetic susceptibility is that it is a material property, independent of quantity. Even if a direct proportionality does not exist between the magnetization and the magnetic field, Eq. (6.61) still defines the instantaneous magnetic susceptibility and can be substituted into Eq. (6.60) to yield

$$B = \mu_0(1 + \chi)H \tag{6.62}$$

The quantity $(1 + \chi)$ is called the *relative magnetic permeability* (also dimensionless), which is also a material property.

Cooperative Learning Exercise 6.5

A coil of wire 0.20 m long and having 200 turns carries a current of 10 A.

Person 1: Calculate the magnetic field strength in vacuum.

Person 2: Look up the magnetic susceptibility for titanium.

A bar of titanium is now placed in the coil. Combine your information to make the following calculations.

Person 1: Calculate the flux density (magnetic induction) with the titanium bar in the coil.

Person 2: Calculate the flux density (magnetic induction) in vacuum (without the titanium bar).

Compare your answers. Does the titanium bar affect the magnetic induction significantly? Why or why not?

Person 1: What is the relative magnetic permeability of titanium?

Person 2: What is the magnetization of titanium inside the coil?

Answers: Equation (6.58) $H = (200)(10)/(0.2) = 10{,}000$ A/m; from Table 6.15, $\chi = 0.182 \times 10^{-3}$. Equation (6.62) $B = 4\pi \times 10^{-7}\ (1 + 0.182 \times 10^{-3})(10.000) = 0.0126\ \text{Wb/m}^2$; $B_0 = 4\pi \times 10^{-7}$ $(10.000) = 0.0125\ \text{Wb/m}^2$; they are essentially the same due to the small susceptibility of titanium. $(1 + \chi) = 1.000182$; Eq. (6.61) $M = (0.000182)(10,\ 000) = 1.82$ A/m.

With this background on magnetic field production and susceptibility in hand, we can return to our description of diamagnetic and paramagnetic behaviors in an applied magnetic field. Substances consisting of atoms or molecules that do not possess permanent magnetic moments exhibit diamagnetism in a magnetic field. That is, they display a negative, but generally quite small, magnetic susceptibility because of the weak, induced opposing magnetic dipole moments. An induced diamagnetic component of magnetization is inherent in all substances but is often masked by larger intrinsic moments that reinforce an applied field. Inert gases and solids like Bi, Cu, MgO, and diamond are examples of diamagnetic materials and have typical magnetic susceptibilities on the order of -10^{-5}. The diamagnetic component of susceptibility is independent of temperature and magnetic field. For the most part, diamagnetic materials are of little engineering significance, with the notable exception of superconductors. Type I superconductors can be perfectly diamagnetic ($\chi = -1$ or $B = 0$) up to the critical magnetic field because induced shielding currents persist indefinitely. Type II superconductors exhibit perfect diamagnetism or partial paramagnetism depending on the applied field (cf. Figure 6.32). The strong diamagnetism of superconductors makes it possible to use such materials for magnetic field shielding purposes.

Substances having magnetic susceptibilities in the range from about 10^{-6} to 10^{-2} are called paramagnetic. As described above, paramagnetism results from net intrinsic magnetic moments associated with electronic orbitals and electron spins. For paramagnetic materials, the interactions between atomic moments and an applied field can be described in terms of thermodynamics. When the individual atomic moments are randomly arrayed, the entropy of the material is high. Countering this tendency is the magnetic interaction energy, which is decreased when the moments align with the field. The final degree of moment alignment, therefore, depends on the competing factors of energy and entropy. Thus, the susceptibility, which reflects the degree of alignment, tends to decrease as temperature increases and entropy becomes more dominant. The results of

Table 6.15 Molar Magnetic Susceptibilities (in mol/cm^3 × 10^6) of Common Paramagnetic Metals at 20°C

IA	IIA	IIIA	IVA	VA	VIA	VIIA		VIII		IB	IIB	IIIB	IVB	VB
Li 14.2	Be Diamagnetic													
Na 16	Mg 13.1	Al 16.5												
K 20.8	Ca 40	Sc 295.2	Ti 151	V 285	Cr 167	Mn 511	Fe Ferromagnetic	Co Ferromagnetic	Ni Ferromagnetic	Cu Diamagnetic	Zn Diamagnetic	Ga Diamagnetic		
Rb 17	Sr 92	Y 187.7	Zr 120	Nb 208	Mo 72	Tc 115	Ru 39	Rh 102	Pd 540	Ag Diamagnetic	Cd Diamagnetic	In Diamagnetic	Sn Diamagnetic	
Cs 29	Ba 20.6	Rare Earths	Hf 71	Ta 154	W 53	Re 67	Os 11	Ir 25	Pt 193	Au Diamagnetic	Hg Diamagnetic	T1 Diamagnetic	Pb Diamagnetic	Bi Diamagnetic

Source: Ref. [62].

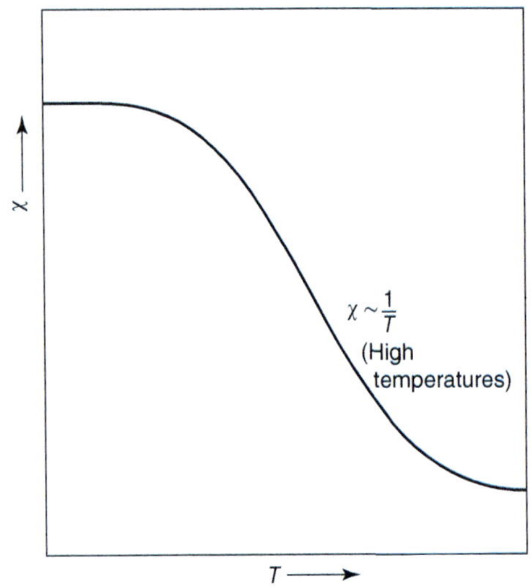

Figure 6.51 Temperature dependence of magnetic susceptibility for a paramagnetic material. Ralls et al. [15]. Copyright © 1976 by John Wiley & Sons, Inc. This material is used by permission John Wiley & Sons, Inc.

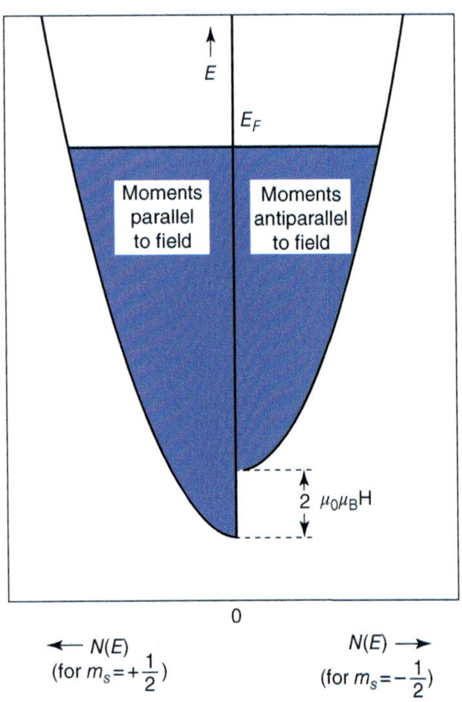

Figure 6.52 Distribution of electrons within the conduction band of a metal in the presence of an applied magnetic field. Ralls et al. [15]. Copyright © 1976 by John Wiley & Sons, Inc. This material is used by permission John Wiley & Sons, Inc.

thermodynamic calculations that minimize the free energy as a function of temperature, and show that the susceptibility decreases with temperature and at intermediate temperatures is given approximately by

$$\chi = \frac{\mu_0 N p_m^2}{3k_B T} \tag{6.63}$$

where p_m is the magnetic dipole moment per atom as given in Eq. (6.57), N is the number of atoms per unit volume, k_B is Boltzmann's constant, T is absolute temperature, and μ_0 is the magnetic permeability in vacuum constant. The dependence of magnetic susceptibility on reciprocal absolute temperature is called the *Curie Law*, and it holds for most paramagnetic nonmetallic materials above room temperature. As shown in Figure 6.51, the Curie Law does not hold at low temperatures, and the susceptibility approaches a constant value due to saturation.

In metals, conduction electrons provide both paramagnetic and diamagnetic moments, with a net paramagnetic moment, in addition to the induced diamagnetic moment resulting from core electrons. The paramagnetic component arising from the conduction electrons usually predominates, and most metals have positive susceptibilities. The origin of this paramagnetism can be explained in terms of the conduction electron energy band viewed as two subbands, one containing electrons with spin up and the other containing valence electrons with spin down. With no applied magnetic field, the subbands have equal populations of electrons and there is no net spin. Application of a magnetic field lowers the potential energy for the subband having spins parallel to the field and raises the potential energy of the other subband. Since the Fermi energy must be at the same level in both subbands when equilibrium exists (i.e., the electronic free energy is constant), a greater number of conduction electrons reinforce the field than oppose it, as illustrated in Figure 6.52. Those electrons with moments aligned parallel to the field ($m_s = +½$) have a magnetic potential energy $2\mu_0\mu_B H$ lower than those electrons with moments antiparallel ($m_s = -½$). Equalization of the Fermi energy within these subbands leads to a greater population of electrons with moments parallel to the field and a resulting net magnetic moment.

It can be shown that the conduction electron net spin susceptibility is proportional to the temperature coefficient of the electronic heat capacity [cf. Eq. (4.42)] and, for free electrons in a single band, having the Fermi energy much lower than any band gap, is given by

$$\chi = \frac{\mu_0 N_v \mu_B^2}{E_F} \tag{6.64}$$

where E_F is the energy of the Fermi level, N_v is the total number of valence electrons per unit volume [see Eq. (6.23)], μ_0 is the magnetic permeability in vacuum constant, and μ_B is the Bohr magneton. Figure 6.53 shows us that only those electrons having energies near the Fermi level are responsible for spin paramagnetism, and this has been accounted for in Eq. (6.64). In general, the spin susceptibility is proportional to the density of states at the Fermi energy, so that transition metals exhibit larger paramagnetic susceptibilities than do nontransition metals (see Table 6.15 for some molar magnetic susceptibilities of paramagnetic elements), just as they exhibit larger electronic heat capacities.

Some metals are diamagnetic because the conduction electron spin susceptibility is smaller than the induced diamagnetic susceptibility component. On the other hand, various rare earth metals display very strong paramagnetism because of unpaired *f* electrons that remain associated with individual atoms rather than entering into energy bands.

Types of Magnetism

In addition to diamagnetic and paramagnetic behavior, there are other types of magnetism, which are important. These include *ferromagnetism*, *antiferromagnetism*, and *ferrimagnetism*. The various types of magnetism will be summarized and compared here, and then described in more detail under the appropriate material-related section. In most cases, the different types of magnetism can be differentiated by their variation of magnetic susceptibility with temperature.

The elements Fe, Co, Ni, Gd, Tb, Dy, Ho, and Tm, as well as a number of alloys and compounds, exhibit very large positive susceptibilities below a critical temperature, θ_c, known as the

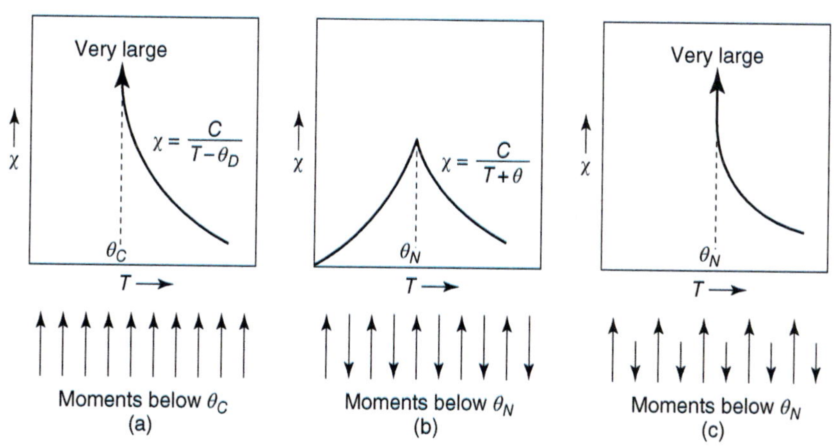

Figure 6.53 The magnetic spin alignments and variation of susceptibility with temperature for (a) ferromagnetic, (b) antiferromagnetic, and (c) ferrimagnetic materials. Ralls et al. [15]. Copyright © 1976 by John Wiley & Sons, Inc. This material is used by permission John Wiley & Sons, Inc.

Table 6.16 Distribution of Electrons and Electronic Magnetic Moments in Some Transition Metals

Element	Fe	Co	Ni	Cu
Total number ($s + d$) valence electrons	8	9	10	11
Average number of electrons in s band	0.2	0.7	0.6	1.0
Average number of electrons in d band	7.8	8.3	9.4	10.0
Number of d electrons with moments up	5.0	5.0	5.0	5.0
Number of d electrons with moments down	2.8	3.3	4.4	5.0
Net moment per atom (units of Bohr magnetons, μ_B)	2.2	1.7	0.6	0

Source: Ralls et al. [15]. Copyright © 1976 by John Wiley & Sons, Inc.

ferromagnetic *Curie temperature*. (We use the Greek lowercase theta and subscript "c" to designate the Curie temperature to differentiate it from the critical transition temperatures of superconductors, which are designated as T_c.) Such substances display spontaneous magnetization, the origin of which arises from an internal interaction that causes a permanent offset in the electronic subbands. Only certain metals and other solids having partially filled d subshells or partially filled f subshells are capable of ferromagnetism (see Table 6.16). In addition to this necessary, but not sufficient, condition for spontaneous permanent magnetization, the interatomic spacing must be such that an internal interaction, termed the *exchange interaction*, occurs. As shown in Table 6.16, the elements Fe, Co, and Ni are magnetized spontaneously, whereas Cu does not. This is because in Fe, Co, and Ni, the average number of electrons in the d band having spin up is not the same as that having spin down, and the difference between the two gives a net magnetic moment. With increasing temperature, the lowering of internal energy because of the exchange interaction is opposed by a greater entropy that results from the disorder of randomized spins. Ferromagnetic materials become paramagnetic above θ_c; susceptibilities can be on the order of 250–100,000 or more and continue to increase with decreasing temperature (see Figure 6.53a). Typically, susceptibilities at $0.9\theta_c$ are approximately half those at absolute zero.

In ferromagnetic materials, the exchange interaction gives rise to a parallel alignment of spins. In some metals like Cr, this exchange interaction, which is extremely sensitive to interatomic spacing and atomic positions, results in an antiparallel alignment of spins. When the strength of antiparallel spin magnetic moments is equal, no net spin moment exists and the resulting susceptibilities are quite small. Such materials are called antiferromagnetic. The most noticeable characteristic of an antiferromagnetic material is that the susceptibility attains a maximum at a critical temperature, θ_N, called the *Néel temperature*.

The maximum susceptibility as a function of temperature can be explained using thermodynamic reasoning. At absolute zero, the spin moments are aligned as antiparallel as possible, yielding a small susceptibility. With increasing temperature, the tendency for disorder or more random alignment of spin moments becomes greater, and a higher net moment results in the presence of an applied magnetic field. This same process accounts for the decrease in net moment of ferromagnetic materials between absolute zero and θ_c. For antiferromagnetic materials, the spin moments become essentially random at θ_N, and above this temperature, paramagnetic behavior with decreasing χ is observed (see Figure 6.53b). Thus, the Néel temperature of an antiferromagnetic temperature is analogous to the Curie temperature of a ferromagnetic material. We will discuss antiferromagnetic materials in further detail in a later section.

Ferrimagnetism is similar to antiferromagnetism in that the spins of different atoms or ions tend to line up antiparallel. In ferrimagnetic materials, however, the spins do not cancel each other out, and a net spin moment exists. Below the Néel temperature, therefore, ferrimagnetic materials behave very much like ferromagnetic materials and are paramagnetic above θ_N (see Figure 6.53c). Ferrites, a particularly important class of ferrimagnetic materials, will be described later.

The different types of magnetism described so far are summarized in terms of the sign and magnitude of their susceptibilities in Table 6.17. A summary of the terms, symbols, and units used in magnetism are provided in Table 6.18, in a manner analogous to the electrical terms in Table 6.5.

Table 6.17 Summary of the Various Types of Magnetism

Type of Magnetic Behavior	Characteristics of Magnetic Susceptibility		Typical Materials
	Sign	Magnitude	
Diamagnetism	Negative	Small[a] χ = constant	Organic materials, superconducting metals, and other metals (e.g., Bi)
Paramagnetism	Positive	Small χ = constant	Alkali and transition metals, rare earth elements
Ferromagnetism	Positive	Large $\chi = f(H)$	Some transition metals (Fe, Ni, and Co) and rare earth metals (Gd)
Antiferromagnetism	Positive	Small χ = constant	Salts of transition elements (MnO)
Ferrimagnetism	Positive	Large $\chi = f(H)$	Ferrites ($MnFe_2O_4$, $ZnFe_2O_4$) and chromites

[a] Diamagnetic susceptibilities for superconducting metals are large.

Table 6.18 Magnetic Symbols, Terms, and SI Units

Quantity	Symbol	SI Units	
		Primary	Derived
Magnetic field strength	H	C/m · s	A/m = 10^{-3} oersted (Oe)[a]
Magnetic induction (magnetic flux density)	B	kg/s · C	V · s/m^2 = Weber (Wb)/m^2 = tesla (T) = 10^4 gauss (G)[a]
Magnetic susceptibility	χ	**(dimensionless)**	**(dimensionless)**
Magnetization	M	C/m · s	A/m
Magnetic dipole moment	p_m	C · m^2/s	A · m^2
Remnant induction	$\boldsymbol{B_r}$	**kg/s · C**	**V · s/m^2 = Weber/m^2 = tesla (T) = 10^4 gauss (G)***
Coercivity (coercive force)	$\boldsymbol{H_c}$	**C/m · s**	**A/m = 1^{-3} Oe (Oe)***
Magnetic permeability in vacuum	μ_0	kg · m/C^2	Henry/m = Wb/A · m
Saturation magnetization	$\boldsymbol{M_s}$	C/m · s	A/m

[a]cgs unit; items in boldface are material properties.

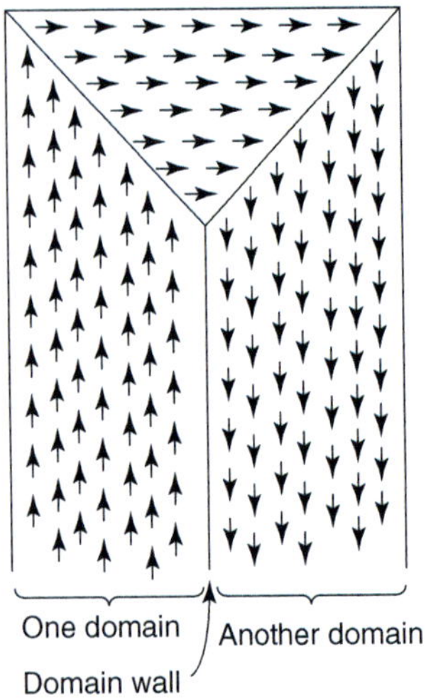

Figure 6.54 Schematic illustration of magnetic domains.

Magnetic Domains and Hysteresis

In addition to susceptibility differences, the different types of magnetism can be distinguished by the structure of the magnetic dipoles in regions called *magnetic domains*. As shown in Figure 6.54, each domain consists of magnetic moments that are aligned, giving rise to a permanent net magnetic moment per domain. The domains are separated by domain boundaries (not the same as the grain boundaries), or domain walls, across which the direction of the spin alignment gradually changes. Typically, the domain wall has a thickness of about 100 nm. This type of arrangement represents the lowest free energy in the absence of an externally applied magnetic field. When the bulk material is unmagnetized, the net magnetization of these domains is zero, because adjacent domains may be oriented randomly in any number of directions, effectively canceling each other out.

The average magnetic induction of a ferromagnetic material is intimately related to the domain structure. When a magnetic field is applied, the domains most nearly parallel to the direction of the applied field grow in size at the expense of the others. This is called *boundary displacement of domains* or *domain growth*. This process continues until only the most favorably oriented domains remain. The spontaneous moments are aligned naturally with respect to certain crystallographic directions; in the case of iron, they are parallel to the [100] direction. When domain growth is completed, a further increase in the magnetic field causes the domains to rotate and align parallel to the applied field. The material reaches the point of saturation magnetization, and no further change will take place on increasing the strength of the magnetic field.

> **Need More Help?**
>
> Try these videos on magnetism and magnetization:
>
> https://www.youtube.com/watch?v=7G3smchVP9U
> https://www.youtube.com/watch?v=MN_XtX7WYYc
> https://www.youtube.com/watch?v=m1PPujngqAw
> https://www.youtube.com/watch?v=hRyxCilntDg
>
> and hysteresis loop:
>
> https://www.youtube.com/watch?v=oyJZ1G8l0eU

This change in magnetization with applied magnetic field is shown schematically in Figure 6.55, in a diagram known as a *hysteresis loop*. Hysteresis is defined as the lag in magnetization change that follows variations in the applied magnetic field. If an unmagnetized specimen, such as iron, is placed in a magnetic field, H, a magnetic induction, B, results, as described earlier. When the field is increased, the magnetic induction will also increase, as shown in curve (a) of Figure 6.55. This is the domain growth region. At sufficiently high magnetic fields, all of the magnetic dipoles become aligned with the external field, and the material becomes magnetically saturated; that is, the magnetization becomes constant and is called the *saturation magnetization*, M_s. There is a corresponding *saturation flux density*, B_s, as the induction reaches a maximum value. Once magnetic saturation has been achieved, a decrease in the applied field back to zero, as illustrated in curve (b) of Figure 6.55, results in a macroscopically permanent induction, called the *remnant induction*, B_r, under which the material is in the magnetized condition, even at zero applied field. At this point, spin orientations within domains have readily rotated back to their favorable crystallographic positions, but the original random domain arrangement is not achieved because domain wall motion is limited and the domain growth process is not

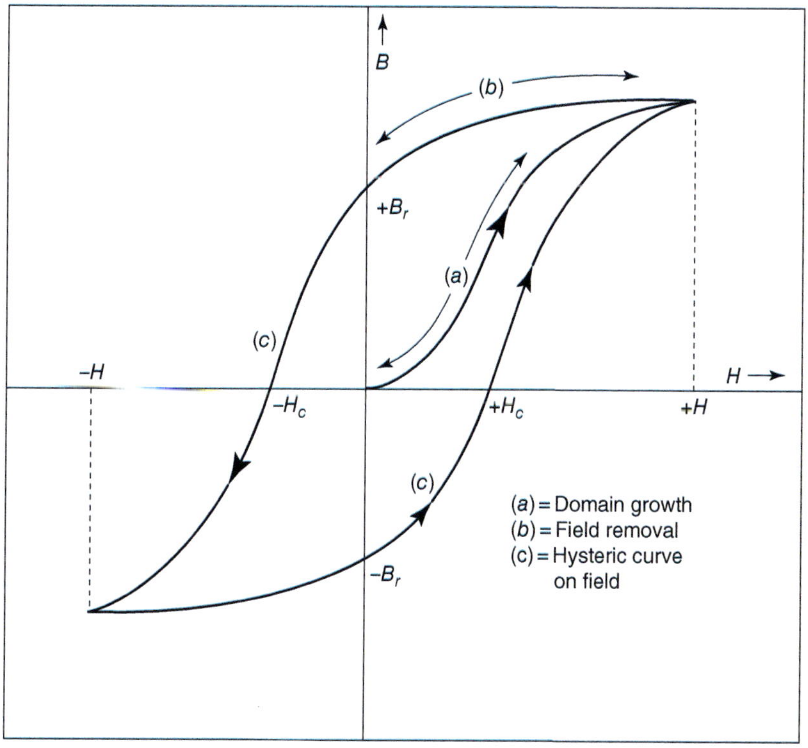

Figure 6.55 A hysteresis loop for a ferromagnetic materials. Ralls et al. [15]. Copyright © 1976 by John Wiley & Sons, Inc. This material is used by permission John Wiley & Sons, Inc.

entirely reversible. To bring the induction back to zero, a field of magnitude H_c, termed the *coercivity*, must be applied antiparallel to the original magnetic field, shown as curve (*c*) in Figure 6.55. A further increase in the magnetic field in the negative direction results in a maximum induction and saturation magnetization but in the opposite direction of the previous maximum achieved under an applied field in the forward direction. The field can once again be reversed, and the field-induction loop can be closed. The area within the hysteresis loop represents the energy loss per unit volume of material for one cycle and, unless suitable cooling is provided, the material heats up. Smaller hysteresis loops result for cycling between applied fields that do not cause saturation magnetization.

Cooperative Learning Exercise 6.6

Recall that the saturation magnetization, M_s, is the maximum possible magnetization in the material. It can be described as the product of the net magnetic moment per atom, p_m, and the number of atoms per unit volume, N:

$$M_s = p_m N$$

The net magnetic moment, in turn, is related to the electronic structure (paired or unpaired electrons), although a number of other factors come into play. Use this information to calculate the saturation magnetization for nickel.

Person 1: Look up the net magnetic moment of nickel in Table 6.16 and convert it from units of Bohr magnetons to $A \cdot m^2$.

Person 2: Look up the density and atomic weight of nickel in Tables 1.12 and 1.3, respectively, and calculate the number of atoms per cubic meter for nickel. (Do not forget Avogadro's number.)

Combine your information to calculate M_s for nickel in units of A/m.

Since the magnetic susceptibilities of ferromagnetic materials are so high, $H << M$, and Eq. (6.60) can be simplified. Use this information and your value of M_s to calculate the saturation flux density, B_s.

Answers: p_m = 0.60 Bohr magnetons = $0.60(9.27 \times 10^{-24}$ $A \cdot m^2 = 5.56 \times 10^{-24}$ $A \cdot m^2$;
$N = \rho_{Ni} N_A/(A.W.)_{Ni} = 98.91)(10^6)(6.02 \times 10^{23})/(58.71) = 9.136 \times 10^{28}$ atoms/m^3.
$M_s = 5.1 \times 10^5$ A/m. Equation (6.61) simplified to $B_s = \mu_0 M_s = 0.638$ Wb/m^2.

When applied magnetic fields cyclically vary at some finite speed, *eddy currents* are formed. Eddy currents are electrical currents that are set up in the material due to the applied field and that lead to additional energy losses. The magnitude of these currents depends on the frequency and flux density imposed by the magnetic field, as well as on the specific resistance and geometry of the material. The effect of eddy currents is to increase the magnetic field, H, for the same induction, causing a much wider hysteresis loop. The sum of the hysteresis loss and eddy current loss is known as the *total core loss*, and it is usually expressed in watts per pound of a standard sample thickness and at a given magnetic induction.

Magnetic materials can be divided into soft and hard magnets, depending on their behavior in the presence of a magnetic field. This is the topic of the next two sections.

Soft Magnets

Soft magnets, sometimes called *permeable magnets*, have high magnetic permeabilities and low coercive forces, H_c; they are easily magnetized and demagnetized. In devices subjected to alternating magnetic fields, such as transformer cores, not only is a high magnetic permeability and high saturation induction desirable, but it is also necessary that the material dissipate as little energy per cycle as possible. In other words, the core material must have a hysteresis loop of small area, as characterized by a low coercive force, as illustrated in the inner loop of Figure 6.56.

In ferromagnetic materials, the saturation magnetization is primarily a function of composition, but the susceptibility, coercivity, and shape of the hysteresis curve are very sensitive to structure. A low coercivity coincides with easy motion of domain walls. Domain wall motion is restricted to structural defects such as nonmagnetic inclusions, voids, and precipitates of a nonmagnetic phase. Domain walls are pinned by these defects because the wall energy is lowered. Consequently, the number of such defects must be minimized in a soft magnetic material. Impurities and dislocation structures resulting from cold deformation also lead to higher hysteresis.

Energy losses in soft magnetic materials arise due to both hysteresis and eddy currents, as described in the previous section. Eddy current losses can be reduced by increasing the electrical resistivity of the magnetic material. This is one reason why solid solution iron–silicon alloys (4% Si) are used at power frequencies of around 60 Hz and why iron–nickel alloys are used at audio frequencies. Some magnetically soft ferrites are very nearly electrical insulators and are thus immune to eddy current losses. Some common soft magnetic materials and their properties are listed in Table 6.19. Soft magnetic alloys are described further in a subsequent section.

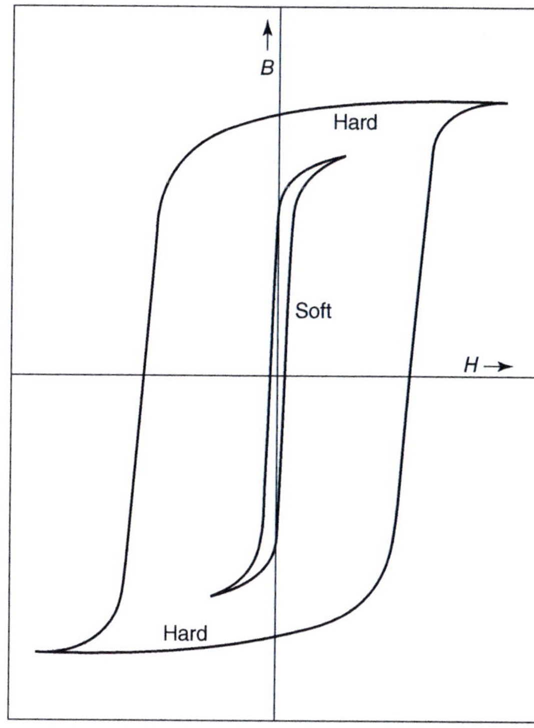

Figure 6.56 Comparison of hysteresis loops for soft and hard magnetic materials. Ralls et al. [15]. Copyright © 1976 by John Wiley & Sons, Inc. This material is used by permission John Wiley & Sons, Inc.

Table 6.19 Selected Properties of Some Soft Magnetic Materials

Material	Initial Susceptibility (χ at $H \sim 0$)	Hysteresis Loss per Cycle (J/m^3)	Saturation Induction (Wb/m^2)
Commercial iron ingot	250	500	2.16
Fe—4% Si, random	500	50–150	1.95
Fe—3% Si, oriented	15,000	35–140	2.0
45 Permalloy (45% Ni—55% Fe)	2700	120	1.6
Mu metal (75% Ni—5% Cu—2% Cr—18% Fe)	30,000	20	0.8
Supermalloy (79% Ni—15% Fe—5% Mo—0.5% Mn)	100,000	2	0.79
Ferroxcube A (Mn, Zn)Fe_2O_4	1200	~40	0.36
Ferroxcube B (Ni, Zn)Fe_2O_4	650	~35	0.29

Source: Ralls et al. [15]. Copyright © 1976 by John Wiley & Sons, Inc.

Hard Magnets

Permanent magnets require materials with high remnant inductions and high coercivities. These materials, called *hard magnets*, also generally exhibit large hysteresis losses. A typical hysteresis loop for a hard magnetic material is shown in the outer loop of Figure 6.56, which is shown relative to a soft material for comparison. The term hard magnet is derived from the mechanical properties of these materials, since increases in mechanical strength in ferromagnetic materials also tend to increase hardness. A relative measure of hardness in magnetic materials is the product of the remnant induction and the coercivity, $B_r \times H_c$, which is roughly twice the energy required to demagnetize a unit volume of material. That is, $(B_r \times H_c)/2$ is roughly twice the area under the B versus H curve between $H = 0$ and $H = -H_c$.

Hard magnetic behavior is intimately related to microstructure. Some permanent magnetic materials, including carbon steel and alloy steels containing combinations of Cr, Co, or W, undergo martensitic transformations upon cooling, resulting in a fine-grained microstructure. Others undergo an order–disorder transformation that results in internal strains. Specific hard magnetic materials will be described further in later sections.

Magnetism in Alloys

Recall from Table 6.15 that the transition metal Ni exhibits spontaneous magnetization, whereas Cu does not. The difference in magnetic behavior between these two transition metals can be explained in terms of band theory—specifically, the overlap of the s conduction band with the d band immediately below in energy. In the transition metals, the d band is not normally filled entirely. One exception is copper, Cu, which has a $3d^{10}4s^1$ outer shell configuration (cf. Table 1.3). The filled $3d$ band of copper has two separate subbands of opposite electron spin orientation, each holding five electrons. With both subbands filled, the net spin, and hence the net magnetization, of the d band is zero. Nickel, which has one less electron than copper, is ferromagnetic and has a saturation magnetic moment of $0.60\mu_B$ per atom at absolute zero. After a small correction for orbital electron motion, the result is that Ni at saturation has an excess of 0.54 electrons per atom with spin preferentially oriented in one direction. At absolute zero, the electrons are distributed as shown in Figure 6.57. We take one $3d$ subband as filled, and we take the other as filled except for a 0.54 hole. The 0.54 excess electron goes into the $4s$ band as shown, and it creates a net spin in the d-band due to the 0.54 hole in one of its subbands. The separation in energy between the $3d$ subbands is a result of a phenomenon called the *exchange interaction energy*, which is a function of temperature. When Ni is heated above θ_c (631 K), the exchange interaction is dominated by thermal effects, the 0.54 hole distributes itself equally in the two $3d$ subbands as shown in Figure 6.58, and the saturation magnetic moment disappears.

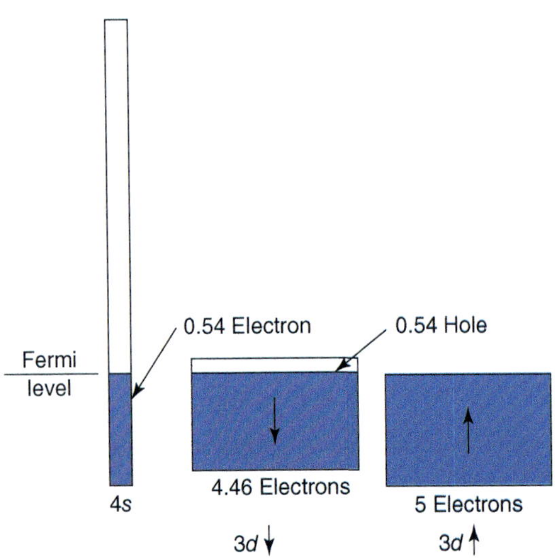

Figure 6.57 Schematic relationship of outer-shell energy bands in nickel at absolute zero.

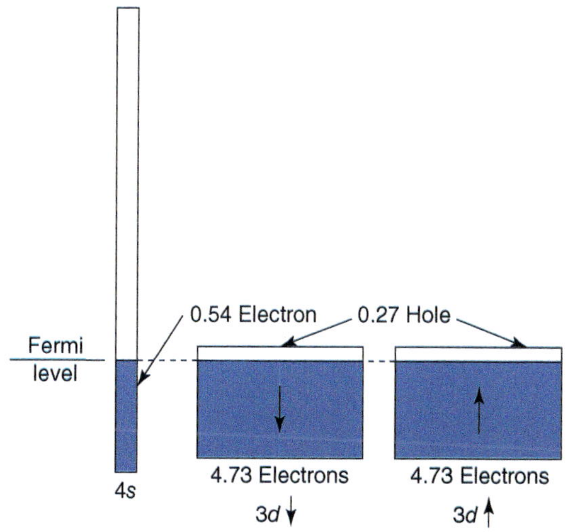

Figure 6.58 Schematic relationship of outer-shell energy bands in nickel over the Curie temperature.

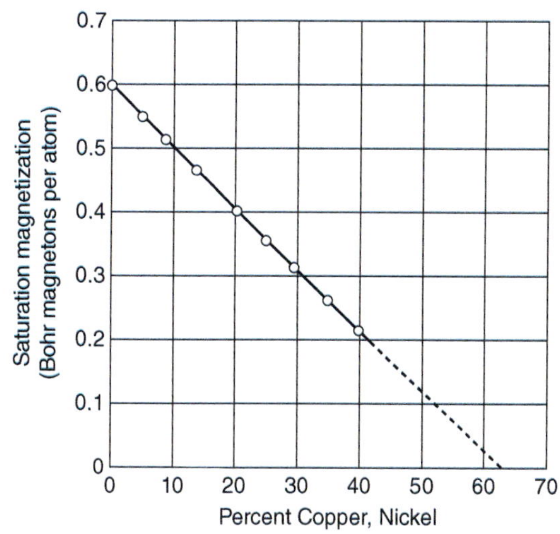

Figure 6.59 Variation in saturation magnetization for a copper–nickel alloy. Kittel [38]/Reproduced with permission of John Wiley & Sons.

When a series of solid solutions between Cu and Ni is formed, it is observed that the spontaneous magnetization decreases linearly with increasing Cu content until no spontaneous magnetization remains at 60 at.% Cu (see Figure 6.59). At this composition, we have added about 0.54 electrons to the d band and about 0.06 electrons to the s band. But the 0.54 electrons added to the d band in Figure 6.58 will fill both d subbands, and the magnetization will be zero. The 10.6 total valence electrons per atom are now equally divided between spin up and spin down, as shown in Figure 6.60.

This simple band model predicts that an alloying metal with $10 + z$ valence electrons outside a filled p shell (rare gas configuration) is expected to decrease the magnetization of nickel by approximately z Bohr magnetons per solute atom (μ_B). As illustrated in Figure 6.61, this simple relationship is obeyed extraordinarily well for small amounts of such atoms as Sn(+4), Al(+3), Zn(+2), Cu(+1), Co(−1), Fe(−2), and Mn(−3).

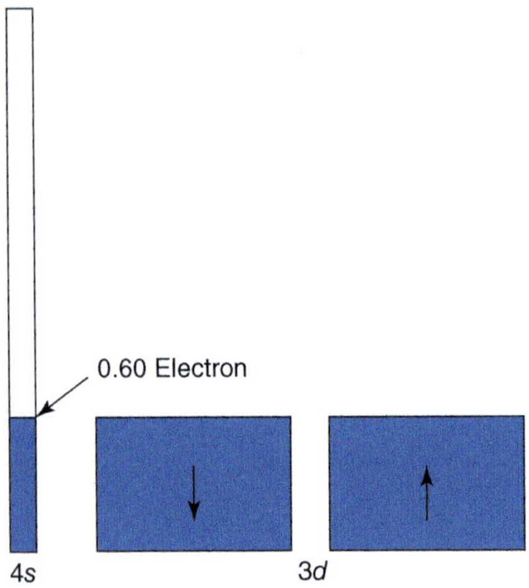

Figure 6.60 Schematic relationship of bands in a 60% Cu and 40% Ni alloy with zero magnetization.

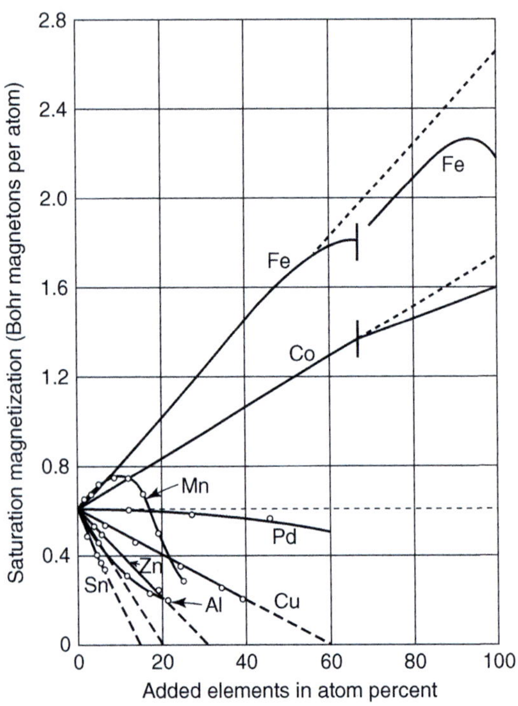

Figure 6.61 Saturation magnetization of some nickel alloys. Kittel [39]/Reproduced with permission of John Wiley & Sons.

Widely used permanent magnetic materials include low-alloy steels, containing between 0.6% and 1.0% carbon, that are hardened by quenching or precipitation hardening. Another important group is iron–nickel–aluminum alloys with a certain amount of cobalt, known as *Alnico alloys*. The permanent-magnet materials tend to be brittle and difficult to shape, so that they must be cast and finished by grinding. More ductile alloys such as copper–nickel–iron and copper–nickel–cobalt are available that can be cold-worked and machined by ordinary methods. All three types of alloys are subject to precipitation hardening, which can be controlled in such a way as to produce a microscopically heterogeneous structure composed of acicular particles oriented in a required direction. The size of these particles is very small and approaches the dimensions of a magnetic domain (about 1 μm). The reversal of magnetization can then occur only by a complete rotation of the magnetic domain in a particle. This, however, would require a very high magnetizing force because of the crystal anisotropy. Consequently, the boundary displacement of the domains is suppressed, resulting in very high coercive forces.

Rare-earth magnets have superior magnetic properties and are particularly suitable for use in compact and lightweight devices due to their coercive forces that are up to 10 times that of ordinary magnets. They are alloys of rare-earth elements, which include elements 57 through 71, so named because of the inherent difficulty in extracting them from ores, not due to their rarity. The most common rare-earth alloys for magnetic applications are samarium–cobalt (Sm–Co) and neodymium–iron–boron (Nd–Fe–B). Sm–Co rare-earth magnets, of which the composition $SmCo_5$ is the most common, are relatively stable to changes in temperature. Nd–Fe–B rare-earth magnets, such as $Nd_2Fe_{14}B$, have magnetic properties superior to those of Sm–Co. However, the magnetic flux variation is very sensitive to changes in temperature, making it necessary to select a magnetic material with properties suited to meet its specific service conditions.

6.1.2 Magnetic Properties of Ceramics and Glasses

Most of the important magnetic ceramics are of the ferrimagnetic class. However, some ceramics do exhibit other types of magnetic behavior. These ceramic materials will be described first, followed by a more thorough description of an important class of ferrimagnetic ceramics called ferrites. Finally, a topic related to the magnetic properties of ceramic superconductors will be introduced.

Diamagnetic, Paramagnetic, and Antiferromagnetic Ceramics and Glasses

In general, diamagnetism is associated with all ceramic materials in which the ions have closed electronic shells—that is, no unpaired electrons. This means that ceramics with no transition metal ions or rare-earth ions are generally diamagnetic. We will return to a special class of diamagnetic ceramics later in this section.

Paramagnetism results from unpaired electrons. As a result, most compounds containing transition, rare-earth, and actinide elements, including oxides, nitrides, carbides, and borides, exhibit paramagnetism. Such ceramics are generally not of importance due to their paramagnetism alone, since they often exhibit other types of magnetism, as well.

Aside from ferrimagnetic ceramics, which is the subject of the next section, the only other type of magnetism exhibited by ceramics is antiferromagnetism. Several of the transition metal monoxides exhibit antiferromagnetic behavior, in which the exchange between the unpaired electrons causes antiparallel spin alignments. These oxides include MnO, FeO, NiO, and CoO, all of which have the rock salt structure. For example, in FeO, the O^{2-} ions possess no net magnetic moment, whereas the Fe^{2+} ions do. Since neutrons have a magnetic moment, neutron diffraction can be employed to determine the spin arrangement, as shown in Figure 6.62. The spins of the *d* electrons of adjacent iron ions are aligned in opposite directions. Furthermore, any ions an a (111) plane have parallel spins, but ions in adjacent (111) planes have antiparallel spins, as illustrated in Figure 6.62. The aligned moments of the ions in the two directions cancel, and the FeO crystal as a whole has no magnetic moment. The ordering of spins in antiferromagnetic substances often introduces a complexity with respect to viewing their crystal structures. Because of the spin ordering, the magnetic unit cell has twice the lattice parameter of the standard unit cell. Some antiferromagnetic ceramics and their magnetic properties are listed in Table 6.20.

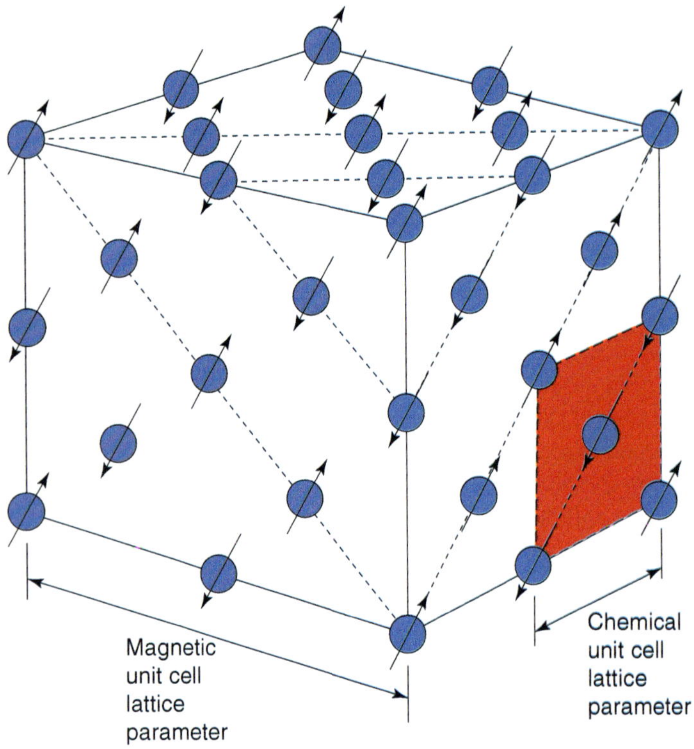

Figure 6.62 Schematic illustration of spin alignments in adjacent (111) planes of FeO. Ralls et al. [15]. Copyright © 1976 by John Wiley & Sons, Inc. This material is used by permission John Wiley & Sons, Inc.

Table 6.20 Selected Magnetic Properties of Some Antiferromagnetic Ceramics

Substance	Néel Temperature, θ_N (K)	χ (0 K)/χ (θ_N)
MnO	116	0.67
MnS	160	0.82
MnF_2	67	0.76
FeF_2	79	0.72
FeO	198	0.80
CrSb	723	~0.25
$FeCO_3$	35	~0.25

Source: Ralls et al. [15]. Copyright © 1976 by John Wiley & Sons, Inc.

Before moving on to ferrites, we should at least mention that magnetic glasses do exist, although any further coverage of them is beyond the scope of this text. Not surprisingly, the magnetic glasses are found chiefly in the group of glasses with high contents of transition elements or rare earths, as well as in the corresponding fluoride glasses but especially in the metallic glasses.

Table 6.21 Some Ferrites and their Magnetic Properties

Composition	Saturation Magnetization (A/m)	Curie Temperature (°C)	Crystal Structure
$MnFe_2O_4$	4.13×10^5	300	Inverse spinel
$FeFe_2O_4$	4.77×10^5	585	Inverse spinel
$NiFe_2O_4$	2.71×10^5	590	Inverse spinel
$CoFe_2O_4$	3.98×10^5	520	Inverse spinel
$CuFe_2O_4$	1.35×10^5	455	Inverse spinel
$BaFe_{12}O_{19}$	3.18×10^5	450	Hexagonal
$SrFe_{12}O_{19}$	3.18×10^5	453	Hexagonal
$MgFe_2O_4$	1.11×10^5	440	Inverse spinel
$Li_{0.5}Fe_{2.5}O_4$	3.10×10^5	670	Inverse spinel
γ-Fe_2O_3	4.13×10^5	575	Inverse spinel
$Y_3Fe_5O_{12}$	1.35×10^5	287	Garnet

Source: Somiya [63]/With permission of Elsevier.

Ferrites

Ferrites are ceramic materials having the general formula $MO \cdot Fe_2O_3$, where M is a bivalent element such as Fe, Cu, Mn, Zn, and Ni. The oldest known ferrite with magnetic properties is magnetite (Fe_3O_4), which may be considered a double oxide composed of ferrous and ferric oxides, $FeO \cdot Fe_2O_3$. Mixed ferrites can also be fabricated in which the divalent cation may be a mixture of ions, such as $Mg_{1-x}Mn_xFe_2O_4$, so that a wide range of composition and magnetic properties are available. The major ferrite solid solutions are Mn–Zn, Ni–Zn, and Cu–Zn. These, along with the Ba and Sr ferrites, form the mainstream of magnetic materials for electronic applications. Table 6.21 presents the most important types of ferrites and their basic properties.

Historical Highlight

Lodestone has long been known to attract iron. Read more about this "magical" ore and the early uses of magnetism here:
https://nationalmaglab.org/magnet-academy/history-of-electricity-magnetism/museum/lode-stone-600-bc

The ferrites generally have one of three crystal structures: inverse spinel, garnet, and hexagonal. The spinel structures (cf. Figure 1.42) have the oxygen ions in a nearly close-packed cubic array. The unit cell contains 32 oxygen ions, with 32 octahedral and 64 tetrahedral sites, of which 16 of the octahedral and 8 of the tetrahedral sites are filled. It is the position of these 24 cations within the unit cell that determines magnetic behavior. The distribution of cations in the sites is specific to the type of cations, and it must be determined experimentally. There are two idealized spinel structures. In the *normal spinel*, all the divalent ions are on the tetrahedral sites, as in $ZnFe_2O_4$. In the *inverse spinel*, the 8 occupied tetrahedral sites are filled with trivalent ions, and the 16 occupied octahedral sites are equally divided between di and trivalent ions (see Figure 6.63). The prototypical inverse spinel ferrite is magnetite, whose structure consists of an FCC oxygen array with Fe^{2+} and Fe^{3+} ions in the interstices. Ferrite compounds with the inverse spinel structure are similar to magnetite, with different ions substituting for the iron atoms. As with FeO (cf. Figure 6.62),

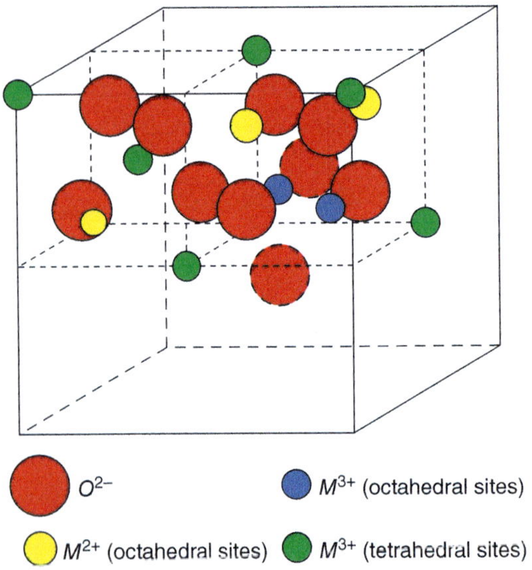

Figure 6.63 A portion of the inverse spinel unit cell found in many ferrites. Ralls et al. [15]. Copyright © 1976 by John Wiley & Sons, Inc. This material is used by permission John Wiley & Sons, Inc.

the oxygen ions have no permanent magnetic moment. Tetrahedral sites in the FCC oxygen array are occupied by half of the trivalent cations, and octahedral sites are occupied equally by divalent cations and the remaining trivalent cations.

The permanent moments of the cations in the octahedral and tetrahedral sites are antiparallel. Thus, the moments of the trivalent cations (Fe^{3+}) cancel out, and the net moment is due to the divalent cations. If the individual divalent cation moments are known, then it is possible to calculate the net moment per formula unit (see Cooperative Learning Exercise 6.7). In a formula unit of magnetite, there is one divalent iron ion that has a total moment of $4\mu_B$ resulting from four unpaired electrons, which is very close to the observed value. Similarly, in $NiFe_2O_4$, the moment per formula unit is $2\mu_B$ identical to that of the divalent nickel ion.

In some systems, and particularly at high temperatures, the cation distribution may be disordered and the cations nearly randomly distributed between tetrahedral and octahedral sites, but generally there is a tendency for individual ions to fit into particular sites so that either the normal or inverse arrangement is preferred. The extent to which inversion of the cations occurs depends on the heat treatment, but, in general, increasing the temperature of a normal spinel causes an excitation of the ions to the inverted position. For example, manganese ferrite is about 80% normal spinel, and this arrangement does not change greatly with heat treatment. Since the Mn^{2+} ion has a moment of $5\mu_B$, the inversion should not affect the net moment of $MnFe_2O_4$. Nickel ferrite has similar properties, and it is 80–90% normal spinel. Magnesium ferrite, however, transforms at high temperatures to the normal spinel structure as the divalent magnesium ions are thermally excited onto the tetrahedral sites. In this case, the magnetization is strongly influenced by the cooling rate. Slow cooling allows the inverse spinel structure to occur because enough thermal energy and time are available to allow the magnesium ions to migrate to the preferred octahedral sites. The saturation moment for a rapidly quenched sample is $2.23\mu_B$; a slow furnace cool results in $1.28\mu_B$. The Mg^{2+} ion has no net moment so the inverse spinel should have zero magnetization and the normal spinel $10\mu_B$. Magnesium ferrite has high resistivity and low magnetic and dielectric losses that make it most suitable for microwave applications.

Cooperative Learning Exercise 6.7

The number of Bohr magnetons contributed by various divalent transition metal ions is summarized below. Since the trivalent ions are equally distributed between half of the occupied octahedral sites (8) and all the occupied tetrahedral sites (8), their moments cancel out, and the net magnetic moment of a ferrite can be predicted from the moment of the divalent ions that occupy the remainder of the octahedral sites (8).

Ion	Magnetic Moment (Bohr magnetons, $\mu_B = 9.27 \times 10^{-24}$ A · m^2)
Mn^{2+}	5
Fe^{2+}	4
Co^{2+}	3
Ni^{2+}	2
Cu^{2+}	1

Person 1: Calculate the net magnetic moment per unit cell for copper ferrite. Remember that there is more than one formula unit ($CuFe_2O_4$) per unit cell in the inverse spinel structure.

Person 2: Calculate the net magnetic moment per unit cell for nickel ferrite. Remember that there is more than one formula unit ($NiFe_2O_4$) per unit cell in the inverse spinel structure.

Exchange information.

Person 1: Calculate the saturation magnetization, M_s, for nickel ferrite, which has a lattice parameter of $a = 0.833$ nm.

Person 2: Calculate the saturation magnetization, M_s, for copper ferrite, which has a lattice parameter of $a = 0.838$ nm.

Answers: p_m (Cu^{2+}) = $8\mu_B$; p_m (Ni^{2+}) = $16\mu_B$. M_s ($NiFe_2O_4$) = $(16)(9.27 \times 10^{-24})/(0.833 \times 10^{-9})^3 = 2.6 \times 10^5$ A/m; M_s ($CuFe_2O_4$) = $(8)(9.27 \times 10^{-24})/(0.838 \times 10^{-9})^3 = 1.3 \times 10^5$ A/m.

Some ferrimagnetic oxides have the garnet structure, which can accommodate large trivalent rare-earth ions with large magnetic moments. The chemical formula for ferrimagnetic garnets is $M_3Fe_5O_{12}$, where M is a rare-earth ion or a yttrium ion. The garnet $Y_3Fe_5O_{12}$, called yttrium–iron–garnet (YIG), has a high electrical resistivity and, correspondingly, a very low hysteresis loss even at microwave frequencies.

The hexagonal ferrites have a structure related to the spinel structure but with hexagonal close-packed oxygen ions and a unit cell made up of two formulae of $MN_{12}O_{19}$, where M is divalent (Ba, Sr, or Pb) and N is trivalent (Al, Ga, Cr, or Fe), which gives a molecular formula of $M^{2+}O \cdot N_2{}^{3+}O_3$. The best-known examples are magnetoplumbite, which has a formula $PbFe_{12}O_{19}$, and barium ferrite, $BaFe_{12}O_{19}$. As in the spinel structures, magnetization of these compounds is the result of unpaired electrons, usually in the Fe ions. Hexagonal ferrites are of interest because of their high magnetocrystalline anisotropy and high coercivity, which make them suitable for permanent magnets.

For all the ferrites, other factors influence the magnetic properties besides structure. These include impurities, oxygen partial pressure, grain size, and porosity. For example, nickel ferrite thin films deposited by reaction sputtering at 0°C are noncrystalline, and the magnetic susceptibility shows the material to behave as a paramagnet. If the deposition temperature is raised to 400°C, a micro polycrystalline film results with grain sizes less than 15 nm, and the resulting film shows superparamagnetic behavior. *Superparamagnetism* is the result of clusters of paramagnetism. If there are n clusters, each with N spins, it can be shown that the magnetic susceptibility is now proportional to N^2/n, where in normal paramagnetic materials the Curie Law [Eq. 6.63] tells us that χ is proportional to N. This causes the magnetic moment to increase by orders of magnitude, provided that $N >> n$. If the size of the clusters gets so big as to form magnetic domains, the advantage is lost.

Ferrite can be used as soft magnets, hard magnets, and a class of materials known as *semihard magnets*. Soft magnetic ferrite is used for core materials for coils and transformers. Hard magnetic ferrite is used in making permanent magnets. Semihard magnetic ferrites have properties between those of soft and hard magnetic ferrites. They are comparatively easy to magnetize, can retain magnetization, and can produce changes in the magnetic flux from this state, as needed. The typical application for semihard magnetic ferrites is in memory cores for computers. In general, the Mn–Mg ferrites and the lithium ferrites are useful for semihard applications. There is a growing list of applications for ferrites, including microwave applications that utilize the unique magnetic properties of these materials.

The Meissner Effect

In Section 6.0.2, we describe the interrelationship between electric current density, magnetic field, and temperature for oxide superconductors, and although a great deal of description was given to upper and lower critical magnetic fields, one particularly important topic related to magnetism was deferred to this section. That topic has to do with an important phenomenon related to superconductivity called the *Meissner Effect*. Like superconductivity itself, the Meissner Effect is relatively old as far as scientific discoveries go. It was found in 1933 by W. Meissner and colleagues, who determined that a superconducting metal expels any magnetic field when it is cooled below the critical transition temperature, T_c, and becomes superconducting. By expelling the field and thus distorting nearby magnetic field lines, as shown in Figure 6.64b, a superconductor will transition from the normal state (Figure 6.64a) and create a strong enough force field to overcome gravity. This expulsion of the magnetic field is to be distinguished from simply preventing the field lines from penetrating—any metal with infinite conductivity

Need More Help?
Try these videos on the Meissner Effect: https://www.youtube.com/watch?v=NVeAmKUbXvA https://www.youtube.com/watch?v=HRLvVkkq5GE

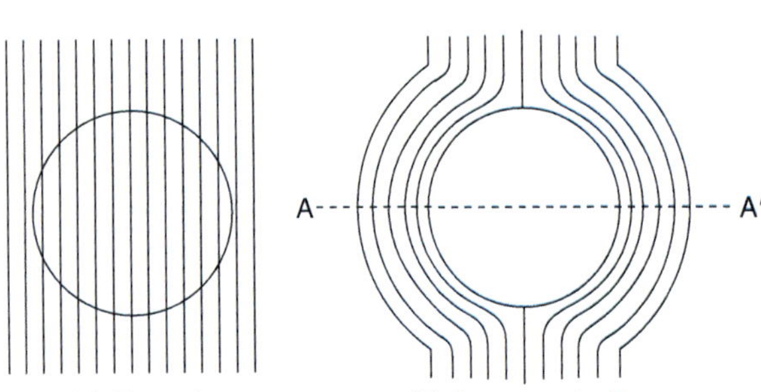

Figure 6.64 Schematic illustration of Meissner Effect in which magnetic flux lines that (a) normally penetrate the material are (b) expelled in the superconducting state.

would do the latter. If a magnetic field is already present, and a substance is cooled below T_c to become a superconductor, the magnetic field is expelled. The significance of this difference cannot be explained merely by infinite conductivity.

The ability of the superconductor to overcome gravitational forces has led to the famous demonstration, typically involving an oxide superconductor and a rare-earth magnet, in which the magnet is levitated above the superconductor when it is cooled with liquid nitrogen. In this experiment, the magnetic field lines from the strong (but small) magnet penetrate the superconductor to a small depth called the *penetration depth*, λ, which sets up circulating currents—called *eddy currents*—in the superconductor. Typically, the penetration depth is less than 0.05 μm, so that on a macroscopic scale, the superconductor is said to exclude the magnetic field. Actually, these eddy currents create their own magnetic field lines that exactly cancel the applied magnetic field from the magnet, thus levitating it.

No superconductor can keep out very strong magnetic fields. As described previously, Type I superconductors have a critical magnetic field, H_c, above which they no longer superconduct. Type II superconductors have a range of critical fields from the upper critical field, H_{c2}, to the lower critical field, H_{c1}. Below these critical fields, the superconductor is perfectly diamagnetic. For a perfect diamagnet, $M = -H$, so that according to Eq. (6.60) the inductance is $B = 0$. Thus, the earlier statement that diamagnetism is not important in ceramic materials must be strongly qualified for the oxide superconductors—diamagnetism is critical to their importance.

At first, the superconducting state was not thought to be a thermodynamic equilibrium state. But, as we know from Chapter 2, any equilibrium state must be the result of a free energy minimization. It can be shown that the superconducting and normal states will be in equilibrium when their free energies are equal and that the free energy difference between the normal state at zero field, G_n $(H = 0)$, and the superconducting state at zero field, G_s $(H = 0)$, is

$$G_n(H = 0) - G_s(H = 0) = \frac{\mu_0 H_c^2}{2} \tag{6.65}$$

The free energies for the respective states can be determined from heat capacity data, and the thermodynamic critical field can be determined from a plot of free energy versus applied field for the two states, as illustrated in Figure 6.65.

The Meissner Effect is a very important characteristic of superconductors. Among the consequences of its linkage to the free energy are the following: (a) the superconducting state is more ordered than the normal state; (b) only a small fraction of the electrons in a solid need participate in superconductivity; (c) the phase transition must be of second order; that is, there is no latent heat of transition in the absence of any applied magnetic field; and (d) superconductivity involves excitations across an energy gap.

6.1.3 Magnetic Properties of Polymers

Much like ceramics, the electrical and magnetic properties of polymers are closely linked. Initially, this statement can be viewed with considerable disappointment, since, with only a few notable exceptions, polymers are poor electrical conductors, and what is known about their potential for magnetization is even less than is known about their electrical conductivity. Nonetheless, there is hope. Just as new classes of polymers are currently being developed with improved electrical conductivities, there is reason to believe that polymers can be developed that will have significant magnetic properties. What a potential benefit this could be for the development of lightweight magnets. We briefly describe here just one topic related to magnetic polymers that is still too "young" in its scientific development to warrant significant coverage in an introductory text: *molecular magnets*.

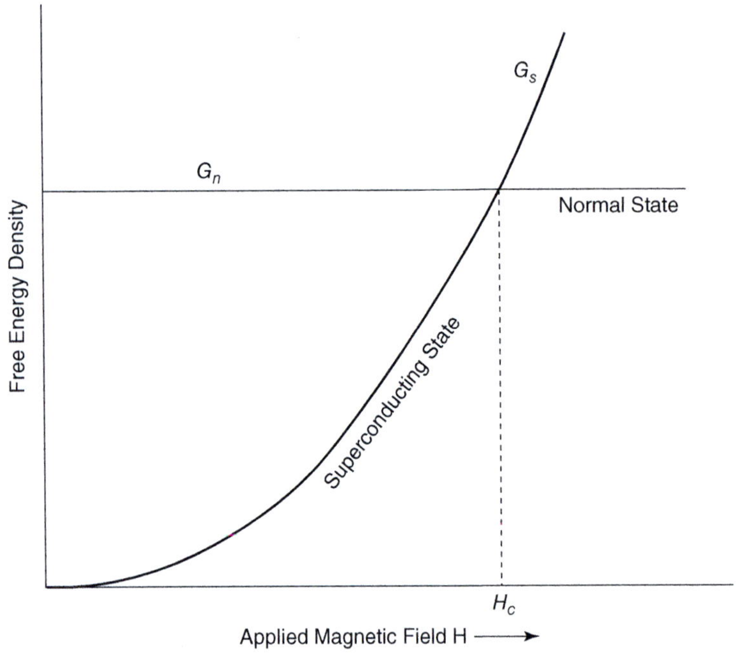

Figure 6.65 Free energy curves for both normal and superconducting states.

The more common use of polymers in magnetic media, that as a matrix for the support of nonpolymeric magnetic particles, is more appropriately addressed in the section on composites.

Molecular Magnets*

Much of the information presented here on molecular magnets is drawn from some excellent, recent review articles [7–10] on this emerging field. Following a brief introduction to the general topic of molecular magnets, a description of organic molecular magnets will be given, with an emphasis on the development of polymeric molecular magnets.

A molecular magnet is a magnet that is molecule-based—that is, composed of molecular or molecular ion entities, and not atoms or atomic ions. This distinction is not new to us, though previously we have not elaborated upon it. Simply compare the crystal structures of such atom-based solids as $CaTiO_3$ in Figure 1.43, in which the positions of the atoms determine the crystal structure, with that of a molecule-based solid such as polyethylene in Figure 1.61, in which the positions of the molecules, in this case, chains determine the crystal structure. Think about this distinction for a moment, and you will realize that most molecular materials are organic in nature. The covalent bonds within the molecules are stronger than the noncovalent bonds that draw the organic compounds into a molecular crystal, such as hydrogen bonding and van der Waals interactions. This is different from metallic and ionic solids, in which any Na–Cl interaction, for example, is essentially the same anywhere in the crystal, so that ordering takes place on the atomic scale.

The magnetic materials described so far in this chapter are all atom-based: their bulk magnetic properties are due to the incorporation of atoms such as Fe or Nd. In these materials, magnetic order arises from cooperative spin–spin interactions between unpaired electrons located in *d* or *f* orbitals of the metal atoms. Molecular-based materials, on the other hand, do not tend to exhibit magnetic properties for some very good reasons. Organic compounds generally have closed-shell structures and are thus diamagnetic. Paramagnetic organic molecules are very rare because the presence of unpaired electrons—which are required for paramagnetism—in the *s* and *p* orbitals gives instability to the molecules, which tend to spontaneously form a covalent bond with another molecule, thus pairing the spins. Even if one or more unpaired electrons are maintained stably in an organic molecule, stabilization of a triple state (parallel alignment of spins) requires that certain quantum mechanical conditions be met which are difficult to satisfy, such that they prefer to couple in an antiferromagnetic manner with no spontaneous magnetic moment. Thus, the development of ferro or ferrimagnetic organic molecules has been elusive. However, recent advances in synthetic molecular chemistry have made possible the formation of organic molecules with sufficient magnetic function that can serve as building blocks for bulk magnetic behavior. Without delving into the theory of magnetism in these materials, we describe some of these building blocks that may one day be the precursors to polymer magnets.

The typical synthetic approach to designing molecular magnets consists of choosing molecular precursors, each of which contains an unpaired spin, as represented by the arrow in Figure 6.66. The precursors are then assembled in such a way that there is no compensation of the spins at the crystal lattice scale. In Figure 6.66, the precursors may contain other functional groups, represented by the star, that can assist in self-assembly, or impart some additional property such as chirality or ferroelectricity. The primary criteria for designing a molecular magnet, then, are that (a) all the molecules in the lattice have unpaired electrons, and (b) the unpaired electrons have spins aligned parallel along a given direction. Since magnetism is a cooperative effect, the spin–spin interaction must extend to all three dimensions of the lattice. These criteria are met in molecules such as the nitronyl nitroxide radical, NITR, shown in Figure 6.67a, where R is an alkyl group. Through

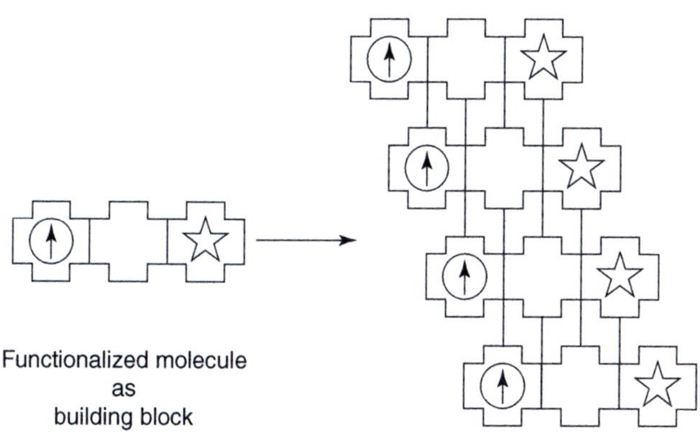

Figure 6.66 Schematic illustration of building blocks used to construct molecular magnets. An arrow indicates an unpaired electron, and a star represents some other desirable functionality. Yakhmi [40]/ Reproduced with permission of Elsevier.

Figure 6.67 Chemical structures of (a) NITR and (b) *p*-NPNN used in molecular magnets. An asterisk, ∗, indicates an unpaired electron. Yakhmi [40]/ Reproduced with permission of Elsevier.

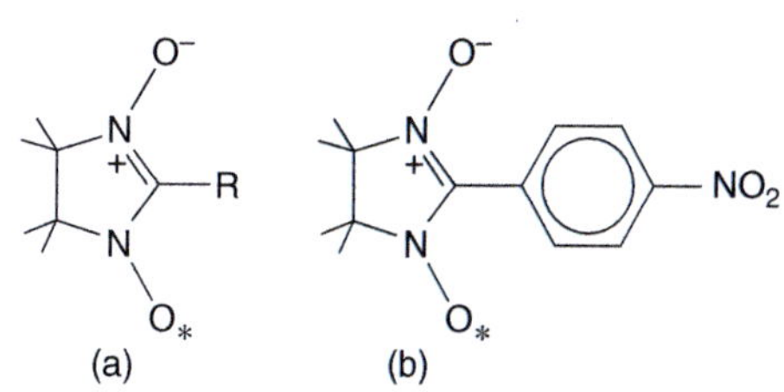

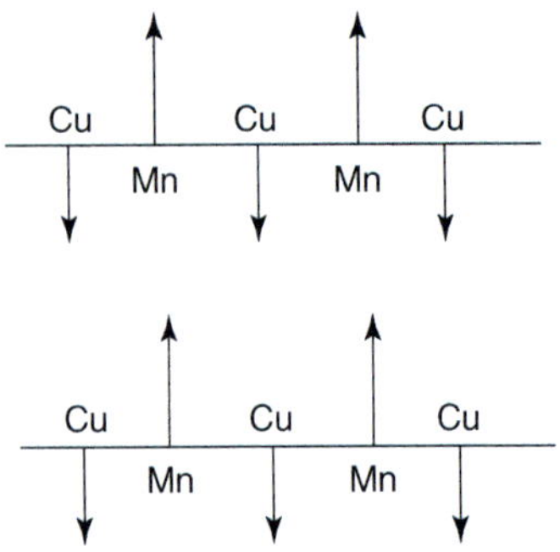

Figure 6.68 Schematic illustration of spin assembly metal-based ferrimagnetic chains. Yakhmi [40]/Reproduced with permission of Elsevier.

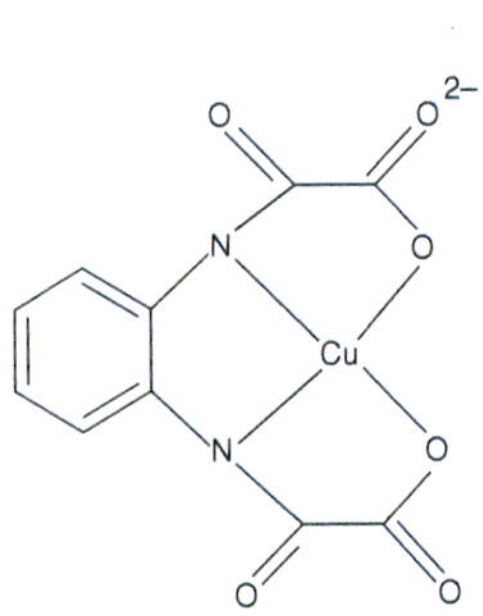

Figure 6.69 Structure of the copper dianion precursor $[Cu(opba)]^{2-}$. Yakhmi [40]/Reproduced with permission of Elsevier.

different substitutions for the alkyl group, the spin delocalization effect can be fine-tuned. An example is when R is a nitrophenyl compound, which results in 4-nitrophenyl nitronyl nitroxide, or *p*-NPNN for short, as shown in Figure 6.67b. Ferromagnetism has been exhibited in this purely organic compound at around 0.6 K. Other similar, organic compounds have Curie temperatures of 16 K and above.

An alternative strategy is to utilize ferrimagnetic chains containing alternating spins of unequal magnitude and assemble them in such a way that there is a net spin. For example, two different spin carriers such as Mn ions, with $m_s = 5/2$, and Cu ions, with $m_s = 1/2$, can be incorporated within the same molecular precursor, as schematically illustrated in Figure 6.68. It is also possible to create different spins using organic radicals instead of metal ions. An example of such a building block is shown in Figure 6.69 for $[Cu(opba)]^{2-}$, where "opba" stands for ortho-phenylenebis(oxamato). This copper precursor can be reacted with a divalent ion such as Mn^{2+}, which results in an amorphous magnet with a spontaneous magnetization below 6.5 K. Similar precursors can be assembled into crosslinked chains with still higher Curie temperatures, as shown in Figure 6.70. Note that even though these structures contain transition metals such as Mn and Cu, it is the entire *molecule*, including the metal ion, that exhibits the spin, and that can be used as a building block for larger structures.

One final example is a class of related materials that undergo reversible magnetic behavior through a mild dehydration–rehydration process. Because these materials show a reversible crossover under dehydration to a polymerized long-range, magnetically ordered state with spontaneous magnetization and transformation back to isolated molecular units under rehydration, they are called *molecular magnetic sponges*. Coercivity values for these sponges are high (see Table 6.22), and some even undergo color changes that coincide with the change in magnetic properties. In addition to the "opba" anion described above, these molecular magnetic sponges are composed of "obbz" and "obze" ligands, where obbz = oxamido bis(benzoato) and obze = oxamido-*N*-benzoato-*N'*-ethanoato.

As this field continues to develop, look for improvements not only in the magnetic properties of these organic polymers but also in the theoretical descriptions of intrinsic ferro and ferrimagnetism in polymers.

6.1.4 Magnetic Properties of Composites

In this section, we describe the properties of two magnetic composites that represent extremes in terms of applications for magnetic materials. The first, flexible magnets illustrate how the incorporation of less-expensive components into a material can retain the functionality of the magnetic material while significantly reducing cost. The second, magnetic storage media illustrates how a simple concept such as a reversible magnetic moment can be utilized to create an entirely new industry and revolutionize the manner in which we conduct our daily lives.

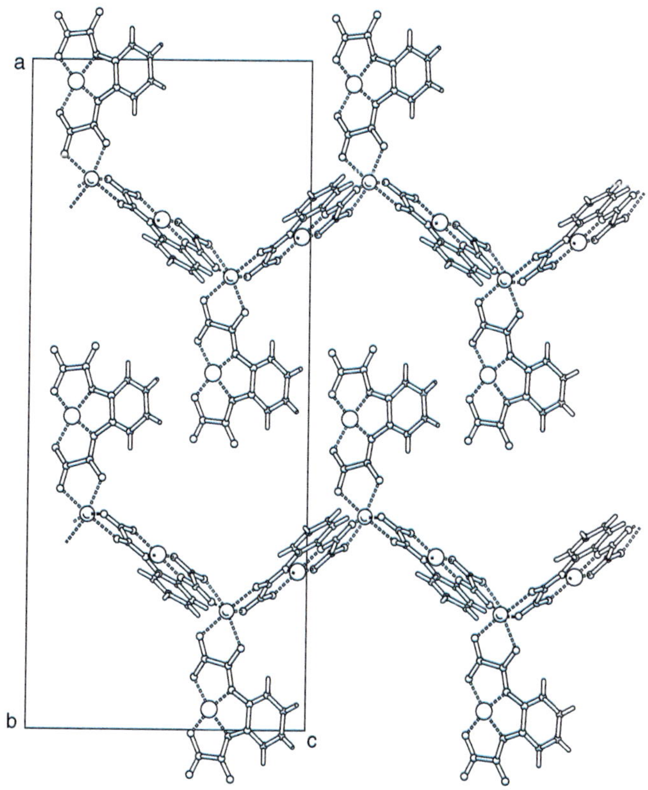

Figure 6.70 Structure of 1:2 Mn(II)Cu(II) compound consisting of a zigzag chain with terminal "opba" groups. Yakhmi [41]/Reproduced with permission of Elsevier.

Table 6.22 Selected Magnetic Properties of Some CoCu-Based Molecular Magnetic Sponges

Compound	T_c (K)	H_c (kOe)
$CoCu(pbaOH)(H_2O)_3 \cdot 2H_2O$	38	5.66
$CoCu(pba)(H_2O)_3 \cdot 2H_2O$	33	3
$CoCu(obbz)(H_2O)_4 \cdot H_2O$	25	1.3
$CoCu(obze)(H_2O)_4 \cdot 2H_2O$	20	1

Source: Yakhmi [64]/With permission of Elsevier.

Flexible Magnets

Flexible magnets are made by embedding ferrimagnetic ceramics in a polymer matrix, such as barium ferrite or strontium ferrite in nylon or poly(phenylene sulfide). In an anisotropic magnet (one-sided) the ferrite and polymer components are mixed and processed, usually by extrusion, into a thin sheet and are then coated or printed on one side. The resulting magnetic field structure (Figure 6.71) is made of a striped pattern of alternating pole alignments on the back side of the flexible magnetic sheet. Typical tape thicknesses are 1–2 mm. To obtain maximum alignment, the ferrite particles are physically rotated in a magnetic field during the molding process. In this way, energy products equivalent to those of cobalt steel are obtained.

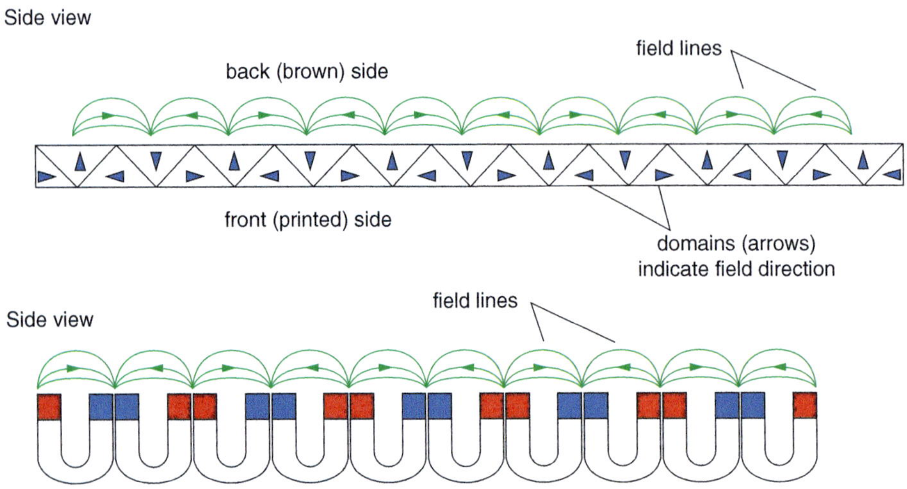

Figure 6.71 Side view of magnetic flux lines in an anisotropic flexible magnet (mrsec.wisc.edu/edetc).

The magnetic field structure (see Figure 6.71) in an anisotropic magnet is made of a striped pattern of alternating pole alignments on the back side of the flexible magnetic sheet. This arrangement also channels the magnetic field to the back of the magnet so that maximum holding power is obtained, which is important for many applications. It is also possible to make isotropic, or two-sided, magnetic tapes and sheets, for which the field lines shown in Figure 6.71 are present on both sides of the sheet.

The magnetic properties of the flexible magnets depend upon the properties of the ferrite, of course, and the relative amounts of ferrite and binder, as well as processing conditions, but generally for strontium ferrite-based materials, residual inductions, B_r, range from 1.6 to 2.6 gauss, coercive forces, H_c, range from 1.3 to 2.3 kOe, and densities range from 3.5 to 3.8 g/cm^3. The magnetic properties degrade linearly with increasing temperature, and there is generally an upper-use temperature of around 100°C at which point the polymeric component begins to melt or flow.

Magnetic Storage Media

Magnetic materials are ideal candidates for information storage media because magnets can be used like switches. They can be made to point in one direction or another in the presence of a magnetic field. Reverse the direction of the field, and the magnet will switch to point in the other direction. This represents a simple way to store binary information. The magnet represents a "1" when pointing in one direction and represents a "0" when pointing in the other. A magnetic memory can be constructed from an entire array of magnets, in which the binary data can be encoded by applying an external magnetic field to each magnet in turn, the direction of which determines whether a "1" or a "0" is written. As we have seen, individual atoms can act as magnets. In a ferromagnetic material such as iron, for example, the magnetic moments in each atom tend to line up so that they all point in the same direction. Furthermore, recall that the magnetic moments tend to align in domains, separated by domain walls, the net magnetic moments of which are randomly oriented to create a bulk material with no net magnetic moment.

The ability to influence the alignment of magnetic moment by the application of a magnetic field forms the basis of magnetic recording. By applying a strong, localized magnetic field to one region of a thin layer of a ferromagnetic material such as magnetite (above the Curie temperature), a domain can be created in which all the magnetic moments are aligned. The magnetized domains, in general, are larger than the magnetic particles. Application of the field in the opposite direction in another part of the magnetic medium creates a domain of opposite alignment. One alignment direction represents the "1" in the binary code, and the opposite direction represents the "0." A *recording head* is used to both record and retrieve the information.

Magnetic tape technology—while still important in limited applications—has been almost completely replaced by two-dimensional disk technology due to the limitations of the one-dimensionality of the tape. To retrieve information from one part of the tape, the entire preceding information must first be trawled through. Similar disadvantages exist for magnetic hard drives in computers where mechanical devices must be used to record and retrieve the digital information. Moreover, the magnetic domains are subject to external magnetic fields and data loss over time. The advent of the MOSFET as a digital storage medium (see page 507) has limited magnetic storage media to archival data storage in corporate environments.

6.1.5 Magnetic Properties of Biologics*

Much as was the case with polymeric materials, most biologics do not possess significant magnetic activity. There is one topic related to the interaction of biological tissues with magnetic fields that is of such technological importance, and related enough to the chemical makeup of biologics, that it is worth describing here. *Nuclear Magnetic Resonance Imaging*, or MRI for short, has revolutionized diagnostic medicine. What is so intriguing about it from a scientific standpoint is that it is based upon the interaction of only one atom with a magnetic field—hydrogen.

Magnetic Resonance Imaging

Virtually all the topics in this textbook deal in some way with the chemical aspects of atoms and molecules—that is, how electrons and the electronic structure dictate interactions between atoms and molecules, and ultimately, the bulk properties of an assembly of atoms or molecules. All the types of magnetism described in this section are related in some way to the pairing of electrons. What we describe in this subsection is not a bulk magnetic property of a material and is not related to any of the types of magnetism described heretofore. *Nuclear magnetic resonance* (NMR), as the name implies, is related to the interaction of an atom's nucleus with an external magnetic field. Let us take a moment to review a bit of nuclear chemistry, then describe the general topic of NMR before completing this section with a description of the application of NMR to imaging of hydrogen in biological tissues—MRI.

Recall from general chemistry that an atom consists of electrons orbiting a central nucleus composed of *nucleons* called protons and neutrons. The protons are positively charged, and the neutrons have no charge associated with them. There are two kinds of spins associated with an atom: (a) the spin of an electron on its own axis and (b) the spin of a nucleus on its own axis. Nuclear spin is the basis upon which NMR and MRI function.

Particles that spin have an associated angular momentum with values that are integral or half-integral multiples of $h/2\pi$, where h is Planck's constant. The maximum spin component for a particular nucleus is its *spin quantum number*, I, and a nucleus has $(2I + 1)$ discrete states. The component of angular momentum for these states in any chosen direction will have values of I, $I - 1$, $I - 2$, ..., $-I$, called the *magnetic quantum number*, m. In the absence of an external magnetic field, the various states have identical energies. The spin number for a proton is ½; thus two spin states are possible: $I = +½$ and $I = -½$. Heavier nuclei have spin numbers that range from 0 to over 9/2. As shown in Table 6.23, the spin number of a nucleus is related to the relative number of protons and neutrons it contains. Since the nucleus contains a charge, its spin gives rise to a magnetic field, much like the field produced when electricity flows through a coil of wire.

Table 6.23 Spin Quantum Number for Various Nuclei

Number of Protons	Number of Neutrons	Spin Quantum Number *I*	Examples
Even	Even	0	^{12}C, ^{16}O, ^{32}S
Odd	Even	$\frac{1}{2}$	^{1}H, ^{19}F, ^{31}P
		$\frac{3}{2}$	^{11}B, ^{79}Br
Even	Odd	$\frac{1}{2}$	^{13}C
		$\frac{3}{2}$	^{127}I
Odd	Odd	1	^{2}H, ^{14}N

The resulting magnetic dipole is oriented along the axis of spin and has a value that is characteristic of each type of nucleus. For a hydrogen atom, whose nucleus contains only one proton, the spin quantum number is $I = ½$, and the magnetic quantum number can have values of $+½$ and $– ½$. These two quantum energies correspond to the two possible orientations of the spin axis: spin up and spin down. Those nuclei with a spin-up orientation are termed low energy, and those with a spin-down orientation are termed high energy. In the absence of a magnetic field, the energies of these two states are identical, despite their names. Consequently, a large assemblage of hydrogen atoms will contain an identical number of nuclei with $m = +½$ and $m = –½$.

When brought into the influence of an applied magnetic field, the spins tend to become oriented parallel to the field, termed spin up, and the lower-energy spin state ($m = +½$) is more highly populated. Some spins do populate the high-energy state, however. In fact, at room temperature, there is only a minute excess (<10 ppm) of nuclei in the lower-energy state due to thermal energies that are several orders of magnitude greater than the magnetic energy differences. This small difference in spin orientations is sufficient, however, to result in a net magnetic moment, as characterized by the *net magnetization vector*, NMV. The orientation of nuclear spins in a magnetic field is shown in Figure 6.72.

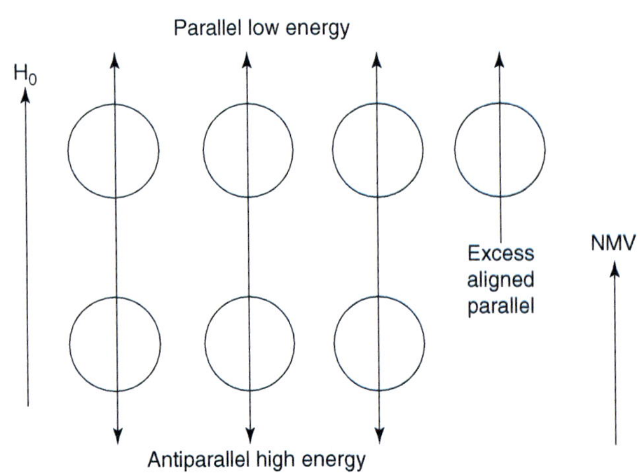

Figure 6.72 Orientation of nuclear spins in the presence of a magnetic field, and the resulting net magnetization vector (NMV) from excess low-energy spins aligned parallel to field. Westbrook and Kaur [42]/John Wiley & Sons.

The application of a magnetic field to the spinning nucleus creates an additional movement in the nucleus, called *precession*. Much like a compass needle that will swing back and forth indefinitely in the presence of an external magnet, a spinning particle will be forced to move in a circular path in the presence of a magnetic field. This is a gyroscopic effect similar to what occurs in a gyroscope when it is displaced from vertical by application of an external force. In this case, the external force is the applied magnetic field, which causes the spinning particle to move in a circular path, or precess, about the magnetic field, as illustrated in Figure 6.73. The *angular velocity of precession*, ω_0 is directly proportional to the applied force and inversely proportional to the angular momentum of the spinning body. The force on a spinning nucleus in a magnetic field is the product of the field strength, H_0, and the magnetic moment of the particle. When combined with the angular momentum, the magnetic moment creates a quantity called the *magnetogyric ratio*, γ, sometimes (inappropriately) called the gyromagnetic ratio. The result is the Larmor equation:

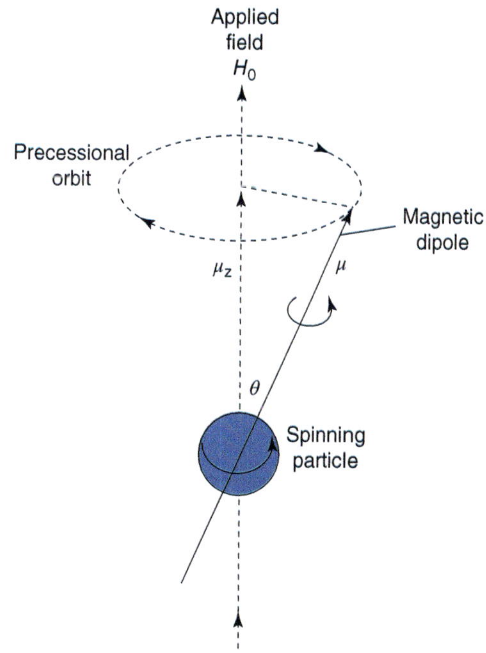

Figure 6.73 Precession of a spinning particle in a magnetic field. Skoog and West [43]/Reproduced with permission of Saunders College.

$$\omega_0 = \gamma H_0 \tag{6.66}$$

The magnetogyric ratio is then the ratio between the magnetic moment and the angular momentum of a rotating particle and has a characteristic value for each type of nucleus:

$$\gamma = \frac{\mu\beta}{I(h/2\pi)} \tag{6.67}$$

where μ is the magnetic moment of the nucleus expressed in nuclear magnetons, β is a constant called the *nuclear magneton* (5.051×10^{-24} erg/G), I is the spin quantum number as described previously, and h is Planck's constant. For reference, the magnetic moment for the proton is 2.7927 nuclear magnetons. The angular velocity of precession given by the Larmor equation can be converted to a frequency of precession, ν_0, by simply dividing by 2π. Thus, as the field strength is increased, the precessional frequency increases.

We turn our attention now to NMR, which forms the basis for MRI. *Resonance* is a phenomenon that occurs when an object is exposed to an oscillating perturbation that has a frequency close to its own natural frequency of oscillation. The nucleus gains energy and resonates if the energy is delivered at exactly its precessional frequency. If energy is delivered at a frequency different from that of the Larmor frequency of the nucleus, resonance does not occur. When this energy, typically in the form of radio-frequency (RF) electromagnetic radiation, is absorbed by the nucleus, *excitation* occurs, and the angle of precession, θ, shown in Figure 6.73, must change. If the frequency of the radio wave is the same as the precessional frequency, absorption, and flipping can occur.

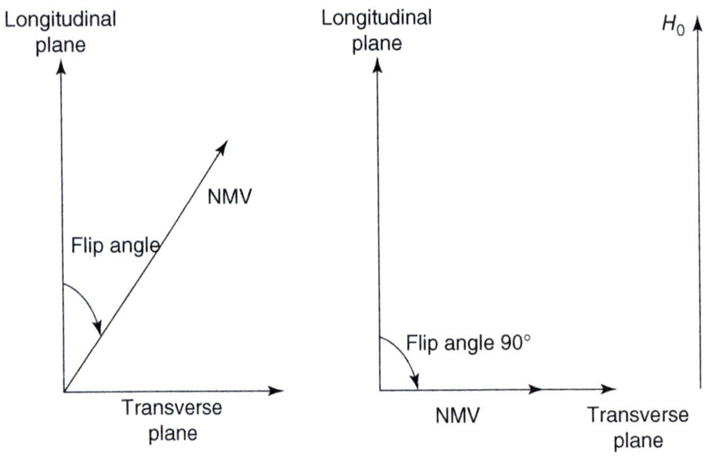

Figure 6.74 The flip angle with and without an applied field. Westbrook and Kaur [44]/John Wiley & Sons.

The first result of resonance is that the NMV moves out of alignment away from the direction of the applied field called the *flip angle*. The magnitude of the flip angle depends upon the amplitude and duration of the electromagnetic pulse. Usually, the flip angle is 90°; that is, the NMV is given enough energy by the pulse to move through 90° relative to H_0 (see Figure 6.74). With a flip angle of 90° the nuclei are given sufficient energy so that the longitudinal NMV is completely transferred into a transverse NMV. This transverse NMV rotates in the transverse plane at the Larmor frequency.

The process is reversible, and the excited particle can thus return to the ground state by reemission of the radiation. This is known as the *relaxation process*. Relaxation following resonance is the basis for NMR spectroscopy. It turns out that there is a low probability of emission of radiation of the exact frequency of resonance. Instead, relaxation processes usually involve radiationless mechanisms. The relaxation processes should be as rapid as possible, so the lifetime of the excited state should be small—about 0.1–1 s. There are two important types of relaxation processes: longitudinal, or *spin–lattice relaxation*, designated T_1; and transverse, or *spin–spin relaxation*, designated T_2. Each T is the average lifetime of the nuclei in the higher energy state.

In addition to depending upon the magnetogyric ratio and the absorbing nuclei, T_1 is strongly affected by the mobility of the lattice. In crystalline solids and viscous liquids, where mobilities are low, T_1 is large. T_2 is caused by nuclei exchanging energy with neighboring nuclei. Values for T_2 are generally so small for crystalline solids and viscous liquids (as low as 10^{-4} s) that these samples are not typically used for high-resolution NMR spectra. The NMR process, then, essentially involves placing a sample in a magnetic field, applying a RF pulse, and determining the decrease in power (attenuation) of the radiation caused by the absorbing sample.

Need More Help?

Try these videos on NMR:

https://www.youtube.com/watch?v=RZLew6Ff-JE
https://www.youtube.com/watch?v=XXoc12uaB6M

and MRI:

https://www.youtube.com/watch?v=Ok9ILIYzmaY
https://www.youtube.com/watch?v=rJ9gV4yFMi8

The principles of NMR can be adapted to the generation of images from biological tissues. MRI relies upon the nuclear magnet response of only one atom: hydrogen. Though there are certainly other magnetically susceptible atoms in the human body such as C^{13}, N^{15}, O^{17}, Na^{23}, and P^{31}, hydrogen is in such abundance and has such a relatively large magnetic moment owing to its single proton in the nucleus that its selection as the nucleus of choice for MRI is an easy one. Hydrogen atoms in different environments within the body, in principle, can be differentiated according to their relaxation times. Practically, this means that the NMV can be separated into the longitudinal and transverse components for differing tissues such as fat, cerebrospinal fluid, and muscle.

A tissue has a large NMR signal if it has a large transverse component of magnetization. If there is a large component of transverse magnetization, the amplitude of the signal received by the coil is large, resulting in a bright area on the MRI image. A tissue returns a low signal if it has a small transverse component of magnetization, resulting in a dark area on the image. In general, the two extremes of image contrast in MRI are the result of hydrogens found in fat (lipids) and water (see Figure 6.75).

The hydrogens in lipids are linked to carbon, whereas the hydrogen in water is linked to electron-withdrawing oxygens. The removal of electrons from around the hydrogens in water causes its nucleus to be less shielded and to interact more strongly with the applied magnetic field. The carbon in lipids is less electron-withdrawing, so lipid hydrogens do not interact with the magnetic field as strongly. Therefore, the Larmor frequency of hydrogen in water is higher than that of hydrogen in lipids. Lipid hydrogens recover more rapidly along the longitudinal axis than water, and they lose transverse magnetization faster than water hydrogens. The result is different imaging characteristics in MRI. Lipid hydrogens, therefore, have a high signal and appear bright on a T_1 contrast image. The opposite is true for water hydrogens, which appear as dark on a T_1 contrast image. These images are called *T_1-weighted images*. Conversely, the magnitude of transverse magnetization in

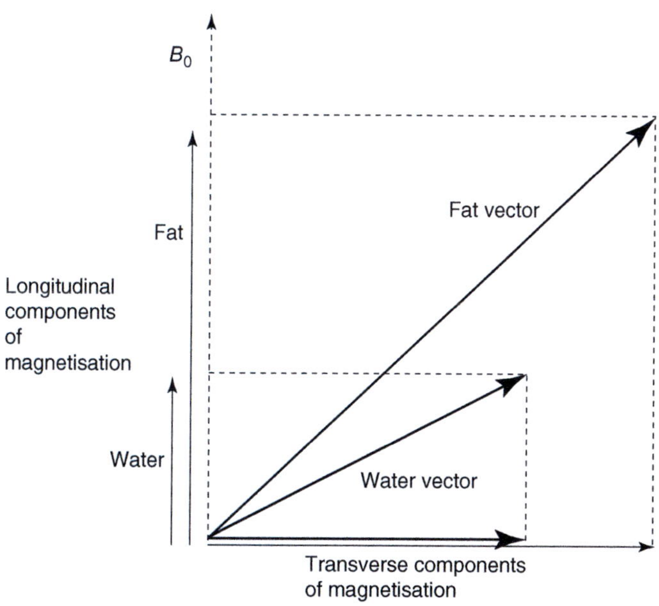

Figure 6.75 Magnitude of transverse and longitudinal magnetization for fat and water. Westbrook and Kaur [45]/Reproduced with permission of John Wiley & Sons.

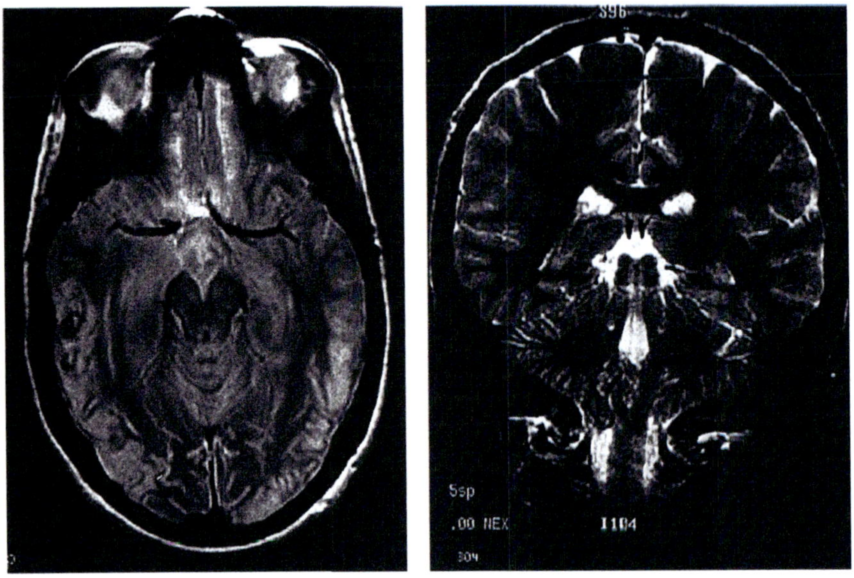

(a) (b)

Figure 6.76 (a) Proton density-weighted axial MRI image of the brain, TE = 20 ms, TR = 2700 ms. (b) T_2 weighted coronal image of the brain, TE = 90 ms, TR = 2700 ms. Reproduced with permission from Westbrook and Kaur [46]. © 1998, John Wiley & Sons, Inc.

water is large, and it has a high signal, thus appearing bright on a T_2 contrast image. Similarly, fat appears dark on a *T_2 contrast image*. The result is astounding detailed images of biological tissue, like that of the human brain shown in Figure 6.76.

The final effect on MRI image contrast is proton density or the relative number of protons per unit volume. To produce contrast due to proton density differences, the transverse component of magnetization must reflect these differences. Tissues with high proton density, such as brain tissue, have a large transverse component of magnetization and therefore have a high signal. A bright *proton density contrast image* results. Tissues with low proton density, such as cortical bone, have small transverse components of magnetization and have low signal strengths, and they appear dark on proton density contrast images.

There are, of course, more topics and techniques associated with MRI imaging, such as spin echo and gradient pulse echo pulse sequences. Advances in magnetic field strength and image processing have led to the development of *functional MRI*, or fMRI, that allows the human brain to be imaged in real time. The societal benefits of materials engineering principles are perhaps nowhere more evident or impactful than in the areas of human cognition and the prediction, treatment, and ultimate elimination of neurological disease.

6.2 Optical Properties of Materials

The final broad topic of this chapter deals with a combination of the previous two topics: electricity and magnetism. The interaction of electromagnetic radiation with materials involves many of the same principles used to describe the electrical and magnetic properties of materials, such as magnetic moments, spins, and electronic configurations. However, we will see that both wave (frequency) and particle (photon) descriptions of light will be of benefit in this chapter. Thus, we will vacillate between the two with little fanfare or warning.

In this section, we are concerned with the interaction of electromagnetic radiation with a material. When electromagnetic radiation (which we will generally call "light" while recognizing that the visible spectrum is but a small portion of all electromagnetic radiation) strikes a surface, three types of processes may take place, either individually or in tandem. The light may be *absorbed*, *reflected*, or *transmitted*. Like many technical terms that find their way into common usage, these three terms are well known to us through their everyday occurrences. We know that dark clothing absorbs energy, that a mirror reflects light, and that a pane of window glass transmits light. What is not always clear is that these three effects may occur simultaneously and that more than just the visible spectrum of light is involved. Since we are dealing primarily with "optical" properties of materials, this implies that we are primarily considering the visible portion of the electromagnetic spectrum. In some instances, radiation of other energies, such as infrared and ultraviolet radiation, will be examined. In these cases, more specificity will be exercised in the description of the type of "light."

In all instances, the total intensity of the incident light striking a surface, I_0, is the sum of the absorbed, reflected, and transmitted intensities, I_A, I_R, and I_T, respectively:

$$I_0 = I_A + I_R + I_T \tag{6.68}$$

The intensity is defined as the number of photons impinging on a surface per unit area per unit time. If we divide Eq. (6.68) by the incident radiation, we end up with an equally useful relationship

$$1 = A + R + T \tag{6.69}$$

where A is the *absorptivity* (I_A/I_0), R is the *reflectivity* (I_R/I_0), and T is the *transmissivity* (I_T/I_0). The utility of Eq. (6.69) is that A, R, and T are material properties. Thus, it is not possible for a material to be simultaneously highly reflective, highly absorptive, and highly transmissive. We have some intuition as to which materials are absorptive, reflective, and transmissive, at least in terms of visible light. Let us utilize this knowledge to examine the fundamentals of how electromagnetic radiation interacts with materials from our four classes and how these interactions can be used in some important applications.

Need More Help?

Try these videos on reflection, absorbance, and transmittance of light:

https://www.youtube.com/watch?v=DOsro2kGjGc
https://www.youtube.com/watch?v=3XKTDRBkLds

6.2.1 Optical Properties of Metals and Alloys

The "shininess" and inability to transmit visible light inherent in most metals indicate that absorptivity and reflectivity are high. In this section, we utilize our knowledge of electronic structure from previous sections to describe why this is so, and we investigate what happens to the energy that is absorbed in these materials.

Consider an incident beam of light interacting with a material as shown in Figure 6.77. Upon initial interaction, a certain fraction of the incident beam is reflected. According to Eq. (6.69), this fraction is R, or I_R/I_0. The associated intensity is I_R, or I_0R. Metals and alloys typically have high values of R, as high as 1.0 in some cases. The light that is not reflected enters the material, the intensity of which is thus $I_0 - I_R$, or $I_0(1 - R)$. Some or all of this remaining light will be absorbed through electronic interactions with the solid, and the result is a continuously decreasing intensity as the light passes through the length of the solid, indicated by l. The fractional change in light intensity,

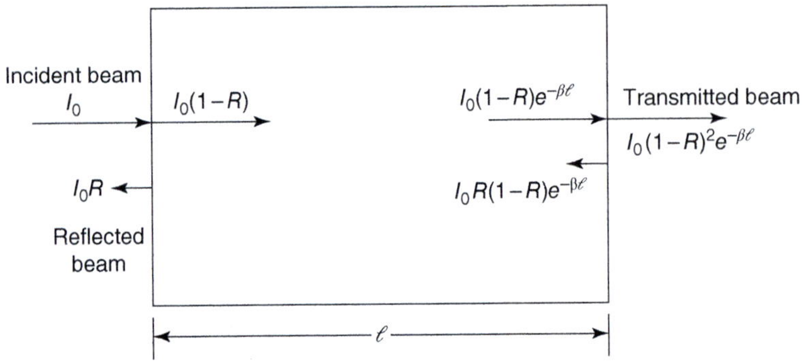

Figure 6.77 Reflection, absorption, and transmission by a solid. Ralls et al. [15]. Copyright © 1976 by John Wiley & Sons, Inc. This material is used by permission John Wiley & Sons, Inc.

dI/I, over a distance x is directly proportional to a material property called the *linear absorption coefficient*, β:

$$\frac{dI}{I} = -\beta\, dx \tag{6.70}$$

> **Cooperative Learning Exercise 6.8**
>
> The reflectivity of silicon at 633 nm is 35% and the absorption coefficient is $3.8 \times 10^5\ \text{m}^{-1}$. Calculate the following quantities for a 10 μm-thick sample of silicon at this wavelength.
>
> *Person 1*: Calculate the fraction of incident light that is absorbed.
>
> *Person 2*: Calculate the fraction of incident light that is transmitted through the sample. Combine your information to check that Eq. (6.69) is satisfied.
>
> *Answers*: If Eq. (6.71) is the fraction of light reaching the back face, then that portion subtracted from the light entering, $I_0(1-R)$ is the fraction absorbed, so $I_A/I_0 = 0.35-(1-0.35)\exp[(-3.8\times10^5)(1\times10^{-5})] = 0.6355$. Equation (6.71) gives $I_T/I_0 = (0.65)^2 \exp[(-3.8\times10^5)(1\times10^{-5})] = 0.00945$. $A+R+T = 0.6355+0.035+0.00945 \approx 1.0$.

where the minus sign indicates a decrease in intensity with distance, since β is taken as a positive quantity. Integration of Eq. (6.70) over the distance l in Figure 6.77 results in

$$I = I_0(1-R)\exp\{-\beta\cdot l\} \tag{6.71}$$

This is the intensity of light reaching the back surface in Figure 6.77. At the back surface, a fraction of the light is once again reflected internally, which is R times the amount given in Eq. (6.71), and the remainder passes out of the material or is transmitted. Thus, the fractional amount of transmitted light, T, is one minus the amount reflected at the back face, or

$$T = (1-R)^2 \exp -\{-\beta\cdot l) \tag{6.72}$$

provided that the same medium, e.g., air is on both the entering and exiting sides of the solid. (We will return to this condition when we discuss refractive index). These relationships indicate that the linear absorption coefficient, β, and reflectivity, R, are the two important parameters in determining how light interacts with a material. Let us investigate these two quantities more carefully, first for metals and alloys.

Reflectance and Color

Recall that in metals and alloys, there are many empty electronic states just above the occupied levels in the conduction bands. When incident radiation strikes a metallic surface, energy can be absorbed by promoting electrons from the occupied to the unoccupied electronic states. This generally occurs over a very short distance in the metal surface, and only very thin films of metals (<0.1 μm) exhibit any transmittance at all. Once the electrons have been excited, they decay back to the lower energy levels, and reemission of light from the metal surface occurs. It is this immediate reemission process that gives rise to reflectivity. As we will see, there are other types of reemission processes, but they differ in time scale and frequency of the emitted light from reflectivity.

The efficiency of the reflection process depends on the frequency of the incident light, ν. For example, as shown in Figure 6.78, silver is highly reflective over the entire visible range. As a result, it has a white metallic *color* and a bright *luster*.* The white color of silver indicates that the reemitted photons cover much of the same frequencies and are comparable in number (intensity) to the incident photons. Metals like copper and gold exhibit characteristic red–orange and yellow colors, respectively, because incident photons above a certain frequency cause excitation of some electrons from a filled *d* electronic band to empty *s* electronic bands. Since this frequency falls in the visible region for Cu and Au (see Figure 6.78), strong absorption of light with higher frequencies occurs, and light with lower frequencies (red and yellow) is strongly reemitted. Many metals, like nickel, iron, and tungsten, have grayish colors because the density of electronic states both for occupied and unoccupied levels exhibit maxima and minima. This, in combination with the fact that the density of electronic states is very large in different parts of the *d* band, leads to relatively strong absorption and thus leads to a relatively low reflectivity in the visible region. We will return to a more quantitative description of color in the section on ceramics and glasses, where there is more variability in color than is commonly found in metals and alloys. Outside of the visible region, most metals become highly reflective for photons having frequencies in the infrared region. This is because, unlike free atoms where discrete frequencies are absorbed and emitted, photons having a virtually continuous band of frequencies are absorbed and emitted by a metal because of the almost continuous variation of energy levels in the conduction band. Of course, some intensity is lost because of phonon generation resulting from collisions suffered by the excited electrons in a metal and because of electronic polarization.

The color and luster of a metal can be affected by other factors, such as surface roughness. As illustrated in Figure 6.79, reflection from a smooth or mirror-like surface is termed *specular reflectance*, whereas reflection from surfaces that are not parallel to the average surface due to roughness is termed *diffuse reflectance*. On a completely rough surface, reflectance can occur at all angles, θ, relative to the incident light, as shown in Figure 6.80. The relative intensity of reflection, I_θ, varies as the cosine of the angle as

$$I_\theta = I_0 \cos\theta \quad (6.73)$$

*The technical definition of *luster* is a bit elusive. In its common use, it is defined as "the quality, condition, or fact of shining by reflected light; gloss; sheen." In fact, the term *metallic luster* is used in mineralogy as a distinguishing feature from *mineral luster*, which is less "shiny" than metallic luster. In this context, luster is defined as "the appearance of a mineral in reflected light." However, this does not help us differentiate between the luster of two metals. So, while luster is commonly used to describe both metals and minerals, there is no absolute or relative scale for luster that makes it a useful quantity for describing metals.

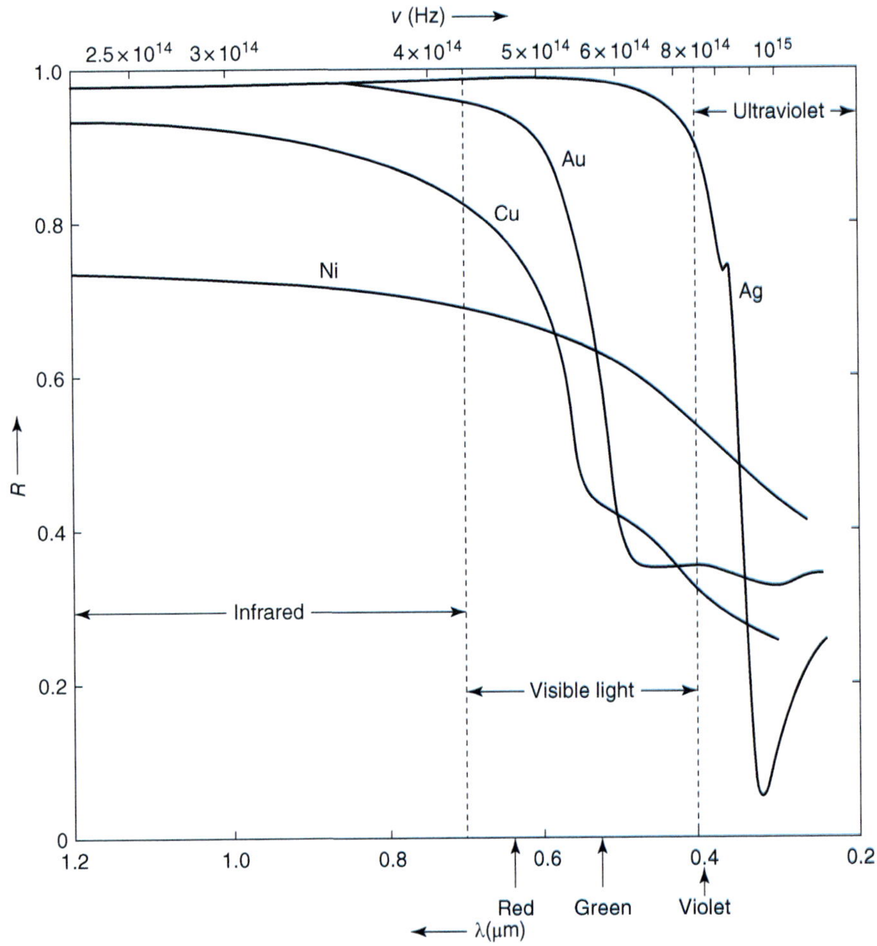

Figure 6.78 The frequency dependence of reflectivity for selected metals. Ralls et al. [15]. Copyright © 1976 by John Wiley & Sons, Inc. This material is used by permission John Wiley & Sons, Inc.

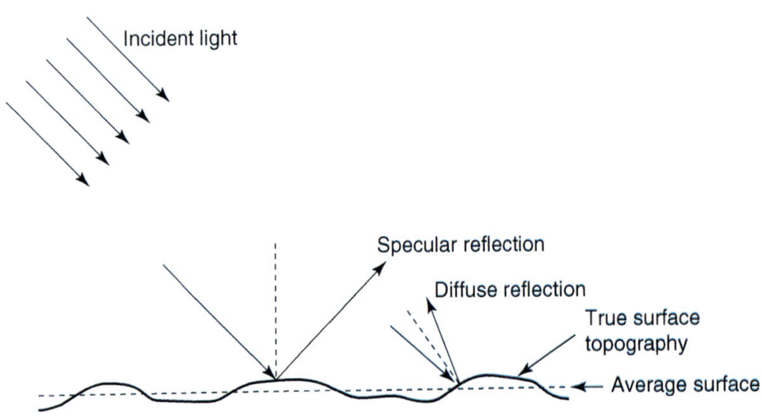

Figure 6.79 Schematic illustration of specular reflection from a flat surface and diffuse reflection from a rough surface, both relative to the average surface. Shackelford [47]/Reproduced with permission of Pearson.

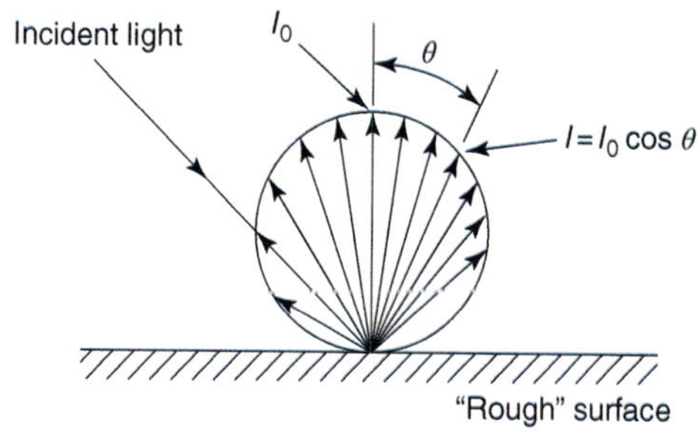

Figure 6.80 Polar diagram illustrating the directional intensity of reflection from a rough surface. Shackelford [48]/Reproduced with permission of Pearson.

where I_0 is the intensity of scattering at $\theta = 0°$. Since any area segment, A_θ, will be reduced when viewed at an angle, the luster of the diffuse surface will be a constant independent of viewing angle:

$$\text{luster} = \frac{I_\theta}{A_\theta} = \frac{I_0 \cos\theta}{A_0 \cos\theta} = \text{constant} \tag{6.74}$$

This, at least in part, helps us understand the concept of luster.

The Photoelectric Effect

A phenomenon that is related to the bombardment of metallic surfaces with light (photons) is the *photoelectric effect*, which is the release of electrons due to the absorption of light energy. The photoelectric effect arises from the fact that the potential energy barrier for electrons is finite at the surface of the metal. This process is illustrated in Figure 6.81. The finite energy barrier at the surface is represented by W. For an electron to escape from the metal, it must possess kinetic energy equal to or greater than W. Since the free electrons in the metal have kinetic energies ranging from nearly zero at the bottom of the potential energy well to E_f at the Fermi level, the electrons must absorb energy between W and $(W - E_f)$ to be ejected from the metal. The work that must be supplied to eject a free electron is called the *work function* and is denoted by the Greek lowercase phi, φ, where:

$$\varphi = W - E_f \tag{6.75}$$

The critical energy required for a photon to remove an electron is then $h\nu_c = \varphi$, where ν_c is the critical frequency of the photon and h is Planck's constant. When the frequency of the incident

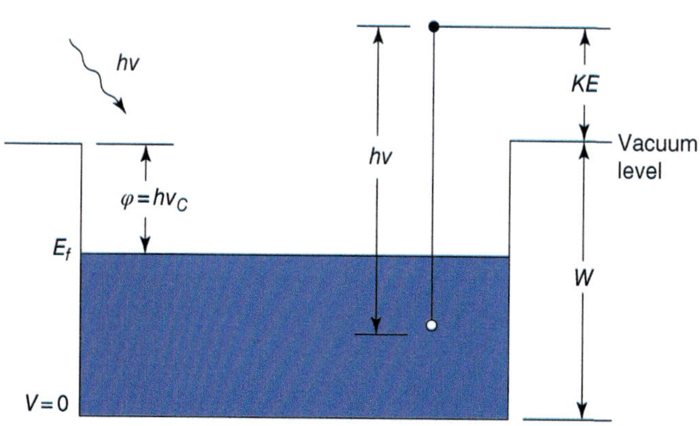

Figure 6.81 Illustration of potential energy well for the surface of a metal, and the energy required by an incident photon to remove an electron from it.

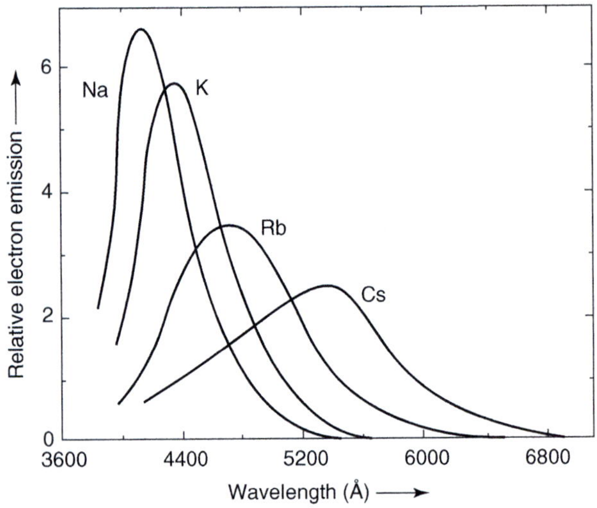

Figure 6.82 Dependence of photoelectric emission on wavelength of incident light for various alkali metals. Barrett et al. [49]/Reproduced with permission of Pearson.

radiation is less than the critical frequency, electrons will not be ejected. Similarly, when the wavelength of the incident radiation is greater than the *critical wavelength*, λ_c, electrons will also not be emitted (recall that $v = c/\lambda$, where c is the speed of light in vacuum). This relationship between photon energy (in terms of wavelength) and electron emission for several alkali metals is shown in Figure 6.82. The decrease in number of emitted electrons at short wavelengths (high energy) is due to a decrease in the efficiency of the electronic excitation. In this way, the work function for each element, in principle, can be determined and can be used to identify a metallic sample.

Historical Highlight

It is somewhat surprising that despite all his contributions to science and engineering, including his work on Brownian motion and the theory of relativity, **Albert Einstein** won his only Nobel Prize "for his services to Theoretical Physics, and especially for his discovery of the law of the photoelectric effect." Read more here: https://www.nobelprize.org/prizes/physics/1921/summary/

We will see that many important optical phenomena are functions of the wavelength of light. This is probably a good time to review the various categories and associated wavelengths (and frequencies) of electromagnetic radiation through the summary in Figure 6.83.

6.2.2 Optical Properties of Ceramics and Glasses

Refractive Index and Dispersion

The velocity of light varies depending on the density of the medium in which it is propagating. In a vacuum, the speed of light is a constant, c, which has a value of 3.08×10^8 m/s. In any other

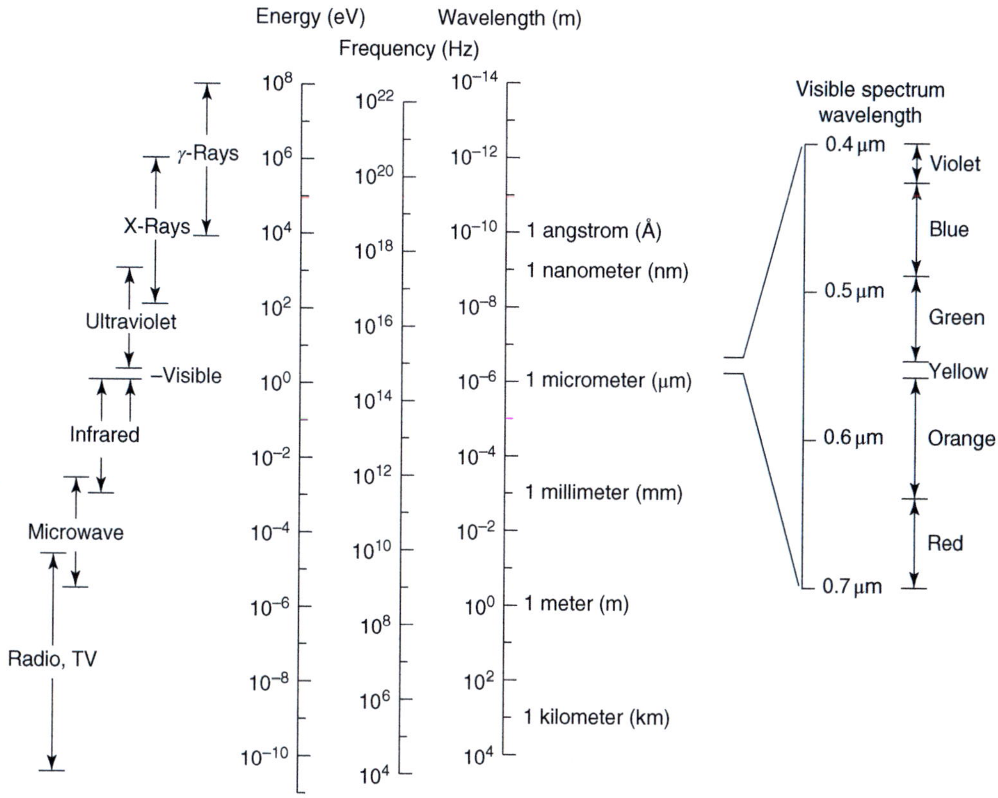

Figure 6.83 Energies, wavelengths, and frequencies of various categories of electromagnetic radiation. Callister [50]/Reproduced with permission of John Wiley & Sons.

medium, such as a gas, liquid, or solid, the velocity is given by the variable, v. The ratio between these two velocities determines the *index of refraction*, n, sometimes called the *refractive index*:

$$n = \frac{c}{v} \tag{6.76}$$

Hence, the refractive index is a dimensionless quantity. The velocity of light in a vacuum can be related to the electric permittivity in a vacuum, ε_0 [cf. Eq. (6.33)], and the magnetic permeability in a vacuum, μ_0 [cf. Eq. (6.59)]:

$$c = \frac{1}{\sqrt{\varepsilon_0 \mu_0}} \tag{6.77}$$

Similarly, the velocity of light in a medium is related to the electric permittivity and magnetic permeabilities in the medium, ε and μ, respectively:

$$v = \frac{1}{\sqrt{\varepsilon \cdot \mu}} \tag{6.78}$$

Thus, we see the initial connection between optical properties and the electrical and magnetic properties from the two previous sections. Substitution of Eqs. (6.77) and (6.78) into (6.76) shows that the refractive index can be expressed in terms of the relative electric permittivity (dielectric constant), ε_r (cf. Table 6.5), and relative magnetic permeability of the medium, $(1 + \chi)$ [cf. Eq. (6.62)], where χ is the magnetic susceptibility:

$$n = \sqrt{\frac{\varepsilon_r}{(1 + \chi)}} \tag{6.79}$$

Since most ceramic substances (and most nonmetallic substances, for that matter) possess small magnetic susceptibilities, the quantity $(1 + \chi)$ is approximately unity, and the refractive index can be approximated as the square root of the dielectric constant:

$$n \approx \sqrt{\varepsilon_r} \tag{6.80}$$

Refractive index values vary from 1.0003 for air to over 2.7 for some solid oxide ceramics. Silicate glasses have a much narrower range of values, from about 1.5 to 1.9. The refractive indices (or indexes) of other materials can be found in Appendix 8.

The relationship between refractive index and the dielectric constant is a logical one and brings into play many of the considerations of electronic structure utilized in Section 6.0.2 to describe the dielectric constant. A dielectric material interacts with electromagnetic radiation because it contains charge carriers that can be displaced. Light waves are retarded in a dielectric; that is, the velocity is decreased because of the interactions of the electromagnetic radiation with the electrons. Recall from Section 6.0.2 that the polarizability, α, is a measure of the average dipole moment per unit field strength [cf. Eq. (6.40)]. The polarizability can be related to the index of refraction for a simple, monatomic gas through the Lorentz–Lorenz equation

$$\alpha = \frac{3\varepsilon_0 M(n^2 - 1)}{N_A \rho(n^2 + 2)} \tag{6.81}$$

where ε_0 is again the electric permittivity in vacuum, M is the atomic weight of the gas, ρ is its density, N_A is Avogadro's number, and n is the refractive index.

Need More Help?

Try these videos on refractive index:

https://www.youtube.com/watch?v=ohQheheySDw
https://www.youtube.com/watch?v=qPO2Eo7Naes
https://www.youtube.com/watch?v=uAcfzqswuQY

In ionic solids, we expect the ionic polarizability to dominate. Ionic polarization should increase with the size of the ion and with the degree of negative charge on isoelectric ions. Since the index of refraction increases with polarizability in accordance with Eq. (6.81), large refractive indices arise with large ions (e.g., $n_{PbS} = 3.912$) and low refractive indices occur with small ions (e.g., $n_{SiCl_4} = 1.412$). However, as we know, the electronic environment of the ions also affects the polarizability and, in turn, the refractive index. Only in glasses and in cubic crystals is the refractive index independent of direction. In other crystal systems, the refractive index is larger in directions that are close-packed. This also follows directly from Eq. (6.81). Similarly, the more open structures of high-temperature polymorphic forms have lower refractive indices than do crystals of the same composition; for example, for SiO_2, $n_{glass} = 1.46$, $n_{tridymite} = 1.47$, $n_{cristobalite} = 1.49$, and $n_{quartz} = 1.55$. The refractive index of a typical soda-lime glass is about 1.5. The addition of large ions such as lead and barium has the effect of increasing the index of refraction. For example, glasses containing 90 wt% PbO have an index of refraction of around 2.1, which is about the upper limit of the obtainable values for practical optical glasses.

In the same way that close-packed directions in a crystal have larger refractive indices, so too can the application of tensile stress to an isotropic glass increase the index of refraction normal to the

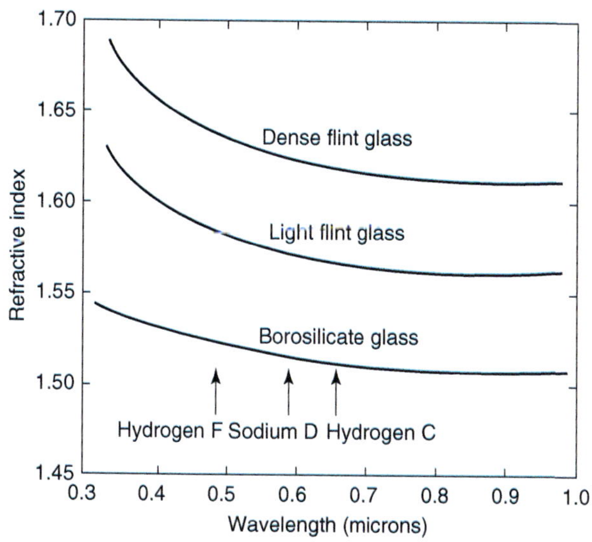

Figure 6.84 Effect of light wavelength on refractive index for some common glasses. Kingery et al. [22].

direction of the applied stress. Uniaxial compression has the reverse effect. The resulting variation in refractive index with direction is called *birefringence*, which can be used as a method of measuring stress.

The refractive index is a function of the frequency of the light, and it normally decreases with increasing wavelength, as illustrated in Figure 6.84 for some typical glasses. The variation in refractive index with wavelength, λ, is called *dispersion*, and it is given in one definition by

$$\text{Dispersion} = \frac{dn}{d\lambda} \tag{6.82}$$

As can be seen in Figure 6.84, the slope of the refractive index versus wavelength—the dispersion—varies with wavelength. However, most practical measurements are made by determining the dispersion at specific wavelengths, such as the hydrogen F line (λ = 486.1 nm), sodium D line (λ = 589.3 nm), and hydrogen C line (λ = 656.3 nm), as indicated in Figure 6.84. The complete dispersion curves for some common inorganic materials are shown in Figure 6.85.

Cooperative Learning Exercise 6.9

It can be shown that for some ionic crystals, such as LiF, both ionic and electronic polarizations can contribute to the overall dielectric constant, ε_r. In such cases, Eq. (6.80) is not entirely correct, and the electronic contribution to the polarizability, α_e, is given by Eq. (6.81), since the refractive index affects only the frequencies in the electronic range, and the number of ions per unit cell, in this case, two must be included in the denominator. The total polarizability, $\alpha = \alpha_e + \alpha_i$, is then given by

$$\alpha = \frac{3\varepsilon_0 M}{N_{A\rho}}\left(\frac{\varepsilon_r - 1}{\varepsilon_r + 2}\right),$$

where ε_r is the dielectric constant as given in Table 6.3 and the remaining quantities are all the same as in Eq. (6.81). The index of refraction for LiF is 1.395, its density is 2.635 × 10^3 kg/m^3, and its molecular weight is 26 × 10^{-3} kg/mol. Recall that ε_0 = 8.854 × 10^{-12} C/V-m.

Person 1: Calculate the total polarizability for LiF.

Person 2: Calculate the electronic contribution to the total polarizability.

Combine your information to calculate the ionic polarizability, α_i.

Answers: α = [3(8.854 × 10^{-12})(26 × 10^{-3})/(6.02 × 10^{23})(2.635 × 10^{3})][(9 − 1)/(9 + 2)] = 3.166 × 10^{-40} F m^2; α_e = [3(8.854 × 10^{-12})(26 × 10^{-3})/2(6.02 × 10^{23})(2.635 × 10^{3})][(1.95 − 1)/(1.95 + 2)] = 5.2 × 10^{-41} F m^2. $\alpha_i = \alpha - \alpha_e$ = 3.16 × 10^{-40} − 5.2 × 10^{-41} = 2.64 × 10−40 F m^2.

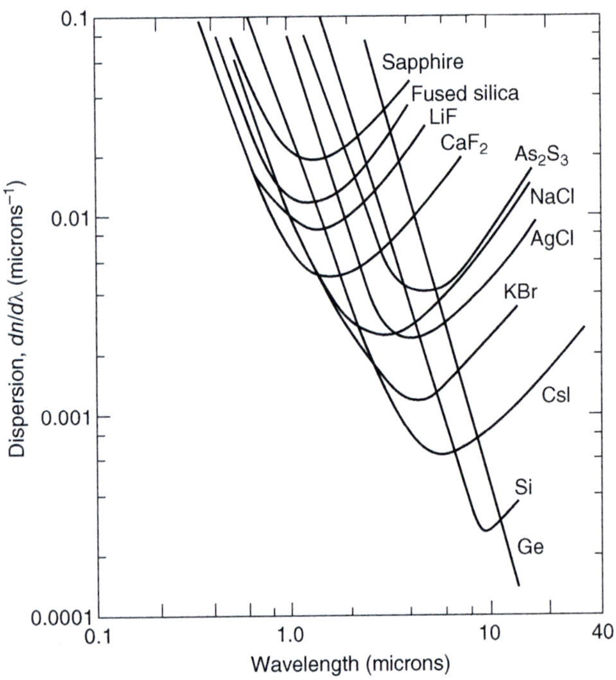

Figure 6.85 Effect of light wavelength on dispersion for some common inorganic materials. Kingery et al. [22]. Copyright © 1976 by John Wiley & Sons, Inc. This material is used by permission of John Wiley & Sons, Inc.

The data shown in Figures 6.84 and 6.85 are a subset of the more general behavior, which is composed of two types of dispersion: *normal dispersion* and *anomalous dispersion*. Normal dispersion arises in the visible range of light, and the index of refraction increases with decreasing wavelength (increasing energy), as in Figure 6.85. In this region, the refractive index can be correlated with wavelength according to the following empirical expression, known as the Cauchy formula:

$$n = A + \frac{B}{\lambda^2} + \frac{C}{\lambda^4} \tag{6.83}$$

where A, B, and C are constants. In the region where refractive index decreases with decreasing wavelength (typically in the ultraviolet), anomalous dispersion occurs. Anomalous dispersion is the result of interactions between the light and the electronic oscillators in the material.

At the so-called *resonant frequencies* of the electronic oscillators, absorption of the electromagnetic radiation can occur, resulting in reinforcement of the natural electronic oscillations. This resonance tends to decrease the refractive index.

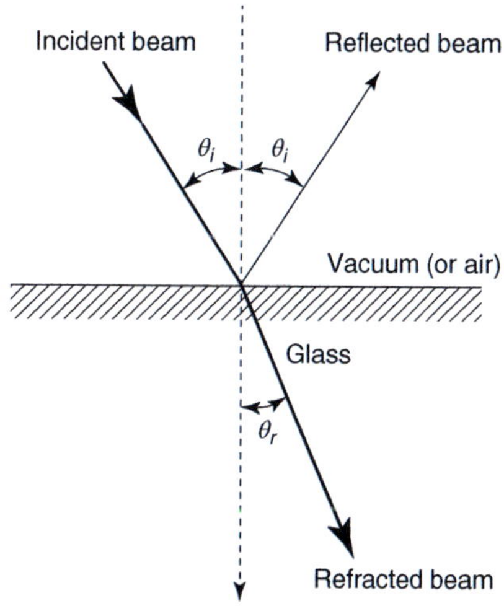

Figure 6.86 Reflection and refraction at a solid surface. Shackelford [47]/ Reproduced with permission of Pearson.

Reflection and Refraction

With the principle of refractive index now in mind, we can return momentarily to the concept of reflection, which was first introduced in Section 6.2.1. Recall that the velocity of light changes as it passes from vacuum to a medium, such as air. The velocity also changes at any interface between two phases with different indices of refraction, such as at an air–solid interface as illustrated in Figure 6.86. The change in velocity causes the light to bend upon passing through the interface. The bending of transmitted light at an interface is more properly termed *refraction*—hence, the term refractive index. We also know that some of the light is reflected. The ratio of reflected to refracted (transmitted) light at the interface is a function of (a) the refractive index difference between the two phases and (b) the angle of the incident light.

If the *angle of incidence* from a normal to the surface is θ_i (see Figure 6.86) and the *angle of refraction* is θ_r, the refractive index of the solid medium, n (provided that the incident light is coming from a phase of low refractive index such as vacuum or air), is given by

$$n = \frac{\sin \theta_i}{\sin \theta_r} \tag{6.84}$$

Recall from Figure 6.83 that the light that is reflected at the same angle of incidence, θ_i, is due to specular reflectance. We can now use the refractive index to show that the reflectivity, R [cf. Eq. (6.69)], at the interface in Figure 6.86 is given by

$$R = \frac{(n-1)^2}{(n+1)^2} \tag{6.85}$$

Equation (6.85) holds for a solid–gas interface where the refractive index of the gas (or vacuum) is approximately unity. It is a simplification of a more general expression for light propagating through a medium of refractive index n_1, striking perpendicular to a surface ($i = 0$) of refractive index n_2, which is presented here without derivation:

$$R = \frac{(n_2 - n_1)^2}{(n_2 + n_1)^2} \tag{6.86}$$

Thus, the higher the refractive index of the solid, the greater the reflectivity. For example, the reflectivity of a typical silicate glass, with $n = 1.5$, is about 0.04. Just as the refractive index varies with wavelength, so too does the reflectivity.

Absorbance and Color

Once the light that is not reflected enters the medium, it can either continue to be transmitted or be absorbed by the medium (cf. Figure 6.77). The absorption process, too, is a function of the energy (wavelength) of the light.

Recall from Eq. (6.71) that the fractional change in light intensity, dI/I, over a distance x within the absorbing medium is directly proportional to the linear absorption coefficient, β:

$$\frac{dI}{I} = -\beta\, dx \tag{6.71}$$

The absorption coefficient is a material property and is a function of the wavelength of light, λ:

$$\beta = \frac{4\pi k}{\lambda} \tag{6.87}$$

where k is called the *index of absorption*.

Light is absorbed by two basic mechanisms: electronic polarization and electronic excitation. Electronic polarization and its effect on refractive index were described in the previous section and will not be elaborated upon here. The process of electronic excitation is an important one, however, and has implications for many optical phenomena such as lasing and luminescence.

Absorption of a photon of light may occur by the promotion or excitation of an electron from the valence band to the conduction band. In doing so, the promoted electron leaves behind a hole in the valence band and is free to move in the conduction band. Absorption and excitation occur only if the photon has energy greater than that of the band gap, E_g:

$$h\nu \geq E_g \tag{6.88}$$

Cooperative Learning Exercise 6.10

Use the data in Figure 6.83 to perform the following calculations.

Person 1: Calculate the maximum band gap energy for which absorption of visible light is possible.

Person 2: Calculate the minimum band gap energy for which absorption of visible light is possible.

Compare your answers. What type of electrical properties will materials within these band gap energy limits possess?

Answers: E_g (max) = hc/λ_{min} = $(4.13 \times 10^{-15}\ \text{eV} \cdot \text{s})(3 \times 10^8\ \text{m/s})/(4 \times 10^{-7}\ \text{m})$ = 3.1 eV; E_g (min) = 1.8 eV

The magnitude of the band gap energy, then, will determine whether the material absorbs no light, and is transparent, or absorbs certain wavelengths of light, thus becoming *opaque*.

The absorption of certain wavelengths of light in the visible spectrum results in *color*. As illustrated in Figure 6.87 for a ruby, the absorption of light centered at 0.42 and 0.6 μm (represented here as decreases in transmittance rather than increases in absorbance) correspond to absorption of blue–violet and yellow–green light, respectively. The result is that red is the most strongly transmitted visible light, giving ruby its characteristic red color. The transmission spectrum for sapphire,

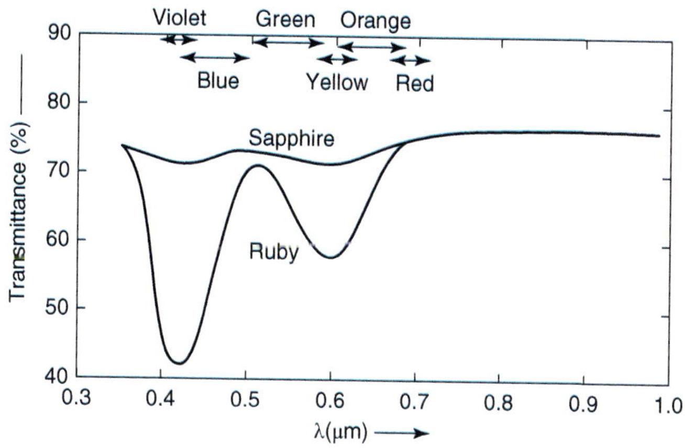

Figure 6.87 Transmission spectrum for sapphire and ruby. Ralls et al. [15]. Copyright © 1976 by John Wiley & Sons, Inc. This material is used by permission John Wiley & Sons, Inc.

which has a slight bluish color, is also shown in Figure 6.87. The interesting fact is that both ruby and sapphire are primarily Al_2O_3, with only trace amounts of Ti^{3+} or Cr^{3+} substituted for Al^{3+} in sapphire and ruby, respectively. In the case of ruby, the Cr^{3+} ions are point defects with associated energy states that lie between the valence and conduction bands of Al_2O_3. The separation of levels is such that blue-violet light can be absorbed and red light emitted. A characteristic of such an absorption–emission process is that the emitted light cannot be of a higher energy than the absorbed light. We will further describe these radiative transitions when we get to the topics of luminescence and phosphorescence.

In general, the most common ions for creating color in ceramics are transition elements with their incomplete *d* shells, such as V, Cr, Mn, Fe, Co, Ni, and Cu, and, to a lesser extent, the rare-earth elements with their incomplete *f* shells. In addition to the individual ion and its oxidation state, absorption phenomena are affected by their ionic environment. The introduction of impurity atoms to affect color is also utilized in glasses. In this case, a single ion such as Co^{2+} can give different colorings in different glasses, since it can have different coordination numbers associated with it (cf. Section 1.2.4). Some common metal ions and their effect on the color of silicate glasses are listed in Table 6.24.

Transmittance, Transparency, and Translucency

The descriptions of absorption processes and their relationship to transmittance and color in the preceding section serve to underscore the relationship first presented by Eq. (6.69)—namely, that the fractions of light reflected, absorbed, and transmitted must sum to unity. In this section, we complete our description of these three processes with some issues related to transmittance—namely, transparency and translucency.

A material is transparent if it transmits a clear image through it. Dielectrics tend to be transparent to visible radiation because they possess such large energy gaps that light cannot cause electronic excitation. Thus, materials such as diamond, sodium chloride, and silicate glasses readily transmit light. When the energy of the incident radiation is sufficiently high, as in ultraviolet radiation, electrons can be excited across the energy gap, and strong absorption occurs, as described in the previous section.

Table 6.24 Selected Metal Ions and their Effect on Color in Silicate Glasses

	In Glass Networks		In Modifier Position	
Ion	**Coordination Number**	**Color**	**Coordination Number**	**Color**
Cr^{2+}				Blue
Cr^{3+}			6	Green
Cr^{6+}	4	Yellow		
Cu^{2+}	4		6	Blue-green
Cu^{+}			8	Colorless
Co^{2+}	4	Blue-purple	6–8	Pink
Ni^{2+}		Purple	6–8	Yellow-green
Mn^{2+}		Colorless	8	Weak orange
Mn^{3+}		Purple	6	
Fe^{2+}			6–8	Blue-green
Fe^{3+}		Deep brown	6	Weak yellow
U^{6+}		Orange	6–10	Weak yellow
V^{3+}			6	Green
V^{4+}			6	Blue
V^{5+}	4	Colorless		

Source: Shackelford [51]. Copyright © 2000 by Prentice-Hall, Inc.

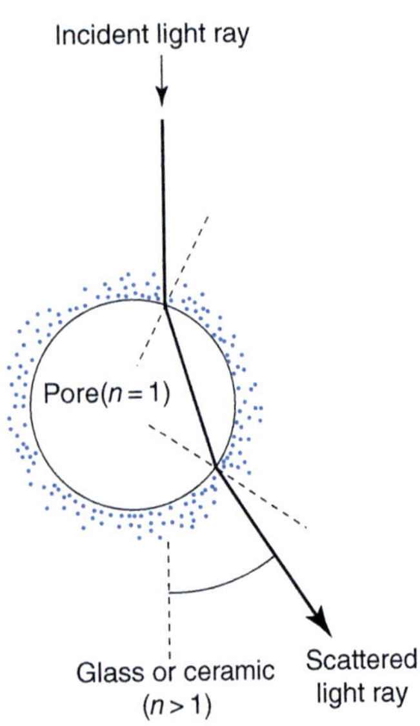

Figure 6.88 Schematic illustration of light scattering in a solid due to a pore. Shackelford [51]/Reproduced with permission of Pearson.

Many inherently transparent materials appear "milky" because of diffusive transmission, also known as *translucency*. Translucence results from multiple internal reflections. In noncubic polycrystalline materials that possess an anisotropic refractive index, light is reflected or scattered at grain boundaries where the index of refraction undergoes a discontinuous change. This effect is similar to that which results in reflection at a free surface. Translucency is slightly different from opacity, which was first described in the previous section, insofar as a diffuse image is transmitted through a translucent medium, whereas a total loss of image occurs in an opaque medium. In both cases, however, a certain fraction of some light of certain wavelengths is transmitted.

In addition to surface roughness and absorbing chemical species that affect transmittance and have already been described, such factors as porosity and particulate inclusions can affect translucence and opacity, mostly through a process called *scattering*. Figure 6.88 illustrates how scattering can occur at a single pore by refraction. The light is refracted as it exits a medium of higher refractive index and enters one of lower refractive index, such as air, inside the pore. This is essentially a miniaturized, reversed situation of the refraction process illustrated in

Figure 6.86. A similar scattering result can be created with second-phase particles such as SnO_2 with a higher index of refraction ($n = 2.0$) than that of the medium, such as glass ($n = 1.5$). Such materials are called *opacifiers*. The degree of scattering and opacification created by the pores or particles depends on their average size and concentration, as well as the mismatch between the refractive indices. If individual pores or particles are significantly smaller than the wavelength of light, they are ineffective scattering centers. Conversely, particles or pores in the 400- to 700-nm size range maximize the scattering effect. This effect is what leads to transparent ceramics, such as glass ceramics (cf. Section 1.2.5) where the crystallite sizes are carefully controlled through nucleation and growth to be less than the wavelength of visible light (cf. Figure 1.48).

Decay Processes: Luminescence and Lasers*

The absorption and reemission of electromagnetic radiation lead to some interesting phenomena that can be utilized for important applications. Both lasers and luminescence are built upon the principle of emission processes and will be described here in brief. The topic of electroluminescence will be reserved for Section 6.2.3.

As we have seen, the absorption of photons by a material can lead to the reemission of other photons of equal or lesser energy. When photon absorption is accompanied by emission of photons in the visible spectrum, the process is called *photoluminescence*, or simply *luminescence*. In fact, luminescence can be used to describe the emission of visible light as a result of absorption of any number of forms of energy, including thermal, mechanical, or chemical energy. The photon emission is the result of excited electrons returning to their ground state. The duration of this relaxation process is used to distinguish between the two main types of luminescence: *fluorescence* and *phosphorescence*. If the relaxation process occurs in less than about 10 ns (10^{-8} s), it is termed fluorescence. The fluorescent process also involves additional internal or external energy conversion, as illustrated by the energy diagrams in Figure 6.89. In this way, the energy of the reemitted photons is less than the energy of the absorbed photons. If the reemission process takes longer than about 10 ns, it is termed phosphorescence.

Luminescence is generally produced through the addition of impurities to compounds. For example, zinc sulfide containing small amounts of excess zinc, or impurities such as gold or manganese, emits light in the visible spectrum after excitation with ultraviolet radiation (see Figure 6.90). Zinc sulfide and a similar compound, Mn-activated zinc orthosilicate, Zn_2SiO_4, are also examples of compounds used as *phosphors*. Phosphors are useful for detecting X-ray or γ-ray radiation because they emit visible light upon exposure. They are used for other, more mundane purposes as well, such as the luminous hands on analog watches and clocks.

Light amplification by stimulated emission of radiation, or *LASER*, is another application of radiative relaxation processes. Unlike most radiative processes, such as luminescence, which produce *incoherent* light (light waves are out of phase with each other), the light produced by laser emission is *coherent* (light waves are all in phase with each other). The principal benefit of coherent light is that the photons do not disperse as they do in an incoherent beam of light. Consequently, the laser beam does not spread out in the same way that an incoherent beam can. Additionally, lasers produce a beam of light that is *monochromatic*—that is, entirely of one wavelength. This is in contrast to most light sources—even phosphors—that produce a range of wavelengths, or polychromatic light. Polychromatic light is not entirely a bad thing; we much prefer white light (all or most of the wavelengths in the visible spectrum) to monochromatic light for our daily and nightly operations. But the monochromatic, coherent, and collimated (photons directed along the same straight line) beam of a laser opens a realm of possible applications. Let us see how this type of beam can be produced.

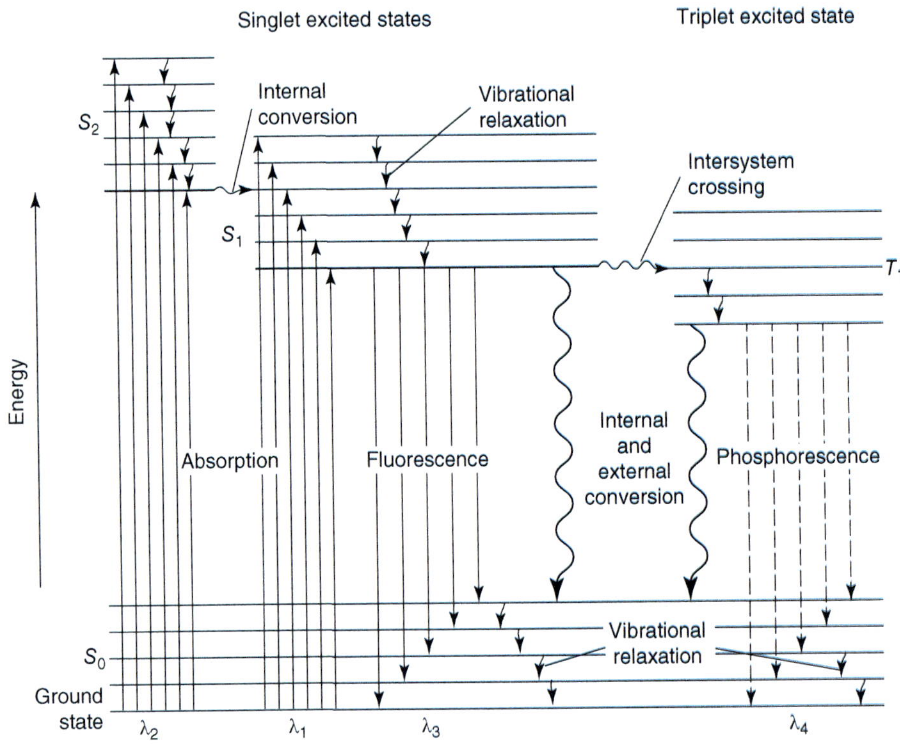

Figure 6.89 Schematic diagram of transitions between electronic energy levels involved in fluorescence and phosphorescence. Skoog and West [52]. Copyright © 1980 by Saunders College.

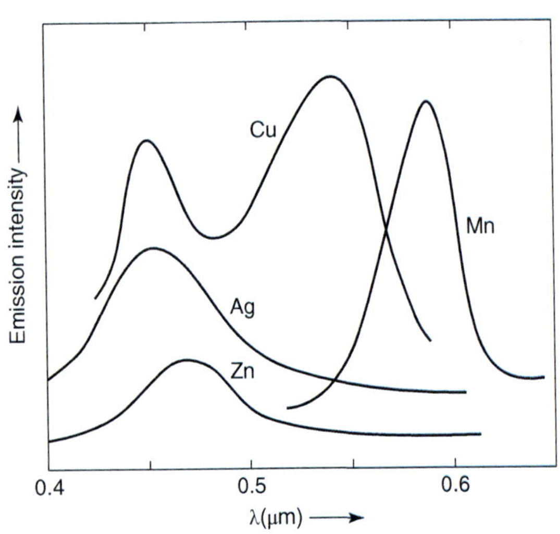

Figure 6.90 Luminescence spectra for Zn doped with Cu, Ag, Mn, and excess Zn. Kingery et al. [22]. Copyright © 1976 by John Wiley & Sons, Inc. This material is used by permission of John Wiley & Sons, Inc.

The radiative transitions of the previous descriptions have all been spontaneous: Relaxation from the excited state to the ground state and emission of photons occur without external aid. In contrast, a stimulated emission occurs when the half-life of the excited state is relatively long, and relaxation can occur only through the aid of a stimulating photon. In stimulated emission, the emitted photon

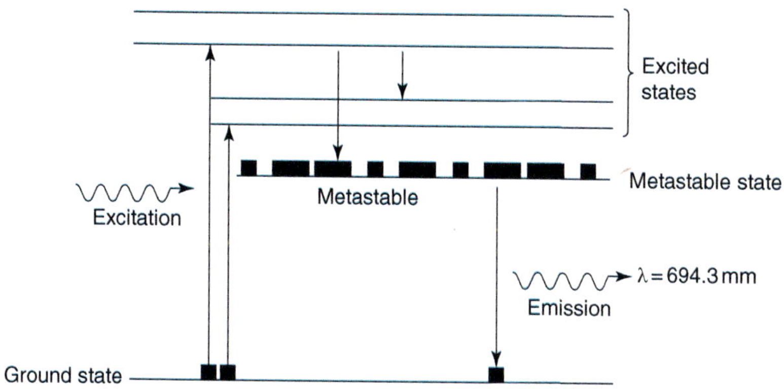

Figure 6.91 Schematic illustration of energy levels in ruby that are used to create a populated metastable electronic state, which can then be stimulated to emit monochromatic, coherent radiation for a laser. Shackelford [53]/Reproduced with permission of Pearson.

has the same direction as, and is in phase with, the stimulating photon. The example of Cr^{3+}-doped Al_2O_3 that we utilized earlier for our description of the color of ruby works equally well for a description of stimulated emission. Recall that the presence of chromium in alumina alters the electronic structure, creating a metastable state between the valence and conduction bands. Absorption of a blue–violet photon results in the excitation of an electron from the ground state to the excited state, as illustrated in Figure 6.91. The excited electron relaxes to a metastable electronic state through a non-radiative process involving the transfer of energy to a phonon. *Spontaneous decay* results in the emission of a photon with wavelength in the red part of the visible spectrum ($\lambda = 694.3$ nm). However, most of the electrons reside in this metastable state for up to 3 ms before decaying to the ground state. This time span is sufficiently long that several electrons can occupy this metastable state simultaneously. This fact is utilized to create artificially large numbers of electrons in the metastable state through a process known as *optical pumping*, in which a xenon flash lamp excites a large number of electrons into the excited state, which decays rapidly (in the nonradiative process) to the metastable state. When spontaneous emission finally occurs due to a few electrons, these electrons stimulate the emission of the remaining electrons in the metastable state, and an avalanche of emission occurs. The photons are of the same energy and are in phase (coherent).

Need More Help?

Try these videos on lasers:

https://www.youtube.com/watch?v=NpePZjTXqRw

(and how to communicate science to different audiences):

https://www.youtube.com/watch?v=3eTogq7rknQ

Historical Highlight

Not until 1954 were the theoretical principles of coherent radiation put into practice when ***Charles Townes*** built the first *maser*, a precursor to the *laser*. Read more about it here: https://physics.aps.org/story/v15/st4

In addition to ruby, lasers can be constructed from gases such as CO_2 or He–Ne and from semiconductors such as GaAs (cf. Section 6.0.2) and InGaAsP. Gas lasers generally produce lower intensities and powers but are more suitable for continuous operation since solid-state lasers generate appreciable amounts of heat. The power in a solid-state laser can range from microwatts to 25 kW, and more than a megawatt in lasers for defense applications. Some common lasers and their associated powers are listed in Table 6.25. Lasers that can selectively produce coherent light of more than one wavelength are called *tunable lasers*.

Semiconducting lasers have become increasingly important in information storage and lithography. In these materials, a voltage is applied across a layered semiconductor (see Figure 6.92), which excites electrons into the conduction band, just as in the ruby laser. However, the holes created in the valence band play an important role in the emission process. It is the recombination of the electron–hole pairs that results in the emission of photons with a specific wavelength. As in the ruby laser, the ends of the semiconducting "sandwich" are fully and partially silvered, and a voltage is continuously applied to ensure that there is a steady source of holes and electrons. These devices are

Table 6.25 Characteristics and Applications of Some Common Lasers

Laser	Type	Common Wavelengths (μm)	Applications
He-Ne	Gas	0.6328, 1.15, 3.39	Barcode scanning, spectroscopy
CO_2	Gas	9.4, 10.6	Material processing, surgery
Argon	Gas ion	0.488, 0.5145	Surgery, lithography
HeCd	Metal vapor	0.441, 0.325	Printing and typesetting
Dye	Liquid	0.39–1.0	Spectroscopy, surgery
Ruby	Solid state	0.694	holography, tattoo removal
Nd-YAG	Solid state	1.064	Material processing
Tm-YAG	Solid state	2.0	Laser ranging (LIDAR)
AlGaInP AlGaAs	Semiconductor	0.63–0.9	Optical discs, laser pointers

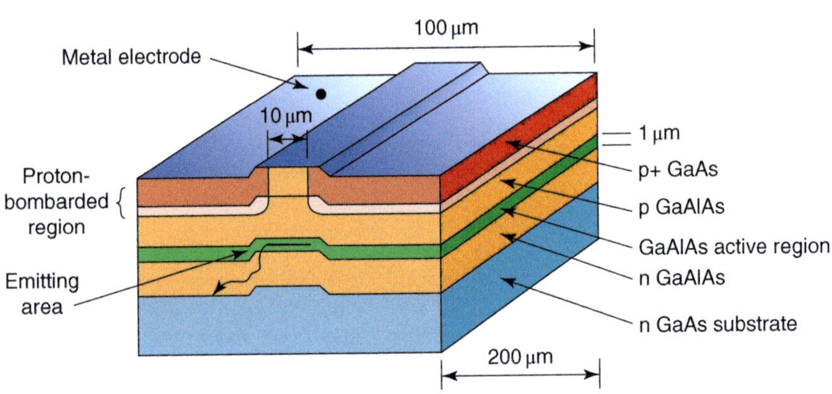

Figure 6.92 Schematic diagram of a layered compound semiconductor laser. www.mtmi.vu.lt/pfk/funkc_dariniai/diod/led.htm.

extremely temperature-sensitive, since the current required to cause lasing, the wavelength of emitted radiation, and the lifetime of the diode all depend on temperature.

The conventional III–V compound semiconductors such as GaAs are limited to infrared or red emitting diode lasers and are not capable of generating shorter wavelength light. Short-wavelength LEDs and laser diodes require a different class of semiconductor materials; principally, the group III nitrides. The growth of high-quality AlGaN single-crystalline layer sequences has been a challenge, and only recent developments in fabrication processes have produced blue lasers.

Photoconductivity*

In Section 6.2.1 we saw how electricity can be generated from the surface of a metal when it is bombarded with photons. A similar process occurs when certain semiconductors absorb photons and create electron–hole pairs that can be used to generate current. This process is called *photoconductivity* and is slightly different from the photoelectric effect, since it is an electron–hole pair that is being generated instead of a free electron, and since the energy of the electron–hole pair is related to the band gap energy, and not the energy of the Fermi level.

When a semiconductor is bombarded with photons equal to or greater in energy than the band gap, an electron–hole pair is formed. The current that results is a direct function of the incident light intensity. Photoconductive devices consist of a *p–n* junction called a *photodiode*, or a *p–i–n* junction commonly used in a photodetector, an example of which is schematically illustrated in Figure 6.93. When light is absorbed in the undoped region (*i* region), the electron–hole pair is created, and the two charge carriers are pulled in opposite directions by the applied electric field to the two poles, thus creating current. Silicon can be used to detect infrared wavelength light, and InGaAs is used

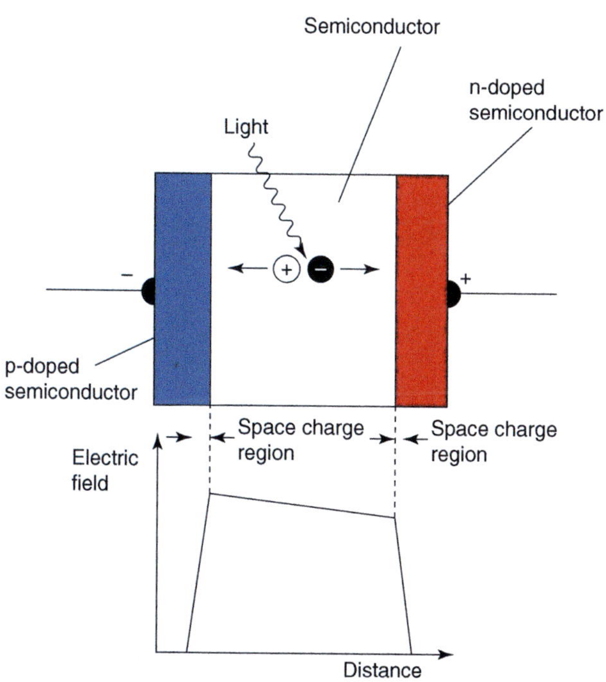

Figure 6.93 Schematic illustration of *p–i–n* junction used in a photodiode. Ball [54]/Reproduced with permission of Princeton University Press.

for longer wavelengths, including visible light. However, optical detectors tend to be relatively slow and have low sensitivity. CdS is widely used for the detection of visible light, as in photographic light meters. Photoconductivity is also the underlying principle of the *photovoltaic cell*, commonly called the *solar cell*, used for the conversion of solar energy into electricity.

Optical Fiber*

With the proliferation of wireless communications (e.g., cellular telephones and satellite-based internet) the use of land-based transmission media such as optical fiber has grown since for every cellular transmission tower, there needs to be the associated land-based components that connect the tower to the communications "backbone." Hence, it is worthwhile to describe how this important materials technology works. One of the benefits that lasers have provided is a method for transmitting information in binary code (cf. Section 6.0.1). This is easily performed by assigning a "one" to a high-intensity pulse (on) and assigning a "zero" to a low-intensity pulse (off). The key to utilizing this technology for telecommunications, however, is the ability to transmit this digital information over long distances. One method for transmitting optical information has been with optical fiber.

Optical fiber operates on the principle of *total internal reflectance*. Recall from Eq. (6.84) that when light encounters a change in refractive index, it is bent at an angle. Consider now the case where light goes through three media, each of differing refractive index, as illustrated in Figure 6.94. The angles of refraction are related to the refractive index change at each of the interfaces, through the more general form of Eq. (6.84) known as *Snell's Law*:

$$\frac{n_2}{n_1} = \frac{\sin\theta_1}{\sin\theta_2} \tag{6.89}$$

and

$$\frac{n_3}{n_2} = \frac{\sin\theta_2}{\sin\theta_3} \tag{6.90}$$

Note that Eq. (6.90) reduces to Eq. (6.86) for the case where medium 1 is air ($n_1 = 1$).

At some angle of θ_2, the transmitted wave will travel parallel to the second interface (i.e., $\theta_3 = 90°$), and all the light is reflected back into medium 2. This value of θ_2 is called the *critical*

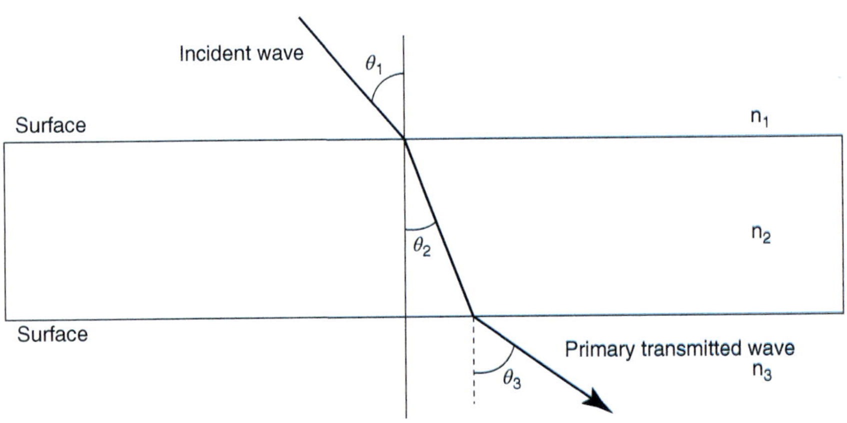

Figure 6.94 Schematic illustration of light propagating through three media of differing refractive indices.

angle for total internal reflectance, θ_c. For example, if medium 3 were air ($n_3 = 1$) and medium 2 were a glass of $n_2 = 1.5$, the condition for which $\theta_3 = 90°$ could be found from Eq. (6.88) to be $\theta_c = 42°$. You should be able to visualize, with the aid of Figure 6.94, that total internal reflectance can also be accomplished by selecting an index of refraction for medium 3 such that $\theta_3 = 90°$. This is called the *core and cladding* structure, where medium 1 is air at the entrance to the optical fiber, medium 2 is the core, and medium 3 is the cladding. The core–cladding structure is the principle behind fiber optics and simply consists of a center core, through which the optical signal is transmitted, surrounded by a cladding of different refractive index. The indices of refraction are selected such that $n_{\text{cladding}} < n_{\text{core}}$. Once the light enters the core from the source, it is reflected internally and propagates along the length of the fiber. Typically, both the core and cladding are made of special types of glass with carefully controlled indices of refraction. A polymeric coating is usually applied to the outside of the fiber for protection only.

The sharp change of refractive index between the core and cladding results in what is called a *step-index optical fiber*, as illustrated in Figure 6.95a. It is also possible, through chemical manipulation of the glass components, to vary the index of refraction in a continuous, parabolic manner, resulting in a *graded-index optical fiber*, as shown in Figure 6.95b. This results in a helical path for the light rays, as opposed to a zigzag path in a step-index fiber. The digital pulse is less distorted in a graded-index fiber, which allows for a higher density of information transmission. Both step- and graded-index fibers are termed *multimode fibers*. A third type of optical fiber is called a *single-mode fiber*, as shown in Figure 6.95c, in which light travels largely parallel to the fiber axis with little

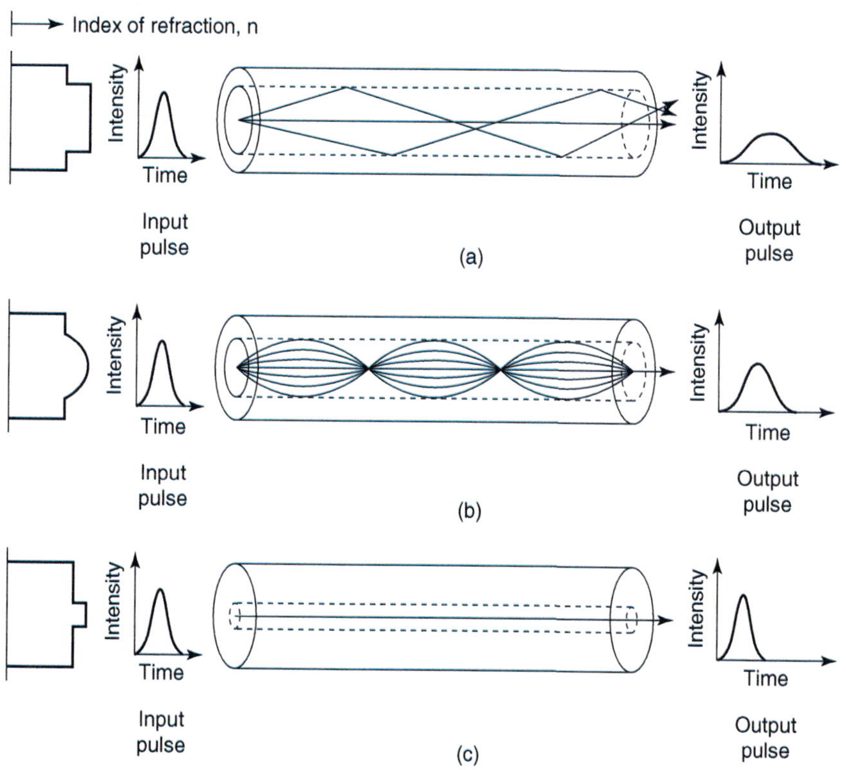

Figure 6.95 Schematic illustration of (a) step-index, (b) graded-index, and (c) single-mode optical fibers. Shackelford [55]/Reproduced with permission of Pearson.

distortion of the digital light pulse. These fibers are used for long transmission lines, in which the incoming signals are combined, or *multiplexed*, into a single signal.

Core and cladding materials are selected not only for their refractive index, but also for their processability, attenuation loss, mechanical properties, and dispersion properties. However, density, ρ, and refractive index, n, are critical and have been correlated for over 200 optical glasses, for which the following formula applies with an accuracy of 2% or better:

$$n = \frac{\rho + 10.4}{8.6} \tag{6.91}$$

Refractive index can also be found from the weighted average of the refractive index of its components, to a first approximation.

Need More Help?

Try these videos on total internal reflection:

https://www.youtube.com/watch?v=bPrCQvF48gc

and optical fiber manufacturing and use:

https://www.youtube.com/watch?v=4DIKVVY5NbA

High-purity silica-based glasses are used as the fiber material, with fiber diameters ranging between about 5 and 100 μm. The fibers are carefully fabricated to be virtually free of flaws and, as a result, are extremely strong and flexible. The production of kilometer-long optical fiber is now routine and is critical to success of virtually all wireless communications.

6.2.3 Optical Properties of Polymers

As with virtually every other material property, if a polymer possesses optical properties comparable to those of a glass or metal, it will be utilized simply on the basis of weight savings, all other factors being equal. As a result, much of the development in the optical properties of polymers is related to creating glass-like transmissivities or metal-like reflectivities. There are also a few unique optical characteristics of polymers that make them useful for applications for which there is no true glass or metal analog. We investigate some of these optical characteristics of polymers in this section.

Polymer Structure, Transmittance, and Birefringence

There is a great deal of variability in the light transmission properties of polymers. Some of this variability is due to structural differences, and some is due to the presence of fillers and colorants that are purposely added to make otherwise translucent polymers appear opaque. We will focus for the moment on structural differences in polymers that lead to differences in the transmission of light.

The primary structural characteristic affecting the transmission of light in polymers is crystallinity. As in ceramic materials, the presence of grain boundaries, pores, and particles that are of the order in size of the wavelength of light will create scattering. Crystallites in polymers act as these scattering centers. An example is found in polyethylene. Low-density polyethylene is semicrystalline, with only a few crystallites to act as scattering centers, so that it is mostly transparent. However, high-density polyethylene, which is highly crystalline, is translucent. In fact, most polymer crystals are of the right size to cause scattering, so a rule of thumb is that crystalline polymers are translucent (or opaque), whereas noncrystalline polymers are transparent. Some examples of noncrystalline, transparent polymers are polycarbonate, acrylics such as PMMA, and polystyrene. There are some transparent, crystalline polymers, such as polyethylene terephthalate (PET), in which the crystallites are smaller than the wavelength of light.

An interesting artifact of polymer crystallinity is the anisotropic nature of the refractive index, which leads to birefringence, as first introduced in Section 6.2.2. When the topic of polymer crystallinity was first introduced back in Chapter 1, we presented a photomicrograph of polyethylene spherulites that was produced under cross-polarized light (cf. Figure 1.58). This image is obtained by placing a sample of the polymer between two quarter-wave plate optical filters, or *polarizers*, which allow the light passing through to vibrate only along one plane. In this way, if the second filter, called the *analyzer*, is rotated 90° from the polarizer, no light will pass through the filter combination, since the plane of vibration allowed through the polarizer is extinguished in the analyzer. However, when a crystalline sample such as polyethylene is placed between the polarizer and analyzer, an interesting pattern is transmitted through the analyzer, which includes some dark regions (no transmission) and light regions (complete transmission). An idealized form of the resulting *Maltese cross* is shown in Figure 6.96, and an actual image of which was shown in Figure 1.58. This image is the result of birefringence, or refractive index differences in the oriented polymer chains. The index of refraction for most polymers is greater parallel to the chain than normal to the molecular axis. As a result, the refractive index in the tangential direction of the spherulite is greater (see Figure 1.57) than that along the spherulite radius. This birefringence, coupled with the spherical geometry of the spherulite, produces light extinction along the axis of each of the filters, hence the 90° angles in the Maltese cross.

Figure 6.96 Schematic of Maltese cross produced by spherulites in cross-polarized filters. Strobl [56]/ Reproduced with permission of Springer Nature.

Electroluminescence

In Section 6.2.2, we saw that some materials—in particular, semiconductors—can reemit radiation after the absorption of light in a process called photoluminescence. A related type of emission process—common in polymer-based semiconductors—is called *electroluminescence*, which results when the electronic excitation necessary for emission is brought about by the application of an electric field rather than by incident photons. The electric field injects electrons into the conduction band, and holes into the valence band, which upon recombination emit light.

As we saw in Section 6.0.3, most electrically conducting polymers contain polyconjugated structures, such as extended double bonds along the polymer backbone. It is not surprising, then, that electrically conducting polymers such as polyacetylene can also be electroluminescent. The first visible-light-emitting polymer was made from poly(phenylene vinylene), or PPV, the structure for which is shown in Figure 6.97a. PPV exhibits luminescence in the yellow part of the spectrum, and it can be fabricated into a *light-emitting diode* (LED) by sandwiching a micron-thick layer of PPV between contacts made of indium–tin-oxide (a transparent electrode material) on the bottom and cadmium metal on the top. These contact materials are selected to match their electronic bands to that of the polymer. When 12 V are applied across the contacts, electrons and holes enter the polymer and recombined to generate photons of yellow light. Other polymers with different band gaps can be used to emit light of various wavelengths, such as polythiophene-based compounds (see Figure 6.97b) for the emission of blue light.

The applications for polymer LEDs (PLEDs) continue to grow, including one class of *organic light-emitting diodes* (OLED) or organic electroluminescent diodes for televisions and flat panel

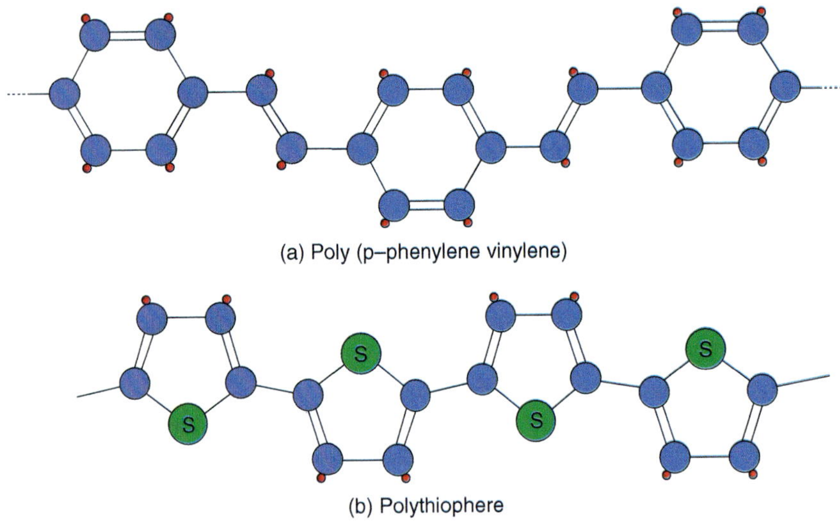

Figure 6.97 Molecular structure of electroluminescent polymers (a) poly(phenylene vinylene) and (b) polythiophene. Ball [57]/Reproduced with permission of Princeton University Press.

displays (small molecules can also be used as OLEDs). These devices utilize a single electroluminescent polymer that is capable of emitting light of variable frequency depending upon the applied voltage. The polymer is typically part of a bilayer structure that is sandwiched between an anode and a cathode. The electronic structure of the conjugated polymer allows for different band gaps between the *highest occupied molecular orbital* (HOMO) and the *lowest unoccupied molecular orbital* (LUMO), much like a semiconductor bandgap discussed previously in Section 6.0.1. Application of a specific voltage across the film causes electrons to flow into the LUMO, thereby creating a corresponding hole in the HOMO. This *excited state* can decay when the voltage is removed; light is generated when electrons and holes undergo *radiative recombination* and emit a photon. We will see in Chapter 7 that *inkjet printing* is an evolving materials processing technique that finds wide application in such areas as the formation of these OLED films and devices.

Need More Help?
Try these videos on OLEDs: https://www.youtube.com/watch?v=uwmQ3oXVV2w

6.2.4 Optical Properties of Composites and Biologics*

Of the three physical properties covered in this chapter, optical properties have the least importance in composite and biological applications. This is not to say that there are no applications of optical properties in composites or biological materials. There are indeed, such as the use of birefringence in the analysis of stress distribution and fiber breakage in fiber–matrix composites [11] and in the development of materials for ophthalmic implants such as intraocular devices [12]. These topics are beyond the scope of this text, however, and introduce no new concepts from a material property standpoint. There are many interesting articles and books on these topics, however, and the interested reader is encouraged to consult them as these—and many more—new uses for the optical properties of materials are applied to composite and biological materials systems.

References

1 Solymar, L., and D. Walsh, *Lectures on the Electrical Properties of Materials*, 5th ed., Oxford University Press, Oxford, 1993, p. 425.

2 Sheahan, T., *Introduction to High-Temperature Superconductivity*, Plenum, New York, 1994, p. 3.

3 Zawodzinski, T. A., C. Derouin, S. Radzinski, R. J. Sherman, V. T. Smith, T. E. Springer, and S. Gottesfeld, *J. Electrochem. Soc.*, **140**, 1041 (1993).

4 Schmidt, C. E., V. R. Shastri, J. P. Vacanti, and R. Langer, Stimulation of neurite growth using an electrically-conducting polymer, *Proc. Natl. Acad. Sci. USA*, **94**, 8948–8953 (1997).

5 Sun, A., H. Xu, Z. Chen, L. Cui, and X. Hai, Research on electrical properties of amphiphilic lipid membranes by means of interdigital electrodes, *Mater. Sci. Eng. C*, **2**(3), 159–163 (1995).

6 Hontsu, S., T. Matsumoto, J. Ishii, M. Nakamori, H. Tabata, and T. Kawai, Electrical properties of hydroxyapatite thin films grown by pulsed laser deposition, *Thin Sol. Films*, **295**, 214–217 (1997).

7 Gatteschi, D., P. Carretta, and A. Lascialfari, Molecular magnets and magnetic nanoparticles: new opportunities fo μSR investigations, *Physica B*, **289–290**, 94–105 (2000).

8 Yakhmi, J. V., Magnetism as a functionality at the molecular level, *Physica B*, **321**, 204–212 (2002).

9 Chavan, S. A., J. V. Yakhmi, and I. K. Gopalakrishnan, Molecular ferromagnets—a review, *Mater. Sci. Eng. C*, **3**, 175–179 (1995).

10 Miller, J. S., and A. J. Epstein, Molecular magnets: an emerging area of materials chemistry, in *Materials Chemistry: An Emerging Discipline*, L. V. Interrante, L. A. Casper, and A. B. Ellis, eds., American Chemical Society, Washington, D.C., 1995, p. 161.

11 Schuster, D. M., and E. Scala, *Trans. AIME*, **230**, 1639 (1964).

12 Obstbaum, S. A., Ophthalmic implantation, in *Biomaterials Science*, B. D. Ratner, A. S. Hoffman, F. J. Schoen, and J. E. Lemons, eds., Academic Press, San Diego, 1996, p. 435.

13 Callister, W., *Materials Science, and Engineering: An Introduction*, 5th ed., John Wiley & Sons, Inc, 2000, p. 607.

14 Jastrzebski, Z., *The Nature and Properties of Engineering Materials*, 2nd ed., John Wiley & Sons, Inc, 1976.

15 Ralls, K. M., T. H. Courtney, and J. Wulff, *Introduction to Materials Science and Engineering*, John Wiley & Sons, Inc, 1976.

16 Solymar, L., and D. Walsh, *Lectures on the Electrical Properties of Materials*, 5th ed., Oxford University Press, 1993, p. 427.

17 Solymar, L., and D. Walsh, *Lectures on the Electrical Properties of Materials*, 5th ed., Oxford University Press, 1993, p. 428.

18 Solymar, L., and D. Walsh, *Lectures on the Electrical Properties of Materials*, 5th ed., Oxford University Press, 1993, p. 429.

19 Callister, W., *Materials Science, and Engineering: An Introduction*, 5th ed., John Wiley & Sons, Inc, 2000, p. 631.

20 Callister, W., *Materials Science, and Engineering: An Introduction*, 5th ed., John Wiley & Sons, Inc, 2000, p. 632.

21 Callister, W., *Materials Science, and Engineering: An Introduction*, 5th ed., John Wiley & Sons, Inc, 2000, p. 633.

22 Kingery, W. D., H. K. Bowen, and D. R. Uhlmann, *Introduction to Ceramics*, John Wiley & Sons, Inc, 1976.

23 Callister, W., *Materials Science, and Engineering: An Introduction*, 5th ed., John Wiley & Sons, Inc, 2000, p. 640.

24 Sheahan, T. P., *Introduction to High-Temperature Superconductivity*, Plenum Press, 1994, p. 121.
25 Callister, W., *Materials Science, and Engineering: An Introduction*, 5th ed., John Wiley & Sons, Inc, 2000, p. 699.
26 Sheahan, T. P., *Introduction to High-Temperature Superconductivity*, Plenum Press, 1994, p. 5.
27 Scholze, H., *Glass*, Springer Nature, 1991, p. 312.
28 Callister, W., *Materials Science, and Engineering: An Introduction*, 5th ed., John Wiley & Sons, Inc, 2000, p. 634.
29 Rodriguez, F., *Principles of Polymer Systems*, 2nd ed., Hemisphere Publishing Corporation, 1982, p. 271.
30 Cowie, J. M. G., *Polymers: Chemistry & Physics of Modern Materials*, 2nd ed., Taylor & Francis, p. 411.
31 Cowie, J. M. G., *Polymers: Chemistry & Physics of Modern Materials*, 2nd ed., Taylor & Francis, p. 417.
32 Cowie, J. M. G., *Polymers: Chemistry & Physics of Modern Materials*, 2nd ed., Taylor & Francis, p. 419.
33 Interrante, L. V., L. A. Casper, and A. B. Ellis, *Materials Chemistry*, American Chemical Society, 1995, p. 109.
34 Interrante, L. V., L. A. Casper, and A. B. Ellis, *Materials Chemistry*, American Chemical Society, 1995, p. 110.
35 Mallick, P. K., *Fiber-Reinforced Composites*, Marcel Dekker, Inc, 1988, p. 365.
36 Moffatt, D. M., J. Runt, W. Huebner, S. Yoshikawa, and R. Nenham, Chpt. 3, "PTC Effects In Conductor-Filled Amorphous Polymer Composites", in *Composite Applications*, T. L. Vigo, and B. J. Kinzig, eds., VCH Publishing, Inc, 1992, p. 52.
37 Moffatt, D. M., J. Runt, W. Huebner, S. Yoshikawa, and R. Nenham, Chpt. 3, "PTC Effects In Conductor-Filled Amorphous Polymer Composites", in *Composite Applications*, T. L. Vigo, and B. J. Kinzig, eds., VCH Publishing, Inc, 1992, p. 56.
38 Kittel, C., *Introduction to Solid State Physics*, John Wiley & Sons, Inc, 1957, p. 333.
39 Kittel, C., *Introduction to Solid State Physics*, John Wiley & Sons, Inc, 1957, p. 335.
40 Yakhmi, J. V., Magnetism as a functionality at the molecular level, *Physica B*, **321**, 206 (2002).
41 Yakhmi, J. V., Magnetism as a functionality at the molecular level, *Physica B*, **321**, 207 (2002).
42 Westbrook, C., and C. Kaur, *MRI in Practice*, 2nd ed., Blackwell Science Ltd, 1998, p. 6.
43 Skoog, D. A., and D. M. West, *Principles of Instrumental Analysis*, 2nd ed., Saunders College, 1980, p. 381.
44 Westbrook, C., and C. Kaur, *MRI in Practice*, 2nd ed., Blackwell Science Ltd, 1998, p. 9.
45 Westbrook, C., and C. Kaur, *MRI in Practice*, 2nd ed., Blackwell Science Ltd, 1998, p. 18.
46 Westbrook, C., and C. Kaur, *MRI in Practice*, 2nd ed., Blackwell Science Ltd, 1998, p. 35.
47 Shackelford, J. F., *Introduction to Materials Science for Engineers*, 5th ed., Prentice-Hall, Inc, 2000, p. 597.
48 Shackelford, J. F., *Introduction to Materials Science for Engineers*, 5th ed., Prentice-Hall, Inc, 2000, p. 598.
49 Barrett, C. R., W. D. Nix, and A. S. Tetelman, *The Principles of Engineering Materials*, Prentice-Hall, Inc, 1973, p. 378.
50 Callister, W., *Materials Science, and Engineering: An Introduction*, 5th ed., John Wiley & Sons, Inc, 2000, p. 709.
51 Shackelford, J. F., *Introduction to Materials Science for Engineers*, 5th ed., Prentice-Hall, Inc, 2000, p. 599.
52 Skoog, D. A., and D. M. West, *Principles of Instrumental Analysis*, 2nd ed., Saunders College, 1980, p. 282.
53 Shackelford, J. F., *Introduction to Materials Science for Engineers*, 5th ed., Prentice-Hall, Inc, 2000, p. 607.
54 Ball, P., *Made to Measure*, Princeton University Press, 1997, p. 46.

55 Shackelford, J. F., *Introduction to Materials Science for Engineers*, 5th ed., Prentice-Hall, Inc, 2000, p. 611.
56 Strobl, G., *The Physics of Polymers*, 2nd ed., Springer-Verlag, 1997, p. 146.
57 Ball, P., *Made to Measure*, Princeton University Press, 1997, p. 381.
58 Rumble, J. R., ed., Selective properties of superconductive elements physical constants of organic compounds, in *CRC Handbook of Chemistry and Physics*, 103rd Edition (Internet Version 2022), CRC Press/Taylor & Francis, Boca Raton, FL.
59 Rumble, J. R., ed., Interactive table: dipole moments of gases and liquids, in *CRC Handbook of Chemistry and Physics*, 102nd Edition (Internet Version 2021), CRC Press/Taylor & Francis, Boca Raton, FL.
60 Lubin, G., ed., *Handbook of Fiberglass and Advanced Plastics Composites*, John Wiley & Sons, 1969.
61 Shackelford, J., and W. Alexander, *The CRC Materials Science and Engineering Handbook*, CRC Press, 1992.
62 Rumble, J. R., ed., Magnetic susceptibility of the elements and inorganic materials, in *CRC Handbook of Chemistry and Physics*, 103rd Edition (Internet Version 2022), CRC Press/Taylor & Francis, Boca Raton, FL.
63 Somiya, S., *Advanced Technical Ceramics*, Academic Press, 1984.
64 Yakhmi, J. V., Magnetism as a functionality at the molecular level, *Physica B*, **321**, 204–212 (2002).

Problems

Level I

6.I.1 What will be the resistance of a copper wire 0.08 in. in diameter and 100 ft long if its resistivity is 1.7 $\mu\Omega \cdot$ cm?

6.I.2 A maximum resistance of 1 Ω is permitted in a copper wire 25 ft long. What is the smallest wire diameter that can be used?

6.I.3 What is the electrical conductivity of iron (a) at room temperature? (b) at 212°F?

6.I.4 Silicon has a density of 2.40 g/cm^3. (a) What is the concentration of silicon atoms per cubic centimeter? (b) Phosphorus is added to the silicon to make it an n-type semiconductor with a conductivity of 1 mho/cm and an electron mobility of 1700 cm^2/V-s. What is the concentration of the conduction electrons per cubic centimeter?

6.I.5 (a) How many silicon atoms are there for each conduction electron in Problem 6.I.4? (b) The lattice constant for silicon is 5.42 Å, and there are eight atoms per unit cell. What is the volume associated with each conduction electron?

6.I.6 Germanium used for transistors has a resistivity of 2 $\Omega \cdot$ cm and an electron "hole" concentration of 1.9×10^{15} holes/cm^3. (a) What is the mobility of the electron holes in the germanium? (b) What impurity element could be added to germanium to create electron holes?

6.I.7 Calculate the mobility of electrons in Cu. The resistivity of Cu is 1.72×10^{-8} $\Omega \cdot$ m at 25°C and its density is 8.9 g/cm^3. Assume each copper atom donates one valence electron to the conduction band.

6.I.8 A coil of wire 0.1 m long and has 15 turns carries a current of 1.0 A. (a) Compute the magnetic induction if the coil is within a vacuum, (b) A bar of molybdenum is now placed in the coil, and the current adjusted to maintain the same magnetic induction as in part (a). Calculate the magnetization. Adapted from Callister, p. 703, Problem 21.2.

6.I.9 Why is there a maximum power loss when the dielectric is changing most rapidly as a function of temperature and frequency?

6.I.10 Crown glass has a refractive index of 1.51 in the visible spectral region. Calculate the reflectivity of the air–glass interface, and determine the transmission of a typical glass window.

6.I.11 Look up the refractive indices for fused silica and dense flint glass and calculate the ratio of their reflectivities. Cite the source of your information.

Level II

6.II.1 At room temperature, the temperature dependence of electron and hole mobilities for intrinsic germanium is found to be proportional to $T^{-3/2}$ for temperature in degrees Kelvin. Thus, a more appropriate form of Eq. (6.31) is

$$\sigma = C''T^{-\frac{3}{2}} exp\left(\frac{-E_g}{2k_BT}\right)$$

where C'' is a temperature-independent constant. (a) Calculate the intrinsic electrical conductivity using both Eq. (6.31) and the equation above for intrinsic germanium at 150°C, if the room temperature electrical conductivity is 2.2 $(\Omega \cdot m)^{-1}$. (b) Compute the number of free electrons and holes for intrinsic germanium at 150°C assuming this $T^{-3/2}$ dependence of electron and hole mobilities. Adapted from Callister, Problem 19.40, p. 654.

6.II.2 Assume there exists some hypothetical metal that exhibits ferromagnetic behavior and that has a simple cubic structure, an atomic radius of 0.153 nm, and a saturation flux density of 0.76 tesla. Determine the number of Bohr magnetons per atom for this material. Callister, Problem 21.10, p. 704.

6.II.3 The formula for yttrium iron garnet ($Y_3Fe_5O_{12}$) may be written in the form $Y_3^cFe_2^aFe_3^dO_{12}$, where the superscripts a, c, and d represent different sites on which the Y^{3+} and Fe^{3+} ions are located. The spin magnetic moments for the Y^{3+} and Fe^{3+} ions positioned in the c and a sites are oriented parallel to one another and antiparallel to the Fe^{3+} ions in d sites. Compute the number of Bohr magnetons associated with each Y^{3+} ion, given the following information: (1) each unit cell consists of eight formula ($Y_3Fe_5O_{12}$) units; (2) the unit cell is cubic with an edge length of 1.2376 nm; (3) the saturation magnetization for this material is 1.0×10^4 A/m; and (4) assume that there are 5 Bohr magnetons associated with each Fe^{3+} ion. Callister, Problem 21.17, p. 704.

6.II.4 The detectors used in optical fiber networks operating at 850 nm are usually made of silicon, which has an absorption coefficient of 1.3×10^5 m^{-1} at this wavelength. The detectors have coating on the front face that makes the reflectivity at the design wavelength negligibly small. Calculate the thickness of the active region of a photodiode designed to absorb 90% of the light.

6.II.5 A thin copper film adhered to a polyimide substrate can undergo strains in excess of 50% without rupturing the copper film. Assuming that the resistivity of the copper film is unchanged during uniaxial deformation, derive a relationship between the final resistance of the copper film, R, its initial resistance, R_0, and the final and initial lengths of the film, L and L_0, respectively. Clearly state your assumptions.

6.II.6 Dielectric generators can be produced by plating both sides of a stretchable dielectric polymer with thin metal films that serve as electrodes. If a voltage is applied across the electrodes and the films is biaxially stretched, the net amount of energy per unit volume that can be extracted during a constant field cycle, u_E, is given by:

$$u_E = \varepsilon_p E_{\max}^2 \, ln(\gamma)$$

where ε_p is the permittivity of the polymer film, $E_{\max}$ is the maximum electric field applied during the cycle, and γ is the area stretch ratio (stretched area/unstretched area). Calculate the energy per unit volume (in J/m^3) for a dielectric elastomer generator made of silicone-based elastomer film with a dielectric constant of 2.8 at a maximum applied electric field of 350×10^6 V/m and an area strain of 64%.

Level III

6.III.1 The *magnetocaloric effect* (*Scientific American*, May 1998, p. 44) relies on the ability of a ferromagnetic material to heat up in the presence of a magnetic field and then cool down once the field is removed. When a ferromagnet is placed in a magnetic field, the magnetic moments of its atoms become aligned, making the material more ordered. But the amount of entropy must be conserved, so the atoms vibrate more rapidly, raising the material's temperature. When the material is taken out of the field, the material cools. Water, or some other heat transfer fluid, can be cooled down by running it through the ferromagnetic material as it cools. (a) The magnitude of the magnetocaloric effect reaches a maximum at the Curie temperature of the ferromagnet. Based upon the above description of the application and the Curie temperature, which of the following three materials will make the best ferromagnetic material for a typical household refrigeration unit? *Justify your selection.*

Material	Saturation Magnetization[a] at Room Temperature	Curie Temperature (K)	Melting Point (K)
Iron	1707	1043	1811
Gadolinium	1090	289	1585
Nickel	485	631	1728

[a]Maximum possible magnetization that results when all magnetic dipoles are aligned with the external field.

(b) The amount of refrigeration is also determined by the strength of the applied magnetic field. Superconducting magnets are used to generate massive magnetic fields that cause large magnetizations in the ferromagnetic inductor. Below are three candidate materials for the superconducting magnet. Based on what you know about superconductors, which of these three materials will make the best superconducting magnet for this application? *Justify your selection.*

Material	T_c (K)	H_c (tesla)
$Bi_2Sr_2Ca_2Cu_3O_{10}$	110	100
Gadolinium	5.9	0.0001
NbTi	9.5	10

6.III.2 A team of chemical engineers [*Proc. Natl. Acad. Sci. USA*, **94**, 8948 (1997)] has shown that oxidized polypyrrole supports the growth of structural support cells needed for nerve regeneration in rats. In particular, the team showed that rat nerve cells responded to electrical stimulus from the polymer by producing extensions called neurites that are twice as long as neurites produced in the absence of the stimulus, (a) Calculate the current generated in a 10-mm-long, 10.0×10^{-6} m-diameter polypyrrole filament that is subjected to a potential difference of 12 V. Assume that the conductivity of oxidized polypyrrole is $1.5 \times 10^{7}\ \Omega^{-1}\ m^{-1}$. (b) One drawback of this application is that polypyrrole is very fragile and not biodegradable. Suggest a method for evaluating the long-term biocompatibility of polypyrrole fibers *in vivo*.

7

Processing of Materials

7.0 Introduction and Objectives

With an understanding of the structure and properties of engineering materials, now firmly in place, we can discuss how these materials can be formed or fabricated into useful products and components. The methods used to shape materials span orders of magnitude across spectrums of dimension, time, and cost. Processes that form water bottles, artificial organs, and trusses for highway bridges are described here—and the available techniques are rapidly evolving. Parts that were once manufactured by the traditional techniques of casting can now be formed by additive processes such as 3D printing. Components with previously impractical geometric complexity and incompatible materials are now possible. The common thread is that all the chemical and physical phenomena needed to introduce these processing techniques have already been described in the previous chapters.

By the end of this chapter, you should be able to

- Differentiate between additive and subtractive manufacturing processes.
- Differentiate between different types of heating techniques.
- Identify and describe different types of metal processing techniques.
- Calculate the power and penetration depth of metals heated by induction heating.
- Identify a proper milling technique for the preparation of metallic and ceramic powders of a desired size.
- Calculate pressures and identify the neutral point in rolling operations.
- Identify and describe different types of ceramic slurry processing techniques.
- Identify and describe different types of continuous and batch polymer processing techniques.
- Use equations of change to derive simple design equations for flow and pressure drop in polymer processing techniques.
- Differentiate between pultrusion, resin transfer molding (RTM), and filament winding methods for fiber-reinforced composite production.
- Describe the benefits of plasma processing to improvement of biocompatibility in biologics.
- Describe the benefits and limitations of 3D printing.

Before describing material-specific processing techniques, it is useful to categorize processing methods into traditional and emergent technologies and describe the common components of each. Related topics such as metal joining (e.g., welding) and surface finishing (e.g., polishing), although important, are beyond the scope of this text.

Materials Engineering and Science: Principles, Properties, and Processes, Second Edition. Brian S. Mitchell.

Companion website: www.wiley.com/go/materialsengineeringandscience2e

7.0.1 Traditional Manufacturing Processes

Many of the oldest and still most common fabrication processes involve the shaping of materials into useful components through a series of steps that may include melting, forming, and solidification in a procedure generally referred to as *casting*. Casting is the process of melting a material and solidifying it in a cavity, called a *mold*, to produce an object whose shape is determined primarily by the mold configuration.

Melting

Melting is the initial step in the casting operation and can be achieved through any number of well-established techniques, including heating elements or open flame. *Resistance heating*, also called *Joule heating* or *Ohmic heating*, is one of the simplest and most common heating techniques. Recall from Section 6.1 that resistivity, ρ, is a property of a material and that the resistance, R, in an object is a function of resistivity and geometry. Electrons flowing through a material can lead to dissipative energy losses which can be utilized to heat the object—typically in wire or ribbon form called a *heating element*—in a controllable way. The power, P, generated in the heating element is given in differential form by

$$\frac{dP}{dV} = \boldsymbol{J} \cdot \boldsymbol{\mathcal{E}} \tag{7.1}$$

where V is voltage, $\boldsymbol{J}$ is the current density, and $\mathcal{E}$ is the electric field gradient. For the simple case of direct current and flow through a perfect resistor in which all the electric power is converted into heat, this equation reduces to the useful relation

$$P = I^2R = \frac{V^2}{R} \tag{7.2}$$

where I is the scalar quantity of the current. Platinum, metallic alloys such as nichrome (80% Ni/20% Cr) and Kanthal (FeCrAl), molybdenum disilicide ($MoSi_2$), ceramics like silicon carbide (SiC) and silicon nitride (Si_3N_4), and even composite materials like positive temperature coefficient (PTC) rubber (carbon nanoparticles embedded in polydimethylsiloxane) are all air-stable resistive heating elements in common use.

Another common heating technique is *arc melting* in which a direct or alternating current is applied across an electrode and the material in the container or *crucible* called the *charge*. The electrode may be nonconsumable when made of carbon or tungsten, or consumable when made of the metal to be melted. An *electric arc* is formed between the electrode and charge, and the energy imparted to the charge causes it to melt. This process is carried out under a vacuum, or in an inert atmosphere such as helium or argon, to prevent contamination with oxygen or nitrogen. Other heating methods more common to specific types of materials are discussed in the following sections.

Shaping and Forming

The molten material can then be shaped by any number of methods. It can be transferred to a mold for subsequent solidification as is commonly the case for metals and ceramic slurries or shaped during solidification through an orifice called a *die* for higher-viscosity materials like polymers. The specifics of these transfer and shaping operations will be discussed in more detail for each material class, but they all have in common the use of a mold or die to impart the primary shape. Mold and die design are complex processes involving not only shaping of the melt but also heat transfer considerations that allow for control of the solidification process.

Solidification

The solidification step is perhaps the most important in determining the final properties of the shaped component. Material-specific considerations of solidification will be given in the following sections, but in general, the solidification time of a cast component, t, is given by *Chvorinov's Rule*:

$$t = B\left(\frac{V}{A}\right)^n \tag{7.3}$$

where V is the mold volume, A is mold surface area, and n and B are mold-specific parameters. A mold with high surface area and small volume will have higher heat transfer rates and a shorter solidification time than one with small surface area and high volume. The mold constant, B, depends on such mold material properties as density, thermal conductivity, and wall thickness, casting material properties such as heat capacity and heat of fusion, and processing parameters like degree of superheat. The value of n often lies between 1.5 and 2.

These three processing steps of melting, shaping, and solidification necessitate that the material to be shaped is in molted, liquid, or solution form. We will consider such traditional shaping operations for all the material classes. But there are some existing and emerging processing methods that can be used for materials in the solid state. We reserve some of these methods like powder compaction to a later section, but now turn our attention to a burgeoning method of shaping materials that is revolutionizing how we think of materials processing: additive manufacturing (AM).

7.0.2 Additive Manufacturing Processes

AM is a term applied to a family of materials processing techniques that produce a component through the addition of materials one step at a time, often in a *layer-by-layer manner*. The term was developed as a contrast to many traditional shaping techniques such as forming operations described above and those involving the removal—or subtraction—of material by machining with cutting tools like lathes and drills from a block of material called a *preform* or *blank* (see Figure 7.1). There are many AM processes and more being added each year. As a result, the terminology is evolving and not always an accurate description of the process steps involved nor the range of materials

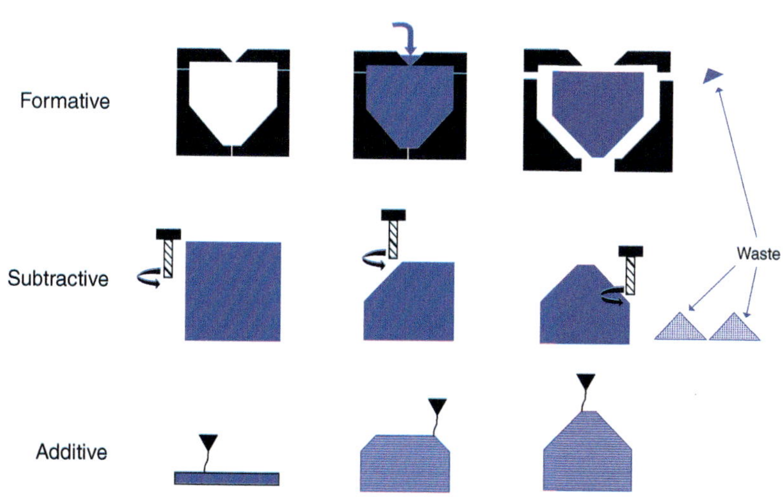

Figure 7.1 Comparison of formative (top), subtractive (middle), and additive (bottom) manufacturing processes.

to which it applies. Every attempt is made here to be up-to-date with the terminology, while also providing an introduction to this exciting approach to the fabrication of technologically relevant, geometrically complex, and materialistically diverse components.

Examples of AM techniques include *Selective Laser Sintering* (SLS) and *powder bed fusion* (PDF) of metal powders; *stereolithography* (SL) and *fused filament fabrication* (FFF) of polymers; and *material jetting* (MJ) of all types of materials. A special type of MJ is called *3D printing*, in which a device much like an inkjet printer uses a *computer-aided design* (CAD) digitized description of an object to place material in a predetermined location on a substrate. Subsequent layers of material produce a three-dimensional piece, which can then be refined further if necessary, through sintering and surface modification.

Take a Peek! Jump to Section 7.3.3 to see how 3D printing of polymers works.

Need More Help?

Try these videos on additive manufacturing:

https://www.youtube.com/watch?v=_K1C_dkZK98
https://www.youtube.com/watch?v=t4S0mKjXtT4
https://www.youtube.com/watch?v=fzBRYsiyxjl
https://www.youtube.com/watch?v=ICjQ0UzE2Ao

laser powder bed fusion:

https://www.youtube.com/watch?v=r8_M995cwAA

selective layer sintering:

https://www.youtube.com/watch?v=9E5MfBAV_tA

and stereolithography:

https://www.youtube.com/watch?v=NM55ct5Kwil

The inherent benefits of AM processes are illustrated in Figure 7.2. The costs of AM equipment, materials, and design are high but relatively constant as the number of components increases. The key benefit of AM is that fabrication costs do not scale with size or component complexity. Traditional machining costs drop with the number of parts produced but increase dramatically with part complexity. This trade-off makes AM more economical for low volume but highly complex components such as those in the aerospace and medical industries.

We will delve into only one example of AM: 3D printing with polymers (Section 7.3.3). Many new applications are being developed every day, and the interested reader is referred to the growing examples of AM available online.

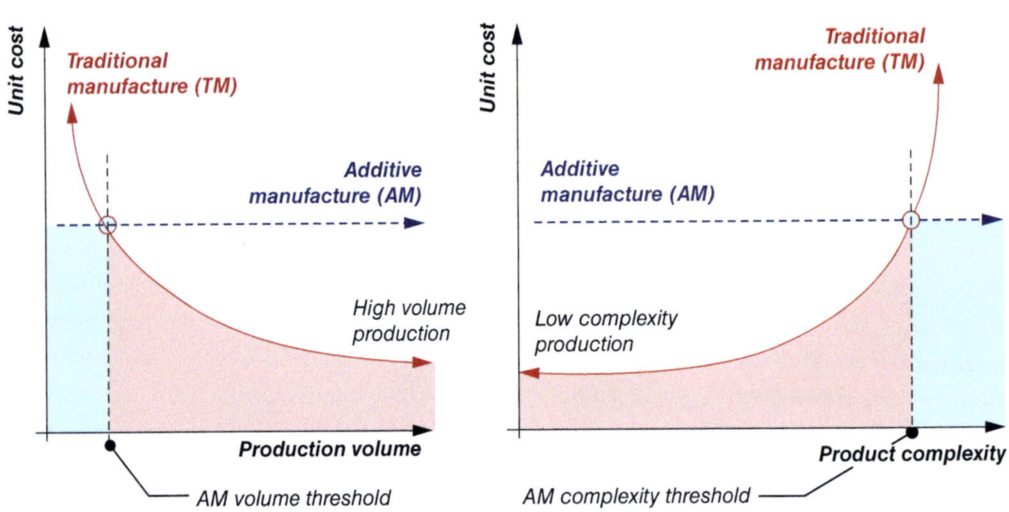

Figure 7.2 Economic trade-off of unit cost with production volume (left) and component complexity (right) for additive and traditional manufacturing. Leary [1]/With permission of Elsevier.

7.1 Processing of Metals and Alloys

Despite the relatively recent development of polymers and composites as structural materials, metals continue to be the dominant group of engineering materials for many applications including the automotive, aerospace, and construction industries. This is due, in part, to their ease in component fabrication or *formability*. In this section, we describe three of the most common and widely utilized metal-processing techniques: casting, wrought processing, and powder metallurgy.

7.1.1 Metal Casting

Almost all metals are cast during some stage of the fabrication process. They can be cast either directly into the shape of the component or as *ingots* that can be subsequently shaped into a desired form using processes that are described later in this section. Casting offers several advantages over other methods of metal forming: it is adaptable to intricate shapes, extremely large pieces, and mass production; and it can provide parts with uniform physical and mechanical properties throughout. From an economic standpoint, it would be desirable to form most metal components directly from casting, since subsequent operations such as forming, extruding, annealing, and joining add additional expense. However, since the growth rate of crystalline phases is much higher in the liquid state than in the solid state (cf. Sections 2.1.2 and 3.2.1), the microstructure of as-cast materials is much coarser than that of heat-treated or annealed metals. Consequently, as-cast parts are usually weaker than wrought or rolled parts. Casting is generally only used as the primary fabrication process when (1) the structural components will not be subjected to high stresses during service; (2) high strength can be obtained by subsequent heat treatment; (3) the component is too large to be easily worked into shape; or (4) the shape of the component is too complicated to be easily worked into shape. Let us more closely examine the important steps in the casting process and see how some properties can be related to the important processing parameters in each step.

Induction Melting

Induction melting utilizes Joule heating in a unique way and can offer distinct advantages for melting in certain metals. The metal is placed in a container within an electrical coil that produces a high-frequency alternating magnetic field. The alternating current flowing through the induction coil generates an alternating magnetic field that produces a current in the metal, called a *workpiece*. Heat is generated within the metal due to the passage of these *secondary currents* induced in the coils by electromagnetic induction (cf. Section 6.2.1). The electrical energy through the coil is thus converted to magnetic energy, which is in turn converted back to heat in the workpiece due to the induced secondary electrical currents called *eddy currents*. The current density in the workpiece is dictated by the *skin effect*, in which most of the eddy currents occur on the workpiece surface. Toward the center, the current density decreases exponentially. The heat develops in the workpiece itself; there is no need for a transmission medium such as air or water, or a mechanical connection. As the outer portions of the metal melt, mixing begins and thermal conduction eventually results in the melting of the entire workpiece.

The power imparted to the workpiece or melt, P, is a function of the magnetic susceptibility of the metal, χ, and the frequency of the applied field, ω:

$$P = kI^2\sqrt{(1+\chi)\rho\omega} \tag{7.4}$$

where P is in watts, k is a constant, I is the current in the induction coil (in amps), ρ is the specific resistivity of the metal in $\Omega{\cdot}\text{mm}^2/\text{m}$, and ω is in Hz. As mentioned above, the induced currents occur at the outer portions of the metal, as given by the *penetration depth*, δ, of the currents (in mm)

$$\delta = 503\sqrt{\frac{\rho}{(1+\chi)\omega}} \tag{7.5}$$

Obviously, the melt or workpiece must be larger in diameter than the penetration depth. The induction process, too, is often carried out under vacuum or inert gas atmosphere. The advantage of induction melting is the ability to melt cleanly and quickly. It is applicable to virtually any metal or alloy, even if they are not inductively suscepting, since it is possible to use the crucible as a susceptor as well. Furthermore, once the metal is molten, the induction lines from the secondary currents stir the melted charge and improve melt homogeneity.

Need More Help?

Try this video on induction melting:

https://www.youtube.com/watch?v=C9gqmATPGf8

Other techniques such as *electron beam melting* and *plasma flame melting* are used for highly refractory metals, such as tantalum. These too are often performed in vacuum. The use of vacuum in melting operations is used not only to prevent reactions such as oxidation but also to remove all dissolved gases in the metals. Most metals dissolve considerable amounts of gases in their liquid state. The solubility of these gases is much less in the solid state than in the liquid state. So, as the metal solidifies gases can escape from the melt, forming *blowholes* and *pinholes* in the metal ingot. This degasification process results in porosity that can adversely affect properties of the metal or alloy.

Historical Highlight

Read more about one of the defining technological advances of the industrial revolution, the Bessemer process for steel manufacturing: https://en.wikipedia.org/wiki/Bessemer_process.

Oxidation is not always to be avoided in the melting process. It is an important step in the making of steel, in which *pig iron*, the raw product of iron production, is remelted in a blast of oxygen. The pig iron contains about 4 wt% carbon, which is converted to carbon monoxide in the presence of oxygen. The furnace used to melt iron and form steel is thus termed a *blast furnace*, due to the blast of oxygen that is provided. The amount of oxygen is carefully controlled, so that not all of the carbon is removed, but just enough to form steel of the desired carbon content (cf. Figure 2.8 and Table 2.4). The oxygen also reacts to form oxides of highly reactive trace impurities in the pig iron, such as silicon, aluminum, and manganese. These oxides float to the top of the melt, since they are less dense than the molten metal, and form a product called *slag* that can be skimmed from the top of the melt. An additional benefit of the oxygen blast is that these reactions with carbon and trace impurities produce an enormous amount of energy that can be used to melt the iron. Many modifications have since been made to this process, and many variations exist, including the formation of pig iron from iron ore (Fe_2O_3). The chemistry of the steel-making operation is quite complex, and the interested reader is referred to metallurgy texts for further information.

Forming Operations

In principle, the molten metal can then be poured into a mold of the desired shape and become solidified to form a component. As mentioned above, and as described further in the next section, solidification from casting operations usually leads to undesirable crystal structures, such that the most common operation after melting is the formation of a large metal "blank" called an *ingot*. Thus, there are two main categories of metal casting processes: *ingot casting* and *casting to shape*.

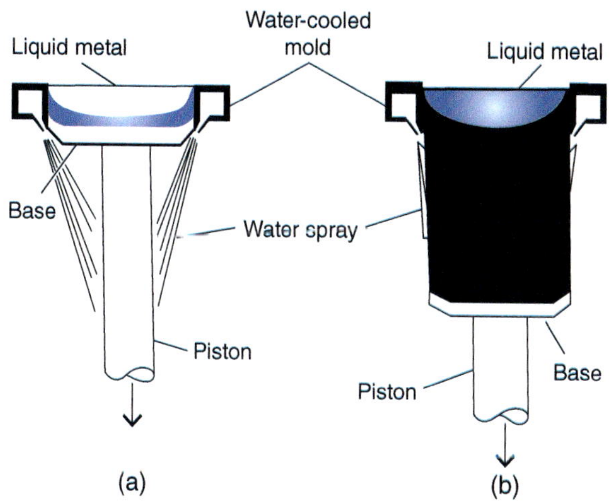

Figure 7.3 Schematic illustration of semicontinuous casting process in which (a) molten metal solidifies in a water-cooled mold with a movable base and (b) the base is moved downward so more metal can be poured into the reservoir. Ref. [2]/Reproduced with permission of McGraw-Hill.

Ingot casting makes up the majority of all metal castings and can be separated into three categories: *static casting*, semicontinuous or *direct-chill casting*, and *continuous casting*. Static ingot casting simply involves pouring molten metal into a permanent, refractory mold with a removable bottom, or plug in the bottom. Ingots have a slight taper toward the bottom of the mold that facilitates their removal and aids in the solidification process. The plug in the bottom of the mold allows the metal to be drained from the mold upon remelting if desired. Semicontinuous ingot casting is primarily employed in the production of cast aluminum. In this process, the molten metal is transferred to a water-cooled permanent mold that has a movable base mounted on a long piston (see Figure 7.3a). As the metal solidifies, the piston is moved downward and more metal continues to fill the reservoir (see Figure 7.3b). The piston is allowed to move its entire length, at which point the process is stopped.

Continuous casting is used in the steel and copper industries. In this process, molten metal is delivered to a permanent mold in much the same way as semicontinuous casting. However, instead of the process ceasing after a certain length of time, the solidified ingot is continually cut into lengths and removed from the process as quickly as it is cast. This method has many economic advantages over the more conventional ingot casting techniques.

Need More Help?

Try these videos on metal casting of:

aluminum ingots
https://www.youtube.com/watch?v=o0RwdHv0GKI

stainless steel parts
https://www.youtube.com/watch?v=KEXANt8lqLI

iron parts
https://www.youtube.com/watch?v=wMzSS3BnMLs

Casting to shape is generally classified according to the molding process, molding material, or method of feeding melt to the mold. There are four basic types of casting-to-shape processes: *sand casting, permanent mold casting, die casting*, and *centrifugal casting*. In sand casting, the mold consists of a mixture of sand grains, water, clay, and other materials, which are compacted around a pattern to 20–80% of their mixture bulk density. The two halves of the mold, called the *cope* and *drag*, are clamped together to form internal cavities, into which the metal is cast. When the molds are made of cast iron, steel, or bronze, the process is called *permanent mold casting* because the molds can be reused. In this case, the mold cavity is formed by machining. This process is used to cast iron and nonferrous alloys, and it has advantages over sand casting such as smoother surface finish, closer tolerances, and higher production rates. A further development of the permanent molding process is die casting. Molten metal is forced through a plate with an opening of a desired geometry known as a *die*, under pressures of 0.7–700 MPa (100–100,000 psi). Two basic types of die-casting machines are *hot-chamber* and *cold-chamber*. In the hot-chamber machine, a portion of the molten metal is forced into a cavity at pressures up to 14 MPa (2000 psi). The process is used for casting low-melting-point alloys such as lead, zinc, and tin. In the cold-chamber process, the molten metal is ladled into the injection cylinder and forced into the cavity under pressures that are about 10 times those in the hot-chamber process. High-melting-point alloys such as aluminum-, magnesium-, and copper-based alloys are used in this process. Die casting has the advantages of high production rates, high quality and strength, excellent surface finish, and close tolerances. Finally, inertial forces of rotation distribute molten metal into the mold cavities during centrifugal casting. The rotational speed in centrifugal casting is selected to give 40–60 gravitational constants of acceleration. Dies may be made of forged steel or cast iron. Other types of casting processes include *evaporative casting* and *zero gravity casting*.

Solidification

When the ingot or casting solidifies, there are three main possible microstructures that form (see Figure 7.4). We will describe here only the final structures; the thermodynamics of the liquid–solid phase transformation was described in Chapter 2. The outside layer of the ingot is called the *chill zone* and consists of a thin layer of equiaxed crystals with random orientation. This is due to the high degree of subcooling of the melt adjacent to the mold wall. Recall from Eq. (3.43) that a high degree of undercooling creates a driving force for a high nucleation rate. Many nuclei are formed at the mold wall and a fine grain size results. The second zone, which constitutes most of the ingot, consists of elongated grains, often of dendritic shape, having great length, and oriented parallel to the heat-flow direction in the ingot. This is called the *columnar zone*. Finally, the *central zone* of the ingot is the last to solidify, resulting in a much lower heat transfer rate and a coarser, equiaxed grain size.

The formation of the central equiaxed zone is caused by a process called *constitutional supercooling*. This enhances nucleation in the melt at the center of the casting prior to the advance of the solid

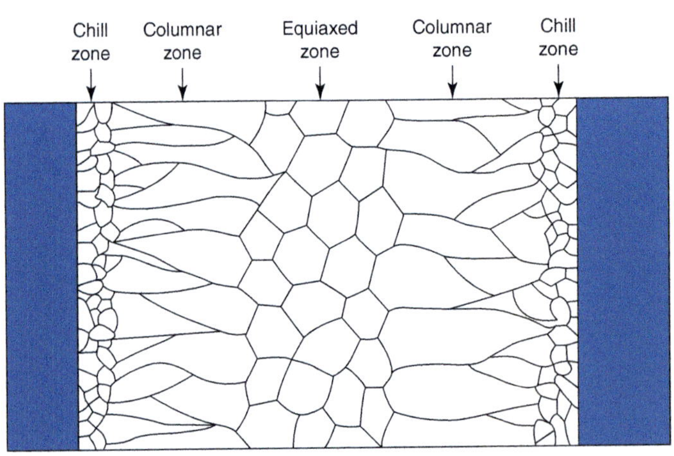

Figure 7.4 Schematic illustration of crystallization zones in a solidified metal ingot. From Jastrzebski [3]. Copyright © 1976 by John Wiley & Sons, Inc. This material is used by permission of John Wiley & Sons, Inc.

interface. The growth of the equiaxed zone can also be promoted by adding some nucleating agents. In general, columnar growth dominates in pure metals, while equiaxed crystallization prevails in solid–solution alloys. With increasing amounts of solute atoms, and with increasing freezing range of the solid alloy, the equiaxed crystal zone increases, and, in some cases, the columnar growth can be eliminated. In the extreme of cooling rates, nucleation and crystallization are suppressed leading to amorphous metals and alloys. This is the topic of the next section.

Some common phenomena that occur during solidification are *segregation* and *shrinkage*. During the solidification process, variations in the concentration can occur in certain regions of alloys that result in segregation. Two main types of segregation can occur: *macrosegregation* and *microsegregation*. In macrosegregation, compositional changes occur over relatively large dimensions as a result of gravity, normal, or inverse segregation. Gravity segregation is caused by density differences in melt components. In normal segregation, the solute is rejected at an advancing solid–liquid interface because of different diffusion rates in solid and in liquid. Inverse segregation is the result of outward movement of the impurity-enriched interdendritic liquid, thereby increasing the average impurity content at the head of the ingot.

Microsegregation refers to compositional variations over relatively small distances. Insoluble foreign particles can be trapped during solidification, resulting in segregation at the grain boundaries.

Need More Help?

Try this video on solidification and microstructure:

https://www.youtube.com/watch?v=G5z9KknF_s8

The relationship between the processing history and the solidification event is illustrated in Figure 7.5 for a cast component. Solidification begins with the formation of dendrites on the mold walls, which grow with time into the liquid and eventually consume the entire liquid volume. In alloys, it is the growth of dendrites that leads to segregation. In addition, the dendrites provide opportunities for gases to become trapped and form porosity. Most or all of these defects can be eliminated by processing in a *microgravity* environment. Macrosegregation, for example, is driven by gravity effects in which solute-rich materials sink in response to density differences.

Rapid Solidification

In Section 6.2.1, we saw the relationship between grain size, magnetic domain size, and magnetic properties in ferrous materials. In the previous section, we saw the effect of cooling rate on grain size in metals. You would expect, then, that there would be a strong relationship between solidification processes and magnetic properties in ferrous alloys, and this is indeed the case. We also saw that increased solidification rates at the walls of a mold can produce a fine-grained crystalline structure due to increased nucleation rates. But in the extreme, even nucleation rates drop off at sufficient subcooling (cf. Figure 3.15), such that it is possible, in theory, to produce *amorphous metals and alloys*, also known as *metallic glasses*, with high enough cooling rates. Such amorphous materials are formed by a class of techniques known collectively as *rapid solidification processing*, in which solidification rates, or *quench rates* as they are called for these processes, are on the order of 10^5 to 10^6 C/s. It is not yet possible to produce amorphous pure metals since the minimum cooling rates through the liquid–solid transition must be at least 10^{14} C/s, so metallic glasses always contain at least two or more different kinds of atoms. The compositions of some common amorphous ferrous alloys produced by rapid solidification techniques are listed in Table 7.1. Other known glass-forming metallic systems include alloys of transition metals or noble metals that contain about 10–30% semimetals, such as the iron–boron systems above, platinum–phosphorus, and niobium–rhodium, alloys containing group II metals such as magnesium–zinc, and alloys of rare-earth metals and transition metals such as yttrium–iron and gadolinium–cobalt. With further solute substitution, the stability and glass-forming ability can be drastically enhanced. Ternary alloy

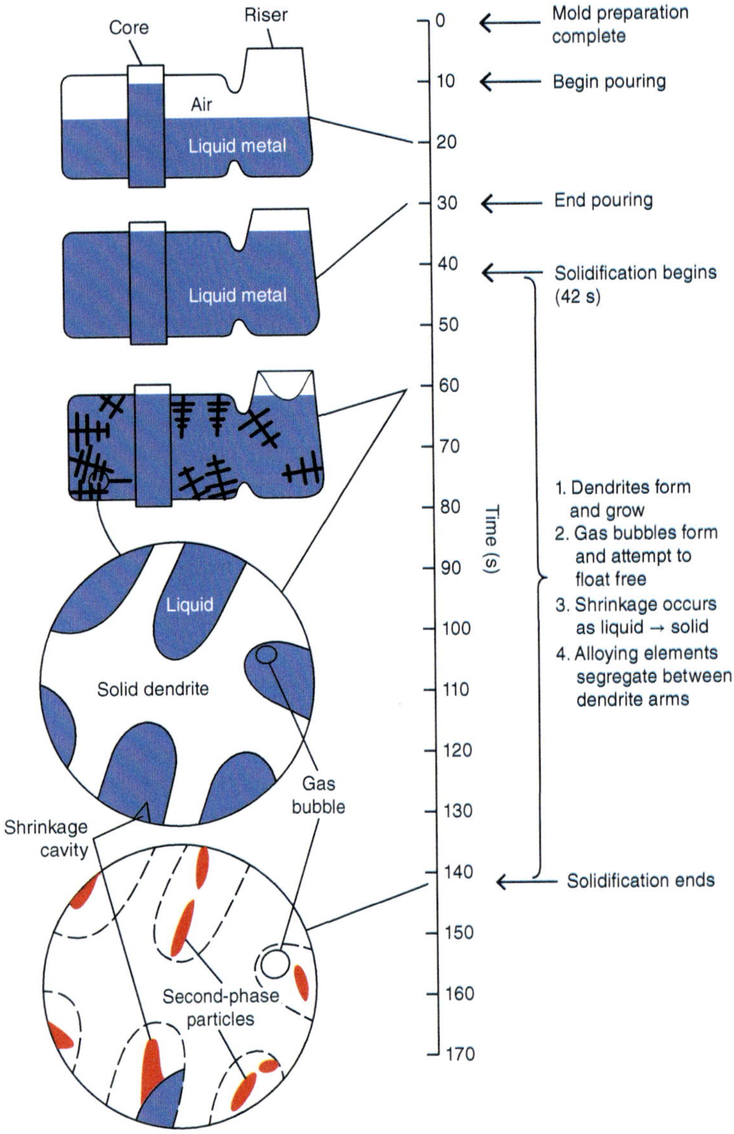

Figure 7.5 Solidification history for a cast-to-shape component. Ref. [4]. Copyright © 1997 by McGraw-Hill.

Table 7.1 Composition of Some Amorphous Ferrous Alloys

Composition (wt%, balance Fe)					
B	**Si**	**Cr**	**Ni**	**Mn**	**P**
20	—	—	—	—	—
10	10	—	—	—	—
28	—	6	—	6	—
6	—	—	40	—	14

glasses such as palladium–copper–silicon and platinum–nickel–phosphorus have been prepared as cylindrical rods of 2.5 mm in diameter at quench rates of 10^2 C/s or less. The X-ray diffraction (XRD) pattern (cf. Section 1.1.2) of an amorphous alloy is very similar to that of the same alloy in the liquid state. From the XRD pattern, it can be determined that the coordination number in metallic glasses is close to 12, just as in the close-packed crystalline structure.

Need More Help?
Try this video on rapid solidification: https://www.youtube.com/watch?v=C9gqmATPGf8

The high quench rates necessary to achieve the noncrystalline structures are primarily achieved through contacting the molten alloys with a surface that has a high thermal conductivity and is maintained at a low temperature. In a process called *splat-quenching*, the molten metal is propelled by means of a shock tube onto copper substrates with high thermal conductivity. Ribbons, or tapes, of metallic glasses can be formed by feeding a continuous jet of liquid alloy on the outside rim of a rapidly rotating cylinder. The molten alloy solidifies rapidly into a thin ribbon, about 50 μm thick, that is ejected tangentially to the rotating cylinder at rates as high as 2000 m/min. The width of the ribbon is limited to a few millimeters. Ribbons up to 15 cm in width can be produced by using a planar flow casting method. By injecting a stream of melt through a nozzle into a rotating drum containing water instead of onto a rotating cylinder, wires with circular cross sections can be formed. The centrifugal force ensures the formation of a water layer on the inner surface of the drum. The molten jet solidifies rapidly into a wire that is collected at the bottom of the drum. The wire has a uniform cross section with diameter ranging from 50 to 200 μm. The metallic glass wires show excellent mechanical and magnetic properties, comparable to or surpassing those of corresponding ribbons.

7.1.2 Wrought Metals and Alloys

The ingots produced by casting can either be used as cast, or further processed into usable geometries such as bars, rods, and sheets. Those metals and alloys that are manufactured by plastic deformation in the solid state are referred to as *wrought metals and alloys*.* The mechanical working of metals during wrought processing improves the properties of the wrought metals over that of the cast products. Plastic deformation, when carried out below the recrystallization temperature (cf. Section 5.1.2), results in work hardening, which improves the mechanical properties of the metal.

In all wrought processes, the flow of metal is caused by application of an external force or pressure that pushes or pulls a piece of metal or alloy through a metal die. The pressure required to produce plastic flow is determined primarily by the yield stress of the material (cf. Section 5.1.4) which, in turn, controls the load capacity of the machinery required to accomplish this desired change in shape. The pressure, P, used to overcome the yield stress and cause plastic deformation is given by

$$P = \sigma_y \varepsilon \tag{7.6}$$

where σ_y is the yield stress of the material and ε is the strain during the deformation. It is often useful to utilize the true strain in this relationship [see Eq. (5.28)] rather than the engineering strain. The pressure calculated in this way is generally much lower than that required in the actual

* There is no participial form of the word "wrought" in common use that is analogous to the term "casting," so the modified gerund *wrought processing*, although awkward, is generally used.

process due to losses caused by friction, work hardening, and inhomogeneous plastic deformation. For example, the pressure predicted in Eq. (7.6) is only 30–55% of the working pressure required for extrusion, 50–70% of that required for drawing, and 80–90% of the pressure required for rolling. The actual work requirement is the sum of (a) the work due to plastic deformation and (b) the losses mentioned above. The mechanical working of metals can be accomplished by various processes, but the methods of forging, rolling, and extrusion will be described here.

Forging

Forging is the process of applying large forces, mainly compressive forces, to cause plastic deformation to occur in the metal. The use of compressive forces means that tensile necking and fracture are generally avoided. Forging can be used to produce objects of irregular shape. There are two main categories of forging: *pressing* and *hammering*.

Need More Help?
Try this video on forging, rolling, and hammering: https://www.youtube.com/watch?v=XTU0Z-FkhtU

Pressing is the relatively slow deformation of metal to the required shape between mating dies. This method is preferred for homogenizing large cast ingots because the deformation zone extends throughout the cross section. All presses are composed of some basic components. These include a *ram* for applying the compressive force, a *drive* to move the ram up and down, power to the drive, a *bed* to rest the workpiece upon, and a *frame* to hold the components. Most press forging is done on upright *hydraulic presses* or *mechanical presses*. Pressures (in tons/in.^{2} of projected area) have been found to be 5–20 for brass, 19–20 for aluminum, 15–30 for steel, and 20–40 for titanium. Presses exceeding 50,000 tons have been built for forging large and complex parts.

Hammering involves more rapid deformation than pressing and is mainly restricted to the surface region of the metal. Large forgings require very large forces and usually involve elevated temperatures. The most common forging hammers are steam or air operated. A related operation is *drop forging*, in which a drop hammer forms parts with impression or cavity dies. Stock metal in the form of bars, slugs, or billets is placed in a cavity in the bottom half of a forging die on the anvil of a drop hammer. The upper half of the die is attached to the hammer, which falls on the stock, shaping it into the desired geometry. Generally, a finished forging cannot be formed in one blow, so most dies have several impressions, and the workpiece is transferred from one to the next between blows until the finished product is attained.

Rolling

In the process of *rolling*, the metal is continuously drawn between rotating rollers by the friction forces between the surfaces of the rolls and the metal, as illustrated in Figure 7.6, where $h_f < h_0$. The process can be run either hot or cold; it is much more economical than forging since it is faster, consumes less power, and produces items of a uniform cross section in a continuous fashion. In *cold rolling*, it is possible to attain production speeds of over 1500 m/min for thin strips of metal. Temperatures in the *hot rolling* process are similar to those in forging, namely 400–450°C for aluminum alloys, 820°C for copper alloys, 930–1260°C for alloys steels, 760–980°C for titanium alloys, and 980–1650°C for refractory metals. Unlike other wrought processes, rolling can be used without a die.

Deformations are restricted to a small volume at any given time, so the loads are relatively low. The metal is subjected to a longitudinal tension force that, combined with the normal force, produces a shear stress, τ, which causes deformation

$$\tau = f \times P \tag{7.7}$$

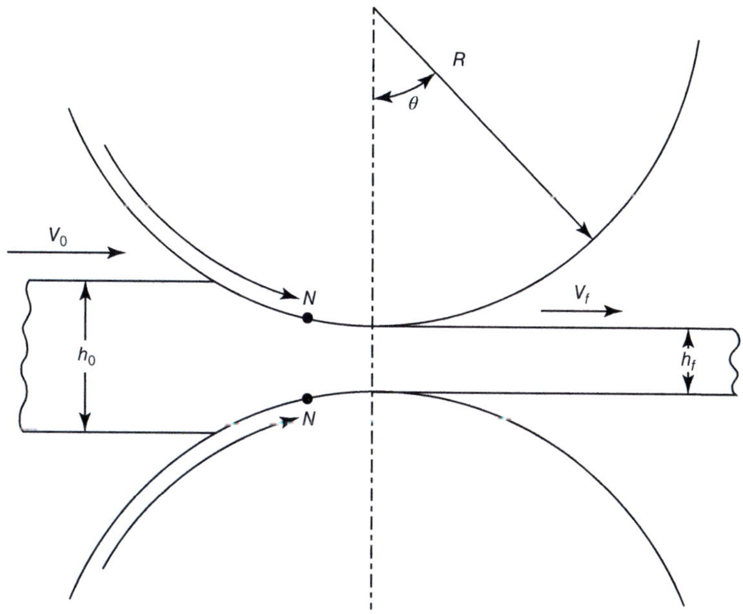

Figure 7.6 Schematic diagram of roll gap region in the rolling of metals.

where P is the normal pressure applied by the roller and f is the coefficient of friction between the metal surface and the roller surface. In hot rolling, f may be as much as 0.7, whereas in cold rolling it generally ranges from 0.02 to 0.3. The normal pressures entering, P_e, and exiting, P_x, the neutral point can be determined at any point of contact along the arc formed between the roller and the sheet by

$$P_e = \sigma' \frac{h}{h_0} \exp[f(H_0 - H)] \tag{7.8}$$

and

$$P_x = \sigma' \frac{h}{h_f} \exp(fH) \tag{7.9}$$

where f is again the coefficient of friction, σ' is the plane strain flow stress at a particular point in the roll gap where the instantaneous thickness is h, h_0 is the entering thickness, h_f is the exiting thickness, and H is a parameter given by

$$H = 2\sqrt{\frac{R}{h_f}} \tan^{-1}\left(\theta\sqrt{\frac{R}{h_f}}\right) \tag{7.10}$$

Example Problem 7.1

The ability of a hammer to deform metal depends on the energy it delivers on impact. Consider a steam hammer that has a falling weight of 2000 lb and a steam bore, d, equal to 12 in. Assume that the mean effective steam pressure, P, is 80 psi and that the stroke is 30 in. If the hammer travels 1/8 in. into the metal after striking it, determine the average force exerted on the workpiece. You should be able to do this without an equation, through the application of some dimensional analysis.

Answer: Steam force = $\left(\frac{\pi d^2}{4}\right) \times P = 9050\ \text{lb}_f$

$$\begin{aligned}\text{Total downward force} &= \text{Steam force} + \text{Hammer weight}\\ &= 9050 + 2000 = 11{,}050\ \text{lb}_f\end{aligned}$$

Energy in *a* blow = Force × Distance = 11.050 lb_f × 30 in. = 331,500 in. · lb_f

Average force exerted = 331,500/0.125 = 2.652,000 lb_f (1326 tons)

where R is the radius of the roller and θ is the angle subtended by the roll center by the vertical and the point of interest at the roll surface. H takes on a value of H_0 when $\theta = \theta_0$, the angle at the plane of entry, which can be approximated by

$$\theta_0 = \sqrt{\frac{h_0 - h_f}{R}} \tag{7.11}$$

The speed at which the metal moves through the rollers must change in order to keep the volume rate of flow constant through the roll gap. Hence, as the thickness decreases, the velocity increases. However, the surface speed of the roller is constant, so there is relative sliding between the roller and the metal. The direction of the relative velocity changes at a point along the contact area called the *neutral* or *no-slip point* (point N in Figure 7.6). At the neutral point, the roller and metal have the same velocity. Prior to the neutral point, the metal moves more slowly than the roller; to the right, it moves faster ($V_f > V_0$ in Figure 7.6). The direction of frictional forces is thus opposite in these two regions. The net frictional force acting on the metal must be in the direction of exit to enable the rolling operation to take place.

Cooperative Learning Exercise 7.1

The neutral plane (point N in Figure 7.6) is the point at which the subtended angle is $\theta = \theta_N$ and can be found in several ways. A precise method is to balance the entry and exit forces pressures; *i.e.*, $P_e = P_x$, and solve for H at the neutral plane, H_N, which can then be substituted into Eq. (7.10) to solve for θ_N. Alternatively, there are empirical relations for θ_N. Consider the cold rolling of an aluminum strip that is rolled from 4 to 3.3 mm in thickness with a roller 500 mm in diameter. The coefficient of friction is 0.06.

Person 1: Use Eq. (7.8) to (7.11) to determine the subtended angle at the neutral point. Hint: First, calculate θ_0 from Eq. (7.11), then calculate H_N by equating Eqs. (7.8) and (7.9) at $H = H_N$ Then use Eq. (7.10) to find θ_N.

Person 2: Use the following empirical relationship to determine the subtended angle at the neutral point

$$\theta_N = \sqrt{\frac{h_0 - h_f}{4R}} - \frac{h_0 - h_f}{4fR}.$$

Then help your partner to finish their calculations. Compare your answers. How similar or different are they?

Answers: θ_N = 0.0141 radians; θ_N = 0.0148 radians

It is sometimes necessary to use multiple rolling steps in sequence, interspersed with annealing steps to improve plastic deformation. The largest product in hot rolling is called a *bloom*. By successive hot- and cold-rolling operations, the bloom is reduced to a *billet*, *slab*, *plate*, *sheet*, *strip*, and *foil*, in decreasing order of thickness and size. The initial breakdown of the ingot by rolling changes the course-grained, brittle, and porous structure into a wrought structure with greater ductility and finer grain size.

Extrusion

Extrusion involves the pressing of a small piece of metal, called a *billet* or *slug*, through a die by the application of force. A billet or slug is differentiated from an ingot primarily by its smaller size. The die in extrusion processing is sometimes called an *orifice* or *nozzle*. The forces necessary to push the metal through the orifice are generated by a ram that pushes on the billet, generating both compressive and shear forces. Because no tensile forces are involved, fracture is avoided. When the billet is forced through the orifice by a ram, the process is called *direct extrusion*. It is also possible to force the die down around the billet to cause the same change in cross-sectional geometry, in a process called *indirect extrusion*. In either case, the efficiency of the extrusion process is measured by the ratio of work necessary for deformation [Eq. (7.6)] to the actual work performed.

> **Need More Help?**
>
> Try this video on metal extrusion:
>
> https://www.youtube.com/watch?v=ELgtjeJyFw8

Extrusion is primarily used for lower-melting point, nonferrous metals (see Table 7.2); but with the aid of lubricants and improvements in presses, the extrusion of steel has become available. Most carbon steels and stainless steels are extruded at temperatures of about 1200°C, in which glass is used as a lubricant. Glass is inexpensive, easy to use, and chemically stable, and it assists in dissolving oxide layers on the metal surface. The extrusion process is commonly used to form tubes and other continuous shapes of complex cross section. We will describe the extrusion process in more detail in Section 7.3 within the context of polymer processing.

7.1.3 Powder Metallurgy

The process of converting metallic powders into ingots or finished components via compaction and sintering is called *powder metallurgy*. Powder metallurgy is used whenever porous parts are needed, whenever the parts have intricate shapes, whenever the alloy or mixture of metals cannot be achieved in any other manner, or whenever the metals have very high melting points. The

Table 7.2 Some Typical Processing Parameters for the Extrusion of Nonferrous Metals

Materials	Minimum section thickness (in.)	Extrusion temperature (°F)	Extrusion pressure (ksi)	Extrusion exit speed (fpm)	Extrusion ratio $\ln(A_0/A_1)^a$
Aluminum alloys	0.04–0.30	550–1050	40–130	4–300	4–7
Copper alloys	0.05–0.30	1200–1650	30–130	80–1000	5–6.5
Magnesium alloys	0.04	570–800	100–130	4–100	4–5.3
Zinc alloys	0.06	400–660	90–110	75–100	4–5

[a] A_0 is the cross-sectional area of the billet; A_1 is the cross-sectional area of the extruded shape.
Source: Altan et al. [5]/ASM International.

technique is suitable for virtually any metal, however. An organic binder is frequently added to the metallic powders during the mixing state to facilitate compaction but volatilizes at low temperatures during the sintering process. The advantages of powder metallurgy included improved microstructures and improved production economies. Although the cost of the powders can be greater than that of ingots and mill products obtained by casting or wrought processing, the relative ease of forming the powder into a final product more than offsets the production costs and metal losses associated with the other two processes.

The powder metallurgy process consists of three basic steps: powder formation; powder compaction; and sintering. Each of the steps in powder metallurgy will be described in more detail.

Powder Formation

Metallic powders can be formed by any number of techniques, including the reduction of corresponding oxides and salts, the thermal dissociation of metal compounds, electrolysis, atomization, gas-phase synthesis or decomposition, or mechanical attrition. The *atomization* method is the most common because it can produce powders from alloys as well as from pure metals. In the atomization process, a molten metal is forced through an orifice, and the stream is broken up with a jet of water or gas. The molten metal forms droplets to minimize the surface area, which solidifies very rapidly. Currently, iron–nickel–molybdenum alloys, stainless steels, tool steels, nickel alloys, titanium alloys, and aluminum alloys, as well as many pure metals, are manufactured by atomization processes.

Chemical powder formation methods include the reduction of oxides, reduction of ions in solution by gases, electrolytic reduction of ions in solution, and thermal decomposition of gaseous molecules containing metal atoms. *Sponge iron* powder, for example, is produced by reacting a mixture of magnetite (Fe_3O_4) ore, coke, and limestone. The product is crushed to control the iron particle size, and the iron is magnetically separated. Iron powder can also be produced by hydrogen reduction of ground mill scale and the reaction of atomized high-carbon steel particles with iron oxide particles. Copper, tungsten, and cobalt powders can be produced by hydrogen reduction of oxide powders. Nickel and copper powders can be produced by hydrogen reduction of their corresponding amine sulfates, and iron and nickel powders of high purity and fine particle size are produced by thermal decomposition of their gaseous carbonyls $Fe(CO)_5$ and $Ni(CO)_4$, respectively.

Mechanical methods of powder production—also known as *mechanical attrition*—typically require a brittle material. There are various types of equipment that perform particle size reduction, termed *comminution*, but they are generally called *mills* and operate on the principles of grinding, crushing, and particle fracture (see Figure 7.7). In a device known as a *ball mill*, spherical milling media are used inside a vial that is agitated via any number of mechanical means to cause ball–ball and ball–vial collisions. The metallic powder is trapped during these collisions. Figure 7.8 shows the process of trapping an incremental volume of metallic powder between two balls. The powder undergoes *compaction* and eventually fracture into smaller particles. In the case of metals, fracture is the result of work hardening and embrittlement (cf. Section 5.1.2). For brittle materials, particle fracture is well described by Griffith theory (cf. Section 5.2.1). Cooling of the metal below the brittle–ductile transition, such as with liquid nitrogen in *cryogenic milling*, can aid in increasing brittle fracture during milling. As fragments decrease in size, the tendency to aggregate increases, and fracture resistance increases. Particle fineness approaches a limit as milling continues and maximum energy is expended. We will see that these milling methods are also useful for the formation of powders and particles from brittle materials such as ceramics.

Compaction and Hot Pressing

The shaping of components from metallic powders is accomplished by compacting the particles under pressure, and sometimes at elevated temperatures. Typical compaction temperatures exceed

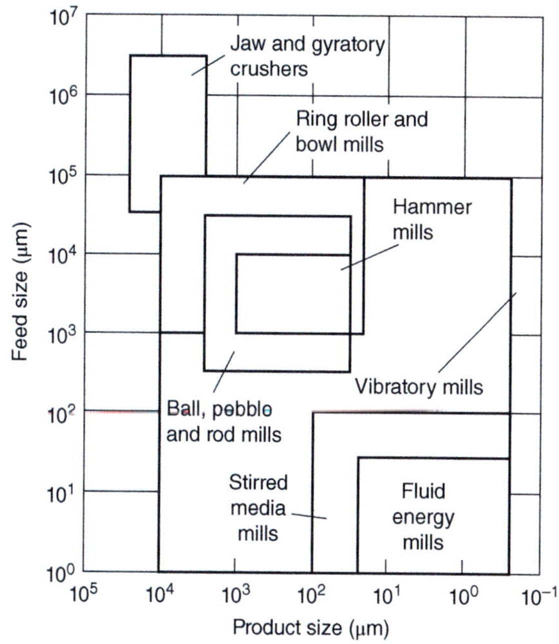

Figure 7.7 Classification of equipment used in mechanical attrition. Reprinted, by permission from Ref. [6].

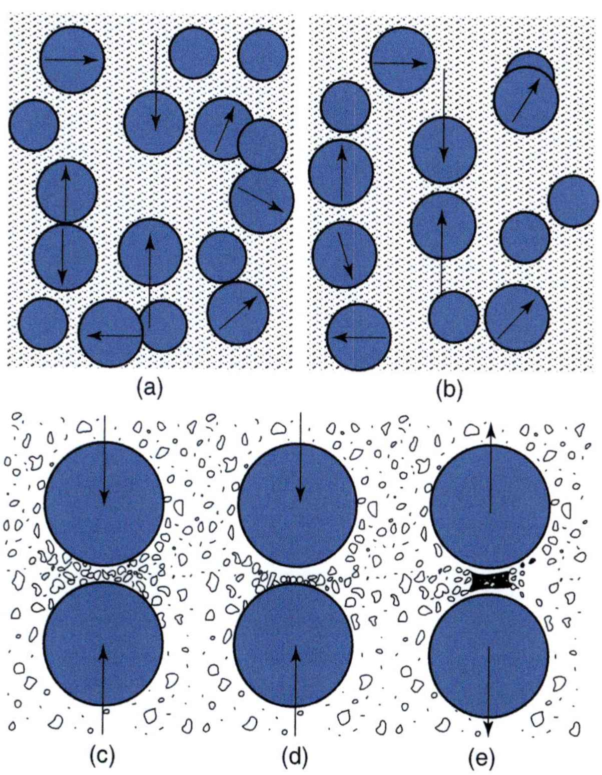

Figure 7.8 Schematic illustration of compaction and fracture processes in the formation of metallic powders via ball milling. Ref. [7]/Reproduced with permission of ASM International.

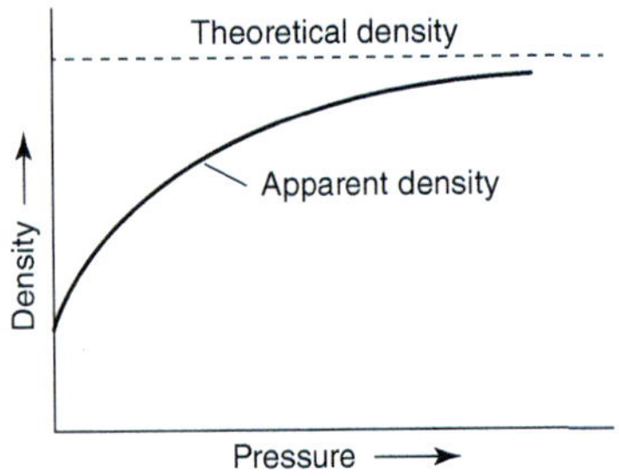

Figure 7.9 Pressure–density relationship in the compaction of metallic powders. Cahn [8]/ Reproduced with permission of Elsevier.

75% of the absolute melting temperature, and compaction times are on the order of 10^3 to 10^4 s. The compaction operation is performed primarily to provide form, structure, and strength, known as *green strength*, necessary to allow for handling before sintering. Compaction at room temperature involves pressures of 70–700 MPa, depending on the powder, and *green densities* (density prior to sintering) of 65–95% of theoretical (powder material) densities are achievable. Pressures for hot pressing are usually lower due to die or pressure chamber limitations.

The density–pressure relationship for powder compaction at room temperature typically increases from the apparent density at zero pressure to values that approach the theoretical density at high pressures, as illustrated in Figure 7.9. A compact with 100% theoretical density would indicate that it contains no porosity. Soft powders are more easily densified than hard powders at a given pressure, and irregularly shaped powders have lower densities than spherical powders in the low-pressure regime.

The mechanism of densification begins with individual particle motion at low pressure before interparticle bonding becomes extensive. At higher pressures, the compact acts as a coherent mass and densifies with decreased interparticle motion. At pressures of about 100–700 MPa, it has been shown that the slope of the density–pressure curve is generally proportional to the pore fraction, $1-\rho_r$:

$$\frac{d\rho_r}{dP} \approx K(1-\rho_r) \tag{7.12}$$

where ρ_r is the ratio of actual compact density to theoretical density (ρ/ρ_t), P is the pressure, and K is a proportionality constant that increases as the powder becomes softer. In general, K is proportional to the reciprocal of the nominal yield strength of the powder material.

Need More Help?
Try this video on a punch and die process: https://www.youtube.com/watch?v=4zXkn9iyC0c

Powders can be compacted at room temperature using a so-called *die and punch* design, in which the powders are compacted in a tool steel or cemented carbide die. The process can be automated so that the die can be filled with powder, the upper and lower portions of the die, called *punches*, are brought together to compact the powder, and the compact is then ejected to prepare for the next cycle. Die compaction of simple shapes can be carried out at elevated temperatures using carbide, superalloy, refractory metal, or graphite dies. An inert gas atmosphere or vacuum is often used to protect the die and/or the powder.

The die-and-punch method of compact imparts an anisotropic pressure to the compact. It is possible to apply pressure equally in all directions using a process called *isostatic pressing*, in which the powder is compacted with a fluid, such as hydraulic fluid, water, or an inert gas. Isostatic pressing is carried out using formed, flexible molds, called *cans*, that contain the powder and allow the pressure from the fluid to be transferred to the powder. The flexible molds are then stripped off the compact after densification. Isostatic pressing eliminates the die-wall friction inherent in die compaction and allows for consolidation of large components. However, isostatic pressing cannot provide the dimensional tolerances and cycle rates of die compaction. Isostatic pressing is useful for

> **Need More Help?**
>
> Try these videos on hot isostatic pressing (HIP):
>
> https://www.youtube.com/watch?v=_mqDqdAkLL0
> https://www.youtube.com/watch?v=BsnzgsEXT_A

forming intricate shapes and for forming parts that are near their final, desired dimensions prior to sintering, in what is termed *near-net-shape processing*.

When the powder is isostatically compacted at elevated temperatures, the process is called *hot isostatic pressing* (HIP). In this case, the flexible dies are often made of thin metals and high-pressure gases, such as argon, are used to heat the part rapidly and reduce thermal losses. Pressure up to 100 MPa and temperatures in excess of 2000°C are possible using hot isostatic pressing (HIP), and parts up to 600 kg can be fabricated. Metals that are processed commercially by HIP include various specialty steels, superalloys, hard metals, refractory alloys, and beryllium. We will see that HIP is also particularly useful for the densification of ceramic components.

7.2 Processing of Ceramics and Glasses

There are perhaps a wider variety of techniques that can be used to form and shape ceramics and glasses than for any of the other materials classes. Much of the variation in processing is due to the end use of the product. Certain techniques are used for structural ceramics, while others are used for semiconductors, superconductors, and magnetic materials. In general, there are three types of processes we will consider: powder-forming, melt processing, and chemical processing. The melt processing of ceramics and glasses is similar enough to the melt processing of polymers that its description will be delayed to Section 7.3. The latter category, which includes vapor phase and sol–gel processing, is considered in *Advanced Topics*. Two topics related to powder processing—firing and sintering—will be described here.

The powder-forming processes are similar in many ways to those used for powder metallurgy described in the previous section. For example, pressing is a common method for processing ceramics; however, ceramic powders can be pressed in either dry or wet form. In wet form, they can also be extruded, just like metals, and cast in a variety of process variations. The nominal forming pressures and shear rates associated with some of these processing methods are summarized in Table 7.3. You may want to refer to this table when each of the various processing techniques is described in more detail.

Before a description of powder processing is initiated, there is a consideration in ceramic-forming that is more prominent than in metal-forming which must be addressed: dimensional tolerance. Post-forming shrinkage is much higher in ceramics processing than in metals and polymer processing because of the large differential between the final density and the as-formed density. *Volume shrinkage* may occur during drying (Sv_d), during the removal of organic binders or reaction of bond phases producing consolidation (Sv_b), and during sintering (Sv_s). The total volume shrinkage Sv_{tot} is then the sum of these contributions:

$$Sv_{tot} = Sv_d + Sv_b + Sv_s \tag{7.13}$$

where each shrinkage is specified using an as-formed volume basis that is $(V_{formed} - V_{final})/V_{formed}$. Therefore, the final volume depends on the as-formed volume and the total shrinkage as follows:

$$V_{final} = (1 - Sv_{tot})V_{formed} \tag{7.14}$$

Table 7.3 Nominal Processing Pressures and Shear Rates in Some Ceramic Forming Processes

Forming Process	Pressure (MPa)	Shear Rate (1/s)
Pressing		
Roll/isostatic	>150	
Metal die	<100	
Plastic forming		
Injection molding	Varies	10–10,000
Extrusion	<40	10–1000 (die)
		100–10,000 (die-land)
Casting		
Slip (mold suction)	<0.2	<10 Pouring/draining
Slip (slurry pressurized)	<10	<100 Pumping
Slip (vacuum on mold)	<0.7	<10 Filling
Gel	<0.1	<10 Filling
Tape	<0.1	10–2000

Source: Reed [9]. Copyright © 1995 by John Wiley & Sons, Inc.

Precise control of the final product dimensions depends on control of the various shrinkages. For dry-pressed and injection-molded ceramics, Sv_d is in the range 3–12%. Binder elimination shrinkage is relatively high for injection-molded and tape-cast parts. The sintering shrinkage is commonly about 25–45% for products having high fired density and may range from 20% to 50%.

The final volume, V_{final}, is larger than the volume of the particulate material, V_{solids}, when the product has *residual porosity*. We define a ratio of these two volumes, D_r, that gives a measure of the residual porosity:

$$D_r = \frac{V_{solids}}{V_{final}} \tag{7.15}$$

When the densified component is pore-free, $V_{final} = V_{solids}$ and $D_r = 1.0$. Anything less than this value indicates the presence of porosity. Control of D_r requires control of both the microstructure of the as-formed part and the firing because both affect the sintering shrinkage.

When shrinkage is isotropic, the *linear shrinkage* ($\Delta L/L_0$) is calculated from the volume shrinkage as:

$$\frac{\Delta L}{L_0}(\%) = \left[1 - (1 - Sv_{tot})^{1/3}\right] \times 100 \tag{7.16}$$

Cooperative Learning Exercise 7.2

The same alumina ceramic product is made using two different hypothetic processes. In both cases, the ultimate density for the product is 3.96 g/cm^3.

Person 1: In process A, the bulk density is 2.43 g/cm^3 green and 3.84 fired. Calculate the volume shrinkage, linear shrinkage, and the densification ratio for this process.

Person 2: In process B, the bulk density is 2.74 g/cm^3 green and 3.90 fired. Calculate the volume shrinkage, linear shrinkage, and the densification ratio for this process.

Compare your answers. Which process produces results in the greatest shrinkage?

Answers:

$$Sv_{s(A)} = 36.7\%; \Delta L/L_{0A} = 14.1\%; D_{r(A)} = 0.97$$

$$Sv_{s(B)} = 29.7\%; \Delta L/L_{0B} = 11.1\%; D_{r(B)} = 0.98$$

The linear shrinkage on drying is typically less than 2% and on binder elimination and sintering less than 15%.

With these general principles of shrinkage and porosity in mind that apply to most ceramic processes, we proceed to the specific processing methods by examining some of the oldest materials processing techniques known, and we progress through time and technical difficulty to current techniques that produce complex components of incredible purity and dimensional tolerances.

7.2.1 Pressing

In much the same way that metal powders can be compacted and sintered to form a densified product, so, too, can many ceramic powders. Recall that pressing is the simultaneous compaction and shaping of a powder within a die or flexible mold. Just as we did for metals processing, we will look at each of the separate steps involved in the pressing operation: powder formation (called granulation for ceramics), die filling, and compaction. We reserve the final step, firing, to Section 7.2.3, since it applies not only to pressing operations but to casting operations as well, which are the topic of Section 7.2.2.

Granulation

Powders of glass and ceramic materials may be formed in many of the same ways as those used to form metal particles. This is particularly true of milling operations, for which Figure 7.7 applies to both metals and ceramics. However, wet milling is much more common with ceramic materials, particularly when finer particle sizes are required. The combination of dry powders with a dispersant such as water is called a *slurry*. Unlike a solution or suspension, the solids contents of slurries are very high, typically 50 wt% solids and above. Depending on the type of milling equipment being used, the solids content can affect the efficiency of the milling operation. In general, a high solids content is desirable to develop a coating of adequate viscosity on the milling media (balls and vial) to prevent escape of the particles from the grinding zone (cf. Figure 7.8), but not completely dissipate the grinding stress. The effect of slurry solids content on milling efficiency, in terms of the percentage of fines (smaller particles) produced, is illustrated in Figure 7.10 for the milling of an alumina slurry.

For the formation of fine particles necessary for most pressing and casting operations, ball and vibratory milling are the most common operations. The increase in specific surface area of the powders called the *grinding rate*, is 10–50% greater in vibratory mills than in ball mills, as illustrated in Figure 7.11 for two types of ceramic materials. Power consumption and media wear are also lower for vibratory mills than for ball mills. Milling efficiency is an important parameter in these

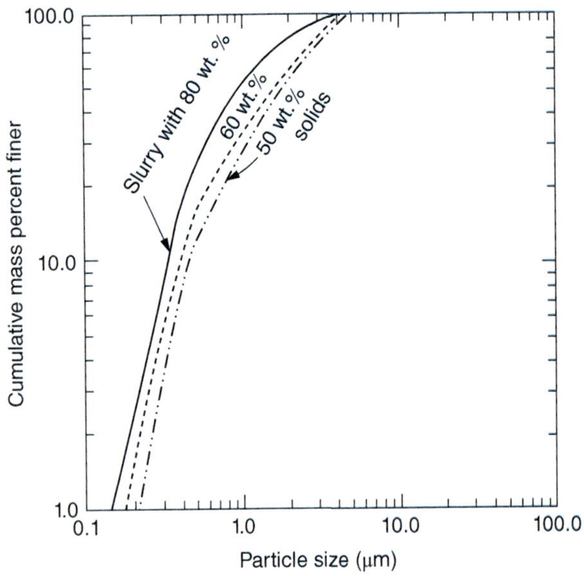

Figure 7.10 Effect of solids content in the milling of an alumina slurry on milling rate, expressed as cumulative mass percent fines. From Reed [9]. Copyright © 1995 by John Wiley & Sons, Inc. This material is used by permission of John Wiley & Sons, Inc.

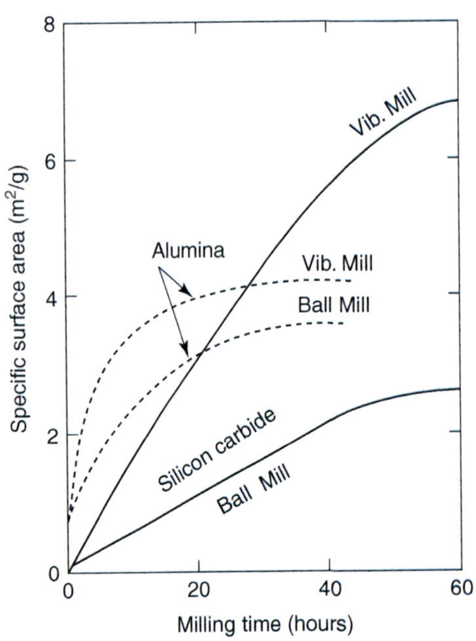

Figure 7.11 Comparison of grinding rates for vibratory and ball mills for two ceramics. From Reed [9]. Copyright © 1995 by John Wiley & Sons, Inc. This material is used by permission of John Wiley & Sons, Inc.

operations. The total energy input used to produce a unit of milled product, U_T, is commonly used as a measure of efficiency. An empirical expression that is used to relate the dependence of U_T in kW to the characteristic mean particle size during milling, $\bar{a}$ (in μm), and the type of mill is

$$U_T = A_c\left(\frac{1}{\bar{a}^m} - \frac{1}{\bar{a}_0^m}\right) \tag{7.17}$$

where A_c is an efficiency constant for a particular milling system, $\bar{a}_0$ is the initial size of the material in μm, and m is a fracture constant for a particular material. Some representative values of m are presented in Table 7.4, and some values of A_c are presented in Table 7.5.

Spray granulation is the formation of granules when a liquid or binder solution is sprayed into a continually agitated powder. The formation of granules occurs in a nucleation and growth process but on a larger size scale than the crystallization processes of Section 3.1.2. The nucleation and growth processes in granulation are illustrated in Figure 7.12. While the powder is stirred, particles roll and slide, and fines become airborne. An

Table 7.4 Size-Dependence Index, *m*, for Several Industrial Raw Materials

Material	*m*
Fused alumina	1.1
Silicon carbide	1.3
Quartz	1.4
Tabular alumina	1.6
Bauxite	2.4
Calcined alumina (porous aggregate)	4.0
Calcined titania (porous aggregate)	4.4

Source: Reed [9]. Copyright © 1995 by John Wiley & Sons, Inc.

Table 7.5 Values of *m* and A_c for Milling to a Submicron Mean Size for Various Types of Mills

Mill	Material	*m*	$A_c \left[\frac{KW(\mu m)^m}{(T)}\right]$	$A_c \left[\frac{KW(\mu m)^m}{(m^3)}\right]$
Ball	Zircon	1.8	650	3350
Vibratory	Zircon	1.8	100	520
Attrition	Quartz	1.8	920	2680
Attrition	Limestone	1.8	500	1500

Multiply A_c by the appropriate milling temperature or mill capacity.
Source: Reed [9]. Copyright © 1995 by John Wiley & Sons, Inc.

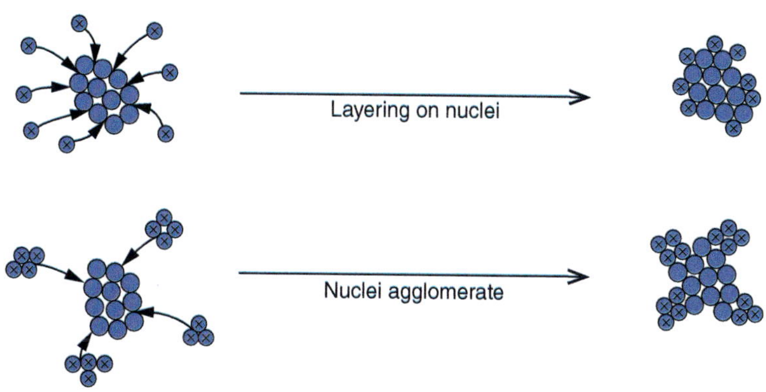

Figure 7.12 Schematic illustration of nucleation and growth process in spray granulation of ≅ 1 μm alumina. From Reed [9]. Copyright © 1995 by John Wiley & Sons, Inc. This material is used by permission of John Wiley & Sons, Inc.

agglomerate nucleus forms when a droplet of liquid or binder hits and is absorbed on the surface of a group of particles. Capillary forces give the nuclei strength. Nuclei are more numerous when the liquid is introduced as a fine mist and the powder is vigorously agitated. The growth of granules by a process called *layering* occurs through the contact and adhesion of individual particles to the nuclei (see the top illustration in Figure 7.12).

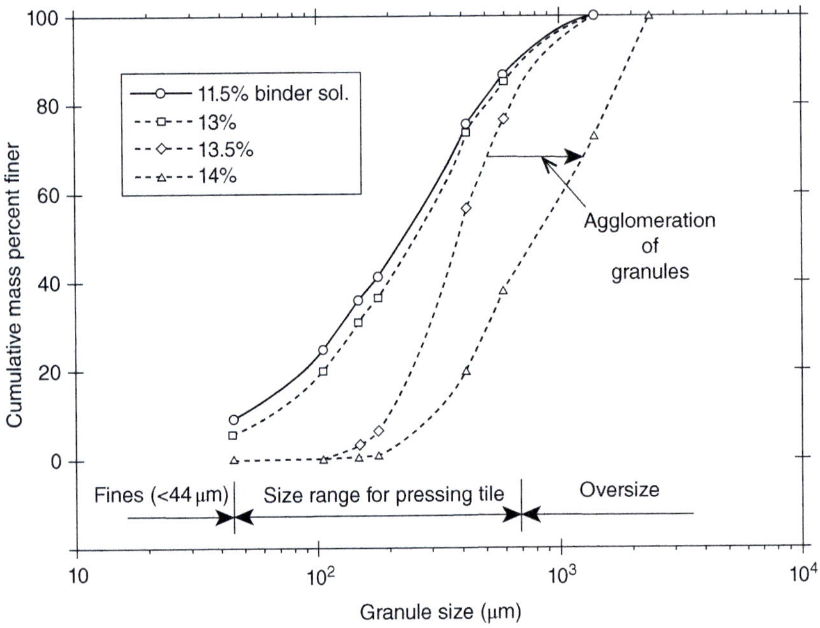

Figure 7.13 Effect of binder concentration on granule size in spray granulation of alumina. From Reed [9]. Copyright © 1995 by John Wiley & Sons, Inc. This material is used by permission of John Wiley & Sons, Inc.

Alternatively, the agglomeration of small nuclei and fracture fragments can also produce a granule, as in the bottom illustration in Figure 7.12. The rate of each mechanism depends on the liquid feed rate, the adsorption of liquid into the agglomerate, and the mixing action. Rubbing between granules during tumbling may cause particle transfer at surfaces and result in surface smoothing. There is a critical range of liquid content for granulation. At low liquid binder contents, granules are eventually formed when the processing energy is sufficiently high. The addition of greater amounts of liquid generally increases the mean granule size, the size distribution, and the porosity of the granules, as illustrated in Figure 7.13 for alumina. The liquid requirement for granulation is greater when the specific surface area of the powder is higher. Common liquid requirements in the range of 20–36 vol% indicate that during granulation the pores in granules are incompletely saturated. Granules produced by spray granulation are more nearly spherical than granules produced by other methods such as compaction and extrusion. Spray-granulated materials are usually partially dried prior to use in pressing operations.

The process of *spray drying* is fundamentally different from spray granulation. In spray drying, a particle slurry is sprayed into a warm drying medium such that a portion of the binder is evaporated, resulting in nearly spherical powder granules that are more homogeneous than those formed in spray granulation. Spray drying is used for preparing granulated pressing feed from powders of ferrites, titanates, alumina, carbides, nitrides, and porcelains.

The granules from spray granulation or spray drying can then be used for pressing, though any one of a number of *additives* is typically used prior to forming. The liquid content of the ceramic and the composition of the processing additive will differ for each type of processing technique, but, in general, additives serve to stabilize the viscosity and provide a ceramic mixture suitable for consistent processing. A list of typical additives and their specific functions

Table 7.6 Some Typical Additives to Ceramic Powders Prior to Processing

Additive	Functions
Deflocculant	Particle charging, aid and maintain dispersion
Coagulant	Uniform agglomeration after dispersion
Binder/flocculant	Modify rheology, retain liquid under pressure, yield strength, green strength, adhesion
Plasticizer	Change viscoelastic properties of binder at forming temperature, reduce T_g of binder
Lubricant	Reduce die friction (external), reduce internal friction, mold release
Wetting agent	Improve particle wetting by liquid, aid dispersion
Antifoam	Eliminate foam
Foam stabilizer	Stabilize foam
Antistatic agent	Charge control of dry powder
Chelating agent/sequestering agent/ precipitant	Inactivate undesirable ions
Antioxidant	Retard oxidative degradation of binder

Source: Reed [9]. Copyright © 1995 by John Wiley & Sons, Inc.

are given in Table 7.6. Some of these additives will be described over the next several sections, as necessary.

Need More Help?

Try this video on granulation, die filling, and compaction:

https://www.youtube.com/watch?v=R4tc3B29dB4

Die Filling

Powder flow rates into the dies can be determined using traditional mass or volume per unit time techniques during processing. An alternative method for estimating the *flowability* of untested powders is to use the *angle of repose* that a powder forms when poured onto a flat surface (see Figure 7.14). The angle of repose is the greatest angle measured from the horizontal to the slope of a cone-shaped pile of free-flowing material. A highly flowable powder forms a relatively flat pile with a low angle of repose, whereas a poorly flowable powder has a higher angle of repose. The flow rate and angle of repose for the pressing of granular powders correlates well, as shown in Figure 7.15. Dense, spherical particles with smooth surfaces and diameters of 20 μm or greater provide the optimal flow behavior. The presence of more than 5% *fines* (particles less than about 20 μm in diameter) may prohibit flow altogether. Binders and other additives can inhibit proper flow when the temperature exceeds the glass transition point of the additive, and relative humidity can affect the flow behavior of water-soluble binders.

The bulk density of the powder inside the die is called the *fill density*, D_{fill}. A high fill density reduces both the air content in the powder and the distance the punch needs to travel to further compress the powder during pressing. Poor powder flowability can adversely affect the fill density and, in turn, the final product density. Mechanical vibration is often used to cause particle settling in the die, which further improves the fill density. A typical fill density for a granulated powder is in the range of 25–35%, as indicated for several representative ceramic powders in Table 7.7.

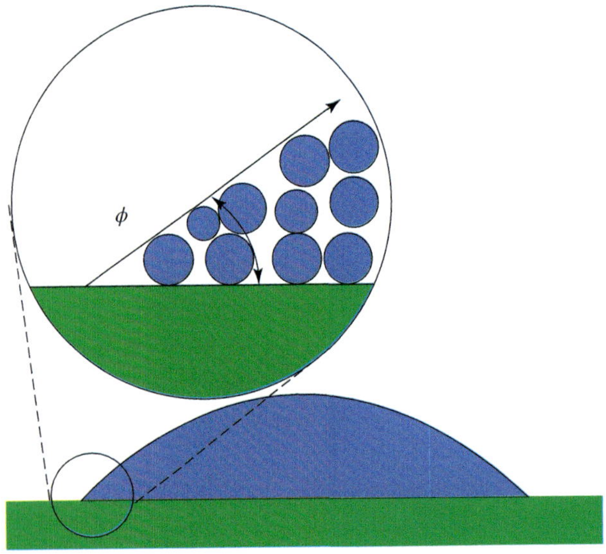

Figure 7.14 Definition of angle of repose for a powder pile.

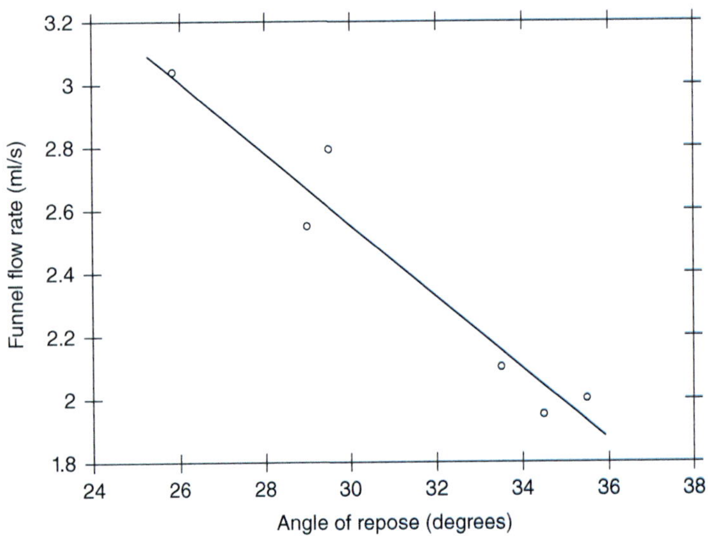

Figure 7.15 Relationship between angle of repose and flowability of a powder. From Reed [9]. Copyright © 1995 by John Wiley & Sons, Inc. This material is used by permission of John Wiley & Sons, Inc.

Compaction

Compaction is the process of granule densification via applied pressure to produce a cohesive part having a particular shape and microstructure. When the pressure is first applied, it is transmitted throughout the powder, or compact, by means of intergranular contact. The rate of initial densification is high but drops off rapidly as the particle compacts. As illustrated in Figure 7.16, relatively little densification occurs above about 50 MPa. In these initial stages, air from the pores is removed

Table 7.7 Particle Size and Bulk Density Characteristics of Some Spray-Dried Ceramic Pressing Powders

	Particle Size (μm)	Organics (vol%)	Granule Size (μm)	Granule Density (%)	Fill Density (%)
Alumina substrate	0.7	3.6	92	54	32
Spark plug alumina	2.0	13.3	186	55	34
Zirconia sensor	1.0	10.5	75	55	37
MnZn ferrite	0.7	10.0	53	55	32
MnZn ferrite	0.2	8.5	56	31	18
Silicon carbide	0.3	20.4	174	45	29

Source: Reed [9]. Copyright © 1995 by John Wiley & Sons, Inc.

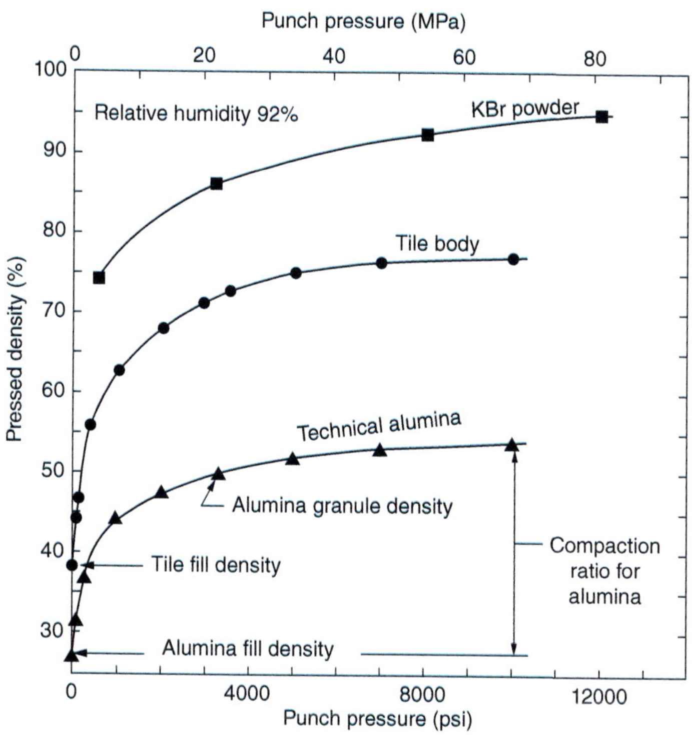

Figure 7.16 Compaction behavior of KBr powder and spray-dried granules of alumina and clay tile as a function of punch pressure. From Reed [9]. Copyright © 1995 by John Wiley & Sons, Inc. This material is used by permission of John Wiley & Sons, Inc.

and exhausted between the punch and die. Most industrial presses operate below 100 MPa for high-performance technical ceramics and below 40 MPa for whiteware and tile compositions.

The final density of the compact is less than the maximum packing fraction of the particles, PF_{max} [cf. Eqs. (4.8) and (4.68)], due to frictional forces at particle contacts that retard particle sliding. The effectiveness of the compaction process is quantified using the *compaction ratio*, *CR*, which is defined as the ratio of the pressed product density to the filled die density (an inverse ratio of volumes also works since the mass stays constant):

$$CR = \frac{V_{fill}}{V_{pressed}} = \frac{D_{pressed}}{D_{fill}} \tag{7.18}$$

For most ceramic pressing, a $CR < 2.0$ is desired since it reduces both the punch displacement and the compressed air in the compact. As indicated in Eq. (7.18), a high fill density leads to a low CR. For comparison, the CR in metal powder pressing is typically much greater than 2.0 due to the ductility of the particles.

There are essentially three stages of densification in the compaction process. In Stage I, a small amount of sliding and granular rearrangement occurs at low pressure when the punch first contacts the powder. The density is only slightly above the fill density at this point. In Stage II, granular deformation into neighboring interstices begins when the pressure exceeds the apparent yield pressure of the granules, P_y. A reduction in the volume and size of the relatively large interstices occurs in Stage II, as illustrated in Figure 7.17. The apparent yield pressure of the granules is related to the packing fraction (PF), the volume of the binder and particles, V_b and V_p, respectively, and the strength of the binder, σ_0, as follows:

$$P_y = C\left(\frac{PF}{1-PF}\right)\left(\frac{V_b}{V_p}\right)\sigma_0 \tag{7.19}$$

where C is an experimentally determined proportionality constant.

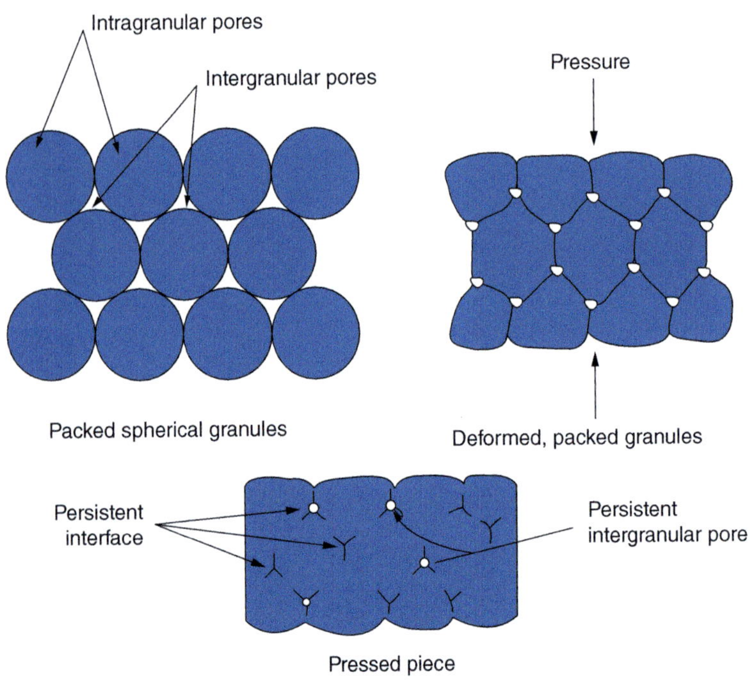

Figure 7.17 Schematic illustration of change in shape and change of pore size distribution during compaction. From Reed [9]. Copyright © 1995 by John Wiley & Sons, Inc. This material is used by permission of John Wiley & Sons, Inc.

Most of the densification occurs in Stage II. Granule deformation is the predominant mechanism of densification. The *compact density*, $D_{compact}$, may be approximated in this stage as

$$D_{compact} = D_{fill} + m \log(P_a/P_y) \tag{7.20}$$

where P_a is the applied pressure, P_y is the apparent yield pressure of the granules, D_{fill} is the fill density, and m is a *compaction constant* that depends upon the deformability and densification of the granules. Powders containing high-density granules may produce a high compact density with a relatively low value of m, but the elimination of all intergranular pores is favored when powders contain granules of somewhat lower density, for which m has intermediate values of 0.07 to 0.10. For each powder, there is an optimum granule density, fill density, and value of m that produces the best compromise of compaction ratio and the elimination of large pores.

Finally, in Stage III of the compaction densification process, densification continues by the sliding and rearrangement of particles into a slightly denser packing configuration. In powders containing granules of varying yield strength and density, the transition between stages does not occur uniformly throughout the compact. Ideally, the large pores between deformed granules have disappeared in Stage II, and interfaces between granules no longer exist. The high pressures can cause particle fracture that blocks further densification. Interstices between large granules and large pores within doughnut-shaped granules may persist into Stage III. These defects are not eliminated during sintering and reduce the mechanical reliability of the product.

As with metals, compaction can be done in an isostatic manner using inert gas pressurization. Isostatic pressing is used for products with one elongated dimension, complex shapes, or large volumes that cannot be easily compacted using dry pressing. Roll pressing can also be applied to ceramic powders, and it is used to produce substrates up to 30 cm in width with a uniform thickness in the range of 0.5–1.5 mm. Once the ceramic powder has been compacted by any of these means, it is ejected from the mold or die and is transferred to the final processing steps, which include sintering and surface finishing.

7.2.2 Casting

The casting of ceramics is similar to the casting of metals insofar as both involve the pouring of liquid-like material into a mold, followed by the development of a solidified product via a transformation that causes the development of yield strength. Unlike metal casting, which involves a molten metal, the pourable substance in ceramic casting is typically a slurry. And unlike metal casting, in which the transformation required to achieve sufficient yield strength in the cast part is invariably a physical transformation (solidification), the transformation in ceramic casting can be either a physical, chemical, thermal transformation, or a combination thereof. Since the fundamental steps of casting have been previously described, let us proceed by examining some of the differences between the more common types of ceramic casting.

Slip Casting

An aqueous slurry of fine clay is called a *slip*. *Slip casting* is then the casting of a slip, typically using a porous gypsum mold. As capillary suction of the water from the slurry into the mold proceeds, the slurry particles coagulate near the mold surface, and the cast is formed. Variations in the slip casting process include the application of pressure (*pressure casting*), vacuum (*vacuum-assisted casting*), or centrifugal force (*centrifugal casting*), all of which serve to increase the casting rate. There are two main types of slip casting: *drain casting* and *solid casting*. In drain casting, the molds are first filled with the slurry. After some initial solidification has taken place, the excess slurry is drained from the molds. This technique is suitable for the formation of objects with hollow centers, such as crucibles or tubes. In solid casting, the entire amount of slurry in the mold is used to form a dense, finished shape. The important steps of the slip casting process that will be elaborated upon here include slurry preparation, mold filling, draining, and partial drying while in the mold. Other steps that will not be described here include (a) separation of the slip from the mold after drying and (b) post-mold trimming and surface finishing prior to the final firing or sintering step.

Need More Help?

Try this video on slip casting:

https://www.youtube.com/watch?v=iLXsThRJTAo

Slurries are prepared from a mixture of liquid (typically water), milled powders or granules, and additives (see Table 7.6). The compositions of two common ceramic slips are given in Table 7.8. Casting slurries are typically formulated to be shear thinning (cf. Section 4.1.2) with an apparent viscosity below 2000 cpoise (2000 MPa·s) at a mold-filling shear rate of between 1 and 10 s^{-1} (see Table 7.3). A lower viscosity (<1000 cpoise) is desirable when the slurry is screened prior to mold filling, or when air bubbles must be quickly removed from the slurry. The density, apparent viscosity, and yield stress of the slurry should be high enough to minimize particle settling, called *deflocculation*, during casting. The yield stress of the slurry, τ_y, can be related to the slurry density, D_{slurry}, and the diameter and apparent density of the suspended particle, a and D_p, respectively:

$$\tau_y = (2/3)a\left(D_p - D_{slurry}\right)g \tag{7.21}$$

Table 7.8 Composition of Some Common Slip Casting Slurries

	Concentration (vol%)	
Material	**Porcelain**	**Alumina**
Alumina		40–50
Silica	10–15	
Feldspar	10–15	
Clay	15–25	
Water	45–60	50–60
	100 vol%	100 vol%
Processing additives (wt%[a])		
Deflocculant		
Na silicate	<0.5	
NH_4 polyacrylate		0.5–2
Na citrate		0.0–0.5
Coagulant		
$CaCO_3$	<0.1	
$BaCO_3$	<0.1	
Binder		
Na carboxymethylcellulose		0.0–0.5

Underline indicates that values above are summed to yield total below.
[a] Based on weight of solids in slurry.
Source: Reed [9]/Reproduced with permission of John Wiley & Sons.

where g is the gravitational constant. When the slurry does not have a yield stress, a high viscosity at the low shear rates during casting can also retard settling of the particles.

The drying process is caused by the removal of water from the slurry through capillaries in the gypsum mold, as illustrated in Figure 7.18. The permeable gypsum molds typically have a porosity of 40–50% and an effective pore size of 1–5 μm. As water is removed from the slip due to capillary action in the mold, a compact layer of slip particles (clay in Figure 7.18) forms at the mold–slip interface. The rate of the process is determined by the transport of water out of the slip and into the capillaries; that is, the rate-limiting step is the flow of water through the compact layer. As the thickness of the compact layer, x, increases, the overall rate of water removal, $J^*_{H_2O}$, decreases. We can determine the molar flux of water using a form of Fick's Law [Eq. (4.4)], in which we replace the concentration gradient driving force with a pressure gradient driving force, dP/dx:

$$J^*_x = -K\frac{dP}{dx} \tag{7.22}$$

where the diffusivity, D, in Eq. (4.4) has been replaced by a related quantity, the *permeation coefficient*, K, to make the units work out correctly. Until the capillaries become filled with water, the pressure difference is effectively constant, $dP = \Delta P = P_s - P_m$, where P_s is the pressure in the slip, which is essentially atmospheric pressure, and P_m is the pressure in the mold. Recall from Section 2.2.2 that the pressure drop across a curved surface, such as at a capillary opening, is given by Eq. (2.57), such that

$$\Delta P = \frac{2\gamma}{R} \tag{2.57}$$

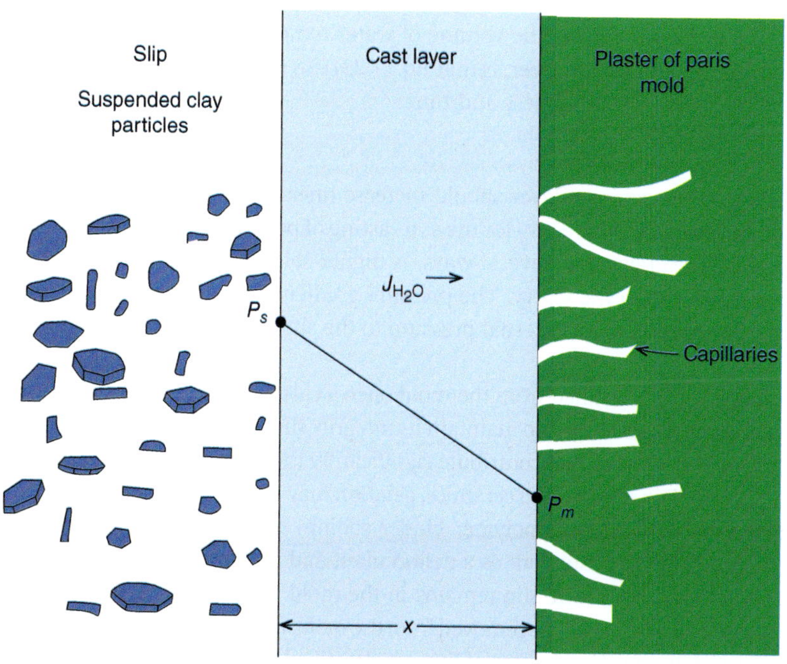

Figure 7.18 Schematic illustration of the formation of a slip-cast layer formed in the extraction of water by capillary action from a mold. From Kingery et al. [10]. Copyright © 1976 by John Wiley & Sons, Inc. This material is used by permission of John Wiley & Sons, Inc.

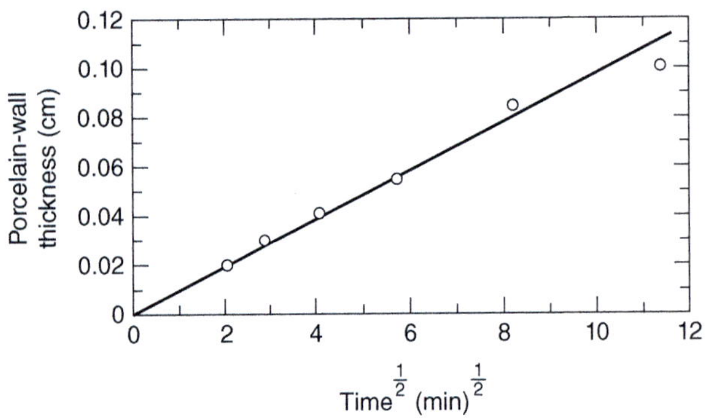

Figure 7.19 Dependence of compact layer thickness with time for a porcelain slip in a gypsum mold. From Kingery et al. [10]. Copyright © 1976 by John Wiley & Sons, Inc. This material is used by permission of John Wiley & Sons, Inc.

where γ is the surface tension of the slip, and R is the capillary radius in the mold. The surface tension will vary with the type of binders and deflocculants added to the slip, as well as with temperature and the chemical composition of the ceramic particles. To a first approximation, then, the molar flux of water is proportional to the change in thickness of the compact layer with time, dx/dt, and the change in thickness of the compact layer, dx, becomes x, which is now a function of time, so Eq. (7.22) becomes

$$J_x^* \propto \frac{dx}{dt} = -K' \frac{2\gamma}{Rx} \tag{7.23}$$

The permeation coefficient, K', has been modified slightly to include some other terms such as the density of the slip and a factor for converting the volume of water removed to the volume of clay particles deposited. The important point, however, is that Eq. (7.23) can now be integrated to give a relationship between the compact layer thickness and time:

$$x = (K't)^{0.5} \tag{7.24}$$

Equation (7.24) predicts that the compact layer should increase linearly with the square root of time. This functionality is illustrated in Figure 7.19 for the slip casting of porcelain in a gypsum mold. The rate of water removal can be affected in several ways. A higher solids content in the slip will increase the water flux, as can heating the slurry. The pressure gradient may also be increased to increase the flux, through application of an external pressure to the slip. This process is known as *pressure casting*.

In drain casting, the excess slurry is drained from the mold once a cast has been formed. The cast must possess adequate yield strength, however, to maintain its integrity through the draining process. In some systems, coagulation forces in the cast contribute significantly to the strength. This coagulation is commonly called *gelation*. For engineering ceramics, gelation may occur from the flocculation of an organic binder in the cast. The anionic polymer binder sodium carboxymethylcellulose (cf. Table 4.5) has been shown to act beneficially both as a deflocculant and as a binder in aqueous slip castings. Of course, in solid casting, the entire slip remains in the mold and must be dried prior to removal of the cast from the mold. Once the cast is removed from the mold, it is subjected to post-casting processes such as trimming and drying. The process of firing is described further in the next section.

Tape Casting

As with slip casting, a tape casting slurry is prepared that contains deflocculated powders or granules, binders and/or plasticizers, and a solvent such as water. Unlike slip casting, however, the slurry can be processed into thin, flexible sheets called *green sheets*, which are then cut and dried. The term "green" refers to the fact that it is a pre-fired or pre-sintered product and has nothing to do with color. The dried tape is rubbery and flexible due to the high binder content and has a very smooth surface. Tape-cast ceramics are used primarily as substrates for electronic conductors, resistors, capacitors, photovoltaic cells, electrical sensors, and solid electrolytes for batteries. As with slip casting, we will briefly investigate the topics of slurry composition, casting, and drying as they relate to this important processing technique.

Need More Help?

Try this video on tape casting:

https://www.youtube.com/watch?v=8eXnqQHk72E

Ceramic powders used for tape casting typically have a maximum particle size of 1–5 μm and a specific surface area of 2–5 m^2/g. Finer powders result in a smoother tape surface. Because the tapes are often used for electronic component fabrication, chemical homogeneity is critical. Particle size and dispersion also affect dimensional stability during the drying and firing processes. Therefore, the liquid–solvent system must dissolve the additives yet permit their adsorption on the ceramic particles. A solvent blend is thus typically used. Organic solvents have a low viscosity, low boiling point, low heat of vaporization, and high vapor pressure, all of which promote short drying times. Some common nonaqueous slurry compositions are listed in Table 7.9. Longer drying times and higher drying temperatures are required when an aqueous system is used.

Table 7.9 Composition of Some Nonaqueous Tape Casting Slurries

Function	Composition	Vol%	Composition	Vol%
Powder	Alumina	27	Titanate	2.8
Solvent	Trichloroethylene	42	Methylethyl ketone	3.3
	Ethyl alcohol	16	Ethyl alcohol	16
Deflocculant	Menhaden oil	1.8	Menhaden oil	1.7
Binder	Polyvinyl butyral	4.4	Acrylic emulsion	6.7
Plasticizer	Polyethylene glycol	4.8	Polyethylene glycol	6.7
	Octyl phthalate	4.0	Butylbenzylphthalate	6.7
Wetting agent			Cyclohexanone	1.2

Source: Reed [9]. Copyright © 1995 by John Wiley & Sons, Inc.

Example Problem 7.2

Tape is cast at a velocity of 1200 mm/min and the blade height is 0.4 mm. Estimate the shear rate (s^{-1}) and compare the value to that in Table 7.3.

Answer: The shear rate for flow in the x direction is calculated as dv/dy which for a planar geometry at steady state is the linear velocity divided by the separation distance between the moving and static surfaces (y direction):

$$\dot{\gamma} = \frac{dv}{dy} = \frac{v}{h_0} = \frac{(1200\ \text{mm/min})(1\ \text{min}/60\ \text{s})}{0.4\ \text{mm/min}} = 50\ \text{s}^{-1}$$

This shear rate is not high when compared to the range 10–2000 s^{-1} given in Table 7.3.

As mentioned in the previous paragraph, binder and plasticizer concentrations for tape casting are much higher than for slip casting and for most other types of slurry processing such as extrusion and pressing. More flexible molecules such as acrylics and vinyls are common (see Table 7.10). These macromolecules orient during casting and impart toughness and strength to the casting. The combination of binder and plasticizer must be carefully controlled to provide the required mechanical properties but permit a high concentration of ceramic particles in the slurry. An excess of binder can increase the particle separation and reduce the PF of the particles in the tape. Other additives include a *wetting agent* to promote spreading of the slurry on the substrate, a *homogenizer* that enhances surface quality, and *antifoaming agents*. As for slip casting, the tape casting slurry should be shear thinning and have a viscosity greater than about 4000 cpoise at the tape casting shear rate (refer to Table 7.3).

In the tape casting process the slurry is cast on a smooth, clean, impervious surface called the *carrier*. The tape is formed when the slurry flows beneath a blade, forming a film on the carrier. The thickness of the tape varies directly with the height of the blade or gate above the carrier, the speed of the carrier, and the drying shrinkage. A flow model that gives the dependence of the tape thickness, H, as a function of the carrier velocity, v, can be obtained using the parameters in Figure 7.20:

$$H = AD_r h_0 \left(1 + \frac{h_0^2 \Delta P}{6\eta_s vL}\right) \tag{7.25}$$

Table 7.10 Some Common Binders and Plasticizers Used in Tape Casting Slurries

	Binder	Plasticizer
Nonaqueous	Polyvinyl butyral	Dioctyl phthalate
	Polymethylmethacrylate	Dibutyl phthalate
	Polyvinyl alcohol	Benzyl butyl phthalate
	Polyethylene	Polyethylene glycol
Aqueous	Acrylics	Glycerine
	Methyl cellulose	Polyethylene glycol
	Polyvinyl alcohol	Dibutyl phthalate

Source: Reed [9]. Copyright © 1995 by John Wiley & Sons, Inc.

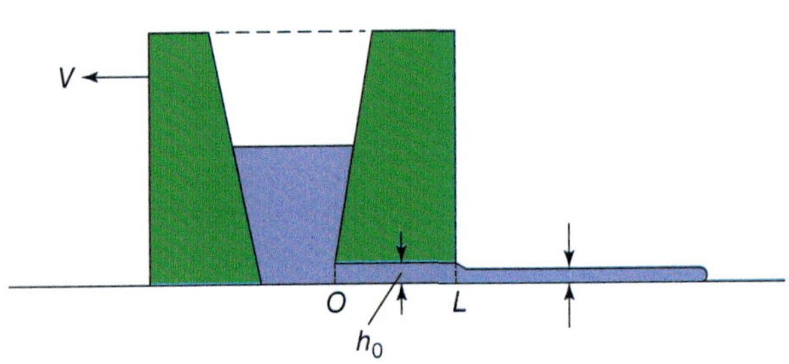

Figure 7.20 Tape casting flow model used for the development of Eq. (7.25). From Reed [9]. Copyright © 1995 by John Wiley & Sons, Inc. This material is used by permission of John Wiley & Sons, Inc.

where A is a constant that depends on the amount of side flow, D_r is the ratio of the density of the slurry to the density of the as-dried tape, h_0 is the cast thickness at the blade, ΔP is the pressure that causes the slurry to flow, η_s is the viscosity of the slurry, and L is the length of the casting. Equation (7.25) indicates that the tape thickness is nearly independent of variations in the carrier speed when the ratio h_0/η_s is small. A high slurry viscosity and a carrier velocity exceeding about 0.5 cm/s are desirable for thickness uniformity, as illustrated in Figure 7.21.

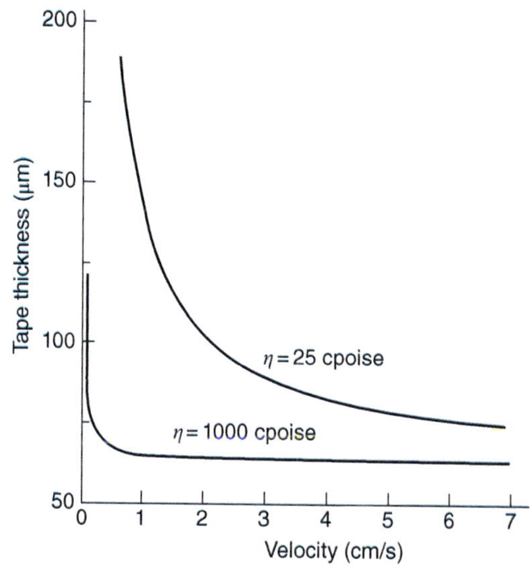

Figure 7.21 Variation of tape thickness with carrier velocity and slurry viscosity. Reed [9]/With permission of John Wiley & Sons.

In the final step, a viscoelastic tape is formed as the casting moves through a drying tunnel and the solvent vaporizes. The drying rate varies directly with the solvent concentration on the surface and the temperature and solvent content in the drying medium, which is typically air. Capillary forces transport the solvent to the drying surface. Initial vaporization is relatively slow because the counterflowing air contains much solvent and is cooler. Shrinkage occurs as solvent is lost and the particles move closer together. Shrinkage occurs primarily along the dimension of tape thickness, and the final thickness is typically about one-half the initial blade height. Some migration of the plasticizer to the drying surface occurs, but migration of higher-molecular-weight binders is minimal. The liquid surface is interrupted; drying by vapor transport begins when the degree of pore saturation (DPS) is <1 (cf. Section 4.1.2). The viscoelastic tape becomes more elastic as drying continues. The PF of the particles in the dried tape is about 55–60%, and the remaining volume is about 35 vol% organics and 15 vol% porosity. The dried tape may go directly to a post-processing step or may be stored on reels.

Other Casting Techniques

As alluded to earlier, there are several modifications to the slip casting technique that give rise to other types of casting processes. Most of these methods have to do with increasing the pressure, ΔP, which causes liquid migration from the slip into the mold, as indicated in Eq. (7.22). *Vacuum casting* is widely used for forming very porous refractory insulation having a complex shape. The slurry typically contains partially deflocculated ceramic powder and chopped refractory fiber. The fiber increases the viscosity and liquid requirement and increases the porosity of the cast. An expendable, permeable preform is coupled to a vacuum line and then submerged in the slurry. After the desired wall thickness has been cast, the preform is withdrawn and removed for drying. Vacuum casting may also be used to describe the process of applying a vacuum to the external surface of the gypsum mold during slip casting to increase the casting rate. *Centrifugal casting* is used for forming advanced ceramics having a complex shape where a very high cast density is required. Centrifuging increases the drive force for the settling of the particles, and liquid is displaced to the top of a centrifuge cell. For a relatively dilute suspension, the settling rate is described by Stoke's Law [cf. Eq. (4.70)], and particles having a greater size or density settle at a faster rate and a structural gradient is produced. The casting rate increases with centrifugal speed and the flow rate of the liquid away from the deposition surface. Casts of complex shapes having a uniform density and sedimentation rates of 0.6 mm/min can be obtained with this process.

There are additional variations to the casting process, and additional post-processing steps such as shaping, trimming, laminating, and printing that are beyond the scope of this text. The interested reader is referred to a ceramic processing textbook for more information.

7.2.3 Firing

Once the green ceramic component has been cast by any of the methods described in the previous section, dried, and surface finished, it is generally subjected to a process known as firing. Firing involves heat treatment of the green ceramic in a high-temperature furnace, known as a *kiln*, to develop the desired microstructure and properties. Kiln design varies widely, depending upon the ceramic to be fired, but they are generally designed to operate in either an intermittent or continuous fashion. Several processes occur during firing, including further drying beyond that performed in the casting process, binder burnout, and, most importantly, sintering. We will briefly touch on each of these topics.

Pre-sintering Processes

Material changes that occur upon firing prior to sintering may include drying, the decomposition of organic additives, the vaporization of hydrolyzed water from the surfaces of particles, the pyrolysis of particulate organic materials, changes in the oxidation state of some of the transition metal and rare-earth ions, and the decomposition of carbonates and sulfates, termed *calcination*, that are introduced either as additives or as raw component constituents.

The decomposition of organic additives, such as binders, called *thermolysis*, is an important step prior to densification on sintering. Incomplete binder removal and uncontrolled thermolysis may reduce yields or impair performance of the product. Ideally, the green ceramic should survive thermolysis without deformation, distortion, the formation of cracks, or expanded pores. Binder burnout is dependent upon the composition of the binder and the composition and flow rate of the kiln atmosphere. It is also very dependent upon the microstructure of the organic, powder, and porosity phases and dynamic changes in the microstructure as the binder is eliminated. Some of the chemical and physical changes that take place during thermolysis are summarized in Table 7.11.

Sintering of Ceramics

The thermodynamics of the sintering process were presented in Section 2.2.2. Recall that there are three primary stages of sintering: initial sintering, intermediate sintering, and final-stage sintering. These three stages, and the accompanying densification that occurs in each stage, can be observed for the sintering of two 0.8-(finer) and 1.3-μm (coarser), magnesia-doped alumina powders in Figure 7.22. For comparison, the densification process for hot pressing of the same powder is shown. Notice that most of the densification takes place in the intermediate stage. The mass transport mechanisms that lead to densification include grain boundary diffusion, lattice diffusion, viscous flow, and plastic flow. Such mass transport processes as surface diffusion and evaporation take place during sintering, but do not contribute to densification.

The shrinkage that accompanies densification is controlled by different mechanisms in the three stages of sintering. In the initial stage of sintering, the linear shrinkage, $\Delta L/L_0$, is described by

$$\frac{\Delta L}{L_0} = \left(\frac{K D_v \gamma_s V_v t}{d^n k_B T}\right)^m \tag{7.26}$$

Table 7.11 Some Chemical and Physical Changes that Occur During Thermolysis

Chemical reactions	Decomposition
	Depolymerization
	Carburization
	Oxidation
Physical changes	Melting/softening (solid–liquid)
	Sublimation (solid–gas)
	Evaporation (liquid–gas)
Reaction zone	External
	Internal
	Uniform displacement
	Nonuniform displacement
Transport	Gas diffusion
	Liquid flow

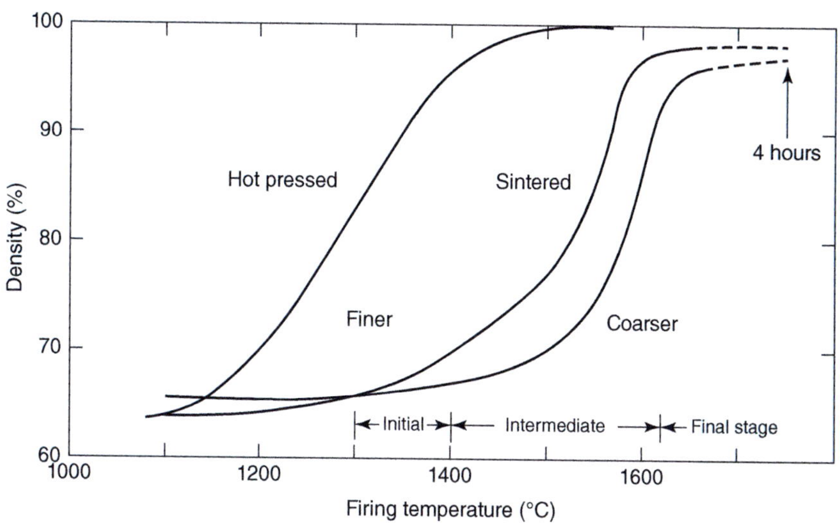

Figure 7.22 Densification behavior of compacts of two alumina powders (0.8- and 1.3-μm average particle size), with the sintering stages indicated for the coarser powder. A hot-pressed powder is shown for comparison. From Reed [9]. Copyright © 1995 by John Wiley & Sons, Inc. This material is used by permission of John Wiley & Sons, Inc.

where D_v is the apparent diffusivity of the vacancies of volume V_v, d is the grain diameter, K is a geometry-dependent constant, γ_s is the solid surface energy [cf. Eq. (2.64)], t is time, k_B is Boltzmann's constant, T is absolute temperature, and m and n are constants depending on the mechanism of mass transport. The value of n is approximately 3, which implies that surface diffusion predominates, and m is in the range of 0.3–0.5. The shrinkage is very temperature-dependent,

Need More Help?
Try this video on sintering kinetics: https://www.youtube.com/watch?v=eQQ54VmTHMU

since in addition to the direct temperature dependence in Eq. (7.26), the diffusivity is also temperature-dependent [cf. Eq. (4.71)].

In the intermediate stage, Eq. (7.26) can still apply, but the situation is complicated by grain growth (i.e., d is no longer constant) and a change in the pore geometry. More than one mass transport mechanism may be contributing to the changes in microstructure. In the final stages of sintering, densification is very dependent upon the association of pores with grain boundaries and the rate and mode of grain growth. Diffusion of atoms across the grain boundary causes the grain boundary to be displaced. The grain boundary mobility, M_b, can also be described by an Arrhenius-type expression, so it is temperature-dependent. When M_b is roughly constant, it can be used to determine the rate of grain size growth with time, t:

$$d_t^n - d_0^n = 2AM_b\gamma_{gb}t \tag{7.27}$$

where n is the dimensionality parameter of Eq. (7.26) with a value of 2–3, d_0 is the original mean grain size, d_t is the mean grain size at time t, A is a constant that depends on geometry, and γ_{gb} is the grain boundary interfacial energy. A log-normal grain size distribution is often observed during regular grain growth.

Pores can also affect grain growth and shrinkage, typically leading to exaggerated grain sizes. The tendency for exaggerated grains is higher when powder aggregates or dense, extremely coarse particles are present and the pore distribution is inhomogeneous. A high sintered density with minimal grain growth occurs when the material contains densely packed, homogeneously distributed, fine particles. A common *grain growth inhibitor* is magnesium oxide, MgO, which inhibits exaggerated grain growth by decreasing surface diffusion. Surface diffusion and pore mobility, M_p, are related by

$$M_p = \frac{KD_s}{Tr_p^4} \tag{7.28}$$

where K is a geometry-dependent constant, D_s is the surface diffusivity, T is absolute temperature, and r_p is the pore radius. Thus, M_p is larger for smaller pores. Pore growth can also occur during sintering, either by the coalescence of pores at grain boundaries or by the diffusion of vacancies from smaller pores to larger pores rather than to grain boundaries. This process of pore growth during heating is referred to as *Ostwald ripening*. A rapid heating schedule may produce densification with a smaller concomitant grain size, owing to the different activation energies for the two processes (see Figure 7.23). Surface diffusion, which predominates at low temperatures, can cause grain coarsening. Fast heating can increase the vacancy diffusivity (D_v) and densification at a faster rate than diffusion that causes grain coarsening.

There are many other aspects of ceramic and glass sintering that are beyond the scope of this text, including vitrification, sintering in the presence of liquid, the sintering of glazes and thick films, and sintering during hot pressing. There are also numerous specialized ceramic and glass processing technique such as chemical vapor deposition, chemical vapor infiltration, and sol–gel processing that lead to many important technological components and devices. The interested reader is directed to one of the many fine references available on these individual topics.

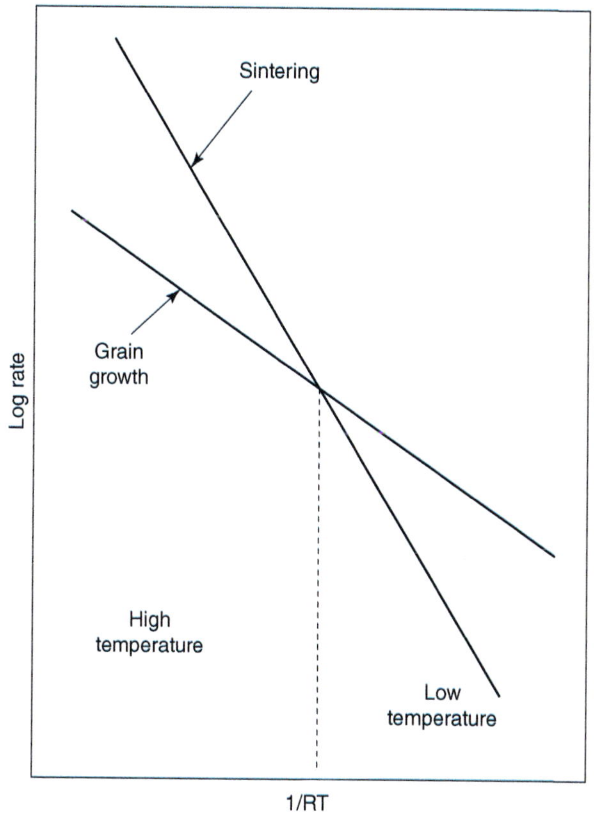

Figure 7.23 Temperature dependence of grain growth and sintering (densification) rates. The slope of each curve represents the activation energy. Reed 8[]/With permission of John Wiley & Sons.

7.3 Processing of Polymers

Just as was the case for a metal, a polymer must have formability to be shaped. Although an exact definition of formability is as difficult to give for polymers as it was for metals, it involves such polymer physical properties as viscosity, crystallinity, and thermal conductivity and such processing parameters as temperature and pressure. Most pure polymers are difficult to process, due in part to their high molecular weights or their cohesive forces. As a result, additives, oftentimes low-molecular-weight species are sometimes added to polymers to improve their formability (or for other reasons as we will see) to form *plastics*. Up to this point, we have made little distinction between polymers and plastics, and we will continue to do so. Be aware, however, of the relationship between the two and recognize that when the term "polymer" is used as an adjective to describe a certain type of processing, it may well be that the term "plastics" is more appropriate. Nonetheless, the principles of each process are fundamentally the same, whether a pure polymer or a plastic is being formed, such that the distinction is not necessary for our purposes.

There are three primary steps to any polymer forming procedure: *pre-shaping, shaping*, and *post-shaping*. Pre-shaping involves the melting, mixing, and conveying of the material to next step. The thermomechanical history in this step can affect properties, and time/temperature effects

Need More Help?

Try this video on an overview of polymer processing methods:

https://www.youtube.com/watch?v=qn16JtE_vLc

are important. In the shaping operation, deformation of the mass after pre-shaping forms the product into the desired shape. In this step, anisotropic properties can result. Finally, post-shaping consists primarily of solidification, although secondary operations such as printing, painting, fastening, and electroplating can occur here, as well. Heat transfer is the primary consideration in solidification, and tolerance of dimensions are dictated by this step. We will concentrate primarily on the shaping step, since it is the one that involves the most engineering principles, but we will also introduce aspects of the other two steps where appropriate. Let us begin by looking at some of the more important shaping methods.

There are two primary polymer processing methods, *continuous processing* and *cyclic processing*. *Die forming* and *calendering* constitute the continuous processes; whereas *molding* and *mold casting* are cyclic processes. We will discuss the methods within each of these categories, as well as *secondary shaping* methods such as *thermoforming*. Finally, we will discuss some of the more specialized types of polymer processing.

7.3.1 Continuous Processing

The two primary types of continuous polymer processing techniques are die forming (extrusion) and calendering. The advantages these methods offer are those of any continuous process: high throughput and product uniformity. The product geometries that can be made with these methods are limited, however. As with metals, the products that can be formed in a continuous fashion are primarily sheets and rods. In addition to extrusion and calendering, we will describe a specialized type of extrusion process known as fiber-spinning, which is used to make synthetic fibers for many important applications. Where appropriate, we will describe the engineering principles that can affect polymer structure and performance in these continuous operations.

Die Forming and Extrusion

Just as in the extrusion of metals, polymer extrusion is the pushing of a polymer melt across a metal die that continuously shapes the melt into a desired form. Recall that a die is a metal flow channel or restriction that serves the purpose of imparting a specific cross-sectional shape to a material—in this case, a stream of polymer melt. Dies are primarily used in extrusion and are used to form tubes, films, sheets, fibers, and complex profiles. They are positioned at the end of the melt generator or conveying equipment and consist of three general regions: the *manifold*, the *approach channel*, and the *die lip* (see Figure 7.24). The manifold distributes the incoming melt stream over a cross-sectional area similar in shape to the desired final product. The approach channel streamlines melt into the final die opening, and the die lip, or final die opening area, gives the proper cross-sectional shape to the product.

The objective of the die is to achieve the desired shape within set limits of dimensional uniformity at the highest possible production rate. However, nonuniformities can arise in the final part and are classified according to their geometry. The first type of nonuniformity is in the machine direction, or direction of flow [z direction, Figure 7.25a]. These arise due to time variations of inlet temperature, pressure, and composition of the melt. The second type of part nonuniformity is in the cross-machine direction [perpendicular to the z direction, Figure 7.25b]. These are generally due to improper die design.

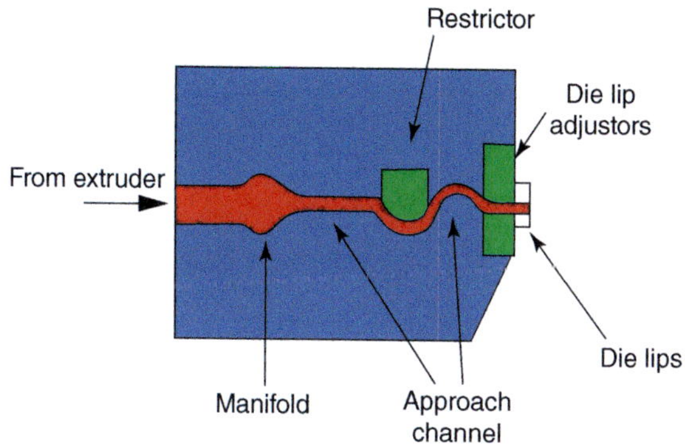

Figure 7.24 Schematic diagram of a polymer die. From Tadmor and Gogos [11]. Copyright © 1979 by John Wiley & Sons, Inc. This material is used by permission of John Wiley & Sons, Inc.

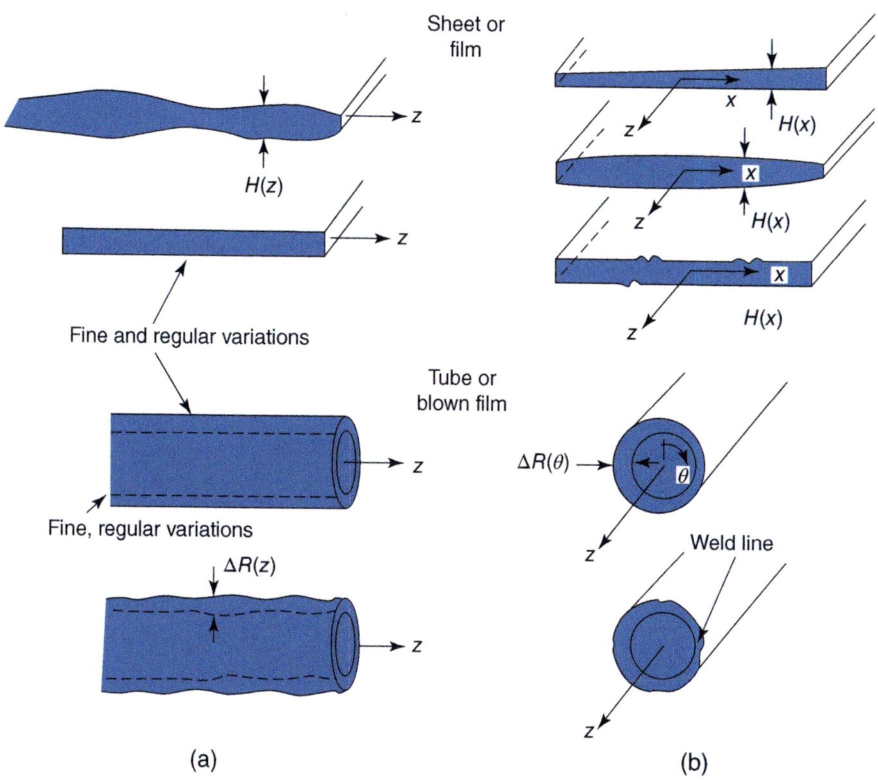

Figure 7.25 Die-formed product nonuniformities (a) in the machine direction and (b) in the cross-machine direction. From Tadmor and Gogos [11]. Copyright © 1979 by John Wiley & Sons, Inc. This material is used by permission of John Wiley & Sons, Inc.

A common phenomenon in die processing of polymers that influences the geometry of the final product is called *extrudate swelling* or *die swelling*. Extrudate swelling refers to the phenomenon observed with polymer melts and solutions that, when extruded, emerge with cross-sectional dimensions appreciably larger than those of the flow conduit. For the cylindrical extrudate

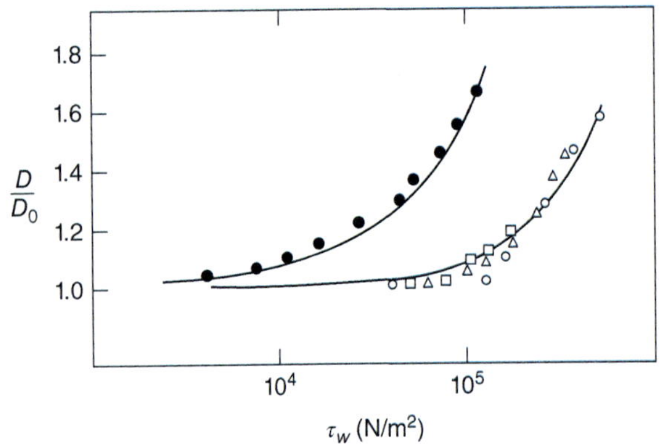

Figure 7.26 Extrudate swelling in polystyrene melts for (●) broad molecular weight distribution and (○, Δ) narrow molecular weight distribution samples. From Tadmor and Gogos [11]. Copyright © 1979 by John Wiley & Sons, Inc. This material is used by permission of John Wiley & Sons, Inc.

geometry, the ratio of the final diameter to that of the die capillary, D/D_0, varies from 1.12 at low shear rates to 0.87 at high shear rates for Newtonian fluids. Non-Newtonian fluids such as polymer melts have similar D/D_0 values at low shear rates, but they swell 2–4 times the extrudate diameter at high shear rates (see Figure 7.26). Experimentally, D/D_0 depends on the following variables: the shear stress at the die wall, τ_w, the molecular weight distribution, and the length-to-diameter ratio of the die capillary. Extrudate swelling is related to the ability of polymer melts and solutions to undergo delayed elastic strain recovery (cf. Section 5.3.1). The more strained and the more entangled the melt is at the die exit, the more it will swell.

How the melt is conveyed to the die is of equal, if not greater, interest when compared to the die forming process, since much of the polymer flow characteristics through the die are determined in the conveying section of the extruder. A typical extrusion apparatus is shown in Figure 7.27. As with all extruders, there is a rotating screw, called an *extrusion screw*, that is used to convey the polymer from the feed, called a *hopper*, to the die. The polymer is typically placed in pellet form into the hopper. From the hopper, the material falls through a hole in the top of the extruder (feed throat) onto the extrusion screw. This screw, which turns inside the *extrusion barrel*, conveys the polymer

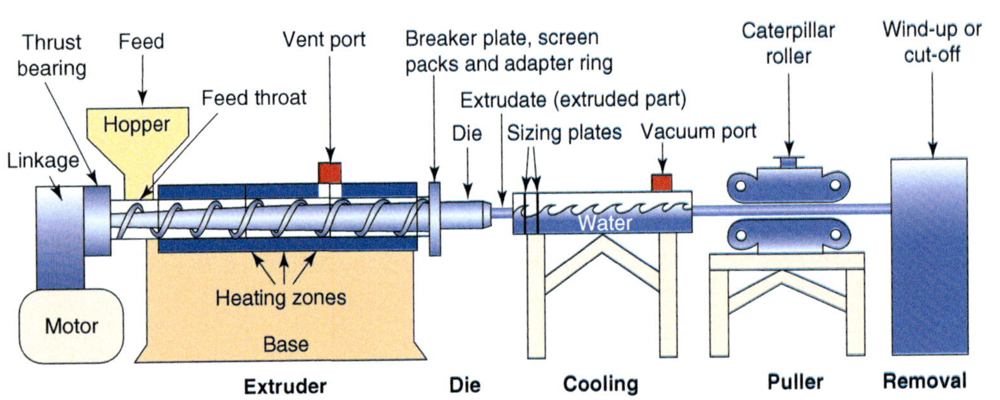

Figure 7.27 Schematic illustration of a typical extrusion apparatus. Strong [12]/Reproduced with permission of Pearson.

forward into a heated region of the barrel where a combination of external heating and frictional heating melts the polymer. The screw moves the molten polymer forward until it exits through the die. The extruded polymer is immediately cooled and solidified, usually in a water tank. The continuous output from the extruder is called the *extrudate*. Auxiliary equipment will cut, form, or coil the extrudate for secondary shaping or processing.

In the simplest type of extruder, a single screw is used to convey the polymer from the feed to the die. Every screw has certain geometric characteristics that can be summarized in Figure 7.28. The *barrel diameter*, D_b, is the diameter of the screw barrel in which the screw resides (not shown). There is a small clearance between the tip of the screw, called the *flight*, and the barrel wall, δ_f, on each side of the screw. The *screw diameter* at the tip of the flight, D_s, is then related to the barrel diameter as $D_b - 2\delta_f$. The *channel depth*, H, is the distance between the barrel wall and surface of the main portion of the screw, called the *screw root*. The channel depth varies along the length of the screw because the screw root diameter increases from the feed to the die. This is because the screw is conveying solids near the hopper, but molten polymer near the die. The ratio of the channel depth at the feed to the channel depth in near the die, called the *metering section*, is called the *compression ratio*. The axial distance of one full turn of the flight is called the *lead*, L_s, and e is the width of a flight. The actual width of the flow channel, W, is related to the lead and flight width:

$$W = L_s \cos\theta - e \tag{7.29}$$

where θ is called the *helix angle*, or *screw pitch*, and is given by

$$\tan\theta = \frac{L_s}{\pi D_s} \tag{7.30}$$

For square pitched screws, $L_s = D_s$, and the flights are all perpendicular to the screw axis. Extruder screws range from 25 to 150 mm in diameter, with screw lengths typically of 25 to 30 screw diameters. They rotate at speeds from 50 to 150 rpm and deliver melt at rates from 10 to 1000 kg/h with pressures up to 40 MN/m^2.

It is important to understand the flow characteristics of the polymer inside the screw. If we "unwind" the screw channel, as illustrated in Figure 7.29, we see that the polymer is conveyed to the die through nothing more than a square channel formed between the screw and the barrel surface. For the purposes of this development, we will assume that the channel has a constant channel depth, H (at least in the metering region). Furthermore, since the screw channel moves relative to the barrel surface, we can model the flow either as in a moving channel against a fixed barrel wall or as a fixed channel with a moving barrel wall. Both give the same result, and we will find it easier to use the latter model. So, assuming that the barrel surface moves at constant velocity, V_b, the

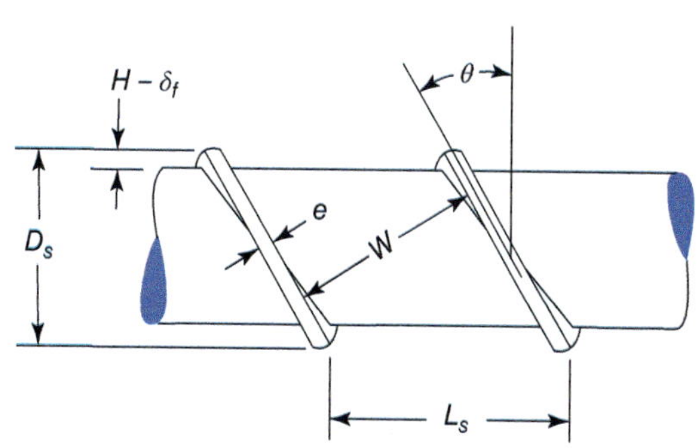

Figure 7.28 Characteristic dimensions of an extrusion screw. From Tadmor and Gogos [11]. Copyright © 1979 by John Wiley & Sons, Inc. This material is used by permission of John Wiley & Sons, Inc.

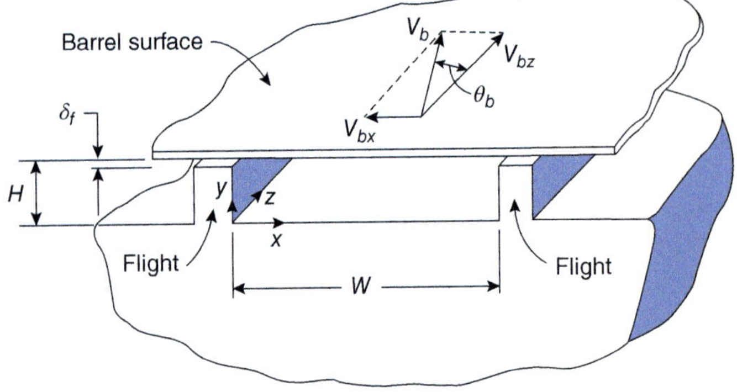

Figure 7.29 Schematic diagram of an "unwound" screw channel. From Tadmor and Gogos [11]. Copyright © 1979 by John Wiley & Sons, Inc. This material is used by permission of John Wiley & Sons, Inc.

surface velocity of the barrel surface can be resolved into components as illustrated in Figure 7.29: the *down channel velocity*, or *drag velocity*, $V_{bz} = V_b \cos\theta$ and the *cross-channel velocity* $V_{bx} = V_b \sin\theta$. The components of the barrel surface motion lead to circulatory flow of the polymer in the channel.

To obtain the velocity profile of the polymer in down channel direction, v_z, we start with the appropriately simplified equation of motion in rectangular coordinates

$$\frac{\partial P}{\partial z} = \mu\left(\frac{\partial^2 v_z}{\partial x^2} + \frac{\partial^2 v_z}{\partial y^2}\right) \tag{7.31}$$

and apply the following boundary (no slip) conditions:

$$v_z = 0 \quad @x = x, y = 0$$
$$v_z = V_{bz} \quad @x = x, y = H$$
$$v_z = 0 \quad @x = 0, y = y$$
$$v_z = 0 \quad @x = W, y = y$$

It can be shown that the pressure gradient is a constant. We then integrate v_z across the channel and perform similar analysis for v_x to get the volumetric flow rate, Q,

$$Q = \frac{V_{bz}WH}{2}F_d + \frac{WH^3}{12\mu}\left(-\frac{\partial P}{\partial z}\right)F_p \tag{7.32}$$

where F_d and F_p are *shape factors* for the drag and pressure flows, respectively. They have values less than 1.0 and depend only on the H/W ratio, as illustrated in Figure 7.30. The shape factors represent the reducing effect of the flights on flow rate between infinite parallel plates, as in our model.

The first term in Eq. (7.32) represents the contribution of *drag flow*, Q_d, and the second term represents the contribution of *pressure flow*, Q_p. The two terms are independent of one another and represent the two ways in which polymer moves down the channel. The ratio of pressure to drag flow, Q_p/Q_d, is a measure of the relative amount of pressure to drag flow and indicates the velocity profile and movement of fluid elements in the flow channel, among other things. For pure drag flow, $\partial P/\partial z = 0$ (no pressure flow) and $Q_p/Q_d = 0$. The polymer flows by drag only. A value of -1 for this ratio indicates a situation of zero net flow rate, or $Q_p = -Q_d$. An increase in the positive pressure gradient will increase the pressure backflow. This decreases the net flow rate, despite an

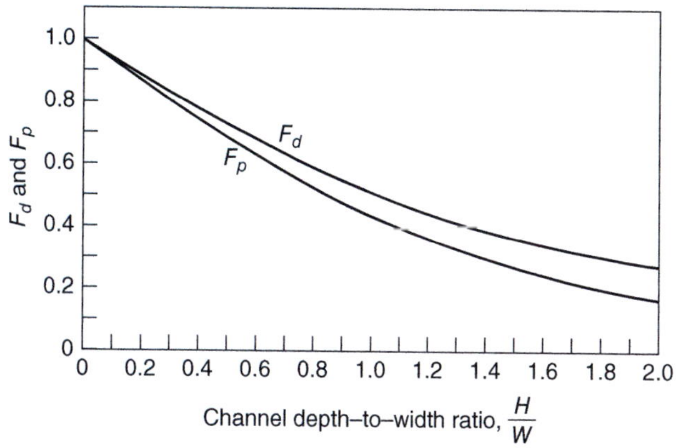

Figure 7.30 Shape factors for polymer extrusion. From Tadmor and Gogos [11]. Copyright © 1979 by John Wiley & Sons, Inc. This material is used by permission of John Wiley & Sons, Inc.

unchanging drag flow. Eventually, the drag and pressure flows are equal, but opposite, and discharge ceases. As the throughput increases from zero, the fluid elements travel along a sort of flattened helical path through the channel, as illustrated in Figure 7.31 for the two extremes cited above and for an intermediate case of $Q_p/Q_d = -0.5$. The closer the flow to pure drag flow, the more open the loops of the helix will be. Thus, polymer particles travel in a path that is a flat helix within a helical channel.

In addition to conveying the polymer toward the die, the screw helps perform two more important functions. The first is melting. In addition to the heat supplied by resistance heating elements along the barrel length, the screw provides frictional heat to the melt to assist in melting. The last function is mixing. In addition to the mixing provided by a combination of drag flow and pressure flow, mixing elements can be added to the screw that accentuate dispersive or distributive mixing action.

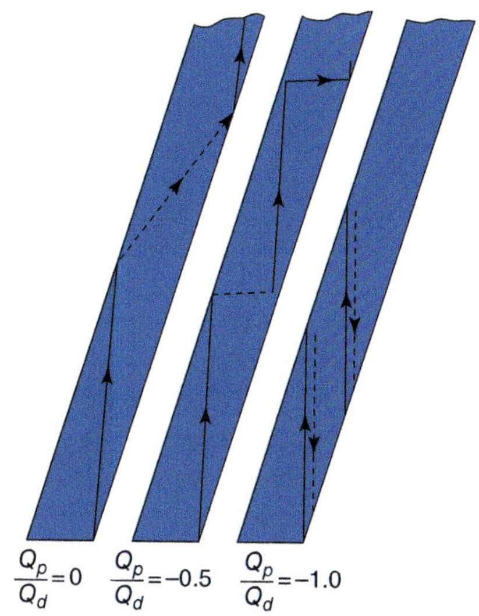

Figure 7.31 The path of fluid elements in the extruder screw channel for various ratios of Q_p/Q_d. From Tadmor and Gogos [11]. Copyright © 1979 by John Wiley & Sons, Inc. This material is used by permission of John Wiley & Sons, Inc.

Need More Help?
Try this video on twin-screw extrusion: https://www.youtube.com/watch?v=WMxlunR1LC0

In addition to single-screw extruders, there are twin-screw and multiscrew extruders that perform essentially the same functions, but with additional benefits. Among these, the *intermeshing twin-screw* extruders are the most important ones. They are used primarily for heat-sensitive resins (such as PVC), which are difficult to process. The intermeshing screws create a relative motion of one flight in another, such that it acts like a paddle to push the material from screw to screw and from flight to flight. Twin-screw extruders have a more positive pumping action than single-screw extruders and can therefore be used more effectively in high-output situations.

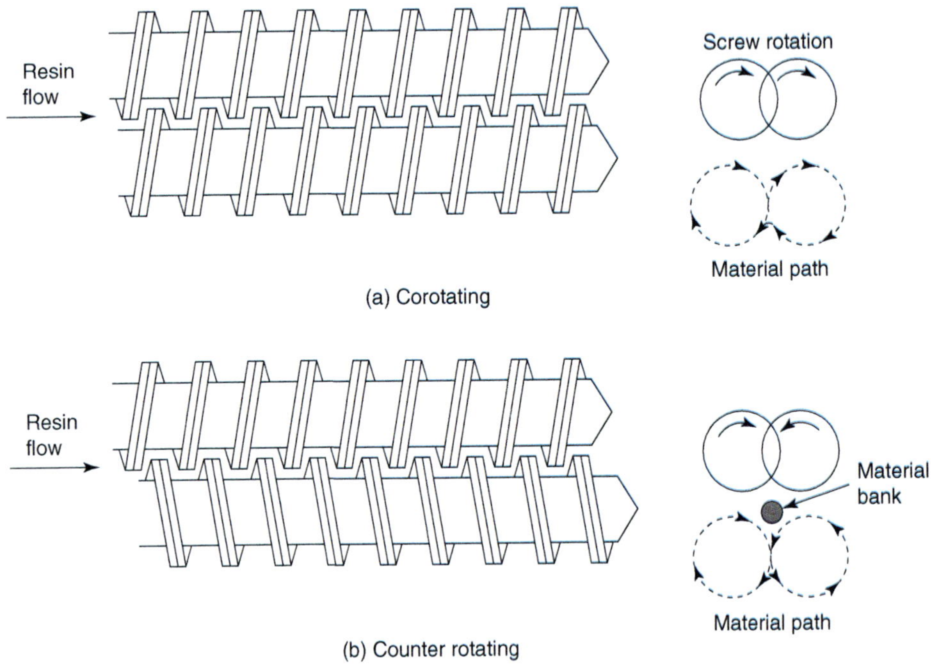

Figure 7.32 Schematic diagram of (a) corotating and (b) counterrotating twin screws. Strong [13]/ Reproduced with permission of Pearson.

The screws can rotate either in the same direction, known as *corotating screws*, or in opposite directions, known as *counterrotating screws*, as shown in Figure 7.32. In the corotating screws system, the material is passed from one screw to another and follows a path over and under the screws. In a counterrotating screws system, the material is brought to the junction of the screws and builds up what is called a *material bank* on the top of the junction. This buildup of material is conveyed along the length of the screws, high shear is created, but shear elsewhere is very low. Only a small amount of material passes between the screws. Therefore, total shear is lower than in single-screw extruders and in corotating twin-screw extruders. A comparison between some characteristics of single, corotating and counterrotating twin screws is given in Table 7.12.

Calendering

Calendering is used for the continuous manufacture of polymer sheet or film. The calendar consists of four horizontal, counterrotating steel rolls, called *bowls* or *nip rollers*. The top two bowls look just like a two-roll mill used in the rolling of metals and are charged with a uniform polymer melt. The melt may come from a melt-conveying operation such as extrusion. The gap between the two rolls is small, and the polymer builds up a small pool on the top of the nip area called the *bank*. The polymer is drawn through the first nip and, adhering to one bowl, is transferred to the next nip where it is reduced in thickness again. The gaps between subsequent rolls get smaller and the temperature of the rolls lower to finally set the dimensions of the part. The thickness reduction in the third nip provides fine control of the sheet and confers the required surface finish.

Table 7.12 Comparison of Single- and Twin-Screw Extruder Screw Characteristics

Type	Single Screw	Corotating Screw: Low-Speed Type	Corotating Screw: High-Speed Type	Counterrotating Twin Screw
Principle	Friction Between Cylinder and Materials and the Same Between Material and Screw	Mainly Depend on the Frictional Action as in the Case of Single-Screw Extruder		Forced Mechanical Conveyance Based on Gear Pump Principle
Conveying efficiency	Low	Medium		High
Mixing efficiency	Low	Medium/high		High
Shearing action	High	Medium	High	Low
Self-cleaning effect	Slight	Medium/high	High	Low
Energy efficiency	Low	Medium/high		High
Heat generation	High	Medium	High	Low
Temp distribution	Wide	Medium	Narrow	Narrow
Max. revolving speed (rpm)	100–300	25–35	250–300	35–45
Max. effective length of screw L/D	30–32	7–18	30–40	10–21

Source: Tadmor and Gogos [11]. Copyright © 1979 by John Wiley & Sons, Inc.

The advantage of calendering over direct sheet extrusion is that a complicated and expensive die is not required. Some additional mixing can be done as part of the calendering process, as well, and a surface finish can be applied by the final roller, if desired. Sheet up to a few millimeters thick and a meter wide can be made using calendering.

Need More Help?

Try this video on calendering:

https://www.youtube.com/watch?v=Tm1maH6taZ8

Fiber Spinning

The production of polymer fibers is important for many applications, including woven fabrics for such things as composite reinforcement and clothing. There are three primary methods of spinning polymeric fibers: *melt spinning; dry solution spinning*; and *wet solution spinning*. A fourth type of fiber spinning, known as *gel spinning*, is a hybrid form of dry-wet spinning.

As the name implies, melt spinning involves the drawing of fibers directly from a polymer melt. A schematic illustration of the apparatus is shown in Figure 7.33a. The polymer is fed and melted in much the same manner as it is for other melt processing techniques (e.g., extrusion), but instead of being forced through a single die, it is pressed through a multiple-orifice plate called a *spinneret*. The spinneret consists of 50–60 holes, each 0.12 mm in diameter. The final diameter of the fiber, or *denier*, is determined by pumping and winding rate—not hole size. Production rates of about 2500 ft per minute are typical for this operation. The multiple fibers are gathered together on a roller in a bundle known as a *tow*. To help keep the group of fibers together, the tow can be twisted slightly, at which point it is called a *yarn*. Melt spinning is used to make nylon, polyester, polyethylene

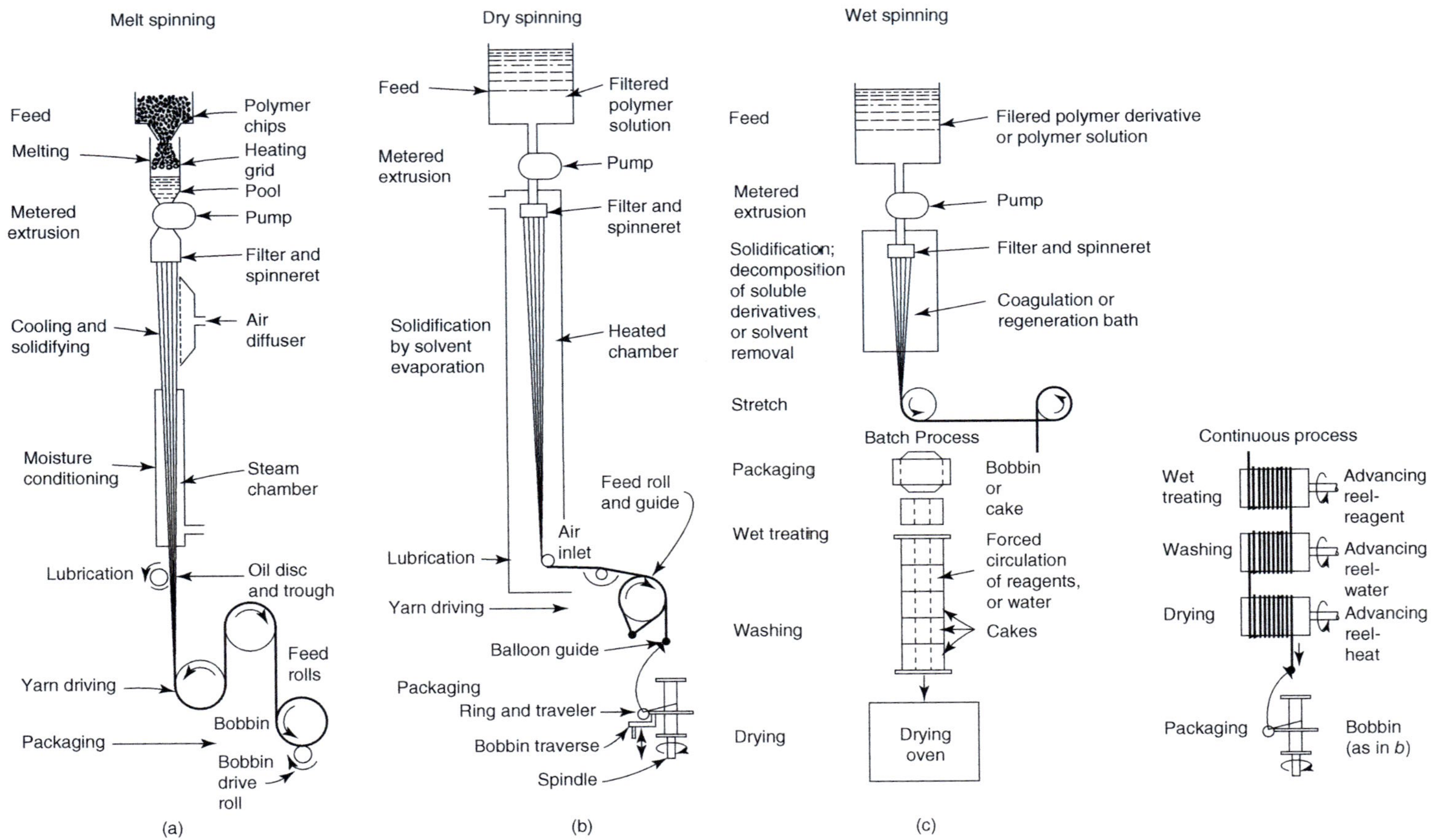

Figure 7.33 Schematic illustration of three types of polymer fiber spinning operations: (a) melt spinning, (b) dry spinning, and (c) wet spinning. From Billmeyer [14]. Copyright © 1984 by John Wiley & Sons, Inc. This material is used by permission of John Wiley & Sons, Inc.

terephthalate (PET), and olefin fibers, among others. The desired range of viscosity at the spinneret is between 1000 and 2000 poise. If the polymer melt is too viscous and would require a dangerously high temperature to reach this viscosity range, it is spun in solution.

Need More Help?

Try this video on an overview of polymer fiber spinning:

https://www.youtube.com/watch?v=uhaeEV5_bso

Although the term "dry solution spinning" may seem like an oxymoron, it simply describes a process by which a polymer is dissolved in an appropriate solvent, and the solution is pumped through a spinneret in much the same manner as in melt spinning (see Figure 7.33b). In this case, however, the fibers enter air when they exit the spinneret and solidify because the solvent is evaporated and recycled. This process is used for the formation of such filaments as cellulose acetate, acrylic, spandex, triacetate, vinyon, and poly acrylonitrile (PAN) fibers, which are the precursors to carbon fibers (cf. Section 1.4.5). In wet solvent spinning, or wet spinning, the polymeric fiber is formed in an analogous fashion to dry spinning, except that the solvent is leached out by another liquid, as illustrated in Figure 7.33c. This process is used to make rayon (cellulosic), acrylic, aramid, modacrylic, and spandex fibers. Gel spinning is used to form some high-strength polyethylene fibers.

We can derive some useful design equations for fiber spinning, just as we did for extrusion and calendering, by using a capillary flow model, as illustrated in Figure 7.34. Though the principles apply to all spinning methods, we will concentrate on melt spinning, since it involves a single phase system. In capillary flow, the polymer melt is contained in a reservoir. In the entrance region, the melt is forced into a converging flow pattern and undergoes large axial accelerations—that is, stretching. As the flow rate is increased, the axial acceleration also increases, and the polymer becomes much more elastic. It is possible for melt fracture to occur in this regime due to the viscoelastic nature of the polymer (cf. Section 5.3.1). Barring any such instability phenomena, a fully developed velocity profile is reached a few diameters after the entrance region. The flow in the capillary, which for pseudoplastic fluid is characterized by a flat velocity profile, imparts a shear strain on the melt near the capillary walls. The core of the melt, if the capillary L/R is large and the flow rate is small, can undergo a partial strain recovery process during its residence in the capillary. At the exit region, the melt is under the influence of no externally applied stresses. It can thus undergo delayed strain recovery, which, together with the velocity profile rearrangement to one that is pluglike, results in the phenomenon of die swelling that was introduced earlier in this section. Let us develop some more quantitative descriptions of these phenomena.

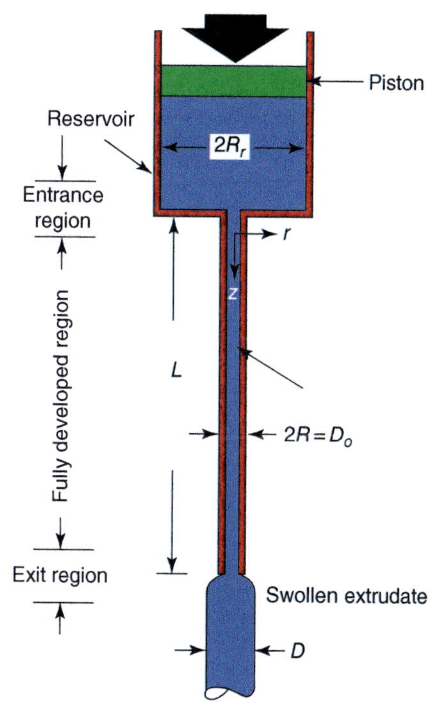

Figure 7.34 Schematic diagram for capillary flow. From Tadmor and Gogos [11]. Copyright © 1979 by John Wiley & Sons, Inc. This material is used by permission of John Wiley & Sons, Inc.

Need More Help?
Try this video on the Hagen–Poiseuille equation: https://www.youtube.com/watch?v=_hvgvZuIZOc

The velocity profile for a Newtonian fluid in a capillary is well-described in most introductory transport texts by the Hagen–Poiseuille equation

$$v_z = \frac{\Delta PR^2}{4\mu L}\left[1 - \left(\frac{r}{R}\right)^2\right] \tag{7.33}$$

with the implicit assumption that the pressure gradient, dP/dz, is constant and given by $\Delta P/L$, where $\Delta P = P_0 - P_L$, and L is the length of the capillary. This assumption is not correct near the capillary entrance, where the converging flow causes extra velocities (e.g., v_r), and velocity gradients to be present. An example of the entrance flow patterns in a molten polymer is given in Figure 7.35. Although the flow pattern is not known precisely, we know that a higher pressure drop is needed to support the additional velocity gradients for any viscous or viscoelastic fluid. Schematically, then, the pressure profile along the length of the capillary can be represented as in curve *a* (bottom curve including the dotted line) in Figure 7.36. Thus,

$$\frac{-dP}{dz} = \frac{\Delta P}{L^*} = \frac{\Delta P}{(L + ND_0)} \tag{7.34}$$

where L^* has been substituted for L to indicate a modified flow length, $N > 0$ is the *entrance loss correction factor*, sometimes called the *Bagley correction*, and is applied to increase the effective length of the capillary by a multiple of capillary diameters, D_0. This has the same effect as reducing the pressure drop by a value of ΔP_{ent}, the *entrance pressure drop*. It turns out that the value of the entrance pressure drop becomes larger than the total capillary pressure drop at higher shear rates. It follows then that in polymer processing, where the length-to-opening ratios are small and shear rates are high, entrance pressure drops must be included in calculations of the die pressure in die design equations. This is done through *Bagley plots*, an example of which is shown in Figure 7.37. If the L/D_0 of the capillary and the volumetric flow rate, Q, are known the pressure

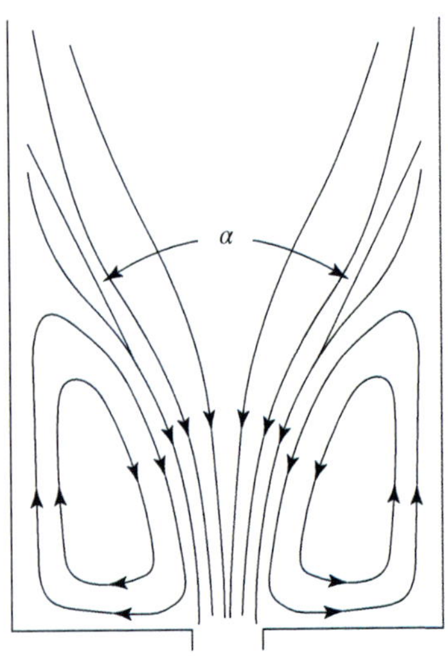

Figure 7.35 Entrance flow patterns in molten polymers. From Tadmor and Gogos [11]. Copyright © 1979 by John Wiley & Sons, Inc. This material is used by permission of John Wiley & Sons, Inc.

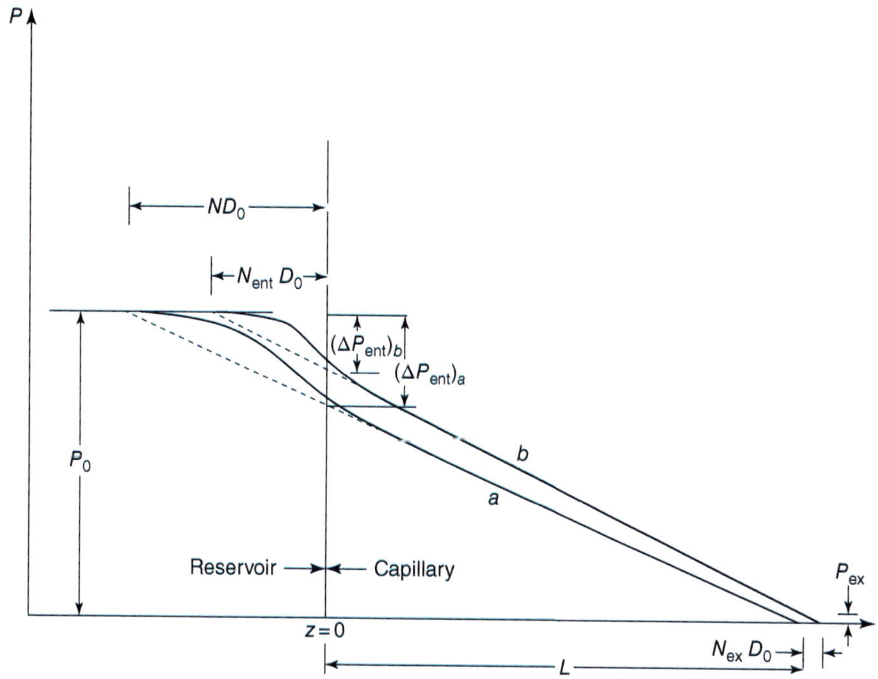

Figure 7.36 Schematic representation of the capillary pressure along its axis: curve *a*, without exit effects; curve *b* with exit effects. From Tadmor and Gogos [11]. Copyright © 1979 by John Wiley & Sons, Inc. This material is used by permission of John Wiley & Sons, Inc.

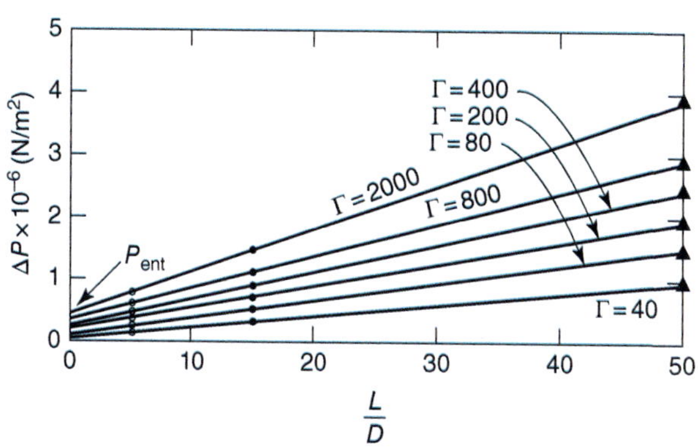

Figure 7.37 Bagley plots for a polystyrene melt at 200°C. From Tadmor and Gogos [11]. Copyright © 1979 by John Wiley & Sons, Inc. This material is used by permission of John Wiley & Sons, Inc.

drop can be found from Figure 7.41, where ΔP_{ent} is the value of the pressure drop at $L/D_0 = 0$. The values of Γ are given by

$$\Gamma = \frac{32Q}{\pi D_0^3} \tag{7.35}$$

Cooperative Learning Exercise 7.3

At right is shown data for the ratio of entrance pressure drop to shear stress at the capillary wall, τ_w, as a function of Newtonian wall shear rate, Γ, for a number of common polymers. Consider the melt spinning of PS at a volumetric flow rate of 4.06×10^{-1} cm^3/s through a spinneret that contains 100 identical holes of radius 1.73×10^{-2} cm and length 3.46×10^{-2} cm. Assume that the molecular weight distribution is broad.

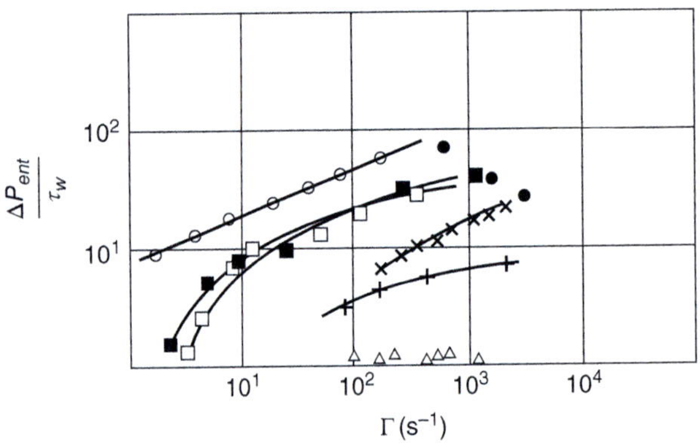

The ratio of entrance pressure drop to shear stress at the capillary wall versus Newtonian wall shear rate, Γ. ■, PP; □, PS; ○, LDPE; +, HDPE; ●, 2.5% PIB in mineral oil; ×, 10% PIB in decalin; Δ, NBS-OB oil. Reprinted, by permission, from Z. Tadmor and C. G. Gogos, *Principles of Polymer Processing*, p. 537. Copyright © 1979 by John Wiley & Sons, Inc.

Person 1: Estimate the entrance region pressure drop, ΔP_{ent}, using the Bagley correction. What can you say about the percentage of the entrance pressure drop relative to the total pressure drop?

Person 2: Estimate the ratio of entrance region pressure drop to shear stress at the capillary wall using the plot above.

Combine your results to obtain a value for the shear stress at the capillary wall, and use that value to estimate the extrudate swell, D/D_0 for polystyrene using Figure 7.26. What effect does molecular weight have on die swell at these shear levels?

Answers: $Q = 4.06 \times 10^{-3}$ cm^3/s; $\Gamma = 10^3$ s^{-1}; $L/D_0 = 1$; $\Delta P_{ent} \approx 0.4 \times 10^6$ N/m^2; the close proximity of the entrance pressure drop to the overall pressure drop in the Bagley plot indicates that most of the pressure drop is due to entrance effects. $\Delta P_{ent}/\tau_w \approx 40$; $\tau_w \approx 10^4$ N/m^2; $D/D_0 \approx$ 1.1. At this stress level, molecular weight does not have a large effect.

Similarly, one must worry about exit pressure effects, or *end corrections*. Experimentally, it has been observed that there is a nonzero gauge pressure at the capillary exit, P_{ex} (curve *b* of Figure 7.36). It has been found that the ratio of exit to entrance effects, $P_{ex}/\Delta P_{ent}$, is between 0.15 and 0.20 and that although ΔP_{ent} does not depend on L/D_0, P_{ex} decreases up to $L/D_0 = 10$ and then remains constant. With these observations in mind, we can rewrite Eq. (7.34) such that L^* now includes both entrance and exit effects:

$$\frac{\Delta P}{L^*} = \frac{\Delta P}{L + N_{ent}D_0 + N_{ex}D_0} \tag{7.36}$$

where N_{ent} and N_{ex} are Bagley corrections for entrance and exit effects, respectively. If $P_{ex}/\Delta P_{ent} \ll 1$, then $N = N_{ent}$. If P_{ex} is included, it is found that $N_{ent} < N$.

Other flow effects include extrudate swelling and *melt fracture*. The phenomenon of extrudate swell was elaborated upon in the context of polymer extrusion (cf. Figure 7.26). We will not describe the phenomenon of melt fracture here. There are many excellent textbooks on polymer processing that deal with subject and more.

7.3.2 Cyclic Processing

In contrast to extrusion, calendering, and fiber spinning, which are continuous processes, molding operations are by their very nature cyclical, or batch, processes. In this section, we describe five types of cyclic molding operations involving polymers: *injection molding, compression molding, blow molding, thermoforming*, and a special type of injection molding called *reaction injection molding (RIM)*. These techniques all have in common the use of molds to shape the polymer into a desired geometry. Although molds are utilized in the batch processing of other materials, they are most widely used in the shaping of polymers, so we will briefly elaborate on the various types of molds before continuing on to the molding operations.

Molds can be of three basic types. Open molds are used to define one surface of a product. They can be of the female (indenting) or male (projecting) type. Two matched female half-molds can be used to make hollow items, and a matched female and male pair is used to form a cavity that is to be completely filled. Shaping of the polymer inside the mold is usually achieved by applying pressure to cause the fluid to conform to one or both mold surfaces. The variation in types of molding is primarily due to the type of mold used and the method of melt delivery to the mold.

The choice of process and of mold type depends greatly on the number of pieces to be made, bearing in mind the lowest possible price per product, and on the tolerances to be attained. In turn, the choice of the process will affect the choice of material and the shape of the product. For example, the price of producing a simple plastic tray is compared in Figure 7.38 for the various molding techniques described in this chapter. The price is determined by such things as raw material, mold, machine, energy, and personnel costs. There are, of course, considerations other than total cost, but injection molding generally provides the lowest cost per part for many polymeric components. So, it is fitting that we begin our review of cyclic polymer processing techniques with injection molding.

Injection Molding

Injection molding is the process of producing identical articles from a hollow mold. Because of their high viscosity, polymers cannot be poured into a mold, or cast, in the same way that metals are, because gravitational forces are not sufficient to produce appreciable flow rates. Thus, the melt must be injected into the mold cavity by the application of large forces from a plunger. Moreover, once the mold is filled with melt and solidification starts, an additional amount of melt must be packed into the mold to offset polymer shrinkage during solidification.

Injection molding involves two distinct processes. The first is melt generation, mixing, and pressurization and flow, which is carried out in the injection unit of the molding machine. The second is product shaping, which takes place in the mold cavity. Injection molders, therefore, have two distinct parts: the injection unit and the mold/clamping unit. The function of the injection unit

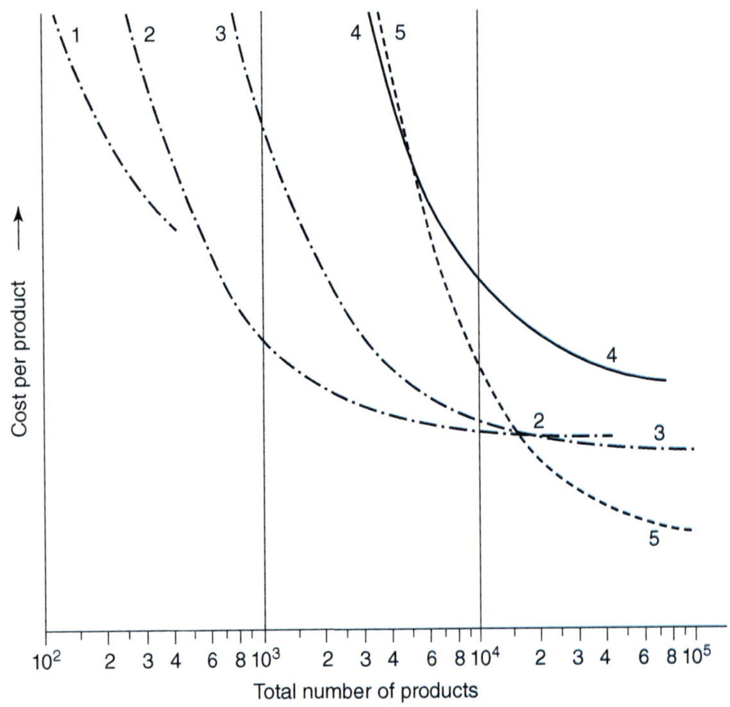

Figure 7.38 Comparison of production costs per product for a polymer tray formed using (1) thermoforming in a wooden mold, (2) thermoforming in an aluminum mold, (3) blow molding with a single mold, (4) compression molding, and (5) injection molding. Powell and Housz [15]/Reproduced with permission of Stanley Thorned Publishers.

is to melt the polymer and inject it into the mold; the function of the clamping unit is to hold the mold, open and close the mold, and eject the finished product.

Two systems have been used in injection molding machines to melt and inject the polymer. The most commonly used types use a *reciprocating screw*, which has many similarities to an extruder screw, but with a unique reciprocating (back and forth) action. The other type of injector system is the *ram injector*. We will concentrate on the reciprocating screw since it offers many advantages over ram injectors, including more uniform melting, improved mixing, lower injection pressures, the formation of larger parts, fewer stresses in parts, and faster cycle time.

The reciprocating screw used for injection molding is similar to an extruder screw but is generally much shorter. Typical length-to-diameter ratios for reciprocating screws are 12:1 to 20:1. The compression ratios are in the range of 2:1 to 5:1, which means that there is less mechanical action added during melting and more thermal energy is needed. As with an extrusion screw, the reciprocating screw has three sections: feed, compression, and metering. The reciprocating action of the screw is illustrated in Figure 7.39. The screw acts like an extrusion screw to melt and meter the polymer toward the mold, forming a pool at the mold entrance. It then moves forward like a piston to inject the polymer into the mold. After mold filling, the screw retracts, and the part is rejected. Let us look more closely at the entire injection molding cycle.

Need More Help?
Try this video on injection molding: https://www.youtube.com/watch?v=xn2HgMjbKf4

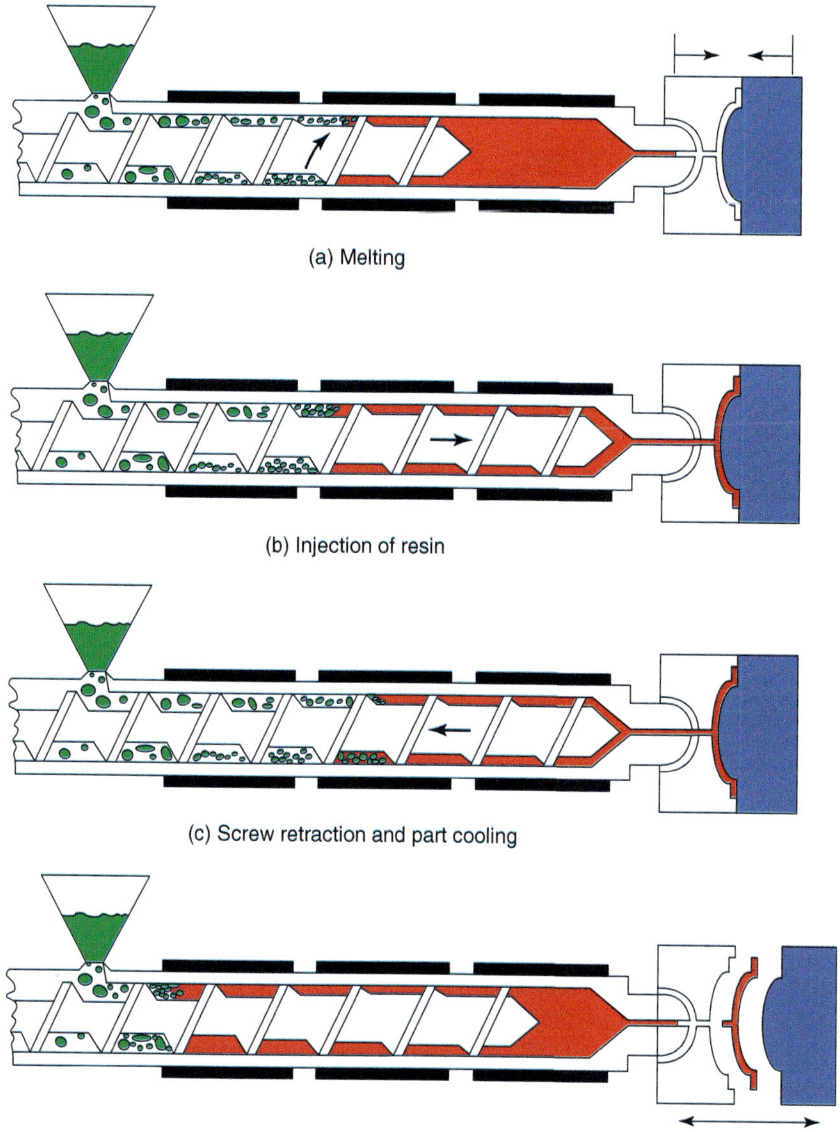

(a) Melting

(b) Injection of resin

(c) Screw retraction and part cooling

(d) Part removal

Figure 7.39 Schematic illustration of reciprocating action of injection molding screw. Strong [16]/ Reproduced with permission of Pearson.

The injection molding cycle is depicted in Figure 7.40. We can begin the cycle at any point we wish but let us start at the point the screw moves forward and fills the mold with polymer melt. The screw moves forward and fills the mold with melt and maintains the injected melt under pressure, during what is called the *hold time*. To ensure that polymer does not flow backward, a check valve is attached to the end of the screw. During the hold time, additional melt is injected, offsetting contraction due to cooling and solidification. As the melt cools and solidifies, the pressure drops. Once the part is cooled, screw rotation commences and the screw moves backward, conveying molten polymer to the front as it retracts. After sufficient melt generation for the next injection, called

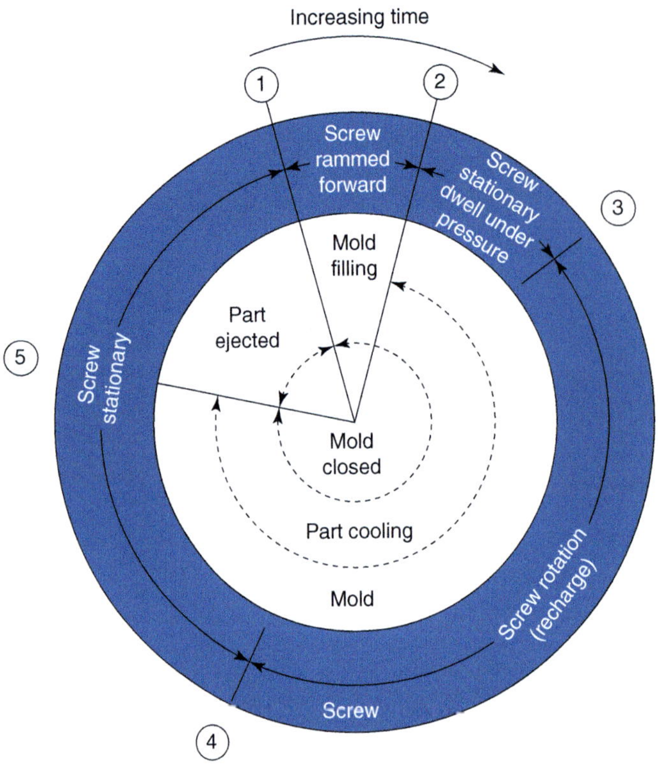

Figure 7.40 The injection molding cycle. From Tadmor and Gogos [11]. Copyright © 1979 by John Wiley & Sons, Inc. This material is used by permission of John Wiley & Sons, Inc.

the *shot*, screw rotation ceases. The polymer on the stationary screw continues to melt by heat conduction from the hot barrel. Meanwhile, the mold opens, and solidified part is ejected from the mold. The mold then closes, and is ready to accept the next shot, thus initiating the cycle once again. All of these operations are automated and computer controlled. Perhaps the most important measure of the size of an injection molding machine is the amount of polymer that can be injected, or shot size. Typical shot sizes range from 20 g to 20 kg. Because the shot size is dependent upon the density of the polymer being injected, polystyrene has been chosen as the standard for the rating of machines.

A typical injection mold is made of at least two parts, one of which is movable so that it can open and close during different parts of the molding cycle. The entire mold is kept at a constant temperature below T_g or T_m of the polymer. The various components of an injection mold are illustrated in Figure 7.41. The *sprue* is the part of mold that joins the mold cavity to the machine nozzle. In Figure 7.41 the polymer, then, would be coming into the page from the injection screw. The *runners* are channels cut in the mold through which molten polymer flows after entering the sprue, and on to the *gates*, which are narrow constrictions between the runner and each individual cavity. The gates are purposely designed to be as narrow as possible to allow the molded part to be easily removed from the runner and sprue while still allowing polymer to flow easily into each mold. It is the material within the gate that solidifies first, due to the more rapid heat transfer from the small part. Once the gate is solidified, the parts at the end of the runners are isolated from the injection pressure, and the part can be safely ejected.

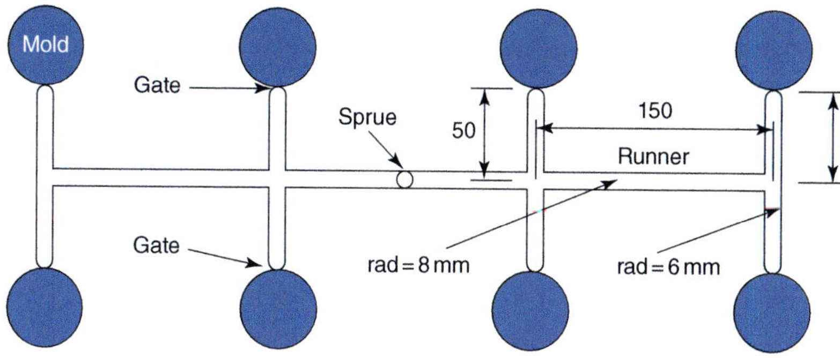

Figure 7.41 Schematic diagram of eight-cavity mold for injection molding operations. The molded parts of interest are indicated by the shaded regions.

The function of the runner system is to transmit the molten polymer to the cavities with a minimum of material and pressure drop. Therefore, the runner length must be kept to a minimum, and the cross section should be optimally set for low pressure drop, low material waste, and relatively slow cooling. Slow cooling is desirable to prevent premature solidification of the polymer in the molds, which causes incomplete filling called *short shot*. The runner dimension is about 1.5 times the characteristic thickness of the molded part, and it is of circular cross section to minimize heat loss.

Short shot can be avoided by proper mold design and control of polymer melt conditions—namely, temperature and injection pressure. This relationship is shown in Figure 7.42. Within the area bounded by the four curves, the specific polymer is moldable in the specific cavity. If the pressure and/or temperature are too low, short shot will result. If the temperature is too high, thermal degradation of the polymer can occur. If the temperature is too low, the polymer will not be molten. If the pressure is too high or the polymer is too fluid, the melt can flow into the gaps of the mold, creating thin webs of polymer attached to the molded article in an undesirable part defect called *flash*. Obviously, mold design is critical, as is the proper matching of mold design to process

Figure 7.42 Schematic illustration of molding area for a given polymer/mold combination. From Tadmor and Gogos [11]. Copyright © 1979 by John Wiley & Sons, Inc. This material is used by permission of John Wiley & Sons, Inc.

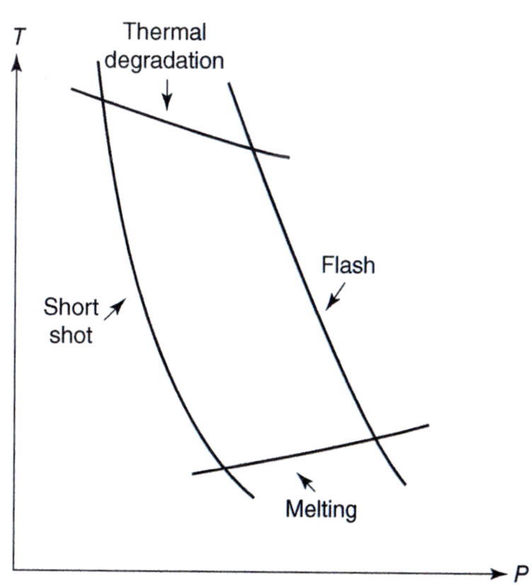

conditions. The most common material for injection molds is tool steel, although beryllium–copper alloys are used when high heat transfer rates are required. Aluminum is also used for molds.

Let us turn our attention now to developing some useful flow equations for injection molding, just as we have done for the other polymer processing techniques. Consider a steady, isothermal, laminar, fully developed pressure flow of an incompressible fluid in a horizontal tube, such as might be found in the runner of an injection mold. For a Newtonian fluid, the analysis is essentially the same as for capillary flow, and the velocity profile can be given by the Hagen–Poiseuille equation, Eq. (7.33). Let us take this opportunity, however, to add an additional level of complexity that is common in polymer processing. As introduced in Section 4.1.2 for ceramic slurries, and later applied to polymer melts and solutions, non-Newtonian behavior can occur when the viscosity, η, is a function of the shear rate, $\dot{\gamma}$. One model that is widely used for non-Newtonian fluids is the power law model [cf. Eqs. (4.15) and (4.17)]. We will dispense with the derivation here, but starting from the continuity equation and the equations of motion in cylindrical coordinates as before, the following velocity profile, $v_z(r)$, for a power law fluid in a horizontal tube can be obtained:

$$v_z(r) = \left(\frac{R}{1 + 1/n}\right)\left[\frac{-R}{2K}\left(\frac{dP}{dz}\right)\right]^{1/n}\left[1 - \left(\frac{r}{R}\right)^{1+1/n}\right] \tag{7.37}$$

where R is the tube radius, dP/dz is the pressure drop, and n and K are the power law exponent and consistency index, respectively, as defined in Eq. (4.15). The corresponding volumetric flow rate, Q, assuming a constant pressure drop over the tube length, L, can be obtained from the velocity profile:

$$Q = \frac{\pi R^3}{3 + 1/n}\left(\frac{-R\Delta P}{2KL}\right)^{1/n} \tag{7.38}$$

Notice that Eq. (7.38) reduces to the Hagen–Poiseuille Law for $n = 1$, where $K = \mu$, the Newtonian viscosity.

Obviously, molds are much too complicated to be adequately represented by simple flow in a tube, even if a power law model is employed. Other important factors to consider include heat transfer rates and shrinkage, which are dependent upon both the material of construction for the mold and the polymer being molded. One of the advantages of injection molding is that parts with very complex shapes can be made. Even hollow parts can be made. However, sharp bends in the part are to be avoided. These are difficult for the viscous resin to fill and can be the origin of stresses in the part. The minimum thickness of the part depends upon the type of resin used. Surface finish in the part is imparted by the surface of the mold. Part imperfections include (a) surface indentations where the ejector pins hit the part and (b) rough edges from mold lines or remaining after the part is separated from the secondary runner.

The most serious imperfection that must be accounted for is *mold shrinkage*, which is the difference in the part dimensions and the dimensions of the mold. The amount of shrinkage is dependent upon several factors, including the temperature of the polymer, the type of polymer, the flow field of the polymer in the mold, injection pressure and hold time, and the presence of additives. Characteristic shrinkages for several common polymers are given in Table 7.13. These values are used by calculating the dimensions of the mold cavity from the following equation:

$$\text{Mold dimension} = \text{Part dimension} \times (1 + \text{Shrinkage value}) \tag{7.39}$$

Table 7.13 Characteristic Shrinkages for Various Plastics

Shrinkage (mm/mm)	Type of Material
0–0.002	Polyester (thermoset)
0.001–0.004	Polycarbonate, 20% fiberglass
0.002–0.008	Acrylic
0.002–0.003	PVC
0.004–0.007	ABS
0.004–0.006	Polystyrene
0.005–0.007	Polycarbonate
0.005–0.008	Polyphenylene oxide
0.008–0.015	Nylon (6/6)
0.010–0.020	Polypropylene
0.018–0.023	Acetal
0.007–0.025	LDPE
0.020–0.040	HDPE

Notice the strong dependence of shrinkage on the presence of a reinforcement such as fiberglass. Polymers without reinforcement have a substantially greater shrinkage than those with reinforcement. In general, crystalline polymers tend to have higher characteristic shrinkages than do amorphous polymers. To a first approximation, shrinkage is isotropic, although nonuniform part thicknesses can result in substantial orientation effects on shrinkage.

Example Problem 7.3

Consider a straight tubular runner of length L. A melt following the power-law model is injected at constant pressure into the runner. The melt front progresses along the runner until it reaches the gate located at its end. Calculate the melt front position, $Z(t)$, and the instantaneous flow rate, $Q(t)$, as a function of time. Assume an incompressible fluid and an isothermal and fully developed flow, and make use of the pseudo-steady-state approximation. For a polymer melt with $K = 2.18 \times 10^4$ N·s^n/m^2 and $n = 0.39$, calculate $Z(t)$ and $Q(t)$ for an applied pressure $P_0 = 20.6$ MN/m^2 in a runner of dimensions $R = 2.54$ mm and $L = 25.4$ cm.

Answer: We begin with the relationship for flow rate of a power law fluid, as given by Eq. (7.38), and recognize that the length the fluid has traveled, L, is now replaced by the melt front position, $Z(t)$, such that

$$Q(t) = \frac{\pi R^3}{3 + \dfrac{1}{n}} \left[\frac{R P_0}{2KZ(t)} \right]^{1/n} \tag{E7.1}$$

The position expression, $Z(t)$, is obtained from a mass balance:

$$Z(t) = \frac{1}{\pi R^2} \int_0^t Q(t)\, dt \tag{E7.2}$$

Differentiation of Eq. (E7.2) with respect to time gives

$$\frac{dZ(t)}{dt} = \frac{Q(t)}{\pi R^2} \tag{E7.3}$$

Substitution of Eq. (E7.1) into (E7.3) gives

$$Z(t) = R\left[\frac{t(1+n)}{1+3n}\right]^{\frac{n}{n+1}}\left(\frac{P_0}{2K}\right)^{\frac{1}{n+1}} \tag{E7.4}$$

Substitution of Eq. (E7.4) into Eq. (E7.1) gives the flow rate expression:

$$Q(t) = \pi R^3\left(\frac{1+n}{1+3n}\right)^{n/(1+n)}\left(\frac{n}{1+n}\right)\left(\frac{P_0}{2K}\right)^{1/(1+n)}\left(\frac{1}{t^{1/(1+n)}}\right) \tag{E7.5}$$

Insertion of the appropriate values for R, n, and K gives the following simplified expressions:

$Z(t) = 0.188t^{0.281}$ and $Q(t) = 1.07 \times 10^{-6}t^{-0.719}$

Values for the flow front and flow rate at various times are listed in the table below. The results clearly indicate that we should expect a very high flow rate and quick runner filling initially, followed by a rapid drop in Q and long filling times for the remainder of the long runner. The first 50% of the runner is filled in 10% of the total runner time. This is the situation with a constant applied pressure and decreasing flow rate with time. In practice, the initial part of the mold filling cycle is one of increasing applied pressure and almost constant flow rate.

t (s)	*Z* (m)	*Q* (m³/s)
0	0	∞
0.5	0.155	1.76×10^{-6}
1	0.188	1.06×10^{-6}
1.5	0.211	8×10^{-7}
2	0.228	6.5×10^{-7}
2.88	0.253	5×10^{-7}

Compression Molding

Whereas the previous polymer processing techniques are used primarily for thermoplastic polymers, *compression molding*, also called *matched die molding*, is used almost exclusively for molding thermoset polymers. In compression molding, a charge of thermosetting resin is placed in the cavity of a matched mold. The mold is closed by bringing the male and female halves together, and pressure is applied to squeeze the resin so that it uniformly fills the mold cavity. While under pressure, the material is heated, causing it to flow, crosslink, and harden. When the material is hard, the mold is opened and the part is removed. As with injection molding, then, this is a cyclic, or batch, operation.

Need More Help?

Try this video on compression molding:

https://www.youtube.com/watch?v=KXxGqNzAry8

The advantages of compression molding over other molding operations are the simplicity of the molds, which makes them less expensive to manufacture and maintain, and the relatively small amount of waste produced. Disadvantages include relatively slow cycle times and geometrical limitations to the moldable parts. Both compression molding and injection molding are used to produce discrete parts of widely differing geometries, but the difference in resin type commonly used between the two processes dictates very different mold conditions. In compression molding, the mold is heated so that the thermoset polymer will cure. In traditional injection molding, the mold is cold so that the molten polymer will freeze (there are heated injection molding designs). Since curing usually takes longer than freezing, the molding cycle for compression molding is longer than for injection molding, typically 1–2 minutes versus 20–60 seconds, respectively. Another consequence of making thermoset polymer parts by compression molding is that reject parts cannot be reprocessed. To overcome some of these limitations, *transfer molding* was developed. Transfer molding is a hybrid between compression molding and injection molding, in which the charge is melted in a separate pot, which is part of the heated mold, and then transferred under pressure by a ram, through runners and gates, into the mold cavity. Transfer molding cycles tend to be shorter than compression molding cycles. We will describe transfer molding in more detail in the context of composites processing (Section 7.4.2), for which it is most widely utilized.

Figure 7.43 represents the various stages of the compression molding cycle from the point of view of the plunger force needed to close the mold at a constant rate. In the first region, the force increases rapidly up to the fill time, t_f, as the preform is squeezed and heated. At t_f, the polymer is in the molten state and is forced to flow and fill the cavity. Filling terminates at t_c, when compression of the polymer melt begins, to compensate for the volume contraction that occurs during polymerization. The bulk of the curing process takes place after t_c. To derive relationships for the flow and pressure characteristics, then, we simply need to consider the time period $t_f < t < t_c$, since prior to t_f the polymer is solid and after t_c it is curing.

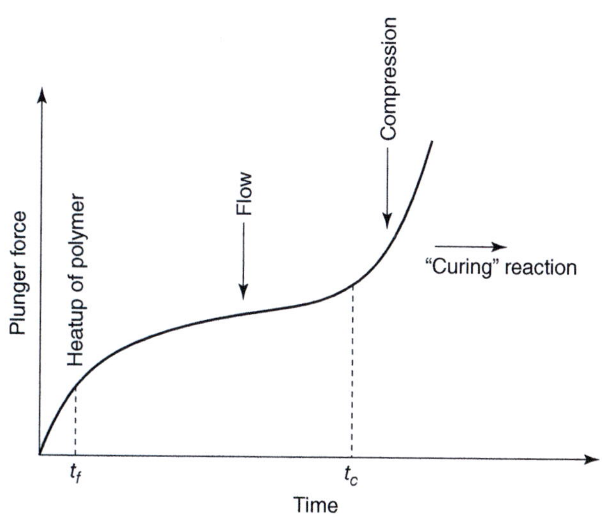

Figure 7.43 Schematic representation of the plunger force applied during compression molding. From Tadmor and Gogos [11]. Copyright © 1979 by John Wiley & Sons, Inc. This material is used by permission of John Wiley & Sons, Inc.

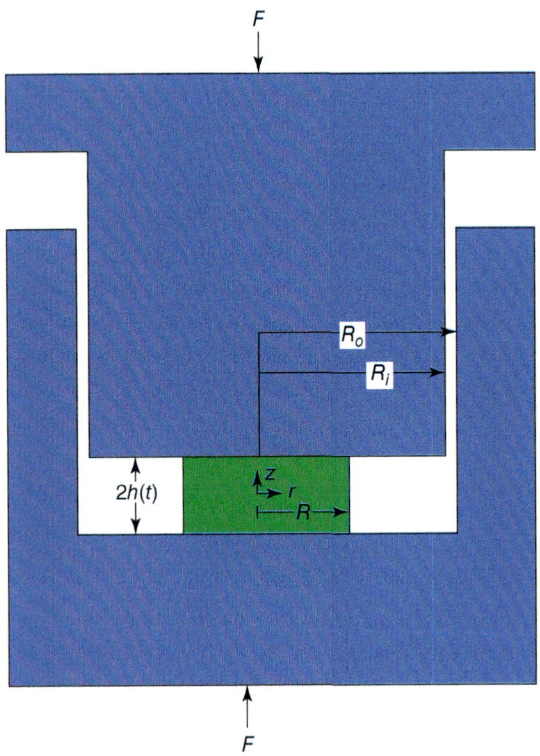

Figure 7.44 Model of compression molding process. From Tadmor and Gogos [11]. Copyright © 1979 by John Wiley & Sons, Inc. This material is used by permission of John Wiley & Sons, Inc.

The flow characteristics can be determined using the model shown in Figure 7.44. During the time interval under consideration, it is assumed that the polymer has been heated to a uniform temperature, T_w, that is equivalent to the mold wall temperature. As long as the preform radius, R, is less than the radius of the outer wall of the mold cavity, R_o, we can treat the problem as an isothermal radial flow of an incompressible power law fluid flowing between two disks that approach each other at a constant rate, $\dot{h}$. In this way, the velocity field, $v_r(z, r, t)$, the pressure distribution in the mold, $P(z, r, t)$, and the plunger force, $F_N(z, r, t)$ can be obtained as follows:

$$v_r = \frac{nh^{\frac{n+1}{n}}}{n+1}\left(\frac{-\partial P}{K\partial r}\right)^{1/n}\left[1-\left(\frac{z}{h}\right)^{\frac{n+1}{n}}\right] \tag{7.40}$$

$$P = P_a + \frac{-\dot{h}KR^{n+1}\left(\frac{2n+1}{n}\right)^n}{2^n(n+1)h^{2n+1}}\left[1-\left(\frac{r}{R}\right)^{n+1}\right] \tag{7.41}$$

$$F_N = \frac{K\pi\left(\frac{2n+1}{n}\right)^n\left(-\dot{h}\right)^n R^{n+3}}{2^n(n+3)h^{2n+1}} \tag{7.42}$$

where K and n are the power law parameters, and all other geometric variables are as defined in Figure 7.44. When the radius of the flowing preform reaches the mold wall (i.e., $R = R_o$), the fluid is forced to flow in the annular space $R_o - R_i$. Equation (7.40) through (7.42) can be modified to account for this flow. The pressure rises substantially during annular flow. The interested reader is referred to the text by Tadmor and Gogos for more details on the addition of annular flow to these equations.

After filling ($t > t_c$), curing takes place, and the analysis switches from one of modeling flow to modeling heat transfer and kinetics. More details on the reaction kinetics will be given when we describe RIM, but for now we simply model the curing process as the reaction of trifunctional A groups, A_3, with bifunctional B groups, B_2, as follows:

$$A_3 + B_2 \rightarrow \text{Crosslinked polymer} \tag{7.43}$$

The result is that the extent of reaction, p (cf. Section 3.3.1), can be related to the weight average molecular weight, $\overline{M}_w$, for a polyurethane system under equal stoichiometry:

$$\overline{M}_w = \frac{\frac{2}{3}(1 + p^2)M_{A_3}^2 + (1 + 2p^2)M_{B_2}^2 + 4pM_{A_3}M_{B_2}}{\left(\frac{2}{3}M_{A_3} + M_{B_2}\right)(1 - 2p^2)} \tag{7.44}$$

where M_{A_3} and M_{B_2} are the molecular weights of species A_3 and B_2, respectively. The *gel point* is defined to be the condition when $\overline{M}_w$ goes to infinity, which occurs at $p = \left(\frac{1}{2}\right)^{\frac{1}{2}}$. It turns out that the center of the curing polymer slab will gel faster than the skin. A knowledge of the temperature, conversion, and molecular weight distribution as a function of thickness and reaction time is essential in determining the required compression mold cycle or the time and temperature in the postcuring step.

Blow Molding

Blow molding is used to form hollow articles such as bottles. The process involves first forming and melting a preshaped sleeve called a *parison*, which is usually produced by extrusion. Hence, the term *extrusion blow molding* is sometimes used for this process. The parison is engaged between two mold halves, into which air is blown, causing the parison to take on the shape of the mold. The polymer quickly solidifies upon contacting the cold mold, and the finished hollow article is ejected. This procedure is schematically illustrated in Figure 7.45.

> **Need More Help?**
>
> Try this video on blow molding:
> https://www.youtube.com/watch?v=oY7hyuMHm68

Blow molding is not the only process from which polymeric hollow articles can be made, but it does offer distinct advantages over other methods such as injection molding and *rotational molding* that makes it ideal for medium-sized hollow articles—for example, parts ranging from a few milliliters to over 500 l in volume. Blow molding can create parts with much lower mold costs than injection molding and can create parts with narrow openings and wide bodies. Inflation pressures are quite modest, typically about 0.2 MN/m^2, so that the clamping forces on the mold are relatively small. The process is suitable for high production rates, and it is typically integrated with automatic polymer feeding systems and part removal. Cycles tend to be very short, so that equipment costs can be borne over many fabrication cycles. There are many variations to the blow molding process, primarily involving different methods of melt delivery and parison production. We will concentrate on continuous extrusion blow molding to illustrate the concepts of the blow molding process.

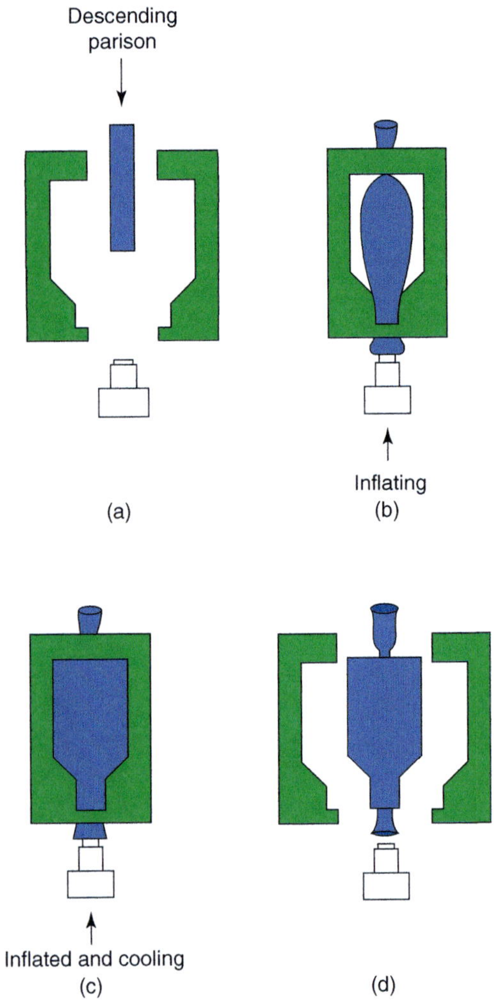

Figure 7.45 Schematic illustration of blow molding process. From Tadmor and Gogos [11]. Copyright © 1979 by John Wiley & Sons, Inc. This material is used by permission of John Wiley & Sons, Inc.

In extrusion blow molding, the parison is formed from an extrusion die. (This process is not to be confused with *blown film extrusion*, which is similar, but is used to form only films.) The molten material flows from the extruder through an adapter that changes the direction of the flow from horizontal to vertical, thus allowing gravity to act uniformly on the polymer melt. The material then enters the die and flows around a mandrel to create a cylindrical extrudate. The die may have a hole down the center of the mandrel for injecting the blown air, or the air is introduced through an inlet in the bottom of the mold. This procedure creates some interesting conflicts, namely, that the output from an extrusion operation is continuous, whereas the blow molding process is a cyclical one. There are methods for producing intermittent polymer flow, such as with the use of a reciprocating screw in injection molding, but the trend in extrusion blow molding is to keep the extrusion process a continuous one and match the cyclical molding process to the polymer flow rate. This type of blow molding is called *continuous extrusion blow molding*.

In continuous extrusion blow molding, the extruder is run continuously and its output matched through the use of multiple molds which seal and blow the parison. For example, if the mold cycle is no more than twice the time to create a parison, a two-mold system can be used. Typically, multiple

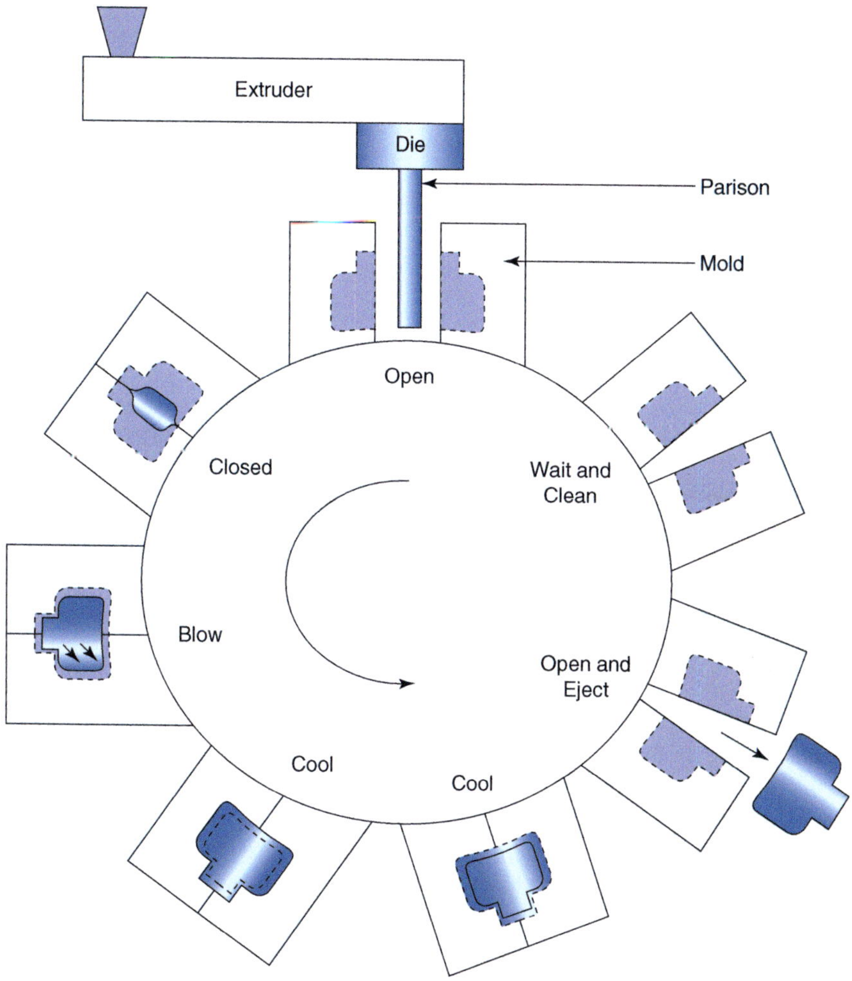

Figure 7.46 Schematic diagram of rotating mold system used in continuous extrusion blow molding. Strong [17]/Reproduced with permission of Pearson.

molds are required to match the extrusion rate of the parisons, so they are mounted on a rotating wheel, as illustrated in Figure 7.46. While one mold is closing to capture the parison, the mold ahead is in position for blowing of the part, and other molds are closed while the part is cooling. Still farther around the wheel the mold opens for part ejection and is prepared to capture another parison. In this system, the rotational speed of the wheel and the number of molds mounted on the wheel are matched to the extrusion rate of the parisons.

From both the product design and economic points of view, the shape and thickness of the parison must be carefully controlled. In general, parison thicknesses are not uniform in the axial direction due to flow rate variations and gravitational effects. These parameters are neither predictable nor controllable, so that empirical relationships must be found. One such method used to quantitatively analyze the parison formation process and relate it to fundamental rheological properties utilizes four parison properties: parison diameter, product weight, severity of melt fracture, and pleating. *Pleating* involves buckling of the parison under its own weight, and it is a function of

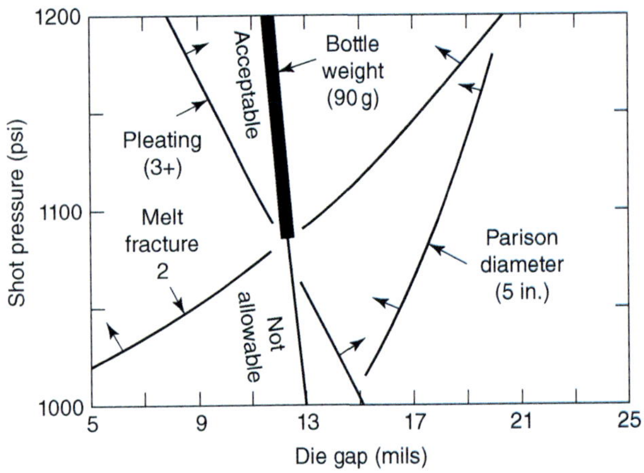

Figure 7.47 Operating diagram for the blow molding of HDPE. Arrows indicate regions of acceptable operation. From Tadmor and Gogos [11]. Copyright © 1979 by John Wiley & Sons, Inc. This material is used by permission of John Wiley & Sons, Inc.

the parison wall thickness and the angle of extrusion. Increased melt strength and higher extrusion rates alleviate pleating. In this method, response surfaces are experimentally obtained, and by specifying minimum acceptable levels for each property, operating lines can be generated. An example of this procedure is shown in Figure 7.47 for high-density polyethylene (HDPE). The heavy line represents the acceptable range of shot pressure and die gap values to produce the specified product. It should be noted that these results apply to HDPE only, and a new curve with new operating lines should be experimentally generated for each polymer under consideration.

Parison inflation is less difficult to model. In general, the parison is inflated very rapidly, and at a predetermined rate such that it does not burst while expanding. An approximate description of the blowing of a cylindrical parison of uniform radius R_i and thickness h_i to that of R_o and h_o can be obtained by assuming that the flow is planar extension, that the flow is isothermal, and that $h/R \ll 1$. In this instance, the parison inflation time, t, at constant inflation pressure, P, for a power-law polymer melt is

$$t = \frac{n}{2CP^{1/n}}\left[\left(\frac{1}{R_i}\right)^{2/n} - \left(\frac{1}{R_o}\right)^{2/n}\right] \tag{7.45}$$

where

$$C = \left(\frac{2\pi L}{VK}\right)^{1/n} \tag{7.46}$$

In Eqs. (7.45) and (7.46), K and n are the power-law parameters, L is the length of the parison, and V is the volume of the blown part, which is a function of both $R(t)$ and $h(t)$.

Parison cooling significantly impacts the cycle time only when the final parison thickness is large. In thin blown articles, the mold is opened when the pinched-off parts have solidified so that they can be easily stripped off. Thus, they are the rate-controlling element in the cooling process. For fast blow molding of even very thin articles, the crystallization rate must be fast. For this reason, HDPE, which crystallizes rapidly, is ideally suited for blow molding, as are amorphous polymers that do not crystallize at all.

Thermoforming

Thermoforming is a process used to shape polymeric sheets and films into discrete parts. (Although there is a distinction between sheets and films based upon thickness, we will use the generic term "sheet" in this section to reflect the forming of both.) As with the other techniques in this section, thermoforming is a cyclic process. It differs from the others, however, insofar as it is technically a secondary shaping operation; that is, it utilizes sheets that must be formed by some primary shaping method such as extrusion or calendering. Other than that, many of the same processing principles apply to thermoforming as they do to other melt processing techniques, even though the polymer in thermoforming is only softened, instead of being melted. Thermoforming is widely used for packing that must conform to the shape of the part it holds.

Need More Help?
Try these videos on polymer thermoforming:
https://www.youtube.com/watch?v=HWX_XxS4zY8 https://www.youtube.com/watch?v=LYfzl8eciG8

In the thermoforming process, the sheet is heated to slightly above T_m or T_g and is placed in a clamp frame that clamps it along the part perimeter. The deformable sheet is then forced to conform to the shape of a mold by means of vacuum, air or gas pressure, or a plunger. There are thus three distinct categories of thermoforming: *vacuum forming, pressure forming*, and *plug forming*. Once the object has been shaped, cooling occurs by conductive heat transfer to the mold wall(s). The main advantage of thermoforming is the low cost of the shaping equipment. The main disadvantage is that thermoforming is limited to simple shapes. Trimming operations are also often required after shaping, which increases cost and the amount of scrap. Cycle times can be long due to the slower heating and cooling rates for thick sheets.

Reaction Injection Molding (RIM)*

RIM is a process used to make polymeric articles by direct polymerization of two reactive monomeric species in a closed mold. It was developed to circumvent two major problems associated with injection molding: (a) the generation of sufficient, homogenized melt in the injection portion of the machine and (b) maintaining sufficient clamping pressure to keep the mold closed during the filling and packing stages of the injection molding cycle. Both issues become problematic as the size of the molded article increases. Thus, RIM is used to produced large, molded parts such as automotive bodies. In the RIM process, two (or more) low-viscosity liquid streams are mixed prior to injection into the mold.

Need More Help?
Try these videos on reaction injection molding:
https://www.youtube.com/watch?v=TWEfbGH-Ewc

In many cases, the components are a polyol, which contain one or more —OH groups, and a diisocyanate, which contains two —NCO groups. These react to form a polyurethane. Other materials are also suitable for RIM, such as epoxies and nylons. The choice of monomers and additives yields a polymer with properties very much tailored to the function of the product, from flexible elastomers to rigid foams. Cycle times are in the range of 2–4 minutes, provided that the polymerization reaction is sufficiently fast; otherwise, there is little advantage offered over injection molding. Foaming agents are often added to (a) ensure that the polymer completely fills the mold and (b) provide porous products with improved thermal insulating properties, for example.

The advantages of RIM, in addition to those cited above, are that the injection pressures are relatively small (1–10 MN/m^2), and homogenization of the polymer is readily accomplished. The primary disadvantages are that there are (a) nonisothermal and transient flow with chemical reaction during the filling stage and (b) conductive heat transfer with heat generation during the polymerization reaction, both of which are difficult to model. We address both issues with a model of the polymerization process.

Consider a long, thin mold being fed at constant temperature with two bifunctional monomers, AA and BB. The feed has a molecular weight of M_0, and the polymerization reaction, which is assumed to be reversible, proceeds by the reaction of A and B functional groups in an idealized step polymerization reaction (cf. Section 3.3.1):

$$\mathrm{AA} + \mathrm{BB} \overset{k_f, k_r}{\rightleftarrows} \mathrm{AA} - \mathrm{BB} \tag{7.47}$$

For simplicity, we assume that the small molecule (such as water) typical of these reactions is not formed. The reaction under consideration, then, Eq. (7.47), is second order in A or B functional groups and first order in A – B bonds. The reaction rate is given by the rate of disappearance of AA or BB, or the rate of formation of AA – BB. This analysis was performed previously in Section 3.3.1 for the case of the forward reaction only. If we include the reverse reaction and assume that the concentrations of the two reactants are equal at all times, [AA] = [BB], the rate expression becomes

$$\frac{d[\mathrm{AA}]}{dt} = -k_f[\mathrm{AA}]^2 + k_r\left([\mathrm{AA}]_0 - [\mathrm{AA}]\right) \tag{7.48}$$

where the substitution $[\mathrm{AA} - \mathrm{BB}] = [\mathrm{AA}]_0 - [\mathrm{AA}]$ has been employed. The forward and reverse reaction rate constants, k_f and k_r, follow the Arrhenius expression, Eq. (3.12).

Recall also from Section 3.3.1 that we introduced the extent of reaction, p, which is related to the degree of polymerization, $\overline{x}_n$. From Eq. (7.48), we can derive the following expressions for the number- and weight-average molecular weights in terms of the extent of reaction:

$$\overline{M}_n = \frac{M_0}{1-p} \tag{7.49}$$

and

$$\overline{M}_w = M_0\left(\frac{1+p}{1-p}\right) \tag{7.50}$$

To solve the filling flow and heat transfer problem with this reacting system, we need to specify the x-direction momentum and energy balances. The x-direction momentum equation during filling is

$$\rho\frac{\partial v_x}{dt} = \frac{-\partial P}{dx} + \eta\frac{\partial^2 v_x}{\partial y^2} + \frac{\partial \eta}{\partial y}\frac{\partial v_x}{\partial y} \tag{7.51}$$

where ρ is the density of the reacting fluid, P is the pressure, v_x is the velocity in the x direction, and η is the non-Newtonian viscosity, which can be described by a relationship such as the Ellis model (see Cooperative Learning Exercise 4.4) or the Carreau model:

$$\eta = \frac{\eta_0}{\left[1 + (\lambda\dot{\gamma})^2\right]^{(1-n)/2}} \tag{7.52}$$

where n is the power-law exponent, η_0 is the zero-shear rate viscosity, $\dot{\gamma}$ is the shear rate, and

$$\lambda = \frac{\lambda_0 \eta_0 \overline{M}_w^{0.75}}{\rho T} \tag{7.53}$$

where λ_0 is a curve-fitting parameter, T is temperature, and all other parameters are defined above. The zero-shear rate viscosity is a function of molecular weight, as well, and can incorporate variables to account for entanglements.

The energy equation for the filling stage is for constant density and thermal conductivity as follows:

$$\rho C_p \left(\frac{\partial T}{\partial t} + v_x \frac{\partial T}{\partial x} \right) = k \frac{\partial^2 T}{\partial y^2} + \eta \left(\frac{\partial v_x}{\partial y} \right)^2 - \Delta H \left(\frac{\partial p}{dt} \right) \tag{7.54}$$

where ρ is the fluid density, C_p is its heat capacity, k is its thermal conductivity, η is its viscosity, T is its temperature, t is time, v_x is the velocity in the x direction, ΔH is the heat of the polymerization reaction [Eq. (7.54)], and p is the extent of the polymerization reaction.

Equations (7.53) and (7.54), along with the rate expression, Eq. (7.48), can be used to analyze the conditions during the filling and polymerization portions of the RIM process. This is normally done by a *finite element analysis*, and not through analytical solutions to the equations. We will not present the details of this analysis here, but rather show some representative results. Figure 7.48 shows some results of the finite element simulation for the filling and polymerization reaction conversion in a typical RIM process with stepwise polymerization. Both temperature and conversion increase with increasing flow direction, which is the result of larger residence times. Drastically different conversion and fill profiles are obtained when the fill time and forward reaction rate constant are changed.

At the end of the filling stage, only heat transfer with chemical reaction occurs, which can be described by the appropriate analogs of the species and energy balance equations, Eqs. (7.48) and (7.54), respectively. We will not present the solution to these equations here, but simply comment that the postfill cure process continues until the thickness-average tensile modulus is high enough at every position along the x direction that the part can be removed.

7.3.3 3D Printing with Polymer Filaments

Need More Help?
Try this video on 3D printing with polymers: https://www.youtube.com/watch?v=tC_z6jznFPY

3D printing was introduced in Section 7.0.2 as an emergent AM method for producing parts with complex shapes (Figure 7.49). The list of materials available for 3D printing operations continues to grow and now includes cement, carbon, and even food-grade materials. Polymers were some of the first materials to be 3D printed, however, and today still constitute the largest category of materials employed in AM. The 3D printing of polymers is also the most mature technologically and enjoys widest availability due to simplicity of the printers and relatively low cost of the polymers. We will focus on the 3D printing of polymers here as an entry point into the emergent domain of AM of engineered materials.

A 3D polymer printer is a type of *fused layer modeling* or FDM device in which a filament is extruded through a heated dye that melts the polymer and places it on a supported substrate. For this reason, the term FFF is often used. The substrate sits atop a movable platform (Figure 7.50). The movement pattern for the platform comes from a computer-generated CAD file (or *STL file*

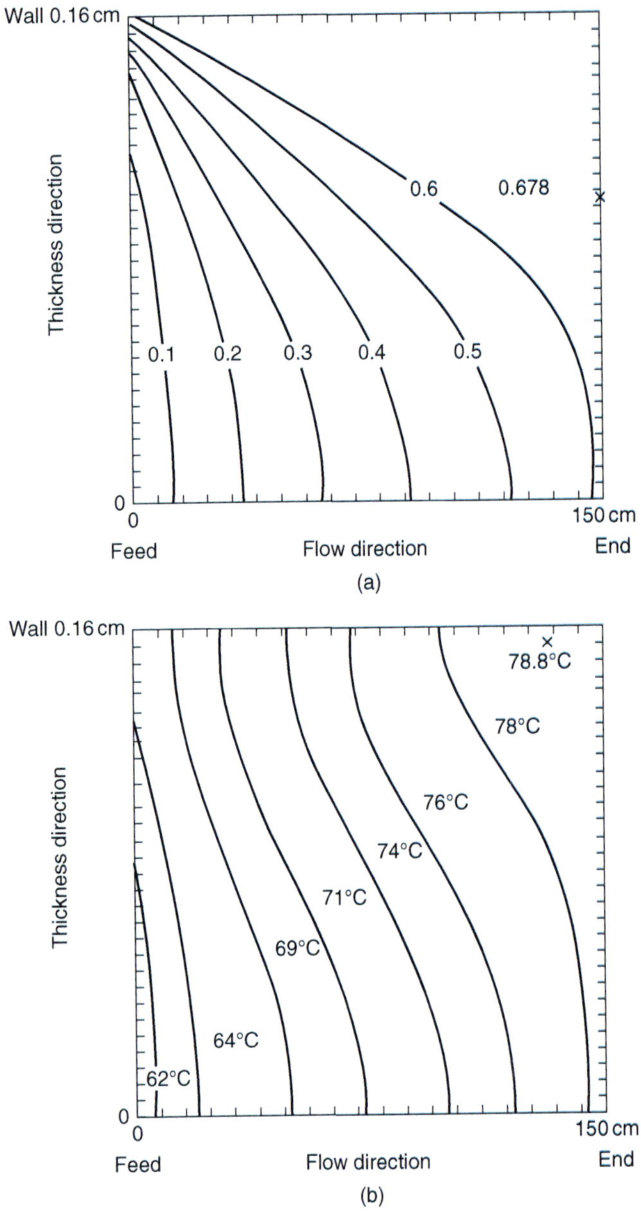

Figure 7.48 Simulation results of the RIM process for $T_0 = T_w = 60°C$, $k_f = 0.51$/mol s, and $t_{fill} = 2.4$ s. Contours in (a) are for conversion; those in (b) are for temperature. From Tadmor and Gogos [11]. Copyright © 1979 by John Wiley & Sons, Inc. This material is used by permission of John Wiley & Sons, Inc.

short for SL) which selectively places the polymer (or not) at the appropriate location in the layer (some devices use a fixed platform and moveable print head). Multiple layers are generated; once the piece solidifies it can be removed from the substrate. The simplest printer then consists of extruder rollers, a heated print head, and a moveable stage to print on. The filament is fed from a coil and through the extruder rollers. In this way, the modeling and analysis of printing with polymer filaments are much like that of die forming and capillary flow, both of which were described

Figure 7.49 Fused deposition modeling (FDM) of polymers into complex shapes from a computer-generated drawing, also known as 3D printing. MIke/Adobe Stock.

in Section 7.3.1. Shear thinning can occur through the die such that the non-Newtonian fluid dynamics of Sections 4.1.2 and 4.1.3 apply. Flow models that allow prediction of printer speed and spatial distribution are now beginning to emerge. One such model is shown in Figure 7.51. Through simultaneous solution of the equations of continuity, moment, and energy, important processing parameters can be estimated, such as the depth of the melt film in the heated extruder head, δ:

$$\delta = \frac{k_m(T_h - T_m)}{\rho_s(-U_{sz})[\lambda + C_s(T_m - T_0)]} \quad (7.55)$$

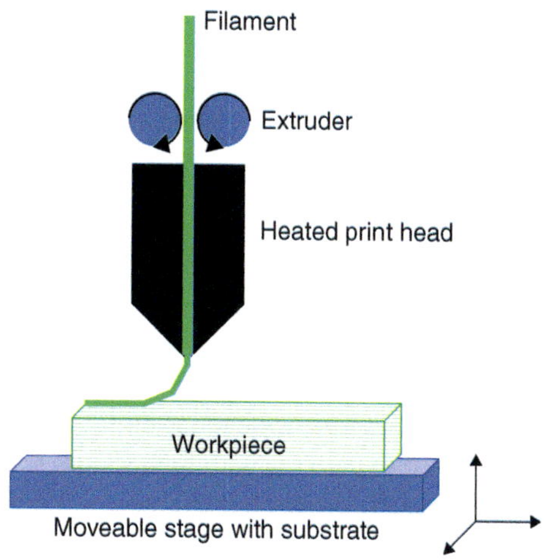

Figure 7.50 Schematic diagram of a 3D printer.

where k_m is the thermal conductivity of the molten polymer, C_s is the specific heat of the solid polymer, ρ_s is the density of the solid polymer, λ is the heat of fusion for the polymer, U_{sz} is the polymer filament velocity, and T_h, T_m, and T_0 are the temperatures at various points in the extruder head as defined in Figure 7.51.

As illustrated by Eq. 7.55, the polymer physical properties play an important role in determining how the 3D printer functions. The polymers available in filament form for 3D printing are numerous with new formulations and compounding agents being added at a rapid pace. Polymers such as poly(lactic acid) (PLA), polyethylene terephthalate glycol (PETG), acrylonitrile butadiene styrene (ABS), and acrylonitrile styrene acrylate (ASA) form the basis for many 3D printing filaments. The relative merits of each polymer from a 3D printing standpoint are beyond the scope of this text but there are many resources online describing the practical aspects of filament selection and the resulting component properties.

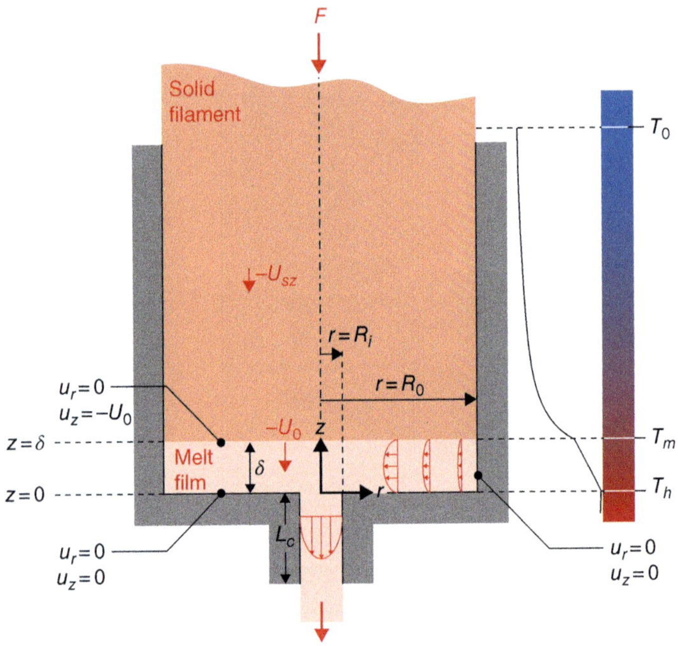

Figure 7.51 Melt flow model for melting of a polymer filament in fused filament fabrication. Osswald et al. [18]/With permission of Elsevier.

Although 3D printing with polymer filaments provides exciting opportunities to produce components with complex shapes and prototypes with unique geometries, the technological limitations of using thermoplastic polymers should be clear. As a result, this field is rapidly evolving to include filaments of metallic, ceramic, composite, and biological materials as are the methods used to deposit them. Devices that utilize more than one class of material are called *multi-material fabricators*. Multi-material fabricators increasingly utilize solution-based stock materials and deposit them in a drop-wise fashion much like an inkjet printer. This process is known as material jetting (MJ) and allows for a wide range of nanoscale particulates suspending in a solvent called *nanoinks* to be printed in patterns with high spatial resolution. The science and engineering of these materials and their 3D processing will be an important part of your future.

7.4 Processing of Composites

In this section, we see how some of the processing methods described in previous sections can be modified to produce composite materials, and how new techniques are being developed to create composites with unique structures and properties. We will concentrate almost exclusively on fiber-reinforced matrix composites, especially those composites based on glass-fiber-reinforced polymers. There are three common methods for forming glass-fiber-reinforced polymer matrix composites: pultrusion, RTM, and filament winding. As with ceramic processing, there are several important vapor-phase-based synthetic methods for forming nonpolymeric matrix composites. We will not cover these topics here, nor the techniques of wet layup (also known as hand layup), autoclave molding, and mandrel wrapping.

7.4.1 Pultrusion

Pultrusion is derived from the terms "pulling" and "extrusion" and reflects a process that involves the pulling of continuous reinforcement fibers through a molten polymer (sometimes called a *resin*), through a die, and into a curing chamber. A schematic illustration of one simple variation on the pultrusion process is shown in Figure 7.52. Unlike extrusion, in which the material is pushed through a die, the material in pultrusion is pulled through the die. This is possible due to the continuous nature of the reinforcing fibers.

Need More Help?
Try these videos on pultrusion: https://www.youtube.com/watch?v=1sH9rIGWNvc

Pultrusion is a continuous process that is ideal for high throughput of constant-cross-section products. As long as the fiber volume passing through the die is constant, excess resin is squeezed out and returned to the resin bath. There is very little waste of the fiber reinforcement, and virtually any length of product is possible. Both solid and hollow cross-sectional profiles are possible, and inserts made of various materials can be encapsulated. Depending on the part complexity, resin viscosity, and cure schedule, the pultrusion line speed may range from 0.05 to 5 m/min. Unlike other molding processes, no external pressure is applied in the pultrusion process, other than the pulling force on the fiber tows.

The major reinforcing phase is typically continuous glass roving, which is a bundle composed of 1000 or more individual filaments. The most common glass roving used in pultrusion is E-glass, although A-glass and graphite filaments are also common (cf. Table 1.32). The total fiber content in a pultruded member is typically very high, over 70% by weight in some cases. In nearly all cases, the fibers are oriented along the axis of the pultrusion member, although mats or woven rings can also be incorporated to improve transverse strength and modulus.

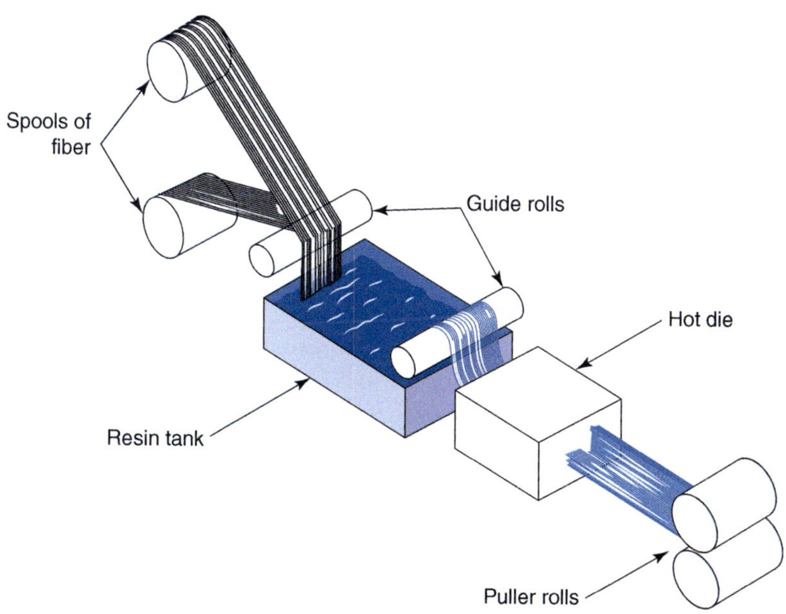

Figure 7.52 Schematic diagram of a typical pultrusion process. Schwartz [19]/Reproduced with permission of McGraw-Hill.

In most applications, polyester and vinyl ester resins are used as matrix materials. Epoxies are also used, although they require longer cure times and do not release easily from the pultrusion dies. Hence, thermosetting resins are often used with pultrusion, although some high-performance thermoplastics such as PEEK and polysulfone can also be accommodated. In addition to the resin, the resin bath may contain a curing agent (initiator, cf. Section 3.3.1), colorants, ultraviolet stabilizer, and fire retardant.

The viscosity of the liquid resin, residence time of the fibers in the resin bath, and mechanical action of the fibers in the resin bath are adjusted to ensure complete wetting of the fibers by the resin, called *wet-out*. The fiber–resin stream is pulled first through a series of preformers and then through a long, preheated die. The preformers distribute the fiber bundles evenly, squeeze out excess resin, and bring the material to its final configuration. Final shaping, compaction, and curing take place in the die, which has a gradually tapering section along its length. The entrance section of the die is usually water-cooled to prevent premature gelling, and the rest of the die is heated in a controlled manner by either oil or electric heaters. Infrared and microwave heating can also be employed to speed the curing process.

The most important factor controlling the mechanical performance of the pultruded members is the fiber wet-out. For a given resin viscosity, the degree of wet-out is improved as the residence time is prolonged by using slower line speeds or longer baths, the resin bath temperature is increased, and/or the degree of mechanical working is increased. The resin viscosity of commercial pultrusion lines may range from 400 to 5000 centipoise. Higher resin viscosities may result in poor fiber wet-out and slower line speed. On the other hand, very low resin viscosities may cause excessive resin draining from the fiber–resin stream after it leaves the resin bath.

Resin penetration takes place through capillary action as well as lateral squeezing between the bundles. The fiber and resin surface energies (see Appendix 3) are also important parameters in improving the amount of resin coating on fiber rovings. Thus, Kevlar 49 fibers, by virtue of their high surface energies, pick up more resin in the resin bath than either E-glass or carbon fibers under similar conditions.

As the fiber resin stream enters the heated die, the resin viscosity first decreases, which aids in the continued wet-out of uncoated fibers. Curing reactions commence a short distance from the die entrance, and soon the resin viscosity rapidly increases, as illustrated in Figure 7.53. The curing reaction continues at an increasing rate as the fiber–resin stream moves toward the die exit. Heat generated by the exothermic curing reaction raises the temperature in the fiber–resin stream. The location of the exothermic peak depends on the pulling speed, as illustrated in Figure 7.54. As the curing reactions finish, the exotherm temperature decreases and cooling begins. The rate of heat transfer from the cured material into the die walls is increased owing to a lower die temperature near the exit zone.

7.4.2 Resin Transfer Molding

Resin Transfer Molding (RTM) is a hybrid of two polymer processing techniques we have already described: transfer molding and RIM. The key difference is that a reinforcement is incorporated during molding to create a composite.

Need More Help?
Try these videos on resin transfer molding: https://www.youtube.com/watch?v=ElN_dR8WTLU

Several layers of dry, continuous-strand mat, woven roving, or cloth called *preform* are placed in the bottom half of a two-part mold, either by a hand-layup or by an automated technique (see Figure 7.55). The

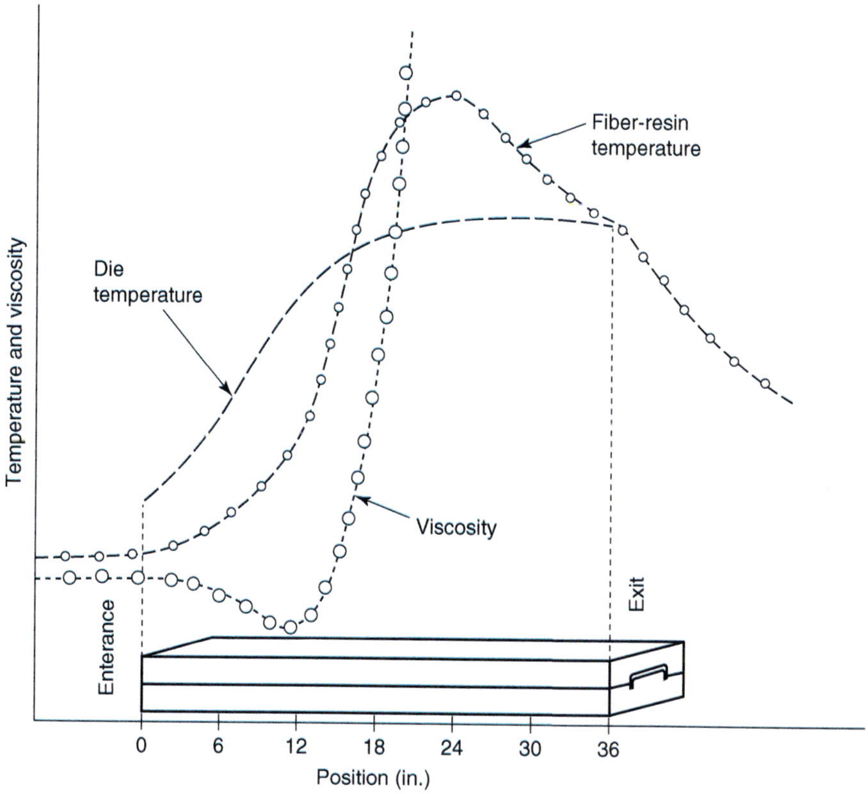

Figure 7.53 Viscosity change of a thermosetting resin in a pultrusion die. Mallick [20]/Reproduced with permission of Taylor & Francis.

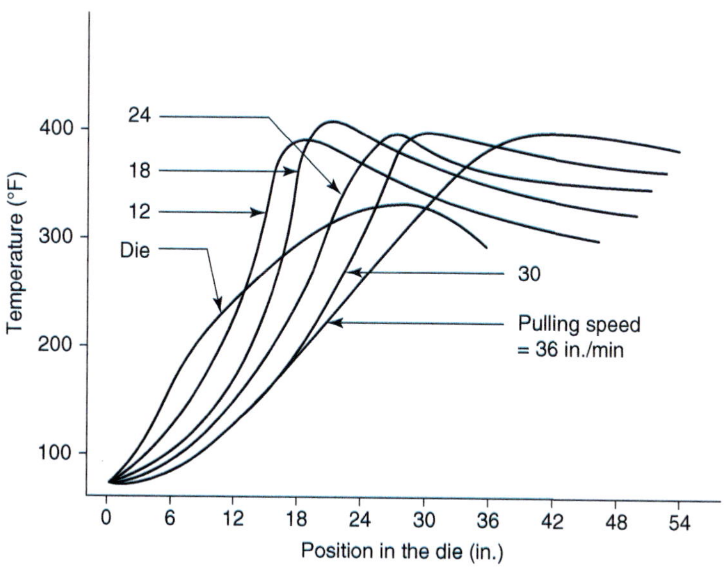

Figure 7.54 Temperature distribution along the length of a pultrusion die. Mallick [21]/Reproduced with permission of Taylor & Francis.

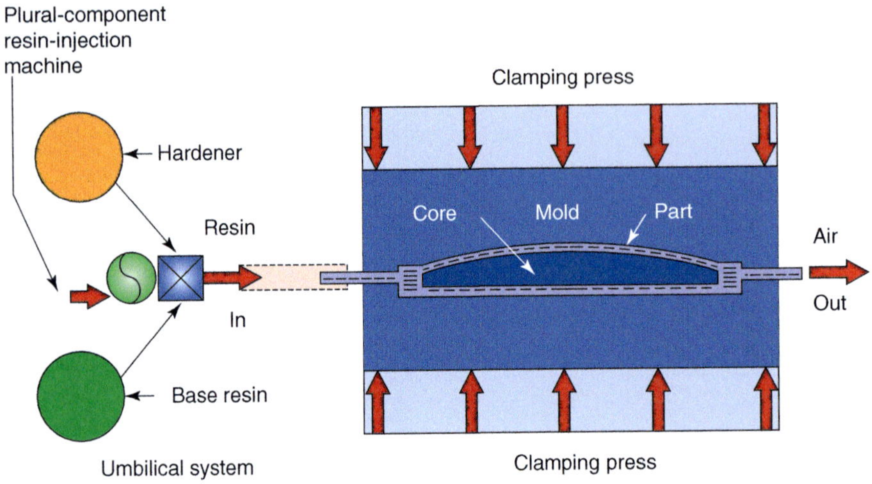

Figure 7.55 Schematic diagram of a typical resin transfer molding (RTM) process. Schwartz [22]/ Reproduced with permission of McGraw-Hill.

mold is closed, and a catalyzed liquid resin is injected into the mold via a centrally located sprue. The resin injection point is usually the lowest point of the mold cavity. The injection pressure is in the range of 70–700 kPa. Care must be taken to inject slowly enough that the fibers in the preform are not moved significantly as the resin fills the mold. As the resin spreads throughout the mold, it displaces the entrapped air through the air vents and impregnates the fibers. The mold can be held under vacuum to assist in air removal and allow inward resin flow. Alternatively, the edges of the mold are vented to allow air to escape as the resin enters. The local speed of the resin will be determined by the local permeability of the fiber material, the viscosity of the resin, and the local pressure decay. The permeability of the fiber material is low if its density is high, so a high-fiber volume fraction can only be realized at the cost of low production rate. By summation, these factors determine the total filling time. When excess resin begins to flow from the vent areas of the mold, the resin flow is stopped and the molded part begins to cure. When cure is completed, which can take from several minutes to several hours, it is removed from the mold and the process can begin again. As with other molding techniques, it is often necessary to trim the part at the outer edges once it is removed from the mold.

The reinforcements amenable to RTM are similar to those used for pultrusion, except that they need not be continuous in nature. Thus, E-glass, S-glass, aramid, and carbon fibers are commonly used, as are discontinuous filaments such as wood fiber and polyesters. Even metal and ceramic fibers can be used in this technique. In one method, the preform is fabricated by spraying 12- to 75-mm-long chopped fiber rovings onto a preshaped screen. A binder sprayed with the fibers keeps them in place and holds the preform shape, which is then placed in the mold.

Although polyurethanes or other reactive resins can be used in RTM, the most common resins are polyesters and epoxies. The two pumping reservoirs in Figure 7.55, then, contain polyester resin and initiator, or epoxy resin and hardener, respectively. Epoxies require relatively long cycle times. When cycle time is critical, low viscosity vinyl ester, acrylamate, or urethanes can be injected very rapidly into the mold. Even the use of thermoplastics is possible.

Compared with compression molding, RTM has very low tooling costs and simple mold clamping requirements. Another advantage of RTM is its ability to encapsulate metal ribs, stiffeners, and inserts within the molded product. Size capability is among the major benefits of RTM, and although RTM is well-suited to molding large objects, part complexity may restrict the use of the process.

A related process is called *structural reaction injection molding* (SRIM). Preform and mold preparation are similar to the RTM process, with changes in mold release and reinforcement sizing made to optimize their chemical characteristics for the SRIM chemistry. Once the mold has been closed, the SRIM resin is rapidly introduced into the mold and reacts quickly to cure fully within a few seconds. Therefore, wet-out and air displacement occur rapidly. No postcure is normally needed. RTM resins are typically two-component low-viscosity liquids in the range of 100–1000 cpoise, require preinjection mixing ratios in the range of 100:1 and can be mixed at low pressure using a static mixer. SRIM resins, on the other hand, are typically two-part, low-viscosity liquids being in the viscosity range of 10–100 cpoise at room temperature. They are highly reactive in comparison to RTM resins and require very fast, high-pressure impingement mixing to achieve thorough mixing before entering the mold. Mix ratios of typical systems are near 1:1, which is desirable for rapid impingement mixing. Preforms for RTM and SRIM are similar in most respects.

7.4.3 Filament Winding

In the *filament winding* process, a band of continuous resin-impregnated rovings or monofilaments is wrapped around a rotating mandrel and cured to produce axisymmetric hollow parts. Among the applications of filament winding are automotive drive shafts, helicopter blades, oxygen tanks, pipes, spherical pressure vessels, and large underground storage tanks. Figure 7.56 shows a schematic diagram of a basic filament winding process. A large number of fiber rovings are pulled from a series of rolls called *creels* into a liquid resin bath containing liquid resin, catalyst, colorants, and UV stabilizers, just as in the pultrusion process. Fiber tension is controlled by fiber guides or scissor bars located between each creel and the resin bath. Just before entering the resin bath, the rovings are usually gathered into a band by passing them through a textile thread board or a stainless steel comb.

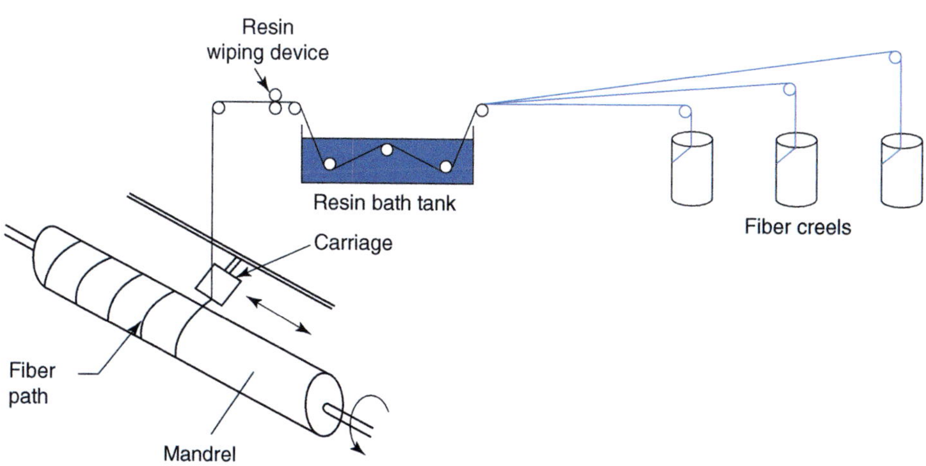

Figure 7.56 Schematic diagram of the filament winding process. Mallick [23]/Reproduced with permission of Taylor & Francis.

> **Need More Help?**
>
> Try these videos on filament winding:
>
> https://www.youtube.com/watch?v=ign6W5ENJAA
> https://www.youtube.com/watch?v=jc52ssQ65cU

At the end of the resin tank, the resin-impregnated rovings are pulled through a wiping device that removes the excess resin from the rovings and controls the resin coating thickness around each roving. Once the rovings have been impregnated and wiped, they are gathered in a flat band and positioned on the mandrel. The band former is usually located on a carriage, which traverses back and forth parallel to the mandrel, like a tool stock in a lathe machine. The traversing speed of the carriage and the winding speeds of the mandrel are controlled to create the desired winding angle patterns. Typical winding speeds are on the order of 100 linear m/min. After winding a number of layers to generate the desired reinforcement thickness, the filament-wound part is generally cured on the mandrel. The mandrel is then extracted from the cured part.

There are several material options available for filament winding. Typical fibers are fiberglass, carbon, and aramid. Fiberglass is still the most common reinforcing fiber, and the choices are like those found in the other composite forming techniques, with E-glass and S-glass rovings being the most common.

The typical resin systems include thermoset polyesters, vinyl esters, epoxies, polyimides, bismaleimide, and phenolics. Thermoplastics are also finding their way into filament winding. Wet thermoset filament winding requires a resin with viscosity in the range of 1000–3000 cpoise. Resin components are chosen based on pot life, winding temperature, viscosity, gel time, and cure time.

Filament winding patterns are designated in one of two primary ways: helical and polar. In *helical winding*, fiber is fed from a horizontal translating delivery head to a rotating mandrel. The angle of the roving band with respect to the mandrel axis is called the *wind angle* or *wrap angle*, as illustrated in Figure 7.57. The wind angle is specified by the ratio of the two relative motions. For a circular mandrel of radius r, rotating with a constant rotational speed of N revolutions per minute and a constant carriage feed of V, the wind angle, θ, is given by

$$\theta = \frac{2\pi Nr}{V} \tag{7.56}$$

Wind angles vary from 20° to 90°, with a 90° wind angle called a *hoop* or *circumferential winding* pattern. In *polar winding*, a delivery unit races around a slowly indexing mandrel, as illustrated in Figure 7.58. Technically speaking, polar winding is a helical winding pattern with a wind angle near

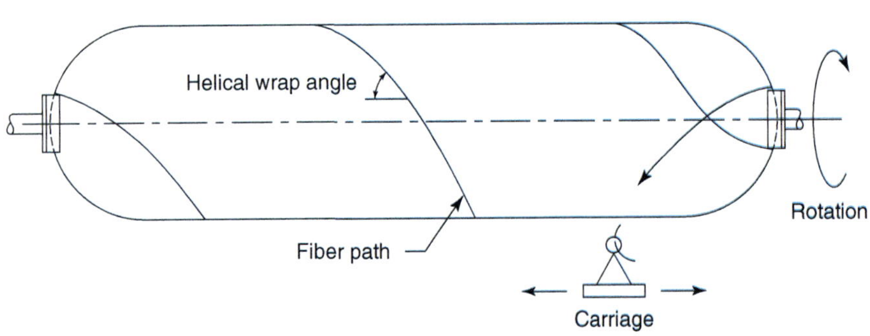

Figure 7.57 A helical winding pattern. Mallick [24]/Reproduced with permission of Taylor & Francis.

0°, typically between 5° and 15°. The principal advantage of polar winding is that it is a simple and rapid winding technique for short geometries with length-to-diameter ratios less than two. The winding pattern is a function of the desired winding angle, the polar port openings, the cylinder length, and the dwell time as the fiber direction is reversed at the end of the dome. The pattern can contain any number of circuits but usually ranges from 5 to 15. The number of patterns to complete a layer requires adjustment; otherwise, a gap or overlap will occur between the first and last pattern. The mechanical properties of the filament wound composite are a complex combination of (a) winding pattern and thickness and (b) resin and filament properties. The mechanical properties of the dome regions in particular require specialized treatment.

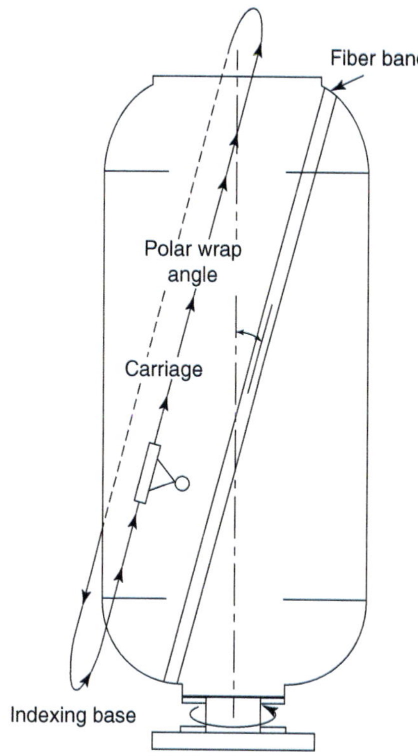

Figure 7.58 A polar winding pattern. Mallick [24]/Reproduced with permission of Taylor & Francis.

7.5 Processing of Biologics

The processing of biologics on a scale comparable to that of the previous material classes is as yet unknown, at least with regard to biomaterials for use in the human body. Most artificial biologics, both hard and soft, serve as scaffolds for the *in situ* manufacture of tissues; that is, the human body is actually the processing center in which the raw materials it provides are used to fabricate some portion of the component (e.g., bone or skin). Even when the component is fabricated almost entirely out of artificial materials, such as in an artificial hip, the device fabrication process is performed almost entirely by hand to meet case-dependent specifications. Furthermore, most of the artificial materials have already been processed into useful shapes using any of the previously described processing techniques, such as casting a titanium rod or extrusion of a polymer sheet.

Despite these observations, there are some common processing techniques that are emerging in the processing of biologics, as self-contained biomedical devices become more widely utilized. We examine two such areas in this section: collagen processing and surface modification.

7.5.1 Processing of Collagen

As described in Section 1.5.3, collagen is one of the most useful and abundant substances in the human body. Both skin and tendon are composed of collagen, and in its purified form, collagen can be used for a wide variety of tissue repair applications. Let us briefly examine the processing of collagen.

There are two distinct ways in which collagen can be isolated and purified. One is a molecular-based technology, and the other is fibrillar technology. The isolation and purification of soluble collagen molecules from a collagenous tissue are achieved by using a proteolytic enzyme such as pepsin to cleave the natural crosslinking sites of collagen called *telopeptides*. This process renders the collagen soluble in aqueous solution. The pepsin-solubilized collagen can be purified by repetitive precipitation with a neutral salt. The collagen molecules may be reconstituted into fibrils of various polymorphisms. However, the reconstitution of the pepsin-solubilized collagen into fibrils of native molecular packing is not as efficient as the intact molecules.

In the second method, collagen fibers are isolated and purified upon the removal of noncollagenous materials from the collagenous tissue. Salt extraction removes the newly synthesized collagen molecules that have not been covalently incorporated into the collagen fibrils. Salt also removes the noncollagenous materials that are soluble in aqueous conditions and are bound to collagen fibrils by nonspecific interactions. Lipids are removed by low-molecular-weight organic solvents such as ethers and alcohols. Acid extraction facilitates the removal of acidic proteins and glycosaminoglycans due to weakening of the interactions between the acidic proteins and collagen fibrils. Alkaline extraction weakens the interaction between the basic proteins and collagen fibrils and thus facilitates the removal of basic proteins. Purified collagen fibers can be obtained through these sequential extractions and enzymatic digestions from the collagen-rich tissues.

The purified collagen obtained from either of these techniques is subjected to additional processing to fabricate the materials into useful devices for specific medical applications. Some of these matrices and their medical applications are shown in Table 7.14 and will be briefly described.

Collagen membranes can be produced by drying a collagen solution or a fibrillar collagen dispersion cast on a nonadhesive surface. The thickness of the membrane is dictated by the concentration and the initial thickness of the casting. In general, membrane thicknesses of up to 0.5 mm can be obtained in this way. Additional chemical crosslinking is required to stabilize the membrane from dissolution. The casting process does not permit manipulation of the pore structure, and the cast membrane is dense and amorphous with minimal permeability to macromolecules.

Porous membranes may be obtained by freeze-drying a casting or by partially compressing a preformed matrix to a predetermined density and pore structure. The freeze-dried porous matrix also requires chemical crosslinking to stabilize the structure. A convenient way to do this is to use a vapor of a volatile crosslinking agent such as formaldehyde or glutaraldehyde. The pore structure depends on the concentration of the collagen in the casting. Other factors that contribute to the pore structure include the rate of freezing, the size of fibers in the dispersion, and the presence or absence of other macromolecules. Apparent densities from 0.05 to 0.3 g/cm^3 can be obtained. The pores in these matrices range in size from 50 to 15,000 μm in diameter.

Table 7.14 Some Examples of Collagen Matrices and Their Medical Applications

Matrix Form	Medical Application
Membrane (sheet, film)	Oral tissue repair; wound dressings; dura repair; patches
Porous (sponge, felt)	Hemostats; wound dressings; cartilage repair; soft tissue augmentation
Gel	Drug delivery; soft- and hard-tissue augmentation
Solution	Drug delivery; soft- and hard-tissue augmentation
Filament	Tendon and ligament repair; sutures
Tubular	Nerve repair; vascular repair
Collagen/synthetic polymer	Vascular repair; skin repair; wound dressings
Collagen/biological polymer	Soft-tissue augmentation; skin repair
Collagen/ceramic	Hard-tissue repair

Source: Bronzino [25]. Copyright © 2000 by CRC Press.

A *gel matrix* is defined as a homogeneous phase between a liquid and a solid. As such, it may vary from a simple viscous liquid to a highly concentrated slurry. Collagen gels may be formed by shifting the pH of a dispersion away from its isoelectric point. Alternatively, the collagen material may be subjected to a chemical modification to change its charge profile to a net positively charged or net negatively charged protein before hydrating the material.

A *collagen solution* is obtained by dissolving the collagen molecules in an aqueous solution. These solutions are prepared as previously described using pepsin. The solubility of collagen depends on the pH, temperature, ionic strength, and molecular weight. Collagen molecules aggregate into fibrils when the temperature of the solution increases to body temperature. Collagen is more soluble at a pH away from the isoelectric point of the protein and is less soluble at higher ionic strength of the solution. The solubility of collagen decreases with increasing size of molecular aggregates. Thus, collagen becomes increasingly less soluble with increasing crosslink density.

Collagen filaments can be produced by extrusion techniques, in which a collagen solution or dispersion in the concentration range 0.5–1.5% weight to volume is first prepared. Collagen is extruded into a coacervation bath containing a high concentration of a salt or into an aqueous solution at a pH of the isoelectric point of the collagen. Tensile strengths of 30 MPa have been obtained for the reconstituted filaments. *Tubular matrices* may be formed by extrusion through a coaxial cylinder or by coating collagen onto a mandrel. Different properties of the tubular membranes can be obtained by controlling the drying properties.

Finally, collagen can form a variety of *collagen composites* with other water-soluble materials. Ions, peptides, proteins, and polysaccharides can all be uniformly incorporated into a collagen matrix. The methods of composite formation include ionic and covalent bonding, entrapment, entanglement, and coprecipitation. A two-phase composite can be formed between collagen, ceramics, and synthetic polymers for specific biomedical applications.

7.5.2 Biologic Surface Modification

The increasing demand for synthetic biomaterials, especially polymers, is mainly due to their availability in a wide variety of chemical compositions and physical properties, their ease of fabrication into complex shapes and structures, and their easily tailored surface chemistries. Although the physical and mechanical performance of most synthetic biomaterials can meet or even exceed that of natural tissue (see Table 5.13), they are often rejected by a number of adverse effects, including the promotion of thrombosis, inflammation, and infection. As described in Section 5.5, biocompatibility is believed to be strongly influenced, if not dictated, by a layer of host proteins and cells spontaneously adsorbed to the surfaces upon their implantation. Thus, surface properties of biomaterials, such as chemistry, wettability, domain structure, and morphology, play an important role in the success of their applications.

Several methodologies have been developed for modifying the surfaces of polymeric materials to manipulate their interactions with biological environments, as summarized in Table 7.15. Biomaterial surfaces can be simply modified by physical deposition or adsorption of amphiphilic molecules, polymers, or proteins. However, such physical coatings are normally unstable. To overcome this limitation, chemical modifications to the surface through liquid- or gas-phase derivatization are usually favored. For example, strong oxidizing acids have been used to introduce carbonyl and carboxylic acid groups to polyolefinic surfaces. The treatment of polyolefins with ozone yields surfaces with hydroxyl, carbonyl, and carboxylic acid groups. Flame and thermal treatments induce radicals and chains scission on polymer surfaces. Also, ion beam treatments are utilized to induce

Table 7.15 Some Surface Modification Methods for Biomaterials

Physical Modification Methods	Chemical Modification Methods	Radiation Modification Methods
Physical adsorption	Oxidation by strong acids	Glow-discharge plasma
Langmuir–Blodgett film	Ozone treatment	Corona discharge
	Chemisorption	UV photoactivation
	Flame treatment	Laser
	Ion beam	Electron beam
		Gamma irradiation

Source: Sheu et al. [26]. Copyright © 1995 by Marcel Dekker, Inc.

chemical and morphological changes of the surface and can be used to incorporate various elements into the polymer surface. Laser treatments, which offer a combination of thermal and ultraviolet effects, have been used to induce chemical and morphological changes, as well. These surface modification methods can also be used to induce further surface functionalization via grafting methods. Radicals on the surfaces, once created, can act as surface-bound initiators for grafting reactions to monomers. For example, acrylic acid has been grafted to PET by radiation or plasmas and has been grafted to polyethylene by electron beam, radiation, or plasma treatment.

Let us briefly examine plasma treatment in more detail as an example of how the surface of these biologics can be modified to improve biocompatibility. Recall that a plasma is a partially ionized complex gas composed of electrons, ions of both charges, free radicals, photons of various energies, and gas atoms and molecules in both the ground and excited electronic states. There are two general types of plasma: a *cold plasma*, also known as a low-temperature plasma and a *hot plasma*. A hot plasma is generated by electrical arcs produced at atmospheric pressure, and it is a highly ionized discharge in thermal equilibrium. Cold plasmas are less ionized and are usually produced by low-pressure glow discharges. They are also called *glow-discharge plasmas*. A typical plasma system consists of a reaction chamber, a vacuum system, a radio-frequency (RF) generator, a gas flow controller, and a process controller. Plasma surface treatment is usually performed in a batch process, although continuous online treatment of fibers, tubing, and films is also common.

The plasmas are in a nonequilibrium state in which the Boltzmann temperature of the electrons is roughly two orders of magnitude higher than that of the gas ions and molecules. This lack of thermal equilibrium makes it possible to obtain a plasma in which the gas temperature is near ambient conditions, while the electrons are sufficiently energetic to cause rupture of the molecular bonds at the material surface. It is this characteristic that makes glow-discharge plasmas well-suited for the promotion of chemical reactions involving thermally sensitive materials. Although the glow discharge is in a thermodynamically nonequilibrium state, the plasma reaction is a kinetically steady-state reaction. In other words, the rate of active species generated is equivalent to the rate of active species consumed on the plasma–solid interface. It is this steady-state condition that makes the plasma surface treatment so valuable.

A broad range of chemical and physical reactions occurs in a plasma. The following are some of the more significant electron–gas and gas–gas reactions in a plasma of a two-atom gas.

Excitation $$A_2 + e^- \rightarrow A_2^* + e^- \quad (7.57a)$$

Dissociation $$A_2 + e^- \rightarrow 2A + e^- \quad (7.57b)$$

Electron attachment	$A_2 + e^- \rightarrow A_2^-$	(7.57c)
Dissociative attachment	$A_2 + e^- \rightarrow A + A^-$	(7.57d)
Ionization	$A_2 + e^- \rightarrow A_2^+ + 2e^-$	(7.57e)
Photoemission	$A_2^* \rightarrow A_2 + h\nu$	(7.57f)
Abstraction	$A + B_2 \rightarrow AB + B$	(7.57g)

During the plasma surface reaction, the plasma and the solid are in physical contact, but electrically isolated. Surfaces in contact with the plasma are bombarded by free radicals, electrons, ions, and photons, as generated by the reactions listed above. The energy transferred to the solid is dissipated within the solid by a variety of chemical and physical processes, as illustrated in Figure 7.59. These processes can change surface wettability (cf. Section 2.2.2), alter molecular weight of polymer surfaces, or create reactive sites on polymers. These effects are summarized in Table 7.16.

The wettability changes of polymers after plasma treatment are possibly due to oxidation, unsaturation effects, electrostatic charging, and surface morphology changes. Improvement of surface wettability is one of the major interests in biomaterials surface modifications because most common polymeric biologics are hydrophobic in nature, such as PE, PP, PET, PMMA, PS, PTFE, PVC, polyurethane, and silicone rubber. Hydrophilic surfaces can be obtained by treating polymers with oxygen, nitrogen, or water plasmas.

Molecular weight changes are due to such surface interactions as scission, branching, and crosslinking. In an inert gas plasma, molecular weight changes are due only to crosslinking or degradation. In addition to molecular weight changes, minor surface oxidation also can be observed

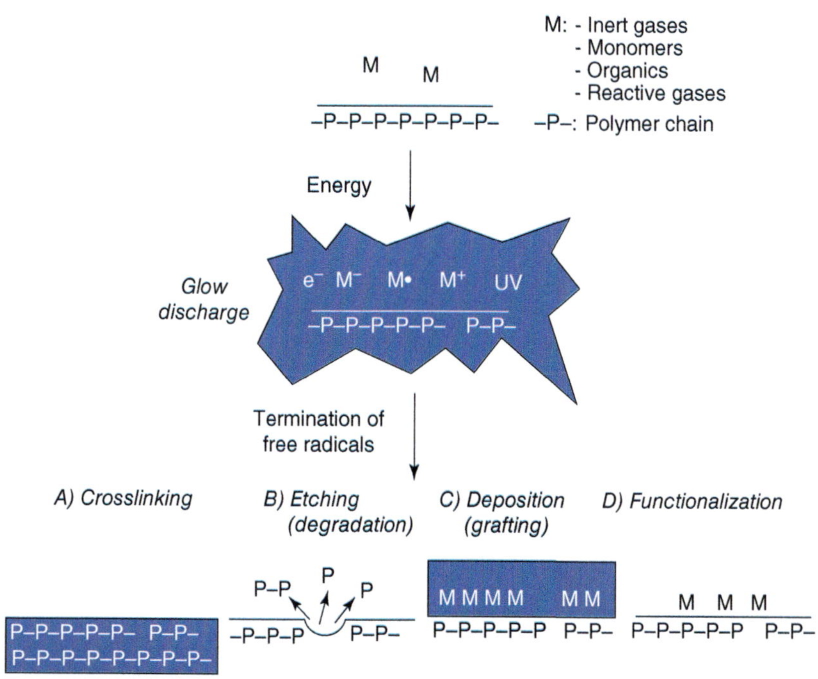

Figure 7.59 Schematic illustration of interactions of glow-discharge plasma with solid surfaces. Sheu et al. [26]/Reproduced with permission of Taylor & Francis.

Table 7.16 Effects of Plasma Treatment on Polymer Surface Modification

Effect	Surface Change	Plasma Gas
Wettability	Oxidation, electrostatic unsaturation	O_2, N_2, H_2O, air (inert gases), NH_3, CO_2, inert gases
Molecular weight	Crosslinking	Inert gases (He, Ar), H_2, N_2
	Degradation	O_2, N_2
	Etching	Ar, CF_4
Functionalization (reactive sites)	—OH	O_2, H_2O, H_2O/H_2O_2
	—C(O)OH	CO_2
	—C—O—O—	Ar (quenching in O_2 or air)
	$-NH_2$	$N_2 + H_2$, NH_3
	C=C	Inert gases

Source: Sheu et al. [26].

on the inert-gas-plasma-treated surface when the surface is exposed to air after the treatment. This oxidation may affect the surface wettability.

Most polymeric biomaterials surfaces are chemically inert. In order to extend their biological performance, there is always great interest in functionalizing their surfaces for further immobilization of bioactive molecules such as enzymes, antibodies, ligands (cf. Section 2.5.2), and drugs. Hydroxyl, carboxylic, and primary amine groups are some of the commonly desired reactive centers that can be generated by plasma treatment, as listed in Table 7.16. However, functional groups generated on the plasma-modified surface are usually not homogeneous due to the presence of undesired side products from various gas fragments in the plasma. Reactive sites, especially peroxide groups, can also be used for surface grafting with other functional polymers, such as hydrogels. Surface peroxides can be obtained using an argon plasma treatment followed by quenching in air or oxygen.

A wide variety of parameters can directly affect the chemical and physical characteristics of a plasma, which in turn affects the surface chemistry obtained by the plasma modification. Some of the more important parameters include electrode geometry, gas type, RF (0–10^{10} Hz), pressure, gas flow rate, power, substrate temperature, and treatment time. The materials and plasmas used for specific biomedical applications are beyond the scope of this text, but the applications include surface modification for cardiovascular, ophthalmological, orthopedic, pharmaceutical, tissue culturing, biosensor, bioseparation, and dental applications.

References

1 Leary, M., *Design for Additive Manufacturing*, Elsevier, 2020.
2 *McGraw-Hill Encyclopedia of Science and Technology*, Vol. **11**, 8th ed., McGraw-Hill, 1997, p. 31.
3 Z. Jastrzebski., *The Nature and Properties of Engineering Materials*, 2nd ed., John Wiley & Sons, Inc, 1976.
4 *McGraw-Hill Encyclopedia of Science and Technology*, Vol. **11**, 8th ed., McGraw-Hill, 1997, p. 35.

5 Altan, T., S.-I. Oh, and H. L. Gegel, *Metal Forming: Fundamentals and Applications*, American Society for Metals, 1983.
6 *ASM Handbook, Vol. 7, Powder Metallurgy*, ASM International, Materials Park, OH 44073-0002, 1984, p. 70.
7 Robert, W., *ASM Handbook, Vol. 7, Powder Metallurgy*, ASM International, Materials Park, OH 44073-0002, 1984, p. 2.
8 Cahn, R. W., *Encyclopedia of Materials Science & Engineering*, Vol. **5**, Pergamon Press, 1986, p. 3874.
9 Reed, J. S., *Principles of Ceramics Processing*, 2nd ed., John Wiley & Sons, Inc, 1995.
10 Kingery, W. D., H. K. Bowen, and D. R. Uhlmann, *Introduction to Ceramics*, John Wiley & Sons, Inc, 1976.
11 Tadmor, Z., and C. G. Gogos, *Principles of Polymer Processing*, John Wiley & Sons, Inc, 1979.
12 Strong, A. B., *Plastics Materials and Processing*, 2nd ed., Prentice Hall, Inc, 2000, p. 352.
13 Strong, A. B., *Plastics Materials and Processing*, 2nd ed., Prentice Hall, Inc, 2000, p. 365.
14 Billmeyer, F. W., *Textbook of Polymer Science*, 3rd ed., John Wiley & Sons, Inc, 1984.
15 Powell, P. C., and A. J. I. Housz, *Engineering with Polymers*, Stanley Thorned Publishers, 1998, p. 58.
16 Strong, A. B., *Plastics Materials and Processing*, 2nd ed., Prentice Hall, Inc, 2000, p. 424.
17 Strong, A. B., *Plastics Materials and Processing*, 2nd ed., Prentice Hall, Inc, 2000, p. 489.
18 Osswald, T. A., J. Puentes, and J. Kattinger, Fused filament fabrication melting model, *Addit. Manuf.*, **22**, 51–59 (2018).
19 Schwartz, M., *Composite Materials Handbook*, 2nd ed., McGraw-Hill, 1992, p. 4.46.
20 Mallick, P. K., *Fiber-Reinforced Composites*, Marcel-Dekker, Inc, 1988, p. 349.
21 Mallick, P. K., *Fiber-Reinforced Composites*, Marcel-Dekker, Inc, 1988, p. 350.
22 Schwartz, M., *Composite Materials Handbook*, 2nd ed., McGraw-Hill, 1992, p. 4.56.
23 Mallick, P. K., *Fiber-Reinforced Composites*, Marcel Dekker, Inc, 1988, p. 352.
24 Mallick, P. K., *Composites Engineering Handbook*, Marcel Dekker, Inc, 1997, p. 184.
25 Bronzino, J. D., ed., *The Biomedical Engineering Handbook*, 2nd ed., CRC Press, 2000.
26 Sheu, M. S., D. M. Hudson, and I. H. Loh, *Encyclopedic Handbook of Biomaterials and Bioengineering, Part A: Materials*, Vol. **1**, Marcel Dekker, Inc, 1995, p. 869.

Problems

Level I

7.I.1 A 1-in. × 36-in. (thickness × width) slab enters a hot-rolling mill at 500 ft per minute. It passes through seven stands and emerges as a strip 0.25 in. × 36 in. What is the exit speed of the strip from the last set of rollers?

7.I.2 In polymer extrusion, show that a square-pitched screw has a helix angle of 17.65°.

7.I.3 In polymer extrusion, show that, for a single-flighted screw, the channel width is related to the inner barrel diameter by $W = \pi D_s \sin\theta$. Clearly state any assumptions you make.

7.I.4 Estimate the shear rate during ceramic tape casting when the substrate velocity is 3 cm/s and the blade height is 50 μm.

Level II

7.II.1 The actual falling weight of a steam hammer is 1200 pounds, the cylinder is 10 in. in diameter, and the stroke is 27 in. The mean average steam pressure is 80 psi. (a) What is the energy of the hammer blow? (b) What is the average force exerted by the hammer if it travels 1/8 in. after striking the workpiece?

7.II.2 Refer to Figure 7.16. (a) Calculate the compact ratio for the granulated tile composition, (b) Estimate the *PF* of the alumina granules in the compact. Assume that the granules pack with $PF = 0.50$ and the particle density is 3.98 mg/m^3.

7.II.3 Look up the relationship for the mold constant, B, in Eq. (7.3) and cite your source. Select a metal to be cast and a material for the mold as well as reasonable temperatures for the "pour temperature" (actual temperature of the molten metal as it is poured into the mold) and the initial temperature of the mold. Calculate the mold constant for your process.

Level III

7.III.1 The following data for a polymer have been collected using a capillary viscometer. The capillary over which the pressure drop has been measured was 1 mm in diameter and 30 mm in length.

Q (cm^3/s)	ΔP (MPa)
0.0196	48.0
0.0441	59.4
0.127	93.6
0.588	144
1.47	198
3.92	288

(a) Using the Hagen–Poiseuille equation, calculate the shear stress at the wall and the apparent shear rate at the wall. Use the ratio of these two values to calculate the apparent viscosity, (b) Calculate the power-law index for the linear portion of the data when plotted on a log–log plot, (c) Does the polymer exhibit power-law behavior over the entire range of shear rates? (d) Look up the equation for something called the Rabinowitsch correction factor in a polymer science or polymer processing reference, and calculate it for this polymer. Cite the source of your information.

Appendix 1

Structure of Some Common Polymers

Chemical Name	Mer Chemical Structure
Epoxy (diglycidyl ether of bisphenol A, DGEPA)	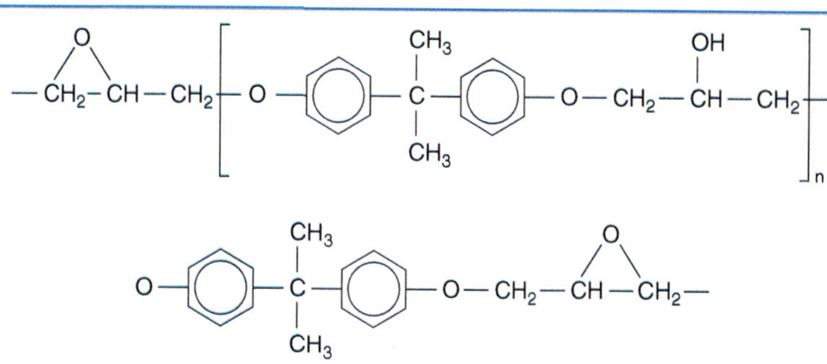
Melamine-formaldehyde (melamine)	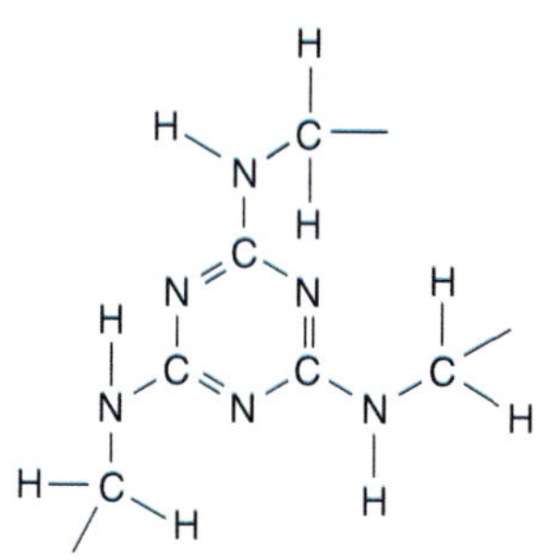
Phenol-formaldehyde (phenolic)	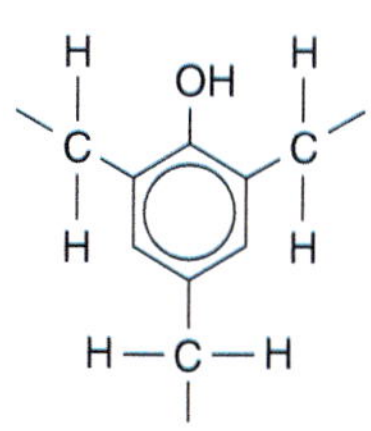
Polyacrylonitrile (PAN)	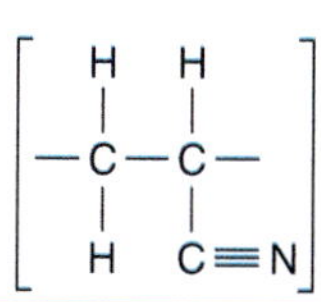

(Continued)

Materials Engineering and Science: Principles, Properties, and Processes, Second Edition. Brian S. Mitchell.

Companion website: www.wiley.com/go/materialsengineeringandscience2e

Chemical Name	Mer Chemical Structure
Polyamide-imide (PAI)	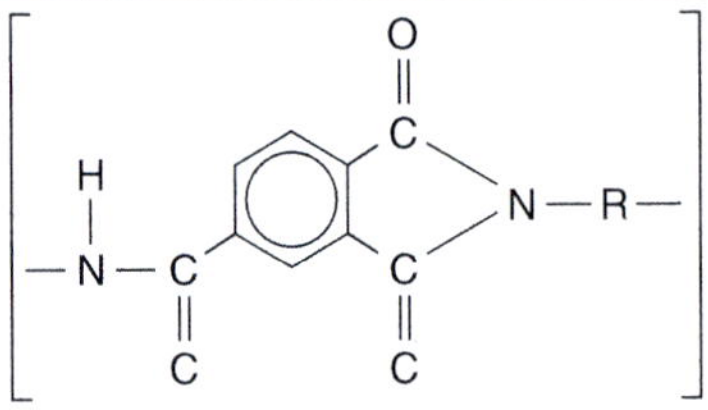
Polybutadiene	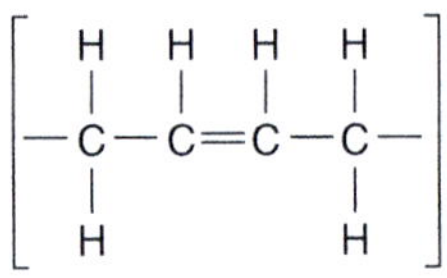
Polybutylene terephthalate (PBT)	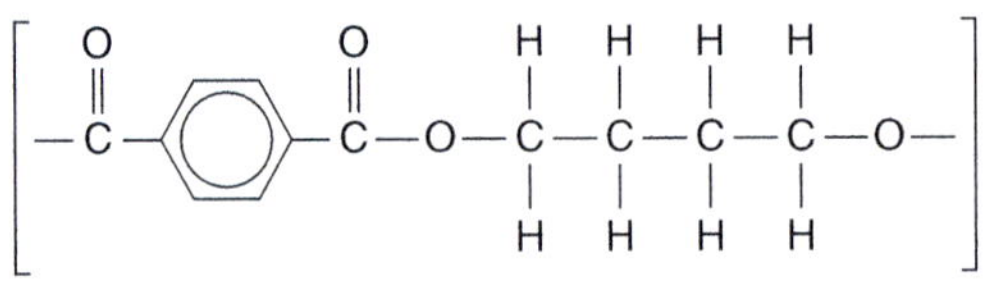
Polycarbonate (PC)	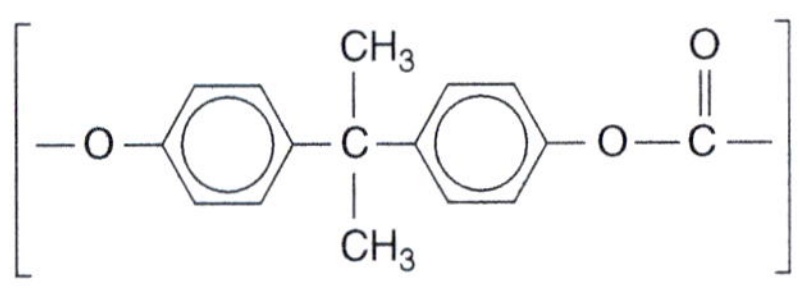
Polychloroprene	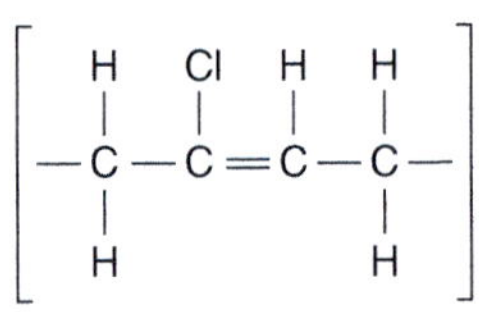
Polychlorotrifluoroethylene	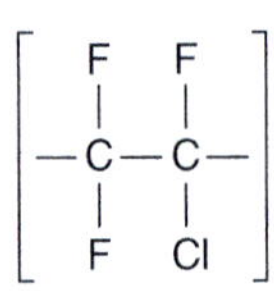
Polydimethyl siloxane (silicone rubber)	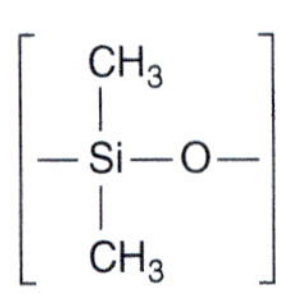
Polyetheretherketone (PEEK)	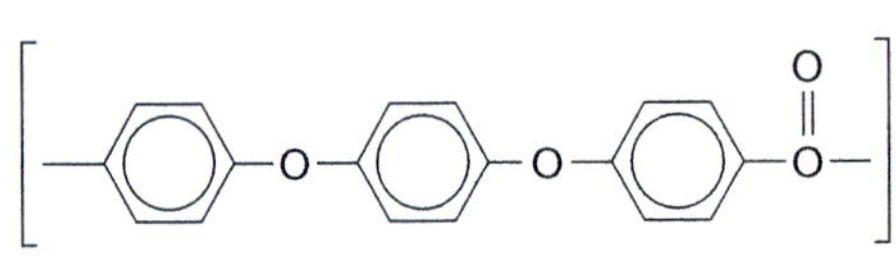
Polyethylene (PE)	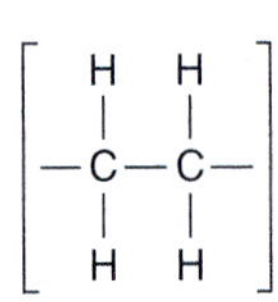

Chemical Name	**Mer Chemical Structure**
Polyethylene terephthalate (PET)	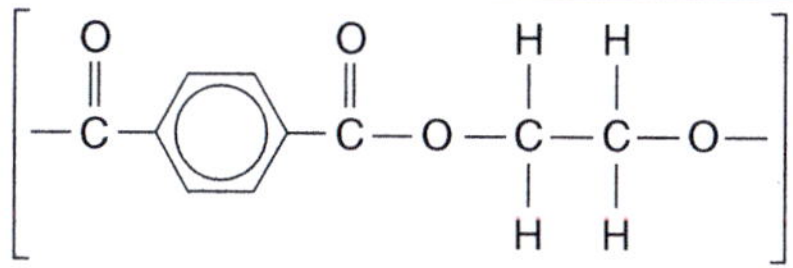
Polyhexamethylene adipamide (nylon 6,6)	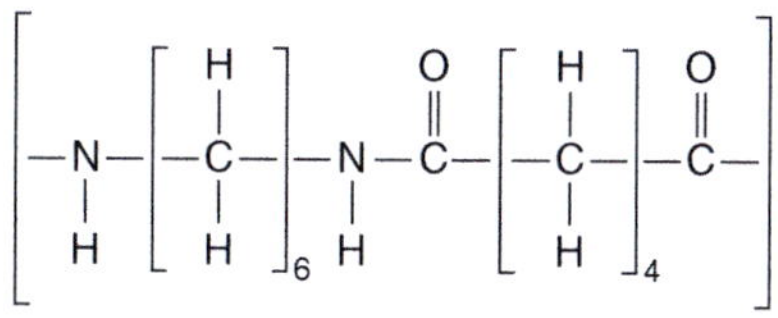
Polyimide	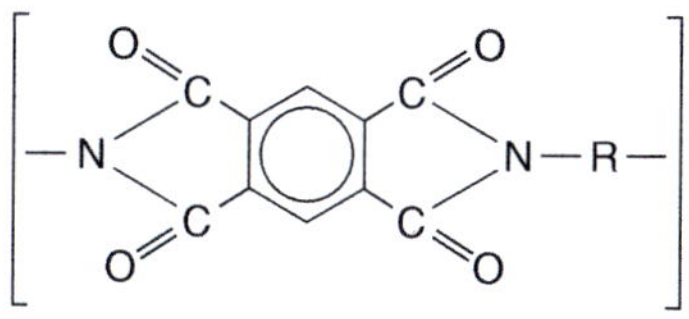
Polyisobutylene	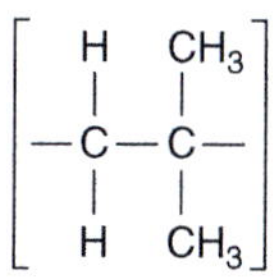
cis-Polyisoprene (natural rubber)	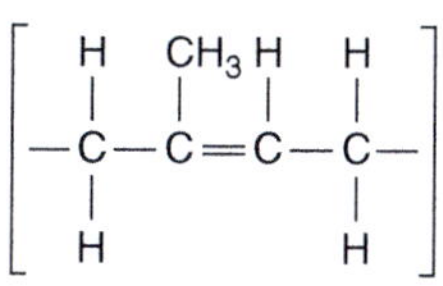
Polymethyl methacrylate (PMMA)	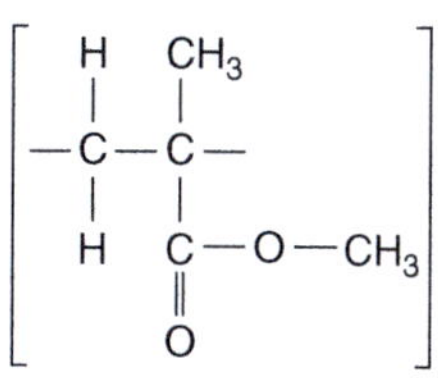
Polyphenylene oxide (PPO)	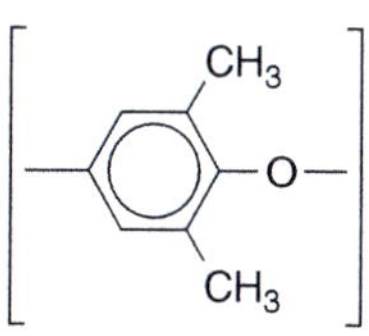
Polyphenylene sulfide (PPS)	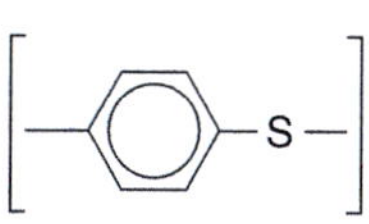
Polyparaphenylene terephthalamide (aramid)	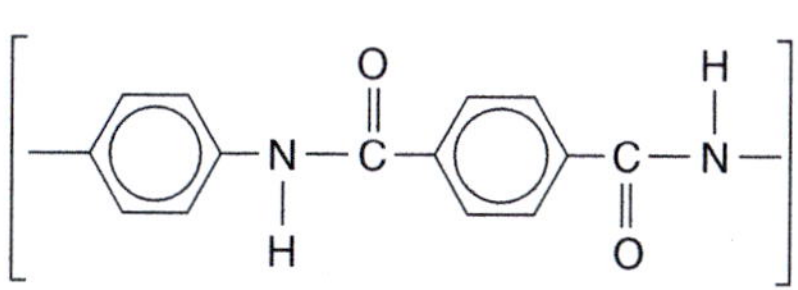

(*Continued*)

Chemical Name	Mer Chemical Structure
Polypropylene (PP)	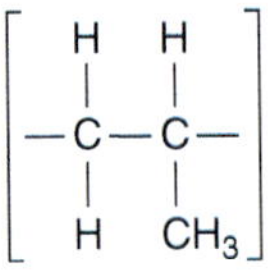
Polystyrene (PS)	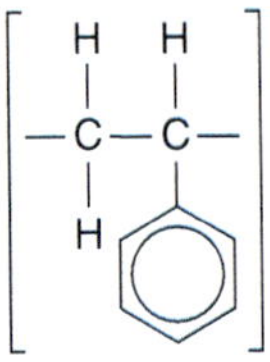
Polytetrafluoroethylene (PTFE)	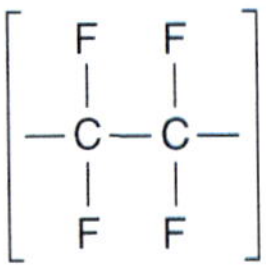
Polyvinyl acetate (PVAc)	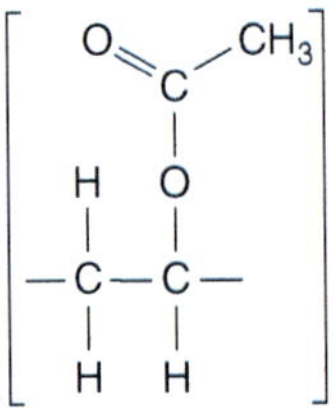
Polyvinyl alcohol (PVA)	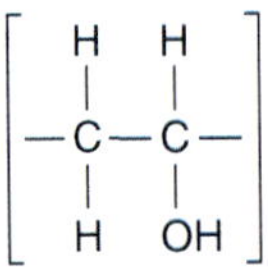
Polyvinyl chloride (PVC)	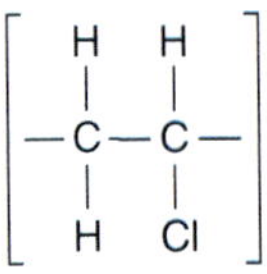
Polyvinyl fluoride (PVF)	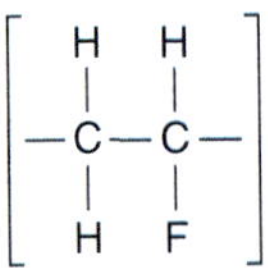
Polyvinylidene chloride (PVDC)	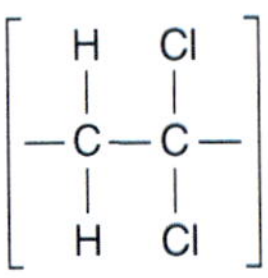
Polyvinylidene fluoride (PVDF)	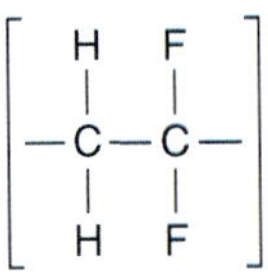

Appendix 2

Composition of Common Alloys

Ferrous Alloys

Standard Stainless and Heat-Resisting Steels

		Composition, %, Balance Iron. Compositions are maximum unless noted								
Type	UNS#	C	Mn	P	S	Si	Cr	Ni	Mo	Other
201	S20100	0.15	5.5–7.5	0.06	0.03	1.00	16–18	3.5–5.5	—	N 0.25
202	S20200	0.15	7.5–10.0	0.06	0.03	1.00	17–19	4.0–6.0	—	N 0.25
205	S20500	0.12–0.25	14–15.5	0.06	0.03	1.00	16.5–18	1–1.75	—	N 0.32–0.4
301	S30100	0.15	2.0	0.045	0.03	1.00	16–18	6–8	—	—
302	S30200	0.15	2.0	0.045	0.03	1.00	17–19	8–10	—	—
302B	S30215	0.15	2.0	0.045	0.03	2.0–3.0	17–19	8–10	—	—
303	S30300	0.15	2.0	0.2	0.15 min	1.00	17–19	8–10	0.6 (opt.)	—
303 Se	S30323	0.15	2.0	0.2	0.06	1.00	17–19	8–10	—	Se 0.15 min
304	S30400	0.08	2.0	0.045	0.03	1.00	18–20	8–10.5	—	—
304 L	S30403	0.03	2.0	0.045	0.03	1.00	18–20	8–12	—	
304 N	S30451	0.08	2.0	0.045	0.03	1.00	18–20	8–10.5	—	N 0.1–0.16
305	S30500	0.12	2.0	0.045	0.03	1.00	17–19	10.5–13	—	—
308	S30800	0.08	2.0	0.045	0.03	1.00	19–21	10–12	—	—
309	S30900	0.20	2.0	0.045	0.03	1.00	22–24	12–15	—	—
309S	S30908	0.08	2.0	0.045	0.03	1.00	22–24	12–15	—	—
310	S31000	0.25	2.0	0.045	0.03	1.5	24–26	19–22	—	—
310S	S31008	0.08	2.0	0.045	0.03	1.5	24–26	19–22	—	—
314	S31400	0.25	2.0	0.045	0.03	1.5–3.0	23–26	19–22	—	—
316	S31600	0.08	2.0	0.045	0.03	1.00	16–18	10–14	2.0–3.0	—
316F	S31620	0.08	2.0	0.2	0.1 min	1.00	16–18	10–14	1.75–2.5	—
316 L	S31603	0.03	2.0	0.045	0.03	1.00	16–18	10–14	2.0–3.0	—
316 N	S31651	0.08	2.0	0.045	0.03	1.00	16–18	10–14	2.0–3.0	N 0.1–0.16

(Continued)

Materials Engineering and Science: Principles, Properties, and Processes, Second Edition. Brian S. Mitchell.

Companion website: www.wiley.com/go/materialsengineeringandscience2e

		Composition, %, Balance Iron. Compositions are maximum unless noted								
Type	UNS#	C	Mn	P	S	Si	Cr	Ni	Mo	Other
317	S31700	0.08	2.0	0.045	0.03	1.00	18–20	11–15	3.0–4.0	—
317 L	S31703	0.03	2.0	0.045	0.03	1.00	18–20	11–15	3.0–4.0	—
321	S31200	0.08	2.0	0.045	0.03	1.00	17–19	9–12	—	Ti5 × C min
329	S32900	0.10	2.0	0.04	0.03	1.00	25–30	3–6	1.0–2.0	—
330	S33000	0.08	2.0	0.04	0.03	0.75–1.5	17–20	34–37	—	—
347	S34700	0.08	2.0	0.045	0.03	1.00	17–19	9–13	—	Cb + Ta 10 × C min
348	S34800	0.08	2.0	0.045	0.03	1.00	17–19	9–13	—	Cb + Ta 10 × C min; Ta 0.10 max, Co 0.20 max
384	S38400	0.08	2.0	0.045	0.03	1.00	15–17	17–19	—	—
403	S40300	0.15	1.0	0.04	0.03	0.5	11.5–13	—	—	—
405	S40500	0.08	1.0	0.04	0.03	1.00	11.5–14.5	—	—	Al 0.1–0.3
409	S40900	0.08	1.0	0.045	0.045	1.00	10.5–11.75	—	—	Ti 6 × C min; 0.75 max
410	S41000	0.15	1.0	0.04	0.03	1.00	11.5–13.5	—	—	—
414	S41400	0.15	1.0	0.04	0.03	1.00	11.5–13.5	1.25–2.5	—	—
416	S41600	0.15	1.25	0.06	0.15 min	1.00	12–14	—	0.6 (opt.)	—
416 Se	S41623	0.15	1.25	0.06	0.06	1.00	12–14	—	—	Se 0.15 min
420	S41000	>0.15	1.0	0.04	0.03	1.00	12–14	—	—	—
420F	S42020	>0.15	1.25	0.06	0.15 min	1.00	12–14	—	0.6 (opt.)	—
422	S42200	0.2–0.25	1.0	0.025	0.025	0.75	11–13	0.5–1.0	0.75–1.25	V 0.15–0.30; W 0.75–1.25
429	S42900	0.12	1.0	0.04	0.03	1.00	14–16	—	—	—
430	S43000	0.12	1.0	0.04	0.03	1.00	16–18	—	—	—
430F	S43020	0.12	1.25	0.06	0.15 min	1.00	16–18	—	0.6 (opt.)	—
430FSe	S43023	0.12	1.25	0.06	0.06	1.00	16–18	—	—	Se 0.15 min
431	S43100	0.20	1.0	0.04	0.03	1.00	15–17	1.25–2.5	—	—
434	S43400	0.12	1.0	0.04	0.03	1.00	16–18	—	0.75–1.25	—
436	S43600	0.12	1.0	0.04	0.03	1.00	16–18		0.75–1.25	Cb + Ta 5 × C min, 0.70 max
440A	S44002	0.6–0.75	1.0	0.04	0.03	1.00	16–18	—	0.75	—
440B	S44003	0.75–0.95	1.0	0.04	0.03	1.00	16–18	—	0.75	—
440C	S44004	0.95–1.2	1.0	0.04	0.03	1.00	16–18	—	0.75	—
442	S44200	0.2	1.0	0.04	0.03	1.00	18–23	—	—	—
446	S44600	0.2	1.5	0.04	0.03	1.00	23–27	—	—	N 0.25

		Composition, %, Balance Iron. Compositions are maximum unless noted								
Type	UNS#	C	Mn	P	S	Si	Cr	Ni	Mo	Other
501	S50100	>0.1	1.0	0.04	0.03	1.00	4–6	—	0.4–0.65	—
502	S50200	0.1	1.0	0.04	0.03	1.00	4–6	—	0.4–0.65	—
503	S50300	0.15	1.0	0.04	0.04	1.00	6–8	—	0.45–0.65	—
504	S50400	0.15	1.0	0.04	0.04	1.00	8–10	—	0.9–1.1	—

Source: Harvey [1].

Iron–Nickel Alloys

	Composition, %, Balance Iron					
Type	Ni	Co	Cr	W	Al	Cu
Invar	36	—	—	—	—	—
Superinvar	31	5	—	—	—	—
Elinvar	32	—	5	2	—	—
Dumet	42	—	—	—	—	—
Kovar	29	17	—	—	—	—
Pllatinite	46	—	—	—	—	—
Permalloy	78.5	—	—	—	—	—
Hipernik	50	—	—	—	—	—
Perminvar	45	25	—	—	—	—
Alnico I	20	5	—	—	12	—
Alnico II	17	12.5	—	—	10	6
Alnico III	25	—	—	—	12	—
Alnico IV	28	5	—	—	12	—

Carbon Steels

		Composition, %, Balance Iron		
AISI-SAE Designation	UNS #	C	Mn	S
1015	G10150	0.12–0.18	0.30–0.60	—
1020	G10200	0.17–0.23	0.30–0.60	—
1022	G10220	0.17–0.23	0.70–1.00	—
1030	G10900	0.27–0.34	0.60–0.90	—
1040	G10400	0.36–0.44	0.60–0.90	—

(Continued)

		Composition, %, Balance Iron		
AISI-SAE Designation	UNS #	C	Mn	S
1050	G10500	0.47–0.55	0.60–0.90	—
1060	G10600	0.55–0.66	0.60–0.90	—
1080	G10800	0.74–0.88	0.60–0.90	—
1095	G10950	0.90–1.04	0.30–0.50	—
1118	G11180	0.14–0.20	1.30–1.60	0.08–0.13
1137	G11370	0.32–0.039	1.35–1.65	0.08–0.13
1141	G11410	0.37–0.45	1.35–1.65	0.08–0.13
1144	G11440	0.40–0.48	1.35–1.65	0.24–0.33

Source: Brady and Clauser [2].

Alloy Steels

		Composition, %, Balance Iron							
AISI-SAE Designation	UNS #	C	Mn	P (max)	S (max)	Si	Cr	Ni	Mo
1330	G13300	0.28–0.33	1.60–1.90	0.035	0.040	0.15–0.30	—	—	—
1340	G13400	0.38–0.43	1.60–1.90	0.035	0.040	0.15–0.30	—	—	—
3140	–	0.38–0.43	0.70–0.90	0.040	0.040	0.20–0.35	0.55–0.75	1.10–1.40	—
4037	G40370	0.35–0.40	0.70–0.90	0.035	0.040	0.15–0.30	0.20–0.30	—	—
4042	G40420	0.40–0.45	0.70–0.90	0.035	0.040	0.15–0.30	0.80–1.10	—	0.15–0.25
4130	G41300	0.28–0.33	0.40–0.60	0.035	0.040	0.15–0.30	0.80–1.10	—	0.15–0.25
4140	G41400	0.38–0.43	0.75–1.00	0.035	0.040	0.15–0.30	0.80–1.10	—	0.15–0.25
4150	G41500	0.48–0.53	0.75–1.00	0.035	0.040	0.15–0.30	0.80–1.10	—	0.20–0.30
4320	G43200	0.17–0.22	0.45–0.65	0.035	0.040	0.15–0.30	0.40–0.60	1.65–2.00	0.20–0.30
4340	G43400	0.38–0.43	0.60–0.80	0.035	0.040	0.15–0.30	0.70–0.90	1.65–2.00	0.20–0.30
4620	G46200	0.17–0.22	0.45–0.65	0.035	0.040	0.15–0.30	1.65–2.00	—	0.20–0.30
4820	G48200	0.18–0.23	0.50–0.70	0.035	0.040	0.15–0.30	3.25–3.75	—	0.20–0.30
5046	G50460	0.43–0.48	0.75–1.00	0.035	0.040	0.15–0.30	0.20–0.35	—	—
50B46	G50461	0.44–0.49	0.75–1.00	0.035	0.040	0.15–0.30	0.20–0.35	—	—
5060	G50600	0.56–0.64	0.75–1.00	0.035	0.040	0.15–0.30	0.40–0.60	—	—
50B60	G50461	0.56–0.64	0.75–1.00	0.035	0.040	0.15–0.30	0.40–0.60	—	—
5130	G51300	0.28–0.33	0.70–0.90	0.035	0.040	0.15–0.30	0.80–1.10	—	—
5140	G51400	0.38–0.43	0.70–0.90	0.035	0.040	0.15–0.30	0.70–0.90	—	—
5150	G51500	0.48–0.53	0.70–0.90	0.035	0.040	0.15–0.30	0.70–0.90	—	—

AISI-SAE Designation	UNS #	Composition, %, Balance Iron							
		C	Mn	P (max)	S (max)	Si	Cr	Ni	Mo
5160	G51600	0.56–0.64	0.75–1.00	0.035	0.040	0.15–0.30	0.70–0.90	—	—
51B60	G51601	0.56–0.64	0.75–1.00	0.035	0.040	0.15–0.30	0.70–0.90	—	—
6150	G61500	0.48–0.53	0.70–0.90	0.035	0.040	0.15–0.30	0.8–1.10	—	—
81B45	G81451	0.43–0.48	0.75–1.00	0.035	0.040	0.15–0.30	0.35–0.55	0.20–0.40	0.08–0.15
8620	G86200	0.18–0.23	0.70–0.90	0.035	0.040	0.15–0.30	0.40–0.60	0.40–0.70	0.15–0.25
8630	G86300	0.28–0.33	0.70–0.90	0.035	0.040	0.15–0.30	0.40–0.60	0.40–0.70	0.15–0.25
8640	G86400	0.38–0.43	0.75–1.00	0.035	0.040	0.15–0.30	0.40–0.60	0.40–0.70	0.15–0.25
86B45	G86451	0.43–0.48	0.75–1.00	0.035	0.040	0.15–0.30	0.40–0.60	0.40–0.70	0.15–0.25
8650	G86500	0.48–0.53	0.75–1.00	0.035	0.040	0.15–0.30	0.40–0.60	0.40–0.70	0.15–0.25
8660	G86600	0.56–0.64	0.75–1.00	0.035	0.040	0.15–0.30	0.40–0.60	0.40–0.70	0.15–0.25
8740	G87400	0.38–0.43	0.75–1.00	0.035	0.040	0.15–0.30	0.40–0.60	0.40–0.70	0.20–0.30
9255	G92550	0.51–0.59	0.70–0.95	0.035	0.040	1.8–2.2	—	—	—
9260	G92600	0.56–0.64	0.75–1.00	0.035	0.040	1.8–2.2	—	—	—
9310	G93106	0.08–0.13	0.45–0.65	0.025	0.025	0.15–0.30	1.00–1.40	3.00–3.50	0.08–0.15
94B30	G94301	0.28–0.33	0.75–1.00	0.035	0.040	0.15–0.30	0.30–0.50	0.30–0.60	0.08–0.15

Source: Brady and Clauser [2].

Gray Cast Irons

UNS #	Composition, %, Balance Iron			
	Mn	Si	P	S
F10004	0.50–0.80	2.3–2.8	0.15	0.15
F10005	0.60–0.90	2.0–2.4	0.12	0.15
F10009	0.60–0.90	1.6–2.1	0.12	0.12
F10006	0.60–0.90	1.9–2.3	0.10	0.16
F10007	0.60–0.90	1.8–2.2	0.08	0.16
F10010	0.60–0.90	1.3–1.8	0.08	0.12
F10011	0.60–0.90	1.3–1.8	0.08	0.12
F10008	0.70–1.00	1.8–2.1	0.07	0.16
F10012	0.60–0.90	1.95–2.4	0.07	0.12

Source: Brady and Clauser [2]

Aluminum Alloys

Type	Composition, %, Balance Aluminum								
	Si	Cu	Mn	Mg	Cr	Zn	Ni	Ti	Other
				Wrought Alloys					
1100	—	0.12	—	—	—	—	—		—
2011	—	5.5	—	—	—	—	—		0.4 Bi; 0.4 Pb
2014	0.8	4.4	0.8	0.5	—	—	—		—
2024	—	4.4	0.6	1.5	—	—	—		—
2219	—	6.3	0.3	—	—	—	—	0.06	0.10 V; 0.18 Zr
2319	—	6.3	0.3	—	—	0.18	—	0.15	0.10 V
2618	0.18	2.3	—	1.6	—	—	1.0	0.07	1.1 Fe
3003	—	—	0.12	1.2	—	—	—	—	—
3004	—	—	1.2	1.0	—	—	—	—	—
3105	—	—	0.55	0.5	—	—	—	—	—
4032	12.2	0.9	—	1.0	—	—	0.9	—	—
4043	5.2	—	—	—	—	—	—	—	—
5005	—	—	—	0.8	—	—	—	—	—
5050	—	—	—	1.4	—	—	—	—	—
5052	—	—	—	2.5	0.25	—	—	—	—
5056	—	—	0.12	5.0	0.12	—	—	—	—
5083	—	—	0.7	4.4	0.15	—	—	—	—
5086	—	—	0.4	4.0	0.15	—	—	—	—
5154	—	—	—	3.5	0.25	—	—	—	—
5182	—	—	0.35	4.5	—	—	—	—	—
5252	—	—	—	2.5	—	—	—	—	—
5254	—	—	—	3.5	0.25	—	—	—	—
5356	—	—	0.12	5.0	0.12	—		0.13	—
5454	—	—	0.8	2.7	0.12			—	—
5456	—	—	0.8	5.1	0.12			—	—
5457	—	—	0.3	1.0	—	—	—	—	—
5652	—	—	—	2.5	0.25	—	—	—	—
5657	—	—	—	0.8	—	—	—	—	—
6005	0.8	—	—	0.5	—	—	—	—	—
6009	0.8	0.35	0.5	0.6	—	—	—	—	—
6010	1.0	0.35	−0.5	0.8	—	—	—	—	—
6061	0.6	0.28	—	1.0	0.2	—	—	—	—
6063	0.4	—	—	0.7	—	—	—	—	—
6066	1.4	1.0	0.8	1.1	—	—	—	—	—
6070	1.4	0.28	0.7	0.8	—	—	—	—	—

	Composition, %, Balance Aluminum								
Type	Si	Cu	Mn	Mg	Cr	Zn	Ni	Ti	Other
6101	0.5	—	—	0.6	—	—	—	—	—
6151	0.9	—	—	0.6	0.25	—	—	—	—
6201	0.7	—	—	0.8	—	—	—	—	—
6205	0.8	—	0.1	0.5	0.1	—	—	—	0.1 Zr
7049	—	1.5	—	2.5	0.15	7.6	—	—	—
7075	—	1.6	—	2.5	0.23	5.6	—	—	—
				Cast Alloys					
201	—	4.6	0.35	0.35	—	—	—	0.25	0.7 Ag
206	0.10 max	4.6	0.35	0.25	—	—	—	0.22	0.15 Fe max
A206	0.05 max	4.6	0.35	0.25	—	—	—	0.22	0.10 Fe max
208	3.0	4.0	—	—	—	—	—	—	
242	—	4.0	—	1.5	—	—	2.0	—	
295	0.8	4.5	—	—	—	—	—	—	
296	2.5	4.5	—	—	—	—	—	—	
308	5.5	4.5	—	—	—	—	—	—	
319	6.0	3.5	—	—	—	—	—	—	
336	12.0	1.0	—	1.0	—	—	2.5	—	
354	9.0	1.8	—	0.50	—	—	—	—	
355	5.0	1.2	0.50 max	0.50	—	0.35 max	—	—	0.6 Fe max
C355	5.0	1.2	0.10 max	0.50	—	0.10 max	—	—	0.2 Fe max
356	7.0	0.25 max	0.35 max	0.32	—	0.35 max	—	—	0.6 Fe max
A356	7.0	0.25 max	0.10 max	0.35	—	0.10 max	—	—	0.2 Fe max
357	7.0	—	—	0.50	—	—	—	—	—
A357	7.0	—	—	0.6	—	—	—	0.15	0.005 Be
359	9.0	—	—	0.6	—	—	—	—	—
360	9.5	—	—	0.5	—	—	—	—	2.0 Fe max
A360	9.5	—	—	0.5	—	—	—	—	1.3 Fe max
380	8.5	3.5	—	—	—	—	—	—	2.0 Fe max
A380	8.5	3.5	—	—	—	—	—	—	1.3 Fe max
383	10.5	2.5	—	—	—	—	—	—	—
384	11.2	3.8	—	—	—	3.0 max	—	—	—
A384	11.2	3.8	—	—	—	1.0 max	—	—	—
390	17.0	4.5	—	0.6	—	1.3 max	—	—	—
A390	17.0	4.5	—	0.6	—	0.5 max	—	—	—
413	12.0	—	—	—	—	—	—	—	2.0 Fe max
A413	12.0	—	—	—	—	—	—	—	1.3 Fe max
443	5.2	0.6 max	—	—	—	—	—	—	—
A443	5.2	0.30 max	—	—	—	—	—	—	—

(Continued)

	Composition, %, Balance Aluminum								
Type	Si	Cu	Mn	Mg	Cr	Zn	Ni	Ti	Other
B443	5.2	0.15 max	—	—	—	—	—	—	—
C443	5.2	0.6 max	—	—	—	—	—	—	2.0 Fe max
514	—	—	—	4.0	—	—	—	—	—
518	—	—	—	8.0	—	—	—	—	—
520	—	—	—	10.0	—	—	—	—	—
535	—	—	0.18	6.8	—	—	—	0.18	—
A535	—	—	0.18	7.0	—	—	—	—	—
B535	—	—	—	7.0	—	—	—	0.18	—
712	—	—	—	0.6	0.5	5.8	—	0.2	—
713	—	0.7	—	0.35	—	7.5	—	—	—
771	—	—	—	0.9	0.13	7.0	—	0.15	—
850	—	1.0	—	—	—	—	1.0	—	6.2 Sn

Source: Committee on Metallurgy [3].

Copper Alloys

		Composition, %, Balance Copper						
Alloy	UNS #	Ni	Zn	Sn	Al	Be	Pb	Other
Brass, naval		—	39.25	0.75	—	—	—	—
Muntz metal		—	40.0	—	—	—	—	—
Bronze, aluminum		—	—	—	8.0	—	—	—
Beryllium copper	C17200	—	—	—	—	1.9	—	Co or Ni
Brass, free-cutting	C36000	—	35.5	—	—	—	2.5	—
Nickel silver, 18%		18	17	—	—	—	—	—
Nickel silver, 13%		12.5	20.0	2.0	—	—	9.0	—
Constantan		45	—	—	—	—	—	—
Brass, red		—	5	5	—	—	5	—
Bronze, silicon		—	4	—	1	—	—	Fe 2.0; Mn 1.0
Bronze, tin		—	4	8	—	—	—	—
Cupronickel	C71500	30	—	—	—	—	—	—

Titanium Alloys

Type	UNS #	Composition, %, Balance Titanium					
		Al	Sn	Mo	V	Fe	Cu
Ti–5 Al–2.5 Sn	R54520	5	2.5				
Ti–8 Al–1 Mo–1 V	R54810	8		1	1		
Ti–6 Al–4 V	R56400	6			4		
Ti–6 Al–6 V–2 Sn	R56620	6	2		6		0.75
Ti–10 V–2 Fe–3 Al	—	3			10	2	

Source: Ref. [4].

Silver Alloys

Type	Composition, %, Balance Silver							
	Au	Cu	Zn	Pt	Pb	Hg	Sn	Pd
Sterling silver	—	7.5	—	—	—	—	—	—
Coin silver (to 1966)	—	10	—	—	—	—	—	—
Brazing alloy	—	28	—	—	—	—	—	—
10-karat green gold	41.7	9	0.3	—	—	—	—	—
Dental amalgam	—	2	0.5	—	—	52	12.5	—
Gold solder	50	17.5	2	—	—	—	—	—
White gold	60	10	1	4	—	—	—	10
14-karat yellow gold	58.3	29.7	2	—	—	—	—	—
Soft solder	—	—	—	—	97.5	—	—	—

Source: Practical Handbook of Materials Science.

Magnesium Alloys

ASTM #	UNS #	Composition, %, Balance Magnesium					
		Al	Zn	Mn	Th	Zr	Rare Earth
AZ80A	M11800	8.5	0.5	0.12	—	—	—
HK31A	M13310	—	—	—	3.0	0.6	—
ZK60A	M16600	—	5.5	—	—	0.45	—
AM60A	M10600	6.0	—	0.13	—	—	—
EZ33A	M12330	—	2.7	—	—	0.6	3.3
AZ91A	M11910	9.0	—	0.13	—	0.7	—

Source: Ref. [4].

References

1 Harvey, P. D., ed., *Engineering Properties of Steel*, ASM, 1982.
2 Brady, G. S., and H. R. Clauser, eds., *Materials Handbook*, McGraw-Hill, New York, 1991.
3 Committee on Metallurgy, *Engineering Metallurgy*, Pitman Publishing, 1957.
4 *ASM Handbook*, American Society for Metals, Metals Park, OH, 1985.

Appendix 3

Surface and Interfacial Energies

Surface Tensions

Material	Temperature (°C)	Surface Tension (ergs/cm^2)
Ag[1]	1000	920
Al_2O_3[1]	2080	700
Al_2O_3[2]	2320	630
BaO[2]	2196	520
B_2O_3[1]	900	80
Bi_2O_3[2]	1098	213
BeO[2]	2843	415
CaO[2]	2860	670
CoO[2]	2078	550
Cu[1]	1120	1270
Cr_2O_3[2]	2573	812
FeO[1]	1420	585
FeO[2]	1641	545
GeO_2[2]	1389	250
H_2O[1]	25	72
K_2O[2]	1400	156
La_2O_3[2]	2573	560
MgO[2]	3073	660
MnO[2]	2058	630
MoO_3[2]	1068	70
Na_2O[2]	1400	308
NaCl[1]	801	114
$Na(SO_4)$[1]	884	196
$NaPO_3$[1]	620	209
Nb_2O_5[2]	1773	279
Pb[1]	350	442

(*Continued*)

Materials Engineering and Science: Principles, Properties, and Processes, Second Edition. Brian S. Mitchell.

Companion website: www.wiley.com/go/materialsengineeringandscience2e

Material	Temperature (°C)	Surface Tension (ergs/cm^2)
PbO[2]	900	132
Pt[1]	1770	1865
P_2O_5[2]	836	60
siO_2[2]	1993	307
Sm_2O_3[2]	2593	815
Sodium silicate[1]	1000	250
Ta_2O_5[2]	2150	280
TiO_2[2]	2143	380
Ti_2O_3[2]	2090	584
V_2O_5[2]	943	80
WO_3[2]	1743	100

Source: Kingery et al. [1]; Eustathopoulos et al. [2].

Solid-Surface Energies

System	Temperature (°C)	Surface Energy (ergs/cm^2)
Ag[1]	1183	1140 ± 90
Al_2O_3[2]	1850	905
Au[1]	1273	1400 ± 50
BaF_2, $N_2(l)$[2]	77 K	300
CaF_2, $N_2(l)$[2]	77 K	500
$CaCO_3$, $N_2(l)$[2]	77 K	300
Co[1]	1678	2282 ± 300
Cu[1]	1243	1650 ± 100
γ-Fe[1]	1648	2150 ± 325
δ-Fe[1]	1723	2220 ± 250
In[1]	420	674 ± 74
LiF, $N_2(l)$[2]	77 K	400
Mica, vacuum[2]	298 K	4500
MgO, $N_2(l)$[2]	77 K	1500
MgO[2]	25	1000
NaCl, $N_2(l)$[2]	77 K	300
Nb[1]	2523	2100 ± 100
Ni[1]	1488	2385 ± 100
Pb[1]	590	620 ± 20
Si, $N_2(l)$[2]	77 K	1800
Sapphire, $(10\bar{1}1)$[2] plane	298 K	6000

System	Temperature (°C)	Surface Energy (ergs/cm^2)
Sapphire, $(10\bar{2}3)^2$ plane	293 K	24,000
Sn[1]	488	685
Ti[1]	1873	1700
TiC[2]	1100	1190
Zn[1]	653	830
Polymers (from critical surface tension)[3]		
Polytetrafluoroethylene		19
Poly(dimethyl siloxane)		24
Poly(vinylidine fluoride)		25
Poly(vinyl fluoride)		28
Polyethylene		31
Polystyrene		33
Poly(hydroxyethyl methacrylate)		37
Poly(vinyl alcohol)		37
Poly(methyl methacrylate)		39
Poly(vinyl chloride)		39
Polycaproamide(Nylon 6)		42
Poly(ethylene oxide)-diol		43
Polyethylene terephthalate		43
Polyacrylonitrile		50

Source: Eustathopoulos et al. [2]; Kingery et al. [3]; Ratner et al. [4].

Interfacial Energies

System	Temperature (°C)	Interfacial Energy (ergs/cm^2)
Al_2O_3(*s*)–silicate glaze (*l*)	1000	<700
Al_2O_3(*s*)–Pb(*l*)	400	1440
Al_2O_3(*s*)–Ag(*l*)	1000	1770
Al_2O_3(*s*)–Fe(*l*)	1570	2300
Ag(*s*)–Na_2SiO_3 (*l*)	900	1040
Cu(*s*)–Na_2SiO_3 (*l*)	900	1500
Cu(*s*)–Cu_2S (*l*)	1131	90
MgO(*s*)–Ag (*l*)	1300	850
MgO(*s*)–Fe (*l*)	1725	1600
SiO_2(glass)–sodium silicate (*l*)	1000	<25
SiO_2(glass)–Cu (*l*)	1120	1370
TiC (*s*)–Cu (*l*)	1200	1225

Source: Kingery et al. [1].

References

1 Kingery, W. D., H. K. Bowen, and D. R. Uhlmann, *Introduction to Ceramics*, John Wiley and Sons, New York, 1976.
2 Eustathopoulos, N., M. G. Nicholas, and B. Drevet, *Wettability at High Temperatures*, Pergamon, Oxford, 1999.
3 Ratner, B. D., A. S. Hoffman, F. J. Schoen, and J. E. Lemons, *Biomaterials Science*, Academic Press, New York, 1996.

Appendix 4

Thermal Conductivities of Selected Materials

All values in units of W/m·K. Divide by 1.73 to obtain values in Btu/hr·ft·F. Divide by 419 to obtain values in cal/cm·K·s. Consult references for values at additional temperatures.

Elements, Metals, and Alloys

	−190°C	0°C	20°C	100°C	1000°C
Source: Geankoplis [1].					
Aluminum		202		—	
Copper		388		377	
Steel		—		45	
Source: Kittel [2].					
Aluminum	418	226			
Cadmium (‖hex. axis)	92.1	83.7			
Cadmium (⊥ hex. axis)	113	105			
Copper	578	394			
Gold	—	306			
Iron	184	92.1			
Magnesium	188	172			
Nickel	113	83.7			
Silver	427	418			
Sodium	155	138			
Source: Gaskell [3].					
1% carbon steel			43		
1% chrome steel			62		
304 stainless steel (18% Cr, 8% Ni)			14		
60% Pt, 40% Rh			46		

(*Continued*)

Materials Engineering and Science: Principles, Properties, and Processes, Second Edition. Brian S. Mitchell.

Companion website: www.wiley.com/go/materialsengineeringandscience2e

	−190°C	0°C	20°C	100°C	1000°C
Al Bronze (90% Cu, 10% Al)			49		
Aluminum			236		
Beryllium			218		
Boron			31.7		
Brass (70% Cu, 30% Zn)			110		
Cadmium			104		
Cast iron			52		
Chromium			95		
Cobalt			104		
Constantan (55% Cu, 45% Ni)			22		
Copper			401		
Germanium			67		
Gold			318		
Iron			83		
Lead			36		
Lithium			79		
Magnesium			157		
Manganese			7.7		
Molybdenum			139		
Nichrome (80% Ni, 10% Cr)			12		
Nickel			94		
Platinum			72		
Rhenium			49		
Rhodium			151		
Silicon			168		
Silver			428		
Sodium			135		
Tin			68		
Titanium			22		
Tungsten			182		
Uranium			27		
Vanadium			31		
Wrought iron			59		
Zinc			122		
Zirconium			23		

Ceramics

	−190°C	0°C	20°C	100°C	1000°C
Source: Kingery [4].					
Al_2O_3				30.2	6.28
BeO				220	20.5
Fire-clay refractory				1.13	1.55
Fused silica glass				2.00	2.51
Graphite				180	62.8
$MgAl_2O_4$				15.1	5.86
MgO				37.7	7.11
Mullite				5.86	3.77
Porcelain				1.67	1.88
Soda–lime–silica glass				1.67	—
ThO_2				10.5	2.93
TiC				25.1	5.86
TiC cermet				33.5	8.37
$UO_{2.0}$				10.0	3.35
ZrO_2 (stabilized)				1.97	2.30
Source: Samsanor [5].					
CaO				15.2	7.79
NiO				12.4	4.48
TiO_2				6.53	3.31
VO_2				9.80	3.41
ZrO_2				1.97	2.0
Source: Gaskell [3].					
Aluminum oxide			40		
Asbestos			0.11		
Beryllium oxide			302		
Bricks					
Chrome			2.2		
Common			0.7		
Fireclay			1.0		
Magnesite			4.0		
Masonry			0.66		
Silica			1.1		

(Continued)

	−190°C	0°C	20°C	100°C	1000°C
Cement mortar			0.9		
Clay earth			1.4		
Coal			0.24		
Concrete			1.0		
Diatomaceous earth			1.3		
Fused quartz			1.3		
Granite			3.0		
Gypsum plaster			0.5		
Limestone			2.0		
Magnesium oxide			53		
Marble			2.7		
Rock wool			0.04		
Sand			0.3		
Sandstone			2.8		

Source: Shackelford and Alexander [6].

Aluminum nitride			30.2	—	—
Aluminum oxide			25.1	—	5.9–6.7
Beryllium oxide			15.9–19.7	—	—
Boron carbide			27.2–28.9	—	—
Boron nitride (‖ to *a* axis)			—	—	12.4
Boron nitride (‖ to *c* axis)			—	—	26.7
Calcium oxide			—	15.5	7.96
Chromium carbide			190	—	—
Chromium diboride			20.5–31.8	—	—
Chromium oxide			10.0–33.0	—	—
Cordierite			3.2	—	—
Hafnium carbide			22.2	—	—
Hafnium diboride			6.29	—	—
Hafnium oxide			11.4	—	—
Magnesium oxide			40.6	—	—
Mullite			—	6.1	3.8
Nickel oxide			12.1	—	—
Silicon carbide			—	—	21.4
Silicon nitride			30.2	—	—
Sillimanite			—	1.8	—
Spinel			—	14.7	—
Tantalum carbide			22.2	—	—
Tantalum diboride			10.9	—	—
Thorium oxide			10.1	8.4	2.9
Titanium carbide			17.1–31.0	—	—

	−190°C	0°C	20°C	100°C	1000°C
Titanium diboride			24.3–26.0	—	—
Titanium nitride			28.9	—	—
Titanium oxide			—	6.7	3.4
Tungsten carbide			84.2	—	—
Uranium dioxide			—	10.5	3.4
Zircon			—	6.1	—
Zirconium carbide			20.5	—	—
Zirconium diboride			23.0–24.3	—	—
Zirconium oxide			—	2.1	—

Glasses

	−190°C	0°C	20°C	100°C	1000°C
Source: Gaskell [3].					
Glass fiber			0.035		
Glass wool			0.038		
Pyroceram			4.1		
Window glass			0.84		

Polymers

	−190°C	0°C	20°C	100°C	1000°C
Source: Geankoplis [7].					
Hard Rubber		0.15			
*******Source:* Shackelford and Alexander [6].					
ABS resins (molded, extruded)					
Heat resistant			0.21–0.35		
High impact			0.21–0.28		
Low-temperature impact			0.13–0.24		
Medium impact			0.13–0.31		
Very high impact			0.02–0.24		
Acrylics (cast, molded, and extruded)			0.21		
Alkyds (molded)			0.35–1.04		
Allyl diglycol carbonate			2.51		
Cellulose acetate (molded, extruded)			0.17–0.33		
Cellulose acetate butyrate (molded, extruded)			0.17–0.33		

(Continued)

	−190°C	0°C	20°C	100°C	1000°C
Cellulose acetate propionate (molded, extruded)			0.17–0.33		
Chlorinated polyether			1.57		
Chlorinated polyvinyl chloride			1.64		
Epoxies (cast, molded)			0.17–0.86		
Fluorinated ethylene propylene (FEP)			0.21		
Nylon 6			2.08–2.92		
Nylon 6–6			2.94		
Nylon 6–10			2.60		
Phenylene oxides			1.9–3.1		
Polyacetal			0.22–0.28		
Polyarylsulfone			1.9		
Polycarbonate			0.19		
Polyester, thermoplastic			0.62–0.95		
Polyester, thermoset			0.17–0.21		
Polyethylene (all densities)			0.33		
Polyimides			6.58–11.7		
Polyphenylene sulfide			3.46		
Polypropylene			2.1–2.35		
Polypropylene, high impact			2.98		
Polystyrene			0.04–0.16		
Polytetrafluoroethylene (PTFE)			0.24		
Polytrifluorochloroethylene (PTFCE)			0.25		
Polyvinyl chloride			0.12–0.17		
Polyvinylidene fluoride (PVDF)			0.24		
Source: Gaskell [3].					
Foam rubber			0.030		
Hard rubber			0.16		
Polystyrene insulation			0.025		
Teflon			0.35		

Composites

	−190°C	0°C	20°C	100°C	1000°C
Source: Gaskell [3].					
Plywood			0.12		
Source: Shackelford and Alexander [6].					
Alkyd, glass-filled			0.35–0.52		
Polycarbonate, 40% glass-fiber-filled			0.22		

	−190°C	0°C	20°C	100°C	1000°C
High-strength epoxy laminate			4.1		
Melamine, glass-fiber-filled			0.48		
Melamine, cellulose-filled			0.29–0.35		
Nylon 6, 30% glass-fiber-filled			2.92–5.66		
Nylon 6-6, glass-fiber-filled			2.60–5.71		
Nylon 6–10, 30% glass-fiber-filled			6.06		
Polyester, high-strength glass-fiber-filled			2.28–2.91		
Polyimide, glass-filled			6.21		
Polyphenylene sulfide, 40% glass-fiber-filled			3.46		
Polystyrene, 30% glass-fiber-filled			0.20		
Polyurea, cellulose-filled			0.29–0.42		
Phenolic, wood-flour-filled			0.17–0.52		
Phenolic, glass-fiber-filled			0.35		

Biologics

	−190°C	0°C	20°C	100°C	1000°C
Source: Gaskell [3].					
Human skin			0.37		
Balsa wood			0.55		
Cypress wood			0.097		
Fir wood			0.11		
Maple/Oak wood			0.17		
White pine wood			0.11		
Yellow pine wood			0.15		
Wool			0.038		
Source: Johnson [8].					
Animal skin			0.50		
Beeswax			0.40		
Bone			0.533		
Cat blood			23.0		
Dental amalgam			0.59		
Enamel			0.82		
Fleece			0.14–0.21		
Animal muscle			0.43–0.50		
Human blood			0.507		

(Continued)

	−190°C	0°C	20°C	100°C	1000°C
Human skin			0.21–0.63		
Human fat			0.21–0.33		
Human muscle			0.41–0.50		
Kidney or liver			0.498		
Porcelain			1.00		
Pork fat			0.187		
Seal blubber			0.190		
Oyster shell			1.95–2.27		
Wool			0.036		
Cork			0.045		
Cotton			0.061		
Cypress wood			0.097		
Fir wood			0.17		
Leaves			0.24–0.50		
Animal coats			0.03–0.15		
Paper			0.13		

References

1 Geankoplis, C. J., *Transport Processes and Unit Operations*, Prentice-Hall, Englewood Cliffs, 1993.
2 Kittel, C., *Introduction to Solid State Physics*, John Wiley and Sons, New York, 1957.
3 Gaskell, D. R., *Transport Phenomena in Materials Engineering*, MacMillan, New York, 1992.
4 Kingery, W. D., H. K. Bowen, and D. R. Uhlmann, *Introduction to Ceramics*, John Wiley and Sons, New York, 1976.
5 Samsanor, G. V., ed., *Oxide Handbook*, Plenum, New York, 1973.
6 Shackelford, J., and W. Alexander, eds., *Materials Science and Engineering Handbook*, CRC Press, Boca Raton, FL, 1992.
7 Geankoplis, C. J., *Transport Processes and Unit Operations*, Prentice-Hall, Englewood Cliffs, 1992.
8 Johnson, A. T., *Biological Process Engineering*, John Wiley and Sons, New York, 1999.

Appendix 5

Diffusivities in Selected Systems

Metals and Alloys

Diffusing Species	System	D_0, m^2/s
H_2	Fe	2.59×10^{-13}
Al	Cu	1.3×10^{-34}

Source: Geankoplis et al. [1].

Ceramics and Glasses

Diffusing Ion	System	D_0, cm^2/s	E_a, kJ/mol (kcal/mol)	Comment
O^{2-}	Al_2O_3	1.9×10^3	36.3 (152)	Single crystal, >1600 °C
O^{2-}	Al_2O_3	2.0×10^{-1}	26.3 (110)	Polycrystal, >1450 °C
O^{2-}	Al_2O_3	6.3×10^{-8}	13.8 (57.6)	Polycrystal, <1600 °C
O^{2-}	MgO	2.5×10^{-6}	14.9 (62.4)	1300–1750 °C
O^{2-}	SiO	1.5×10^{-2}	17.0 (71.2)	Vitreous, 925–1225 °C
O^{2-}	ZrO_2 (Ca-stabilized)	1.0×10^{-2}	6.7 (28.1)	15 mol% CaO, 700–1100 °C
Al^{3+}	Al_2O_3	1.0×10^{-2}	27.2 (114)	Polycrystal, 1670–1905 °C
Ca^{2+}	ZrO_2 (Ca-stabilized)	2.8×10^0 (?)	26.1 (109)	16 mol% CaO
Mg^{2+}	MgO	2.3×10^{-1}	18.8 (78.7)	1400–1600 ° C
Mg^{2+}	$MgAl_2O_4$	2.0×10^2	18.6 (78)	–
Zr^{4+}	ZrO_2 (Ca-stabilized)	1.97	26.1 (109)	16 mol% CaO

Diffusing Species	System	D_0, m^2/s
He	SiO_2	$2.4–5.5 \times 10^{-14}$

*Source: McCauley [2]; Geankoplis [1].

Materials Engineering and Science: Principles, Properties, and Processes, Second Edition. Brian S. Mitchell.

Companion website: www.wiley.com/go/materialsengineeringandscience2e

Polymers

Diffusing Species	System	D_0, m^2/s
O_2	Vulcanized rubber	2.1×10^{-10}
N_2	Vulcanized rubber	1.5×10^{-10}
CO_2	Vulcanized rubber	1.1×10^{-10}
H_2	Vulcanized Rubber	8.5×10^{-10}

Source: Geankoplis [1].

Biologics

Diffusing Species	System	D_0, m^2/s
H_2O	Human skin	1.2×10^{-13}

Source: Johnson [3].

References

1 Geankoplis, C. J., *Transport Processes and Unit Operations*, 3rd ed., Prentice-Hall, 1993.
2 McCauley, R. A., *Corrosion of Ceramics*, Taylor & Francis, New York, 1995.
3 Johnson, A. T., *Biological Process Engineering*, John Wiley and Sons, New York, 1999.

Appendix 6

Mechanical Properties of Selected Materials

Metals and Alloys

Material; Form or Treatment	Modulus of Elasticity (GPa)	Yield Strength (MPa)	Tensile Strength (MPa)	Poisson's Ratio
Wrought Aluminum Alloys; see Appendix 2 for Composition				
1100		34–150	90–165	
2011		295–310	380–405	
2014		97–415	185–485	
2024		76–395	185–495	
2219		76–395	170–475	
2618		370	440	
3003		42–185	110–200	
3004		69–250	180–285	
3105		55–195	115–215	
4032		315	380	
4043		69–270	145–285	
5005		41–195	125–200	
5050		55–200	145–220	
5052		90–255	195–290	
5056		150–405	290–435	
5083		145–285	290–345	
5086		115–255	260–325	
5154		115–270	240–330	
5182		140–395	275–420	
5252		170–240	235–285	
5254		115–270	240–330	
5454		115–310	250–370	

(*Continued*)

Materials Engineering and Science: Principles, Properties, and Processes, Second Edition. Brian S. Mitchell.

Companion website: www.wiley.com/go/materialsengineeringandscience2e

Material; Form or Treatment	Modulus of Elasticity (GPa)	Yield Strength (MPa)	Tensile Strength (MPa)	Poisson's Ratio
5456		160–255	310–350	
5457		48–185	130–205	
5652		90–255	195–290	
5657		140–165	160–195	
6005		105–240	170–260	
6009		130–325	235–345	
6010		170	255	
6061		55–275	125–310	
6063		48–270	90–290	
6066		83–360	150–395	
6070		69–350	145–380	
6101		76	97	
6151		195	220	
6201		300–310	330	
6205		140–290	260–310	
7075		105–505	230–570	
Cast Aluminum Alloys; see Appendix 2 for Compositions				
201		215–415	365–485	
206		345	435	
A206		345	435	
208		97	145	
242		125–290	185–325	
295		110–220	220–285	
296		130–180	255–275	
308		110	195	
319		125–185	185–280	
336		195–295	250–325	
354		285	380	
355		160–280	195–310	
356		140–210	175–265	
357		290	360	
A357		290	360	
359		255–290	330–345	
360		170	325	
A360		165	320	
380		165	330	
383		150	310	
384		165	330	
A384		165	330	

Material; Form or Treatment	Modulus of Elasticity (GPa)	Yield Strength (MPa)	Tensile Strength (MPa)	Poisson's Ratio
390		240–260	280–330	
A390		180–310	180–310	
413		140	300	
A413		130	290	
443		55	130	
B443		62	159	
C443		110	228	
514		85	170	
518		190	310	
520		180	330	
535		140	275	
712		170	240	
713		150	210–220	
771		275	345	
850		75	160	
Magnesium Alloys; see Appendix 2 for Compositions				
M11800		250	340	
M13310		200	255	
M16600		285	350	
M10600		130	220	
M12330		110	160	
M11910		150	230	
Titanium Alloys; see Appendix 2 for Compositions				
R54520		807	862	
R54810		951	1000	
R56400		924	993	
R56620		1000	1069	
Ti–10V–2Fe–3Al		1200	1276	
Copper Alloys; see Appendix 2 for Compositions				
Brass, naval		152–276	386–448	
Muntz metal		138	372	
Bronze, aluminum		172–448	483–724	
Beryllium copper		221–717	483–1310	
Brass, free-cutting		303	483	
Nickel silver, 18%		172–483	400–724	
Nickel silver, 13%		124	241	
Constantan		207–448	414–586	
Brass, red		117	241	

(Continued)

Material; Form or Treatment	Modulus of Elasticity (GPa)	Yield Strength (MPa)	Tensile Strength (MPa)	Poisson's Ratio
	Commercially Pure Metals and Elements			
C, natural diamond	700–1200			0.10–0.30
C, synthetic diamond	800–925			0.20
C, extruded graphite	11			
C, isostatically molded graphite	11.7			
C, PAN carbon fiber	230–400			
Titanium	107	170	240	0.34
Tantalum	185	165	205	0.35
Tungsten	400	760	960	0.28
Molybdenum	320	500	630	0.32
Gold	77	205	130–220	0.42
Silver	74		170–296	0.37
Platinum	171	<13.8	125–240	0.39
Tin	44.3	11		0.33
Zinc	104.5		134–186	0.25
Lead	18			0.40
Zirconium	94			
Nickel (cold drawn)	214			0.30
Copper	110			0.36
Aluminum	69			0.33

Ceramics and Glasses

Material; Form or Treatment	Modulus of Elasticity (GPa)	Flexural Strength (MP)	Compressive Strength (MPa)	Poisson's Ratio	Density (g/cm^3)
Al_2O_3	343–392	294–392	274–343	0.32 [1]	3.98
Glass, E-glass fiber	72.5				
BeO	392	147–196		0.34–0.38 [1]	3.02
CeO_2	294			0.27–0.31 [3]	7.13
Cr_2O_3					5.21
MgO	196–294	157–274	490–588	0.36 [1]	3.58
SiO_2, cristobalite					2.32
SiO_2, quartz	98		1960		2.65
SiO_2, silica glass	69	49–98	686–1862		2.20
TiO_2, rutile	98–196	69–167	274–823	0.28 [3]	4.24
ZrO_2, stabilized	147–196	176–784	980–2940	0.23–0.32 [3]	6.27

Material; Form or Treatment	Modulus of Elasticity (GPa)	Flexural Strength (MP)	Compressive Strength (MPa)	Poisson's Ratio	Density (g/cm^3)
ZrO_2, monoclinic	245	176–784	980–2940		5.56
ZrO_2, high-strength	196	980–1470			5.7–6.1
$Al_6Si_2O_{13}$, mullite	49–147	108–186	392–588	0.238 [3]	3.16
$MgAl_2O_4$, spinel	255	147–167	1667	0.294 [3]	3.58
$Mg_2Al_4Si_5O_{20}$, cordierite	147	118	343–666	0.17–0.26 [3]	2.0–2.5
Sic, α	392–588	441–784	588–4116	0.16–0.17 [2]	3.22
TiC	294–392		745–1352	0.187–0.189 [3]	4.94
B_4C				0.207 [3]	2.52
WC	686	343–833	2646–3528	0.24 [3]	15.6
BN, hexagonal	49–78	49–78	69–98		2.27
Si_3N_4	294–392	490–980	490–784	0.22–0.30 [2]	3.17
GaAs, <100> orientation				0.30 [2]	
Glass, borosilicate (Pyrex)				0.20 [2]	
Glass, soda-line				0.23 [2]	
$MoSi_2$				0.158–0.172 [3]	

Source: Somiya; [1] *Oxide Handbook*; [2] *Callister*; [3] *Materials Handbook*.

Polymers

Material; Form or Treatment	Modulus of Elasticity (GPa)	Yield Strength (MPa)	Tensile Strength (MPa)	Poisson's Ratio
Aramid, Kevlar 49 fiber	131	—	3600–4100	
Epoxy	2.41	—	27.6–90.0	
Nylon 66	1.59–3.79	55.1–82.8	94.5	0.39
Polybutylene terephthalate	1.93–3.00	56.6–60.0	56.6–60.0	
Polycarbonate	2.38	62.1	62.8–72.4	0.36
Polyester (thermoset)	2.06–4.41	—	41.4–89.7	
Rubber, nitrile	0.0034	—	6.9–24.1	
Rubber, styrene butadiene	0.002–0.010	—	12.4–20.7	
Rubber, silicone		—	10.3	
Phenol–formaldehyde	2.76–4.83	—	34.5–62.1	
Polyetheretherketone	1.10	91	70.3–103	
Polyethylene, low density	0.172–0.282	9.0–14.5	8.3–31.4	

(Continued)

Material; Form or Treatment	Modulus of Elasticity (GPa)	Yield Strength (MPa)	Tensile Strength (MPa)	Poisson's Ratio
Polyethylene, high density	1.08	26.2–33.1	22.1–31.0	
Polyethylene, ultrahigh molecular weight	0.69	21.4–27.6	38.6–48.3	
Polyethylene terephthalate	2.76–4.14	59.3	48.3–72.4	
Polymethyl methacrylate	2.24–3.24	53.8–73.1	48.3–72.4	
Polypropylene	1.14–1.55	31.0–37.2	31.0–41.4	
Polystyrene	2.28–3.28	—	35.9–51.7	0.33
Polytetrafluoroethylene	0.40–0.55	—	20.7–34.5	0.46
Polyvinyl chloride	2.41–4.14	40.7–44.8	40.7–51.7	0.38

Composites

Material; Form or Treatment	Modulus of Elasticity (GPa)	Tensile Strength (MPa)	Density (g/cm^3)
Nylon 66/30% milled carbon fiber	8.76	109	1.30
Nylon 66/30% 0.25″ carbon fiber	10.07	241	1.28
Nylon 66/25% milled carbon fiber, 20% 0.25″ glass	11.79	114	1.42
SMC/33% glass fiber	11.7	103	1.67
SMC/35% carbon fiber	26.9	103	1.66
SMC/24% glass fiber, 26% carbon fiber	24	117	1.65
Epoxy/PAN-carbon fiber fabric (8-hamess satin)	689	621	
Epoxy/PAN-carbon fiber fabric (plain weave)	66	552	
Epoxy/pitch-carbon fiber fabric (8-hamess satin)	40	379	
Epoxy/pitch-carbon fiber fabric (5-harness satin)	90–96.5	331–345	
Aluminum(220)/35% PAN-carbon fiber (unidirectional)		483 (longitudinal)	
Aluminum(6061)/35% PAN-carbon fiber (cross-ply)		221 (longitudinal) 103 (transverse)	
Aluminum/boron fiber (unidirectional)	0.22	1207 (longitudinal) 10 (transverse)	
Aluminum/boron fiber (cross-ply, 0/90)	0.14	52 (longitudinal) 52 (transverse)	
Glass/short-fiber-reinforced	50–150		
Glass/continuous-fiber-reinforced	1600		

Source: Schwartz, *Matrix/Reinforcement.*

Biologics

Material; Form or Treatment	Modulus of Elasticity (MPa)	Tensile Strength (MPa)	Poisson's Ratio	Density (g/cm^3)
Arterial wall	1.0	0.5–1.72		
Hyaline cartilage	0.4–19	1.3–18		
Skin	6–40	2.5–16		
Tendon/ligament	65–2500	30–300		
Cortical bone	16,000–20,000	30–211		
Cancellous bone	4600–15,000	51–93		
Hydroxyapatite	19,000	600		
Dentin, mineralized	14,700		0.31	2.14–2.18
Dentin, demineralized	260	29,600		
Enamel	84,000–130,000	10.3	0.33	2.95–2.97

Source: Silver, *Biomaterials Science and Biocompatibility*.

Appendix 7

Electrical Conductivity of Selected Materials

Resistivities and conductivities are the numerical inverses of one another, unless both are listed, indicating two separate sources for the information. Both are highly temperature sensitive. Values listed here are nominally at room temperature. Consult Section 6.1 for further information on compositional and temperature dependences.

Metals and Alloys

All elemental metals are commercially pure unless otherwise indicated.

Material	Conductivity, mho/m	Resistivity, ohm-m	Source
Silver	6.3×10^{7}		[1]
Copper	5.85×10^{7}		[1]
Gold	4.25×10^{7}		[1]
Aluminum	3.45×10^{7}		[1]
Al–1.2% Mn alloy	2.95×10^{7}		[1]
Sodium	2.1×10^{7}		[1]
Zinc		62.0×10^{-7}	[2]
Zirconium, grade 702		3.97×10^{-7}	[2]
Solder (60Sn–40Pb)		1.50×10^{-7}	[2]
Tin		1.11×10^{-7}	[2]
Lead		2.06×10^{-7}	[2]
Stainless steel, 301	0.14×10^{7}		[1]
Nichrome (80% Ni, 20% Cr)	0.093×10^{7}		[1]
Carbon, graphite	10^{5}		[1]
Germanium	2.2		[1]
Silicon (intrinsic)	4.3×10^{-4}		[1]
Carbon, diamond (natural)	$<10^{-14}$		[1]
Molybdenum		5.2×10^{-8}	[2]
Tantalum		13.5×10^{-8}	[2]

(Continued)

Materials Engineering and Science: Principles, Properties, and Processes, Second Edition. Brian S. Mitchell.

Companion website: www.wiley.com/go/materialsengineeringandscience2e

Material	Conductivity, mho/m	Resistivity, ohm-m	Source
Tungsten		5.3×10^{-8}	[2]
Platinum		10.60×10^{-8}	[2]
Titanium		4.2×10^{-7}–5.2×10^{-7}	[2]
Titanium alloy Ti–6Al–4 V		17.1×10^{-7}	[2]
Titanium alloy Ti–5Al–2.5Sn		15.7×10^{-7}	[2]
Carbon, diamond (synthetic)		1.5×10^{-2}	[2]
Steel alloy, A36, 1040, 1020		1.6×10^{-7}	[2]
Steel alloy, 4140		2.2×10^{-7}	[2]
Steel alloy, 4340		2.48×10^{-7}	[2]
Stainless steel, 304		7.2×10^{-7}	[2]
Stainless steel, 316		7.4×10^{-7}	[2]
Stainless steel, 405, 440A		6.0×10^{-7}	[2]
Stainless steel, 17-7PH		8.3×10^{-7}	[2]
Iron, gray, G1800		15×10^{-7}	[2]
Iron, gray, G3000		9.5×10^{-7}	[2]
Iron, gray, G4000		8.5×10^{-7}	[2]
Iron, 60-40-18		5.5×10^{-7}	[2]
Iron, 80-55-06, 120-90-02		6.2×10^{-7}	[2]
Aluminum alloy, 1100		2.9×10^{-8}	[2]
Aluminum alloy, 2024		3.4×10^{-8}	[2]
Aluminum alloy, 6061		3.7×10^{-8}	[2]
Aluminum alloy, 7075		5.22×10^{-8}	[2]
Aluminum alloy, 356.0		4.42×10^{-8}	[2]
Copper alloy, C11000		1.72×10^{-8}	[2]
Copper alloy, C172000 (beryllium–copper)		5.7×10^{-8}–1.15×10^{-7}	[2]
Copper alloy, C26000		6.2×10^{-8}	[2]
Copper alloy, C36000		6.6×10^{-8}	[2]
Copper alloy, C71500 (copper–30% nickel)		37.5×10^{-8}	[2]
Copper alloy, C93200		14.4×10^{-8}	[2]
Magnesium alloy, AZ31B		9.2×10^{-8}	[2]
Magnesium alloy, AZ91D		17×10^{-8}	[2]
Nickel 200		0.95×10^{-7}	[2]
Inconel 625		12.9×10^{-7}	[2]
Monel 400		5.47×10^{-7}	[2]
Invar		8.2×10^{-7}	[2]
Haynes alloy 25		8.9×10^{-7}	[2]
Kovar		4.9×10^{-7}	[2]
Mercury	1.04×10^{6}		[3]

Ceramics and Glasses

Material	Conductivity, mho/m	Resistivity, ohm-m	Source
Al_2O_3	10^{-10}–10^{-12}		[1]
$Al_6Si_2O_{13}$ (mullite)		$>10^{12}$	[4]
AlN		2×10^{9}–10^{11}	[4]
B_4C		3×10^{-3}–8×10^{-3}	[4]
BeO	10^{-12}–10^{-15}		[1]
BN		1.7×10^{-11}	[4]
CrB_2		21×10^{-8}	[4]
E-glass, fiber		4×10^{14}	[2]
GaAs (intrinsic)		10^{6}	[2]
Glass, soda-lime		10^{10}–10^{11}	[2]
Glass, window	$<10^{-10}$		[1]
Glass, silica	$<10^{-16}$		[1]
Glass–ceramic (Pyroceram)		2×10^{14}	[2]
HfB_2		10^{-7}	[4]
HfC		6×10^{-7}	[4]
InSb	17×10^{3}		[3]
$Mg_2Al_4Si_5O_{18}$ (Cordierite)		10^{-2}	[4]
MgO		1.3×10^{13}	[4]
Mica	10^{-11}–10^{-15}		[1]
$MoSi_2$		21.5×10^{-8}	[4]
PbS	38.4		[3]
Si_3N_4		$>10^{12}$	[2]
SiC (highly structurally dependent)		1.0–10^{10}	[4]
SiO_2, fused		$>10^{18}$	[2]
TaB_2		68×10^{-8}	[4]
TaC		25×10^{-8}	[4]
TiB_2		6.6×10^{-8}–28.4×10^{-8}	[4]
TiC		0.3×10^{-2}–0.8×10^{-2}	[4]
TiC	0.17×10^{7}		[1]
TiN		11×10^{-8}–13×10^{-7}	[4]
TiO_2		$>10^{11}$	[5]
WC		5×10^{-7}	[5]
WSi_2		33.4×10^{-8}–54.9×10^{-8}	[4]
ZrB_2		9.2×10^{-8}	[4]
ZrC		61×10^{-8}–64×10^{-8}	[4]
ZrO_2, 3% Y_2O_3		10^{10}	[2]
ZrO2, monoclinic		$>10^{11}$	[5]

Polymers

Material	Conductivity, mho/m	Resistivity, ohm-m	Source
Polymethylmethacrylate	$<10^{-12}$		[1]
Polyethylene	$<10^{-14}$		[1]
Polystyrene	$<10^{-14}$		[1]
Polytetrafluoroethylene	$<10^{-16}$		[1]
Phenol-formaldehyde	10^{-7}–10^{-11}		[1]
Polyethylene, low and high density		10^{15}–5×10^{16}	[2]
Polyethylene, ultrahigh molecular weight		$>5\times10^{14}$	[2]
Polypropylene		$>10^{14}$	[2]
Rubber, nitrile		3.5×10^{8}	[2]
Rubber, styrene butadiene		6×10^{11}	[2]
Rubber, silicone		10^{13}	[2]
Epoxy		10^{10}–10^{33}	[2]
Polyamide (Nylon 66)		10^{12}–10^{13}	[2]
Polybutylene terephthalate		4×10^{14}	[2]
Polycarbonate		2×10^{14}	[2]
Polyester		10^{13}	[2]
Polyetheretherketone		6×10^{14}	[2]
Polyethylene terephthalate		10^{12}	[2]
Polyvinyl chloride		$>10^{14}$	[2]
Cellulose acetate		10^{8}–10^{13}	[4]
Polyimide		4×10^{13}	[4]
Polyarylsulfone		3.2×10^{14}–7.71×10^{14}	[4]

Composites

Material	Conductivity, mho/m	Resistivity, ohm-m	Source
Wood		10^{14}–10^{16}	[2]
Concrete, dry		10^{9}	[2]
Polyimide, glass-fiber-reinforced		9.2×10^{13}	[4]
Nylon, glass-fiber-reinforced		2.6×10^{13}–5.5×10^{13}	[4]
Polyester, glass-fiber-reinforced		3.2×10^{14}–3.3×10^{14}	[4]
Phenylene oxide, glass-fiber-reinforced		10^{15}	[4]
Polyphenylene sulfide, 40% glass-fiber-reinforced		4.5×10^{12}	[4]
Polystyrene, 30% glass-fiber-reinforced		3.6×10^{14}	[4]
Polyacetal, 20% glass-fiber-reinforced		5×10^{12}	[4]

Biologics

Material	Conductivity, mho/cm	Resistivity, ohm-m	Source
Cortical bone, dry	10^{-9}		[6]
Cortical bone, wet (10% moisture)	10^{-3}	$3–4 \times 10^3$	[6]
Tissues, soft		5–15	[6]

References

1 Ralls, K. M., T. H. Courtney, and J. Wolff, *Introduction to Materials Science and Engineering*, John Wiley and Sons, New York, 1976.
2 Callister, W., *Materials Science and Engineering. An Introduction*, 5th ed., John Wiley & Sons, New York, 2000.
3 Shackelford, J. F., *Introduction to Materials Science for Engineers*, 5th ed., McGraw-Hill, New York, 2000.
4 Shackelford, J., and W. Alexander, eds., *Materials Science & Engineering Handbook*, CRL Press, Boca Raton, FL, 1992.
5 Somiya, S., *Advanced Technical Ceramics*, Academic Press, Oxford, 1984.
6 Brighton, C. T., J. Black, and S. R. Pollack, *Electrical Properties of Bone and Cartilage*, Grune and Stratton, New York, 1979.

Appendix 8

Refractive Index of Selected Materials

Ceramics, Glasses, and Ionic Solids

Material	Average Refractive Index	Source
Al_2O_3 (corundum, sapphire)	1.76	[1]
$Al_6Si_2O_{13}$ (mullite)	1.64	[1]
As_2S_3	2.66	[6]
BaF_2	1.48	[8]
$BaTiO_3$	2.40	[6]
C (diamond)	2.417	[2]
$CaCO_3$ (calcite)	1.65	[6]
CaF_2	1.43	[2]
CaO (lime)	1.84	[3]
CdTe	2.74	[6]
CuO	2.84	[3]
Glass, flint	1.65	[2]
Glass, soda–lime–silica	1.51–1.52	[1]
Glass from orthoclase	1.51	[1]
Glass, silica	1.458	[1]
Glass from albite	1.49	[1]
Glass, borosilicate (Pyrex)	1.47	[1]
$H_2O(l)$	1.33	[2]
$H_2O(s)$	1.30	[2]
$KAlSi_3O_8$ (orthoclase)	1.525	[1]
KBr	1.56	[8]
KCl	1.49	[2]
KI	1.673	[8]
LiF	1.392	[6]
$MgAl_2O_4$ (spinel)	1.72	[1]

(Continued)

Materials Engineering and Science: Principles, Properties, and Processes, Second Edition. Brian S. Mitchell.

Companion website: www.wiley.com/go/materialsengineeringandscience2e

Material	Average Refractive Index	Source
MgF_2	1.38	[8]
MgO (periclase)	1.74	[1]
MnO (manganosite)	1.22	[3]
$NaAlSi_3O_8$ (albite)	1.529	[1]
NaCl	1.54	[2]
NaF	1.326	[6]
PbO (litharge)	2.67	[3]
PbS (galena)	3.91	[3]
Si	3.29	[6]
SiC	2.68	[3]
SiO_2 (quartz)	1.55	[1]
SrO	1.87	[2]
SrTiOs	2.49	[6]
TiO_2 (rutile)	2.68	[6]
TiO_2 (anatase)	2.55	[3]
Y_2O_3	1.92	[6]
$ZiSiO_4$ (zircon)	1.95	[6]
ZnO	2.00	[3]
ZnSe	2.62	[6]
ZrO_2	2.1–2.2	[3]

Polymers

Material	Refractive Index	Source
Cellulosics (cellulose acetate)	1.46–1.50	[1]
Epoxies	1.55–1.60	[1]
Phenol-formaldehyde	1.47–1.50	[1]
Polyamides; e.g., Nylon 66	1.53	[1]
Polyarylsulfone	1.651	[4]
Polybutadiene/polystyrene copolymer	1.53	[1]
Polycarbonate	1.586	[4]
Polychloroprene	1.55–1.56	[1]
Polyester	1.52–1.57	[7]
Polyethylene, low density	1.51	[1]
Polyethylene terephthalate	1.64	[7]
Polyethylene, high density	1.545	[1]
Polyisoprene	1.52	[1]

Material	Refractive Index	Source
Polymethylmethacrylate	1.49	[5]
Polypropylene	1.47	[1]
Polystyrene	1.59	[1]
Polytetrafluoroethylene	1.35–1.38	[1]
Polyvinyl chloride	1.54–1.55	[1]
Polyvinylidene fluoride	1.42	[4]
Urethanes	1.5–1.6	[1]

References

1 Shackelford, J. F., *Introduction to Materials Science for Engineers*, 5th ed., McGraw-Hill, New York, 2000.
2 Ralls, K. M., T. H. Courtney, and J. Wolff, *Introduction to Materials Science and Engineering*, John Wiley and Sons, New York, 1976.
3 Samsanor, G. V., ed., *Oxide Handbook*, Plenum, New York, 1973.
4 Shackelford, J., and W. Alexander, eds., *Materials Science & Engineering Handbook*, CRL Press, Boca Raton, FL, 1992.
5 Callister, W., *Materials Science and Engineering. An Introduction*, 5th ed., John Wiley & Sons, New York, 2000.
6 Kingery, W. D., H. K. Bowen, and D. R. Uhlmann, *Ceramics*, John Wiley and Sons, New York, 1976.
7 Billmeyer, F. W., *Textbook of Polymer Science*, 3rd ed., John Wiley and Sons, New York, 1984.
8 Fox, M., *Optical Properties of Solids*, Oxford University Press, New York, 2001.

Index

Materials Engineering and Science: Principles, Properties, and Processes, Second Edition. Brian S. Mitchell.

Companion website: www.wiley.com/go/materialsengineeringandscience2e

d

j

k

n

O

t

Z

Name	Symbol	Atomic number	Atomic weight
Thallium	Tl	81	204.3833 (2)
Thorium	Th	90	232.0381 (1)
Thulium	Tm	69	168.93421 (2)
Tin	Sn	50	118.71 (7)
Titanium	Ti	22	47.867 (1)
Tungsten	W	74	183.84 (1)
Ununbium	Uub	112	[277]
Ununnilium	Uun	110	[269]
Unununium	Uuu	111	[272]
Uranium	U	92	238.0289 (1)
Vanadium	V	23	50.9415 (1)
Xenon	Xe	54	131.29 (2)
Ytterbium	Yb	70	173.04 (3)
Yttrium	Y	39	88.90585 (2)
Zinc	Zn	30	65.39 (2)
Zirconium	Zr	40	91.224 (2)

The number in parentheses following the atomic weight value gives the uncertainty in the last digit. An entry in [square brackets] indicates the mass number of the longest-lived isotope of an element that has no stable isotopes and for which a standard atomic weight cannot be defined because of wide variability in isotopic composition (or complete absence) in nature.

Frequently used physical constants

Quantity	Symbol	Value	Unit
	UNIVERSAL		
Standard acceleration of gravity	g	9.80665	m/s^2
Speed of light in vacuum	c, c_0	299 792 458	m/s
Magnetic permeability	μ_0	$4\pi \times 10^{-7}$	N/A^2
Electric permittivity	ε_0	$8.854187817 \ldots \times 10^{-12}$	F/m
Planck constant	h	$6.62606876(52) \times 10^{-34}$	J s
Elementary charge	e	$1.602176462(63) \times 10^{-19}$	C
Bohr magneton	μ_B	$9.27400899(37) \times 10^{-24}$	$A\,m^2$
	PHYSICO-CHEMICAL		
Avogadro constant	N_A	$6.02214199(47) \times 10^{23}$	mol^{-1}
Faraday constant	F	96 485.3415(39)	C/mol
Molar gas constant	R	8.314 472(15)	J/mol·K
Boltzmann constant R/N_A	k_B	$1.380\,6503(24) \times 10^{-23}$	J/K

Source: Peter J. Mohr and Barry N. Taylor, CODATA Recommended Values of the Fundamental Physical Constants: 1998, *Journal of physical and Chemical Reference Data*, Vol. 28, No. 6, 1999 and *Reviews of Modern Physics*, Vol. 72, No. 2, 2000.

Conversion factors

Name	To convert from	To	Multiply by	Divide by
Acceleration	ft/s^2	m/s^2	0.3048	3.2810
Density	lb/ft^3	kg/m^3	16.02	6.243×10^{-2}
Energy	BTU	J	1055	9.478×10^{-4}
Energy	cal	J	4.1859	0.2389
Energy	erg	J	1.000×10^{-7}	1.000×10^{7}
Energy	ev	J	1.602×10^{-19}	6.242×10^{18}
Energy	$ft \cdot lb_f$	J	1.3557	0.7376
Force	dyne	N	1.000×10^{-5}	1.000×10^{5}
Force	lb_f	N	4.4484	0.2248
Heat capacity	BTU/lb·°F	J/kg·°C	4188	2.388×10^{-4}
Heat transfer coefficient	$BTU/hr \cdot ft^2 \cdot °F$	$W/m^2 \cdot °C$	5.6786	0.1761
Length	ft	m	0.3048	3.2810
Mass	amu	kg	1.661×10^{-27}	6.022×10^{26}
Mass	lb	kg	0.4535	2.2050
Power	BTU/hr	W	0.2931	3.4120
Power	hp	W	745.71	1.341×10^{-3}
Pressure	bar	Pa	1.000×10^{5}	1.000×10^{-5}
Pressure	$dyne/cm^2$	Pa	0.1000	10.0000
Pressure	in Hg	Pa	3377	2.961×10^{-4}
Pressure	in water	Pa	248.82	4.019×10^{-3}
Pressure	lb_f/in^2 (psi)	Pa	6897	1.450×10^{-4}
Pressure	mbar	Pa	100.00	1.000×10^{-2}
Pressure	mm Hg	Pa	133.3	7.501×10^{-3}
Pressure	std atm	Pa	1.013×10^{5}	9.869×10^{-6}
Specific heat	BTU/lb·°F	J/kg·°C	4186	2.389×10^{-4}
Specific heat	cal/g·°C	J/kg·°C	4186	2.389×10^{-4}
Thermal conductivity	BTU/hr·ft·°F	W/m·°C	1.7307	0.5778
Thermal conductivity	$BTU\ in/hr \cdot ft^2 \cdot °F$	W/m·°C	0.1442	6.9340
Thermal conductivity	cal/cm·s·°C	W/m·°C	418.60	2.389×10^{-3}
Thermal conductivity	cal/ft·hr·°F	W/m·°C	6.867×10^{-3}	145.62
Torque	$ft \cdot lb_f$	Nm	1.3557	0.7376
Viscosity—absolute	centipoise	$N\ s/m^2$	1.000×10^{-3}	1000
Viscosity—absolute	g/cm·s	$N\ s/m^2$	0.1000	10
Viscosity—absolute	$lb_f/ft \cdot s$	$N\ s/m^2$	47.87	2.089×10^{-2}
Viscosity—absolute	lb/ft·s	$N\ s/m^2$	1.4881	0.6720
Volume	U.S. gallons	m^3	3.785×10^{-3}	264.20

Printed and bound by CPI Group (UK) Ltd, Croydon, CR0 4YY

17/09/2026

14964258-0007